T0185447

Lecture Notes in Artificial Intelligence   11670

Subseries of Lecture Notes in Computer Science

Series Editors

Randy Goebel
University of Alberta, Edmonton, Canada
Yuzuru Tanaka
Hokkaido University, Sapporo, Japan
Wolfgang Wahlster
DFKI and Saarland University, Saarbrücken, Germany

Founding Editor

Jörg Siekmann
DFKI and Saarland University, Saarbrücken, Germany

More information about this series at http://www.springer.com/series/1244

Abhaya C. Nayak · Alok Sharma (Eds.)

PRICAI 2019:
Trends in
Artificial Intelligence

16th Pacific Rim
International Conference on Artificial Intelligence
Cuvu, Yanuca Island, Fiji, August 26–30, 2019
Proceedings, Part I

 Springer

Editors
Abhaya C. Nayak (ID)
Department of Computing
Macquarie University
Sydney, NSW, Australia

Alok Sharma (ID)
RIKEN Center for Integrative
Medical Sciences
Yokohama, Japan

ISSN 0302-9743 ISSN 1611-3349 (electronic)
Lecture Notes in Artificial Intelligence
ISBN 978-3-030-29907-1 ISBN 978-3-030-29908-8 (eBook)
https://doi.org/10.1007/978-3-030-29908-8

LNCS Sublibrary: SL7 – Artificial Intelligence

© Springer Nature Switzerland AG 2019
This work is subject to copyright. All rights are reserved by the Publisher, whether the whole or part of the material is concerned, specifically the rights of translation, reprinting, reuse of illustrations, recitation, broadcasting, reproduction on microfilms or in any other physical way, and transmission or information storage and retrieval, electronic adaptation, computer software, or by similar or dissimilar methodology now known or hereafter developed.
The use of general descriptive names, registered names, trademarks, service marks, etc. in this publication does not imply, even in the absence of a specific statement, that such names are exempt from the relevant protective laws and regulations and therefore free for general use.
The publisher, the authors and the editors are safe to assume that the advice and information in this book are believed to be true and accurate at the date of publication. Neither the publisher nor the authors or the editors give a warranty, expressed or implied, with respect to the material contained herein or for any errors or omissions that may have been made. The publisher remains neutral with regard to jurisdictional claims in published maps and institutional affiliations.

This Springer imprint is published by the registered company Springer Nature Switzerland AG
The registered company address is: Gewerbestrasse 11, 6330 Cham, Switzerland

Preface

These proceedings in three volumes contain the papers presented at the 16th Pacific Rim International Conference on Artificial Intelligence (PRICAI 2019) held during August 26–30, 2019, in Yanuca Island, Fiji. PRICAI started as a biennial conference inaugurated in Tokyo in 1990. It provides a common forum for researchers and practitioners in various branches of artificial intelligence (AI) to exchange new ideas and share experience and expertise. Over the past years the conference has grown, both in participation and scope, to be a premier international AI event for all major Pacific Rim nations as well as countries from further afield. Indeed, the growth has merited holding PRICAI on an annual basis starting this year.

Submissions to PRICAI 2019 were received through two different routes: (1) some papers were directly submitted to PRICAI as in earlier years, and (2) in a special arrangement with IJCAI 2019, authors of submissions that narrowly missed out being accepted were encouraged to resubmit to PRICAI, along with the reviews and meta-reviews they received. The submissions of the first category underwent a double-blind review process, and were reviewed by the PRICAI Program Committee (PC) members and external reviewers against criteria such as significance, technical soundness, and clarity of presentation. Every paper received at least two, and in most cases three, reviews. Submissions of the second category were not subjected to further review, keeping in mind the workload of the reviewers in the community.

Altogether we received 311 high-quality submissions (with 265 submissions being of the first category) from 34 countries, which was impressive considering that for the first time PRICAI was being held in consecutive years. The program co-chairs read the reviews, the original papers, and called for additional reviews if necessary to make final decisions. The entire review team (PC members, external reviewers, and co-chairs) expended tremendous effort to ensure fairness and consistency in the paper selection process. Of the 265 submissions under the first category, 105 (39.6%) were accepted as full papers for the main-track, and 6 as full papers for the industry-track. A small number of papers were also accepted as short papers for the main-track (6), short papers for the industry-track (7), and as posters (6) – with the understanding that papers in the last category will not be included in these proceedings. The papers are organized in three volumes, under three broad (and naturally overlapping) themes, "Cognition", "Investigation", and "Application."

The technical program consisted of two workshops, five tutorials, and the main conference program. The workshops and tutorials covered important and thriving topics in AI. The workshops included the Pacific Rim Knowledge Acquisition Workshop (PKAW 2019) and the Knowledge Representation Conventicle (2019). The former was co-chaired by Prof. Kouzou Ohara and Dr. Quan Bai, while the latter was organized by Dr. Jake Chandler. The tutorials focused on hot topics including Big Data in bioinformatics, Data Science, Cognitive Logics, and Identity Management. All papers at the main conference were orally presented over the three days in parallel, and in thematically organized sessions. The authors of the posters were also offered the opportunity to give short talks to introduce their work.

It was our great honor to have four outstanding keynote/invited speakers, whose contributions have pushed boundaries of AI across various aspects: Prof. Hiroaki Kitano (Sony Computer Science Laboratories Inc. and The System Biology Institute, Japan), Prof. Grigoris Antoniou (University of Huddersfield, UK), Prof. Mary-Anne Williams (University of Technology Sydney, Australia), and Prof. Byoung-Tak Zhang (Seoul National University, South Korea). We are grateful to them for sharing their insights on their latest research with us.

The success of PRICAI 2019 would not have been possible without the effort and support of numerous people from all over the world. First of all, we would like to thank the PC members and external reviewers for their engagements in providing rigorous and timely reviews. It was because of them that the quality of the papers in this volume is maintained at a high level. We wish to thank the general co-chairs, Professors Abdul Sattar and MGM Khan for their continued support and guidance, and Dr. Sankalp Khanna for his tireless effort toward the overall coordination of PRICAI 2019. We are also thankful to various chairs and co-chairs, namely the industry co-chairs, workshop co-chairs, the tutorial co-chairs, the web and publicity co-chairs, the sponsorship chair, and the local organization chair, without whose support and hard work PRICAI 2019 could not have been successful. We also acknowledge the willing help of Kinzang Chhogyal, Jandson S. Ribeiro, and Hijab Alavi toward the preparation of these proceedings.

We gratefully acknowledge the financial and/or organizational support of a number of institutions including the University of the South Pacific (Fiji), Griffith University (Australia), Macquarie University (Australia), Fiji National University (Fiji), RIKEN Center for Integrative Medical Sciences (Japan), University of Western Australia (Australia), Australian Computer Society (ACS), and Springer Nature. Special thanks to EasyChair, whose paper submission platform we used to organize reviews and collate the files for these proceedings. We are also grateful to Alfred Hofmann and Anna Kramer from Springer for their assistance in publishing the PRICAI 2019 proceedings in the *Lecture Notes in Artificial Intelligence* series, as well as sponsoring the best paper awards.

We thank the Program Chair and the Conference Chair of IJCAI 2019, Professors Sarit Kraus and Thomas Eiter, for encouraging the resubmission of many IJCAI submissions to PRICAI 2019. Last but not least, we thank all authors and all conference participants for their contribution and support. We hope all the participants took this valuable opportunity to share and exchange their ideas and thoughts with one another and enjoyed their time at PRICAI 2019.

August 2019   Abhaya C. Nayak
                                                                                           Alok Sharma

Organization

Steering Committee

Tru Hoang Cao	Ho Chi Minh City University of Technology, Vietnam
Xin Geng	Southeast University, China
Guido Governatori	Data61, Australia
Takayuki Ito	Nagoya Institute of Technology, Japan
Byeong-Ho Kang	University of Tasmania, Australia
Sankalp Khanna	CSIRO, Australia
Dickson Lukose	GCS Agile Pty Ltd., Australia
Hideyuki Nakashima	Sapporo City University, Japan
Seong-Bae Park	Kyung Hee University, South Korea
Abdul Sattar	Griffith University, Australia
Zhi-Hua Zhou	Nanjing University, China

Honorary Members

Randy Goebel	University of Alberta, Canada
Tu-Bao Ho	JAIST, Japan
Mitsuru Ishizuka	University of Tokyo, Japan
Hiroshi Motoda	Osaka University, Japan
Geoff Webb	Monash University, Australia
Wai K. Yeap	Auckland University of Technology, New Zealand
Byoung-Tak Zhang	Seoul National University, South Korea
Chengqi Zhang	University of Technology Sydney, Australia

Organizing Committee

General Co-chairs

Abdul Sattar	Griffith University, Australia
M. G. M. Khan	University of the South Pacific, Fiji

Program Co-chairs

Abhaya C. Nayak	Macquarie University, Australia
Alok Sharma	RIKEN Center for Integrative Medical Sciences, Japan

Local Co-chairs

Salsabil Nusair	University of the South Pacific, Fiji
A. B. M. Shawkat Ali	Fiji National University, Fiji

Workshop Co-chairs

Nasser Sabar   La Trobe University, Australia
Anurag Sharma   University of the South Pacific, Fiji

Tutorial Co-chairs

Min-Ling Zhang   Southeast University, China
Yi Mei   Victoria University of Wellington, New Zealand

Industry Co-chairs

Duc Nghia Pham   MIMOS Berhad, Malaysia
Sankalp Khanna   CSIRO, Australia

Sponsorship Co-chairs

Andy Song   Royal Melbourne Institute of Technology, Australia
Sabiha Khan   Fiji National University, Fiji

Web and Publicity Co-chairs

Mahmood Rashid   Victoria University, Australia
Benjamin Cowley   Griffith University, Australia

Local Arrangements

Priynka Sharma   University of the South Pacific, Fiji
Gavin Khan   University of the South Pacific, Fiji
Wafaa Wardha   University of the South Pacific, Fiji
Goel Aman Lal   University of the South Pacific, Fiji

Program Committee

Eriko Aiba   University of Electro-Communications, Japan
Patricia Anthony   Lincoln University, New Zealand
Quan Bai   Auckland University of Technology, New Zealand
Yun Bai   University of Western Sydney, Australia
Blai Bonet   Universidad Simón Bolívar, Venezuela
Richard Booth   Cardiff University, UK
Zied Bouraoui   CRIL – CNRS, Université d'Artois, France
Arina Britz   CAIR, Stellenbosch University, South Africa
Rafael Cabredo   De La Salle University, Philippines
Longbing Cao   University of Technology Sydney, Australia
Lawrence Cavedon   RMIT University, Australia
Siqi Chen   Tianjin University, China
Songcan Chen   Nanjing University of Aeronautics
                             and Astronautics, China
Wu Chen   Southwest University, China
Yingke Chen   Sichuan University, China
Wai Khuen Cheng   Universiti Tunku Abdul Rahman, Malaysia

Krisana Chinnasarn	Burapha University, Thailand
Phatthanaphong Chomphuwiset	Mahasarakham University, Thailand
Dan Corbett	Optimodal Technologies, USA
Célia da Costa Pereira	Université Côte d'Azur, France
Jirapun Daengdej	Assumption University, Thailand
Xuan-Hong Dang	IBM T.J. Watson, USA
Abdollah Dehzangi	Morgan State University, USA
Clare Dixon	University of Liverpool, UK
Shyamala Doraisamy	Universiti Putra Malaysia, Malaysia
Atilla Elci	Aksaray University, Turkey
Vlad Estivill-Castro	Griffith University, Australia
Eduardo Fermé	Universidade da Madeira, Portugal
Christian Freksa	University of Bremen, Germany
Katsuhide Fujita	Tokyo University of Agriculture and Technology, Japan
Naoki Fukuta	Shizuoka University, Japan
Marcus Gallagher	University of Queensland, Australia
Dragan Gamberger	Ruđer Bošković Institute, Croatia
Wei Gao	Nanjing University, China
Xiaoying Gao	Victoria University of Wellington, New Zealand
Yang Gao	Nanjing University, China
Xin Geng	Southeast University, China
Manolis Gergatsoulis	Ionian University, Greece
Guido Governatori	CSIRO, Australia
Alban Grastien	Data61, Australia
Fikret Gürgen	Boğaziçi University, Turkey
Peter Haddawy	Mahidol University, Thailand
Bing Han	Xidian University, China
Choochart Haruechaiyasak	NECTEC, Thailand
Kiyota Hashimoto	Prince of Songkla University, Thailand
Tessai Hayama	Nagaoka University of Technology, Japan
Jose Hernandez-Orallo	Universitat Politècnica de València, Spain
Juhua Hu	University of Washington, USA
Sheng-Jun Huang	Nanjing University of Aeronautics and Astronautics, China
Xiaodi Huang	Charles Sturt University, Australia
Van Nam Huynh	JAIST, Japan
Masashi Inoue	Tohoku Institute of Technology, Japan
Sanjay Jain	National University of Singapore, Singapore
Jianmin Ji	University of Science and Technology of China, China
Liangxiao Jiang	China University of Geosciences, China
Yichuan Jiang	Southeast University, China
Hideaki Kanai	JAIST, Japan
Ryo Kanamori	Nagoya University, Japan
Byeong-Ho Kang	University of Tasmania, Australia

C. Maria Keet	University of Cape Town, South Africa
Gabriele Kern-Isberner	Technische Universität Dortmund, Germany
Sankalp Khanna	CSIRO, Australia
Frank Klawonn	Ostfalia University of Applied Sciences, Germany
Sébastien Konieczny	CRIL - CNRS, France
Alfred Krzywicki	University of New South Wales, Australia
Young-Bin Kwon	Chung-Ang University, South Korea
Ho-Pun Lam	CSIRO, Australia
Jérôme Lang	CNRS, LAMSADE, University Paris-Dauphine, France
Roberto Legaspi	RIKEN Center for Brain Science, Japan
Gang Li	Deakin University, Australia
Guangliang Li	University of Amsterdam, The Netherlands
Li Li	Southwest University, China
Ming Li	Nanjing University, China
Tianrui Li	Southwest Jiaotong University, China
Yu-Feng Li	Nanjing University, China
Beishui Liao	Zhejiang University, China
Jiamou Liu	University of Auckland, New Zealand
Qing Liu	CSIRO, Australia
Michael Maher	Reasoning Research Institute, Australia
Xinjun Mao	National University of Defense Technology, China
Eric Martin	University of New South Wales, Australia
Maria Vanina Martinez	Universidad de Buenos Aires, Argentina
Sanparith Marukatat	NECTEC, Thailand
Michael Mayo	University of Waikato, New Zealand
Brendan Mccane	University of Otago, New Zealand
Thomas Meyer	University of Cape Town and CAIR, South Africa
James Montgomery	University of Tasmania, Australia
Abhaya Nayak	Macquarie University, Australia
Richi Nayak	QUT, Australia
Kourosh Neshatian	University of Canterbury, New Zealand
M. A. Hakim Newton	Griffith University, Australia
Shahrul Azman Noah	Universiti Kebangsaan Malaysia, Malaysia
Masayuki Numao	Osaka University, Japan
Kouzou Ohara	Aoyama Gakuin University, Japan
Hayato Ohwada	Tokyo University of Science, Japan
Mehmet Orgun	Macquarie University, Australia
Noriko Otani	Tokyo City University, Japan
Lionel Ott	University of Sydney, Australia
Maurice Pagnucco	University of New South Wales, Australia
Hye-Young Paik	University of New South Wales, Australia
Laurent Perrussel	IRIT, Université de Toulouse, France
Bernhard Pfahringer	University of Waikato, New Zealand
Duc Nghia Pham	MIMOS Berhad, Malaysia
Jantima Polpinij	Mahasarakham University, Thailand

Mikhail Prokopenko	University of Sydney, Australia
Chao Qian	University of Science and Technology of China, China
Yuhua Qian	Shanxi University, China
Joël Quinqueton	LIRMM, France
Fenghui Ren	University of Wollongong, Australia
Mark Reynolds	University of Western Australia, Australia
Ji Ruan	Auckland University of Technology, New Zealand
Kazumi Saito	University of Shizuoka, Japan
Chiaki Sakama	Wakayama University, Japan
Ken Satoh	National Institute of Informatics and Sokendai, Japan
Abdul Sattar	Griffith University, Australia
Torsten Schaub	University of Potsdam, Germany
Nicolas Schwind	National Institute of Advanced Industrial Science and Technology, Japan
Nazha Selmaoui-Folcher	University of New Caledonia, New Caledonia
Lin Shang	Nanjing University, China
Alok Sharma	RIKEN Center for Integrative Medical Sciences, Japan
Chuan Shi	Beijing University of Posts and Telecommunications, China
Zhenwei Shi	Beihang University, China
Daichi Shigemizu	NCGG, Japan
Yanfeng Shu	CSIRO, Australia
Guillermo R. Simari	Universidad del Sur in Bahia Blanca, Argentina
Tony Smith	University of Waikato, New Zealand
Chattrakul Sombattheera	Mahasarakham University, Thailand
Andy Song	RMIT University, Australia
Markus Stumptner	University of South Australia, Australia
Xing Su	Beijing University of Technology, China
Merlin Teodosia Suarez	De La Salle University, Philippines
Thepchai Supnithi	NECTEC, Thailand
Michael Thielscher	University of New South Wales, Australia
Shikui Tu	Shanghai Jiao Tong University, China
Miroslav Velev	Aries Design Automation, USA
Serena Villata	CNRS, France
Toby Walsh	University of New South Wales, Australia
Kewen Wang	Griffith University, Australia
Qi Wang	Northwestern Polytechnical University, China
Wei Wang	NJU, China
Paul Weng	UM-SJTU Joint Institute, China
Peter Whigham	University of Otago, New Zealand
Wayne Wobcke	University of New South Wales, Australia
Brendon J. Woodford	University of Otago, New Zealand
Chang Xu	University of Sydney, Australia
Guandong Xu	University of Technology Sydney, Australia
Ming Xu	Xi'an Jiaotong-Liverpool University, China
Shuxiang Xu	University of Tasmania, Australia

Xin-Shun Xu	Shandong University, China
Bing Xue	Victoria University of Wellington, New Zealand
Hui Xue	Southeast University, China
Bo Yang	Jilin University, China
Ming Yang	Nanjing Normal University, China
Roland Yap	National University of Singapore, Singapore
Kenichi Yoshida	University of Tsukuba, Japan
Chao Yu	University of Wollongong, Australia
Yang Yu	Nanjing University, China
Takaya Yuizono	JAIST, Japan
Yifeng Zeng	Teesside University, UK
Chengqi Zhang	University of Technology Sydney, Australia
Dongmo Zhang	Western Sydney University, Australia
Du Zhang	Macau University of Science and Technology, China
Min-Ling Zhang	Southeast University, China
Minjie Zhang	University of Wollongong, Australia
Qieshi Zhang	Chinese Academy of Sciences, Australia
Rui Zhang	University of Melbourne, Australia
Shichao Zhang	Guangxi Normal University, China
Wen Zhang	Beijing University of Technology, China
Yu Zhang	Hong Kong University of Science and Technology, SAR China
Zhao Zhang	Hefei University of Technology, China
Zili Zhang	Deakin University, Australia
Zongzhang Zhang	Soochow University, China
Li Zhao	MSRA, China
Yanchang Zhao	CSIRO, Australia
Shuigeng Zhou	Fudan University, China
Zhi-Hua Zhou	Nanjing University, China
Xiaofeng Zhu	Guangxi Normal University, China
Xingquan Zhu	Florida Atlantic University, USA
Fuzhen Zhuang	Chinese Academy of Sciences, China

Additional Reviewers

Shintaro Akiyama
Yuya Asanomi
Mansour Assaf
Weiling Cai
Rohitash Chandra
Jairui Chen
Armin Chitizadeh
Laurenz A. Cornelissen
Emon Dey
Duy Tai Dinh

Bayu Distiawan
Shaokang Dong
Steve Edwards
Suhendry Effendy
Jorge Fandinno
Zaiwen Feng
Chuanxin Geng
Sayuri Higaki
Jin B. Hong
Yuxuan Hu

Paul Salvador Inventado
Abdul Karim
Karamjit Kaur
Sunil Lal
Ang Li
Haopeng Li
Weikai Li
Yun Li
Yuyu Li
Shenglan Liao
Shaowu Liu
Yuxin Liu
Wolfgang Mayer
Kingshuk Mazumdar
Nguyen Le Minh
Risa Mitsumori
Taiki Mori
Majid Namaazi
Courtney Ngo
Aaron Nicolson
Lifan Pan
Asanga Ranasinghe
Vahid Riahi
Maria AF Rodriguez
Manou Rosenberg
Matt Selway
Cong Shang

Swakkhar Shatabda
Manisha Sirsat
Fengyi Song
Yixin Su
Adam Svahn
Trung Huynh Thanh
Yanlling Tian
Qing Tian
Jannai Tokotoko
Nhi N. Y. Vo
Guodong Wang
Jing Wang
Xiaojie Wang
Yi Wang
Yuchen Wang
Yunyun Wang
Shiqing Wu
Peng Xiao
Yi Xu
Wanqi Yang
Heng Yao
Dayong Ye
Jun Yin
Zhao Zhang
Zhu Zhirui
Zili Zhou
Yunkai Zhuang

Contents – Part I

Knowledge Handling

Image Recognition and Manipulation

Language and Speech

Knowledge Representation and Reasoning

Multi-Agent Systems

Contents – Part II

Learning Models

Neural Networks

Optimization

Traffic and Vehicular Automation

Social and Information Networks

Contents – Part III

Robotics, IOT and Traffic Automation

Biometrics and Bioinformatics

Other Applications

Learning

Explaining Black-Box Models Using Interpretable Surrogates

Deepthi Praveenlal Kuttichira[(⊠)], Sunil Gupta, Cheng Li, Santu Rana, and Svetha Venkatesh

Applied Artificial Intelligence Institute, Deakin University, Geelong, Australia
{dkuttichira,sunil.gupta,cheng.li,santu.rana,
svetha.venkatesh}@deakin.edu.au

Abstract. Explaining black-box machine learning models is important for their successful applicability to many real world problems. Existing approaches to model explanation either focus on explaining a particular decision instance or are applicable only to specific models. In this paper, we address these limitations by proposing a new model-agnostic mechanism to black-box model explainability. Our approach can be utilised to explain the predictions of any black-box machine learning model. Our work uses interpretable surrogate models (e.g. a decision tree) to extract global rules to describe the preditions of a model. We develop an optimization procedure, which helps a decision tree to mimic a black-box model, by efficiently retraining the decision tree in a sequential manner, using the data labeled by the black-box model. We demonstrate the usefulness of our proposed framework using three applications: two classification models, one built using iris dataset, other using synthetic dataset and a regression model built for bike sharing dataset.

Keywords: Explainability · Bayesian optimisation · Gaussian process

1 Introduction

Application of artificial intelligence (AI) can be found in almost any field ranging from medical diagnosis to making million dollar decisions. AI algorithms are popular due to their excellent generalization capabilities. Complex machine learning models make accurate predictions and classifications. However, there is a trade-off between performance and explainability [7]. Simple models e.g. logistic regression are explainable, but have lower predictive power compared to more complex models like deep neural network, Support Vector Machine (SVM) etc. Usually more complex a model gets, the less interpretable it becomes. With the gaining popularity of complex models, lack of interpretability is posing a problem. For any model to be effective, both accuracy and explainability are important. If the decisions made by the model cannot be understood, the model might not be deployed, especially in areas where providing explanation is crucial, like medical diagnosis, law making and finance [1,4]. Recently European Union

© Springer Nature Switzerland AG 2019
A. C. Nayak and A. Sharma (Eds.): PRICAI 2019, LNAI 11670, pp. 3–15, 2019.
https://doi.org/10.1007/978-3-030-29908-8_1

proposed in its regulations that people affected by algorithmic decisions have the right to explanation [6]. With these developments providing explanations to a black-box algorithm has become an urgent step.

Researchers have been trying to come up with explanations for black-box models. The term explanation, is not a monolith [11]. It could mean different things. Explanations can be given in terms of relationship between input features and output. It could also mean algorithmic transparency, *i.e.*, the mechanism of the decision making of an algorithm. Since the term explainability is not precisely defined, multiple approaches have been made to address this problem. These approaches can be broadly classified into two categories. The first one is to provide *post-hoc explanation*, i.e., explaining the output of a black-box model while the other is *algorithmic transparency*, i.e., explaining the internal working of an algorithm. Post-hoc explanations aim to provide explanation for an output through another similar input. Deep learning techniques perform well in image and video classifications. Providing descriptions for image and videos through the outputs of other similar images or videos is a form of post-hoc explanation [8,9]. Algorithmic transparency aims to come up with explanations for existing models [5,13]. Explaining existing complex models is challenging. Most of the existing methods try to extract rules by perturbing the test points and observing how these affect the outputs [5]. Others try to learn rules from neural network by training it with a subset of inputs, and observing which neurons get activated for the dominant feature in the subset. A recent work has been done to approximate a complex model using a surrogate model [13].

Although the explainability of black-box model has been tried to tackle in many ways, it still remains an open problem. Post-hoc explanations do not provide any insight into the working of the model. So the underlying model still remains a black-box. Also the mapping of descriptions to outputs remain opaque. The accuracy of novel interpretable algorithms depends a lot on good prior knowledge. Also these algorithms have to be tailor made for each problem. One of the key reasons for the popularity of neural networks is that it can work across different datasets with slight tuning of the hyperparameters. It would be ideal if complex machine learning models that have been successfully deployed could also be explained. Even better if the mechanism of explanation could be agnostic to model type, that is, if explanation mechanism remains same irrespective of whether it is used to explain a neural network, random forest or a gradient boosting model. Such an explanation scheme, then could be implemented as an abstraction layer over the black-box models and can be applied in a wide range of problems. In this paper, we take up this problem and aim to develop a solution. Our work focuses on extracting global rules to describe internal working of a model, while remaining model agnostic. Decision trees are interpretable models as they make decisions based on a hierarchy of decisions. If decision trees can be modified to mimic a given black-box model, we can use them as a surrogate for the complex black-box model. We make this possible by starting with a black-box model and a decision tree both trained using a training set. Next, we find the region in the input space where the difference between the black-box model and the decision tree model is the highest. We sample a small set of data points in this region and train the decision tree with this data labeled by the black-box model. This process is repeated until the difference between the two models becomes small. After this training process, decision tree rules can be used to explain the internal working of the black-box model. In this paper, we have used a neural network to represent the black-box model. We demonstrate our proposed framework using three applications: a classification model built using synthetic dataset, a classification model built using iris dataset, and a regression model built for bike sharing dataset.

Our contributions can be summarized as below:

- We introduce a novel model-agnostic mechanism for black-box model explainability.
- We develop an optimization procedure to retrain the explainable surrogate model (decision tree) in a sequential manner.
- We demonstrate the usefulness of our proposed framework using two applications: a classification model built using iris dataset and a regression model built for bike sharing dataset.

2 Framework

Complex models are usually black-box. There are many black-box models like neural networks, SVM etc. In our work we have chosen neural network as the black-box model. Decision tree is capable of modelling non-linear functions. The prediction made by the decision tree can be explained by the hierarchy of decisions. For this reason, we have chosen decision tree as our surrogate model. The prediction technique of neural network and decision tree are briefly summarized in the Subsects. 2.1 and 2.2. Both the models are trained with a training set $D^{train} = \{x_i, y_i\}_{i=1}^{N}$,where $x \in \mathbb{R}^d$ and $y = f(x_i)$ is realization from an unknown and smooth function $f : \mathbb{R}^d \to \mathbb{R}$.

The goal of training any model is to learn f as accurately as possible. We denote f^n as the function learned by neural network and f^t as the function learned by decision tree. We expect f^n to model the input data more accurately than f^t. This is usually true as the prediction accuracy of neural network is more than that of decision tree. Our goal is to approximate the function f^t to that of f^n, as they both are likely to be different as shown in Fig. 1(a). For this, we train the decision tree iteratively using the neural network, to make f^t close to f^n, as desirable in Figure 1(b) (c) (d). In each iteration of retraining, we train the surrogate model with the data points sampled around the point at which the difference between the functions is maximum. Finding the point of maximum difference is an optimization problem. Since the data point at which the functions differ maximum might be virtual, we denote this point as $x^v \in \mathbb{R}^d$. Then our task has reduced to iteratively optimizing the following objective function:

$$x^v = \underset{x \in \mathcal{X}}{\operatorname{argmax}} |f^n(x) - f^t(x)| \tag{1}$$

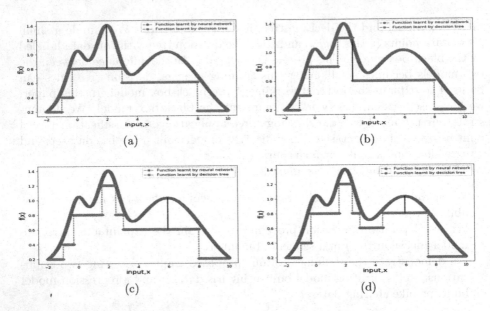

Fig. 1. The illustration of function differences before and after introducing new data. (a) The functions learnt from neural network and decision tree with only the initial training data. (b), (c), (d) The functions learnt by neural network and decision tree after introducing new data iteratively by following our method.

where $\mathcal{X}$ is the search space for both the training data and the new data. Note that both f^n and f^t have no tractable form and thus the objective function is a black-box function. We can use Bayesian optimization [3] - an efficient and popular unknown function optimizer to solve the Eq. 1.

We illustrate our framework in Fig. 2. The first step is to training a neural network and decision tree based on the initial training data (Step 1). The neural network learns a function f^n and decision tree learns f^t. The function f^n by neural network is fixed and we use it only to label a new data. We run Bayesian optimization for several iterations to suggest a new data x^v, where function differences (Eq. 1) are maximum (Step 2). In order to approximate the function of decision tree to that of neural network at the point x^v, we uniformly sample points from $[(1-\delta)x^v, (1+\delta)x^v]$ with the radius coefficient δ. These new samples are labeled using the prediction of the trained neural network (Step 3). We next augment the training data with these samples and retrain the decision tree (Step 4). The steps 2, 3, and 4 are repeated until the performance of the decision tree converges. The algorithm can be found in Algorithm 1. We give a detailed explanation of the key steps below.

2.1   Training Neural Network

Neural network is a framework describing the relationship between input and output by connected neurons in the hidden layers. A neuron is a decision unit

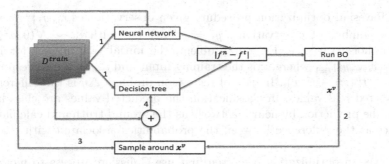

Fig. 2. The diagram of our algorithm using decision tree to interpret neural network. We illustrate the detail steps in texts.

based on the input vector x and the weight vector ω. The output y of a neuron can be formalized as $y = f(\omega^{\mathsf{T}} x + b)$, where b is the bias vector and f is the activation function. The activation function has several choices. The popular ones include rectified linear units (ReLU), tanh and sigmoid function.

The training for neural network aims to learn the weights between connected neurons. A reasonable choice is to compute the gradients of the output with respect to the weights and then we can use any gradient based optimizer. Backpropagation makes it possible by transmitting the gradients using chain rules [2]. The difficulty of the interpretability of neural network is attributed to its highly interconnected hidden layers.

2.2 Training Decision Tree

A decision tree is a predictive model, which starts with a single node and then splits into different branches according to the information gain and finally forms a tree structure. Decision tree can be a classification or regression tree. At each node the feature that can maximize the purity of the classes is used to split the node. The metrics for the purity can be Gini index or information gain.

From the tree structure, we can easily interpret the decision process. Using decision tree as a surrogate model, we expect the function learned by decision tree to approximate the function learned by neural network. As we stated before, the existing training data might be not sufficient to have the decision tree function closely approximate the neural network function. We next need to find additional data to refine the function of decision tree.

2.3 Bayesian Optimization to Search New Data

Having a trained decision tree and neural network from the original training set, our goal is to minimize the Eq. 1 to find the region in the input space, where the two functions are the most different. We choose the technique of Bayesian optimization to accomplish it, since Bayesian optimization (BO) has demonstrated superior power on a black-box function optimization. To simplify, we describe a generic Bayesian optimization procedure given observations $\{x_j, y_j\}_{j=1}^J$, where J is the number of observations, $y_j = f(x_i) + \varepsilon$ with $\varepsilon \sim \mathcal{N}(0, \sigma^2)$ and σ^2 is the noise variance. For our problem, the initial training set for BO is $D_0^{BO} = \{x_i, \triangle y_i\}_{i=1}^n$ where, x is the training input and $\triangle y$ is the corresponding difference $|f^n(x) - f^t(x)|$. In case of regression problem $\triangle y$ is the difference of the predicted real values. In classification the predicted values are class labels. We take the prediction by neural network as the ground truth and calculate the difference as the difference between the probabilities associated with the class labels.

In Bayesian optimization, we often first use Gaussian process to model the latent function based on observations. Then we can construct an acquisition function to query the next point. Specifically, Gaussian process is a stochastic process where the joint distribution of any point in the domain space is still a Gaussian distribution. Therefore, for a predicted point x^*, its predictive posterior distribution is a Gaussian distribution $\mathcal{N}(\mu(x^*), \sigma^2(x^*))$. With a typical zero-mean assumption for GP mean function, we can write the mean and variance

$$\mu(x^*) = k_{*J}^T (K_{JJ} + \sigma^2 I)^{-1} y_{1:J}$$

$$\sigma^2(x^*) = k_{**} - k_{*J}^T (K_{JJ} + \sigma^2 I)^{-1} k_{*J}$$

where k_{**} is the kernel function, $k_{*J} = [k(x^*, x_1), \cdots, k(x_{t+1}, x_t)]$ and K_{JJ} is the Gram matrix between $x_{1:J}$. Note that k is the kernel function representation of the smoothness of the latent function. The popular choice includes the SE-kernel and Matern kernel [14]. The assumptions we make about the data such as the function that models it is smooth etc is incorporated by choosing the kernel function appropriately. In out framework, we have used squared exponential kernel $exp(-\frac{1}{2l^2}||x_i - x_j||^2)$. The hyperparameter l decided the width of then kernel function.

Next based on the built GP before, we want an acquisition function to suggest the next evaluation point. A natural choice is to use a function to measure the possible improvement over the best observation so far (minimal or maximal). The popular ones such as probability of improvement (PI), expected improvement (EI) [3] have been derived. In our work we opted for EI, although other acquisition functions can also be used. We can maximize the acquisition function to obtain the next point and then update Gaussian process and these steps will be repeated. The BO will return the maxima of the difference function. It is a new data point in the function space that maximizes our objective function. We sample around this new data point in a window of $[(1-\delta)x^v, (1+\delta)x^v]$. The size of δ vary with the complexity of data. It is ideally small, so as to limit sampling close to the optima. This new set of points form X_{new}. The target values Y_{new} for these data points are found using neural networks as $Y_{new} = f^n(X_{new})$. The training data is augmented with these new data points. The steps are repeated until the performance of the decision tree converges. BO has also been proposed to work in high dimensions [10,12], although we have used only low dimensional data for our work.

Algorithm 1. The proposed decision tree retraining algorithm.

---

1. **Input:** Training set $D^{train} = \{X, Y\}$ with $X = \{x_i\}_{i=1}^n$ and $Y = \{y_i\}_{i=1}^n$, the number of rounds K, the number of iterations M at each round, a small radius δ.
2. Train Neural network and Decision tree functions $f^n(x)$ and $f^t(x)$ using D^{train}.
3. Initialize a set for BO $D_0^{BO} = \{X, \Delta Y\}$ with $\Delta Y = \{\Delta y_i\}_{i=1}^n$, where $\Delta y_i = f^n(x_i) - f^t(x_i)$.
4. **For** $k = 1 : K$ **Do**
5.   Build up a GP using D^{BO}
6.   **For** $m = 1 : M$ **Do**
7.   Recommend x_m using EI acquisition function and compute $\Delta y_m = f^n(x_m) - f^t(x_m)$
8.   Update the GP based on the data $D^{BO} \bigcup \{x_i, \Delta y_i\}_{i=1}^m$
9.   **end For**
10.   Obtain the point x^v corresponding to the maximum function difference $x^v = \operatorname{argmax}_{x_m \in x_{1:M}} |f^n(x_m) - f^t(x_m)|$
11.   Obtain a new data set $X_{new} = \{x^v \cup x_{1:J}\}$, where $x_{1:J} \sim \mathcal{U}((1-\delta)x^v, (1+\delta)x^v)$ is uniformly sampled.
12.   Use Neural network to label $Y_{new}^n = f^n(X_{new})$
13.   Augment the training data with the new data $D^{train} = D^{train} \bigcup \{X_{new}, Y_{new}^n\}$
14.   Update f^t via retraining decision tree on D^{train} and recompute ΔY for all points in $X = X \bigcup X_{new}$
15.   Reset the data for BO $D_0^{BO} = \{X, \Delta Y\}$
16. **end For**

---

3 Experiments

In this section we show the results obtained by our framework on different datasets. The black-box model we have chosen for our experiments is a simple neural network and the surrogate used to explain it, is a decision tree. We first show the results on synthetic dataset. We demonstrate how the decision boundary of the decision tree changes after it has been trained by neural network. We have also used our framework on regression and classification data. In both these data, our framework helps in the improvement of accuracy of the prediction.

Experiments with Synthetic Dataset

The synthetic data was generated using the function $f(x_1, x_2) = x_1^2 - x_2^2$. If the function value was greater than zero, then the data point was assigned to class 1, else to class 2. The decision space between -15 and 15 was considered for the experiments. The neural network used for the synthetic data was a simple neural network with 10 input nodes, 10 nodes in the hidden layer and two nodes in the output layer. The decision tree used has a maximum depth of 6. Initially a set of 250 data points was generated. The initial points were randomly split into 75% training data and 25% testing data.

The decision tree was iteratively trained for 100 rounds. At each round a set of 60 data points was appended to the original training set. These 60 datapoints were sampled around the point, where the difference between the decision made by neural network and decision tree was found to be maximum. The decision of both models are represented by the class probability associated to datapoints. The difference is calculated as the difference in the class probability predicted by neural network and decision tree. This difference in the function is denoted as Δy. Bayesian optimization was used to find the point of maximum difference. New data points are sampled around this point within a window of $\delta = 0.01$. In Fig. 3 the decision boundary of the neural network and the decision boundary of the surrogate before and after the training are shown. After the training the decision boundary of the decision tree resembles more of that of the neural network.

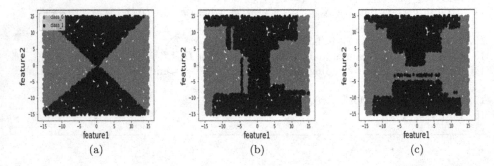

(a)   (b)   (c)

Fig. 3. Decision space. (a) shows the decision boundary of neural network, (b) shows the initial decision boundary of decision tree and the blue point marks the point at which the decision of the two models differs maximum. (c) shows the decision boundary of decision tree after iterative training of 100 rounds. (Color figure online)

As the boundaries are more similar, the predictions of the surrogate is more similar to that of the black-box model. This is further indicated by the improvement in the accuracy of the decision tree and also by the decrease in the difference of the decision probabilities of the black-box model and the surrogate. This is illustrated in Fig. 4.

Experiments with Bike Sharing Dataset

It is a regression dataset from UCI repository. It contains the hourly and daily count of rental bikes with the corresponding weather and seasonal information. We use only the hourly count of rental bikes as the dependent variable. We have selected the most related 5 dependent variables out of 14 based on the correlation between input and output variables. We later use these 5 selected variables ('season', 'hour', 'holiday', 'actual temperature' and 'apparent temperature') to predict the daily count of rental bikes. We randomly split the total 17389 data points into 75% training data and 25% test data.

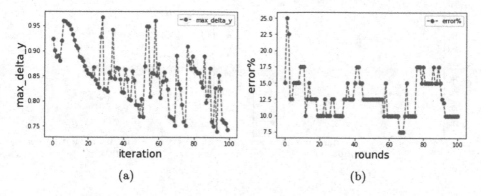

Fig. 4. Performance of the decision tree at each round of training. (a) shows the maximum difference between the functions of 2 models at each round and (b) shows the percentage error of decision tree at each round.

We train a fully-connected neural network with 1 hidden layer consisting of 30 neurons. We also train a decision tree, for which the appropriate depth is estimated via a validation set as 8. We run our method for 20 rounds. For each round, we run Bayesian optimization with 60 iterations (approximately 12 iterations per feature) to recommend a new point. And we sample 30 new data around the recommended one (within a radius of $\delta = 0.01$). Along with the original training data, we retrain the decision tree with these 30 samples. We show how the maximum difference (Δy) between the functions of neural network and decision tree changes at each round in Fig. 5(a). We can see it is decreasing with the number of rounds, that is to say, the function trained by decision tree would be close to that of neural network. The improvement in the prediction of the decision tree is illustrated by the decreasing trend in the RMSE at each iteration as shown in Fig. 5(b). The initial RMSE of test data in decision tree is 96.3 and becomes 95.7 after 20 rounds. We also extracted rules from the decision tree we have obtained finally. We show an example of a subtree and rules extracted from it in Fig. 6 and Table 1 respectively.

Table 1. Rules extracted from the subtree in Fig. 6.

Rule 1 (Blue)	If between 1 am and 5 am, **then** number of bikes rented is 15.92
Rule 2 (Red)	If between 12 midnight and 1 am, **and if** spring or summer **then** number of bikes rented is 38.34, **if** fall **or** winter then 22.24
Rule 3 (Green)	If after 5 am and temperature less than 13 Celsius **then** number of bikes rented is 13.68

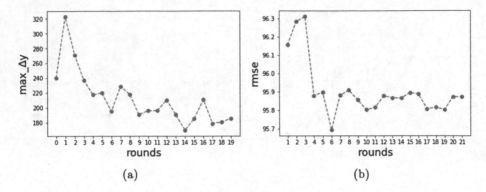

(a)   (b)

Fig. 5. Performance of the decision tree at each round of training. (a) shows the maximum difference between the functions of 2 models at each round and (b) shows the RMSE of decision tree at each round.

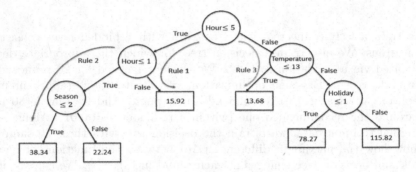

Fig. 6. The examples of the extracted rules from a subtree learnt from bike sharing data. We explain the rules in Table 1. (Color figure online)

Experiments with IRIS Dataset

This data has been used for our classification task, where there are 4 attributes ('sepal length', 'sepal width', 'petal length', 'petal width') and 3 classes of iris flower. We use the same percentage for training data and test data as the bike sharing data. The neural network here is a 1 hidden layer with 10 neurons. We set the depth of decision tree as 3. We run 500 rounds and 150 Bayesian optimization iterations at each round. Since only the class label is available in the classification problem, we use the difference of the class probability from neural network and decision tree as the objective value. Other settings are similar to the bike sharing dataset.

We run our method for 500 rounds. For each round, we run Bayesian optimization with 150 iterations to recommend a new point. As per the suggestion of BO, we again sample 50 new data points around the recommended one (within a radius of $\delta = 0.001$). Along with the original training data, we retrain the decision tree with these additional 50 points. We show how the maximum difference

(Δy) between the functions of neural network and decision tree changes at each round in Fig. 7(a). We can see that after a few round of training, the maximum difference between the decision of the neural network and the decision tree has reduced to a low value. In Fig. 7(b), we show the decrease in the error percentage of the decision tree. It demonstrates that the iterative training of the decision tree reduces the difference in the predictions about a data point. Also, it implies that the decision made by the surrogate are more similar to the neural network model.

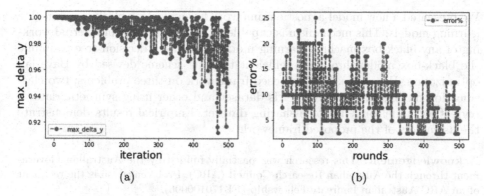

(a)   (b)

Fig. 7. Performance of the decision tree at each round of training. (a) shows the maximum difference between the functions of 2 models at each round and (b) shows percentage error of decision tree at each round.

Figure 8 and Table 2 show some rules that have been learned using our decision tree for iris data.

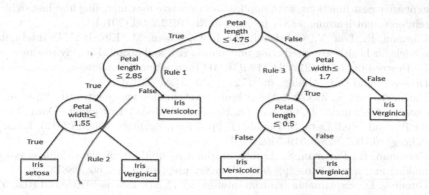

Fig. 8. The examples of the extracted rules from a subtree learnt from iris dataset. We explain the rules in Table 2. (Color figure online)

Table 2. Rules extracted from the subtree in Fig. 8.

Rule 1 (Blue)	**If** petal length is between 4.75 and 2.85, **then** Iris Verisicolor
Rule 2 (Red)	**If** petal length less than 2.85 and width less than 1.55 **then** Iris setosa
Rule 2 (Green)	**If** petal length greater than 4.75 and petal width greater than 1.7 **then** Iris Verginica

4   Conclusion

We presented a new model-agnostic framework for explaining black-box machine learning models. This mechanism is capable of extracting rules of internal working of any black-box machine learning models. We use a decision tree to mimic the black-box model through an efficient training scheme devised by Bayesian optimization. We applied our proposed framework on three problems: two classification model, one built using iris dataset and other using synthetic data, a regression model built for bike sharing dataset. Empirical results demonstrate the usefulness of the proposed framework.

Acknowledgement. This research was partially funded by the Australian Government through the Australian Research Council (ARC). Prof Venkatesh is the recipient of an ARC Australian Laureate Fellowship (FL170100006).

References

1. Ballard, D.I., Naik, A.S.: Algorithms, artificial intelligence, and joint conduct. Antitrust Chronicle **2**, 29 (2017)
2. Bishop, C.: Pattern Recognition and Machine Learning, vol. 16, pp. 461–517. Springer, New York (2006)
3. Brochu, E., Cora, V.M., De Freitas, N.: A tutorial on Bayesian optimization of expensive cost functions, with application to active user modeling and hierarchical reinforcement learning. arXiv preprint arXiv:1012.2599 (2010)
4. Caruana, R., Lou, Y., Gehrke, J., Koch, P., Sturm, M., Elhadad, N.: Intelligible models for healthcare: predicting pneumonia risk and hospital 30-day readmission. In: Proceedings of the 21st ACM SIGKDD International Conference on Knowledge Discovery and Data Mining, pp. 1721–1730. ACM (2015)
5. Datta, A., Sen, S., Zick, Y.: Algorithmic transparency via quantitative input influence. In: Cerquitelli, T., Quercia, D., Pasquale, F. (eds.) Transparent Data Mining for Big and Small Data. SBD, vol. 11, pp. 71–94. Springer, Cham (2017). https://doi.org/10.1007/978-3-319-54024-5_4
6. Goodman, B., Flaxman, S.: European union regulations on algorithmic decision-making and a "right to explanation". arXiv preprint arXiv:1606.08813 (2016)
7. Gunning, D.: Explainable artificial intelligence (XAI). Defense Advanced Research Projects Agency (DARPA), nd Web (2017)
8. Hendricks, L.A., Akata, Z., Rohrbach, M., Donahue, J., Schiele, B., Darrell, T.: Generating visual explanations. In: Leibe, B., Matas, J., Sebe, N., Welling, M. (eds.) ECCV 2016. LNCS, vol. 9908, pp. 3–19. Springer, Cham (2016). https://doi.org/10.1007/978-3-319-46493-0_1

9. Karpathy, A., Fei-Fei, L.: Deep visual-semantic alignments for generating image descriptions, pp. 3128–3137 (2015)
10. Li, C., Gupta, S., Rana, S., Nguyen, V., Venkatesh, S., Shilton, A.: High dimensional Bayesian optimization using dropout. arXiv preprint arXiv:1802.05400 (2018)
11. Lipton, Z.C.: The mythos of model interpretability. arXiv preprint arXiv:1606.03490 (2016)
12. Rana, S., Li, C., Gupta, S., Nguyen, V., Venkatesh, S.: High dimensional Bayesian optimization with elastic Gaussian process. In: Proceedings of the 34th International Conference on Machine Learning, vol. 70, pp. 2883–2891. JMLR.org (2017)
13. Ribeiro, M.T., Singh, S., Guestrin, C.: Why should i trust you?: explaining the predictions of any classifier, pp. 1135–1144 (2016)
14. Williams, C.K., Rasmussen, C.E.: Gaussian Processes for Machine Learning, vol. 2, no. 3, p. 4. The MIT Press, Cambridge (2006)

Classifier Learning from Imbalanced Corpus by Autoencoded Over-Sampling

Eunkyung Park[1], Raymond K. Wong[1(✉)], and Victor W. Chu[2]

[1] University of New South Wales, Sydney, Australia
wong@cse.unsw.edu.au
[2] Nanyang Technological University, Singapore, Singapore

Abstract. Class imbalance is a common problem in classifier learning but it is difficult to solve. Textual data are ubiquitous and their analytics have great potential in many applications. In this paper, we propose a solution to build accurate sentiment classifiers from imbalanced textual data. We first establish topic vectors to capture local and global patterns from a corpus. Synthetic minority over-sampling technique is then used to balance the data while avoiding overfitting. However, we found that residue overfitting is still prominent. To address this problem, we propose an autoencoded oversampling framework to reconstruct balanced datasets. Our extensive experiments on different datasets with various imbalanced ratios and number of classes have found that our approach is sound and effective.

Keywords: Imbalanced learning · Sentiment analysis · Over-sampling · Autoencoding

1 Introduction

Class imbalance problem happens when there are more instances of some classes than the others in a dataset and it often occurs in classification tasks. In such settings, traditional machine learning classifiers are likely to be mainly trained by the large classes than the small ones. This problem is very prevalent in many real-world applications [10]. Among various domains, we would like to focus on textual data. Our goal is to build a unified classifier to predict the sentiment of unseen documents when the class labels in the training data from each class exhibit different distributions.

Traditional supervised learning models are designed to optimize overall accuracy without considering different distributions of different classes. As a result, it is easy to come up with models that overfit observations of majority classes as well as underfit minority classes [10,18]. Researchers have tried many approaches to address this problem, such as algorithmic modification [18,28], cost-sensitive learning [27], data sampling [5], etc. The algorithms modification method modifies learning process to deal with imbalanced data cases [18]. The cost-sensitive learning method applies a penalty cost when training major classes so as to learn

© Springer Nature Switzerland AG 2019
A. C. Nayak and A. Sharma (Eds.): PRICAI 2019, LNAI 11670, pp. 16–29, 2019.
https://doi.org/10.1007/978-3-030-29908-8_2

minority classes properly [27]. Data sampling alleviates imbalance ratio among classes to make fully balanced training dataset [18].

Among those many approaches, data sampling has been widely studied. Since under-sampling can lose information by dropping data of majority class as many as the number of minority class, we do not consider it as a solution. Alternately, over-sampling has been regarded to be appropriate for imbalanced data. However, with Bag-of-Words (BOW) representation, over-sampling does not work properly because BOW representation suffers from high-dimensionality, sparsity and small disjunct problems [10,27,30]. Classification of sparse data is hard [1] and small disjunct problem is likely to create overfitting problem by generating more decision regions [3].

To tackle the above issues, we use topic vector for sentiment classification on class imbalanced textual data. Topic embedding is a hybrid method to construct feature vectors for plain texts by topic modeling and word embeddings. Topic modeling maps documents onto low-dimensional topic space by utilizing global word collocation patterns in a corpus whereas embedding space by exploiting local word collocation patterns in a small context window [15]. Therefore, they can be regarded as complementary. By combining them together, we can obtain more elaborate models for imbalanced textual data. Word embedding approach is better than tradition BOW representation in higher quality of textual data.

Next, we apply SMOTE to the training data. A lot of variants of SMOTE have been published but most of them mainly focus on numerical data and recently. With SMOTE, we can achieve balanced training datasets. However, we found that residue overfitting is still prominent. To address this problem, we propose autoencoded oversampling to reconstruct balanced datasets. Autoencoder is a shallow neural network to reconstruct input at the output layer. Recently, many variants of autoencoder have been published; however, most of them focus on image data [6]. Zhai and Zhang [31] pointed that difficulties of autoencoder for texts came from their complex properties, e.g., high-dimensionality, sparsity and power-law word distribution. These properties are found to be resurfacing after embedding and sampling in empirical studies. So far autoencoder has been proven to be effective in extracting meaningful features from input data. From our experiments, we found that it can also be applied to imbalanced textual data in topic-vector representation. By autoencoded over-sampling, we have successfully reconstructed more meaningful feature vectors and obtained better results in our experiments.

The contributions of this paper are summarized as follows:

- We first apply topic embedding to provide a better representation of our data that combines topic modeling with word embedding
- Next, we address the class imbalanced problem by proposing a novel autoencoded over-sampling framework
- Our findings can potentially be reapplied to other imbalanced training problems

2 Related Work

In this section, we summarize related work which are related to our methodology for sentiment classification on imbalanced textual data.

Topic models and word embeddings are mainstream approaches for text representation. Latent Dirichlet Allocation (LDA) [19] is a well-known topic model to represent documents by distributions of topics and their words. On the other hand, word embedding is to estimate parameters for a word and its context words to learn semantic regularities between the words [16]. While Word2vec uses softmax function as a link function, there exists other extended models which use various link functions [7,14,16,23]. Recently, there have been some research combining topic models and word embeddings. Larochelle and Lauly assigned each word with a unique topic vector summarizing the context of the current word [12]. Paragraph vector assumes that each text has a latent paragraph vector influencing the distributions of all words in this text [13]. Topical word embedding is a hybrid approach of word embedding and LDA by concatenating word embedding and the topical word embedding of a word [17]. Li et al. proposed a link function to combine LDA and word embeddings [15].

On the other hand, data sampling is a common approach to address data imbalance problem that includes under-sampling, over-sampling and other hybrid methods that use both together [18]. Under-sampling reduces the number of examples of majority class as same as the number of the minority to create fully balanced dataset [4]. This method has not been used for opinion analysis and emotion classification because it can lose information by deleting training data [30]. A simple and easy over-sampling method is to duplicate examples of minority randomly to make the number of samples similar to the number of majority class. However, this can cause overfitting. Chawla et al. [4] proposed a synthetic over-sampling method, namely SMOTE, which generates additional samples based on their nearest neighbors. There are many variants of SMOTE. Borderline-SMOTE generates new samples only among the borderline samples of minority class [8]. On the other hand, Safe-Level-SMOTE only generates new samples for the central instances of a minority class [2]. DBSMOTE [3] is a density-based approach by combining DBSCAN clustering [24] and SMOTE. Adaptive synthetic sampling approach for imbalanced learning (ADASYN) [9] uses a weighted distribution for different minority class examples according to their learning difficulty. There have been many variants of SMOTE published in recent time, but most of the works and experiments have been done mainly for numerical data rather than textual data.

Lately, many successes in data reconstruction by autoencoder have been published. The denoising autoencoder (DAE) [29] inputs a corrupted version of data and the contractive autoencoder (CAE) [25] adopts the Frobenius norm of the Jacobian matrix of the encoder activations into the regularization term. The variational autoencoder (VAE) [22] is a generative model by adopting variational inference and the k-sparse autoencoder (KSAE) [20] explicitly enforces sparsity by only keeping the k highest activities in the feedforward phase. So far, autoencoder has been developed for image data and hardly studied on textual data due

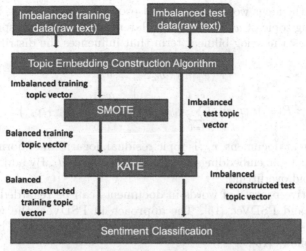

Fig. 1. Our proposed autoencoded over-sampling framework. SMOTE is used to make balanced training dataset and KATE aims to reconstruct topic vector.

to its compounding properties. Zhai and Zhang [31] proposed a semi-supervised autoencoder to overcome some of those difficulties by a weighted loss function. Kumar and D'Haro [11] introduced a new approach adding sparsity and selectivity penalty terms to decrease sparsity because they found that all the topics extracted from the autoencoder were dominated by the most frequent words due to sparsity. Chen and Zaki [6] presented an effective autoencoder for text by adding competition between the neurons in hidden layer, namely k-competitive autoencoder for text (KATE). By doing so, each neuron becomes specialized in recognizing specific data patterns and finally the model can learn meaningful representations of textual data.

3 Our Proposed Autoencoded Over-Sampling

Figure 1 is a sketch of our proposed autoencoded over-sampling solution. First, we construct Topic Vector from raw text corpora. Secondly, we apply SMOTE to make balanced training datasets. We can expect it will improve our model's performance by making distribution of classes equal. Thirdly, we apply KATE to address residue overfitting by reconstructing feature vectors.

3.1 Topic Vector for Document Representation

First, we adopt topic embedding which is a combination of topic models (i.e., LDA) and word embeddings, (i.e., PSDVec).

The main algorithm of most word embedding methods is a link function that connects the embeddings of a focus word with its context words. To define the distribution of the focus word, Li et al. [15] proposed the following link function by incorporating topic of w_c in a way like latent word, and corresponding topic embedding t_{ik} as a new log-bilinear term that influences the distribution of w_c. That is

$$P(w_c|w_0 : w_{c-1}, z_c, d_i)$$

$$\approx P(w_c)exp\{v_{w_c}^T(\sum_{l=0}^{c-1} v_{wl} + t_{z_c}) + \sum_{l=0}^{c-1} a_{w_l w_c} + r_{z_c}\},$$

where d_i is current document, r_{z_c} is topic residual, logarithm of normalizing constant. Note that topic embeddings t_{z_c} may be specific to d_i. By adding constrain r_{z_c}, we can avoid overfitting.

The generative process of words in documents can be regarded as a hybrid between LDA and PSDVec [15]. The approach in PSDVec, the word embedding v_{s_i} and residual $a_{s_i s_j}$ are drawn from respective Gaussians. The generative process is as follows [15]:

1. For the k-th topic, draw a topic embedding uniformly from a hyperball of radius γ, i.e. $t_k \sim \text{Unif}(B_\gamma)$;
2. For each document d_i :
   (a) Draw the mixing proportions ϕ_i from the Dirichlet prior $\text{Dir}(\alpha)$;
   (b) For the j-th word:
      i. Draw topic assignment z_{ij} from the categorical distribution $\text{Cat}(\phi_i)$
      ii. Draw word w_{ij} from S according to $P(w_{ij}|w_{i,j-c} : w_{i,j-1}, z_{ij}, d_i)$

3.2   SMOTE for Topic Vector

Secondly, we apply SMOTE on topic vectors. Although most of variants of SMOTE have been proven on numerical data, WEC-MOTE [30] present success in text domain.

We extend WEC-MOTE by altering topic vectors as an input feature vectors. WEC-MOTE uses sentence vectors as inputs of SMOTE and it has been proven effective to improve the quality of the representation of imbalanced textual data. However, it has limitation that it can only handle up to 25 dimensions due to the complexity of recursive neural tensor network [26], which is used to construct sentence vectors. Also, as WEC-MOTE uses sentence vectors, it cannot convey the meaning of entire text like topic. Differently, our proposed model has less limitation on the dimension of data and also we can capture both local and global word patterns. It is because topic vector algorithm is base on both topic modeling and word embedding.

To make fully balanced training datasets, we generate additional examples of minority classes before classification phase as follows. For the test dataset, we do not apply SMOTE.

For a given minority class where each real sample S corresponds to a topic vector V_t, new synthetic samples V_{new} are generated by

$$V_{new} = V_t + R_{0-1} \times V_{t_n}$$

Table 1. Results for finding best vector normalization parameter using multiple-class dataset

	Precision	Recall	F1
Base (no KATE)	71.9	63.4	65.8
No scaling	70.3	64.1	65.8
$\epsilon = 1$	66.5	68.5	67.3
$\epsilon = 10^{-10}$	68.6	67.8	**68.2**
$V/max\ (V)$	56.9	49.2	47.9

where V_{t_n} is one of the k-nearest neighbors of V_t, R_{0-1} is a real number between zero and one. This algorithm can generate many V_{new} as needed [4].

3.3 K-competitive Autoencoder for Topic Vector

Thirdly, we apply autoencoder KATE to address the residue overfitting of SMOTE and also extract more meaningful patterns from texts.

KATE uses a log-normalized word count vector, but there are some differences between the data of KATE originally designed to handle and our topic vector. While count vector has non-negative integers and zero, topic embeddings include negative value and very small float numbers. Therefore, we enhance the model by introducing a new vector normalization step.

Let $x \in R^d$ d-dimensional input vector,

$$x_i = \frac{log(V_i + \epsilon)}{max_{i \in C} log(V_i + \epsilon)}, \begin{cases} \epsilon = 1, & \text{if } min_{i \in C} V_i > -1 \\ \epsilon = |min_{i \in C} V_i| + 10^{-10} & \text{otherwise,} \end{cases}$$

where V_i is topic vector, C is a set of the entire corpora. The epsilon is decided empirically from the results of below using multiple-class dataset in the experiments (Table 1).

Algorithm 1 represents the competitions among neurons to capture important patterns by adding constraints using mutual competition among neurons. In the hidden layer, each neuron has competition by followings the original KATE autoencoder [15]. Only competitive k neurons that have the largest absolute activation values are used.

3.4 Rationalization of Autoencoded Over-Sampling

Chawla et al. [4] proposed SMOTE algorithm for class-imbalanced classification, which is an elegant and effective over-sampling method. It assumes that there exist a virtual minority sample between two minority samples that are near to each other. Therefore, SMOTE generates new minority samples artificially between the two real minority samples that are near to each other. Compared to over-sampling, the SMOTE algorithm produces new minority samples that are

---

Algorithm 1. K-competitive Layer

1: **function** K-COMPETITIVE-LAYER(z)
2:     sort positive neurons in ascending order $z_1^+...z_p^+$
3:     sort negative neurons in descending order $z_1^-...z_n^-$
4:     **if** $P - \lceil k/2 \rceil > 0$ **then**
5:         $E_{pos} = \sum_{i=1}^{P-\lceil k/2 \rceil} z_i^+$
6:         **for** $i = P - \lceil k/2 \rceil + 1, ..., P$ **do**
7:             $z_i^+ := z_i^+ + \alpha \cdot E_{pos}$
8:         **end for**
9:         **for** $i = 1, ..., P - \lceil k/2 \rceil$ **do**
10:             $z_i^+ := 0$
11:         **end for**
12:     **end if**
13:     **if** $N - \lfloor k/2 \rfloor > 0$ **then**
14:         $E_{neg} := \sum_{i=1}^{N-\lfloor k/2 \rfloor} z_i^-$
15:         **for** $i = N - \lfloor k/2 \rfloor + 1, ..., N$ **do**
16:             $z_i^- := z_i^- + \alpha \cdot E_{neg}$
17:         **end for**
18:         **for** $i = 1, ..., N - \lfloor k/2 \rfloor$ **do**
19:             $z_i^- := 0$
20:         **end for**
21:     **end if**
22:     **return** updated $z_1^+...z_P^+, z_1^-...z_N^-$
23: **end function**

---

different from each other and does not copy the original minority samples at all. This can effectively avoid the overfitting problem caused the over-sampling. At the same time, it is not very complex. However, we found that residue overfitting is still prominent in our dataset. In our experiments, we use python imbalanced-learn library. We set type of SMOTE algorithm as 'svm' because it outperformed other options in our sample tests conducted prior to our main experiments. We varies the number of nearest neighbours to construct synthetic samples from 5 to 120 to find its best match.

We adopt KATE for imbalanced text training because KATE's competitive hidden layer can better capture topic vector of imbalanced text. KATE has been proven to be better representing textual data than other text representation models, including various types of autoencoders such as denosing, constractive, variational, and k-sparse autoencoders. Also, KATE outperforms deep generative models, probabilistic topic models and even word representation models such as Word2Vec [6]. In their experiments, KATE outperforms not only on balanced but also on imbalanced textual data. In our experiments, we set the number of topics (hidden neurons) same as the number of features of topic vector, and varies the number of winners k from 1 to 300 to find its best match to the given number of topics. We set the energy amplification parameter α as default value (i.e., 6.26 in the original paper [6]).

Table 2. Twitter sentiment corpus created by Sanders Analytics - positive, negative, neutral and irrelevant

Exp1	# of classes : 4			
	Positive	Negative	Neutral	Irrelevant
Train (80%)	415	458	1866	1351
Test (20%)	104	114	467	338
Ratio (%)	10.15	11.19	45.63	33.03

In our proposed autoencoded over-sampling framework, we apply SMOTE first to make a balanced training set, and then use KATE to avoid overfitting. Note that Chen and Zaki suggested that KATE can be regarded as a regularized autoencoder in a sense that a positive adder and negative adder constrain the energy [6]. KATE selects the k highest activation neurons as winners in the hidden layer, and reallocates the amplified energy (aggregate activation potential) from the losers [6]. By doing so, recognition of important patterns from texts is made possible.

4 Empirical Evaluation

In this section, we evaluate our proposed model on various imbalanced ratio data with multiple- and binary-class datasets. To examine the effect of our hybrid model, we compare our proposed model with the baseline (i.e., Topic Vector) and other combinations of model components.

4.1 Datasets

Multiple-Class Dataset. The twitter sentiment corpus created by Sanders Analytics has 5,113 hand-classified tweets. Each tweet has a class label (Positive, Negative, Neutral and Irrelevant). Table 2 provides statistics of a multiple-class dataset. We split the entire data into a training (80%) set and a test (20%) set. Note that the original dataset is available from https://github.com/zfz/twitter_corpus.

Binary-Class Dataset. Large Movie Review Data is intended for binary (positive and negative) sentiment classification task. It contains 50,000 reviews splits evenly 25,000 train and 25,000 test sets. The entire distribution of classes is completely balanced (25,000 positive and 25,000 negative). However, we intentionally adjust the number of examples to make imbalanced experiment settings (19:1, 9:1 and 4:1). Table 3 shows the statistic of original dataset and Table 4 shows information of manipulated dataset for imbalanced settings. Note that the original dataset is available from http://andrew-maas.net/data/sentiment.

Table 3. Original data of large movie review data

Class	Positive	Negative	Total
Train	12,500	12,500	25,000
Test	12,500	12,500	25,000

4.2 Experiment Setting

Construction Topic Vector. Li et al. presented two best models using their topic vector algorithm in their paper [15].

- TopicVec: the topic proportions learned by TopicVec
- TV+WV: the topic proportions concatenated with the mean word embedding of the document

TV+WV shows better results for both multiple- and binary-data classifications (with 10 topics for each category) in our experiments as shown in Tables 5 and 6. Therefore, we develop a model base on this representation. To increase readability, we will notate it as just 'Topic Vector' in the rest of this paper. l-1 regularized linear SVM one-vs-all is used for sentiment document classification. The learning process is implemented by Python scikit-learn library.

When data is highly imbalanced, traditional empirical measures including accuracy are no longer appropriate [18]. It is because correctly classifying all examples corresponding to the majority class will achieve a high accuracy rate despite misclassifying minority classes. Considering our datasets are imbalanced, we use macro-averaged precision, recall and F1 (harmonic mean of precision and recall) as our evaluation metrics. They are fundamentally designed for imbalanced data whereas normal accuracy does not consider imbalanced data setting.

To avoid the average results being dominated by the performance of majority class, we adopt macro-averaged evaluation metrics. Macro-averaged method gives equal weight to each class, whereas micro-averaged method gives equal weight to each per-document classification decision. Because the F1 ignores true negatives and its magnitude is mostly determined by the number of true positives, large classes dominate small classes in micro-averaged. Therefore, micro-averaged results are indeed an accurate measure of effectiveness on the large classes in a test collection, and macro-averaged results are effective for small classes [21].

4.3 Experiment Results

Tables 7 and 8 summarize our experiment results. Our proposed model Topic Vector+SMOTE+KATE shows best F1 scores over the multiple- and binary-class datasets with diverse imbalance ratios. Throughout our experiments, both SMOTE and KATE work well for imbalanced textual data. However, we found that SMOTE+KATE can create a synergy effect as KATE alleviates

Table 4. Manipulated data of large movie review data

Exp2	positive:negative = 19:1		
	Positive	Negative	Total
Train	11,875	625	12,500
Test	11,875	625	12,500
Exp3	positive:negative = 9:1		
	Positive	Negative	Total
Train	11,250	1,250	12,500
Test	11,250	1,250	12,500
Exp4	positive:negative = 4:1		
	Positive	Negative	Total
Train	10,000	2,500	12,500
Test	10,000	2,500	12,500

Table 5. Comparing TopicVec and TV+WV with multiple-class dataset

	Precision	Recall	F1
TopicVec	67.9	52.3	54.2
TV+WV	71.9	63.4	**65.8**

Table 6. Comparing TopicVec and TV+WV with binary-class dataset

	Precision	Recall	F1
TopicVec	80.2	57.8	60.8
TV+WV	85.1	79.3	**81.7**

overfitting of SMOTE. For example, in Table 7, SMOTE or KATE alone shows an improvement in F1 and SMOTE+KATE shows the highest F1 value. However, we can observe that using only SMOTE can result in very high F1 in training dataset but with poor result in test data. It is a clear signal of overfitting. Comparing only KATE and KATE+SMOTE, the training result of KATE+SMOTE is much higher than only KATE. However, the test result shows opposite result. Therefore, we can conclude that SMOTE can cause overfitting. In addition, by comparing SMOTE and SMOTE+KATE, we found that KATE is well suited in alleviating the residue overfitting from SMOTE. While only SMOTE shows overfitting, SMOTE+KATE shows better test result.

We can confirm similar trends in the result of the binary-class dataset. When applying only SMOTE, F1 of training data is always higher than 90%. However, applying KATE after SMOTE results less F1 in train data, but shows better F1 in test data. This means that employing KATE after SMOTE can correct overfitting of SMOTE and significantly increase the performance on the test

Table 7. Results from multiple-class dataset

Exp1	Training set			Test set		
	Precision	Recall	F1	Precision	Recall	F1
Topic Vector	86.7	78.2	81.4	71.9	63.4	65.8
Topic Vector+SMOTE	92.1	92.1	92.1	67.4	68.3	67.7
Topic Vector+KATE	81.1	59.4	61.8	68.6	67.8	68.2
Topic Vector+KATE+SMOTE	85.1	85.0	85.0	67.5	68.2	65.1
Topic Vector+SMOTE+KATE	84.8	84.7	84.7	67.8	71.6	**68.9**

Table 8. Results from binary-class dataset

Exp2	positive:negative = 19:1					
	Training set			Test set		
	Precision	Recall	F1	Precision	Recall	F1
Topic Vector	90.1	62.9	68.9	85.0	60.1	64.9
Topic Vector+SMOTE	92.2	92.1	92.2	65.9	76.6	69.5
Topic Vector+KATE	91.0	58.0	62.5	78.2	61.7	66.1
Topic Vector+KATE+SMOTE	94.3	94.3	94.3	62.1	79.1	65.7
Topic Vector+SMOTE+KATE	90.3	88.6	89.4	70.3	73.6	**71.8**
Exp3	positive:negative = 9:1					
	Training set			Test set		
	Precision	Recall	F1	Precision	Recall	F1
Topic Vector	89.3	74.2	79.5	85.3	70.2	75.1
Topic Vector+SMOTE	93.8	93.8	93.8	73.7	80.6	76.5
Topic Vector+KATE	92.5	67.1	72.8	77.9	75.0	76.3
Topic Vector+KATE+SMOTE	92.5	92.5	92.5	62.8	80.5	63.4
Topic Vector+SMOTE+KATE	92.5	92.4	92.4	77.0	78.0	**77.5**
Exp4	positive:negative = 4:1					
	Training set			Test set		
	Precision	Recall	F1	Precision	Recall	F1
Topic Vector	88.5	82.2	84.8	85.1	79.3	81.7
Topic Vector+SMOTE	91.2	91.2	91.2	79.5	84	81.4
Topic Vector+KATE	86.7	77.3	80.7	81.8	82	81.9
Topic Vector+KATE+SMOTE	89.5	89.5	89.5	77.6	83.5	79.8
Topic Vector+SMOTE+KATE	89.7	89.7	89.7	82.1	81.7	**81.9**

data (8.8% to 36.3%). On the other hand, ironically applying KATE and then SMOTE later shows worst results from all cases. F1 of test data is always worse than using one of them separately. In other words, the results show that using only SMOTE and KATE each works a little for class imbalance tasks. However, when using them together it can reach best results. In addition, the order in applying SMOTE and KATE is indeed very important. Our experiment results indicate that KATE first and SMOTE later is the worst combination because SMOTE turns out maximizing overfitting.

5 Conclusion

Class imbalance is a common problem in classifier learning. With the potential in many applications, it is inevitable to exploit class imbalance problem on textual data. In this paper, we have presented a solution to this tricky problem in sentiment analysis. We first establish topic vectors to capture local and global patterns from a corpus. SMOTE is then used to balance the data while avoiding overfitting. However, the residue overfitting is still prominent. To address this problem, we have propose an autoencoded oversampling framework to reconstruct balanced datasets, which is found to be effective from our extensive experiments. Our future work is to investigate how to adapt our proposed approach to more extremely imbalanced data, such as imbalance ratios of 100:1, 1000:1 or 10000:1. Moreover, we also interested in applying our proposed approach to address data imbalance problem in recommender systems and ranking systems.

References

1. Bengio, Y., Ducharme, R., Vincent, P., Janvin, C.: A neural probabilistic language model. J. Mach. Learn. Res. (JMLR) **3**, 1137–1155 (2003)
2. Bunkhumpornpat, C., Sinapiromsaran, K., Lursinsap, C.: Safe-level-SMOTE: safe-level-synthetic minority over-sampling technique for handling the class imbalanced problem. In: Theeramunkong, T., Kijsirikul, B., Cercone, N., Ho, T.-B. (eds.) PAKDD 2009. LNCS (LNAI), vol. 5476, pp. 475–482. Springer, Heidelberg (2009). https://doi.org/10.1007/978-3-642-01307-2_43
3. Bunkhumpornpat, C., Sinapiromsaran, K., Lursinsap, C.: DBSMOTE: density-based synthetic minority over-sampling technique. Appl. Intell. **36**, 664–684 (2011)
4. Chawla, N., Bowyer, K., Hall, L.O., Kegelmeyer, W.P.: SMOTE: synthetic minority over-sampling technique. J. Artif. Intell. Res. (JAIR) **16**, 321–357 (2002)
5. Chawla, N.V., Japkowicz, N., Kotcz, A.: Editorial: special issue on learning from imbalanced data sets. ACM SIGKDD Explor. Newsl. **6**(1), 1–6 (2004)
6. Chen, Y., Zaki, M.J.: Kate: K-competitive autoencoder for text. In: Proceedings of the 23rd ACM SIGKDD International Conference on Knowledge Discovery and Data Mining (KDD 2017), pp. 85–94 (2017)
7. Faruqui, M., Tsvetkov, Y., Yogatama, D., Dyer, C., Smith, N.A.: Sparse overcomplete word vector representations. In: Proceedings of the 53rd Annual Meeting of the Association for Computational Linguistics (ACL 2015) and the 7th International Joint Conference on Natural Language Processing (IJCNLP 2015), vol. 1, pp. 1491–1500 (2015)
8. Han, H., Wang, W.-Y., Mao, B.-H.: Borderline-SMOTE: a new over-sampling method in imbalanced data sets learning. In: Huang, D.-S., Zhang, X.-P., Huang, G.-B. (eds.) ICIC 2005. LNCS, vol. 3644, pp. 878–887. Springer, Heidelberg (2005). https://doi.org/10.1007/11538059_91

9. He, H., Bai, Y., A. Garcia, E., Li, S.: ADASYN: adaptive synthetic sampling approach for imbalanced learning. In: 2008 IEEE International Joint Conference on Neural Networks (IJCNN 2008), pp. 1322–1328 (2008)
10. Jo, T., Japkowicz, N.: Class imbalances versus small disjuncts. ACM SIGKDD Explor. Newsl. **6**(1), 40–49 (2004)
11. Kumar, G., D'Haro, L.F.: Deep autoencoder topic model for short texts. In: International Workshop on Embeddings and Semantics (2015)
12. Larochelle, H., Lauly, S.: A neural autoregressive topic model. In: Proceedings of the 26th Annual Conference on Advances in Neural Information Processing Systems (NIPS 2012), pp. 2708–2716 (2012)
13. Le, Q., Mikolov, T.: Distributed representations of sentences and documents. In: Proceedings of the 31st International Conference on International Conference on Machine Learning (ICML 2014) (2014)
14. Levy, O., Goldberg, Y.: Linguistic regularities in sparse and explicit word representations. In: Proceedings of the Eighteenth Conference on Computational Natural Language Learning (CoNLL 2014), pp. 171–180 (2014)
15. Li, S., Chua, T.S., Zhu, J., Miao, C.: Generative topic embedding: a continuous representation of documents. In: Proceedings of the 54th Annual Meeting of the Association for Computational Linguistics (ACL 2016), pp. 666–675 (2016)
16. Li, S., Zhu, J., Miao, C.: A generative word embedding model and its low rank positive semidefinite solution. In: Proceedings of the 2015 Conference on Empirical Methods in Natural Language Processing (EMNLP 2015), pp. 1599–1609 (2015)
17. Liu, Y., Liu, Z., Chua, T.S., Sun, M.: Topical word embeddings. In: Proceedings of the Twenty-Ninth AAAI Conference on Artificial Intelligence (AAAI 2015), pp. 2418–2424 (2015)
18. López, V., Fernández, A., García, S., Palade, V., Herrera, F.: An insight into classification with imbalanced data: empirical results and current trends on using data intrinsic characteristics. Inf. Sci. **250**, 113–141 (2013)
19. Blei, D.M., Ng, A.Y., Jordan, M.: Latent dirichlet allocation. J. Mach. Learn. Res. (JMLR) **3**, 993–1022 (2003)
20. Makhzani, A., Frey, B.: k-sparse autoencoders. In: Proceedings of the International Conference on Learning Representations (ICLR 2014) (2014)
21. Manning, C.D., Raghavan, P., Schütze, H.: Introduction to Information Retrieval. Cambridge University Press, New York (2008)
22. Kingma, D.P., Welling, M.: Auto-encoding variational bayes. In: Proceedings of the 2nd International Conference on Learning Representations (ICLR 2014) (2014)
23. Pennington, J., Socher, R., Manning, C.: Glove: global vectors for word representation. In: Proceedings of the 2014 Conference on Empirical Methods in Natural Language Processing (EMNLP 2014), pp. 1532–1543 (2014)
24. Ram, A., Sunita, J., Jalal, A., Manoj, K.: A density based algorithm for discovering density varied clusters in large spatial databases. Int. J. Comput. Appl. **3**, 1–4 (2010)
25. Rifai, S., Vincent, P., Muller, X., Glorot, X., Bengio, Y.: Contractive auto-encoders: explicit invariance during feature extraction. In: Proceedings of the 28th International Conference on International Conference on Machine Learning (ICML 2011), pp. 833–840 (2011)
26. Socher, R., et al.: Recursive deep models for semantic compositionality over a sentiment treebank. In: Proceedings of the 2013 Conference on Empirical Methods in Natural Language Processing (EMNLP 2013), pp. 1631–1642 (2013)
27. Sun, Y., Kamel, M.S., Wong, A.K., Wang, Y.: Cost-sensitive boosting for classification of imbalanced data. Pattern Recogn. **40**, 3358–3378 (2007)

28. Tang, Y., Zhang, Y., Chawla, N.V., Krasser, S.: SVMs modeling for highly imbalanced classification. IEEE Trans. Syst. Man Cybern. Part B (Cybern.) **39**, 281–288 (2009)

29. Vincent, P., Larochelle, H., Lajoie, I., Bengio, Y., Manzagol, P.A.: Stacked denoising autoencoders: learning useful representations in a deep network with a local denoising criterion. J. Mach. Learn. Res. (JMLR) **11**, 3371–3408 (2010)

30. Xu, R., Chen, T., Xia, Y., Lu, Q., Liu, B., Wang, X.: Word embedding composition for data imbalances in sentiment and emotion classification. Cogn. Comput. **7**, 226–240 (2015)

31. Zhai, S., Zhang, Z.M.: Semisupervised autoencoder for sentiment analysis. In: Proceedings of the Thirtieth AAAI Conference on Artificial Intelligence (AAAI 2016), pp. 1394–1400 (2016)

Semisupervised Cross-Media Retrieval by Distance-Preserving Correlation Learning and Multi-modal Manifold Regularization

Ting Wang[1,2], Hong Zhang[1,3(✉)], Bo Li[1,3], and Xin Xu[1,3]

[1] College of Computer Science and Technology,
Wuhan University of Science and Technology, Wuhan 430065,
People's Republic of China
zhanghong_wust@163.com
[2] Library, Wuhan Business University, Wuhan 430056,
People's Republic of China
[3] Hubei Province Key Laboratory of Intelligent Information Processing
and Real-Time Industrial System, Wuhan 430065, People's Republic of China

Abstract. Due to the heterogeneous representation and incongruous distribution of cross-media data, like text, image, audio, video, and 3D model, how to capture the correlations of heterogeneous data for cross-media retrieval is a challenging problem. In order to handle with multiple media types, this paper proposes a novel distance-preserving correlation learning and multi-modal manifold regularization (DCLMM) approach to exploit the common representation of heterogeneous data. The method mines the distance-preserving correlation by minimizing (maximizing) the distances between media samples with positive (negative) semantic correlations, while most existing methods only focus on positive correlations of pairwise media types. DCLMM also utilizes an intrinsic multi-modal manifold to well describe the geometry distribution of both labeled and unlabeled heterogeneous cross-media data. Moreover, DCLMM incorporates the distance-preserving correlation and multi-modal manifold into a kernel based regularization framework to explore more rich complementary information from high dimensional space. Extensive experimental results on two widely-used cross-media datasets with up to five media types demonstrate the effectiveness of DCLMM for cross-media retrieval, compared with the state-of-the-art methods.

Keywords: Cross-media retrieval · Distance-preserving correlation · Multi-modal manifold regularization

1 Introduction

Cross-media retrieval is a new search paradigm desired by users in big data era. By using it, people can submit any media type at hand and get relevant results with various media types conveniently. However, there is a challenging issue that how to mine the semantic correlation between low-level features and high-level concepts. Researchers consider that heterogeneous media data with the same semantics have latent

© Springer Nature Switzerland AG 2019
A. C. Nayak and A. Sharma (Eds.): PRICAI 2019, LNAI 11670, pp. 30–42, 2019.
https://doi.org/10.1007/978-3-030-29908-8_3

correlations, it is possible to construct a common representation space where the similarities of different media types can be measured easily. Many approaches have been proposed to model the low-level feature correlation or semantic information [1–4]. However, these methods only deal with two media types.

Modeling multiple media types simultaneously can boost each other, so cross-media retrieval needs diversified media types. Recently, graph regularization methods are successfully introduced for naturally extending to multiple media types [5–8]. For example, joint representation learning (JRL) [5] applies graph regularization to jointly model the correlation and semantic information on XMedia dataset, which is the first dataset containing five media types. Joint graph regularized heterogeneous metric learning (JGRHML) [7] utilizes heterogeneous metric and graph regularization to learn a high-level semantic metric through label propagation. Whereas the methods only exploit positive semantic correlations, the negative correlations between media samples of different semantic categories are neglected. Accordingly, a semisupervised regularization and correlation learning (SSRCL) approach is proposed to exploit both semantic similarity and dissimilarity by graph regularization [8]. However, the performances of these methods are limited by linear projections, which make them powerless when low-level features of cross-media data are high nonlinear.

A positive semantic correlation reflects the concurrence information and a negative semantic correlation supplies the exclusive information. For example, a piece of text description about the "sport" category may have a strong positive correlation with the image of basketball player Jordan. But it also has a negative correlation with the image about a bird sitting on a branch. Therefore, we hold that negative correlation should also be excavated. Inspired by the main idea of metric learning [9] that a sample's good neighbors should lie closer than its bad neighbors, we learn cross-media features by minimizing (maximizing) the distances between media data with positive (negative) correlations. We name it distance-preserving correlation learning.

Moreover, we construct a multi-modal manifold to capture the underlying geometry distribution information of both intra-modality and inter-modality, motived by [10]. In this paper, the distance-preserving correlation learning and multi-modal manifold (DCLMM) are incorporated in a kernel based regularization method to exploit complicated cross-media features. Our main idea is to find a nonlinear projection function of each media type that projects heterogeneous data into a common space for cross-media similarity measure.

The distinct contributions of our approach are summarized as follows.

- DCLMM jointly models not only low-level feature correlations, but also the positive semantic and negative semantic information of all media types in a unified framework, which takes full use of the potential correlations among cross-media data. The retrieval accuracy can be boosted by optimizing them simultaneously.
- DCLMM proposes a multi-modal manifold to better exploit the geometry distribution of cross-media data, which integrates both labeled and unlabeled samples from all modalities for describing diversified relations. While traditional semisupervised methods only exploit the distributional information of single-media data.
- DCLMM utilizes a kernel based method to explore more rich complementary information from high dimensional ambient space, and two regularization terms are

incorporated. One imposes smoothness conditions on possible solutions in ambient space, the other along the intrinsic multi-modal manifold.

Experimental results on two widely-used datasets with up to five media types are reported to show the effectiveness and superiority of our approach. The rest of this paper is organized as follows. Section 2 introduces the overview of our framework. Section 3 describes the objective function and optimization solution of DCLMM. Then experimental results and analyses are shown in Sect. 4. Finally, Sect. 5 draws our conclusions.

2 Overview of Our Framework

We first present the formulated definition of cross-media retrieval problem. Let $\mathcal{D} = \left\{ \mathcal{D}^{(1)}, \ldots, \mathcal{D}^{(s)} \right\}$ be the labeled cross-media dataset, where s is the number of media types, and $\mathcal{D}^{(i)} = \left\{ x_p^{(i)}, y_p^{(i)} \right\}_{p=1}^{l^{(i)}}$, $x_p^{(i)} \in \mathbb{R}^{d^{(i)}}$ is the p-th data of i-th media, $d^{(i)}$ is the dimension of feature vector of i-th media, $l^{(i)}$ is the number of labeled data of i-th media, $y_p^{(i)} \in \mathbb{R}^c$ is the corresponding label of sample $x_p^{(i)}$, and c is the number of semantic category. Let $\mathcal{D}^* = \left\{ \mathcal{D}^{(1)*}, \ldots, \mathcal{D}^{(s)*} \right\}$ be the unlabeled cross-media dataset, where $\mathcal{D}^{(i)*} = \left\{ x_p^{(i)} \right\}_{p=l^{(i)}+1}^{N^{(i)}}$, $N^{(i)} = l^{(i)} + u^{(i)}$, $u^{(i)}$ is the number of unlabeled data of i-th media. Given a query of any media type, the goal of cross-media retrieval is to retrieve the most related results of all media types.

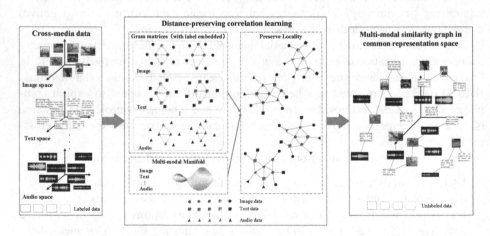

Fig. 1. Framework of proposed DCLMM algorithm.

The framework of our DCLMM is shown in Fig. 1. As we can see from the figure, the proposed approach consists of two main phases. Firstly, we construct s Gram matrices (with label correlations embedded) by kernel transformation, and a multi-modal manifold. Then we exploit a cross-media similarity graph in a common representation space by distance-preserving correlation learning on the Gram matrices and multi-modal manifold. Our approach focuses on finding a nonlinear function of each media type that projects all media objects into common feature space. In projected common space, the similarity of samples can be measured by traditional methods.

3 Semisupervised Distance-Preserving Correlation Learning and Multi-modal Manifold Regularization

3.1 Objective Function of DCLMM Algorithm

Now we begin with the cost function according to the initial label information on training dataset. The loss of label consistency is set to be the Frobenius norm least squares, which is defined as

$$loss(label) = \sum_{i=1}^{s} ||Y^{(i)} - F_l^{(i)}||_F^2 \tag{1}$$

where $F_l^{(i)} = (f^{(i)}(x_1^{(i)}), \ldots, f^{(i)}(x_{l^{(i)}}^{(i)}), \mathbf{0}, \ldots, \mathbf{0}) \in \mathbb{R}^{c \times N^{(i)}}$ is the mapped feature matrix of $l^{(i)}$ labeled samples from i-th media, $\mathbf{0} \in \mathbb{R}^c$ denotes zero vector for an unlabeled sample, $f^{(i)}(x_p^{(i)})$ represents for the nonlinear mapping function of sample $x_p^{(i)}$. $Y^{(i)} = (y_1^{(i)}, \ldots, y_{l(i)}^{(i)}, \mathbf{0}, \ldots, \mathbf{0}) \in \mathbb{R}^{c \times N^{(i)}}$ represents for the label matrix comprising $l^{(i)}$ known tags of labeled samples with $y_p^{(i)} = \mathbf{0}$, for all $p > l^{(i)}$. $|| \cdot ||_F$ is the Frobenius norm. This loss function is the reconstruction error, which restricts media samples are in accordance with their initial semantic labels.

For comprehensively mining the semantic correlation of inter-modality, we make full use of the concurrence information (positive correlation) and exclusive information (negative correlation) among different media types by distance-preserving correlation learning. This idea is achieved by minimizing (maximizing) the distances of similar (dissimilar) media samples. Here similar media samples refer to those that are close to each other in common representation space, or from the same semantic category; otherwise, the dissimilar media samples. Then, a contrastive loss for labeled samples to capture the distance-preserving correlation is designed as follows:

$$loss(dp\_cor) = \lambda_{mij} \sum_{i=1}^{s} \sum_{j=i+1}^{s} ||F_{mij}^{(i)} - F_{mij}^{(j)}||_F^2 - \lambda_{dij} \sum_{i=1}^{s} \sum_{j=i+1}^{s} ||F_{dij}^{(i)} - F_{dij}^{(j)}||_F^2 \tag{2}$$

where $F_{mij}^{(i)}$ and $F_{mij}^{(j)}$ represent for two mapped feature matrices of media samples from i-th media and j-th media with the same semantic labels. Conversely, $F_{dij}^{(i)}$ and $F_{dij}^{(j)}$ denote two mapped feature matrices of media samples with different semantic labels.

Considering the complexity of cross-media low-level features, we use the idea surrounding regularization in vector-valued reproducing kernel Hilbert space (RKHS), and corresponding norm is defined as

$$norm_K(\boldsymbol{F}^{(1)}, \ldots, \boldsymbol{F}^{(s)}) = \gamma_K \sum_{i=1}^{s} ||\boldsymbol{F}^{(i)}||_K^2 \tag{3}$$

where γ_K controls the complexity of the function in ambient space. This term ensures that the solution is smooth with respect to the ambient space.

Next, we will show how to incorporate the geometry knowledge of distribution of both labeled and unlabeled cross-media samples. The support of probability distribution is a compact submanifold of the ambient space, which can be empirically represented by an undirected adjacency graph. We construct a multi-modal manifold by a multi-modal similarity graph $G = (V, E)$, where vertex set V contain total media samples of all modalities. The edge set E connect samples with symmetric, nonnegative weight matrix $\boldsymbol{W}$, which is defined as follows:

$$\boldsymbol{W} = \begin{pmatrix} \boldsymbol{W}^{11} & \cdots & \alpha \boldsymbol{W}^{1s} \\ \vdots & \ddots & \vdots \\ \alpha \boldsymbol{W}^{s1} & \cdots & \boldsymbol{W}^{ss} \end{pmatrix} \tag{4}$$

where α is a trade-off parameter for the similarity effect between intra-modality and inter-modality. And the similarity of samples $\boldsymbol{x}_p^{(i)}$ and $\boldsymbol{x}_q^{(j)}$ is defined as

$$W_{pq}^{ij} = \begin{cases} \exp\left(-||\boldsymbol{x}_p^{(i)} - \boldsymbol{x}_q^{(j)}||^2/\xi^2\right) & \text{if } i = j \text{ and } (\boldsymbol{x}_p^{(i)} \in \mathcal{N}(\boldsymbol{x}_q^{(j)}) \text{ or } \boldsymbol{x}_q^{(j)} \in \mathcal{N}(\boldsymbol{x}_p^{(i)})) \\ 1 & \text{if } i \neq j \text{ and } \boldsymbol{y}_p^{(i)} = \boldsymbol{y}_q^{(j)} \\ 0 & \text{otherwise} \end{cases} \tag{5}$$

where $\mathcal{N}(\boldsymbol{x}_p^{(i)})$ is a set consisting of the k-nearest neighbors of $\boldsymbol{x}_p^{(i)}$. If samples $\boldsymbol{x}_p^{(i)}$ and $\boldsymbol{x}_q^{(j)}$ share a same modality where $i = j$, the similarity between them is measured according to their neighborhood relationship. And if samples $\boldsymbol{x}_p^{(i)}$ and $\boldsymbol{x}_q^{(j)}$ are from different modalities, and correspond to a same semantic tag, the similarity is set to be 1. Laplacian $\boldsymbol{L}$ of graph G is defined as $\boldsymbol{L} = \boldsymbol{D} - \boldsymbol{W}$, where $\boldsymbol{D}$ is a diagonal matrix with $D_{pp} = \sum_q W_{pq}$. Then, an intrinsic norm of data-driven smoothness is given by

$$norm_{\mathcal{M}}(\boldsymbol{F}^{(1)}, \ldots, \boldsymbol{F}^{(s)}) = \frac{\lambda_{\mathcal{M}}}{\hat{N}^2} \sum_{p=1}^{\hat{N}} \sum_{q=1}^{\hat{N}} W_{pq} ||f_p - f_q||^2 = \frac{\lambda_{\mathcal{M}}}{\hat{N}^2} tr(\boldsymbol{F}\boldsymbol{L}\boldsymbol{F}^T) \tag{6}$$

where $\hat{N}$ is the number of the total samples from all modalities, $\frac{1}{\hat{N}^2}$ is the natural scale factor for the empirical estimate of Laplace operator. $tr(\boldsymbol{Z})$ is trace of a matrix $\boldsymbol{Z}$. $\boldsymbol{F} = (\boldsymbol{F}^{(1)}, \ldots, \boldsymbol{F}^{(s)})$ denotes the projected data of all modalities in the common space.

Finally, we obtain our objective function by adding function (1), (2), (3) and (6) as

$$
\arg\min_{F^{(i)}\in\mathbb{R}^{c\times N^{(i)}}} \sum_{i=1}^{s}||Y^{(i)}-F_l^{(i)}||_F^2 + \lambda_{mij}\sum_{i=1}^{s}\sum_{j=i+1}^{s}||F_{mij}^{(i)}-F_{mij}^{(j)}||_F^2 - \lambda_{dij}\sum_{i=1}^{s}\sum_{j=i+1}^{s}||F_{dij}^{(i)}-F_{dij}^{(j)}||_F^2
$$
$$
+\gamma_K\sum_{i=1}^{s}||F^{(i)}||_K^2 + \frac{\lambda_M}{\hat{N}^2}tr(FLF^T)
$$

$$(7)$$

3.2 Optimization Solution

We move to the detail of solving this optimization problem in this section. The classical Representer Theorem states that the minimizer $f^{(i)*}$ of optimization problem (7) is a linear combination of kernel functions, which can be defined as

$$
f^{(i)*}(x^{(i)}) = \sum_{t=1}^{N^{(i)}} \beta_t^{(i)} k(x_t^{(i)}, x^{(i)})
$$

$$(8)$$

where $\beta_t^{(i)} = [\beta_{1t}^{(i)}, \beta_{2t}^{(i)}, \ldots, \beta_{ct}^{(i)}]^T \in \mathbb{R}^c$, $1 \le t \le N^{(i)}$ is the vector to be estimated. The matrix form of target function is that

$$
F^{(i)*} = B^{(i)}K^{(i)}
$$

$$(9)$$

where $B^{(i)} = (\beta_1^{(i)}, \ldots, \beta_{N^{(i)}}^{(i)}) \in \mathbb{R}^{c\times N^{(i)}}$ is weight coefficient matrix, Gram matrix $K^{(i)} = (k(x_p^{(i)}, x_q^{(i)})) \in \mathbb{R}^{N^{(i)}\times N^{(i)}}$ consists of both labeled and unlabeled samples of i-th media. We choose a Gaussian kernel as $k(x_p^{(i)}, x_q^{(i)})$ in our paper, which is given by

$$
k(x_p^{(i)}, x_q^{(i)}) = \exp\left(-||x_p^{(i)} - x_q^{(i)}||^2/\xi^2\right)
$$

$$(10)$$

and we have $F = (B^{(1)}K^{(1)}, \ldots, B^{(s)}K^{(s)})$.

Substituting (9) in the optimization problem (7), we arrive at the following differentiable objective function of variable $B^{(i)}$

$$
\arg\min_{B^{(i)}\in\mathbb{R}^{c\times N^{(i)}}} \sum_{i=1}^{s}||Y^{(i)}-B^{(i)}K^{(i)}J^{(i)}||_F^2 + \lambda_{mij}\sum_{i=1}^{s}\sum_{j=i+1}^{s}||B^{(i)}K_{mij}^{(i)}J_{inter}^{(i)}-B^{(j)}K_{mij}^{(j)}J^{(j)}||_F^2
$$
$$
-\lambda_{dij}\sum_{i=1}^{s}\sum_{j=i+1}^{s}||B^{(i)}K_{dij}^{(i)}J_{inter}^{(i)}-B^{(j)}K_{dij}^{(j)}J^{(j)}||_F^2 + \gamma_K\sum_{i=1}^{s}tr(B^{(i)}K^{(i)}B^{(i)^T})
$$
$$
+\frac{\gamma_M}{\hat{N}^2}\sum_{i=1}^{s}\sum_{j=1}^{s}tr(B^{(i)}K^{(i)}L^{ij}K^{(j)}B^{(j)^T})
$$

$$(11)$$

where $J^{(i)} = diag(1,\ldots,1,0,\ldots,0)$ is the $N^{(i)} \times N^{(i)}$ diagonal matrix where the first $l^{(i)}$ diagonal positions are 1, and the rest 0. $K_{mij}^{(i)}$ and $K_{mij}^{(j)}$ denote two Gram matrices of media objects from i-th media and j-th media with the same labels, whereas $K_{dij}^{(i)}$ and $K_{dij}^{(j)}$ represent for two Gram matrices of that with different labels. $J_{inter}^{(i)}$ is the matrix for coordinating different sample sizes of i-th media and j-th media.

Now, the minimization problem of (7) is reduced to optimizing over s finite dimensional coefficient matrices $B^{(i)}$, $i = 1,\ldots,s$ in function (11). We utilize an alternating optimization method to solve function (11). Differentiating function (11) with respect to $B^{(i)}$ and comparing to zero, we have

$$(B^{(i)}K^{(i)}J^{(i)} - Y^{(i)})J^{(i)}K^{(i)} + \lambda_{mij} \sum_{j=i+1}^{s} (B^{(i)}K_{mij}^{(i)}J_{inter}^{(i)} - B^{(j)}K_{mij}^{(j)}J^{(j)})J_{inter}^{(i)^T}K_{mij}^{(i)}$$

$$- \lambda_{dij} \sum_{j=i+1}^{s} (B^{(i)}K_{dij}^{(i)}J_{inter}^{(i)} - B^{(j)}K_{dij}^{(j)}J^{(j)})J_{inter}^{(i)^T}K_{dij}^{(i)} \qquad (12)$$

$$+ \gamma_K B^{(i)}K^{(i)} + \frac{\gamma_M}{\hat{N}^2}B^{(i)}K^{(i)}L^{ii}K^{(i)} + \frac{\gamma_M}{\hat{N}^2}\sum_{j\neq i}B^{(j)}K^{(j)}L^{ji}K^{(i)} = 0$$

which can be rewritten as

$$B^{(i)} = \left(\begin{array}{c} \lambda_{mij} \sum\limits_{j=i+1}^{s} B^{(j)}K_{mij}^{(j)}J^{(j)}J_{inter}^{(i)^T}K_{mij}^{(i)} - \lambda_{dij} \sum\limits_{j=i+1}^{s} B^{(j)}K_{dij}^{(j)}J^{(j)}J_{inter}^{(i)^T}K_{dij}^{(i)} \\ + Y^{(i)}J^{(i)}K^{(i)} - \frac{\gamma_M}{\hat{N}^2}\sum\limits_{j\neq i}B^{(j)}K^{(j)}L^{ji}K^{(i)} \end{array} \right)$$

$$\left(\lambda_{mij} \sum_{j=i+1}^{s} K_{mij}^{(i)}J_{inter}^{(i)}K_{mij}^{(i)} - \lambda_{dij} \sum_{j=i+1}^{s} K_{dij}^{(i)}J_{inter}^{(i)}K_{dij}^{(i)} + K^{(i)}J^{(i)}K^{(i)} + \gamma_K K^{(i)} + \frac{\gamma_M}{\hat{N}^2}K^{(i)}L^{ii}K^{(i)} \right)^{-1}$$

$$\qquad (13)$$

In view of (13), we propose an iterative optimization approach to minimize the objective function. First, we initialize each weight matrix $B_0^{(i)}$ with a random value. Then, in each iteration, we calculate $\left\{ B_{t+1}^{(1)}, \ldots, B_{t+1}^{(s)} \right\}$ under the condition of given $\left\{ B_t^{(1)}, \ldots, B_t^{(s)} \right\}$. The iteration stops when the ratio change of loss value of (11) between two iterations is less than 1%. We summarize DCLMM algorithm in Algorithm 1. The algorithm requires $O(N^{(i)^3})$ time complexity involving an inversion and several multiplications of $N^{(i)} \times N^{(i)}$ matrix in each iteration. In our experiments, we obtain satisfactory accuracy by only a few iterations, so the time cost is acceptable.

So far, we have learnt s weight matrices $B^{(i)}$, $i = 1,\ldots,s$ for different media types, and each sample $x_p^{(i)}$ is successfully projected into the common feature space $o_p^{(i)} = \sum_{t=1}^{N^{(i)}} \beta_t^{(i)}k(x_t^{(i)},x_p^{(i)}) \in \mathbb{R}^c$. In our common feature space, we use k-nearest neighbors (KNN) classifier to measure the cross-media similarity for cross-media retrieval as [5, 8, 11], where the similarity of sample $o_p^{(i)}$ and $o_q^{(j)}$ is defined as the marginal probability.

Algorithm 1: Common Representation Learning with DCLMM

Input:

The matrix of $N^{(i)} = l^{(i)} + u^{(i)}$ samples $X^{(i)} \in \mathbb{R}^{N^{(i)} \times d^{(i)}}$

The matrix of labels $Y^{(i)} \in \mathbb{R}^{c \times l^{(i)}}$

Parameters $\lambda_{mij}, \lambda_{dij}, \gamma_K, \gamma_M$

Output: Common feature space

1. Initialize each projection matrix $B_0^{(i)}$ with a random value and set $t = 0$;

2. Calculate the Gram matrix $K^{(i)}$ by (10), and calculate the weight matrix W by (4);

3. Compute the Laplacian matrix of multi-modal graph $L \in \mathbb{R}^{\hat{N} \times N}$;

4. **repeat**

5. Compute the loss value according to (11);

6. Update $B_{t+1}^{(i)}$ according to following equation:

$$
B_{t+1}^{(i)} = \left(\lambda_{mij} \sum_{j=i+1}^{s} B_t^{(j)} K_{mij}^{(j)} J^{(j)} J_{inter}^{(i)T} K_{mij}^{(i)} - \lambda_{dij} \sum_{j=i+1}^{s} B_t^{(j)} K_{dij}^{(j)} J^{(j)} J_{inter}^{(i)T} K_{dij}^{(i)} \right.
$$
$$
\left. + Y^{(i)} J^{(i)} K^{(i)} - \frac{\gamma_M}{\hat{N}^2} \sum_{j \neq i} B_t^{(i)} K^{(i)} L^{ii} K^{(i)} \right) \tag{14}
$$
$$
\left(\lambda_{mij} \sum_{j=i+1}^{s} K_{mij}^{(i)} J_{inter}^{(i)} K_{mij}^{(i)} - \lambda_{dij} \sum_{j=i+1}^{s} K_{dij}^{(i)} J_{inter}^{(i)} K_{dij}^{(i)} + K^{(i)} J^{(i)} K^{(i)} + \gamma_K K^{(i)} + \frac{\gamma_M}{\hat{N}^2} K^{(i)} L^{ii} K^{(i)} \right)^{-1}
$$

7. $t = t + 1$;

8. **until** Convergence

9. Compute common feature space according to (9).

---

4 Experimental Results

This section describes our experiments on two real-world datasets to validate the efficiency of DCLMM. We compare our results with five state-of-the-art methods, specifically, JGRHML [7], CMCP [4], HSNN [11], JRL [5] and SSRCL [8].

4.1 Datasets and Evaluation Metric

Wikipedia dataset [12] is the most widely-used dataset for cross-media retrieval. All articles are organized into 29 categories, and 10 most populated categories are preserved at last. Wikipedia dataset consists of 2866 image/text pairs, and is randomly split into 2173 pairs for training and 693 pairs for testing.

XMedia dataset [7] consists of 5000 texts, 5000 images, 1140 videos, 1000 audios, and 500 3D models in our experiments. They are organized into 20 categories, and is randomly divided into 10112 media data for training and 2528 media data for testing. The random division is performed on each media type with the ratio of training set to testing set being 4:1.

We extract 4096 dimensional convolutional neural network (CNN) feature from the highest layer (called FC7) of VGGNet [13] for each image, 3000 dimensional bag-of-words (BoW) feature for each text. And an audio clip is represented by 29 dimensional MFCC feature, a video clip is represented by 4096 dimensional CNN feature, a 3D model is represented by the concatenated 4700 dimensional vector of a set of Light-Field descriptors [14]. All experimental methods adopt the same features in our experiments for fair comparison.

We evaluate the retrieval results by precision-recall (PR) curves and mean average precision (MAP), which are widely used in information retrieval area. In the experiments, we set $\lambda_{mij} = 0.01, \lambda_{dij} = 0.0009, \gamma_K = 0.0006, \gamma_M = 7, K = 150$ and $\alpha = 0.05$ for the two datasets according to the five-fold cross validation result on training set.

4.2 Experimental Results and Analyses

We conduct two cross-modal retrieval experiments on Wikipedia and XMedia datasets: bi-modal retrieval and multi-modal retrieval. Bi-modal retrieval is retrieving one modality in testing set using a query of another modality, and multi-modal retrieval is retrieving multiple modalities in testing set using a query of any modality.

Table 1 lists the MAP scores of bi-modal retrieval on Wikipedia dataset with our DCLMM and five state-of-the-art methods. The "Image → Text" task means we use an image to retrieval texts. Comparing with the state-of-the-art method, DCLMM improves the average MAP from 48.28% to 49.71% on bi-modal retrieval tasks. Table 2 shows the MAP scores of bi-modal retrieval on XMedia dataset with all approaches. Compared with the five methods, our DCLMM improves the average MAP from 49.32% to 55.49%. We also observe that the MAP scores of our method improve the most for the retrieval tasks including Audio or 3D model with small sample size. For example, in "Text → Audio" task, comparing with SSRCL, our map score increases by 15.59%. It suggests that our DCLMM is attractive in practical applications when training samples are insufficient. DCLMM performs better than other methods so far, which is due to that our method learn the distance-preserving correlation in a kernel based regularization framework. This strategy exploits not only positive but also negative semantic correlation from high dimensional space, and more rich complementary information of media samples can be explored.

Table 1. MAP scores of bi-modal retrieval on Wikipedia dataset with different methods

Dataset	Task	JGRHML	CMCP	HSNN	JRL	SSRCL	DCLMM
Wikipedia dataset	Image → Text	0.2633	0.3879	0.4740	0.5150	0.5180	**0.5270**
	Text → Image	0.2074	0.3508	0.4239	0.4382	0.4476	**0.4672**
Average		0.2354	0.3694	0.4490	0.4766	0.4828	**0.4971**

Table 2. MAP scores of bi-modal retrieval on XMedia dataset with different methods

Dataset	Task	JGRHML	CMCP	HSNN	JRL	SSRCL	DCLMM
XMedia dataset	Image→Text	0.4606	0.7099	0.7734	0.8837	0.9014	**0.9045**
	Image→Audio	0.1627	0.3703	0.3814	0.2417	0.3480	**0.4271**
	Image→Video	0.2978	0.4735	0.4840	0.4688	0.5223	**0.559**
	Image→3D	0.4508	0.5410	0.5242	0.5501	0.5912	**0.7127**
	Text→Image	0.3626	0.7269	0.7969	0.8801	0.8951	**0.9022**
	Text→Audio	0.1766	0.3057	0.3150	0.3283	0.3725	**0.5284**
	Text→Video	0.3945	0.4030	0.4193	0.5125	0.5497	**0.5496**
	Text→3D	0.4958	0.4631	0.4848	0.5343	0.5969	**0.7311**
	Audio→Image	0.1108	0.3790	0.3663	0.3204	0.3947	**0.469**
	Audio→Text	0.1346	0.3370	0.2953	0.3612	0.4237	**0.5431**
	Audio→Video	0.1313	0.1442	0.2046	0.2080	0.2372	**0.3156**
	Audio→3D	0.1761	0.2093	0.2380	0.2889	0.3312	**0.4035**
	Video→Image	0.2450	0.4334	0.4856	0.4493	0.5537	**0.5586**
	Video→Text	0.3260	0.3525	0.3842	0.5081	0.5723	**0.5776**
	Video→Audio	0.1443	0.1533	0.1650	0.2065	0.2355	**0.305**
	Video→3D	0.2327	0.3251	0.3344	0.3462	0.4088	**0.4711**
	3D→Image	0.4205	0.5086	0.5069	0.5273	0.6038	**0.6542**
	3D→Text	0.4961	0.4168	0.4087	0.5267	0.6236	**0.6811**
	3D→Audio	0.1531	0.1426	0.1669	0.2390	0.2989	**0.3552**
	3D→Video	0.2170	0.3476	0.3220	0.3067	0.4026	**0.4434**
Average		0.2794	0.3871	0.4028	0.4344	0.4932	**0.5549**

Figure 2 shows the PR curves of bi-modal retrieval with all methods on Wikipedia dataset. For the limitation of pages, we just shows part PR curves of bi-modal retrieval on XMedia dataset in Fig. 3. It can be seen that the results of PR curves are consistent with MAP scores, where our DCLMM achieves the highest precision at most levels of recall rate on two datasets, outperforming the five compared approaches. It should be owning to that our DCLMM models the geometry distribution of all heterogeneous data in a multi-manifold, which integrates the incongruous distribution information of media data from different modalities to describe the diversified relations. Based on multi-manifold regularization, DCLMM imposes smoothness constraints of possible solutions to boost retrieval precision.

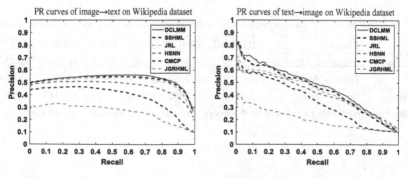

Fig. 2. Precision-recall curves of bi-modal retrieval on Wikipedia dataset.

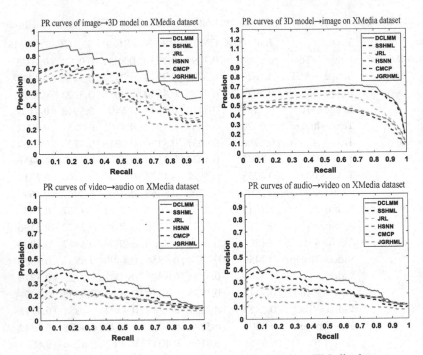

Fig. 3. Precision-recall curves of bi-modal retrieval on XMedia dataset.

Table 3 lists the MAP scores of multi-modal retrieval on Wikipedia dataset with all methods. Comparing with the state-of-the-art method, DCLMM improves the average MAP from 53.17% to 54.97%. It further demonstrates the feasibility and effectiveness of our proposed method for cross-media retrieval. For challenging XMedia dataset, we use one media type to retrieve multiple media types. For example, we use an image to retrieve "images", "images and texts", "images, texts and audios", and so on. Figure 4 shows the PR curves of retrieval tasks where query types are image and video for length limitation. We find when image or text are added into the retrieved media types, the precision increases, and when video, audio and 3D model are added into the retrieved media types, the precision decreases. This is probably due to the large sample size and relatively good low-level features of image and text; on the other hand, the difficulty of retrieval increases with the increase of retrieved media types.

Table 3. MAP scores of multi-modal retrieval on Wikipedia dataset with different methods

Dataset	Task	JGRHML	CMCP	HSNN	JRL	SSRCL	DCLMM
Wikipedia	Image → Multiple*	0.2133	0.3193	0.4020	0.4259	0.4310	**0.4436**
dataset	Text → Multiple	0.3219	0.4787	0.5914	0.6298	0.6323	**0.6558**
Average		0.2676	0.3990	0.4967	0.5279	0.5317	**0.5497**

*Q → Multiple means that Q serves as the query and the results include images and texts

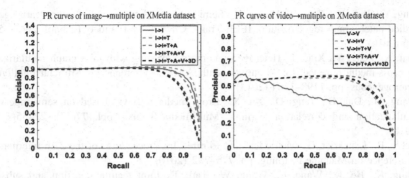

Fig. 4. Precision-recall curves of multi-modal retrieval on XMedia dataset by DCLMM. $Q \rightarrow$ Multiple means that media Q serves as the query and results include multiple media types. I denotes image, T denotes text, A means audio, V means video, and 3D is 3D model.

5 Conclusion

A semisupervised cross-media retrieval approach with distance-preserving correlation learning and multi-modal manifold regularization (DCLMM) has been proposed in this paper. The proposed method jointly models not only the low-level feature correlation, but also the positive semantic and negative semantic information of all media types in a unified kernel based framework, which takes full advantage of the potential information. Besides, DCLMM constructs a multi-modal manifold to better exploit the geometry distribution of all cross-media samples, which further improves the accuracy. Extensive experiments are conducted on two cross-media datasets show the effectiveness of our approach. In the future, we will further optimize the algorithm by modeling more kinds of correlations, such as fine-grained information.

Acknowledgments. This research is supported by the National Natural Science Foundation of China (No. 61373109).

References

1. Ranjan, V., Rasiwasia, N., Jawahar, C.V.: Multi-label cross-modal retrieval. In: IEEE International Conference on Computer Vision, pp. 4094–4102. IEEE Computer Society (2015)
2. Zhang, H., Zhang, W., Liu, W., et al.: Multiple kernel visual-auditory representation learning for retrieval. Multimedia Tools Appl. **75**, 9169–9184 (2016)
3. Peng, Y., Qi, J., Huang, X., Yuan, Y.: CCL: cross-modal correlation learning with multigrained fusion by hierarchical network. IEEE Trans. Multimedia **20**(2), 405–420 (2018)
4. Zhai, X., Peng, Y., Xiao, J.: Cross-modality correlation propagation for cross-media retrieval. In: IEEE International Conference on Acoustics, Speech and Signal Processing (ICASSP), pp. 2337–2340 (2012)
5. Zhai, X., Peng, Y., Xiao, J.: Learning cross-media joint representation with sparse and semisupervised regularization. IEEE Trans. Circuits Syst. Video Technol. **24**(6), 965–978 (2014)

6. Peng, Y., Zhai, X., Zhao, Y., Huang, X.: Semi-supervised cross-media feature learning with unified patch graph regularization. IEEE Trans. Circuits Syst. Video Technol. **26**(3), 583–596 (2016)

7. Zhai, X., Peng, Y., Xiao, J.: Heterogeneous metric learning with joint graph regularization for cross-media retrieval. In: Twenty-Seventh AAAI Conference on Artificial Intelligence Heterogeneous, pp. 1198–1204 (2013)

8. Zhang, H., Dai, G., Tang, D., Xu, X.: Cross-media retrieval based on semi-supervised regularization and correlation learning. Multimedia Tools Appl. **77**(17), 22455–22473 (2018)

9. Mcfee, B., Lanckriet, G.: Metric learning to rank. In: Proceedings of the 27th International Conference on Machine Learning, pp. 775–782 (2010)

10. Wang, K., He, R., Wang, L., Wang, W., Tan, T.: Joint feature selection and subspace learning for cross-modal retrieval. IEEE Trans. Pattern Anal. Mach. Intell. **38**(10), 2010–2023 (2016)

11. Zhai, X., Peng, Y., Xiao, J.: Effective heterogeneous similarity measure with nearest neighbors for cross-media retrieval. In: Schoeffmann, K., Merialdo, B., Hauptmann, A.G., Ngo, C.-W., Andreopoulos, Y., Breiteneder, C. (eds.) MMM 2012. LNCS, vol. 7131, pp. 312–322. Springer, Heidelberg (2012). https://doi.org/10.1007/978-3-642-27355-1_30

12. Rasiwasia, N., Pereira, J.C., Coviello, E., et al.: A new approach to cross-modal multimedia retrieval. In: Proceedings of the 18th ACM International Conference on Multimedia, pp. 251–260. ACM, New York (2010)

13. Simon, M., Rodner, E., Denzler, J.: ImageNet pre-trained models with batch normalization. arXiv:1612.01452v2 [cs.CV] (2016)

14. Chen, D.Y., Tian, X.P., Shen, Y.T., Ouhyoung, M.: On visual similarity based 3D model retrieval. Comput. Graph. Forum **22**(3), 223–232 (2003)

Explaining Deep Learning Models with Constrained Adversarial Examples

Jonathan Moore[1]([✉]), Nils Hammerla[1], and Chris Watkins[2]

[1] Babylon Health, London SW3 3DD, UK
{jonathan.moore,nils.hammerla}@babylonhealth.com
[2] Royal Holloway University of London, Egham, UK
c.j.watkins@rhul.ac.uk

Abstract. Machine learning algorithms generally suffer from a problem of explainability. Given a classification result from a model, it is typically hard to determine what caused the decision to be made, and to give an informative explanation. We explore a new method of generating counterfactual explanations, which instead of explaining why a particular classification was made explain how a different outcome can be achieved. This gives the recipients of the explanation a better way to understand the outcome, and provides an actionable suggestion. We show that the introduced method of Constrained Adversarial Examples (CADEX) can be used in real world applications, and yields explanations which incorporate business or domain constraints such as handling categorical attributes and range constraints.

Keywords: Explainable AI · Adversarial examples · Counerfactual explanations

1 Introduction

The recent explosion in the popularity of machine learning methods has led to their wide adoption in various domains, outside the technology sector. Machine learning algorithms are used to predict how likely convicted felons are to recidivate, which candidates should be interviewed for a job, and which bank customers are likely to default on a given loan. These algorithms assist human decision making, and in some cases may even replace it altogether. When humans are responsible for a decision, we can ask them to explain their thought process and give a reason for the decision (although often that is not done). Asking a machine learning algorithm to explain itself is a challenging problem, especially in the case of deep neural networks.

Throughout this work, we will refer to the following scenario. Assume that a bank has trained a deep learning model to predict which of its customers should be eligible for a loan. The input is a vector that represents the customer, using attributes such as age, employment history, credit score, etc. The output is a label which says whether said customer is likely to repay a loan or default. Now

© Springer Nature Switzerland AG 2019
A. C. Nayak and A. Sharma (Eds.): PRICAI 2019, LNAI 11670, pp. 43–56, 2019.
https://doi.org/10.1007/978-3-030-29908-8_4

suppose that a customer requests a loan, and is denied based on the decision of the algorithm. The customer would obviously like to know why he or she was rejected, and what prompted the decision. The bank, on the other hand, is faced with two problems:

- The bank has difficulty giving a meaningful explanation. Various explanation methods exist, but it is hard to determine which ones give valuable feedback to the customer.
- The bank doesn't want to expose its algorithm, or even the full set of features it uses for classification. Credit scoring and loan qualification mechanisms are typically closely guarded by most banks.

The Constrained Adversarial Examples (CADEX) method presented here aims to answer both problems. Instead of directly explaining why a model classified the input to a particular class, it finds an alternate version of the input which receives a different classification. In the bank scenario it produces an alternate version of the customer, which would get the loan. The customer can act on this explanation in order to receive a loan in the future, without the bank revealing the inner working of its algorithms. Such explanations are referred to as *Counterfactual Explanations*. As shown in a recent study of AI explainability from the perspective of social sciences by [9], people tend to prefer contrastive explanations over detailed facts leading to an outcome, and that they find them more understandable. In fact, when people explain why an event occurred, they tend to explain it in comparison to another event which did not occur.

The CADEX method offers several improvements over current techniques for finding counterfactual explanations:

- It supports directly limiting the number of changed attributes to a predetermined amount.
- It allows specifying constraints on the search process, such as the direction attributes are allowed to change.
- It fully supports categorical one-hot encoded attributes and ordinal attributes.
- It surpasses current explainability methods by providing better, more understandable explanations.

2 Related Work

2.1 Explainability

As machine learning models become increasingly complex and have a large number of internal weights and dependencies, it becomes more and more challenging to explain how they work, and why they produce the predictions they make. Explainability of machine learning models has been an active topic of research recently, and multiple methods and techniques have been developed to try and address these difficulties from several points of view.

Some methods attempt to explain what a model has learned in its training phase. Such methods examine the weights of the trained model and present them in an interpretable way. These methods are particularly common for CNNs, and so the explanations have a highly visual nature. A recent survey [15] mentions many such techniques, which include visualizing the patterns learned in each layer and generating images which correspond to feature maps learned by the network. However, these methods are all specifically tailored to work on CNNs and don't generalize to any black box model.

Other methods seek to explain the output of a classifier for a specific given input. These methods aim to answer the question: "why did the model predict this class?", by assigning a weight or significance score to the individual features of the input. Most notable in this category are LIME [11] and SHAP [8].

LIME takes a given input, and creates different versions of it by zeroing various attributes (or super pixels in the case of images), and then builds a local linear model while weighting the inputs by their distance to the original. The model is trained to minimize the number of non-zero coefficients by using a method such as LASSO. This results in an explainable linear model where the model's coefficients act as the explanation, and describe the contribution of each attribute (or super pixel) to the resulting classification.

SHAP attempts to unify several explanation methods such as LIME and DeepLift [12], in a way that the feature contributions are given in Shapley values from game theory, which have a better theoretic grounding than those produced by LIME. For tabular data the method is called Kernel SHAP, which improves LIME by replacing the heuristically chosen loss function and weighting kernel with ones that yield Shapley values.

Both methods produce an output that highlights which attributes contributed most to the classification, and which reduced the probability of classification. There are several drawbacks to this approach. First, it typically requires a domain expert to understand the significance of the output, and what the values mean for the model. Second, the explanation they provide is not actionable. In the bank scenario, they can tell the user, for example, that she didn't get the loan because of her salary and age. They won't say what she needs to do to get the loan in the future - should she wait until she's older? How much older? Or can she change another attribute such as education level and get the loan?

Some recent methods try to provide such explanations by looking for *counterfactuals*. A counterfactual explanation answers the question: "Why was the outcome Y observed instead of Y^*?". The more specific formulation for machine learning models is: "If X had the values of X^*, the outcome Y^* would have been observed instead of Y", where X represents the input to the model. By observing the difference between X and X^*, we can provide a "what-if" scenario which is actionable to the end user. In the bank scenario the explanation could be, for example, "If you had \$5000 instead of \$4000 in your account you would have gotten the loan".

A naïve way of finding counterfactuals would be to simply find the nearest training set instance to the input, which receives a different classification.

The limitation in that approach is that it is limited by the size and quality of the training set. It cannot find a counterfactual that isn't explicitly in the set. Additionally, showing the user a counterfactual which represents the details of another user may not even be legal considering data protection rights and confidentiality.

[7] finds synthetic counterfactual explanations by sampling from a sphere around the input in a growing radius, until one is found which classifies differently than the original. Then, the number of changed attributes is constrained by iteratively setting them to the value of the original as long as the same contrastive classification is kept. However, the method is sensitive to hyperparameter choices which affect how close the found counterfactuals will be to the original, and the paper doesn't suggest how to determine their optimal value.

[14] generates counterfactuals by optimizing a loss function, which factors the distance to the desired classification as well as a distance measure to the original input. The distance measure is used to limit the number of attributes changed via regularization, but the process of finding the counterfactual requires iteration over various coefficient values, and doesn't allow a hard limit on the number of changed attributes. In addition, it doesn't have a facility to handle one-hot encoded categorical attributes.

2.2   Adversarial Examples

Adversarial examples were discovered by [13], who showed that given a trained image classifier, one could take a correctly classified image and perturb its pixels by a small amount which is indistinguishable to the human eye, and yet causes the image to receive a completely different label by the classifier. [4] developed an efficient way of finding adversarial examples called FGSM, which uses the neural network's loss function's gradient to find the direction where adversarial examples can be found. [6] improved the technique and enabled it to target a specific desired classification, as well as using a more iterative approach to find the adversarial examples.

Most of the discussion around adversarial examples has been in the context of security and attacks against models deployed for real world applications. [10] demonstrate an attack against an online black-box classifier, by training a different classifier on a synthetic dataset and showing that adversarial examples found on that classifier also fooled the online one. They also show that multiple types of classifiers can be attacked that way, such as linear regression, decision trees, SVMs, and nearest neighbours. Others show that adversarial examples can carry over to the real world by printing or 3D printing them, and fooling camera based classifiers [1,2,6].

CADEX uses adversarial examples to facilitate an understanding of the model instead of attacking or compromising it, by finding counterfactual explanations close to the original input. The search process is constrained to enforce domain or business constraints on the desired explanation.

3   Generating Explanations

We present the CADEX method for generating explanations for deep learning models. Let $f(x) = \hat{y}$, where f is the model, x is a specific input sample, and $\hat{y}$ is the output class. The method aims to find x^* for which $f(x^*) = y^*$, where $\hat{y} \neq y^*$, and x^* is as close as possible to x while satisfying a number of constraints. This allows us to present the user a "what if" scenario. In the case of the bank loan application, the user can be told: "if you had the attributes of x^*, you would get the loan". That is, in that scenario $\hat{y} =$ REJECT and $y^* =$ APPROVE. The full algorithm is listed as Algorithm 1. The code used to implement the method and perform the evaluation can be found at https://github.com/spore1/cadex.

3.1   Finding Adversarial Examples

The main motivation in CADEX is to find the explanations through adversarial examples. Adversarial examples work in a very similar sense, by changing the model input with a minimal perturbation so that it receives a different classification.

Given the original input x, we can calculate the loss of the model between the actual output $\hat{y}$, and the desired target classification y^*. This is typically the cross entropy loss between the predicted class probabilities and the desired one. Then, we take the gradient of the loss with respect to the input.

$$\nabla Loss = \frac{\partial}{\partial x} Loss(\hat{y}, y^*) \tag{1}$$

We then follow the gradient in input space using an optimizer such as Adam [5] or RMSProp, until $f(x^*) = y^*$, which is our target classification. This typically results in an input that lies right on the decision boundary between the classes.

3.2   Constraining the Number of Changed Attributes

Following the method above will indeed find adversarial examples that are close to the original sample. However, since we calculate the gradient in input space and don't constrain it, any number of attributes in x may change. Typically, the gradient is nonzero for all attributes, meaning that the resulting x^* is different than x in all attributes. The issue with this approach is that there could be dozens or hundreds of different attributes, and showing the user an explanation which is different in so many attributes is hardly useful, and doesn't constitute an explanation the user can act on. Ideally, we would like to limit the changed attributes to a small number, so that it is perceived as actionable by a human.

Previous work such as [14] has attempted to limit the number of changed attributes by adding a form of L1 regularization to the loss function. However, this approach cannot guarantee the number of changed attributes will in fact be under an acceptable amount.

We limit the number of changed attributes by applying a mask to the gradient. The mask is used to zero the gradient in all attributes except the ones

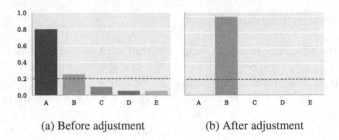

(a) Before adjustment   (b) After adjustment

Fig. 1. Visual explanation of categorical attribute adjustment. On the left, the internal state of the algorithm has category A with the highest value, followed by B. We assume that A is the categorical value of the original sample. Category B is above the predefined threshold of 0.2, so its attribute is set to one and the rest are zeroed.

we wish to allow to change. When this gradient is applied to the input, only the selected attributes will be modified. The decision of where to zero out the gradient is performed as follows. First, get the gradient of the loss function with respect to the input as in Eq. 1. Then, sort the gradient attributes by their absolute value from large to small, and take the top n_{change} attributes, where n_{change} is the number of desired attributes to change (e.g. 3 or 5). Then, prepare a mask which is set to 1 for the top n_{change} attributes and zero elsewhere. At each iteration of gradient descent, after getting the gradient but before applying it to the weights by the optimizer, multiply the gradient by the mask. Then proceed as usual to update the weights.

3.3   Constraining the Direction of the Gradient

In addition to the number of changed attributes, we may want to place another constraint on the search process. The gradient may change each attribute in any direction - positive or negative - which depending on context may not be acceptable. Consider that in the bank scenario, the input may contain attributes such as "age" or "number of children". The algorithm may suggest that the user would get the loan if she were younger, or if she had one less child. For obvious reasons, no bank would ever want to make such a suggestion. We therefore wish to constrain the direction that some of the attributes would be permitted to change in.

We introduce a new parameter C which is used to build a mask to further constrain the gradient. This parameter is a vector of the same dimensions as the model's input, and is defined to be positive for each attribute that may only increase in value, negative for those that may only decrease, and 0 where the value may go in any direction. We assume that this will be defined by a domain expert, who understands each attribute in the data and the implications of changing it. Then, we build the following mask, for each attribute i in the input vector:

Algorithm 1. CADEX - counterfactual explanation for a given input

---

Input: x: original input sample
$\quad\quad$ $f(x)$: trained model
$\quad\quad$ $target$: desired output class for input x
$\quad\quad$ max_epochs: maximum number of epochs to allow
$\quad\quad$ n_{change}: maximum number of changed attributes
$\quad\quad$ C: directional constraints
$\quad\quad$ n_{skip}: number of attributes to skip from the top
$\quad\quad$ t_{flip}: threshold to flip categorical attributes
Output: x^*: modified input sample x which classifies as $target$

1 $\quad x^* \leftarrow x$
2 $\quad \nabla_0 \leftarrow \frac{\partial}{\partial x} Loss(f(x), target)$
3 $\quad \nabla_0 \leftarrow \nabla_0 * C_{mask}$ // See eq 2
4 $\quad i \leftarrow$ ARGSORT(∇_0) in descending order
5 $\quad mask \leftarrow 0$
6 $\quad mask\left[i[n_{skip}..n_{skip} + n_{change}]\right] \leftarrow 1$
7 $\quad result \leftarrow \emptyset; \ epoch \leftarrow 0$
8 $\quad$ **while** $epoch < max\_epochs$ **and** $result = \emptyset$ **do**
9 $\quad\quad \nabla_{epoch} \leftarrow \frac{\partial}{\partial x} Loss(f(x), target) * mask$
10 $\quad\quad x^* \leftarrow x^* -$ADAM$(\nabla_{epoch})$
11 $\quad\quad x^* \leftarrow$ FLIPCATEGORICAL(x^*, t_{flip})
12 $\quad\quad x_{adjusted} \leftarrow$ APPLYCONSTRAINTS(x^*)
13 $\quad\quad$ **if** $f(x_{adjusted}) = target$ **then**
14 $\quad\quad\quad result \leftarrow x_{adjusted}$
15 $\quad\quad$ **end**
16 $\quad\quad epoch \leftarrow epoch + 1$
17 $\quad$ **end**
18 $\quad$ **return** $result$

$$C_{mask_i} = \begin{cases} 1 & \text{if } C_i > 0 \text{ and } \nabla Loss_i < 0 \text{ or} \\ & \quad C_i < 0 \text{ and } \nabla Loss_i > 0 \\ 0 & \text{else} \end{cases} \tag{2}$$

Note that if C is *positive*, we allow only a *negative* gradient, since the gradient is subtracted from the current input at each step of gradient descent, and vice versa when C is negative.

The resulting mask is used during training similar to the process described in Sect. 3.2. In fact, the two techniques can be used together, by first using the directional constraint function to zero the gradient where needed, followed by selecting the top n_{change} attributes. This way, the selected attributes are those which change in the allowed direction.

Algorithm 2. FLIPCATEGORICAL

Input: x^*: modified input sample
      t_{flip}: threshold
Output: x^* with flipped attributes where neccessary
1 $result \leftarrow x^*$
2 **foreach** $attr\_set$ **in** categorical attributes of x^* **do**
3     $i \leftarrow$ ARGSORT($x^*[attr\_set]$) in decreasing order
4   **if** $x^*[i[1]] > t_{flip}$ **then**
5         $result[attr\_set] \leftarrow 0$
6         $result[i[1]] \leftarrow 1$
7   **end**
8 **end**
9 **return** $result$

Algorithm 3. APPLYCONSTRAINTS

Input: x^*: modified input sample
Output: x^* with adjusted attributes
1 $result \leftarrow x^*$
2 **foreach** $attr\_set$ **in** categorical attributes of x^* **do**
3     $i \leftarrow$ ARGMAX($x^*[attr\_set]$)
4     $result[attr\_set] \leftarrow 0$
5     $result[i] \leftarrow 1$
6 **end**
7 **foreach** $attr$ **in** ordinal attributes of x^* **do**
8     $result[attr] \leftarrow$ ROUND($result[attr]$)
9 **end**
10 **return** $result$

3.4 Handling Categorical and Ordinal Attributes

Categorical attributes are frequently found in many real world datasets. They pose a challenge to the algorithm, which relies on changing the attributes gradually by following the gradient. Categorical attributes are typically one-hot encoded, which means that each attribute may only be set to 0 or 1, and only one attribute per attribute set must be set to 1 at any given time. By naïvely following the gradient, the algorithm will easily violate these constraints.

We use the following method to deal with categorical attributes. During training, we continue to treat the categorical attributes as any other attribute, in the sense that they are allowed to change gradually by the gradient. Internally, there could be a moment where the representation of the modified input sample violates the rules of one-hot encoded attributes, but that is acceptable as long as we don't return this as the final result. At each epoch, two extra steps are performed. First, a check is made to determine whether certain categorical attributes need to be "flipped", that is to set the value 1 to a different category than that of the original. Assuming that an "attribute set" is defined to be the set of attributes that represent a one-hot encoded categorical value, then for each attribute set we find the second highest valued attribute, and if it's above a threshold t_{flip} we set it to 1 and zero the rest. The reasoning behind this is that the highest attribute would be that which was equal to one in the original sample, and the second highest is the one that has been most affected by the gradient. This is illustrated visually in Fig. 1 and described in Algorithm 2. The threshold is a hyper-parameter which tunes how quickly the algorithm choses to change the categorical attributes.

Additionally, at each epoch we need to determine whether the stopping condition has been met, which is that the modified observation is classified as the desired label. We test this against an adjusted version of the observation, where the highest attribute in each attribute set to one and the rest to zero. This means we're testing against the valid version of the observation, where attribute values can only have values of 1 or 0.

Ordinal attributes - which must hold integer values - are handled in a similar fashion. During training they are allowed to have any fractional value, but when evaluating the stopping condition we round them to the nearest integer. The adjustment process is described in Algorithm 3.

3.5 Finding Alternate Explanations

In some cases, it would be useful to be able to present more than one adversarial example and show user multiple alternate scenarios with the desired classification.

As described in Sect. 3.2, the method sorts the gradient in descending order of the absolute value of the attributes, and selects the top n_{change} attributes. By skipping the first top n_{skip} attributes, the method chooses a different set of attributes to change and will arrive at a different solution. Thus, by trying various values for n_{skip}, we can generate multiple alternate adversarial examples.

4   Evaluation

We evaluate CADEX with several different approaches. First, we train a feed-forward neural network on the German loan dataset [3], which contains 1000 observations of people who applied for a loan and has a range of numeric, categorical and ordinal attributes. Every categorical attribute was one-hot encoded, and assigned a readable label from the data dictionary supplied with the data. Numerical attributes were normalized to have zero mean and unit standard deviation. The dataset was split to 80% training and 20% validation. The model had one hidden layer with 15 neurons with ReLU activations, and one classification layer with two output neurons with softmax activations to represent the classification labels of APPROVED and REJECTED. The model was trained using the Adam optimizer with early stopping when the validation loss started to increase.

Then for each training set sample, we ran CADEX to find 10 explanations by varying n_{skip} from 0 to 9. We used ordinal constraints on the attributes EXISTING_CREDITS and PEOPLE_MAINTAINED, and directional constraints to allow only positive changed on AGE and PEOPLE_MAINTAINED. t_{flip} was set to 0.2 for all experiments. We repeated the above process for $n_{changed} = (5, 7, 10)$.

4.1   Sample Explanation

Table 1 Illustrates three explanations found for one particular validation set sample, who was refused a loan. The explanations are clear and concise, and can be immediately understood by non-domain experts. They also provide an interesting insight into the inner workings of the model and what it has managed to learn. We can see that the individual would have been given a loan if she were older or had a longer employment history. We can also see it would have been better for her not to have a checking account at all rather than have a negative balance. Finally, we learn that had she been a male instead of a female, she would have gotten the loan which indicates a possible bias of the model to prefer men over women.

Upon investigation, we found that for all of the women which the model classified as REJECT, we were able to produce a counterfactual that changed the sex attribute to male, and therefore the model is in fact biased. We conclude that in addition to providing actionable explanations to an end user, CADEX is also a valuable method to aid in the understanding the inner workings of the model.

4.2   Number of Solutions Found

It is possible that for a particular configuration of CADEX parameters, the method will not converge on an adversarial example. Since we're zeroing many of the gradient's elements, it may get stuck in a local minimum or simply not point at the right direction to cross the decision boundary. To see how significant this is, we plot histograms of how many solutions were found per training set item, for the 3 values of $n_{changed}$. As can be seen in Fig. 2, for most samples CADEX finds at least 3 or 4 explanations which should be enough for any real world use case.

4.3   Comparison to Training Set Counterfactuals

We compared CADEX to the method of finding the counterfactuals directly from the training set. For each item in the validation set which was denied a loan, we find nearest training set sample using L2 distance which receives a different classification, without limiting the number of attributes that are allowed to change. We plot the cumulative distribution of the distances compared with those found using CADEX. As can be seen from Fig. 3, CADEX generates counterfactual explanations that are much closer to the original.

Table 1. Sample explanations for one refused loan candidate

Attribute	Original	Explanation 1	Explanation 2	Explanation 3
Duration	24	21.74	-	-
Credit	3123	2563.85	-	-
Installment percent	4	3.77	3.59	-
Age	27	29.31	31.09	-
Account status	<0 DM	No checking account	-	-
Sex status	Female	-	Male single	-
Property	Building society savings agreement	-	Real estate	-
Employment	<1 year	-	4..7 years	4..7 years
Purpose	Car (new)	-	-	Car (used)

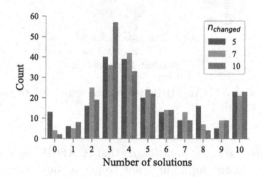

Fig. 2. Number of solutions found by $n_{changed}$

4.4 Comparison to SHAP

We compare CADEX to the well known SHAP method [8] mentioned in Sect. 2.1. SHAP does not directly seek to find counterfactual explanations, but instead explain the effect of each input attribute on the resulting classification. Positive SHAP values are interpreted as increasing the likelihood of the observed classification, and vice versa for the negative values.

When CADEX produces a counterfactual explanation by modifying some attributes in the original input, we expect SHAP to have non-zero coefficients for the same attributes, since they are clearly important to the resulting classification. We have, however, observed that often that is not the case. We perform the comparison as follows. For each CADEX explanation found, we find the attributes which were modified, and count how many of them are zero in the SHAP coefficients of the original input. We used the SHAP implementation on github[1], and used the kernel explainer with the training set as the background dataset. From the results in Fig. 4 we see that in over **93%** of the cases, at least one attribute modified by CADEX had a zero SHAP coefficient.

---

[1] https://github.com/slundberg/shap.

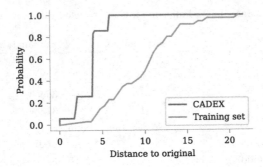

Fig. 3. Cummulative distribution of distances found using CADEX vs. training set

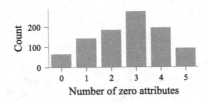

Fig. 4. Distribution of zero SHAP attributes, which were used to produce counterfactual explanations by CADEX

From the comparison we can learn that CADEX can find meaningful attributes to change in the input in order to get a counterfactual explanation, which are undetected and unexplained by SHAP.

4.5   Transferability

We have assumed so far that the bank in our scenario has used a neural network to assign the loan classification to its customers. We now consider the case where the bank has instead used another classifier, which is not a neural network. As shown by [10,13], adversarial examples found using one model can transfer to another one trained on different but similar data, even if that model is not a neural network such as SVM, decision tree and logistic regression. We examined the transferability of CADEX explanations by training a random forest classifier on the same training set with 100 trees and the default scikit-learn parameters. Then, for each validation set item where the classifications of the neural net model and random forest model agreed, we checked how many explanations were indeed adversarial on the random forest model. We repeated the experiment 100 times with different random seeds. We found that on average, in **95.2%** of the cases at least one CADEX explanation was adversarial on the random forest model, and in **87.6%** of the time at least two. In total, **86.1%** of all generated CADEX explanations were found to be adversarial on the random forest model.

This shows that the explanations are largely transferable. For future work, we can consider training more than one neural network model on the data, and to search all of them until a transferable explanation is found.

5 Conclusion

We have shown that CADEX is a robust method to produce counterfactual explanations. Such explanations are by nature highly understandable and actionable by people who receive them. We have demonstrated that CADEX is relatively easy to compute, and can be used to impose various domain and business constraints on the search process.

Going back to the bank scenario, we have shown how the hypothetical bank would benefit from having a way to generate such explanations to its customers. It can use the technique to allow a form of transparency where none exists today, without compromising itself. We believe that such approaches become crucial as machine learning models take a more active part in our daily lives, when we wish to be able to establish trust between the algorithm and the people it serves, and as the public demand for explainability increases.

References

1. Athalye, A., Engstrom, L., Ilyas, A., Kwok, K.: Synthesizing robust adversarial examples. CoRR abs/1707.07397 (2017). http://arxiv.org/abs/1707.07397
2. Brown, T.B., Mané, D., Roy, A., Abadi, M., Gilmer, J.: Adversarial patch. CoRR abs/1712.09665 (2017). http://arxiv.org/abs/1712.09665
3. Dua, D., Karra Taniskidou, E.: UCI machine learning repository (2017). http://archive.ics.uci.edu/ml
4. Goodfellow, I., Shlens, J., Szegedy, C.: Explaining and harnessing adversarial examples. In: International Conference on Learning Representations (2015). http://arxiv.org/abs/1412.6572
5. Kingma, D.P., Ba, J.: Adam: a method for stochastic optimization. CoRR abs/1412.6980 (2014). http://arxiv.org/abs/1412.6980
6. Kurakin, A., Goodfellow, I.J., Bengio, S.: Adversarial examples in the physical world. CoRR abs/1607.02533 (2016)
7. Laugel, T., Lesot, M.-J., Marsala, C., Renard, X., Detyniecki, M.: Comparison-based inverse classification for interpretability in machine learning. In: Medina, J., et al. (eds.) IPMU 2018. CCIS, vol. 853, pp. 100–111. Springer, Cham (2018). https://doi.org/10.1007/978-3-319-91473-2_9
8. Lundberg, S.M., Lee, S.I.: A unified approach to interpreting model predictions. In: Guyon, I., et al. (eds.) Advances in Neural Information Processing Systems 30, pp. 4765–4774. Curran Associates, Inc. (2017). http://papers.nips.cc/paper/7062-a-unified-approach-to-interpreting-model-predictions.pdf
9. Miller, T.: Explanation in artificial intelligence: insights from the social sciences. Artif. Intell. **267**, 1–38 (2019)
10. Papernot, N., McDaniel, P., Goodfellow, I., Jha, S., Celik, Z.B., Swami, A.: Practical black-box attacks against machine learning. In: Proceedings of the 2017 ACM on Asia Conference on Computer and Communications Security, ASIA CCS 2017, pp. 506–519. ACM, New York (2017). https://doi.org/10.1145/3052973.3053009. http://doi.acm.org/10.1145/3052973.3053009

11. Ribeiro, M.T., Singh, S., Guestrin, C.: "Why should i trust you?": explaining the predictions of any classifier. In: Proceedings of the 22nd ACM SIGKDD International Conference on Knowledge Discovery and Data Mining, KDD 2016, pp. 1135–1144. ACM, New York (2016). https://doi.org/10.1145/2939672.2939778. http://doi.acm.org/10.1145/2939672.2939778

12. Shrikumar, A., Greenside, P., Kundaje, A.: Learning important features through propagating activation differences. In: Precup, D., Teh, Y.W. (eds.) Proceedings of the 34th International Conference on Machine Learning. Proceedings of Machine Learning Research, International Convention Centre, Sydney, Australia, 06–11 August 2017, vol. 70, pp. 3145–3153. PMLR (2017). http://proceedings.mlr.press/v70/shrikumar17a.html

13. Szegedy, C., et al.: Intriguing properties of neural networks. In: International Conference on Learning Representations (2014). http://arxiv.org/abs/1312.6199

14. Wachter, S., Mittelstadt, B., Russell, C.: Counterfactual explanations without opening the black box: automated decisions and the GDPR. Harvard J. Law Technol. **31**(2), 841–887 (2018)

15. Zhang, Q.S., Zhu, S.C.: Visual interpretability for deep learning: a survey. Front. Inf. Technol. Electron. Eng. **19**(1), 27–39 (2018). https://doi.org/10.1631/FITEE.1700808

Time-Guided High-Order Attention Model of Longitudinal Heterogeneous Healthcare Data

Yi Huang[1,2], Xiaoshan Yang[1,2], and Changsheng Xu[1,2(✉)]

[1] Institute of Automation, Chinese Academy of Sciences, Beijing, China
[2] University of Chinese Academy of Sciences, Beijing, China
{yi.huang,xiaoshan.yang,csxu}@nlpr.ia.ac.cn

Abstract. Due to potential applications in chronic disease management and personalized healthcare, the EHRs data analysis has attracted much attentions of both researchers and practitioners. There are three main challenges in modeling longitudinal and heterogeneous EHRs data: heterogeneity, irregular temporality and interpretability. A series of deep learning methods have made remarkable progress in resolving these challenges. Nevertheless, most of existing attention models rely on capturing the 1-order temporal dependencies or 2-order multimodal relationships among feature elements. In this paper, we propose a time-guided high-order attention (TGHOA) model. The proposed method has three major advantages. (1) It can model longitudinal heterogeneous EHRs data via capturing the 3-order correlations of different modalities and the irregular temporal impact of historical events. (2) It can be used to identify the potential concerns of medical features to explain the reasoning process of healthcare model. (3) It can be easily expanded into cases with more modalities and flexibly applied in different prediction tasks. We evaluate the proposed method in two tasks of mortality prediction and disease ranking on two real world EHRs datasets. Extensive experimental results show the effectiveness of the proposed model.

1 Introduction

With the wide use of digital devices and information systems in hospital, a large volume of Electronic Health Records (EHRs) data have been accumulated during the patients' admissions to the hospital. EHRs consist of sequential records such as diagnoses, physical test indicators and medication prescriptions. Due to potential applications in chronic disease management and personalized healthcare, such EHRs data have attracted remarkable attentions of both researchers and practitioners. Deep learning based methods are widely used to model EHRs data in healthcare tasks, including disease detection [11,16,23], medical concept embedding [2,5], computational phenotyping [1,3,22] and clinical event prediction [4,11,25]. However, it is still challenging to improve the quality and efficiency of the healthcare/disease management by mining large-scale heterogeneous EHRs data, where the treatment records provided by senior doctors

© Springer Nature Switzerland AG 2019
A. C. Nayak and A. Sharma (Eds.): PRICAI 2019, LNAI 11670, pp. 57–70, 2019.
https://doi.org/10.1007/978-3-030-29908-8_5

and physical examination results monitored during hospital staying always have different formats with various recording frequencies.

There are three challenges in modeling the vast amount of longitudinal heterogeneous EHRs data: (1) *Heterogeneity*: EHRs data are collected from multiple devices and monitors. Multiple data streams are recorded for different destinations in different forms. For example, during a patient's hospital stay, primary diagnostic codes are recorded by doctors for developing treatment plan, while some physical examination results are recorded by medical instruments for monitoring and evaluating the patient's conditions. (2) *Irregular temporality*: On the one hand, the diagnostic codes and physical indicators are always sampled at different frequencies (e.g., ECG sampled dozens per second and vital signs sampled minutely). Moreover, the varying length of hospital staying also leads to the different length of the record sequence in different hospital visits. On the other hand, for a patient with multiple hospital visits, the time interval between two consecutive visits can vary from days to months. (3) *Interpretability*: It is important to improve the interpretability of the healthcare analysis model in addition to the prediction performance on EHRs sequence data. To help doctors and patients with a lot of complex EHRs data, a natural requirement is to identify the supporting evidences for the conclusions.

Over the past few years, a series of deep learning methods have made remarkable progress in resolving these challenges. Existing models often make efforts on improving the prediction performance by capturing the sequential manner of the EHRs data [1,8], or representing the recorded medical concepts [5,17]. In order to getting interpretable results, attention-based models geared towards a specific form of input for a particular task. [6] learns medical concepts with external knowledge. [7,25] learn to selectively attend on different medical features. Most of these models rely on aggregated features via capturing the 1-order temporal dependencies or 2-order multimodal relationships among feature elements of the EHRs data.

In fact, when evaluating patients' health condition, a doctor would comprehensively review both the past medical records and the current reports to find correlation factors, then focus on specific medical features, and make their decisions finally. This kind of reasoning process simultaneously explores the correlations of multiple data sources, such as medical diagnoses, lab indicators and the history medical events. Since most of existing attention models in healthcare only consider 1-order or 2-order relationships, the opportunity is likely to derive from learning high-order correlations (3-order and above) among feature elements. Learning these correlations effectively directs the appropriate attention to the relevant elements in different data modalities and at different time steps that are required to jointly solve the prediction task.

In this paper, we propose a Time-Guided High-Order Attention (TGHOA) model for analyzing the heterogeneous and irregular temporal longitudinal EHRs data. The proposed TGHOA jointly models the correlations of different types of longitudinal medical records and the irregular temporal impact of historical events. Specifically, we compute the one-hot medical diagnose feature by embedding scheme. The uniform representations of physical indicators with different recording frequencies and lengths are computed by convolution kernels. The diagnose features, physical indicator features and historical event features are comprehensively used to compute a relationship matrix which is further transformed to attention scores. Considering that a larger time interval between the previous visit and the current visit leads to less impact of the historical event feature, the time gap is used as an important factor to guide the attention computing. Finally, the attended features are combined together to predict patient's health. Figure 1 shows the framework of the proposed method.

To summarize, the main contributions of this paper are as follows:

- The proposed method can model longitudinal heterogeneous EHRs data via an efficient 3-order attention mechanism, to simultaneously capture the correlations of different modalities and the irregular temporal impact of historical events.
- The proposed high-order attention module can be used to identify the potential concerns of medical features to explain the reasoning process of healthcare model.
- Due to the efficient computation formula of the proposed higher-order attention mechanism, it can be easily expanded into cases with more modalities and flexibly applied in different prediction tasks. In our work, we evaluate the proposed method in two tasks of mortality prediction and disease ranking on two real world EHRs datasets.

2   Related Work

Traditional health analysis system often depends on labor intensive efforts, such as expert-defined phenotyping [18,20] and manual feature engineering [24]. We briefly review the three kinds of deep learning based methods mostly related to our work.

Deep Learning on Longitudinal EHRs Data. [15] shows that RNN models, which can capture the dynamic relationships in sequential data, perform pretty good in large historical data of EHRs. In addition, [3] found that the irregularity of longitudinal EHRs data would affect model performance and used Dynamic Time Warping (DTW) to match irregular temporal patterns in data sequences. [1] proposed a novel LSTM architecture, which performs a subspace decomposition module and a time-decaying memory module followed by the standard gated architecture of LSTM, to handle time irregularities in sequences. These methods do not consider hidden inter-correlation between different medical variables in heterogeneous EHRs data and lack of interpretability.

Deep Learning on Heterogeneous EHRs Data. [12] designed a heterogeneous LSTM structure to explore multiple inter-correlations of different medical sequences with different lengths and record frequencies. [25] proposed an efficient multi-channel attention model of multimodal EHRs time series. However, these models only focus on an instance encounter and do not consider the longitudinally historical records of patients.

Attention-Based Interpretable Deep Methods. RETAIN [7] used two RNNs to model visit-level and variable-level attention mechanisms. Thus it could determine which visit and which medical variable it should pay attention while doing predicting. GRAM [6] used a graph-based attention model in two sequential diagnoses prediction tasks and one heart failure (HF) prediction task. This method could learn robust representations of medical code via a knowledge graph which describes medical ontology relationships. RAIM [25] proposed a recurrent attentive and intensive model for analyzing the multimodal EHR time series. RAIM uses an efficient multi-channel attention on continuous monitored data, which is guided by discrete clinical data. Different from these works, we design a high-order attention module to jointly handle the irregular temporality and heterogeneity of the EHRs data.

3   Methods

In this section, we first define the notations describing the original EHRs events sequence, followed by representation methods of two types of heterogeneous sequential data. Then we describe the details of the proposed time-guided high-order attention module. Finally, we introduce the decision-making process based on the attended features. Figure 1 shows an overview of our method.

3.1   Notations

To reduce clutter, we will introduce our method for a single patient. We define a patient's t-th visit to hospital as one EHRs event $\mathcal{E}^t$, and multiple visits are denoted as a EHRs event sequence $\mathcal{P} = \{\mathcal{E}^1, \mathcal{E}^2, \ldots, \mathcal{E}^t, \ldots, \mathcal{E}^T\}$ where T is the number of this patient's all visits. Each visit $\mathcal{E}^t = \{D^t, W^t\}$ where D^t is an integrated set of discrete diagnoses data indicating what diseases are the patient suffering from. W^t is a set of lab results, such as *saturation of pulse* O_2 and *arterial blood pressure*. y is patient's groundtruth health evaluation after the T visits. In the experiment, y is death rate in the mortality estimation task and the grade of diseases in the disease ranking task.

Patients would be diagnosed with different diseases, so the number of elements in D^t is varying in different visits. We denote $D^t = \{d_1^t, d_2^t, \ldots, d_{n_u}^t\}$, where $d_n^t \in \mathbb{R}^{|\mathcal{D}|}$ is a one-hot representing of patient's n-th disease in t-th visit. The $\mathcal{D}$ denotes the medical code set. The $|\mathcal{D}|$ is the number of unique medical codes of diseases. The n_u is the number of diseases that the patient is suffering from. The lab indicator set $W^t = \{w_1^t, w_2^t, \ldots, w_{n_v}^t\}$, where w_i^t denotes i-th lab indicator at the t-th visit of the patient.

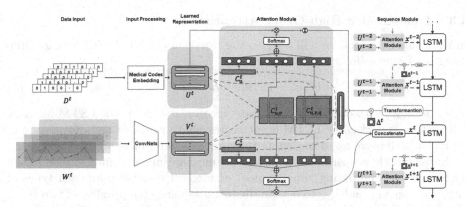

Fig. 1. An overview of the proposed model.

3.2 EHRs Data Representation

In this section, we introduce how to to represent two types of heterogeneous data respectively. The expressive data representations are very important for capturing their correlation patterns.

Diagnose Embedding. Given a medical code representation $\boldsymbol{d}_n^t \in \mathbb{R}^{|\mathcal{D}|}$, we can obtain its embedding representation $\boldsymbol{u}_n^t \in \mathbb{R}^{d_u}$ as follows:

$$\boldsymbol{u}_n^t = \Theta_e \boldsymbol{d}_n^t \tag{1}$$

where $\Theta_e \in \mathbb{R}^{d_u \times |\mathcal{D}|}$ is a learned embedding matrix and d_u is the dimension of the embedding vector. Thus the diagnostic information D^t is transformed to $U^t = \{\boldsymbol{u}_1^t, \boldsymbol{u}_2^t, \ldots, \boldsymbol{u}_{n_u}^t\}$.

Lab Indicator Feature Extracting. As mentioned in Sect. 3.1 that lengths of lab indicator waveforms are different within a single visit. Besides, a lab indicator has different length in multiple visits. To uniformly represent these indicator waveforms, we design a one-dimension convolutional neural network to extract the fixed length features:

$$\boldsymbol{v}_i^t = f_i(\boldsymbol{w}_i^t) \tag{2}$$

where $f_i(\cdot)$ is a two-layer convolutional neural network. The first convolutional block consists of a convolution layer, a max-pooling layer and a activation function ReLu. The second convolutional block consists of a convolution layer and a max-overtime pooling layer [14], which is applied to naturally deals with variable waveform lengths. So we could get the feature representation $\boldsymbol{v}_i^t \in \mathbb{R}^{d_v}$ of the lab indicator $\boldsymbol{w}_i^t$ with a fixed-length. For different lab indicators, we initialize different network parameters of $f_i(\cdot)$ to compute their features respectively. And the network parameters are shared among different visits for the same lab indicator. Then the final feature of n_v lab indicators in t-th visit are represented as $V^t = \{\boldsymbol{v}_1^t, \boldsymbol{v}_2^t, \ldots, \boldsymbol{v}_{n_v}^t\}$.

3.3   Time-Guided High-Order Attention

In the following parts, we will refer to the iteration of LSTM with a single step using notations as follows:

$$h^t, c^t = \text{LSTM}(x^t, h^{t-1}, c^{t-1}) \tag{3}$$

where $h^t \in \mathbb{R}^d$ is the LSTM hidden state vecotor, $c^t \in \mathbb{R}^d$ is the LSTM memory cell vecotor and x^t is the LSTM input vector which contains the information of U^t and V^t. Here we use d to denote the dimensionality of hidden vectors.

Subsequently we consider the attention mechanism as an importance model with each part computing "importance" of medical variable from each types of data. We use $\lambda_{u,q}$ and $\lambda_{v,q}$ to denote the intra-sequence temporality of two types of sequential data. $\lambda_{u,v}$ expresses inter-sequence correlation between two data sequences. $\lambda_{u,v,q}$ captures third-order correlation among two types of sequential data and the history event feature. We compute the importance scores α_u and α_v of the medical diagnose representations and the lab indicator features by combination of intra-sequence irregular temporality unit, inter-sequence correlation unit, third-order correlation unit:

$$\begin{aligned}
\alpha_u^t(i_u) &= \sigma\big(\eta_1 \lambda_{u,q}^t(i_u) + \eta_2 \lambda_{u,v}^t(i_u) + \eta_3 \lambda_{u,v,q}^t(i_u) + \eta_4\big) \\
\alpha_v^t(i_v) &= \sigma\big(\varepsilon_1 \lambda_{v,q}^t(i_v) + \varepsilon_2 \lambda_{u,v}^t(i_v) + \varepsilon_3 \lambda_{u,v,q}^t(i_v) + \varepsilon_4\big)
\end{aligned} \tag{4}$$

here, η_i and ε_i are learned parameters and $\sigma(\cdot)$ refers to the Softmax operation over $i_u \in \{1, \ldots, n_u\}$ and $i_v \in \{1, \ldots, n_v\}$ respectively. Such a linear combination of units provides extra flexibility for the model, since it can learn the reliability of the unit from the data.

Intra-sequence Irregular Temporality. The intra-sequence irregular temporality unit is designed to calculate the importance of medical factors from intra-sequence data based on the historical event feature. We first define attention query q^t as the nonlinearly transformed feature of the previous memory c^{t-1} by a one-layer neural network. What's more, considering that the reference value of historical records would change over time, we use a decaying function $g(\Delta^t) = 1/\log(e + \Delta^t)$ [1,19] as time guidance to adjust impact of historical memory. Δ^t is an irregular time interval between two neighborhood visits. So the memory query is obtained via:

$$q^t = g(\Delta^t) \tanh(\Theta_d c^{t-1} + b_d) \tag{5}$$

where Θ_d and b_d are learned parameters.

The intra-sequence irregular temporality attention weights are formally formulated as:

$$\begin{aligned}
\lambda_{u,q}^t(i_u) &= \tanh\big((\Theta_{u_1} u_{i_u}^t)^{\mathrm{T}} \Theta_{u,q} q^t\big) \\
\lambda_{v,q}^t(i_v) &= \tanh\big((\Theta_{v_1} v_{i_v}^t)^{\mathrm{T}} \Theta_{v,q} q^t\big)
\end{aligned} \tag{6}$$

where $\Theta_{u_1} \in \mathbb{R}^{d \times d_u}$, $\Theta_{v_1} \in \mathbb{R}^{d \times d_v}$, $\Theta_{u,q}$ and $\Theta_{v,q} \in \mathbb{R}^{d \times d}$ are trainable parameters.

Inter-sequence Correlation. Besides the mentioned temporal dependencies of each data sequence, we now introduce a inter-sequence correlation unit, which is able to learn the correlation between the representations of two data sequences. We use a relationship matrix $C_{u,v}$ between data sequences U^t and V^t, where each entry is calculated as follows:

$$C_{u,v}^t(i_u, i_v) = (\Theta_{u_2} u_{i_u}^t)^T \Theta_{v_2} v_{i_v}^t. \tag{7}$$

The $\Theta_{u_2} \in \mathbb{R}^{d \times d_u}$ and $\Theta_{v_2} \in \mathbb{R}^{d \times d_v}$ are trainable parameters. $C_{u,v}^t(i_u, i_v)$ measures the correlation between the i_u-th diagnostic code and the i_v-th lab indicator. Therefore, to retrieve the attention for a specific diagnostic code or lab indicator, we convolve the matrix along the corresponding feature dimension using a 1×1 dimensional kernel. Specifically,

$$\lambda_{u,v}^t(i_u) = \tanh\left(\sum_{i_v=1}^{n_v} \theta_{v_2}(i_v) C_{u,v}^t(i_u, i_v)\right)$$

$$\lambda_{u,v}^t(i_v) = \tanh\left(\sum_{i_u=1}^{n_u} \theta_{u_2}(i_u) C_{u,v}^t(i_u, i_v)\right) \tag{8}$$

where $\theta_{v_2} \in \mathbb{R}^{n_v}$ and $\theta_{u_2} \in \mathbb{R}^{n_u}$ are trainable parameters.

Time-Guided Inter-sequence Correlation. We formulate the high-order correlation between historical records and all data sequences as follows:

$$C_{u,v,q}^t(i_u, i_v) = (\Theta_{u_3} u_{i_u}^t \odot \Theta_q q^t)^T \Theta_{v_3} v_{i_v}^t \tag{9}$$

where $\Theta_{u_3} \in \mathbb{R}^{d \times d_u}$, $\Theta_{v_3} \in \mathbb{R}^{d \times d_v}$ and $\Theta_q \in \mathbb{R}^{d \times d}$ are trainable parameters. Similar to the inter-sequence correlation unit, we use the relationship matrix $C_{u,v,q}^t(i_u, i_v)$ to compute correlated attention scores for each data sequence:

$$\lambda_{u,v,q}^t(i_u) = \tanh\left(\sum_{i_v} \theta_{v_3}(i_v) C_{u,v,q}^t(i_u, i_v)\right)$$

$$\lambda_{u,v,q}^t(i_v) = \tanh\left(\sum_{i_u} \theta_{u_3}(i_u) C_{u,v,q}^t(i_u, i_v)\right) \tag{10}$$

where $\theta_{v_3} \in \mathbb{R}^{n_v}$ and $\theta_{u_3} \in \mathbb{R}^{n_u}$ are trainable parameters.

3.4   Prediction Model

After obtaining attention scores $\alpha_u(i_u)$ and $\alpha_v(i_v)$ for medical diagnoses and lab indicators, the attended features of different data sequences can be calculated respectively. We obtain the final representation of medical codes via attentive mean-pooling as following:

$$\hat{u}^t = \sum_{i_u=1}^{n_u} \alpha_u(i_u) u_{i_u}^t \tag{11}$$

For all features of lab indicators, we concatenate them with attention weights:

$$\hat{v}^t = \alpha_v(1)v_1^t \oplus \alpha_v(2)v_2^t \oplus \ldots \oplus \alpha_v(n_v)v_{n_v}^t. \tag{12}$$

We further concatenate the attended medical diagnose feature $\hat{u}^t$ and lab indicator feature $\hat{v}^t$ and get $x^t = [\hat{u}^t, \hat{v}^t]$. Then, we feed x^t as input into the LSTM sequence model described in Eq. (3). After obtaining the final state h^T, the estimated distribution over possible patient's health evaluation y is given by:

$$\hat{y} = \text{Softmax}(\Theta_o f_o(h^T) + b_o) \tag{13}$$

where $f_o(\cdot)$ a fully-connected layer followed by activation function ReLu. The Θ_o and b_o are learnable parameters of the output layer.

The parameters of all modules are trained end-to-end together by minimizing the following cross entropy loss: $\mathcal{L} = -y^T \log \hat{y} + (1 - y)^T \log(1 - \hat{y})$.

4   Experiments

4.1   Data

In our experiment, we adopt two real world EHRs datasets, namely MIMIC-III [13] and PPMI [9]. For the MIMIC-III dataset, the proposed high-order attention model is applied to a binary classification task of predicting whether the patient would die or survive in ICU. For the PPMI dataset, the proposed attention model is applied in prediction of disease ranking.

MIMIC-III Dataset. Medical Information Mart for Intensive Care III (MIMIC-III) is a publicly available multimodal EHRs dataset comprising deidentified health data associated with critical care patients in Beth Israel Deaconess Medical Center over 11 years [13]. The data contains vital signs, laboratory measurements, diagnostic codes, survival data of 46,520 patients. In the mortality prediction task, we only consider a subset of this dataset. We extract data of patients who have more than two hospital visits. In order to acquire better generalization ability, we choose 1,629 diagnostic codes, whose total frequency of occurrence is greater than 95% in the dataset. For the lab indicators, we choose *heart rate, saturation of pulse O_2, blood glucose* and *arterial blood pressure* from the CHARTEVENTS table as primary physical examination data. We finally get 9,171 records of 2,348 patients. We randomly split the dataset into training and testing sets with a ratio of 4:1. The groundtruth mortality rate in the pre-processed dataset is about 22.7%.

PPMI Dataset. Parkinson's Progression Markers Initiative (PPMI) is an observational clinical and longitudinal study comprising evaluations of people with Parkinson's disease (PD), those people with high risk, and those who are healthy [9]. We refer to [3] for data pre-processing. In our experiments, we use

Table 1. Performance comparison of models on prediction task

Model	MIMIC-III			PPMI		
	Accuracy	AUC-PR	AUC-ROC	Accuracy	AUC-PR	AUC-ROC
LSTM	0.7790	0.8520	0.8555	0.8319	0.8669	0.9595
LSTM-Att	0.7811	0.8710	0.8766	0.8319	0.9180	0.9747
T-LSTM	0.7854	0.8643	0.8643	0.8230	0.8998	0.9660
RETAIN	0.8047	0.8704	0.8772	0.8584	0.9213	0.9755
LSTM+TGA	0.7961	0.8769	0.8829	0.8407	0.9185	0.9764
LSTM+CoA	0.7876	0.8602	0.8664	0.8496	0.9312	0.9808
TGCoA	0.8062	0.8878	0.8867	0.8673	0.9408	0.9837
TGHOA	**0.8155**	**0.9091**	**0.9071**	**0.8938**	**0.9581**	**0.9883**

medication prescriptions as medical codes and choose 318 physical examination features as lab indicators according to [21]. As a result, we get 13,768 records of 586 patients. We randomly split the dataset into training and testing sets with a ratio of 4:1. For the groundtruth labels, we use Hoehn and Yahr (NHY) scale scores [10] which describe how the motor functions of PD patients deteriorate.

4.2 Implementation

All the model parameters introduced in Sect. 3 are randomly initialized and trained in an end-to-end form. We use RMSProp optimizer with gradient descent to train the model. Instead of padding the sequences to the same length, we use the sequences with same number of visits to form a training batch. The learning rate is set to 0.001. Dimension of the medical code embedding is 64. The dimension of the LSTM hidden layer is set to 128. The unit of Δ^t is set to *year* on the MIMIC-III dataset and *day* on the PPMI dataset respectively.

To evaluate the performance of the proposed model, we compare it with the following baseline models:

- **LSTM**: We use basic LSTM as a simple baseline model. Without considering the irregular temporal impact and inter-correlations of EHRs data, we feed the mean-pooled feature $\bar{u}^t$ and mean-concatenated feature $\bar{v}^t$ into the LSTM instead of the attended feature $\hat{u}^t$ and $\hat{v}^t$.
- **LSTM+Att**: This model uses LSTM with attention mechanism which only considers the intra-sequence temporal unit without time-guided query.
- **T-LSTM** [1]: T-LSTM uses a decaying function of time interval to adjust previous memory cell c_{t-1} which affects current output in LSTM. We set T-LSTM as a baseline model which considers the characteristic of varying time intervals in EHRs sequences.
- **RETAIN** [7]: RETAIN uses two RNNs to model visit-level and variable-level attention. It could detect influential past visits and clinical variables.

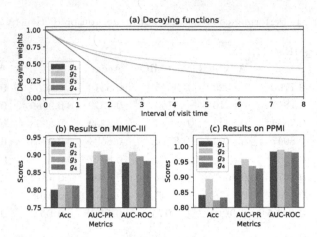

Fig. 2. The effects of time-guided strategies.

- **LSTM+TGA**: This model uses LSTM with interactive attention mechanism which only considers the intra-sequence irregular temporality item $\lambda_{u,q}^t$ and $\lambda_{v,q}^t$ in Eq. (4).
- **LSTM+CoA**: This model uses LSTM with interactive attention mechanism which considers only the inter-sequence correlation item $\lambda_{u,v}^t$ in Eq. (4).
- **TGCoA**: This model uses LSTM with attention mechanism which considers both the intra-sequence irregular temporality item $\lambda_{u,q}^t$ and $\lambda_{v,q}^t$, and the inter-sequence correlation item $\lambda_{u,v}^t$ in Eq. (4).
- **TGHOA**: This is the proposed time-guided high-order attention model which considers all attention items as shown in Eq. (4).

4.3   Result Analysis

The prediction results obtained by all baselines are measured by three evaluation metrics including Accuracy, AUC-PR and AUC-ROC. Table 1 shows the experimental results on both MIMIC-III and PPMI datasets. As shown, The proposed TGHOA outperforms all other models on both datasets.

For the mortality prediction task on the MIMIC-III dataset, LSTM+TGA performs better than LSTM+Att. It indicates intra-sequence irregular temporality unit could better capture the irregular temporal impact than LSTM+Att which do not consider time intervals of sequential data. We also get better performance than RETAIN Besides, LSTM+TGA has better performance than T-LSTM. It shows that considering irregular temporal impact with time-guided attention is more effective. LSTM+CoA has higher scores compared to LSTM model. It indicates that considering the inter-correlation between two types of EHRs data via attention mechanism is helpful. The model LSTM+CoA that incorporates the intra-sequence irregular temporality unit and the inter-source correlation unit further improves the performance. Lastly, the proposed model

Table 2. Diagnoses ranked according to attention scores.

Model	Diagnoses (ICD-9 Code)
TGHOA	Acidosis (276.2); History of kidney neoplasm (V10.52)
	Urinary complications (997.5); Atrial fibrillation (427.31)
	Other noninfectious disorders of lymphatic channels (457.8)
	Other iatrogenic hypotension (458.29)
TGCoA	Other noninfectious disorders of lymphatic channels (457.8)
	Acidosis (276.2); History of kidney neoplasm (V10.52)
	Atrial fibrillation (427.31); Urinary complications (997.5)
	Other iatrogenic hypotension (458.29)
LSTM+CoA	Atrial fibrillation (427.31); Other iatrogenic hypotension (458.29)
	Urinary complications (997.5); History of kidney neoplasm (V10.52)
	Acidosis (276.2)
	Other noninfectious disorders of lymphatic channels (457.8)
LSTM+TGA	Urinary complications (997.5); Other iatrogenic hypotension (458.29)
	Atrial fibrillation (427.31); Acidosis (276.2)
	History of kidney neoplasm (V10.52)
	Other noninfectious disorders of lymphatic channels (457.8)

TGHOA that considers the time-guided high-order correlations obtains the best performance. For the parkinson ranking task on the PPMI dataset, the proposed TGHOA has similar performance improvements over baseline models.

4.4 Effects of Time-Guided Strategy

In the proposed method, the high-order attention module jointly considers the correlation between different modalities and the irregular temporal impact of historical memory. To further analysis the time-guided attention scheme, we investigate the effects of different time-guided functions to the performance of TGHOA. We compare four kinds of decaying functions including $g_1(\Delta^t) = 1$ without any decaying, $g_2(\Delta^t) = 1/\log(\Delta^t + e)$, $g_3(\Delta^t) = e/(\Delta^t + e)$ and $g_4(\Delta^t) = \max\{0, 1 - \Delta^t/e\}$. Here, the g_2 is the adopted decaying function of the proposed method as introduced in Sect. 3.3. Figure 2(a) shows four function curves. Note that the unit of Δ^t is *year* on the MIMIC-III dataset and *day* on the PPMI dataset respectively. Figure 2(b) shows results obtained by our method with four decaying functions.

When using g_1 as a guided function without time decaying, our model obtains worst performance. This further demonstrates that the time-guided attention scheme works well for modeling longitudinal EHRs data. What's more, the decaying function g_2 performs better than g_3 and g_4. It indicates that if the attention model forgets the history feature too quickly, we can only make a suboptimal health assessment, especially obvious on the PPMI dataset.

4.5   Case Study

A key advantage of our model is its interpretability. We conduct a case study of an unseen patient in the testing set of the MIMIC-III dataset.

In Table 2, we rank the diagnostic codes according to their attention scores. We could see that *Acidosis (276.2)* and *History of kidney neoplasm (V10.52)*, which have high fatality rate, have got high attention scores in TGHOA. While other diseases are complications which would not directly cause death. This results demonstrate that proposed attention mechanism gives reasonable cues of the medical features for the mortality prediction. On the other hand, TGHOA and TGCoA generate very different attention scores of the medical feature *Other noninfectious disorders of lymphatic channels (457.8)* while other diagnoses have similar rank. The LSTM+CoA and LSTM+TGA have distinctly different diagnostic attention ranks. It shows that neither LSTM+CoA nor LSTM+TGA has modeled the complete correlation information of the EHRs data.

5   Conclusions

In this paper, we proposed a time-guided high-order attention (TGHOA) model for analyzing the heterogeneous and irregular temporal longitudinal EHRs data. The diagnose features, physical indicator features and historical event features were comprehensively used to compute a relationship matrix which was further transformed to attention scores. The irregular time interval was used as an important factor to guide the attention computing. The proposed high-order attention model was evaluated on the MIMIC-III and PPMI datasets. Extensive experimental results demonstrated the effectiveness and interpretability of the proposed method.

Acknowledgments. This work was supported in part by National Key Research and Development Program of China (No. 2017YFB1002804), National Natural Science Foundation of China (No. 61702511, 61720106006, 1711530243, 61620106003, 61432019, 61632007, U1705262, U1836220) and Key Research Program of Frontier Sciences, CAS, Grant NO. QYZDJSSWJSC039. This work was also supported by Research Program of National Laboratory of Pattern Recognition (No. Z-2018007) and CCF-Tencent Open Fund.

References

1. Baytas, I.M., Xiao, C., Zhang, X., Wang, F., Jain, A.K., Zhou, J.: Patient subtyping via time-aware LSTM networks. In: SIGKDD, pp. 65–74. ACM (2017)
2. Cai, X., Gao, J., Ngiam, K.Y., Ooi, B.C., Zhang, Y., Yuan, X.: Medical concept embedding with time-aware attention. In: IJCAI, pp. 3984–3990 (2018)

3. Che, C., Xiao, C., Liang, J., Jin, B., Zho, J., Wang, F.: An RNN architecture with dynamic temporal matching for personalized predictions of Parkinson's disease. In: SDM, pp. 198–206. SIAM (2017)
4. Choi, E., Bahadori, M.T., Schuetz, A., Stewart, W.F., Sun, J.: Doctor AI: predicting clinical events via recurrent neural networks. In: MLHC, pp. 301–318 (2016)
5. Choi, E., et al.: Multi-layer representation learning for medical concepts. In: SIGKDD, pp. 1495–1504. ACM (2016)
6. Choi, E., Bahadori, M.T., Song, L., Stewart, W.F., Sun, J.: Gram: graph-based attention model for healthcare representation learning. In: SIGKDD. pp. 787–795. ACM (2017)
7. Choi, E., Bahadori, M.T., Sun, J., Kulas, J., Schuetz, A., Stewart, W.: Retain: an interpretable predictive model for healthcare using reverse time attention mechanism. In: NIPS, pp. 3504–3512 (2016)
8. Choi, E., Schuetz, A., Stewart, W.F., Sun, J.: Using recurrent neural network models for early detection of heart failure onset. J. Am. Med. Inform. Assoc. **24**(2), 361–370 (2016)
9. Dinov, I.D., et al.: Predictive big data analytics: a study of Parkinson's disease using large, complex, heterogeneous, incongruent, multi-source and incomplete observations. PLoS ONE **11**(8), e0157077 (2016)
10. Hoehn, M.M., Yahr, M.D., et al.: Parkinsonism: onset, progression, and mortality. Neurology **50**(2), 318–318 (1998)
11. Jagannatha, A.N., Yu, H.: Structured prediction models for RNN based sequence labeling in clinical text. In: EMNLP, vol. 2016, p. 856. NIH Public Access (2016)
12. Jin, B., Yang, H., Sun, L., Liu, C., Qu, Y., Tong, J.: A treatment engine by predicting next-period prescriptions. In: SIGKDD, pp. 1608–1616. ACM (2018)
13. Johnson, A.E., et al.: MIMIC-III, a freely accessible critical care database. Sci. Data **3**, 160035 (2016)
14. Kim, Y.: Convolutional neural networks for sentence classification. arXiv preprint arXiv:1408.5882 (2014)
15. Lipton, Z.C., Kale, D.C., Elkan, C., Wetzel, R.: Learning to diagnose with LSTM recurrent neural networks. arXiv preprint arXiv:1511.03677 (2015)
16. Ma, F., Chitta, R., Zhou, J., You, Q., Sun, T., Gao, J.: Dipole: diagnosis prediction in healthcare via attention-based bidirectional recurrent neural networks. In: SIGKDD, pp. 1903–1911. ACM (2017)
17. Miotto, R., Li, L., Kidd, B.A., Dudley, J.T.: Deep patient: an unsupervised representation to predict the future of patients from the electronic health records. Sci. Rep. **6**, 26094 (2016)
18. Pathak, J., Kho, A.N., Denny, J.C.: Electronic health records-driven phenotyping: challenges, recent advances, and perspectives (2013)
19. Pham, T., Tran, T., Phung, D., Venkatesh, S.: DeepCare: a deep dynamic memory model for predictive medicine. In: Bailey, J., Khan, L., Washio, T., Dobbie, G., Huang, J.Z., Wang, R. (eds.) PAKDD 2016. LNCS (LNAI), vol. 9652, pp. 30–41. Springer, Cham (2016). https://doi.org/10.1007/978-3-319-31750-2_3
20. Richesson, R.L., Sun, J., Pathak, J., Kho, A.N., Denny, J.C.: Clinical phenotyping in selected national networks: demonstrating the need for high-throughput, portable, and computational methods. AIM **71**, 57–61 (2016)
21. van Rooden, S.M., et al.: Clinical subtypes of Parkinson's disease. Mov. Disord. **26**(1), 51–58 (2011)
22. Suresh, H., Szolovits, P., Ghassemi, M.: The use of autoencoders for discovering patient phenotypes. arXiv preprint arXiv:1703.07004 (2017)

23. Thodoroff, P., Pineau, J., Lim, A.: Learning robust features using deep learning for automatic seizure detection. In: MLHC, pp. 178–190 (2016)
24. Xu, Y., Hong, K., Tsujii, J., Chang, E.I.C.: Feature engineering combined with machine learning and rule-based methods for structured information extraction from narrative clinical discharge summaries. J. Am. Med. Inform. Assoc. **19**(5), 824–832 (2012)
25. Xu, Y., Biswal, S., Deshpande, S.R., Maher, K.O., Sun, J.: RAIM: recurrent attentive and intensive model of multimodal patient monitoring data. In: SIGKDD, pp. 2565–2573. ACM (2018)

Towards Understanding Classification and Identification

Mattia Fumagalli$^{(\boxtimes)}$, Gábor Bella$^{(\boxtimes)}$, and Fausto Giunchiglia$^{(\boxtimes)}$

DISI - Department of Information Engineering and Computer Science,
University of Trento, Trento, Italy
{mattia.fumagalli,gabor.bella,fausto.giunchiglia}@unitn.it

Abstract. The paper focuses on two pivotal cognitive functions of both natural and AI agents, namely *classification* and *identification*. Inspired from the theory of teleosemantics, itself based on neuroscientific results, we show that these two functions are complementary and rely on distinct forms of knowledge representation. We provide a new perspective on well-known AI techniques by categorising them as either classificational or identificational. Our proposed *Teleo-KR architecture* provides a high-level framework for combining the two functions within a single AI system. As validation and demonstration on a concrete application, we provide experiments on the large-scale reuse of classificational (ontological) knowledge for the purposes of learning-based schema identification.

Keywords: Classification · Identification · Teleosemantics ·
Cognitive architecture · Knowledge representation

1 Introduction

Class and *classification* are powerful notions in computer science and AI, yet the terms hide a variety of interpretations. Library classifications, for instance, are a traditional form of *knowledge organisation* that apply principled methods to structuring written human knowledge. The notion of class as used in *ontologies* by the Semantic Web community, while also a form of knowledge organisation, is different as it is defined through formal logic and it aims to cater to computational applications such as reasoning or data integration. The *machine learning* community also heavily relies on the notion of classification, understanding it as the sorting of a discrete number of input elements into a discrete number of output categories, *classes* or *clusters*, in a supervised or unsupervised manner.

Our paper looks behind the diverse uses of these notions by various AI communities to find that they are not merely the result of different procedural approaches towards similar goals. Rather, they are complementary and serve tasks with markedly different purposes and representational needs.

This paper was partly supported by the *InteropEHRate* project, co-funded by the European Union (EU) Horizon 2020 programme under grant number 826106.

© Springer Nature Switzerland AG 2019
A. C. Nayak and A. Sharma (Eds.): PRICAI 2019, LNAI 11670, pp. 71–84, 2019.
https://doi.org/10.1007/978-3-030-29908-8_6

The theoretical underpinning of our work is the philosophical theory of *teleosemantics* (also known as *biosemantics* or the *teleological theory of mental content*), and in particular Ruth Millikan's results. Teleosemantics is one of the most popular naturalistic explanations of mental representations: it binds together models of cognition, such as the *classical* and *connectionist* models, and has yielded results in fields such as communication theory or genetics [18]. Based on neuroscientific evidence from animals and humans—and thus formulated in total independence from results in AI or computer science—teleosemantics states that *classification* and *identification* are two distinct tasks that are performed using separate devices of the brain that rely on separate representations of knowledge [23].

The paper offers four main contributions. (1) Based on a teleosemantic perspective, we interpret the notions of classification and identification and clarify the difference between the two. Our goal is not to redefine terminology already in use in various fields of AI, but rather to propose a both theoretically and practically useful distinction between kinds of functions that are often conflated into the same task. (2) We categorise a wide range of AI solutions as based on either of the two or their combination, shedding light on why 'classificational' and 'identificational' tasks need different representations of knowledge in order to be efficient. (3) We introduce a novel *Teleo-KR architecture* that bridges these two fundamental cognitive functions and combines them into a unified AI agent. This high-level theoretical framework may serve, in our view, as a blueprint for future hybrid AI solutions for learning to map between different kinds of representations. (4) Finally, we demonstrate the application of the framework on the AI task of matching data schemas via a combined use of the two kinds of knowledge. We implement the setup as a series of experiments on large sets of data schemas and interpret the results.

In Sect. 2, we define and describe classification and identification based on results from teleosemantics. In Sect. 3, we situate well-known AI tasks with respect to these two functions. Section 4 presents the *Teleo-KR architecture* that models cognitive abilities of artificial agents. Section 5 presents our case study on schema identification. Finally, in Sect. 6 we look at the significance of our results and possible future work.

2   Classification and Identification

Teleosemantics considers biological *perceptual-cognitive systems* (PCS)—i.e., what is able to perceive the external environment, to organize sensory information and to know—to be composed of *devices* having specific *functions*. A device corresponds to a biological component of the brain while the notion of function, as used in neurobiology, describes the role fulfilled by the device. Devices perform tasks with specific goals, in relation to other devices or to the external environment.

In a classic clarifying example [21], bees can be considered as PCSs, i.e., *sender/receiver representational systems*, having a device whose function is to accumulate information about a portion of the environment, such as the location where nectar can be found, as well as a device to *communicate* it to other bees, e.g., through the *bee dance*.

'Communication' and 'accumulation' can be generalised as pivotal applications of *classification* and *identification*, respectively. Classificational representations are views over the stream of diverse data sources collected by the representational system over time. Different individuals may have different classificational representations for the same world state (e.g., a car dealer and a mechanic may classify cars differently), and even the same individual may describe the same world state differently according to context and pragmatic requirements. Nevertheless, classes within individual classifications aim to remain consistent and unequivocal.

Identification, in turn, is required to make learning possible: its purpose is to keep track of things over time, to understand whether they were previously encountered or not, and to focus on new incoming information. In contrast to classification, identification relies on an open and adaptive space for diverse, potentially fuzzy, or contradicting information. Identificational representations afford non-invariant knowledge, adapting to changes in how one perceives things over multiple encounters [3,23].

The device implementing identification relies on knowledge necessary to recognise what is encountered through sensory experience (directly observing the world through seeing, hearing, etc.) and to gather information about it. The device implementing classification builds unequivocal and shareable knowledge from the stream of diverse data collected over experience. Accordingly, a central statement of teleosemantics-which this paper applies to AI as a key contribution-is that *devices may provide their own distinct representations of the world, rather than sharing one common representation. In particular, classificational and identificational representations of knowledge are distinct and are organised in different ways* [20].

Applying these insights to computational agents, we model *identificational representation* (KR^I) as follows:

$$KR^I = \langle S, C^I, \{(s, c^I)\} \rangle \tag{1}$$

where s is a formalization of a *perceptual state*, i.e., a cognitive representation posterior to perception, also called *neural state* in [3]. A perceptual state is the initial cognitive encoding of an object encountered by the agent in the external environment. S is a set of all such perceptual states represented within KR^I. c^I is the representational unit of KR^I that [20] calls substance concept and defines as *'nodes that help in storing knowledge and information arriving at the sensory surfaces'* [23]. For our purposes, c^I is a symbol in KR^I that groups perceptual states together as being from the same object in the external environment [11]. C^I is the set of such substance concepts in KR^I. While the simple formalization above suits the purposes of our paper, in practice we expect KR^I to be more complex and fine-grained both for biological and artificial agents.

We model *classificational representation* (KR^C) as follows:

$$KR^C = \langle S, C^C, \{(i, c^C)\} \rangle \tag{2}$$

where $i \in I$ are *instances*, i.e., *representations of occurrences* of a *given object* [14], and $c^C \in C^C$ are *classes*. Here we commit on the classical definition of class provided in [1], taken as a *set of instances*. A true classificational KR may be a superset of this minimal modelling, e.g., based on first-order logic.

It is important to notice that the difference between instance and class (e.g., *'cat'* and *'my cat Misty'*), pivotal in classificational knowledge, does not occur in the identificational knowledge: both always map into a substance concept [14].

A teleosemantic cognitive device can be modelled as a pair consisting of a knowledge representation and a cognitive function: $D = \langle KR^D, f^D \rangle$. Accordingly, the classification and identification devices are composed, respectively, as $D^C = \langle KR^C, f^C \rangle$ and $D^I = \langle KR^I, f^I \rangle$, where f^C and f^I correspond to the cognitive functions of classification and identification.

We model *identification* (f^I) as the function:

$$f^I : \langle S, KR^I \rangle \rightarrow C^I \tag{3}$$

that assigns perceptual states resulting from an encounter to a given substance concept. For example, recognising a black shape on a photo as *'a cat'* or *'my cat Misty'* is an act of identification.

We model *classification* (f^C) as the function:

$$f^C : \langle I, KR^C \rangle \rightarrow C^C \tag{4}$$

that assigns the instances of a given classification to a given class. The statement *'cats are mammals'*, where the *mammal* is applied to *cat*, both defined within KR^C, is an example of classification.

The representation of identificational knowledge via the *substance concept* strongly relates, in our view, to what in cognitive linguistics is called *basic level category*. As shown in Eleanor Rosch's experiments, the power of identifying something (such as *a cat* or *my cat*) depends highly *on the ability to mirror the structure of information perceived in the world* [27], and this key indicator can be tuned through the accumulation of new information.

Despite the fundamental differences, classification and identification heavily rely on each other. On the one hand, teleosemantics states that the act of recognising is necessarily prior to the act of classifying [20]. On the other hand, the means employed in identification are often heavily influenced by organised classificational knowledge. For instance, the phrase *'lynxes are large-sized wild cats'* may help someone in correctly recognising a cat-like creature in the forest as a lynx. In this particular example, natural language is used to vehicle classificational knowledge that the receiver can use to improve their identification abilities. Language, for humans, is on par with other perceptual ways of acquiring information, such as vision or hearing [21,22]. We adopt this point of view for artificial agents in our case study, where we process semi-formal language as a particular form of perceptual input.

3   Classification and Identification in AI

The findings of teleosemantics bear a high relevance to computational models of intelligence. While AI communities have not always been defining the terms *classification* and *identification* in exactly the same manner as above, the respective functions do have AI equivalents. In this section, we map a few important existing AI approaches and tasks to either of these two functions, explaining their differences in the light of teleosemantics. We also show examples of complementary use of identificational and classificational knowledge in existing AI solutions.

In computational systems, classifications are crucial for reasoning, the sharing of knowledge, standardisation, and are generally widely used as vehicles of semantic interoperability, e.g., for data integration. In AI, and in particular in the field of KR, several kinds of representational systems were developed to model classifications as formal and machine-readable grid or tree structures: semantic networks such as in KL-ONE [6] or top-level and domain ontologies (e.g., Dolce [12] or FOAF [7], respectively).

Identification being such a crucial function in processing sensory input in living beings, it is no surprise to find it playing a central role in AI as well. Machine learning (ML) has proven successful for identificational tasks, especially on unstructured 'sensory-like' input such as images or spoken or written natural language [16,25,28]. ML classifiers expect such input to be pre-processed ('perceived') as *features* (that map to S in Eq. 1) and produce *classes* or *clusters* as output (that map to C^I) [5]. ML models, that map to KR^I, are built through the accumulation of input associated to hypotheses ('training'), as foreseen by teleosemantics for identificational representations, instead of the clear-cut classes of classificational KR.

ML is far from being the sole example of identification in AI. *Schema/ontology matching or entity matching*, crucial tasks in practical applications such as data integration, involve identification that maps one or more incoming structures to a set of reference structures. While the inputs of these matching tasks are typically classificational and not perceptual, most matchers analyse them using techniques common for unstructured input, e.g., the extraction of 'features' from ontology labels via NLP [4,26] and then perform a similarity-based (but not necessarily learning-based) analysis of such features. Note that our teleosemantic model of identification considers the matching of schemas/classes on the one hand and instances on the other hand as essentially the same task over data of different levels of granularity, as opposed to state-of-the-art approaches that regard them as distinct tasks [10]. The need for unifying these tasks has already been recognised in AI in the field of *Structured Machine Learning* [9].

There have been efforts in AI for the mutual reuse of classificational KRs for identificational purposes and vice versa. *Statistical Relational Learning* [13] applies ML to classificational structures. In OntoClean [15], a lot of work has been devoted to defining *identifying* (i.e., *rigid*) *properties* for instances of a certain class (e.g., for an instance of the class *Person*, the *birth date* is an identifying property while *profession* is not as people can change their jobs). In this approach, identificational knowledge is fixed by design as part of classificational knowledge, instead of being derived by gradual accumulation of information. In ontology matching, relying on classificational background knowledge is a common technique for improving precision and recall [10]. Likewise, reusing symbolic knowledge in learning-based (e.g., neural) applications has been a challenging research topic in AI [1,2,17,29,30]. Giunchiglia and Fumagalli [14] motivate the need for two distinct data layers for the two kinds of knowledge in a context of an ontology built for recognition.

The other direction, namely using identification for building classifications, is manifest in *ontology learning* from unstructured, e.g., textual input [8], ontology matching combined with *repair* [19], and *Inductive Logic Programming* [24]. The latter constructs classificational knowledge by learning from examples, without, however, the use of separate identificational knowledge.

4   The Teleo-KR Architecture

This section aims to formalise the principles of teleosemantics, presented in Sect. 2, as a high-level *Teleo-KR architecture*. We intend the architecture as a frame of reference for interpreting and structuring AI solutions that combine the two essential—classificational and identificational—functions of cognition. The approach is demonstrated in a concrete AI use case in Sect. 5.

Figure 1 shows a high-level schema of the architecture. Rounded boxes correspond to teleological devices, and arrows represent the flow of information. Devices fall into one of three general functional areas or *layers*, modelled within a classic perceptual-cognitive paradigm:

- the *perceptual layer* contains devices that take various forms of input from the outside world: sensory, structured data, unstructured text, etc.;
- the *cognitive layer* with devices that collect and organise information about the world;
- and the *behavioural layer* with devices that act upon the world: moving the agent, communicating with other agents, etc.

The contributions of this paper mostly concern the cognitive layer. As shown in Fig. 1, the two pivotal devices of teleological representational systems, namely classification and identification, play the role of connecting environmental inputs to behavioural outputs. (Other devices may also be part of this layer, such as one for linguistic reasoning, but they are out of scope for our paper.) Perceptual input first enters into the cognitive layer through the identification device. This design choice encodes the teleosemantic hypothesis that *identification precedes classification* and, more generally, other cognitive and behavioural acts. The fact that in our architecture identification acts as a bridge between perception and other cognitive functions reflects neuroscientific evidence on the complex transition between perception and cognition and is in line with combined perceptual–conceptual theories of knowledge [3].

On the other hand, as shown by the architecture, both forms of knowledge play a role in controlling the agent's behaviour. For instance, a communicative act may either be the direct result of instinctive recognition (e.g., shouting upon seeing something frightening) or the vehicle of knowledge in an organised manner.

In this model, classificational knowledge is generated through a process of formalization, which we model as a function $f^F : KR^I \rightarrow KR^C$ by which classificational knowledge is derived from identificational knowledge.

Formalisation synthesizes information coming from the external environment, collected through identification during encounters, obtained through various perceptual devices, into a theory about the world. For example, a biologist may observe a living organism from diverse points of view, using an array of sensory inputs (his or her own eyes and hearing, the image provided by a microscope, etc.), before concluding on having discovered an individual of a new species. In the knowledge representation community, this process is known as ontological commitment.

Classifications and deductive thought processes may, in turn, play a role in revising the hypotheses within identificational knowledge, as in the example of the lynx in Sect. 2. Accordingly, we model revision as a function $f^R : KR^C \rightarrow KR^I$ by which classificational knowledge is used to update identificational knowledge.

The two processes that interconnect the two forms of reasoning—*formalisation* and *revision*—hide deep open questions about both biological and artificial cognitive systems. Formalisation, i.e., converting a set of incomplete and potentially contradictory hypotheses into a representation of formal classes and relations, amounts to 'making sense' of identification results in a conscious and fine-grained manner. In the context of AI, it is an instance of the *semantic gap problem* that remains only partially solved, especially in the case of deep learning approaches to identification. Likewise, the process of revision, i.e., controlling the inductive process of identification using formally organised rational knowledge, remains ill-understood: one of the major challenges in current AI research is to find efficient ways for plugging in formal knowledge into learning-based systems. These two functions within the Teleo-KR architecture map to an important set of open problems in AI that will remain subject to extensive research in the near future.

5 Case Study

The goals of our case study are: (1) to demonstrate the conceptual power of the Teleo-KR architecture by applying it to a well-known AI task, showing how the latter can be solved through combining classificational and identificational knowledge; and (2) to propose and test a novel idea on the large-scale reuse of existing classificational knowledge for identificational purposes.

The underlying scenario can be described as the *identification of data schemas*: given a set of input *attributes* (or *properties*), find the schema that matches them best. It is a sub-problem of the well-known *schema alignment problem*, used in applications of semantic interoperability, such as data integration or dynamic data matching.

We map this problem onto the Teleo-KR architecture by building a 'teleological AI agent'. We use this agent to simulate a 'cognitive cycle' that starts from perception, identifies the input, builds identificational and classificational knowledge through accumulation and formalisation, respectively, and finally performs a revision of its identificational knowledge to optimise its abilities. We cover the entire cycle through four successive experiments.

Input. As input classificational knowledge we used schemas collected from 15 resources from *Linked Open Vocabularies*[1] (details will be given in each experiment). A major role of such vocabularies, as explained in Sect. 2, is to communicate conventions for interoperability. It thus makes sense to consider them as natural language input received by an intelligent agent through perception, also considering the commitment of teleosemantics on language being on par with other forms of perceptual input (see Sect. 2).

Perceptual Preprocessing. We consider the preprocessing of linguistic input as part of perception before identification. Its goal is to generate the *perceptual states* (see Sect. 2 above) that constitute the input of identification. We filtered the input classificational knowledge to retain only (Schema, $attribute_1$, ..., $attribute_k$) relations of labels, e.g., *Person* or *dateOfBirth*. We did not consider attributes inherited from ancestors in order not to bias results by the inheritance hierarchy. Perceiving attribute names as natural language text, we converted them to lowercase, and discarded frequent or meaningless stop words, e.g., $dateOfBirth \rightarrow \{date, birth\}$. The goal was to eliminate surface variations related to orthography, word order, etc. The final output was, for each schema, a bag-of-words vector representation of its corresponding attribute words $Schema_i \rightarrow (w_1^{attr}, w_2^{attr}, \ldots, w_n^{attr})$. In machine learning terms, we consider the words in attributes names as the *features* used by the subsequent identification function. While we could just as well have used a different set of features, optimising this aspect of the setup any further was irrelevant with respect to our experiments.

Table 1. Accuracies of identificational devices trained (down) and tested (across) on three schema resources.

		TEST		
		Schema.org	DBpedia	SUMO
TRAINING	Schema.org	63.77%	2.11%	1.46%
	DBpedia	4.19%	94.42%	7.42%
	SUMO	1.90%	5.82%	93.38%

---

[1] https://lov.linkeddata.es.

Identification. We modelled identification essentially as a machine-learning-based supervised document classification task, KR^I being the trained learning model and f^I the learning algorithm. We pre-evaluated multiple algorithms, such as *maximum entropy* or *decision trees*; however, our tests showed that, while the results changed in absolute terms, there was no effect on the overall trends and insights gained. The optimisation of f^I not being of concern to this paper, we finally settled for a decision-tree-based implementation. The training and test sets were all based on the perceived input as described above, with schemas corresponding to output classes and bags of attributes words being the input.

Formalisation. We provide examples to show how acquired identificational knowledge can enrich classificational knowledge. As a form of *ontology learning*, we formalised similarities found among class definitions of various resources by converting them into ontological knowledge of *class equivalence, subsumption,* or *semantic similarity.*

Revision. To close the loop, we reused the newly created classificational knowledge to revise identificational knowledge through the optimisation of training data, and thus improve identification results.

Experiment 1: Identification Ability

In our first experiment we trained three identificational devices using three well-known top-level classificational KR resources: $SUMO^2$ (*178 schemas, 755 attributes*), *Schema.org*[3] (*608, 877*), and *DBpedia*[4] (*775, 2861*). We then evaluated each device with respect to their ability to identify schemas, both over themselves (using the same data as for training) and over each other. These evaluations, shown in Table 1, quantify the ability of each resource to serve as identificational knowledge.

While identification did perform much better, as expected, when the training and test sets were identical, it is also clear that there can be major differences between resources in this respect. Schema.org thus fared much worse for identifying its own schemas. On close analysis, this was due to major overlaps between attribute sets of different schemas, such as the schemas *TVSeries* and *RadioSeries* whose attributes sets were almost identical. The very weak results across resources are, in turn, explained by the relatively low overlap among the schemas and their names (e.g., *Film* in *DBpedia* and *Movie* in *Schema.org* are considered as distinct schemas). This experiment suggests the possibility of a practical tool that evaluates the potential performance of an ontology or a set of schemas in matching tasks. The results may be used, e.g., to finetune schemas in an open-world data integration scenario.

---

[2] http://www.adampease.org/OP/.
[3] https://schema.org/.
[4] https://wiki.dbpedia.org/.

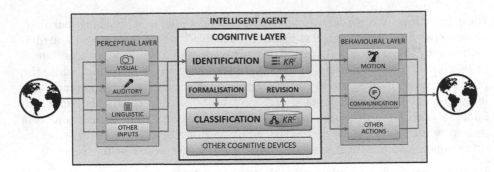

Fig. 1. The Teleo-KR architecture, showing the two pivotal teleosemantic cognitive devices and their relation to perception and behaviour. The schema does not aim an exhaustive description of intelligent agents, hence the inclusion of 'other' devices.

Experiment 2: Knowledge Accumulation

This experiment investigates the effect of *accumulation* of training information on identification results. We increased the size of training sets by merging the three resources from the previous experiment: (A) Schema.org alone; (B) Schema.org + DBpedia; (C) Schema.org + DBpedia + SUMO. We tested the resulting models on a new, more heterogeneous test set consisting of the fusion of 12 vocabularies, some general and some domain-specific, retrieved once again from LOV: *Proton, Bibo,* the *Semantic Web for Research Communities, SwetoD-blp,* the *Comic Book Ontology, Linked Earth, DNB Metadata Terms, Ontology Design Patterns, PREMIS, EBU, Bio,* and *FOAF.* We restricted the evaluation to top-level schemas that were shared by most resources: *Action, Event, Place, Organization, Person, Vehicle, CreativeWork,* and *Product.*

Results can be seen in Fig. 2. While accumulation improves the identification of *Action, Event,* and *Place,* the improvement is only partial for *Organization* and *Person,* and a deterioration is observed for *Vehicle, CreativeWork,* and *Product.* The most salient observation we can make is one well known to the machine learning community: more training data does not systematically lead to higher accuracy. The latter greatly depends on a number of other factors such as input data quality and relevance with respect to the task, how features are defined, the learning algorithm, or the structure of the hypothesis space. In our case, we attribute the low overall scores and the lack of salient improvement of results after accumulation to the high level of heterogeneity of input KRs with respect to the amount of training data.

In conclusion, in a scenario of sparse and heterogeneous identificational knowledge, alternative ways to improve f^I need to be considered beyond the accumulation of more evidence. The Teleo-KR architecture suggests us the improvement of perception (e.g., through feature engineering) but also the cyclic revision of KR^I using knowledge from KR^C. Our two last experiments illustrate the latter process.

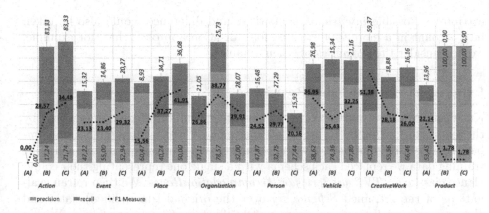

Fig. 2. The effect of accumulation of training data on precision, recall, and F1 for eight core types.

Table 2. Formalisation results: equivalence classes of schemas from Schema.org, derived from identificational similarity scores.

Similar schemas	Similarity
Apartment, SingleFamilyResidence	1.00
Accommodation, House	1.00
Authorize-, Donate-, Give-, Pay-, Return-, TipAction	1.00
Inform-, Invite-, Join-, LeaveAction	1.00
Insert-, Move-, TransferAction	1.00
Comment-, Order-, Reply-, TrackAction	1.00
PropertyValue, QuantitativeValue	0.98
TvSeries, RadioSeries	0.96

Experiment 3: Knowledge Formalisation

This experiment demonstrates formalisation by reusing the output of identification to enrich classificational knowledge. This operation is analogous to the *ontology repair* or *ontology learning* step that is a regular post-processing feature of many ontology matchers [19].

Table 2 shows sets of schemas from *Schema.org* that were found to be identical or very similar by f^I due to overlapping attributes. The high number of shared attributes found across schemas (the table only shows the tip of the iceberg, as we used a similarity cutoff of 0.95) explains the relatively low identificational power of *Schema.org* in experiment 1. Formalisation converts these observed similarities into acquired classificational knowledge of equivalence, e.g., $TVSeries \equiv RadioSeries$. With a larger-scale analysis that includes property set containment, subsumption relations could also be discovered. Note that in this experiment we only consider *extensional* similarity based on shared attributes. Possible *intensional* similarities and differences could also be taken into account in a more sophisticated formalisation approach that, for example, would consider the semantics of schema names.

Experiment 4: Knowledge Revision

Knowledge revision updates KR^I by classificational knowledge, in our case by the axioms formalised in the previous experiment. We re-trained the *Schema.org*-based model of experiment 1 with schemas found equivalent in step 3. In the training data we replaced each schema with a single one representing their equivalence class, e.g., *TVRadioSeries* or *AccommodationHouse*. We then re-ran evaluations of the retrained *Schema.org* over the original (unmodified) data, and obtained an overall accuracy increase of 2.61%, from 63.77% to 66.38%. With a more aggressive approach to formalisation that does not stop at the similarity threshold of 0.95, accuracy could be increased up to 96.52%. This demonstrates the importance of the formalisation–revision cycle as a means to improve the overall cognitive abilities of the artificial agent.

6   Conclusion and Perspectives

Our paper aimed to reframe a range of tasks and open issues of AI with respect to the functions of *classification* and *identification*. Building on the results of teleosemantics, we defined the two notions, clarified their difference based on analogous functions of natural agents, and demonstrated their pivotal role in AI. Our Teleo-KR architecture proposed a schematic model for AI agents based on the combination of these two functions through *formalisation* and *revision*. We demonstrated the use of the architecture on a set of AI tasks inspired from the well-known problems of *schema identification* and *repair*. The case study also introduced a novel idea for the large-scale reuse of symbolic knowledge resources for identification and learning tasks in general.

Among the potential paths of research opened up by our results, we now present two areas of future work. A first perspective concerns the testing of Rosch's seminal hypotheses on the relatedness of identification with *basic level categories*. We plan to verify her results through computational experimentation using the Teleo-KR framework. A second perspective concerns the notion of reward, a central tool of teleosemantics for the evolution and stabilisation of accumulated knowledge. We wish to formalise reward within the Teleo-KR architecture and investigate parallels with results in AI on reinforcement learning.

References

1. Baader, F.: The Description Logic Handbook: Theory, Implementation and Applications. Cambridge University Press, Cambridge (2003)
2. Bader, S., Hitzler, P.: Dimensions of Neural-Symbolic Integration–a Structured Survey. arXiv preprint cs/0511042 (2005)

3. Barsalou, L.W.: Perceptual symbol systems. Behav. Brain Sci. **22**, 577–660 (1999)
4. Bella, G., et al.: Language and domain aware lightweight ontology matching. J. Web Semant. **43**, 1–17 (2017)
5. Bishop, C.M.: Pattern Recognition and Machine Learning. Springer, New York (2006)
6. Brachman, R.J., Schmolze, J.G.: An overview of the KL-ONE knowledge representation system. In: Readings in Artificial Intelligence and Databases, pp. 207–230. Morgan Kaufmann, Burlington (1989)
7. Brickley, D., Miller, L.: FOAF vocabulary specification 0.91 (2010)
8. Buitelaar, P., Cimiano, P., Magnini, B.: Ontology learning from text: an overview. In: Ontology Learning from Text: Methods, Evaluation and Applications, vol. 123, pp. 3–12 (2005)
9. Domingos, P.: Structured machine learning: ten problems for the next ten years. In: Proceedings of the Annual International Conference on Inductive Logic Programming (2007)
10. Euzenat, J., Shvaiko, P.: Ontology Matching, vol. 18. Springer, Heidelberg (2007). https://doi.org/10.1007/978-3-540-49612-0
11. Evans, G., McDowell, J.: The Varieties of Reference. Clarendon Press (1982)
12. Gangemi, A., Guarino, N., Masolo, C., Oltramari, A., Schneider, L.: Sweetening ontologies with DOLCE. In: Gómez-Pérez, A., Benjamins, V.R. (eds.) EKAW 2002. LNCS (LNAI), vol. 2473, pp. 166–181. Springer, Heidelberg (2002). https://doi.org/10.1007/3-540-45810-7_18
13. Getoor, L., Taskar, B.: Introduction to Statistical Relational Learning. MIT Press, Cambridge (2007)
14. Giunchiglia, F., Fumagalli, M.: Concepts as (recognition) abilities. In: FOIS (2016)
15. Guarino, N., Welty, C.: Evaluating ontological decisions with OntoClean. Commun. ACM **45**(2), 61–65 (2002)
16. Hinton, G.: Learning multiple layers of representation. Trends Cogn. Sci. **11**(10), 428–434 (2007)
17. da Cunha Lamb, L.: The grand challenges and myths of neural-symbolic computation. In: Dagstuhl Seminar 08041. Recurrent Neural Networks-Models, Capacities, and Applications. Schloss Dagstuhl, Dagstuhl (2008)
18. Macdonald, G., Papineau, D., et al.: Teleosemantics. Oxford University Press, Oxford (2006)
19. McNeill, F., Bundy, A.: Dynamic, automatic, first-order ontology repair by diagnosis of failed plan execution. IJSWIS **3**(3), 1–35 (2007)
20. Millikan, R.G.: On Clear and Confused Ideas: An Essay About Substance Concepts. Cambridge University Press, Cambridge (2000)
21. Millikan, R.G.: Language: A Biological Model. Oxford University Press, Oxford (2005)
22. Millikan, R.G.: Learning Language. Published in German translation by Alex Burri as 'Spracherwerb'. Biosemantik. Sprachphilosophische Aufsätze, pp. 85–115. Surkamp, Berlin (2012)
23. Millikan, R.G.: Beyond Concepts: Unicepts, Language, and Natural Information. Oxford University Press, Oxford (2017)
24. Muggleton, S., De Raedt, L.: Inductive logic programming: theory and methods. J. Logic Program. **19**, 629–679 (1994)
25. O'Reilly, R.C.: Six principles for biologically based computational models of cortical cognition. Trends Cogn. Sci. **2**(11), 455–462 (1998)
26. Ritze, D., et al.: Linguistic analysis for complex ontology matching. In: CEUR Workshop Proceedings, vol. 689. RWTH (2010)

27. Rosch, E.: Principles of categorization. In: Concepts: Core Readings, vol. 189 (1999)
28. Smith, L.B.: Learning to recognize objects. Psychol. Sci. **14**(3), 244–250 (2003)
29. Sarker, M.K., et al.: Explaining Trained Neural Networks with Semantic Web Technologies: First Steps. arXiv preprint. arXiv: 1710.04324 (2017)
30. Sleeman, J., et al.: Entity type recognition for heterogeneous semantic graphs. AI Mag. **36**(1), 75–86 (2015)

What Prize Is Right? How to Learn the Optimal Structure for Crowdsourcing Contests

Nhat Van-Quoc Truong[1]([⊠]), Sebastian Stein[1], Long Tran-Thanh[1],
and Nicholas R. Jennings[2]

[1] Electronics and Computer Science, University of Southampton, Southampton, UK
{n.truong,s.stein,l.tran-thanh}@soton.ac.uk
[2] Department of Computing, Department of Electrical and Electronic Engineering,
Imperial College, London, UK
n.jennings@imperial.ac.uk

Abstract. In crowdsourcing, one effective method for encouraging participants to perform tasks is to run contests where participants compete against each other for rewards. However, there are numerous ways to implement such contests in specific projects. They could vary in their structure (e.g., performance evaluation and the number of prizes) and parameters (e.g., the maximum number of participants and the amount of prize money). Additionally, with a given budget and a time limit, choosing incentives (i.e., contest structures with specific parameter values) that maximise the overall utility is not trivial, as their respective effectiveness in a specific project is usually unknown a priori. Thus, in this paper, we propose a novel algorithm, *BOIS* (Bayesian-optimisation-based incentive selection), to learn the optimal structure and tune its parameters effectively. In detail, the learning and tuning problems are solved simultaneously by using online learning in combination with Bayesian optimisation. The results of our extensive simulations show that the performance of our algorithm is up to 85% of the optimal and up to 63% better than state-of-the-art benchmarks.

Keywords: Incentive · Crowdsourcing · Bayesian optimisation

1 Introduction

Crowdsourcing has emerged as an efficient approach for obtaining solutions to a wide variety of problems by engaging a large number of Internet users from many places in the world (Ghezzi et al. 2018; Doan et al. 2011). However, the success of crowdsourcing projects relies critically on a crowd to contribute

© Springer Nature Switzerland AG 2019
A. C. Nayak and A. Sharma (Eds.): PRICAI 2019, LNAI 11670, pp. 85–97, 2019.
https://doi.org/10.1007/978-3-030-29908-8_7

(Simula 2013; Doan et al. 2011). Given this, contests[1] have been shown to be an effective approach in these projects, as they are effective and cheap. In particular, by rewarding participants in a contest, task requesters do not necessarily have to pay for every task completed as in other types of financial rewarding schemes, such as paying for performance (Mason and Watts 2010) or using bonuses (Yin and Chen 2015). Indeed, they have to pay only for a certain number of participants, e.g., the top two who have completed the most tasks or the top participant who has completed the tasks with the highest quality. 99designs (www.99designs. com), TopCoder (www.topcoder.com), and Taskcn (www.taskcn.com) are some well-known crowdsourcing platforms that use contests to attract participants.

Much work has taken a game-theoretic approach to investigate the optimal (or efficient) design of contests in general and crowdsourcing contests in particular. It tries to answer the questions of how to distribute the prizes (number of prizes and their values) in contests (Luo et al. 2015; Cavallo and Jain 2012; Moldovanu and Sela 2001). Yet, applying this body of research in building efficient contests for real-world crowdsourcing projects is still challenging. This is because these studies assume rational participants, whereas real participants in crowdsourcing might be partly rational or irrational, as they might lack information, knowledge, or time. Also, these studies do not consider other factors related to the participants' intrinsic motivation that might affect their behaviour such as the project purpose (e.g., collecting data for scientific studies, for government agencies or for companies) or the task nature (e.g., interesting or boring) (Rogstadius et al. 2011; Frey and Jegen 2001).

Furthermore, currently on many crowdsourcing platforms such as Amazon Mechanical Turk (www.mturk.com) and Figure Eight (www.figure-eight.com), the requesters can create tasks and get the submissions in an autonomous manner using programmable Application Programming Interfaces (APIs). This makes it possible to build autonomous agents to monitor and adaptively switch contest structures (e.g., performance evaluation and the number of prizes) and parameters (e.g., the maximum number of participants and the amount of prize money) when appropriate. We refer to a contest structure with specific values of the parameters as an *incentive*[2]. Indeed, it is inconvenient or almost impossible in many cases to switch between incentives manually to identify the best one.

Therefore, another direction for dealing with the incentive problem is to design incentives that are likely to be effective based on previous studies and then empirically select the most effective one. In detail, the above-mentioned studies can be used to design several contest structures with specific ranges of

---

[1] We use the term "contest" in a broad sense to refer to any situation in which participants exert effort to submit tasks for prizes, which are provided based on relative performance. The prizes can be tangible rewards, points, or positions on a leaderboard. Thus, all-pay auctions, lotteries, and leaderboards are considered as contests for the purpose of this paper.

[2] Although the incentives focused on in this paper relate to contests, the problem stated and the algorithms discussed can be used with any other types of incentive in the literature, such as pay for performance or bonuses. Thus, to keep the problem general, we use the term "incentives" instead of "contest structures".

their parameters which are referred to as *candidate incentives*. Then, based on the proposed candidates, an adaptive approach could be used to identify the most effective candidate efficiently. Hence, finding an appropriate way for an autonomous agent (i.e., a computer programme) to select an effective incentive in a crowdsourcing project is a key problem. We refer to this as the *incentive selection problem* (ISP) (Truong et al. 2018).

To identify the most effective incentive to utilise (i.e., exploit), the agent has to try each incentive several times to evaluate its respective effectiveness (i.e., explore). Given this need to balance exploitation and exploration, budgeted multi-armed bandits (MABs) are a promising approach for the ISP. Specifically, they model the problem as a machine with N arms (corresponding to N incentives), pulling an arm (offering the corresponding incentive to a group of participants) incurs a fixed cost (attached to the arm) and delivers a random utility (e.g., the number of tasks completed) drawn from an unknown distribution. The objective in an MAB problem is to find a policy that maximises the total utility within a given budget (e.g., £500) before a deadline (e.g., in the next two weeks).

A number of studies about budgeted MABs have been conducted, such as Badanidiyuru et al. (2018), Ho et al. (2016), and Tran-Thanh et al. (2010). But these studies cannot be applied directly to the ISP as they cannot deal with the tuning problem (i.e., choosing appropriate parameter values for a contest structure) effectively. This is because they do not take advantage of the possible correlations between the arms (i.e., the incentives in a contest structure). Many-armed bandits work well with many or even an infinite number of arms (Li and Xia 2017; Trovo et al. 2016; Bubeck et al. 2011). Yet, none of them can be used to solve the ISP. Actually, they do not consider all important characteristics of the ISP, such as the budget constraints, multidimensional structure of the incentives (i.e., a contest structure has a certain number of parameters), correlations between the arms, and the group-based nature of the arm. Bayesian optimisation (BO) is shown to be an efficient alternative (Snoek et al. 2012). Indeed, BO is designed to find the global optima of functions in as few steps (i.e., function evaluations) as possible. This fits the ISP as applying an incentive incurs a cost. Also, as BO incorporates prior beliefs, if we have some prior knowledge about user performance in the current crowsourcing project, BO can make use of this to find the global optimum more quickly.

Therefore, in this paper we combine the two (online learning with MABs and tuning with BO) to deal with the ISP. By so doing, we decouple a complicated problem (with both learning the best structure and tuning its parameters) into two simple problems and deal with these in a learning process). The ultimate purpose of this work is to build an autonomous agent that can automatically and effectively select the right incentives, so that we can easily deploy projects on crowdsourcing platforms by using the provided APIs. To this end, our main contributions are:

(1) We formalise the ISP and then introduce BOIS, a novel algorithm to solve the ISP effectively by combining an MAB approach to learn the contest structures and BO to tune the parameters of the structures.

(2) We empirically demonstrate that BOIS is generally more effective compared to the state-of-the-art approaches in an extensive series of simulations.

2   The Incentive Selection Problem

Suppose a requester wants to run a crowdsourcing project. The objective is typically to maximise the requester's overall utility with a given financial budget B and time budget T. We can include task quantity, task quality, task completion time, or some subset of them in the utility function. For example, Yin and Chen (2015) consider the quantity and quality of the tasks. To achieve this objective, the requester spends the available budget on providing incentives to encourage participants (referred to as *users*) to perform tasks. For a better presentation, we group the incentives with the same structure in a cluster, which is referred to as incentive cluster (or *cluster* for short). We assume that there are correlations between different incentives in a cluster. Figure 1 shows possible correlations between the incentives in a cluster. Specifically, Fig. 1a shows the effectiveness of the incentives, measured by utility per cost unit[3], in a cluster when there is only one parameter (group size). This figure depicts that the utility initially increases with increasing group size. However, when it is larger than 20, the effectiveness starts decreasing. Figure 1b shows another example in a cluster with two parameters, the group size and the amount of prize money for the best user. We are interested in finding an effective means of selecting the incentives (i.e., exploring their effectiveness and exploiting the most effective one) in order to maximise the requester's overall utility. This is referred to as the ISP.

Formally, let C denote the number of clusters that are being considered in a crowdsourcing project. Cluster i (or C_i for short) has K_i parameters. An incentive a in C_i corresponds to a structure vector $v_a = (v_a^{(1)}, \ldots, v_a^{(K_i)})$, where $v_a^{(k)}$ is the value corresponding to the kth parameter and $v_a^{(k)} \in [\mathrm{v}_{min,i}^{(k)}, \mathrm{v}_{max,i}^{(k)}]$ ($\mathrm{v}_{min,i}^{(k)}, \mathrm{v}_{max,i}^{(k)} \in \mathbb{R}$). Let g_a be the group size of a and c_a denote the cost of applying incentive a once. The expected utility of a is μ_a which is unknown a priori. Let $\mathcal{N} = \{n_a^{(t)} \mid t = 1, \ldots, T; a \in C_i; i = 1, \ldots, C\}$ denote a policy, where $n_a^{(t)}$ is the number of times incentive a is applied in period t, i.e., incentive a is offered to $n_a^{(t)}$ different groups. Let $u_a^{(t)}$ be the total utility of applying this incentive $n_a^{(t)}$ times in period t. The objective is to find a policy that maximises the overall utility:

$$\max \sum_{t=1}^{T} \sum_{i=1}^{C} \sum_{a \in C_i} u_a^{(t)} \quad \text{subject to} \quad \sum_{t=1}^{T} \sum_{i=1}^{C} \sum_{a \in C_i} n_a^{(t)} c_a \leq B. \tag{1}$$

---

[3] The measurement of an incentive's effectiveness will be discussed in Subsect. 3.1.

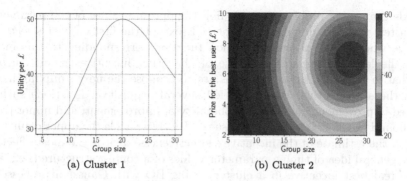

Fig. 1. Illustrative examples of correlations between the incentives in a cluster when it has one (a) and two (b) parameters.

3 The BOIS Algorithm

In this section, we introduce BOIS (which stands for BO-based Incentive Selection), a novel algorithm for the ISP. However, we first describe how the algorithm and the benchmarks measure the effectiveness of the incentives (Subsect. 3.1). We then give an overview of the algorithm (Subsect. 3.2). Finally, we detail how BOIS splits the learning and tuning process into steps and how it acts in these steps (Subsects. 3.3–3.5).

3.1 Measuring the Effectiveness of the Incentives

To measure the effectiveness of the incentives, we use the utility-cost ratio[4], as it reflects the average utility per cost unit. The effectiveness of incentive a is defined as $\delta_a = \mu_a/c_a$. However, as the real effectiveness of the incentives are unknown in advance, we have to estimate them. Right after period t, the estimate of incentive a's effectiveness is:

$$d_a^{(t)} = \hat{\mu}_a^{(t)}/c_a, \tag{2}$$

where $\hat{\mu}_a^{(t)} = \left(1/m_a^{(t)}\right)\sum_{\tau=1}^{t} u_a^{(\tau)}$ is the current estimate of incentive a's expected utility $\left(m_a^{(t)}\right.$ is the number of times incentive a has been applied until the end of period t). To keep the presentation simple, we use *the best incentive* to denote the incentive with the highest estimate, as opposed to *the real best incentive*.

3.2 Algorithm Overview

The idea of BOIS is using an MAB approach to deal with the learning problem (i.e., identifying the best cluster) and using BO[5] with Gaussian processes

---

[4] This ratio is called "density" in Tran-Thanh et al. (2010).

[5] See Snoek et al. (2012) for more information about the method.

to tackle the tuning problem (i.e., finding the optimal values of the parameters of a cluster). In more detail, in each period (except the first one), it selects the incentive whose value of the acquisition function corresponding to this incentive is the largest compared to those of the other incentives in all clusters. Note that in BO, acquisition functions are to propose the next sampling incentive in the search space. We have tried several acquisition functions such as expected improvement, maximum probability of improvement, and upper confidence bound (UCB). However, we chose the UCB (which is the upper confidence bound of the estimate of the incentive's effectiveness) as it is the most effective.

The general idea of tuning parameter values of a contest structure (i.e., finding the real best incentive in a cluster) using BO with Gaussian processes is the following. In each period, based on the incentives sampled in the previous periods, BOIS estimates the mean utilities of the incentives in the cluster using Gaussian process regression (GPR). Then, it calculates the UCBs of the incentives. After that, the incentive with the highest UCB will be the candidate to be applied next in the cluster. BOIS will then choose the candidate incentive in the cluster which has the highest UCB to be applied in that period. In order for the algorithm to use BO, it must have initial estimates of the incentives in each cluster. Therefore, in the first period (i.e., period 1), it samples several incentives, in order to obtain good estimates of the incentives. This step is referred to as the *sampling step*. Then, in each of the next periods (except the last one), it applies the most promising incentive (a), i.e., the incentive with the largest UCB. After that, it updates the UCBs of the incentives in the same cluster (i.e., C_i if $a \in C_i$). We refer to this step as the *stepped exploitation step*. Finally, in the last period it applies the best incentive with the remaining budget. This step is called the *pure exploitation step*, as it simply exploits the best incentive after exploring in the previous periods.

Regarding the UCBs, to select an incentive in a period $t + 1$, at the end of the previous period (t), BOIS uses GPR to estimate the mean utilities of all incentives in each cluster. The results of the estimation are $\hat{\mu}_a^{(t)}$ and $\hat{\sigma}_a^{(t)}$ $\forall a \in C_i; i = 1, \ldots, C$. Then, it calculates the potential effectiveness of all incentives:

$$d_a^{*(t)} = \frac{1}{c_a} \left(\hat{\mu}_a^{(t)} + z^{(t)} \frac{\hat{\sigma}_a^{(t)}}{\sqrt{m_a^{(t)} g_a}} \right). \tag{3}$$

In Eq. 3, $z^{(t)} = Z\left(1 - \frac{t-1}{T-2}\right)$, where Z is the critical value (e.g., 1.96) corresponding to the initial confidence level (e.g., 95%) of the estimates. In more detail, as in the first periods we are not confident about the estimates of the incentives, the confidence intervals should be large to make sure that the algorithm does not leave out the real best incentive. That means at first, it is better to focus on exploration. Then in the next periods, the intervals should become smaller gradually. By so doing, it not only solves the learning and tuning problems simultaneously, but also it performs a smooth transition from exploration to exploitation. Literally, the first period ($t = 1$), $z^{(t)} = Z$ means that it focuses more on exploration. Then, its value gradually decreases as time goes by. And finally, when $t = T$, $z^{(t)} = 0$ means that it focuses only on exploitation.

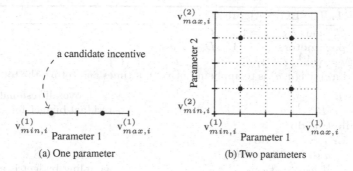

Fig. 2. An illustration of candidate incentives in a cluster in the Sampling step when the cluster has one (a) and two (b) parameters

Additionally, the denominator, $\sqrt{m_a^{(t)} g_a}$, signifies that the exploration level is inversely propotional to the number of sampled users.

In the next subsections, details of the steps will be discussed. The explanations will be linked to the corresponding parts of the pseudocode of BOIS shown in Algorithm 1.

3.3 The Sampling Step

As mentioned above, the purpose of this step (Lines 2–10) is to obtain initial estimates of the incentives in each cluster, which are then used for the regression in the next step. BOIS uses the miniMax distance design (Johnson et al. 1990) to sample the incentives in each cluster to ensure that all other incentives in the cluster are not too far from the sampled ones. An illustration of this space-filling design is shown in Fig. 2. In more detail, for the kth parameter of cluster i, BOIS chooses two values, one in the first quarter and the other in the third quarter of its range, i.e., $v_{min,i}^{(k)} + 0.25\Delta_i^{(k)}$ and $v_{min,i}^{(k)} + 0.75\Delta_i^{(k)}$, where $\Delta_i^{(k)} = v_{max,i}^{(k)} - v_{min,i}^{(k)}$ (Fig. 2a). From these values, we have a set of 2^{K_i} candidate incentives to be sampled. Figure 2b shows four candidate incentives in a cluster which has two parameters.

One issue is that the financial budget is limited and we also want to spend it on further exploration and exploitation. So, BOIS only uses $\epsilon_1 B$ (e.g., $0.2B$) for sampling. This amount might not be enough to sample all the above-mentioned 2^{K_i} candidate incentives ($\forall i = 1, \ldots, C$). Therefore, BOIS simply iterates over the clusters (Line 4) and at each cluster it chooses a random (without repetition) incentive from this set. This is conducted by the NextSample() function (Line 5). Once an incentive is chosen, it will be applied several times so that it has about $\mathbf{U}_1$ (e.g., 20) sampled users, which is calculated by rounding the division $\mathbf{U}_1/g_a$ to the nearest integer (Lines 8–9). By so doing, it guarantees to have enough sampled users if the group size of the incentive is small (e.g., 2). Note that $\lfloor b_1/c_a \rfloor$ in Line 8 is to guarantee the budget being used in this step does not exceed $\epsilon_1 B$. BOIS stops sampling when the budget for sampling is exceeded (Lines 6–7).

Algorithm 1. The BOIS Algorithm

Input: $B, T, C,$ and $K_i \ \forall i = 1, \ldots, C$
Predefined parameters: $\epsilon_1, \epsilon_2, \mathbf{U}_1, D_{min},$ and Z
Output: $u, \mathcal{N} = \{n_a^{(t)} \mid t = 1, \ldots, T; a \in C_i; i = 1, \ldots, C\}$
Note: ApplyIncentive(a, n) is to apply incentive a n times and return the total utility.

Sampling
01: $b \leftarrow B;$ ▷ overall residual budget
02: $b_1 \leftarrow \epsilon_1 B;$ ▷ residual budget for sampling
03: **while** true **do**
04:      **for** $i = 1 \rightarrow C$ **do**
05:         $a \leftarrow$ NextSample(C_i);
06:         **if** $b_1 < c_a$ **then**   ▷ sampling budget is exceeded
07:            Stop the for and while loops;
08:         $n_a^{(1)} \leftarrow \max\{1, \min\{\lceil \mathbf{U}_1/g_a \rceil, \lfloor * \rfloor b_1/c_a\}\};$
09:         $u_a^{(1)} \leftarrow$ ApplyIncentive($a, n_a^{(1)}$); $b_1 \leftarrow b_1 - n_a^{(1)} c_a;$ $b \leftarrow b - n_a^{(1)} c_a;$
10:      UpdateEstimates($C_i, 1, Z$) $\forall i = 1, \ldots, C;$

Stepped Exploitation
11: $b_2 \leftarrow \epsilon_2 b;$ ▷ residual budget for exploration
12: **for** $t = 2 \rightarrow T - 1$ **do**
13:      $a \leftarrow \arg\max_{a' \in C_i; \ i=1,\ldots,C} \{d_{a'}^{*(t-1)}\};$
14:      **if** $d_a^{*(t-1)} < D_{min}$ **then**   ▷ a is too bad
15:         $i \leftarrow$ a random cluster; $a \leftarrow$ a random incentive in $C_i;$
16:      **if** $b_2 < c_a$ **then**   ▷ budget for exploration is exceeded
17:         Stop the for loop;
18:      $n_a^{(t)} \leftarrow \max\{1, \min\{\lceil \mathbf{U}_1/g_a \rceil, \lfloor * \rfloor b_2/c_a\}\};$
19:      $u_a^{(t)} \leftarrow$ ApplyIncentive($a, n_a^{(t)}$); $b_2 \leftarrow b_2 - n_a^{(t)} c_a;$ $b \leftarrow b - n_a^{(t)} c_a;$
20:      UpdateEstimates($C_i, t, Z,$);

Pure Expl.
21: $a \leftarrow \arg\max_{a' \in C_i; \ i=1,\ldots,C} \{d_{a'}^{*(T-1)}\};$
22: $n_a^{(T)} \leftarrow \max\{1, \lfloor * \rfloor b/c_a\};$
23: $u_a^{(T)} \leftarrow$ ApplyIncentive($a, n_a^{(T)}$);

24: $u \leftarrow \sum_{t=1}^{T} \sum_{i=1}^{C} \sum_{a \in C_i} u_a^{(t)};$ ▷ overall utility
25: **return** $u, \mathcal{N};$

3.4 The Stepped Exploitation Step

At first, BOIS sets the budget for stepped exploitation, a specific portion of the residual budget which is identified by ϵ_2, e.g., 0.5 (Line 11). Then, in each period (t) before the deadline, it will choose the incentive (a) with the highest potential effectiveness (Line 13). The incentives are chosen based on their UCBs which contain both the estimates of the incentives' effectiveness so far and the certainty of the estimates. Thus, this step can be considered as both exploiting (choosing the incentives whose estimates are high) and exploring (choosing the incentives whose potential to be the real best one are high).

In some cases, the potential effectiveness of this incentive $(d_a^{*(t-1)})$ can be very low since the sampled incentives so far in this cluster (C_i) had very low utilities. To prevent it from falling into the trap of exploring ineffective incentives,

---

Algorithm 2. The UpdateEstimates() Function

Input: C_i, t and Z
Output: C_i with updated $d_a^{*(t)}$ $\forall a \in C_i$
  1: Use Gaussian process regression to estimate $\hat{\mu}_a^{(t)}$ and $\hat{\sigma}_a^{(t)}$ $\forall a \in C_i$;
  2: Calculate $d_a^{*(t)}$ based on Equation 3 $\forall a \in C_i$;

---

if $d_a^{*(t-1)}$ is less than some lower bound (D_{min}), BOIS will randomly choose another incentive (Lines 14–15). It is not difficult to determine a value for D_{min}. For example, if the utility is measured by the number of tasks completed and we expect an acceptable incentive to have about 20 completed tasks per £, then we can set D_{min} to this value or even 10 if we are not quite sure about this number. Yet, it should be larger than the possible minimum number of tasks, e.g., 0. As in the sampling step, after having an incentive, BOIS will apply the incentive several times so that it obtains about $\mathbf{U}_1$ sampled users (Lines 18–19). This step stops when the budget for exploration is exceeded (Lines 16–17).

3.5   The Pure Exploitation Step

This step (Lines 21–23) simply applies the best incentive with the residual budget. Indeed, from Eq. 3 we can see that in this period the factor $z^{(T)}$ is zero. That means it does not explore anymore but totally exploits the incentive with the highest estimate of the effectiveness.

4   Experimental Evaluation

To systematically evaluate the performance of BOIS, we run simulations in a wide range of settings. It would be infeasible to undertake this evaluation in a real crowdsourcing project as we have to deploy the project multiple times with different financial budgets, time budgets, and numbers of clusters, as well as different values of the parameters of each cluster. Even then, we could not guarantee we have explored the main cases in a comprehensive fashion. In the following, we present the benchmarks (Subsect. 4.1), the experimental settings (Subsect. 4.2), and then discuss the corresponding results (Subsect. 4.3).

4.1   Benchmarks

As the state-of-the-art algorithms are not specifically designed to deal with choosing the best cluster together with tuning its parameter values, we make a number of modifications for them to perform well with the ISP.

(1) **ε-first:** This algorithm spends ϵB (where ϵ is specified in advance, e.g., 0.1) in the first period to explore by sequentially applying a random incentive in each cluster until this budget is exceeded (Tran-Thanh et al. 2010). With a chosen incentive a, it applies this incentive $\max\{1, \lceil \mathbf{U}_1/c_a \rceil\}$ times to obtain

about U_1 (e.g., 20) sampled users. In the second period, it uses GPR to estimate the best incentive. Then it spends the subsequent period purely exploiting the best incentive explored in the first period with the remaining budget, i.e., $(1 - \epsilon)B$.

(2) **Decaying ϵ-greedy** (or ϵ-greedy for short): It spreads the budget B over T periods. In each period, with the given budget, it applies the best incentive with probability $(1 - \epsilon)$ and a random incentive in a random cluster with probability ϵ, where $\epsilon = (T - t)/(T - 1)$. It totally explores when $t = 1$ (i.e., $\epsilon = 1$). When t increases, ϵ gradually decreases. And when $t = T$, it completely exploits the best incentive (i.e., $\epsilon = 0$). At the end of period t $(1 < t < T)$, it uses GPR to estimate the best incentive for period $t + 1$.

(3) **Random**: It spreads the budget B over T periods. Then in each period, it applies a random incentive in a random cluster with the given budget.

(4) **Optimal Solution**: It simply applies the real best incentive all the time. To have this optimality, we have to know the values μ_a $(\forall a \in C_i; \forall i = 1, \ldots, C)$ in advance, which is typically impossible in practice. Thus, this approach represents an upper bound of what any algorithm could achieve.

4.2 Simulation Settings

To evaluate the performance of the algorithms, we run simulations in three different settings where the independent variables are time budget, financial budget, and number of clusters. In the simulations of each setting, the related quantities, i.e., utility, B, T, C, group size, and the amount of prize money for the best user (except the corresponding independent variable) are drawn uniformly from specific ranges. The ranges are chosen to represent realistic settings in real crowdsourcing projects. Specifically, C is generated randomly from 1 to 10. The group sizes (g_a) are from 1 to 50. The amount of prize money for the best user is between £1 and £25. T is between 2 and 30. And B is from 10 to 200 times the round cost. Here, round cost is the cost of applying all the clusters, where in each cluster the incentive which the highest cost is applied once. This is to guarantee B is not too small compared to the generated values of C and g_a, so that we can carry out a meaningful performance comparison.

For each value of the independent variables, we run 2,000 simulations to achieve statistically significant results at the 99% confidence level. Error bars of the line graphs in Fig. 3 represent the confidence intervals. We run the algorithms with different values of the predefined parameters and then choose appropriate values for the parameters. For instance, with ϵ of ϵ-first, we first run this algorithm with different values (such as 0.05, 0.1, 0.2, 0.3, and 0.4). Then we choose one value that helps ϵ-first perform well in different settings. A similar process is used for the other predefined parameters such as ϵ_1 and ϵ_2 of BOIS. As changing these values slightly does not result in a significant difference (i.e., the trends of the algorithms' performance are broadly the same), in Subsect. 4.3 we only present the results on the simulations with the following values of the algorithms' predefined parameters. With BOIS, $\epsilon_1 = 0.1$, $\epsilon_2 = 0.5$, $U_1 = 20$, and $Z = 1.96$.

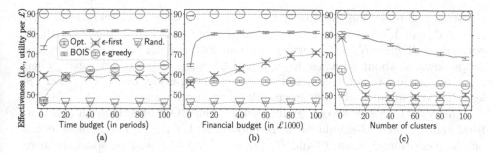

Fig. 3. Results of the simulations

With ϵ-first, $\epsilon = 0.1$ and $\mathbf{U}_1 = 20$. In the simulations, we assume that the performance of a group (i.e., the total utility of all users in the group) is linearly proportional to the group size. This means the more users there are in a group, the better the performance of the whole group. This assumption is based on an empirical study conducted by Araujo (2013).

4.3   Results

In general, BOIS performs best in most cases (Fig. 3). With a looser deadline, the algorithm performs better, especially when T is greater than 15 (Fig. 3a). This is mainly because of the miniMax space-filling design and the BO. Specifically, if $T = 2$ (i.e., no exploration) the performance of BOIS is good enough (which is a utility of about 70 per £). And if $T = 15$, its performance increases clearly (up to about 79 per £). Note that the time budget is used to learn all the clusters. This confirms that BO can quickly approach a global optimum (i.e., the real best incentive). ϵ-greedy also performs better with a larger T, since it has more time to explore. Yet, its performance is far below that of BOIS. Whereas, different values of T does not affect the performance of ϵ-first as it always uses two periods. Nonetheless, with a larger financial budget, ϵ-first performs better, as there is more budget for exploration (Fig. 3b). As the way it explores is inflexible (i.e., always ϵB), when B is small, the budget for exploration is not enough, so that the GPR conducted in the second period does not have enough samples to identify one of the best incentives.

Figure 3b suggests that B should be large enough (e.g., at least £5000 as in the simulations) for BOIS to achieve a good performance. A larger B helps improve its performance slightly. Actually, it needs enough budget to sample all 2^{K_i} candidate incentives $\forall i = 1, \ldots, C$. And with a larger B, the amount of the added budget will be used for exploiting. In Fig. 3c, the performance of BOIS drops significantly when C becomes larger. This is easy to understand, as with a fixed B and a larger C, $\epsilon_1 B$ is not enough to sample all the candidate incentives in all clusters. Similarly, ϵ-first's performance drops more quickly than that of BOIS. The reason is that it does not make use of the time budget to conduct further exploration. Regarding the number of parameters, as BOIS does not scale well to settings with large values of K_i, we only ran experiments with $K_i = 2$, 3, and 4. These results have a similar trend as in Fig. 3c, i.e., that BOIS performs well when $K_i = 2$ (a utility of about 82 per £). Then, its performance drops down to about 69 when $K_i = 3$ and 62 when $K_i = 4$. Also, even when $K_i = 4$, the time to run the algorithm (the whole episode, i.e., $t = 1, \ldots, T$) is less than one minute, which is acceptable in practice.

The results suggest several guidlines for using BOIS effectively in practice. First, both T and B should be large enough and a larger T has more effect on the algorithm. Second, C and K_i ($\forall i = 1, \ldots, C$) should be small. If there are many (e.g., 15) candidate clusters to choose from, it is better to continue using related studies from psychology, sociology, or computer science to filter out clusters which are not actually promising. A similar process should be done with the parameters.

5   Conclusions and Future Work

We have discussed the incentive selection problem (ISP) and introduced an algorithm (BOIS) to solve the ISP effectively. Our algorithm performs efficiently in a wide range of different cases without the need to tune its predefined parameters. It is shown to outperform the state-of-the-art approaches in simulations. Even though BOIS is specifically designed for incentives in the form of contests, it can also be used with other types of incentives where the group size is 1 (i.e., there are no contests, such as pay for performance or using bonuses). Although BOIS is an important initial step towards solving the ISP, there are some areas of further work. First, we assume that time steps are homogeneous and a new incentive can be started only when all previous ones have completed. Addressing this limitation would shorten waiting times and thereby the total time used by the algorithm. Additionally, this could improve the overall performance as the algorithm has more time to conduct exploring, especially when the time budget is limited. Second, we also asume that the cost of applying an incentive is the same at all times. This may be limiting in more general settings. For example, some incentives are inherently designed with variable payment such as pay for performance or using bonuses. Third, the model of user performance used in the simulations is rather simple, while it might be more complicated in different projects. Thus, running experiments might help us better understand how people behave in different cases. Hence, we can design better algorithms to solve the ISP more efficiently. Regarding other applications of our work, the model and the algorithm developed can be applied in other domains with a group-based nature such as in schools, companies, or organisations (i.e., finding the most effective groups of students or employees to work or study together).

Acknowledgments. This research was sponsored by the U.S. Army Research Laboratory and the U.K. Ministry of Defence under Agreement Number W911NF-16-3-0001. The views and conclusions contained in this document are those of the authors and should not be interpreted as representing the official policies, either expressed or implied, of the U.S. Army Research Laboratory, the U.S. Government, the U.K. Ministry of Defence or the U.K. Government. The U.S. and U.K. Governments are authorised to reproduce and distribute reprints for Government purposes notwithstanding any copyright notation hereon.

References

Araujo, R.M.: 99designs: an analysis of creative competition in crowdsourced design. In: HCOMP, pp. 17–24 (2013)

Badanidiyuru, A., Kleinberg, R., Slivkins, A.: Bandits with knapsacks. JACM **65**(3), 1–55 (2018)

Bubeck, S., Stoltz, G., Szepesvári, C., Munos, R.: X-armed bandits. JMLR **12**, 1655–1695 (2011)

Cavallo, R., Jain, S.: Efficient crowdsourcing contests. In: AAMAS, vol. 2, pp. 677–686 (2012)

Doan, A., Ramakrishnan, R., Halevy, A.Y.: Crowdsourcing systems on the world-wide web. CACM **54**(4), 86–96 (2011)

Frey, B.S., Jegen, R.: Motivation crowding theory. J. Econ. Surv. **15**(5), 589–611 (2001)

Ghezzi, A., Gabelloni, D., Martini, A., Natalicchio, A.: Crowdsourcing: a review and suggestions for future research. IJMR **20**(2), 343–363 (2018)

Ho, C.J., Slivkins, A., Vaughan, J.W.: Adaptive contract design for crowdsourcing markets: bandit algorithms for repeated principal-agent problems. JAIR **55**, 317–359 (2016)

Johnson, M., Moore, L., Ylvisaker, D.: Minimax and maximin distance designs. JSPI **26**(2), 131–148 (1990)

Li, H., Xia, Y.: Infinitely many-armed bandits with budget constraints. In: AAAI, pp. 2182–2188 (2017)

Luo, T., Kanhere, S.S., Tan, H.P., Wu, F., Wu, H.: Crowdsourcing with tullock contests: a new perspective. In: INFOCOM, pp. 2515–2523 (2015)

Mason, W., Watts, D.J.: Financial incentives and the "performance of crowds". ACM SigKDD Explor. Newsl. **11**(2), 100–108 (2010)

Moldovanu, B., Sela, A.: The optimal allocation of prizes in contests. AER **91**(3), 542–558 (2001)

Rogstadius, J., Kostakos, V., Kittur, A., Smus, B., Laredo, J., Vukovic, M.: An assessment of intrinsic and extrinsic motivation on task performance in crowdsourcing markets. In: ICWSM, pp. 321–328 (2011)

Simula, H.: The rise and fall of crowdsourcing? In: HICSS, pp. 2783–2791 (2013)

Snoek, J., Larochelle, H., Adams, R.P.: Practical Bayesian optimization of machine learning algorithms. In: NIPS, p. 9 (2012)

Tran-Thanh, L., Chapman, A., De Cote, E.M., Rogers, A., Jennings, N.R.: Epsilon-first policies for budget-limited multi-armed bandits. In: AAAI, pp. 1211–1216 (2010)

Trovo, F., Paladino, S., Restelli, M., Gatti, N.: Budgeted multi-armed bandit in continuous action space. In: ECAI, pp. 560–568 (2016)

Truong, N.V.Q., Stein, S., Tran-Thanh, L., Jennings, N.R.: Adaptive incentive selection for crowdsourcing contests. In: AAMAS, pp. 2100–2102 (2018)

Yin, M., Chen, Y.: Bonus or not? Learn to reward in crowdsourcing. In: IJCAI, pp. 201–207 (2015)

Simple Is Better: A Global Semantic Consistency Based End-to-End Framework for Effective Zero-Shot Learning

Fan Wu[1], Shuigeng Zhou[1(✉)], Kang Wang[1], Yi Xu[1], Jihong Guan[2], and Jun Huan[3]

[1] Shanghai Key Lab of Intelligent Information, and School of Computer Science, Fudan University, Shanghai, China
{fanwu15,sgzhou,kangwang17,yxu17}@fudan.edu.cn
[2] Department of Computer Science and Technology, Tongji University, Shanghai, China
jhguan@tongji.edu.cn
[3] Big Data Lab, Baidu Research, Beijing, China
huanjun@baidu.com

Abstract. In image recognition, there are many cases where training samples cannot cover all target classes. Zero-shot learning (ZSL) addresses such cases by classifying the samples of unseen categories that have no corresponding samples contained in the training set via class semantic information. In this paper, we propose a novel and simple end-to-end framework, called Global Semantic Consistency Network (GSC-Net for short), which makes complete use of the semantic information of both seen and unseen classes to support effective zero-shot learning. We also employ a soft label embedding loss to further exploit the semantic relationships among classes and use a seen-class weight regularization to balance attribute learning. Moreover, to adapt GSC-Net to the setting of Generalized Zero-shot Learning (GZSL), we introduce a parametric novelty detection mechanism. Experiments on all the three widely-used ZSL datasets show that GSC-Net performs better than most existing methods under both ZSL and GZSL settings. Especially, GSC-Net achieves the state of the art performance on two datasets (AWA2 and CUB). We explain the effectiveness of GSC-Net from the perspectives of class attribute learning and visual feature learning, and discover that the validation accuracy of seen classes can serve as an indicator of ZSL performance.

Keywords: Zero-shot learning · Global semantic consistency · Label embedding loss

© Springer Nature Switzerland AG 2019
A. C. Nayak and A. Sharma (Eds.): PRICAI 2019, LNAI 11670, pp. 98–112, 2019.
https://doi.org/10.1007/978-3-030-29908-8_8

1 Introduction

In some real computer vision applications, such as species classification [3], activity recognition and anomaly detection [22], labeled training samples cannot cover all target classes. Zero-shot Learning (ZSL) [1] provides a systematic way to address this type of problems by utilizing the semantic information of classes. Such class semantic information, including annotated attributes [9], label word vectors [16] *etc.*, can be uniformly encoded in attribute vectors [20,32]. This process is also referred to as class embedding or (label) semantic embedding.

ZSL uses the samples of the seen classes (those having training samples) for training and tests on the samples of the unseen classes (those having no training samples). The semantic embeddings of both seen and unseen classes are used as the bridge connecting them. The essence of ZSL is to learn the association between the visual features of samples (images) and the class embeddings, which is then transferred to the samples of unseen classes.

In the test stage, ZSL considers only classifying new images of unseen classes. However, in some real-world applications, an image classification system usually needs to recognize new images from both seen and unseen classes of the application domain. This is addressed by the so-called *generalized zero-shot learning* (GZSL). Figure 1 illustrates both ZSL and GZSL tasks. Most of the existing ZSL methods [29] can be grouped into three types:

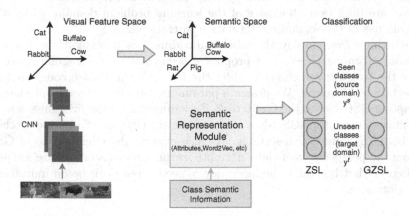

Fig. 1. Illustration of ZSL and GZSL tasks. Available data are labeled images of the seen classes (source domain, $\mathcal{Y}^s$) and semantic information of both seen and unseen classes ($\mathcal{Y}^{s+t}$). In essence, both ZSL and GZSL learn the mapping or compatibility between visual feature space and semantic space, then apply it to unseen classes (target domain, $\mathcal{Y}^t$). At the test stage, ZSL model is only evaluated on unseen classes ($\mathcal{Y}^t$) whereas GZSL recognizes images from both seen and unseen classes ($\mathcal{Y}^{s+t}$).

The 1st-type of works includes these that learn a compatibility function between the image features and the class embeddings, and treat ZSL classification as a compatibility score ranking problem [2,10,26]. However, these methods suffer from the following drawbacks: the attribute annotations are pointwise rather than pairwise, compatibility scores are unbounded, and ranking may fail to learn some semantic structures due to the fixed margin [4].

Methods of the 2nd-type project the visual features and semantic embeddings into a shared space and treat ZSL training as ridge regression. The shared space can be visual space, semantic space or a common space of visual features and semantic embeddings. The prediction process of these methods is a nearest neighbor search in the shared space, which may cause *hubness problems* [19].

Most of recent works fall into the 3rd-type. They either employ deep neural networks [5,15,30], or use generative models [13,28,33], to pursue better performance. For example, Morgado *et al.* [17] adopted a semantically consistent regularization of the last fully-connected (FC) layer's weights of the neural network in end-to-end training, based on the attribute matrix of the seen classes. These methods are usually complex and hard to be deployed in general situation, and very time-consuming to be trained.

To overcome the limitations of existing ZSL methods, in this paper we propose a novel and simple end-to-end framework, called *global semantic consistency network* (GSC-Net) to exploit the semantic embeddings of both seen and unseen classes while preserving the global semantic consistency. By treating the global semantic consistency layer as a fully-connected (FC) layer with fixed weights, we can easily employ all kinds of CNN techniques such as the dropout policy, sigmoid activation, and cross entropy loss. The softmax layer and loss layer in GSC-Net are both over all classes of the learning problem domain, which thus makes full use of the semantic information in training.

Furthermore, we employ the label embedding loss to exploit the semantic relationships among classes and propose a seen-class weight regularization to balance the training, which thus guides the net to learn a more comprehensive representation. Moreover, We design a parametric novelty detection mechanism for adapting GSC-Net to the GZSL task. Experimental results over three widely-used datasets show that GSC-Net performs better than most existing methods under both ZSL and GZSL settings. We also explain the effectiveness of GSC-Net from the perspectives of class attribute learning and visual feature learning, and discover that the validation accuracy of seen classes can be an indicator of ZSL performance.

2   Method

2.1   Problem Formulation

Assume there are n_s seen classes (denoted by set $\mathcal{Y}^s$) and n_t unseen classes (denoted by set $\mathcal{Y}^t$) in a problem domain, where seen classes and unseen classes are disjoint, *i.e.*, $\mathcal{Y}^s \cap \mathcal{Y}^t = \varnothing$. So the number of total classes $n_c = n_s + n_t$. In the seen class space $\mathcal{Y}^s$, given a dataset with N_s labeled samples, $\mathcal{D}_s = \{(\mathbf{I}_i, y_i), i = 1, \ldots, N_s\}$ where $\mathbf{I}_i$ is the i-th training image, and $y_i \in \mathcal{Y}_s$ is the label of $\mathbf{I}_i$. Given the class attribute matrix $\mathbf{A} = [\mathbf{A}^s, \mathbf{A}^t]$ where $\mathbf{A}^s \in \mathbb{R}^{L \times n_s}$ corresponds

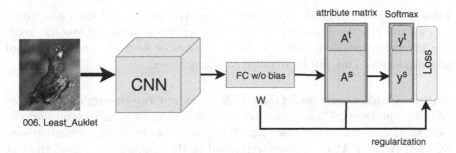

Fig. 2. The GSC-Net architecture. The class attribute matrix $\mathbf{A} = [\mathbf{A}^s, \mathbf{A}^t]$ where $\mathbf{A}^s$ is for the seen/training classes and $\mathbf{A}^t$ is for the unseen/test classes. Though no training images belong to the unseen classes, the Global Semantic Consistency (GSC) Layer, softmax layer and loss layer are designed for all classes $\mathcal{Y}^{s+t}$.

to the seen classes, $\mathbf{A}^t \in \mathbb{R}^{L \times n_t}$ corresponds to the unseen classes, L is the attribute dimension.

Now, given a new test image $\mathbf{I}_j$, the goal of ZSL is to predict its label $\hat{y}_j$ just among the unseen classes, *i.e.*, $\hat{y}_j \in \mathcal{Y}^s$, while the goal of GZSL is to predict its label $\hat{y}_j$ among all classes, *i.e.*, $\hat{y}_j \in \mathcal{Y}^{s+t}$ where $\mathcal{Y}^{s+t} = \mathcal{Y}^s \cup \mathcal{Y}^t$.

2.2 Global Semantic Consistency Network

Architecture. To exploit the semantic attributes of both seen and unseen classes for training, we propose a novel, simple yet effective end-to-end approach, called *Global Semantic Consistency Network* (GSC-Net for short) for the ZSL task, and adapt it to the GZSL task later. Figure 2 is the architecture of GSC-Net, which consists of four major components as follows:

1. **CNN block**: $x = CNN(\mathbf{I})$. In this paper, we use the pretrained resnet50 [11] as the CNN by default. The pretrained CNN acts as a feature extractor, with the original last fully-connected (FC) layer being dropped. For fast end-to-end training, we freeze this block's parameters in the first 5 epochs.
2. **FC w/o bias**: $x_a = \mathbf{W}x$. This FC layer (its weight matrix is $\mathbf{W}$ and bias is 0) maps the CNN features into a L-dimensional space. Its output can be interpreted as the image embedding in attribute space.
3. **Global Semantic Consistency (GSC) Layer**: $y^{out} = \mathbf{A}x_a$. Here, $\mathbf{A}$ is the class attribute matrix (it can also be label word embeddings). [32] discussed how to fuse multiple semantic vectors together. If the auxiliary information needs a neural encoding layer, then we can include this layer in end-to-end co-training. Since the semantic information is usually about classes and can be fixed for different samples, like the class attribute matrix, we can freeze it in the net, which thus makes it equivalent to a fully connected network with no bias. In this framework, the prediction process can be almost the same in both the training stage and the test stage by just taking the class with the maximum score.

4. **Loss**: First, we normalize the output score vector to $[0, 1]$ with a softmax $\hat{y} = softmax(y^{out})$. Then, we adopt a *global attribute balancing loss* to handle the imbalance problem between the attributes of seen classes and that of unseen classes. This will be detailed in the next section.

Semantic Consistency *vs*. Global Semantic Consistency. In order to investigate whether GSC can give a better supervision on both seen and unseen classes, we also design a *semantic consistency network* (SC-Net) for comparison. In SC-Net, $\mathbf{A}^s$ and $\mathbf{A}^t$ are respectively used in the training stage and the test stage, which means the semantic manifold formed by seen classes ($\mathbf{A}^s$) is not aware of the unseen class information ($\mathbf{A}^t$).

In GSC-Net, as we use unseen class information ($\mathbf{A}^t$) in the training stage, though unseen class images are not input to the net, we can still use the global softmax training to form a more comprehensive discriminant space. Intuitively, this can improve performance not only on the ZSL task, but also on the GZSL task that recognizes both training and test classes ($\mathcal{Y}^{s+t}$) at the same time. Furthermore, the softmax and cross entropy loss are also applied to the ($n_s + n_t$)-dimension output vector $\hat{y}$. Therefore, GSC-Net pays more attention to the attributes mainly owned by unseen classes, which can make the learned features more discriminative among the unseen classes.

2.3   Global Attribute Balancing Loss

Since the class attribute matrix is given as the only term connecting seen and unseen classes, the key of ZSL or GZSL is to make the net learn a suitable embedding x_a on the L-dimension (attribute) space. In GSC-Net, there are two major reasons that may leave the embedding x_a extremely imbalanced on different attributes: (1) only the seen classes are supervised positively in GSC-Net; (2) there may be domain shift between seen and unseen classes. Taking these into account, we propose a *global attribute balancing loss* (GAB-loss) for GSC-Net as follows:

$$L_{GAB} = \alpha L_{CE} + (1 - \alpha)L_{SLE} + \lambda||\mathbf{W}\mathbf{A}^s||_2^2 + \beta||\mathbf{W}||_2^2 \tag{1}$$

where the 1st term is the standard one-hot target cross entropy loss L_{CE}, and the 4th term is a simple weight decay on $\mathbf{W}$ for better generalization. Our contributions lie in the 2nd term and the 3rd term. Concretely, the 2nd term is the soft label embedding loss L_{SLE}, and the 3rd term is a L_2 regularization to constrain the weights of seen classes, where $\mathbf{A}^s$ is the seen class attribute matrix. It is actually to balance the attributes of seen and unseen classes, so we call it *attribute balancing regularization*, or *AB-regularization* for short. In what follows, we give detailed explanations on these terms.

Cross Entropy Loss L_{CE}. Formally, it is

$$L_{CE} = q(\hat{y}, y^{true}) \tag{2}$$

where $q(\cdot)$ is a typical cross entropy loss function, $\hat{y}$ is the output vector of the net, y^{true} is the one-hot vector of the target label. Here, we do not use weighted approximate ranking loss [1] because the class semantic matrix used in experiments is point-wisely labeled and cross entropy loss performs better in various experiments.

Soft Label Embedding Loss L_{SLE}. With the GSC-Net, less seen class images will be misclassified into unseen classes in GZSL, but more unseen class images will be misclassified into seen classes. This is because the training samples all fall into seen classes y^s, making the weights corresponding to y^s larger and larger than those corresponding to y^t during training process.

As the one-hot supervision will cause the net to 'lazily' learn a smaller weight for these attributes on which unseen classes have high scores (in the class attribute matrix), so we add a soft label guide to the original cross entropy loss as in [23]:

$$L_{SLE} = q(\hat{y}, Y_{emb}^l) \tag{3}$$

where l is the true label index and Y_{emb}^l is the l-th row of soft label embedding matrix Y_{emb}. We have to utilize the semantic information again to generate the soft label embedding matrix Y_{emb} for all classes $\mathcal{Y}^{s+t}$. Inspired by label propagation, we use the class attribute matrix $\mathbf{A}$ to build a label graph, and employ the adaptive scale policy [31] to compute the class similarity. The similarity between two classes is

$$S_{ij} = \begin{cases} e^{-\eta \frac{||A_i - A_j||^2}{h(A_i)h(A_j)}}, & A_j \in \mathcal{N}(A_i); \\ 0, & \text{otherwise.} \end{cases} \tag{4}$$

$\mathcal{N}(A_i)$ is the neighbor set of A_i, which can be evaluated by setting a distance threshold to reduce the computation cost. We can also directly replace the values of relatively small S_{ij} with 0. The **local scale function** $h(x)$ is defined as

$$h(x) = ||x - x^{(k)}|| \tag{5}$$

where $x^{(k)}$ is the k-th nearest neighbor of point x. In experiments, we find that it is good enough to set k to 1 or 2.

η in Eq. (4) is a hyperparameter to control the centralization degree of S. The larger η is, the farther a node is away from its neighbors, then Y_{emb} will degenerate to the naive one-hot label. Since the local scale function $h(x)$ actually normalizes the numerator term of Eq. (4), it can be easy to set η to get an appropriate similarity.

Normalizing S by row, then we get the normalized class embedding matrix $Y_{emb} \in \mathbb{R}^{n_c \times L}$, each row can be viewed as the soft label.

Attribute Balancing Regularization. We have two L_2 regularizations (the third and the fourth terms) in Eq. (1). The two terms can be derived as follows:

Inspired by [12], we can minimize the reconstruction error and the regression term as follows:

$$\min_{X_i, W} ||W^T X_i - A_i||^2 + \lambda ||W A_i - X_i||^2 + \beta ||W||^2 \tag{6}$$

where X_i is the CNN feature of the i-th sample while A_i is the attribute vector of the sample's corresponding class. Since A_i is fixed, this formula can be rewritten as:

$$\min_{X_i, W} -2(1 + \lambda)A_i^T W^T X_i + ||W^T X_i||^2 + ||X_i||^2 \\ +\lambda ||WA_i||^2 + \beta ||W||^2 \tag{7}$$

Through simple deduction, the optimization directions of the first two terms in Eq. (7) are consistent with L_{CE}. We can approximately replace the first two terms in Eq. (7) with L_{CE}. Furthermore, the third term $||X_i||^2$ is restricted by batch normalization. Since only seen class samples are put into the training pipeline, the regularization on A_i can be generalized into A^s. Then the target function turns out to be:

$$\min_{X, y^*} L_{CE} + \lambda ||WA^s||^2 + \beta ||W||^2 \tag{8}$$

which matches the GAB-loss in Eq. (1).

Overall, GAB-loss can be applied to many problems with unbalancing training data. In the GAB-loss of Eq. (1), α is a hyperparameter falling in $[0, 1]$. A large α will degenerate the loss to a standard cross entropy. We set it around 0.5 if no prior knowledge. If $\alpha = 0$, L_{SLE} dominates GAB-loss. If the FC layers are randomly initialized at the beginning, the projection on each class is almost the same, so L_{SLE} will make the learning process slow at the starting stage. By increasing the value of α, we can make training faster and get higher accuracy for seen classes. Since the training samples all belong to seen classes, L_{GAB} puts more positive supervision to the unseen class attributes.

Relationship to Existing Deep ZSL Models. Many methods [1,14,32] map the visual features and the label semantic vectors into a shared space, then do classification by computing the nearest label embedding vector:

$$c = \arg\min_c ||\theta(x) - \mathbf{A}_y^c||^2 \tag{9}$$

where $\mathbf{A}_y^c$ is the embedding vector of the c-th class. This nearest search method can be clearly visualized and easy to interpret. However, the mean square error is less effective than cross entropy loss in end-to-end training. So we actually transform the search into a softmax classification. Since $\theta(x)$ is independent of classification, Eq. (9) can be written as

$$c = \arg\min_c -\theta(x)^T \mathbf{A}_y^c + \frac{1}{2}||\mathbf{A}_y^c||^2. \tag{10}$$

Since $\mathbf{A}_y^c$ is set statistically equal for each class, Eq. (10) can be simplified to

$$c = \arg\max_c \theta(x)^T \mathbf{A}_y^c \tag{11}$$

where $\theta(x)^T \mathbf{A}_y^c$ can be seen as expression score on class c. Equation (11) is equivalent to the last FC layer with no bias in GSC-Net. This maximization process can be integrated into a softmax layer and trained with cross entropy loss.

Table 1. Details of the ZSL datasets with the proposed splits

Dataset	No. of attributes	No. of seen classes	No. of unseen classes	No. of sample	No.of samples (Train)	No. of samples from unseen classes (Test)	No. of samples from seen classes (Test)
SUN [18]	102	645	72	14340	10320	1440	2580
AWA2 [27]	85	40	10	37322	23527	7913	5882
CUB [25]	312	150	50	11788	7057	2967	1764

2.4 Parametric Novelty Detection for GZSL

Here we adapt our model for the generalized zero-shot learning (GZSL) task by adding a parametric novelty detection (PND) mechanism. In GSC-Net, unseen class images still have relatively high scores on seen classes, which means in most cases $y^{Seen} > y^{Unseen}$ in the output vector. Therefore, we set a hyperparameter γ to control the novelty detection as in [7]. When

$$\max_i y_i^{Seen} < \gamma \cdot \max_j y_j^{Unseen}, \tag{12}$$

we say an unseen class image is detected, and take the maximum y^{Unseen} term as the predicted class. So the prediction method with controllable novelty detection goes as follows:

$$c = \begin{cases} \operatorname{argmax}_i y_i^{Seen}, & \max_i y_i^{Seen} \geq \gamma \cdot (\max_j y_j^{Unseen}); \\ \operatorname{argmax}_j y_j^{Unseen}, & \text{otherwise.} \end{cases} \tag{13}$$

In experiments, γ must be larger than 1. The larger the γ value is, the higher the accuracy on unseen classes is. Our PND mechanism can be easily applied to a typical deep ZSL model. When applied to a certain method, we just append '*' to the method's name for notation.

3 Performance Evaluation

3.1 Datasets and Experimental Settings

Datasets: Xian *et al.* [27] gave a comprehensive evaluation on the existing ZSL methods on several widely used datasets, and proposed an adapted dataset Animals with Attributes 2 (**AWA2**) as well as some suggestions on dataset splits for these ZSL datasets. Since our target is to develop a unified end-to-end ZSL framework, we choose 3 datatsets that have open original images and class attribute annotations: **AWA2** [27], CUB-200-2011 (**CUB**) [25] and Scene UNderstanding (**SUN**) [18]. Table 1 shows more details about these datasets.

In order to make our approach more practical and applicable to more scenarios, we utilize only the class attribute annotations rather than individual samples' attributes. It is common in the datasets that the numbers of images in some classes are much larger than that in the other classes. Therefore, we use the average per-class accuracy to present our results.

Settings: The 2-stage methods use the 2048-D ResNet101 [11] features provided by [27] for all the datasets. To show that our framework can get better results on even smaller CNN base models, we use pretrained ResNet50 [11] as our CNN module, which also outputs 2048-D vectors. In the beginning epochs, since CNN is well pretrained on ImageNet, we can freeze the CNN parameters and train the FC layers only.

Training Policy: We use AdaGrad optimizer [8] with a learning rate 10^{-3}. Regularization ratios λ and β are set to 0.1 and 0.005. L_{CE} ratio α is set to 0.5 by default. For our 3 datasets, to avoid tuning parameters according to test results, we set the affinity factor $\eta = 1.4$, and novelty factor $\gamma = 1.4$. Since we use local function, $\eta \in [1.2, 1.8]$ is suitable enough. In real applications, γ can be set to meet different requirements. If the number of training samples per class is large, which means the seen classes overwhelm unseen classes, γ needs to be large. If α is small, the target label will be soft, then small γ is considered. We run experiments on Titan Xp GPUs with early stopping policy.

3.2   Ablation Study

To testify the benefit of each component in GSC-Net, we consider 3 comparison cases: SC-Net, GSC-Net without L_{SLE} (setting $\lambda = 0.1$ and $\alpha = 1.0$) and GSC-Net without attribute balancing regularization (setting $\lambda = 0$ and $\alpha = 0.5$).

ZSL Results. The results are presented in Table 2. The upper part shows the 2-stage methods whose results were reported in [27]. ALE [2] is simple but effective on all datasets. These methods all use 2048-D ResNet101 features. The lower part stands for end-to-end approaches. Under the same protocol, we implemented Deep-SCoRe, DEM, and our models SC-Net and GSC-Net on ResNet50. The result of S^2GA [30] is directly cited from the original paper where it was evaluated only on CUB.

On the basis of SC-Net, GSC-Net improves performance a lot by making full use of the total class attribute matrix and boosting the feature learning for unseen classes. With L_{SLE}, GSC-Net further lifts the performance. Overall, GSC-Net surpasses the existing methods and achieves the state-of-the-art performance on all the three datasets.

Comparing the end-to-end (E2E) methods and 2-stage (2S) methods, we can easily discover that E2E methods exceed 2 S methods significantly on AWA2 and CUB, but hit a draw on SUN. The reasons may be: (1) there are only 16 images per seen class in SUN, which does not contribute much to CNN finetuning. (2) There are 717 classes but only 102 attributes annotated in SUN. Note that the dimension of the class attribute matrix W, i.e., the last FC weights, is 717×102, therefore the feature dimensionality of 102 is not large enough for 717-way classification.

Table 2. Average per-class accuracy (top-1 in %) for the ZSL task

Method	SUN	AWA2	CUB
LATEM [26]	55.3	55.8	49.3
ALE [2]	58.1	62.5	54.9
DEVISE [10]	56.5	59.7	52.0
SJE [3]	53.7	61.9	53.9
ESZSL [21]	54.5	58.6	53.9
SYNC [6]	56.3	46.6	55.6
SAE [12]	40.3	54.1	33.3
Deep-SCoRe [17] (Resnet50)	51.7	69.5	61.0
DEM [32] (Resnet50)	51.1	68.7	60.1
RELATION NET [24] (GoogleNet)	-	-	62.0
S^2GA [30]	-	-	68.9
SC-Net (baseline, Resnet50)	52.7	71.2	61.4
GSC-Net without L_{SLE} (Resnet50)	56.9	73.7	65.1
GSC-Net without AB-regularization (Resnet50)	58.1	74.5	68.2
GSC-Net (Resnet50)	**58.3**	**75.4**	**69.2**

Table 3. Results on the GZSL task. '*' refers to employing our novelty detection mechanism.

Method	SUN			AWA2			CUB		
	ts	tr	H	ts	tr	H	ts	tr	H
LATEM [26]	14.7	28.8	19.5	11.5	77.3	20.0	15.2	57.3	24.0
ALE [2]	21.8	33.1	26.3	14.0	81.8	23.9	23.7	62.8	34.4
DEVISE [10]	16.9	27.4	20.9	17.1	74.7	27.8	23.8	53.0	32.8
SJE [3]	14.7	30.5	19.8	8.0	73.9	14.4	23.5	59.2	33.6
ESZSL [21]	11.0	27.9	15.8	5.9	77.8	11.0	12.6	63.8	21.0
SYNC [6]	7.9	**43.3**	13.4	10.0	90.5	18.0	11.5	**70.9**	19.8
SAE [12]	8.8	18.0	11.8	1.1	82.2	2.2	7.8	54.0	13.6
DeepSCoRe* [17]	17.3	30.8	22.2	8.8	91.1	16.0	20.3	65.8	31.0
f-CLSWGAN with softmax [28]	**42.6**	**36.6**	**39.4**	-	-	-	43.7	57.7	49.7
SC-Net* (baseline)	10.3	33.4	15.8	3.8	**93.4**	7.3	15.0	**70.1**	24.7
GSC-Net* without L_{SLE}	35.3	30.1	32.5	27.0	72.9	39.4	51.9	59.7	59.1
GSC-Net* without AB-regularization	30.7	**35.3**	32.8	21.3	90.8	34.5	50.4	61.3	55.3
GSC-Net*	**37.5**	31.5	**34.2**	40.2	80.5	**53.7**	**53.6**	68.9	**60.3**

GZSL Results. In GZSL setting, the search space contains both the seen classes and the unseen classes. We use the same evaluation protocol as in [27]. Let **ts** be GZSL accuracy on unseen classes and **tr** GZSL accuracy on seen classes. **H** is the harmonic mean between **ts** and **tr**. **H** pays attention to the smaller one

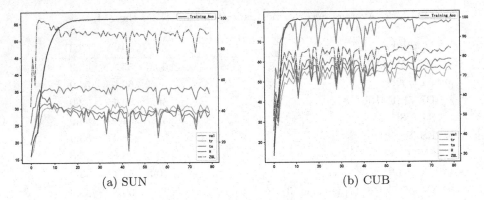

(a) SUN   (b) CUB

Fig. 3. GSC-Net ($\alpha = 0.5$) training processes for ZSL task and GZSL task on (a) SUN and (b) CUB respectively. The X-axis is the number of training epochs. The left Y-axis means ZSL (GZSL) accuracy while the right Y-axis is training accuracy. The blue and purple lines indicate training accuracy and validation accuracy on seen classes. (Color figure online)

between **tr** and **ts**, it is a balanced evaluation for the GZSL task. Table 3 reports the results of GZSL on the three datasets. Some results of existing approaches are obtained from [27]. In the upper part, we can see that most existing ZSL methods perform very poorly on GZSL task in terms of **H** and **ts**. Comparing with these methods, our method can effectively boost the **H** accuracy on all 3 datasets by a large margin.

For the three datasets, GSC-Net improves performance most significantly on CUB, with **H** increasing from 24.7% to 60.3%, mainly due to better attribute balancing between seen and unseen classes. For SUN, there are too many classes and only 16 images per training seen class, which makes it a challenging problem to get high accuracy on both **ts** and **tr**, since the small number of images per class in SUN cannot support end-to-end finetuning well on this setting. It is worthy to notice that [28] uses pretrained ResNet101 features, so it gets better results on SUN. On the other hand, AWA2 faces an extremely unbalancing situation: the number of images in each seen class is quite large, which may make many test images of unseen classes be classified into seen classes in GZSL. Nevertheless, our method still significantly improves the performance on AWA2, lifting **H** from 7.3% (baseline) to 53.7%.

Figure 3 shows the training processes of GSC-Net ($\alpha = 0.5$) for ZSL task and GZSL task on SUN and CUB respectively. We can see that **ts** for unseen classes in GZSL is much lower than ZSL accuracy for seen classes, which shows that GZSL is a much harder task than ZSL.

The model reaches a high accuracy in less than 20 epochs and then oscillates irregularly, so we save the earlier models with early stopping policy. Figure 3 also shows that *ZSL/GZSL accuracy fluctuates with the validation accuracy val* (purple line in Fig. 3) *almost in the same pace*. This obviously reveals that better feature learning gives better ZSL/GZSL prediction. Therefore, we can refer to

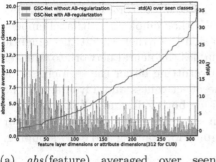
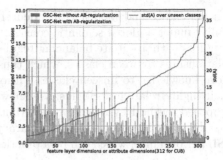

(a) *abs*(feature) averaged over seen classes in CUB

(b) *abs*(feature) averaged over unseen classes in CUB

Fig. 4. Attribute analysis on CUB. Attributes (dimensions) are sorted by std (A). (Color figure online)

the *validation accuracy for seen classes* to select the saved models in real scenarios. This can effectively alleviate the situation that the previous deep learning methods of ZSL have to leave some of seen classes as unseen validation set. So this discovery can help exploit the full power of training data.

3.3 Effectiveness of the AB-Regularization

Here, we investigate how GSC-Net and AB-regularization work on the CUB dataset. First, we get the feature vectors (x_a layer) for validation images of the 200 CUB classes and compute the average features for each class. Thus, we can get a $200*312$ matrix $\mathbf{X}_a$ by concatenating these 200 vectors, which can be compared with the class attribute matrix $\mathbf{A}$. Then, we compute the standard deviation of $\mathbf{A}$, i.e. $std(A)$, for both seen and unseen classes. As shown in Fig. 4, the attributes are sorted in ascending order by $std(A)$. From left to right, $std(A)$ goes bigger, which in some extent means that the classes are more distinguishable on those attributes of the *right* part in Fig. 4(a) and (b). From Fig. 4, we can see that GSC-Net with AB-regularization tends to learn balanced features, rather than biased to a small part of attributes when learning on samples from seen classes. The AB-regularization makes the network tend to utilize more attributes, and the features more balanced and effective, with larger values on the *right* part attributes of unseen classes, as shown by the orange histograms in Fig. 4(b). Moreover, for seen and unseen classes, both the attribute feature distributions and $std(\mathbf{A})$ in corresponding positions are nearly similar, which can explain why ZSL works well on CUB.

4   Conclusion

In this work, we try to make full use of the global class semantic information to improve the classification performance of ZSL and GZSL. We first propose a novel end-to-end model with a neural weighted unit to increase the learning ability under a global semantic constraint. We then employ a soft label embedding loss with attribute balancing regularization to further exploit the semantic relationships between classes, which thus enables the neural network to transfer more knowledge to unseen classes without overfitting either the seen classes or their highly related attributes. We show the effectiveness and advantage of the proposed method by extensive experiments for both ZSL and GZSL tasks. We also discover that the validation accuracy on seen classes can be an indicator for ZSL performance, which can be a practical guide for training and early stopping.

Acknowledgement. This work was supported by the Science and Technology on Complex System Control and Intelligent Agent Cooperation Laboratory.

References

1. Akata, Z., Perronnin, F., Harchaoui, Z., Schmid, C.: Label-embedding for attribute-based classification. In: 2013 IEEE Conference on Computer Vision and Pattern Recognition (CVPR), pp. 819–826. IEEE (2013)
2. Akata, Z., Perronnin, F., Harchaoui, Z., Schmid, C.: Label-embedding for image classification. IEEE Trans. Pattern Anal. Mach. Intell. **38**(7), 1425–1438 (2016)
3. Akata, Z., Reed, S., Walter, D., Lee, H., Schiele, B.: Evaluation of output embeddings for fine-grained image classification. In: Proceedings of the IEEE Conference on Computer Vision and Pattern Recognition (CVPR), pp. 2927–2936. IEEE (2015)
4. Annadani, Y., Biswas, S.: Preserving semantic relations for zero-shot learning. arXiv preprint arXiv:1803.03049 (2018)
5. Annadani, Y., Biswas, S.: Preserving semantic relations for zero-shot learning. In: The IEEE Conference on Computer Vision and Pattern Recognition (CVPR), pp. 7603–7612, June 2018
6. Changpinyo, S., Chao, W.L., Gong, B., Sha, F.: Synthesized classifiers for zero-shot learning. In: 2016 IEEE Conference on Computer Vision and Pattern Recognition (CVPR), pp. 5327–5336 (2016)
7. Chao, W.-L., Changpinyo, S., Gong, B., Sha, F.: An empirical study and analysis of generalized zero-shot learning for object recognition in the wild. In: Leibe, B., Matas, J., Sebe, N., Welling, M. (eds.) ECCV 2016. LNCS, vol. 9906, pp. 52–68. Springer, Cham (2016). https://doi.org/10.1007/978-3-319-46475-6_4
8. Duchi, J., Hazan, E., Singer, Y.: Adaptive subgradient methods for online learning and stochastic optimization. J. Mach. Learn. Res. **12**(Jul), 2121–2159 (2011)
9. Farhadi, A., Endres, I., Hoiem, D., Forsyth, D.: Describing objects by their attributes. In: Proceedings of the IEEE Conference on Computer Vision and Pattern Recognition (CVPR), pp. 1778–1785. IEEE (2009)
10. Frome, A., Corrado, G.S., Shlens, J., Bengio, S., Dean, J., Mikolov, T., et al.: Devise: a deep visual-semantic embedding model. In: Advances in Neural Information Processing Systems (NIPS), pp. 2121–2129 (2013)
11. He, K., Zhang, X., Ren, S., Sun, J.: Deep residual learning for image recognition. In: 2016 IEEE Conference on Computer Vision and Pattern Recognition (CVPR), pp. 770–778 (2016)

12. Kodirov, E., Xiang, T., Gong, S.: Semantic autoencoder for zero-shot learning. In: The IEEE Conference on Computer Vision and Pattern Recognition (CVPR), pp. 4447–4456, July 2017
13. Kumar Verma, V., Arora, G., Mishra, A., Rai, P.: Generalized zero-shot learning via synthesized examples. In: The IEEE Conference on Computer Vision and Pattern Recognition (CVPR), pp. 4281–4289, June 2018
14. Lampert, C.H., Nickisch, H., Harmeling, S.: Attribute-based classification for zero-shot visual object categorization. IEEE Trans. Pattern Anal. Mach. Intell. (T-PAMI) **36**(3), 453–465 (2014)
15. Li, Y., Zhang, J., Zhang, J., Huang, K.: Discriminative learning of latent features for zero-shot recognition. In: The IEEE Conference on Computer Vision and Pattern Recognition (CVPR), pp. 7463–7471, June 2018
16. Mikolov, T., Sutskever, I., Chen, K., Corrado, G.S., Dean, J.: Distributed representations of words and phrases and their compositionality. In: Advances in Neural Information Processing Systems (NIPS), pp. 3111–3119 (2013)
17. Morgado, P., Vasconcelos, N.: Semantically consistent regularization for zero-shot recognition. In: Proceedings of the IEEE Conference on Computer Vision and Pattern Recognition (CVPR), vol. 9, pp. 2037–2046 (2017)
18. Patterson, G., Hays, J.: Sun attribute database: discovering, annotating, and recognizing scene attributes. In: Proceedings of the IEEE Conference on Computer Vision and Pattern Recognition (CVPR), pp. 2751–2758. IEEE (2012)
19. Radovanović, M., Nanopoulos, A., Ivanović, M.: Hubs in space: popular nearest neighbors in high-dimensional data. J. Mach. Learn. Res. **11**(Sep), 2487–2531 (2010)
20. Reed, S., Akata, Z., Lee, H., Schiele, B.: Learning deep representations of fine-grained visual descriptions. In: 2016 IEEE Conference on Computer Vision and Pattern Recognition, pp. 49–58 (2016)
21. Romera-Paredes, B., Torr, P.: An embarrassingly simple approach to zero-shot learning. In: International Conference on Machine Learning (ICML), pp. 2152–2161 (2015)
22. Socher, R., Ganjoo, M., Manning, C.D., Ng, A.: Zero-shot learning through cross-modal transfer. In: Advances in Neural Information Processing Systems (NIPS), pp. 935–943 (2013)
23. Sun, X., Wei, B., Ren, X., Ma, S.: Label embedding network: Learning label representation for soft training of deep networks. arXiv preprint arXiv:1710.10393 (2017)
24. Sung, F., Yang, Y., Zhang, L., Xiang, T., Torr, P.H., Hospedales, T.M.: Learning to compare: relation network for few-shot learning. arXiv preprint arXiv:1711.06025 (2017)
25. Wah, C., Branson, S., Welinder, P., Perona, P., Belongie, S.: The caltech-UCSD birds200-2011 dataset. California Institute of Technology (2011)
26. Xian, Y., Akata, Z., Sharma, G., Nguyen, Q., Hein, M., Schiele, B.: Latent embeddings for zero-shot classification. In: Proceedings of the IEEE Conference on Computer Vision and Pattern Recognition (CVPR), pp. 69–77 (2016)
27. Xian, Y., Lampert, C.H., Schiele, B., Akata, Z.: Zero-shot learning-a comprehensive evaluation of the good, the bad and the ugly. IEEE Trans. Pattern Anal. Mach. Intell. **41**, 2251–2265 (2018)
28. Xian, Y., Lorenz, T., Schiele, B., Akata, Z.: Feature generating networks for zero-shot learning. In: The IEEE Conference on Computer Vision and Pattern Recognition (CVPR), pp. 5542–5551, June 2018

29. Xian, Y., Schiele, B., Akata, Z.: Zero-shot learning - the good, the bad and the ugly. In: IEEE Computer Vision and Pattern Recognition (CVPR), pp. 2251–2265 (2017)
30. Yu, Y., Ji, Z., Fu, Y., Guo, J., Pang, Y., Zhang, Z.: Stacked semantic-guided attention model for fine-grained zero-shot learning. arXiv preprint arXiv:1805.08113 (2018)
31. Zelnik-Manor, L., Perona, P.: Self-tuning spectral clustering. In: Advances in Neural Information Processing Systems (NIPS), pp. 1601–1608 (2005)
32. Zhang, L., Xiang, T., Gong, S.: Learning a deep embedding model for zero-shot learning. In: 2017 IEEE Conference on Computer Vision and Pattern Recognition (CVPR), pp. 3010–3019 (2016)
33. Zhu, Y., Elhoseiny, M., Liu, B., Peng, X., Elgammal, A.: A generative adversarial approach for zero-shot learning from noisy texts. In: The IEEE Conference on Computer Vision and Pattern Recognition (CVPR), pp. 1004–1013, June 2018

A Reinforcement Learning Approach to Gaining Social Capital with Partial Observation

He Zhao[1], Hongyi Su[1], Yang Chen[2](✉) ⓘD, Jiamou Liu[2] ⓘD,
Hong Zheng[1], and Bo Yan[1]

[1] Beijing Lab of Intelligence Information Technology, School of Computer Science,
Beijing Institute of Technology, Beijing, China
{2120171104,henrysu,hongzheng,yanbo}@bit.edu.cn
[2] School of Computer Science, The University of Auckland, Auckland, New Zealand
{yang.chen,jiamou.liu}@auckland.ac.nz

Abstract. Social capital brings individuals benefits and advantages in societies. In this paper, we formalize two types of social capital: *bonding capital* refers to links to neighbours, while *bridging capital* refers to brokerages between others. We ask the questions: How would a marginal individual gain social capital with imperfect information of the society? We formalize this issue as the *partially observable network building* problem and propose two reinforcement learning algorithms: one guarantees the convergence to optimal values in theory, while the other is efficient in practice. We conduct simulations over a real-world dataset, and experimental results coincide with our theoretical analysis.

Keywords: Social capital · Network building · Reinforcement learning

1 Introduction

Social networks grant individuals with both tangible benefits, e.g., economic resources and human resources, and impalpable advantages such as social support, information control, and social influence. The concept of *social capital* epitomizes various incarnations of benefits obtained via engaging in and maintaining social relationships [4]. Prominent differences of origins divide researches on social capital into two dimensions: *bonding capital* depicts the aggregate social benefits that an individual draws from its neighbours [2], while *bridging capital* rises from brokering diverse communities that captures an individual's potentials over acquiring opportunities and information [3]. Human societies can be seen as the product of interactions among all participating individuals regarding gaining social capital. A question naturally arises from this scenario that *how would an individual take a tactic to gain social capital?* The *reward theory of attraction* in social psychology – indicating that people like those whose behaviour is rewarding to them or whom they associate with rewarding events – points

© Springer Nature Switzerland AG 2019
A. C. Nayak and A. Sharma (Eds.): PRICAI 2019, LNAI 11670, pp. 113–117, 2019.
https://doi.org/10.1007/978-3-030-29908-8_9

out approaches to this question on a fundamental level [7]. That is, regarding social capital as rewards and accessing others through building interpersonal ties. Several assumptions are naturally derived from this theory: (1) Since rewards are typical of hindsight, an individual demands learning from experience via trial-and-error. (2) An individual has limited abilities to establish and maintain relationships. Since if otherwise, trivially she can link to all others to maximize rewards. (3) An individual gives priorities to surrounding persons on creating links, as it is difficult to access to and gain rewards from remote parts of society.

Motivated by concerns and assumptions above, we model this issue as *partially observable network building process* (PONB) problem, which involves a network and an agent within. The agent has a partial observation of the network and can create a limited number of ties. To capture that the agent learns a strategy for gaining social via trial-and-error, we tackle PONB problem using the reinforcement learning method. We propose two Q-learning algorithms. One is optimal in theory but impractical, while the other is efficient in practice.

Related Works. Pioneering works of sociologists advanced the research on social capital, in which homophily and weak ties are the sources of two types of social capital, respectively [2–4]. Game-based research on *network formation* focuses on equilibria among rational agents [5], where behaviours of agents are subject to restricted predefined rules. Our work surpasses theirs as the learning process captures initiatives of agents. Algorithmic research on *network building* problem asks for integrating a newcomer to the center of an existing static or dynamic network via establishing a minimum number of links [6,11]. Our work differs from theirs as: (1) we assume an agent does not have global knowledge of the network which is more realistic; (2) we solve PONB using reinforcement learning instead of heuristic-based algorithms.

2   Problem Setup

A social network is considered as an undirected graph $G = (V, E)$, where V is a set of nodes social actors (agents) and $E \subseteq V^2 \setminus \{uu \mid u \in V\}$ is a set of edges. A (k-length) *path* is a sequence of nodes $u_1 u_2, \ldots u_{k+1}$ where $u_i u_{i+1} \in E, \forall 1 \leq i \leq k$. Denote by $\mathsf{dist}_G(u, v)$, the distance between u and v is the length of a shortest path between u and v. If $\mathsf{dist}_G(u, v) = d$, then we say that u, v are *d-hop neighbors*. The *d-hop neighbor set* of $v \in V$ is $N_G^d(v) := \{u \in V \mid \mathsf{dist}_G(u, v) = d\}$. From an individualistic perspective, the social surrounding of a node $v \in V$ contains all ties that v maintains and perceives. This can be captured using the *2-level ego network*, which is the subgraph O_G^v of G consisting of v, v's 1- and 2-hop neighbors and edges between, i.e., the nodes set of O_G^v is $V_G = \{u \in V \mid \mathsf{dist}_G(u, v) \leq 2\}$ and the edges of O_G^v is $E_G^v = \{uw \in E \mid u, w \in V_G\}$.

Bonding Capital. Bonding social capital is inclined to the collective resources generated by the strong relationship between group members. This kind of collective resource is generated by the social cohesion of the agent. We measure such kind of cohesion using *personalized PageRank index*, which is adapted from PageRank that evaluates structural proximity between nodes through predicting the likelihood of edges between any pairs of nodes [8]. It takes as input a starting node s, and assigns a score to every node u that captures the likelihood of a random walk from s to u [10]. More formally, let a_u be the column vector in the adjacency matrix of G corresponding to node u. The personalized PageRank vector **pr** is defined by

$$\mathsf{pr}_u = (1 - \beta)r_u + \beta(\mathbf{pr} \cdot a_u / |N_G^1(u)|),$$

where $\beta \in (0, 1)$ is the restart probability, and $r_u = 1$ if $u = s$ and $r_u = 0$ otherwise. pr_u is used as the link prediction score between s and u. In this work, we use pr_u to evaluate likelihood between s and u. Denote by $\mathsf{bo}_G(v)$, the *bonding capital* of v is defined by summing personalized PageRank indices between v and v's neighbors, namely, $\mathsf{bo}_G(v) := \sum_{u \in N_G^1(v)} \mathsf{pr}_u$.

Bridging Capital. Occupying a central position to act as a gateway for information exchange brings an individual bridging capital [3]. Following the work [1], we formalize bridging capital using betweenness centrality. Formally, $\mathsf{br}_G(v) := \sum_{s \neq v \neq t \in V} \sigma_{st}(v)/\sigma_{st}$, where σ_{st} is the number of shortest paths between nodes s and t, and $\sigma_{st}(v)$ is the number of shortest paths passing v.

Mixed Capital. An individual may show different preferences to two types of capital. To cope with this, we employee a *preference weight* $w \in [0, 1]$. The *mixed capital* induced by w is defined as $\mathsf{mix}_G^w(v) := w\mathsf{bo}_G(v) + (1 - w)\mathsf{br}_G(v)$.

Assuming an inside individual that aims to gain social capital through building interpersonal ties. We model this interaction as the (ℓ-round) *network building process* in discrete time steps $\tau = 0, 1, \ldots, \ell - 1$, which involves a network G and an agent ν within. In each time step τ, ν chooses a node u_τ from its 2-hop neighbor set to create a link, resulting in a sequence of networks $G_0, G_1, \ldots, G_{\ell-1}, G_\ell$ where $G_0 = G$ and $G_{\tau+1}$ is obtained by adding edge νu_τ to G_τ. The tactic for agent v to select the node is called a *network building strategy*. More formally, a network building strategy is a function φ that takes as input a 2-level ego network O_G^ν and outputs a node $u \in N_G^2(\nu)$. Any NB process is said to be *consistent* with a strategy φ if $\forall 0 \leq \tau \leq \ell - 1 : u_\tau = \varphi(O_{G_\tau}^\nu)$.

We are now ready to formally define *partially observable network building* (PONB) problem. A PONB problem is a tuple $\langle G, \nu, \ell, w \rangle$, where $G = (V, E)$ is a graph, $\nu \in V$ is the learner, $\ell \in \mathbb{N}^+$ and $w \in [0, 1]$. The problem asks for an NB strategy φ^* so as to maximize the mixed capital $\mathsf{mix}_{G_\ell}^w(\nu)$ via an ℓ-round NB process consistent with φ^*.

3 Reinforcement Learning Algorithms for PONB

To cope with the inherent high computational complexity and incomplete knowledge of the environment, we tackle PONB using reinforcement learning approaches, which also captures an individuals' learning process in reality.

Algorithm 1 (OQL). We first investigate the optimal Q-learning algorithm (OQL) that uses 2-level ego networks as states and nodes as actions. The reward is defined as the difference of mixed capital between two time steps. More formally, for a PONB problem $\langle G, \nu, \ell, w \rangle$, the Q-value update rule of OQL is

$$Q_{t+1}\left(O_{G_\tau}^\nu, a_\tau\right) = (1 - \alpha_t)Q_t\left(O_{G_\tau}^\nu, a_\tau\right) + \alpha_t\left[r_{\tau+1} + \max_{a'} Q_t\left(O_{G_{\tau+1}}^\nu, a'\right)\right], \quad (1)$$

where $\alpha_t \in [0, 1)$ is the *learning rate*, $a_\tau \in N_{G_\tau}^2(\nu)$, $r_{\tau+1} = \text{mix}_{G_{\tau+1}}^w(\nu) - \text{mix}_{G_\tau}^w(\nu)$ and $a' \in N_{G_{\tau+1}}^2(\nu)$. It is clearly that the underlying process of OQL is an Markov decision process. Therefore, the optimality and convergence of OQL hold. Hence, after convergence an optimal strategy is obtained by setting $\forall 0 \leq \tau < \ell$: $\varphi(O_{G_\tau}^\nu) = \arg\max_{a_\tau} Q(O_{G_\tau}^\nu, a_\tau)$.

Algorithm 2 (FQL). However, OQL is impractical for most of the time as the space of 2-level ego networks is typically too large to learn from. Therefore we propose a fast Q-learning algorithm (FQL) that directly use time steps, $\tau = 0, 1, \ldots, \ell - 1$, as states. Hence, the Q-value update rule of FQL is

$$Q_{t+1}(\tau, a_\tau) = (1 - \alpha_t)Q_t(\tau, a_\tau) + \alpha_t\left[r_{\tau+1} + \max_{a'} Q_t(\tau + 1, a')\right]. \quad (2)$$

The reward distribution is not stationary in FQL as the underlying network $G_{\tau+1}$ induced by a same a_τ may vary, which implies that the cost of a fast speed is losing the guarantee of optimality. However, surprisingly, we observe that FQL achieves a good trade-off between time and accuracy in practice (see Sect. 4).

4   Experiments and Discussion

We test two algorithms on a real-world dataset, which represents American football games between Division IA colleges during regular season Fall 2000. Nodes and edges represent teams and matches, respectively [9]. We set learning rate $\alpha = 0.1$ and use ε-greedy for exploration with $\varepsilon = 0.3$. We set the length of NB processes $\ell = 5$ and the preference weight $w = 0.5$. We select two possibly encountered 2-level ego networks (denoted by o_1 and o_2) at round $\tau = 1$, and a node a (see Fig. 1). We execute 10 independent runs for this experiment. The learning curves of $Q(o_1, a)$ and $Q(o_2, a)$ in OQL, $Q(\tau = 1, a)$ in FQL, and tendencies of social capital are plotted in Fig. 2.

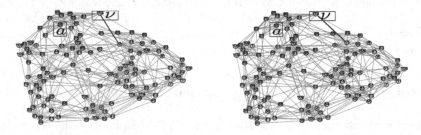

Fig. 1. The agent ν and action a are set as node 113 and 56, respectively. Red nodes represent ν's initial 2-hop neighbors. Green nodes in the left figure represents ν's new 2-hop neighbors after linking to node 0 at round 1. Blue nodes in the right figure represents 2-hop neighbors after linking to node 11 at round 1. o_1 is induced by red and green nodes (left figure), and o_2 is induced by red and blue nodes (right figure). (Color figure onlone)

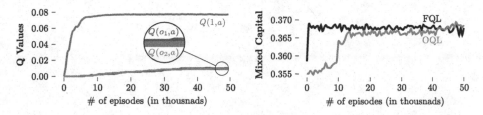

Fig. 2. Results of applying two Q-learing algorithms on football network. Left: Q-values by OQL ($Q(o_1, a)$ and $Q(o_2, a)$) and FQL ($Q(1, a)$). The line shows the median over 10 independent runs. Right: Results of mixed capital by OQL and FQL with preference weight $w = 0.5$.

We make two discussions: (1) FQL standouts as $Q(1, a)$ stabilizes considerably faster than two Q-values in OQL, though to a non-optimal value. This coincides with our theoretical results. (2) Thanks to the fast stabilization, FQL surpasses OQL in the speed of enhancing social capital. Therefore, it is explicitly that FQL successfully achieves a trade-off between efficiency and accuracy.

References

1. Alaa, A.M., Ahuja, K., van der Schaar, M.: A micro-foundation of social capital in evolving social networks. IEEE Trans. Netw. Sci. Eng. **5**(1), 14–31 (2018)
2. Bourdieu, P.: The forms of capital. In: Handbook of Theory and Research for the Sociology of Education (1986)
3. Burt, R.S.: Structural holes and good ideas. Am. J. Sociol. **110**(2), 349–399 (2004)
4. Coleman, J.S.: Social capital in the creation of human capital. Am. J. Sociol. **94**, S95–S120 (1988)
5. Jackson, M.O.: A survey of network formation models: stability and efficiency. Group Formation Econ. Netw. Clubs Coalitions **664**, 11–49 (2005)
6. Moskvina, A., Liu, J.: How to build your network? A structural analysis. In: Proceedings of the Twenty-Fifth International Joint Conference on Artificial Intelligence, pp. 2597–2603. AAAI Press (2016)
7. Myers, D.: Relationship rewards. In: Social Psychology, pp. 392–439 (2010)
8. Page, L., Brin, S., Motwani, R., Winograd, T.: The pagerank citation ranking: bringing order to the web. Technical report, Stanford InfoLab (1999)
9. Rossi, R.A., Ahmed, N.K.: The network data repository with interactive graph analytics and visualization. In: Proceedings of the Twenty-Ninth AAAI Conference on Artificial Intelligence (2015)
10. Tong, H., Faloutsos, C., Pan, J.Y.: Fast random walk with restart and its applications. In: Sixth International Conference on Data Mining, pp. 613–622. IEEE (2006)
11. Yan, B., Liu, Y., Liu, J., Cai, Y., Su, H., Zheng, H.: From the periphery to the center: information brokerage in an evolving network. In: Proceedings of the Twenty-Seventh International Joint Conference on Artificial Intelligence, pp. 3912–3918. AAAI Press (2018)

Knowledge Handling

Emotion Recognition from Music Enhanced by Domain Knowledge

Yangyang Shu and Guandong Xu[✉]

Advanced Analytic Institute, University of Technology Sydney, Sydney, Australia
Yangyang.Shu@student.uts.edu.au, guandong.xu@uts.edu.au

Abstract. Music elements have been widely used to influence the audiences' emotional experience by its music grammar. However, these domain knowledge, has not been thoroughly explored as music grammar for music emotion analyses in previous work. In this paper, we propose a novel method to analyze music emotion via utilizing the domain knowledge of music elements. Specifically, we first summarize the domain knowledge of music elements and infer probabilistic dependencies between different main musical elements and emotions from the summarized music theory. Then, we transfer the domain knowledge to constraints, and formulate affective music analysis as a constrained optimization problem. Experimental results on the Music in 2015 database and the AMG1608 database demonstrate that the proposed music content analyses method outperforms the state-of-the-art performance prediction methods.

Keywords: Music emotion recognition · Domain knowledge · Probabilistic dependencies

1 Introduction

We are surrounded by digital music collections due to the popularity of the Internet and the proliferation of user friendly MP3 players. Since almost every piece of music is created to convey emotion, naturally, music emotion recognition has attracted increasing attention in recent years. Automatic emotion recognition from music pieces has wide potential application in both music creation and music distribution.

The framework of current research into music emotion recognition mainly consists of feature extraction and classification. First, various features, including timbre, rhythm and harmony, are extracted from music pieces. Then, a classifier, such as support vector machine, is used to classify music pieces into several discrete emotion categories, or a regressor, such as support vector regression, is adopted to predict continuous emotional dimensions, such as valence and arousal. An extensive review of emotion recognition from music can be found in [18].

© Springer Nature Switzerland AG 2019
A. C. Nayak and A. Sharma (Eds.): PRICAI 2019, LNAI 11670, pp. 121–134, 2019.
https://doi.org/10.1007/978-3-030-29908-8_10

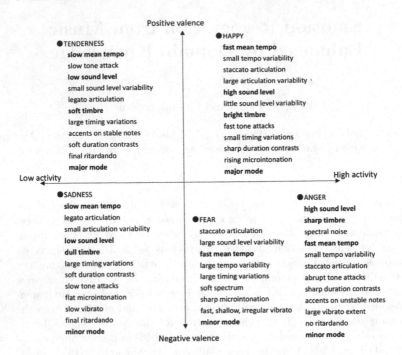

Fig. 1. Music elements used by composers to communicate emotions to audiences

Although various discriminative features and classifiers have been developed, automatic emotion recognition from music pieces is still a very challenging task due to the complexity and subjectivity of human emotions, and the rich variety of music content.

Almost all the current work on music emotion recognition focuses on developing discriminative features and classifiers. This kind of data-driven approach does not successfully exploit the domain knowledge of emotion and music, i.e. the inherent psychological relationship between human emotion and music, which carries crucial information for music emotion recognition.

Specifically, main musical dimensions, i.e., rhythm, tonality, timbre, dynamics and pitch are often used to affect users' emotional experience. The tempo, mode, brightness, loudness and pitch can represent the five main musical dimensions respectively [10]. Figure 1 [15] summarized the relations between music elements and emotions. From Fig. 1, we can find that fast tempo is usually utilized to result in the arousal atmosphere, while the slow tempo is utilized to generate quiet environment [5–8]. Major mode can be used to induce happiness and excitement, and minor mode can create a more tense and sad music [13]. Brightness is related to arousal [6]. Higher brightness can be used to induce excitement and astonishment, while lower brightness can be used to induce sadness and softness. As for loudness, higher loudness can be used to induce anger, fear and excitement, and lower loudness can create a more relaxed and quiet music [6]. High pitch may lead to happiness, anger and fear, while low pitch may induce sadness [6]. Such inherent dependencies between music elements and emotions can be leveraged for emotion recognition from music, but have not been explored yet.

Therefore, in this paper, we propose a novel method to analyze musical emotion through exploring domain knowledge. As a primary study to explore music theory for music emotion analysis, this paper utilize main musical dimensional elements to demonstrate the superiority of the proposed music emotion analyses method enhanced via domain knowledge. Specifically, summarized in music theory, we first infer probabilistic relations between main musical dimensional elements and emotions. Then we transfer this probabilistic dependencies based on domain knowledge as a constrained optimization problem. In order to demonstrate the superiority of the proposed method, we conduct our experiments on two benchmark databases.

2   Domain Knowledge

Rhythm, tonality, timbre, dynamics are often used by composers to invoke emotions to audiences, which constitute musical main dimensions, audiences [9]. In this section, we introduce the dependencies between musical elements and emotions from the summarized music theory.

2.1   Rhythm

Tempo is one of the most import musical elements to affect the mood of audiences [5–8]. Generally, tempo express the rhythm and fluency of the music [9]. By changing the tempo, the composers can fully put the emotions into their music. Especially, as mentioned in [6], *fast tempo* is usually used to generate the exaggerated mood, which result in high arousal atmospheres. On the other hand, the *slow tempo* is utilized to generate quiet mood, which result in low arousal atmospheres.

From the perspective of the audiences, the perceived music tempo can well influence their feelings. Specifically, when receiving fast music tempo signals, people will intuitively associate with activity/excitement, happiness/joy/pleasantness, potency, surprise, flippancy, anger, uneasiness and fear. However, when perceiving slow-tempo music, people may be associated with calmness/serenity, peace, sadness, dignity/solemnity, tenderness, longing, boredom and disgust [6]. Thus the fast tempo music can induce high arousal from audiences, while slow tempo music can invoke low arousal from audiences.

Normally, the tempo are categorized into Largo (40–60 bpm), Adagio (66–76 bpm), Andante (76–10 bpm), Moderato (108–120 bpm) and Allegro (120–168) as shown in Table 1 [13]. Since the difference between slow tempo and intermediate tempo are not obvious, we adopt the 108 bpm as the threshold and categorize the tempo as fast tempo and slow tempo.

Table 1. The correspondence between common music speed terms and speed values

	tempo mark	bpm
slow tempo	Grava	40
	Largo	46
	Lento	52
	Adagio	56
medium tempo	Larghetto	60
	Andante	66
	Andantino	69
	Moderato	88
	Allegretto	108
fast tempo	Allegro	132
	Vivace	160
	Presto	184
	Prestissimo	208

Table 2. The dependencies between four music elements (tempo, mode, brightness and loudness) and emotions. Note that the $\sqrt{}$ demonstrates great dependencies between emotion and the music elements. Details are discussed in Sect. 2

	high Arousal	low Arousal	high Valence	low Valence
fast tempo	$\sqrt{}$			
slow tempo		$\sqrt{}$		
major mode			$\sqrt{}$	
minor mode				$\sqrt{}$
high brightness	$\sqrt{}$			
low brightness		$\sqrt{}$		
high loudness	$\sqrt{}$			
low loudness		$\sqrt{}$		

2.2 Tonality

In the composers' perspective, musical tonality is one of the most important musical elements for music presentation. Since mode is a system of musical tonality involving a type of scale coupled with a set of characteristic melodic behaviors [13], composers control the musical tonality by adjusting the mode. Generally, the mode is scaled in to a heptationic scale, in which the first, third, and fifth scale degrees play important roles. As stated in [13], the mode is categorized into two groups: major mode and minor mode. Specifically, the major mode is constructed by adjusting the first, third, and fifth scale degrees with a major triad, while the minor mode is constructed by adjusting the first, third, and fifth scale degrees with a minor triad. In hands of the composers, they intend to present an audio of grace with major mode, while present an audio of anxiety and sad with minor mode. Thus, major mode is used for invoking high valence from the audiences, and minor mode is used to induce low valence. From the perspective of the audiences, studies also show that major mode is strongly correlated to grace, serene, and solemn, while minor mode is strongly correlated to dreamy, dignified, tension, disgust, and anger [8].

In this paper, we extract the mode features with the MIR toolbox, which ranges between -1 and $+1$. After obtaining the mode features, we adopt the median mode value as the threshold and categorized the audio clips into major mode and minor mode. Specifically, audio clips whose mode values are above the median are assigned as major mode while audio clips whose mode values are below the median are assigned as minor mode.

2.3 Timbre

Musical timbre denoted as brightness [17] is a powerful component in constructing the music piece. The composer usually express their music emotion with the audiences via different musical instruments and equalizer. Normally, in order to show joyful, angry or thrill atmosphere, the composer compose the music using a bright tone, while with depression or tender, compose the music with dull sound [6]. Thus, dull sound is utilized to deliver low valence while bright sound is utilized for delivering high valence. From the perspective of the psychology, while perceiving bright sound, the audience is more likely to feel the arousal emotion [16], e.g. excitement and astonishment. On the other hand, they feel sadness or softness after hearing dull sound.

In this paper, we extract the brightness features via measures the proportion of high frequency (over 1500 HZ) in the music piece. The formulation is shown as below:

$$Brightness = X_{above}/X_{total} * 100\% \qquad (1)$$

where X_{above} represents component whose the energy above 1500 Hz, and X_{total} represents the total energy of the music.

2.4 Dynamics

In music, the dynamics of a piece is the variation in loudness between notes or phrases, which usually used by musicians to deliver dynamic in a music piece instrumentalist and Singers playing a song to express their emotion and create atmosphere through the loudness. As stated in [6], the volume of the music can strongly influence arousal. Specifically, great joy, surprise, excitement and anger are often invoked by Loud sound while peaceful mood, tender and sadness associated with soft sound [6]. Thus, the loudness can well influence the audiences. Specifically, audiences tend to feel high arousal mood while listening high loudness songs, and tend to feel low arousal mood while listening low loudness songs.

In this paper, we use root-mean-square amplitude (rms) to calculate loudness. Rms can be extacted with the MIR toolbox with its values from 0 to 1. We calculate the median value as the threshold to categorize the rms and divide loudness into *high loudness* and *llow loudness*. In conclusion, the dependencies between emotions and main musical dimensions including rhythm, tonality, timbre and dynamics discussed above are shown in Table 2.

3   Proposed Method

3.1   Problem Statement

Denote three tuple $S = \{(x_i, h_i, y_i) | i = 1, ..., N\}$, where x_i represents D-dimensional features, $h_i = (h_i^t, h_i^m, h_i^b, h_i^l) \in \{0, 1\}$ represents the binarized tempo values, mode values, brightness values and loudness values respectively, $y_i \in \{y_i^v, y_i^a | -1 \leq y_i^v, y_i^a \leq 1\}$ represents continuous valence and arousal values, and N is the number of training samples. The goal is to learn a classifier f(x, w) as follows:

$$\min_w \sum_{i=1}^{N} \alpha \ell(f_\theta(x_i), y_i) + \sum_{i=1}^{N} \beta L(x_i, h_i, y_i) \tag{2}$$

where α and β are the coefficients, $\ell(f_\theta(x_i), y_i)$ represents the basic loss function, and $L(x_i, h_i, y_i)$ captures the domain knowledge between music elements h and the emotion values y. The first section denotes the basic loss function. The second section denotes the regularization term associating domain knowledge.

For the first term, any loss function can be used. In this paper, we adopt the support vector regression as the basic loss function:

$$\ell(f_\theta(x_i), y_i) = \frac{1}{2}||\mathbf{w}||^2 + \alpha \sum_{i=1}^{N} \ell_\epsilon(f(x_i, w) - y_i) \tag{3}$$

where the function $\ell_\epsilon(z)$ satisfy the below:

$$\ell_\epsilon(z) = \begin{cases} 0, & if \ |z| \leq \epsilon \\ |z| - \epsilon, & otherwise. \end{cases} \tag{4}$$

where ϵ is a constant which defines the maximum deviation allowed for a prediction to be considered as correct; α is used as a trade-off between the model complexity and regression loss.

As for the second term, the relations between music elements and emotions as domain knowledge, can be exploited to build better emotion classifiers from music. In this paper, domain knowledge of four music elements, i.e., tempo, mode brightness and loudness are discussed, with respect to dynamic, rhythm, timbre and tonality of the music dimension.

3.2   Representation of Domain Knowledge

Domain Knowledge in Arousal Space. From Table 2, tempo, brightness and loudness have the strong relationship with musical emotion in the arousal space. Fast tempo features, high brightness and high loudness are more possible to express high arousal mood of audiences, while the slow tempo features, low brightness and low loudness are more likely to deliver the low arousal of the audiences. Thus the probabilistic dependencies between tempo and arousal emotion shown as:

$$p(\hat{y}^a \geq 0|h^{\{t,b,l\}} = 1) > p(\hat{y}^a < 0|h^{\{t,b,l\}} = 1)$$
$$p(\hat{y}^a < 0|h^{\{t,b,l\}} = 0) > p(\hat{y}^a \geq 0|h^{\{t,b,l\}} = 0)$$

(5)

where $p(\hat{y}^a \geq 0|h^{\{t,b,l\}} = 1)$ and $p(\hat{y}^a < 0|h^{\{t,b,l\}} = 1)$ indicate the probabilities of high arousal and low arousal respectively, when observing fast tempo, high brightness and loudness. $p(\hat{y}^a < 0|h^{\{t,b,l\}} = 0)$ and $p(\hat{y}^a \geq 0|h^{\{t,b,l\}} = 0)$ show the probabilities of low arousal and high arousal respectively, when given slow tempo, low brightness and low loudness.

ReLU function is adopt in our method to penalize the samples violating the domain knowledge. The corresponding penalty $l_i^{\{ta,ba,la\}}(x_i, h_i, \hat{y}_i)$ from the domain knowledge according to Eq. 5 is encoded as below:

$$\ell_i^{\{ta,ba,la\}}(x_i, h_i, \hat{y}_i) = h_i^{\{t,b,l\}} * [p(\hat{y}^a < 0|h^{\{t,b,l\}} = 1) - p(\hat{y}^a \geq 0|h^{\{t,b,l\}} = 1)]_+ +$$
$$(1 - h_i^{\{t,b,l\}}) * [p(\hat{y}^a \geq 0|h^{\{t,b,l\}} = 0) - p(\hat{y}^a < 0|h^{\{t,b,l\}} = 0)]_+$$
$$= h_i^{\{t,b,l\}} * [1 - 2 * p(\hat{y}^a \geq 0|h^{\{t,b,l\}} = 1)]_+$$
$$+ (1 - h_i^{\{t,b,l\}}) * [2 * p(\hat{y}^a \geq 0|h^{\{t,b,l\}} = 0) - 1]_+$$

(6)

where $[\cdot] = max(\cdot, 0)$.

Since there is no obvious relationship between mode and arousal, we treat the major mode and minor equal important. In other words, major mode and minor mode have equal chances to invoke low arousal mood or high arousal mood from audiences. Hence, mode information is not used in arousal space.

Domain Knowledge in Valence Space. From Table 2, major mode (high-value mode) features are more possible to invoke high valence mood from audiences, while the minor mode (low-value mode) features are more likely to invoke the low valence of the audiences in the valence space. Thus we can infer the probabilistic dependencies between mode and valence emotion as:

$$p(\hat{y}^v \geq 0|h^m = 1) > p(\hat{y}^v < 0|h^m = 1)$$
$$p(\hat{y}^v < 0|h^m = 0) > p(\hat{y}^v \geq 0|h^m = 0)$$

(7)

Thus the corresponding constraint $l_i^{mv}(x_i, h_i, \hat{y}_i)$ for valence according to Eq. 7 is encoded as below:

$$\begin{aligned}
\ell_i^{mv}(x_i, h_i, \hat{y}_i) &= h_i^m * [p(\hat{y^v} < 0|h^m = 1) - p(\hat{y^v} \geq 0|h^m = 1)]_+ \\
&+ (1 - h_i^m) * [p(\hat{y^v} \geq 0|h^m = 0) - p(\hat{y^v} < 0|h^m = 0)]_+ \\
&= h_i^m * [1 - 2 * p(\hat{y^v} \geq 0|h^m = 1)]_+ \\
&+ (1 - h_i^m) * [2 * p(\hat{y^v} \geq 0|h^m = 0) - 1]_+
\end{aligned} \tag{8}$$

Since there is no obvious relationship between valence and another elements, e.g. tempo, brightness, loudness, the information of tempo, brightness and loudness is not used in valence space.

3.3 Proposed Model

We propose to learn classifier with the objectives as below:

$$\begin{aligned}
F^{\{a,v\}} &= \frac{1}{2}w^T w + \alpha \sum_{i=1}^{N} \ell_\epsilon(f(x_i, w) - y_i) + \\
\beta^t \sum_{i=1}^{N} \ell_i^{\{ta\}}(x_i, h_i^t, \hat{y}_i) &+ \beta^m \sum_{i=1}^{N} \ell_i^{\{mv\}}(x_i, h_i^m, \hat{y}_i) + \\
\beta^b \sum_{i=1}^{N} \ell_i^{\{ba\}}(x_i, h_i^b, \hat{y}_i) &+ \beta^l \sum_{i=1}^{N} \ell_i^{\{la\}}(x_i, h_i^l, \hat{y}_i)
\end{aligned} \tag{9}$$

where w is the parameter of the classifier, α, β^t, β^m, β^b and β^l are coefficients. We use $f(x, w) = w \cdot \phi(x)$ as our function where $\phi(x)$ maps the features space into the kernel space. According to the property of logistic regression, we apply sigmoid function to replace the probabilistic dependencies between audio elements and emotion labels as follow:

$$\begin{aligned}
p(\hat{y} > 0|h) &= sigmoid(f(x, w)) \\
p(\hat{y} \leq 0|h) &= 1 - sigmoid(f(x, w))
\end{aligned} \tag{10}$$

where $sigmoid(x) = \frac{1}{1+e^{-x}}$.

In order to solve the optimization we adopt the stochastic gradient descent (SGD) to solve the problem. The updating rule is shown as follows:

$$w^{(t+1)} = w^{(t)} - \eta^{(t)} \frac{\partial F^{\{a,v\}}}{\partial w} \tag{11}$$

where t and η are the number of iterations and the learning rate differently.

The gradient of loss function to the weight can be computed as below:

$$\frac{\partial F^{\{a,v\}}}{\partial w} = w + \alpha \sum_{i=1}^{N} \frac{\partial \ell_i(f(x_i, w) - y_i)}{\partial w}$$

$$+ \beta^t \sum_{i=1}^{N} \frac{\partial \ell_i^{\{ta\}}(f(x_i, h_i^t, \hat{y}_i)}{\partial w} + \beta^m \sum_{i=1}^{N} \frac{\partial \ell_i^{\{mv\}}(f(x_i, h_i^m, \hat{y}_i))}{\partial w} + \quad (12)$$

$$\beta^b \sum_{i=1}^{N} \frac{\partial \ell_i^{\{ba\}}(f(x_i, h_i^b, \hat{y}_i))}{\partial w} + \beta^l \sum_{i=1}^{N} \frac{\partial \ell_i^{\{la\}}(f(x_i, h_i^l, \hat{y}_i))}{\partial w}$$

where the specific gradient of loss function to the weight is computed as:

$$\frac{\partial \ell_i(f(x_i, w) - y_i)}{\partial w} = \begin{cases} 0, & if \ |f(x_i) - y_i| \le \epsilon \\ \phi(x), & otherwise. \end{cases} \quad (13)$$

$$\frac{\partial \ell_i^{ta}(f(x_i, h_i^t, \hat{y}_i))}{\partial w} = \begin{cases} -2sigmoid(f(x_i, w))[1 - sigmoid(f(x_i, w))]\phi(x_i), \\ if \ h_i^t = 1 \ and \ 1 - 2sigmoid(f(x_i, w)) \ge 0 \\ 2sigmoid(f(x_i, w))[1 - sigmoid(f(x_i, w))]\phi(x_i), \\ if \ h_i^t = 0 \ and \ 2sigmoid(f(x_i, w)) - 1 \ge 0 \\ 0, \qquad otherwise. \end{cases}$$

$$(14)$$

Gradients of ℓ_i^{ta}, ℓ_i^{mv}, ℓ_i^{ba} and ℓ_i^{la} can be computed as Eq. 14 similarly.

The learning algorithm is shown in Algorithm 1.

Algorithm 1. Training algorithm of the proposed model

Input:

  training samples(x_i, h_i, y_i),

  coefficient α, β^t, β^m, β^b and β^l learning rate η

Output: Model parameters w

  Randomly initialize w;

  repeat

    for each training sample (x_i, h_i, y_i) **do**

      Calculate the probabilistic dependencies $p(\hat{y} > 0|h)$

      and $p(\hat{y} \le 0|h)$ as Eq. 10;

      Calculate the specific gradient as Eq. 13 and Eq. 14;

    end for

    Calculate $\frac{\partial F^{\{a,v\}}}{\partial w}$ as Eq. 12

    $w \leftarrow w - \eta(\frac{\partial F^{\{a,v\}}}{\partial w})$

  until

  Converges

  return w

After training, the proposed approach can evaluate the predicted emotion value for testing samples according to function $f(x, w)$.

4   Experiments

4.1   Experimental Conditions

We conduct experiments on two benchmark databases: the Music Emotion in 2015 database [1] and the All Music Guide 1608 database (AMG1608) [3].

The Music Emotion in 2015 database consists of royalty-free music, with diverse genres of rock, classical, pop, jazz, country, folk, rap etc. [2]. The database is divided into two subsets: the development set and the test set. Specifically, the development set consists of 430 clips of 45 s, and the test set is comprised of 58 complete music pieces with an average duration of 234 ± 105.7 s. We use 260 low-level feature set provided by [1], which are extracted using openSMILE features. The 260 dimensional feature set represent the music from 65 dimensional mean deviation, 65 dimensional standard deviation, and their first-order derivatives from acoustic descriptors. We also extract tempo, mode, brightness, loudness with MIR toolbox.

The AMG1608 database consists of 1608 preview clips of Western songs, collected from a popular music stream service named 7 digit. Each preview clips is 30-second long. For experiments, we adopt the four-fold cross-validation on the database. We use the public feature set provided by [3], including MFCC, Tonal, Spectral and Temporal. We also extract tempo, mode, brightness, loudness with MIR toolbox.

To further demonstrate the effectiveness of domain knowledge, we conduct the following experiments in the arousal space: music audio emotion analysis ignoring all domain knowledge (**none**), music audio emotion analysis only exploiting single domain knowledge (**tempo, brightness, loudness**), music audio emotion analysis exploiting two of domain knowledge(**tempo+brightness, tempo+loudness, brightness+loudness**) and music audio emotion analysis exploiting all domain knowledge (**tempo+brightness+loudness**). In the valence space, since mode is the only musical elements that affects the valence, we conduct experiments as: music audio emotion analysis ignoring all domain knowledge (**none**), and music audio emotion analysis exploiting mode (**mode**). We also conduct experiments using music audio emotion analysis fusing the musical elements as features (**fusion**).

Root-Mean-Square Error (RMSE) and Pearson Correlation (R) is adopted to evaluate the effectiveness of the proposed method.

During training phrase, we use grid search to select our hyper parameter. Specifically, we first initialize small random number as weights, then through grid search, we choose the hyper parameter α, β^t, β^m, β^b and β^l ranging from $\{0.1, 1, 10, 20, 50\}$. On the Music Emotion in 2015 database, a fixed split of training/validation/testing 400/30/58 is adapted. On the AMG1608 database, we adopt 4-fold cross-validation.

Table 3. Music emotion analyses results on the music in 2015 database and the AMG1608 database in valence space

	Music in 2015 database		AMG1608 database	
	RMSE	R	RMSE	R^2
none	0.357	0.012	0.275	0.064
fusion	0.351	0.019	0.272	0.063
mode	**0.318**	**0.044**	**0.254**	**0.140**

Table 4. Music emotion analyses results on the music in 2015 database and the AMG1608 database in arousal space

	Music in 2015 database		AMG1608 database	
	RMSE	R	RMSE	R^2
none	0.270	0.3740	0.2670	0.5680
fusion	0.270	0.377	0.262	0.589
tempo	0.2626	0.4649	0.265	0.5975
brightness	0.2650	0.4887	0.266	0.6257
loudness	0.2618	0.4759	0.252	0.6068
tempo+brightness	0.2454	0.5185	0.264	0.6395
tempo+loudness	0.2550	0.5417	0.246	0.6162
brightness+loudness	0.2566	0.5782	0.244	0.6461
tempo+brightness+loudness	**0.2340**	**0.5970**	**0.240**	**0.669**

4.2 Experimental Results and Analysis

Tables 3 and 4 show the music audio analyses results on the Music Emotion in 2015 database and the AMG1608 database in the valence space and arousal space. From Tables 3 and 4, we observe as follows:

First, the proposed method exploiting all domain knowledge has the best performance among all methods with the lowest RMSE and highest Pearson correlation. Specifically, compared with music audio analyses ignoring all domain knowledge, the proposed method achieves 0.039 and 0.021 decrement of RMSE, and 0.032 and 0.076 increment of Pearson correlation, with respect to the Music Emotion in 2015 database and the AMG1608 database in the valence space. In the arousal space, the proposed method decrease the RMSE of 0.036 and 0.027, and increase the Pearson correlation of 0.223 and 0.101 on the Music Emotion in 2015 database and the AMG1608 database respectively. The method ignoring domain knowledge is totally data-driven method, which only learns the mapping from the extracted features to the predictions and it ignores the well-established music knowledge. On the contrary, our method capture the relations between domain knowledge and training data, and thus achieves better performance.

Second, the methods utilizing more domain knowledge have better performance than that using less domain knowledge. Specifically, in the arousal space, the methods with one domain knowledge is worse than the methods leveraging two domain knowledge. Since temp, brightness, and loudness describes the music from different aspects, the effects of these musical elements on the music emotion analyses are complementary. Thus, the methods using more domain knowledge can build more relations between music elements and emotion, and achieves better prediction.

4.3 Comparison with Related Work

In this section, we aim to evaluate the effectiveness of the proposed method. We compared the proposed method with the state-of-the-art methods.

On the Music Emotion in 2015 database, we compare the proposed method with Aljanaki's [1], Liu's [11], Chin's [4], Markor's [12], and Patra's [14]. Specifically, Aljanaki *et al.* provided the baseline for MediaEval 2015. Liu *et al.* proposed Arousal-Valence Similarity Preserving Embedding (AV-SPE) to extract the intrinsic features embedded in music signal, and train the SVR which takes the extracted features as the input and the emotion values as labels; Chin *et al.* adopted deep recurrent neural network to predict the valence and arousal for each moment of a song; Markor *et al.* used Kernel Bayes Filter (KBF) for predicting the valence and arousal. Patra *et al.* proposed the music emotion recognition system consisting of feed-forward neural networks, which predicts the dynamic valence and arousal values continuously. The comparisons are given in Table 5. As we can see from the table, we conclude that:

Compared with the others' works, our method achieves best performance in most cases. The state-of-the-art method only learns the maps from the features, and makes prediction of the music emotion. On the contrary, the proposed method not only learns the mapping from the features, but also captures the dependencies between musical elements and emotions through domain knowledge. Thus the proposed capture more information, and achieves better performance.

Rare work is conducted on the AMG1608 database. Thus, we only compare the proposed method with the baseline methods provided in [3]. In [3], Chen *et al.* adopted the Music emotion recognition (MER) system to recognize music emotion on the AMG1608 database. We adapted the Average Euclidean Distance (AED) and Pearson correlation as evaluation. The comparison is shown in Table 4. From the table, we observe as follows:

Compared with baseline method, the proposed method achieve better performance of AED and Pearson correlation. Since the proposed method captures the more information by constraints of domain knowledge, it is reasonable that the proposed method achieves better performance.

Taking the comparisons above into consideration, the proposed method has an excellent generalization ability with respect to affective audio music analy-

Table 5. Comparison with related works on Music Emotion 2015 and AMG1608 database

Database models	Music emotion in 2015				AMG1608			
	Arousal		Valence		Arousal		Valence	
	RMSE	R	RMSE	R	AED	R^2	AED	R^2
Our model	**0.234**	**0.597**	**0.318**	**0.044**	**0.240**	**0.669**	**0.254**	**0.140**
Baseline	0.27	0.36	0.37	0.01	0.288	0.651	0.288	0.120
Liu et al.'s	0.2377	0.5610	0.3834	−0.0217				
Chin et al.'s	0.2555	0.3417	0.3359	−0.0103				
Markov et al.'s	0.419	0.498	0.620	−0.035				
Patra et al.'s	0.2689	0.4678	0.3538	−0.0082				

sis. This demonstrates our approach successfully achieves higher music emotion prediction supported by domain knowledge.

5 Conclusion

This paper has proposed to analyze music emotion recognition by exploring domain knowledge. Probabilistic dependencies is used for music emotion recognition between emotions and music elements, i.e., tempo, mode, brightness and loudness. Then we model such probabilistic dependencies to the domain knowledge constraints in order to regularize our objective function. Experimental results on the Music emotion in 2015 database and the AMG1608 database demonstrate that our model outperforms the state-of-the-art approaches. This further demonstrates the importance of the domain knowledge to music emotion recognition.

References

1. Aljanaki, A., Yang, Y.H., Soleymani, M.: Emotion in music task at mediaeval 2015. In: Working Notes Proceedings of the MediaEval 2015 Workshop (2015)
2. Bittner, R.M., Salamon, J., Tierney, M., Mauch, M., Cannam, C., Bello, J.P.: MedleyDB: a multitrack dataset for annotation-intensive MIR research. In: ISMIR, pp. 155–160 (2014)
3. Chen, Y.A., Yang, Y.H., Wang, J.C., Chen, H.: The AMG1608 dataset for music emotion recognition. In: 2015 IEEE International Conference on Acoustics, Speech and Signal Processing (ICASSP), pp. 693–697. IEEE (2015)
4. Chin, Y.H., Wang, J.C.: Mediaeval 2015: recurrent neural network approach to emotion in music tack. In: Working Notes Proceedings of the MediaEval 2015 Workshop (2015)
5. Fernández-Sotos, A., Fernández-Caballero, A., Latorre, J.M.: Influence of tempo and rhythmic unit in musical emotion regulation. Front. Comput. Neurosci. **10**, 80 (2016)

6. Gabrielsson, A., Lindström, E.: The role of structure in the musical expression of emotions. In: Handbook of Music and Emotion: Theory, Research, Applications, pp. 367–400 (2010)
7. Gomez, P., Danuser, B.: Relationships between musical structure and psychophysiological measures of emotion. Emotion **7**(2), 377–387 (2007)
8. Husain, G., Thompson, W.F., Schellenberg, E.G.: Effects of musical tempo and mode on arousal, mood, and spatial abilities. Music Percept.: Interdisc. J. **20**(2), 151–171 (2002)
9. Lartillot, O.: Mirtoolbox 1.3. 4 user's manual. Finnish Centre of Excellence in Interdisciplinary Music Research, University of Jyväskylä, Finland (2011)
10. Lartillot, O., Toiviainen, P.: A Matlab toolbox for musical feature extraction from audio. In: International Conference on Digital Audio Effects, pp. 237–244 (2007)
11. Liu, Y., Liu, Y., Gu, Z.: Affective feature extraction for music emotion prediction (2015)
12. Markov, K., Matsui, T.: Dynamic music emotion recognition using kernel Bayes' filter (2015)
13. Miller, M.: The Complete Idiot's Guide to Music Theory. Penguin, New York (2005)
14. Patra, B.G., Maitra, P., Das, D., Bandyopadhyay, S.: Mediaeval 2015: music emotion recognition based on feed-forward neural network. In: MediaEval (2015)
15. Sloboda, J.: Handbook of Music and Emotion: Theory, Research, Applications. Oxford University Press, Oxford (2011)
16. Trochidis, K., Lui, S.: Modeling affective responses to music using audio signal analysis and physiology. In: Kronland-Martinet, R., Aramaki, M., Ystad, S. (eds.) CMMR 2015. LNCS, vol. 9617, pp. 346–357. Springer, Cham (2016). https://doi.org/10.1007/978-3-319-46282-0_22
17. Wessel, D.L.: Timbre space as a musical control structure. Comput. Music J. **3**, 45–52 (1979)
18. Yang, Y.H., Chen, H.H.: Machine recognition of music emotion: a review. ACM Trans. Intell. Syst. Technol. (TIST) **3**(3), 40 (2012)

A Better Understanding of the Interaction Between Users and Items by Knowledge Graph Learning for Temporal Recommendation

Chunjing Xiao[✉], Cong Xie, Shuyan Cao, Yuxiang Zhang,
Wei Fan, and Hongjun Heng

School of Computer Science and Technology, Civil Aviation University of China,
Tianjin, China
chunjingxiao@163.com

Abstract. Recently the knowledge graph (KG) as extra auxiliary information is widely used to improve recommendation. Existing methods usually treat knowledge representation as characteristic information for addressing data sparsity and cold start issues. However, they ignore the implicit and explicit interaction between users and items, which may be gained by the relation extraction and knowledge reasoning, to lead to suboptimal performance. Thus, we believe that it is crucial to incorporate both relations and attributes of users and items into recommender system. That can better capture the extent that a user prefer to an item. In this paper, we propose a novel knowledge graph-based temporal recommendation (KGTR) model. Firstly, we design a lightweight KG on the basis of a single independent domains knowledge without extra supplement. We define three relationships to express interactions within/between users and items, including the interaction of a user browsing an item, the social relation of two users browsing one item, and the behavior of a user browsing items in the meantime. Different from previous knowledge translation-based recommendation methods, we embed interactions by adding them to the transformation from one entity to another in KG. Extensive experiments on real world dataset show that our KGTR outperforms several state-of-the-art recommendation methods.

Keywords: Knowledge graph · Implicit interaction · Explicit interaction · Temporal recommendation

1 Introduction

The various facts from different domains interlink with each other and store in a complex heterogeneous graph called knowledge Graph (KG). The entities, such as people, books, musics, movies, are treated as nodes in KG and the relations between entities are denoted as edges. Owing to the connection of

© Springer Nature Switzerland AG 2019
A. C. Nayak and A. Sharma (Eds.): PRICAI 2019, LNAI 11670, pp. 135–147, 2019.
https://doi.org/10.1007/978-3-030-29908-8_11

various information from different topic domains in KG, knowledge exploration can develop insights on problems. That are difficult to determine on a single domain data. Over the past years, KG has been widely adopted in many fields, including dialogue system, Web search, and recommendation system.

The rich information of KG has recently shown great potential to enhance accuracy and explainability of recommendation [17]. For example, Zhang et al. [19] extract items' semantic representations from structural content, textual content and visual content by considering the heterogeneity of both nodes and relationships in KG. Sun et al. [15] employ recurrent networks learning semantic representations of both entities and paths for characterizing user preferences to improve recommendation. In these cases, the semantic representations of user and item or user's preference have accurately been obtained. However, the potential of the KG may still fail to be exploited since they suffer from the following limitations: (1) relying on a large-scale knowledge graph and extra knowledge base to extract features by heavy feature engineering process. (2) only utilizing the semantic representations into recommender system while ignoring the implicit and explicit interaction between users and items. For instance, two users are likely to have interaction when they both connect one item.

To address the above issues, we propose a novel knowledge graph-based temporal recommendation (KGTR) model, which captures the joint effects of users and items interactions information. We design a lightweight KG by only utilizing the facts in one domain as the knowledge, meanwhile, extra auxiliary data is lack. Three categories relationships are defined to exploit user-item interaction, including user relationship, item relationship and rating activity. User social relationship implies that two different users browse one item simultaneously, and can be called as user relationship. Item relationship means that one user browses various items. They are considered as implicit interaction in recommendation. Rating activity expresses that a user has rated the item, which is regard as explicit interaction in recommendation. Then representations of users' and items' static feature are obtained by TransE [1] in the light of three kind of relationships separately. Meanwhile, embeddings of users' and items' various attributes are learned by the KR-EAR [11] on the basis of former static representations, which serve as explicit information of user and item.

Considering the important effect of temporal context, we hold that prevailing items at the previous moment, similar to users' preference changing, have affected in recommendation result. Therefore, different from traditional temporary recommendation, we aggregate long-term and short-term features of users and items in recommender process. The attributes features and static feature learned by above procedure constitute the user's long-term features. The item's long-term features are similar. The user's short-term features are learned by LSTM [9] with the user's interaction data in a short period of time (such as hourly, weekly). The item's short-term features are learned by attention machine [16] according to all users' behavior at the latest moment. The personalized recommendation process applies implicit and explicit interactions of users and items to long-term and short-term of users and items features.

We summarize our main contributions as follows:

- We design a lightweight KG based on one topic domain without extra aux-iliary data, and explore the nature of interaction under less information by relation extraction and knowledge reasoning.
- We learn implicit and explicit interaction of users and items by TransE accord-ing to the second-order proximity between the entities, which is determined by the shared neighborhood structures of the entities.
- KGTR considers freshness and popularity of items in recommendation, and learning the items' short-term feature by attention machine with all users behavior at previous moment.

2 Related Work

2.1 Knowledge Representation Learning

User/item clustering or matrix factorization techniques only represent single relation between the connectivity entities. Most existing methods have been designed to learn multi-relations from latent attributes [5,18]. Making use of multi-relational KG in recommender systems has been found to be effective in recent years.

TransRec [6] represented a user as a relation vector to capture the transition from the previous item to the next item in large sequences. A user's previous preference is important for predicting the next item in sequence recommenda-tions, but the social relationships among users cannot be overlooked in context recommendations. TransTL [13] took both time and location into consideration with a translation-based model, which captured the joint effects of spatial and temporal information. Cao et al. [3] jointly learned the recommendation model and KG, which utilized the facts in KG to augment the user-item interaction. These models are not general for arbitrary recommendation scenarios, and ignore structures relationships among entities. Recent studies for KG focus on learning low-dimensional representations of entities and relations, and structural informa-tion of the graph is preserved. For completing knowledge graph and extracting relation from text, TransE learned a continuous vector space to preserve certain information of the graph, regarded relations as a translation between entities. TransE and its extensions TransH [21] and TransR [12] promoted prediction accuracy and computational efficiency by modeling multi-relational data. The most related work to ours is KR-EAR model, the method distinguished existing KG-relations into attributes and relations.

The entities embeddings were learned by building translation between enti-ties according to relations, and attribute values embeddings were learned based on entity embeddings. We extend the KR-EAR to learn user's and item's repre-sentations. Three relations are defined according to the second-order proximity between the entities, which is determined by the shared neighborhood structures of the entities in KG.

2.2   Implicit and Explicit Interaction

Due to the significant impact for the quality of recommendations, many works have made great effort on gaining variety information. Cao et al. [2] described heterogeneity exists between users and between items, meanwhile detected the various coupling relationships to essentially disclose why a user liked an item. Methods [20] incorporated explicit and implicit couplings about users-items interactions and attributes' inter-coupled interactions. A classic work was the CoupledCF model [20], which integrated the explicit user-item couplings within/between user's and item's attributes and the implicit user-item couplings. The model were trained by deep learning. Different from these models, the explicit and implicit information of our model is more substantial by adding user's and item's interactions in various relationships.

User/item information has been increasingly involved into CF. NCF [7] can express and generalize matrix factorization by replacing the inner product with a neural architecture. NCF model may be supercharged with non-linearities, a multi-layer perceptron to learn the user-item interaction function. Wide&Deep [4] trained wide linear models by using cross-product feature transformations and deep neural networks to generalize recommendation. Unlike Wide&Deep model, we treat raw features as input by knowledge representation learning.

3   Our Proposed Model

In this section, we introduce our Model. Suppose there is a sparse user-item rating matrix that consists of users, items, and the rating. The rating is represented by numerical values from 1 to 5, where the higher value indicates the user has more interest in an item. Meanwhile, there are various attributes of users and items, such as gender and profession, which are important additional information for recommendation result. Given a dataset with user-item rating matrix and explicit attributes, we aim to build temporal personalized recommendation model for a user, and recommend a ranked list of items that are of interest to her/him accordingly.

As shown in Fig. 1, long-short term features of users and items are jointed in recommender process. The attributes features and static features capturing by knowledge representations learning are considered as the long-term features of users and items, that is explained in Sect. 3.1. The static features of items browsed by user previously are treated as input to LSTM. The fashionable items are interacted by users at the latest moment, and their attributes features and static features are served as input to attention machine. The implicit and explicit interactions of users and items are blended into long-short term of users and items features to recommendation.

3.1   Knowledge Representation Learning for Interaction

We design a lightweight KG with information of the dataset. As shown in Fig. 2(a), the users and items are treated as entities in KG. When the user has

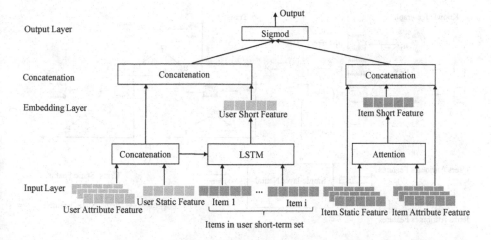

Fig. 1. The framework of knowledge graph-based temporal recommendation model

rated the item, there is a edge between the user and the item. The attributes of users and items are linked with corresponding entities. According to the extracted neighborhood structures of the entities in KG, we express first-order and second-order proximity [10] as three relationship definitions for learning static features.

We define the attribute triple additionally, for the purpose of learning attributes features based on former relationships representations. Our objective is to learn embeddings of users, items, and attributes preserving the structures information and semantic relations. The static feature belongs to the implicit interactions within/between user and item. The attribute feature is part of the explicit information.

Definition 1. *Rating Activity. As shown in up of Fig. 2(b), a rating activity is a triple (u, r, v), which means user u give a rating to item v.*

Definition 2. *Users Relationship. As shown in left of Fig. 2(b), a triple (u_i, v, u_j) represent users relationship, which implies both user u_i and u_j give ratings to item v.*

Definition 3. *Items Relationship. As shown in right of Fig. 2(b), Item relationship is a triple (v_i, u, v_j), which shows user u give rating to item v_i and v_j.*

Definition 4. *Attribute Triple. An attribute triple of user or item is a triple $(u/v, a, e)$, which indicates the attribute a of user u or item v with values e, such as $(u_1, gender, female)$ illustrates the gender of user u_1 is female.*

We aim to embed users and items to capture the implicit and explicit correlations between them. We usually optimize the probability $P(u, r, v)$, $P(u_i, v, u_j)$ and $P(v_i, u, v_j)$ for learning from relational triples. In this paper, we adopt TransE

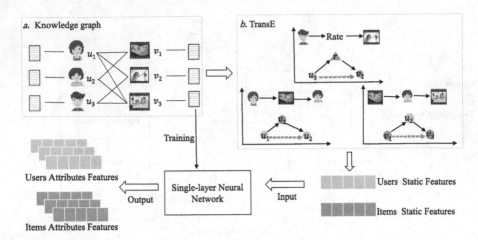

Fig. 2. Knowledge representation learning for implicit and explicit interaction

to encode relational triples. So the probability $P(u, r, v)$ is formalized as follows:

$$P(u,r,v) = \sum_{(u,r,v^+)\in KG} \sum_{(u,r,v^-)\in KG^-} \sigma(g(u,r,v^+) - g(u,r,v^-)) \qquad (1)$$

where $\sigma(x) = 1/(1 + exp(x))$ is sigmoid function, $g(\cdot)$ is the energy function which indicates the correlation of rating r and entity pair (u, v). The KG and the KG^- are the positive and negative instances set, respectively. KG^- contains incorrect triplets constructed by replacing tail entity in a valid triplet randomly. The probabilities of $P(u_i, v, u_j)$ and $P(v_i, u, v_j)$ are similar. Here, we can follow TransE to define the function $g(u, r, v)$ as Eq. 2:

$$g(u,r,v) = ||u + r - v||_{L_1/L_2} + b_1 \qquad (2)$$

where b_1 is a bias constant. A classification model is used for capturing the correlations between entities and their attributes. Hence, we consider the probability $P(u, a, e)$ for each triple (u, a, e) and $P(v, a, e)$ for each triple (v, a, e), and formalize the probability $P(u, a, e)$ for example, it is formalized as follows:

$$P(u,a,e) = \sum_{(u,a,e^+)\in KG} \sum_{(u,a,e^-)\in KG^-} \sigma(h(u,a,e^+) - h(u,a,e^-)) \qquad (3)$$

where $h(\cdot)$ is the scoring function for each attribute value of a given entity. The function $h(\cdot)$ is described in Eq. 4. We first transform entity embedding into the attribute space by a single-layer neural network. For training attribute embedding, we calculate the semantic distance between the transformed embedding and it, as shown in Eq. 4:

$$h(u,a,e) = ||f(uW_a + b_a) - e_{ae}||_{L_1/L_2} + b_2 \qquad (4)$$

where $f(\cdot)$ is a nonlinear function such as $tanh$, W_a is the parameters by learning, e_{ae} is the embedding of attribute value a and b_2 is a bias constant.

The final embeddings of users' and items' static features propagating information between triples are recorded as U_r, V_r, respectively. And the attribute embeddings of users' and items' are written as U_a, V_a, respectively. The static features based on knowledge representation learning remain relatively stable over time, while a user's preference is affected by current prevalence. The freshness and temporal dynamics of the items are more likely to improve recommendation. Therefore, we extend our model by including user-item temporal information. Different from existing models, the item's short-term features are also discussed.

3.2   Temporal Recommendation

Users Preference. The users' short-term features are learned by recurrent neural networks (RNN). Instead of modeling the user history sequence using RNN which is difficult to calculate, our model combines users' static features and attribute features as pre-train input. This can make neural network training .faster and more effective. The key issue of dynamic preferences is to choose the granularity of each input time spot t. Using smaller time spans can capture more fine-grained interest changes, but the feature space is very sparse and learning process is difficult [14]. Having large time spans may lead to sufficient content at each time spot, but makes the model less adaptive for capturing users' dynamics change. Unlike the previous model, we order 16 items for one user according to the latest browsing records. That can sure the enough context in behavior sequence to train user preference. To this end, we propose leveraging LSTM in capturing sequential patterns, and use it to model user's recent interaction trail. The output of LSTM U_S is took as the users' short-term features.

Items Preference. The items' popularity are changing over time, and the features of most fashionable items currently have a greater impact on user preference. Here, we apply attention to obtain items' short-term characteristics. Attention can keep the contextual sequential information and capture the relationships between elements in the sequence. The items viewed by all users in the latest hour are considered as the items sequence. That is matched with items of the whole training datas C to refine representation. The input of attention consists of items' attribute features and static features. The output is a weighted sum of the items, where the weight matrix T^t is determined by similarity. Similar to [16], the attention vector are calculated at each output time t over the input items $(1, \ldots, I)$ with Eq. 5.

$$T_i^t = z^T tanh(W_c c_t + W_y y_i)$$
$$S_i^t = softmax(T_i^t) \tag{5}$$
$$V_s' = \sum S_i^t y_i$$

The vector z and matrices W_c, W_y are learnable parameters, c_t is the train item at time t and y_i is i-th item of input sequence. The i-th item of vector $T^t \in R^I$

indicates the similarity between y_i and the training datas. The attention weight matrix S^t is created by normalizing similarity scores with softmax. Lastly, we concatenate c_t with V_s', which is regarded as c_{t+1} at the next time step. Here, the final attentive output V_s can be viewed as item's short-term feature.

3.3   Model Learning

Objective Function. Our task is to predict the item which the user will inter-act at next time, according to given long-short term preferences of users and items. A straightforward solution is to combine the outputs of their charac-teristics. So $U_{s,a,r}$ are the concatenation of U_s, U_a, U_r, and $V_{s,a,r}$ are the con-catenation of V_s, V_a, V_r. Similar with the NCF, hidden layers are added on the concatenated vector by using a standard multi-layer perceptron (MLP) to learn the long-short term features. Specifically, the model can be formulated as

$$q_1 = \Phi_1(U_{s,a,r}, V_{s,a,r}) = \begin{bmatrix} U_{s,a,r} \\ V_{s,a,r} \end{bmatrix}$$

$$\Phi_2(q_1) = \alpha_2(w_2^T q_1 + b_2)$$
$$\cdots \qquad\qquad (6)$$

$$\Phi_l(q_{l-1}) = \alpha_l(w_l^T q_{l-1} + b_l)$$
$$\hat{y}_{uv} = \sigma(h^T \Phi_l(q_{l-1}))$$

where w_x, b_x and α_x denote the weight matrix, bias vector, and ReLU activation function for the x-th layer's perceptron, respectively. $\hat{y}_{uv}$ indicates whether the user u is likely to interact with the item v.

Considering implicit feedback of interaction, we treat the value of y_{uv} as a label. 1 means user u has browsed item v, and 0 otherwise. The prediction score $\hat{y}_{uv}$ represents how likely u interacts with v. We limit the output $\hat{y}_{uv}$ in the range of $[0,1]$, thus, the output is achieved by using a probabilistic function as the activation function. Finally, we define the likelihood function as

$$p(y, y^- | \Theta_f) = \prod_{(u,v)\in y} \hat{y}_{uv} \prod_{(u,v)\in y^-} (1 - \hat{y}_{uv}) \qquad (7)$$

Taking the negative logarithm of the likelihood, we gain the objective function to minimize for KGTR in Eq. 8.

$$L = -\sum_{(u,v)\in y} \log \hat{y}_{uv} - \sum_{(u,v)\in y^-} \log(1 - \hat{y}_{uv})$$
$$= -\sum_{(u,v)\in y\cup y^-} y_{uv} \log \hat{y}_{uv} + (1 - y_{uv})\log(1 - \hat{y}_{uv}) \qquad (8)$$

For the negative instances y^-, we uniformly sample them from unobserved inter-actions in each iteration and control the sampling ratio about the number of observed interactions. The sigmoid function restricts each neuron to be in $[0,1]$, where neurons stop learning when their output is near either 0 or 1.

We optimize the proposed approach with adaptive gradient algorithm which could adapt the step size automatically. Hence it reduces the efforts in learning rate tuning. In the recommendation stage, candidate items are ranked in ascending order based on the recommendation score computed by Eq. 8, and the top ranked items are recommended to users.

4 Experiments

In this section, we evaluate our proposed framework for movie recommendation scenarios. We test our methods against the related baselines for recommendations items to users. The experimental results demonstrate that our method better than many competitive baselines.

4.1 Experimental Settings

Dataset Description. We used MovieLens-1M[1] dataset in our experiments. The dataset consists of one million ratings from 6,040 users and 3,952 items, user auxiliary information (Gender, Age, Occupation and Zip code) and some item attributes (Genres, Title and release dates). We transformed the original rating matrix scaled from $R \in \{1, 2, ..., 5\}$ into a binarized preference matrix $R \in \{0, 1\}$. Each rating was expressed as either 0 or 1, where 1 indicates an interaction between a user and an item, otherwise 0. Then we sampled four negative instances per positive instance.

For each user, we sorted the user-item interactions by the time stamps at first. Then we took her/his latest interaction as the test positive instance and utilized the remaining data for training positive instance. Finally we randomly sampled 99 items that are not interacted by the user as the test negative instance and randomly sampled four negative instances for per positive instance.

Evaluation Metrics. Similar to [4], we ranked the test item among the 100 items and used Hit Ratio (HR) and Normalized Discounted Cumulative Gain (NDCG) to evaluate the performance of a ranked list [8]. The HR intuitively measures whether the test item is included in top-K list. The NDCG measures the position of the hit on top-K list. The higher NDCG scores show that the test item hits at top ranks. We calculated both metrics for each test user and reported the average score.

Baseline Methods. We evaluated our framework from in three versions based on the different input.

– KGTR_user: for every user, we used the sequence of items recently watched by user as the input vector of LSTM and the sequence of items recently viewed by all users as the input of attention. The input vector of LSTM was learned by translation representation learning and the input vector of attention was one-hot encoding of items;

---

[1] https://grouplens.org/datasets/movielens/.

- KGTR_item: different from KGTR_user, the input vector of LSTM is one-hot encoding of items and the input vector of attention was learned by translation representation learning.
- KGTR_NCF:different from KGTR_user and KGTR_item, both the input vector of LSTM and attention were learned by translation representation learning.

The following relevant and representative state-of-the-art methods were used as the baselines to evaluate our methods.

- NCF [7]: It presents a neural architecture replacing the inner product and proposes to leverage a multi-layer perceptron to learn the user-item interaction function.
- CoupledCF [20]: This model proposes a neural user-item coupling learning for collaborative filtering, which jointly learns explicit and implicit couplings within/between users and items.
- Wide&Deep [4]: It combines memorization and generalization for recommendation, which involves feature engineering (such as cross-product features) of the input to the wide network.

Parameter Settings. The configurations of TransE are $k = 100, b_1 = 7, b_2 = -2$, and taking L_1 as distance metric. The KGTR model was implemented in Python based on the Keras framework. We selected 16 items recently watched by every user as the input vector of LSTM, and the items viewed by the users in the latest hour were selected as input to the attention.

To determine hyper-parameters of KGTR, we randomly sampled one interaction for each user as the validation data and tuned hyper-parameters on it. All KGTR models were learnt by optimizing the log loss of Eq. 8. We used mini-batch Adam as the optimizer for our model. We initialized the embedding matrix with a random normal distribution (the mean and standard deviation are 0 and 0.01 respectively). All biases are initialized with zero. We tested all combinations of the batch size (S = {128, 256, 512, 1024}) and the learning rate (R = {0.0001, 0.0005, 0.001, 0.005}) that S = 256 and R = 0.001 was the best combination. From Fig. 3, we could see that HR@10 and NDCG@10 increased firstly. When S = 256 and R = 0.001, the performance of KGTR was best, and then HR@10 and NDCG@10 decreased or stabilized with the increase of batch size and learning rate. So we set S = 256 and R = 0.001 as the optimal parameters.

4.2   Results and Analysis

First, the performances of the KGTR and baselines were shown in Table 1 for TOP@10 recommendation. From Table 1, A number of interesting observations could be noted. Our method KGTR_item and CoupleCF were superior to NCF and Wide&Deep in both HR@10 and NDCG@10. The reason is that the user preference and item popularity contributed to improve the recommendation performance. Both the KGTR_item and CoupleCF integrated item popularity, user-item interaction and implicit user-item interaction to gain the best performance.

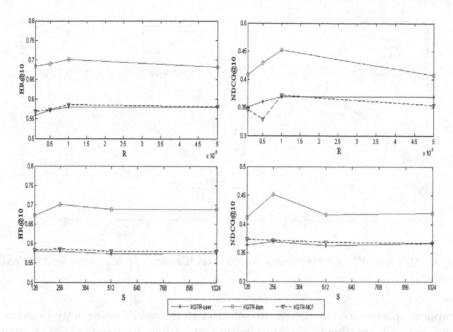

Fig. 3. Performance of KGTR models w.r.t the learning rate (bach size = 256) and the bach size s(learning rate = 0.001)

The results confirmed that the interactions between users and items were useful. Beside, KGTR_item is slightly worse than CoupleCF. That may because the CoupleCF used the rating values from the users and the items, our method only used the rating relationship between the users and the items.

Second, we also tested the top@K item recommendations in Fig. 4. As previously introduced, KGTR models were customized to three versions: KGTR_user, KGTR_item and KGTR_NCF. KGTR_item was compared with KGTR_user, KGTR_NCF and all the baselines. Figure 4 shows the performance of the top@K recommendation, where K ranges from 1 to 10. As shown in Fig. 4, all the baselines and KGTR_item highly outperformed KGTR_user and KGTR_NCF. That were mainly because KGTR_user and KGTR_NCF were personalized recommendation method via learning individual user's preference. When learning user

Table 1. HR@10 and NDCG@10 for Top-10 item recommendation

	#HR@10	#NDCG@10
NCF	0.6947	0.4149
CoupledCF	0.7310	0.4819
Wide&Deep	0.6864	0.4082
KGTR_user	0.5801	0.3691
KGTR_item	0.7012	0.4518
KGTR_NCF	0.5861	0.3719

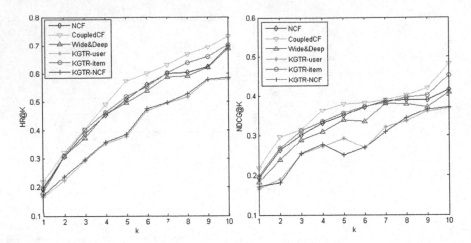

Fig. 4. HR@K and NDCG@K results comparison between our framework and related baselines

temporary preferences by LSTM, we selected 16 items viewed recently by every user, and did not divide the time period according to the traditional time spots (minutes, hours or days). Therefore, recommendation performances in KGTR_user and KGTR_NCF were not improved by adding user preferences.

5   Conclusions

In this paper, we proposed a knowledge graph-based temporal recommendation (KGTR) model to explore implicit/explicit and latest interactions between users and items. Firstly, we designed a lightweight KG based on one domain knowledge without extra information. We defined three categories relationships by TransE, according to the implicit/explicit interactions of users' and items'. That could capture global structural dependencies in the historic behavior and united information between triples. Taking the different impact of long-short term interest into account, our model was trained by deep learning. Specially, the popular features of items were jointed in the course of learning dynamic preferences. The experimental results showed significant improvement has obtained over state-of-the-art baselines on large dataset. In future, we will study how to incorporate short-term users' preferences and the ratings for better recommendation.

Acknowledgement. This work was supported by the Fundamental Research Funds for the Central Universities (No. ZXH2012P009).

References

1. Bordes, A., Usunier, N., Garcia-Duran, A., Weston, J., Yakhnenko, O.: Translating embeddings for modeling multi-relational data. In: Advances in Neural Information Processing Systems, pp. 2787–2795 (2013)

2. Cao, L.: Non-IID recommender systems: a review and framework of recommendation paradigm shifting. Engineering **2**(2), 212–224 (2016)
3. Cao, Y., Wang, X., He, X., Chua, T.S., et al.: Unifying knowledge graph learning and recommendation: Towards a better understanding of user preferences. arXiv preprint arXiv:1902.06236 (2019)
4. Cheng, H.T., et al.: Wide & deep learning for recommender systems (2016)
5. Han, X., Shi, C., Wang, S., Philip, S.Y., Song, L.: Aspect-level deep collaborative filtering via heterogeneous information networks. In: IJCAI, pp. 3393–3399 (2018)
6. He, R., Kang, W.C., McAuley, J.: Translation-based recommendation. In: Proceedings of the Eleventh ACM Conference on Recommender Systems, pp. 161–169. ACM (2017)
7. He, X., Liao, L., Zhang, H., Nie, L., Hu, X., Chua, T.S.: Neural collaborative filtering. In: Proceedings of WWW, pp. 173–182. WWW (2017)
8. He, X., Tao, C., Kan, M.Y., Xiao, C.: Trirank: review-aware explainable recommendation by modeling aspects (2015)
9. Hochreiter, S., Schmidhuber, J.: Long short-term memory. Neural Comput. **9**(8), 1735–1780 (1997)
10. Jian, T., Meng, Q., Wang, M., Ming, Z., Yan, J., Mei, Q.: Line: large-scale information network embedding. In: 24th International Conference on World Wide Web, WWW 2015 (2015)
11. Lin, Y., Liu, Z., Sun, M.: Knowledge representation learning with entities, attributes and relations. Ethnicity **1**, 41–52 (2016)
12. Lin, Y., Liu, Z., Sun, M., Liu, Y., Zhu, X.: Learning entity and relation embeddings for knowledge graph completion. In: Proceedings of AAAI (2015)
13. Qian, T.Y., Liu, B., Hong, L., You, Z.N.: Time and location aware points of interest recommendation in location-based social networks. J. Comput. Sci. Technol. **33**(6), 1219–1230 (2018)
14. Song, Y., Elkahky, A.M., He, X.: Multi-rate deep learning for temporal recommendation. In: Proceedings of ACM SIGIR, pp. 909–912. ACM (2016)
15. Sun, Z., Yang, J., Zhang, J., Bozzon, A., Huang, L.K., Xu, C.: Recurrent knowledge graph embedding for effective recommendation. In: Proceedings of the 12th ACM Conference on Recommender Systems, pp. 297–305. ACM (2018)
16. Vinyals, O., Kaiser, L., Koo, T., Petrov, S., Sutskever, I., Hinton, G.: Grammar as a foreign language. Eprint Arxiv, pp. 2773–2781 (2015)
17. Wang, H., Zhang, F., Xie, X., Guo, M.: DKN: deep knowledge-aware network for news recommendation. In: Proceedings of WWW, pp. 1835–1844. WWW (2018)
18. Wang, X., Peng, Z., Wang, S., Yu, P.S., Fu, W., Hong, X.: Cross-domain recommendation for cold-start users via neighborhood based feature mapping. In: Pei, J., Manolopoulos, Y., Sadiq, S., Li, J. (eds.) DASFAA 2018. LNCS, vol. 10827, pp. 158–165. Springer, Cham (2018). https://doi.org/10.1007/978-3-319-91452-7_11
19. Zhang, F., Yuan, N.J., Lian, D., Xie, X., Ma, W.Y.: Collaborative knowledge base embedding for recommender systems. In: Proceedings of the 22nd ACM SIGKDD, pp. 353–362. ACM (2016)
20. Zhang, Q., Cao, L., Zhu, C., Li, Z., Sun, J.: CoupledCF: learning explicit and implicit user-item couplings in recommendation for deep collaborative filtering. In: IJCAI, pp. 3662–3668 (2018)
21. Zhen, W., Zhang, J., Feng, J., Zheng, C.: Knowledge graph embedding by translating on hyperplanes. In: Proceedings of 28th AAAI (2014)

Knowledge-Aware and Retrieval-Based Models for Distantly Supervised Relation Extraction

Xuemiao Zhang[1], Kejun Deng[2], Leilei Zhang[1], Zhouxing Tan[1],
and Junfei Liu[3(✉)]

[1] School of Software and Microelectronics, Peking University, Beijing 100871, China
{zhangxuemiao,zhang_leilei,tzhx}@pku.edu.cn
[2] School of Electronics Engineering and Computer Science, Peking University,
Beijing 100871, China
kejund@pku.edu.cn
[3] National Engineering Research Center for Software Engineering, Peking University,
Beijing 100871, China
liujunfei@pku.edu.cn

Abstract. Distantly supervised relation extraction (RE) has been an effective way to find novel relational facts from text without a large amount of well-labeled training data. However, distant supervision always suffers from wrong labelling problem. Many neural approaches have been proposed to alleviate this problem recently, but none of them can make use of the rich semantic knowledge in the knowledge bases (KBs). In this paper, we propose a knowledge-aware attention model, which can leverage the semantic knowledge in the KB to select the valid sentences. Furthermore, based on knowledge representation learning (KRL), we formalize distantly supervised RE as relation retrieval instead of relation classification to leverage the semantic knowledge further. Experimental results on widely used datasets show that our approaches significantly outperform the popular benchmark methods.

Keywords: Distantly supervised relation extraction ·
Knowledge-aware attention · Relation retrieval

1 Introduction

Relation extraction (RE), aiming at extracting semantic relations between entities, is a fundamental task in natural language processing (NLP). It can augment current knowledge bases (KBs) by adding new relational facts to them, which are widely used in NLP tasks, such as question answering [12,24]. Formally, given a pair of annotated head entity h and tail entity t, the goal of RE is to predict the relation between h and t. Most supervised methods of RE are limited by a large amount of well-labeled training data. Distant supervision [15] is proposed to solve this challenge by automatically generate a large amount of training data.

© Springer Nature Switzerland AG 2019
A. C. Nayak and A. Sharma (Eds.): PRICAI 2019, LNAI 11670, pp. 148–161, 2019.
https://doi.org/10.1007/978-3-030-29908-8_12

Its intuition is that any sentence containing a pair of entities that participate in a KB will express the relation in some way. However, it suffers from wrong labelling problem inevitably. For example, as shown in Fig. 1, distant supervision believes that sentences S_1, S_2, S_3 that mentioning *(Steve_Jobs, Apple)* express of the relation *Found*. However, S_2, S_3 do not express *Found*.

Recent studies have explored multi-instance learning [5] paradigm to solve this problem. In this paradigm, all instances (sentences) aligned by a triplet (h, t, r) constitute a bag, for example, in Fig. 1, S_1, S_2, S_3 can constitute a bag corresponding to *(Steve_Jobs, Found, Apple)*. And distantly supervised RE learns a relation extractor from the bags in training data to predict the relation of an unseen bag towards the entity pair (h, t). In this paradigm, Zeng et al. [25] proposed piecewise convolutional neural networks (PCNNs) to extract the features of a sentence. Lin et al. [11] and Ji et al. [8] proposed two different attention mechanisms to alleviate the wrong labelling problem. The former uses selective attention model to filter out meaningless sentences (S_2 and S_3), the latter (denoted by APCNN) uses sentence-level attention model based on PCNNs to select valid sentences (S_1) in a bag. APCNN assumes that vector ($e_1 - e_2$) represents the relation, where e_1 and e_2 are word embeddings of the two annotated entities. Although APCNN achieves significant improvements, it suffers from the following flaws: (1) The semantics of the word embeddings of most entities are not rich because of their low frequencies of occurrence; (2) Reusing their word embeddings can not employ auxiliary features, since PCNNs have taken into account various features between them when extracting the sentence features.

In this paper, we propose a novel knowledge-aware attention mechanism based on PCNNs (denoted by KBPCNN) to leverage semantic knowledge in KBs and text semantics of the bag to help select valid sentences in a bag by assigning higher weights to the valid instances and lower weighs to the invalid ones, as shown in Fig. 2, inspired by the success of [23] which leverages KBs in long and short time memory networks (KB-LSTM). Specifically, for each triplet (h, r, t), the translation model TransE [3] in knowledge representation learning (KRL) task treat the relation vector l_r as the translation from the head entity vector l_h to the tail entity vector l_t, i.e., $l_h + l_r \approx l_t$. We define the vector $l_a = l_t - l_h$ as an abstract representation of all the relations between the two entities. As shown in Fig. 1, $l_a = l_{Apple} - l_{Steve\_Jobs}$ is an abstract representation of the two relations *Found* and *WorkIn*. Vector l_a is full of semantic knowledge, which helps neural networks predict whether *Steve_Jobs* has relation *Manage* with *Apple*, since the founder of a company still working in it is likely to manage it. In addition, as shown in Fig. 2, our model can rationally weight text semantics and semantic knowledge by calculating a generation probability p_{gen} [18], which is used to assess the importance of semantic knowledge relative to text semantics. KBPCNN can also use p_{gen} to handle the cases that annotated entities are unseen in the KB by assigning a extremely low value to the semantic knowledge part.

We further improve KBPCNN by introducing an innovative knowledge relation retrieval module (denoted by KBPCNN+R). Based on the relation embeddings trained by the TransE model, KBPCNN+R innovatively formalizes distantly supervised RE as relation retrieval instead of relation classification which is the practice of previous works, as shown in *for Retrieval* channel of Fig. 2. Different from treating relations as isolated categories, which usually represented by onehot embedding ignoring semantic knowledge in the relations, relation retrieval can introduce features from informative relation embeddings to the networks by forcing the bag embedding to be more and more relevant to the target relation embeddings.

In general, our contributions can be summarized as follows:

- We propose a novel knowledge-aware attention model, which can leverage the semantic knowledge in KBs to better select the valid sentences in a bag.
- We further improve the model by creatively treating RE as relation retrieval instead of relation classification.
- We conduct experiments on real-world dataset to prove the effectiveness of the proposed method, and provide some in-depth analysis.

2 Methodology

In this section, we will introduce the models in three main parts: Sentence Encoder, Knowledge-aware Attention Module and Relation Retrieve Module.

2.1 Sentence Encoder

Given an instance S from a bag and two entities h and t corresponding to the bag, the sentence encoder uses PCNNs to compute a representation vector s for S.

In order to fit the input of the neural networks, we should convert raw words in sentences into low-dimensional vectors. Similar to [8], in our methods, each word vector is obtained by concatenating the word embedding and position embedding of the word. We look up the word embedding matrix pre-trained by the word2vec [14] model to transform words to word embeddings. Position Embedding is defined as the combination of the relative distances from the current word to h and t, and each word has two relative distances [8]. For example, in the sentence *Steve Jobs is the founder of Apple*, the relative distance from the word *founder* to head entity *Steve Jobs* is 3 and tail entity *Apple* is 2. Assuming that the dimension of the word embedding is k_w and the dimension of the position embedding is k_d, then the dimension of the word vector of a word is $k = k_w + 2k_d$.

Given an input sentence sequence $S = \{v_1, v_2, \ldots, v_{|S|}\}$, where v_i represents the word vector of the i-th word, We use piecewise convolutional neural networks (PCNNs) to calculate the distributed representation s of the sentence.

$$s = PCNNs(S) \tag{1}$$

where $s \in \mathbb{R}^{3n}$, PCNNs define convolution operations and piecewise max-pooling operations, n is the number of convolution kernel. In this paper, we strictly follow [25] to implement PCNNs, and we recommend readers to read their works carefully.

2.2   Knowledge-Aware Attention

In order to reduce the impact of the invalid instances in a bag, we introduce the knowledge-aware attention mechanism, and hope that the attention model can learn the effective attention weight distribution by leveraging the semantic knowledge in the KB. That is, the model can assign higher weights for valid instances and lower weights for invalid ones. Then the model calculates the bag features by using all the sentence feature vectors in the bag computed by PCNNs according to the attention weights.

KB relational facts $\begin{cases} \text{(Steve\_Jobs, Found, Apple)} \\ \text{(Steve\_Jobs, WorkIn, Apple)} \end{cases}$

Sentence	Latent Label
S₁:Apple was founded by Steve Jobs, Steve Wozniak, and Ronald Wayne in April 1976.	Found
S₂:During the development of Apple, Steve Jobs did not work in the company for some time.	WorkIn
S₃:Steve Jobs was a competent manager of Apple.	Manage

Fig. 1. Samples of relational fact and sentence containing the entity pair *(Steve_Jobs, Apple)*.

Knowledge Abstract Relation. We leverage semantic knowledge in KBs by using their knowledge embeddings trained by a translation model. In this paper, we use the TransE model. Better methods can be chosen, while it is not the focus of this paper. Specifically, for each triplet (h, r, t), TransE assumes that their corresponding knowledge embeddings l_h, l_r and l_t satisfy $l_h + l_r \approx l_t$. But there may be m relations $\{r_1, r_2, \ldots, r_i, \ldots, r_m\}$ $(m \in \{0, 1, 2, \ldots\})$ between two entities in a KB, all the m relations should satisfy $l_h + l_{r_i} \approx l_t$, i.e., $l_{r_i} \approx l_t - l_h$. Let $l_a = l_t - l_h$, that is, l_a is related to any relation embedding l_{r_i}. We argue that l_a is a more abstract representation of all relations between two entities, called abstract relation between the two entities. It contains a wealth of semantic knowledge that can improve our model.

Attention Mechanism. Our method weighs semantic knowledge in the KB and text semantics of the bag to generate attention weight distribution, as shown in Fig. 2. We look up the knowledge embedding matrices trained by TransE

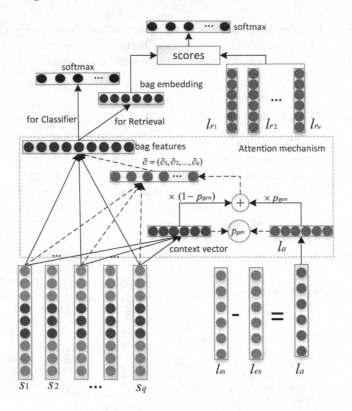

Fig. 2. The architecture of the Knowledge-aware Attention Module with the Knowledge Relation Retrieval Module. *for Classifier* and *for Retrieval* channel are softmax classifier in KBPCNN and relation retrieval module in KBPCNN+R. $s_1, s_2, \ldots, s_q$ are instance feature vectors computed by PCNNs, $l_{e_t}, l_{e_h}, l_{r_i}$ $(1 \leq i \leq u)$ are knowledge embeddings of tail entity, head entity and target relation r_i respectively, and l_a is the embedding of the knowledge abstract relation.

model to transform entities to entity embeddings, relations to relation embeddings. In Fig. 2, $s_1, s_2, \ldots, s_q$ are the feature vectors (computed by PCNNs) of all instances in a bag, the context vector c_t which represents the text semantics of the bag can be computed as follows:

$$c_t = W_c \left(\frac{1}{q} \sum_{i=1}^{q} s_i\right)^T \tag{2}$$

where $W_c \in \mathbb{R}^{k_e \times 3n}$ is a transformation matrix, and k_e is the dimension of knowledge embeddings. We use a generation probability p_{gen} [18] to integrate text semantics and semantic knowledge and to address the problem that the annotated entities are not in the KB. In our method, we assign the generation probability $p_{gen} \in [0, 1]$ to the abstract relation vector l_a to determine its importance relative to the context vector c_t. Further, when the two entities are unseen in the KB, that is, l_a is filled with 0, we hope the model to assign a extremely low value, even 0, to p_{gen} to reduce noise. In our method, p_{gen} is calculated as follows:

$$p_{gen} = \delta(l_a W_g c_t + b_g) \tag{3}$$

where $W_g \in \mathbb{R}^{k_e \times k_e}$ is a intermediate matrix, b_g is a bias, and $\delta(\bullet)$ is the sigmoid function. Next, the alignment vector can be calculated as follows:

$$v_{align} = p_{gen} l_a + (1 - p_{gen}) c_t^T \tag{4}$$

Then, we use the general score method [13] to calculate the attention weight between each instance feature vector and alignment vector, which reflects the similarity or relevance between them. The formula is as follows:

$$\alpha_i = \frac{exp(e_i)}{\sum_{j=1}^q exp(e_j)} \tag{5}$$
$$e_i = v_{align} W_a s_i^T + b_a$$

where $1 \le i \le q$, $W_a \in \mathbb{R}^{k_e \times 3n}$ is an intermediate matrix and b_a is an offset value. And weight vector $\boldsymbol{\alpha} = [\alpha_1, \alpha_2, \ldots, \alpha_q]$ is the attention weight distribution of all instances in the bag. Then the bag features can be calculated as follows:

$$b = \sum_{i=1}^q a_i s_i \tag{6}$$

2.3 Relation Retrieval

Based on the bag feature vector b, we have two ways to calculate the conditional probability of relation r_i, as shown in Fig. 2. One in *for Classifier* channel (KBPCNN) is to treat the distantly supervision RE as a classification task, which is the practice of previous works, and the other in *for Retrieval* channel (KBPCNN+R) is to treat it as a relation retrieval task, which is first proposed in this paper.

Softmax Classifier. The final output scores of the neural networks, which are associated to all relations, can be calculated as follows:

$$o = W_o b^T + d_o \tag{7}$$

where $o \in \mathbb{R}^u$, $W_o \in \mathbb{R}^{u \times 3n}$ is the weight matrix and $d_o \in \mathbb{R}^u$ is the bias vector. Then the conditional probability of relation r_i can be defined as:

$$p(r_i|B, \theta_1) = \frac{exp(o_i)}{\sum_{j=1}^u exp(o_j)} \tag{8}$$

where θ_1 indicates all the parameters of the KBPCNN model.

Relation Retrieval Ranking. As shown in *for Retrieval* channel of Fig. 2, knowledge relation retrieval module of KBPCNN+R can further leverage the semantic knowledge in the KB. As a neural retrieval model [16,22], we calculate the relevance score between the bag B and each relation r_i of u target relations in distantly supervised RE task by measuring the cosine similarity between the bag feature vector $\boldsymbol{b}$ and the relation embedding $\boldsymbol{l}_{r_i}$. The relevance score $R(B, r_i)$ can be computed as follows:

$$R(B, r_i) = cosine(\boldsymbol{l}_{r_i}, \boldsymbol{W}_b \boldsymbol{b}^T) = \frac{\boldsymbol{l}_{r_i}(\boldsymbol{W}_b \boldsymbol{b}^T)}{\|\boldsymbol{l}_{r_i}\| \|\boldsymbol{W}_b \boldsymbol{b}^T\|} \tag{9}$$

where $0 \leq i \leq u$, $\boldsymbol{W}_b \in \mathbb{R}^{k_e \times 3n}$ is a transition matrix. In practice, given a bag, the u target relations are ranked by their semantic relevance scores, and the most relevant one is considered to be the relation between the two entities.

Given a bag B, we calculate the conditional probability of relation r_i by converting the semantic relevance score between the bag and the relation to the posterior probability through softmax:

$$p(r_i|B, \theta_2) = \frac{exp(\gamma R(B, r_i))}{\sum_{j=1}^{u} exp(\gamma R(B, r_j))} \tag{10}$$

where γ is a smoothing factor in the softmax function, which is set empirically in our experiment, θ_2 indicates all parameters of KBPCNN+R model.

Optimization. Here we introduce the details of the learning and optimization of our models. Assume there are N bags in the training set $\{B_1, B_2, \ldots, B_N\}$, and their corresponding labels are relations $\{r_1, r_2, \ldots, r_N\}$. We define the training objective function using cross-entropy as follows:

$$J(\theta) = \sum_{i=1}^{N} p(r_i|B_i, \theta) \tag{11}$$

where $\theta \in \{\theta_1, \theta_2\}$. To train our models, we use stochastic gradient descent (SGD) to minimize the objective function, and employ dropout strategy [6] to prevent overfitting.

3   Experiments

3.1   Dataset and Evaluation Metrics

Dataset. We evaluate our models on the dataset generated by aligning Freebase relations with the New York Times (NYT) corpus, which is developed by [17] and has been widely used in distantly supervised RE [7,8]. And sentences from the years 2005–2006 of the NYT corpus used for training and sentences from the year 2007 used for testing. There are 53 possible target relations including a special NA relation. There are 570,088 sentences, 63,428 entities and 19,601 relational facts (excluding NA) in training data, 172,448 sentences, 16,705 entities and 1,950 relational facts (excluding NA) in testing data.

We choose the available mini dataset from Freebase, i.e., FB15k used in [10], as a part of our external KBs. In order to make the results of the model more convincing, we first removed the triplets that appeared in the testing set from FB15k. As shown in Table 1, we build three knowledge bases with different completeness based on FB15k and training set, named ZeroKB, HalfKB and AllKB. AllKB is obtained by taking unions of the relation sets, entity sets and triplet sets in FB15k and training data respectively. We randomly select half of the entity set in the training data to form HalfE, then HalfKB is the remaining set after filtering out the entities and the triplets that contain the entities of HalfE from AllKB. And ZeroKB is the remaining set after filtering out the entities and the triplets that contain the entities in the entity set in the training data from AllKB.

Table 1. Details of various KBs with different completeness. The above part shows the numbers of entities e and triplets t in each KB. The following shows the details of the intersections of each KB and the training set s_{tr}.

Items		FB15k	ZeroKB	HalfKB	AllKB
e		14951	9310	41558	73007
t		483142	193415	437131	1052915
s_{tr}	$e(x/63428)$	5640	0	31848	63428
	$t(x/293143)$	2261	0	71454	293143

Evaluation Metrics. Following [8,15], we evaluate our models in two ways: held-out evaluation and manual evaluation. The held-out evaluation automatically compares the extracted relation instances from bags against Freebase relation data, i.e., labels of the bags, and reports the precision/recall curves of the experiments. Due to the incomplete nature of Freebase, the held-out method marks all extracted relation instances which are unseen in Freebase as incorrect, and it needs manual evaluation to find those relation instances that are essentially correct from them. And we use P@N metric to report the results.

3.2 Experimental Settings

Knowledge and Word Embeddings. We use the *word2vec* tool to train word embeddings on NYT corpus and keep the words which occur more than 100 times in the corpus as vocabulary [11]. We use the open source tool *KB2E*, which has implemented the TransE model, to train knowledge embeddings for these three KBs: ZeroKB, HalfKB, and AllKB. Following [3], we set latent dimension $k_e = 50$, learning rate $\lambda_e = 0.01$, margin $\gamma_e = 1$.

Parameter Settings. We tune our models using three-fold validation on training set, and use a grid search to determine the optimal parameters. We select smoothing factor γ among $\{0.5, 1, 2, 5\}$, the learning rate λ among $\{0.1, 0.01, 0.001, 0.0001\}$ and the bath size b_s among $\{30, 60, 120\}$. Following [8,11], we set the dimension of word embedding k_w to 50, the dimension of position embedding k_d to 5, the window size l to 3, the number of convolution kernels n to 200. And the optimal settings of experimental parameter are $\lambda = 0.001$, $\gamma = 1$, and $b_s = 60$. Following [6], the dropout rate is set to 0.5.

3.3   Experimental Results and Analysis

Results and Comparisons. We compare the proposed models with the following seven previous work. (1) **Mintz** [15] represents the distant supervision model. (2) **MultiR** [7] is a method based on multi-instance learning. (3) **MIML** [21] is a multi-instance and multi-relation model. (4) **PCNN+ONE** [25] uses PCNNs to extract feature vectors of sentences and select the most valid one to represent the bag. (5) **PCNN+ATT** [11] proposes an inner selective attention model to select valid sentences based on PCNNs. (6) **APCNN** [8] use features from word embeddings of the entities to align sentence feature vectors to generate attention weights. (7) **APCNN+D** [8] uses external description information based on APCNN. We implement the APCNN model ourselves, and the other models with the source codes provided by the authors.

Held-Out Evaluation. Figure 3 shows the aggregate precision/recall curves of our models (KBPCNN and KBPCNN+R) and other baseline methods. From Fig. 3 we can find out that our models significantly outperform all the baselines. We can find that: (1) Neural models based on PCNNs perform much better than feature-based methods, indicating that the human-designed features can not effectively express the semantics of sentences, and the error propagation brought by NLP tools will hurt the performances of the extractors; (2) In all the neural models, the attention-based models achieve further improvement, because the attention models can select the valid sentences and alleviate the wrong labelling problem in the distantly supervised RE; (3) Among all the attention mechanisms, the proposed knowledge-aware attention models perform better, indicating that employing the semantic knowledge in the external KB with text semantics of the bag can select valid instances more effectively; (4) Compared to KBPCNN, the KBPCNN+R using the knowledge relation retrieval module achieves higher precision in most range of recalls, indicating that further knowledge of the KB can be used to further improve model, and for the cases where the precision of both ends of the curve is lower than KBPCNN, we believe that this is due to the incompleteness of the KB, which results in insufficient representation of semantic knowledge.

Manual Evaluation. We further use manual evaluation because the held-out evaluation suffers from false negative in Freebase [8]. We conduct manual evaluation by two PhD and one master students whose research directions are Information Extraction. Table 2 presents the precisions of the manual evaluation on

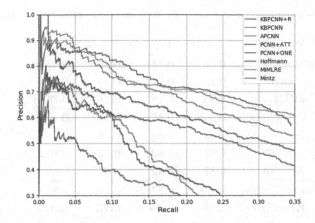

Fig. 3. Aggregate precision/recall curves of various methods.

the top 100, top 200, and top 500 extracted instances. The results show the KBPCNN achieves higher precision, which indicates that our knowledge-aware attention mechanism can more effectively select valid instances in a bag. And KBPCNN+R achieves the best performance, illustrating that employing semantic knowledge of KBs in neural models is effective.

Effect of KB's Completeness. Figure 4 shows the effect of KB's completeness on model KBPCNN+R. AllKB, HalfKB and ZeroKB are three KBs with different completeness. All, Half, and Zero are the models trained by using KBPCNN+R method based on these KBs respectively. From Fig. 4, we can find out that KB's completeness dose have a significant effect on the method. With Zero as the baseline, Half performs slightly better than Zero. This is because the HalfKB contains relatively few triplet facts, so the knowledge Half can leverage is limited. But All achieves a significant improvement over the Zero and Half, because the AllKB contains more triplet facts. In general, the more complete the KB is, the better the model performs.

Cases Study. Table 3 shows examples of generation probability p_{gen}. From Table 3, we can find that when the annotated entity pair does not appear in the KB, that is, the model can not use semantic knowledge, the attention model assigns a low weight (2.92e−10) to the semantic knowledge part, and leverages only the text semantics of the bag to generate the attention distribution almost, as the 1st and 2nd cases. Otherwise, the model will consider both semantic knowledge and text semantics, as the 4th case; sometimes assigns a high value (0.99933) to p_{gen} to leverage almost only semantic knowledge, as the 3rd case.

Table 4 shows an example of knowledge-aware attentions from the testing data. The bag contains 5 instances, where S_4 is an invalid instance. Our models assign S_4 low weights (0.08699 and 0.07850) and assign higher ones to other

Table 2. Precision values for the top 100, 200, and 500 extracted relation instances.

Methods	Top 100	Top 200	Top 500	Average
Mintz	0.77	0.71	0.55	0.676
MultiR	0.83	0.74	0.59	0.720
MIML	0.85	0.75	0.61	0.737
PCNN+ONE	0.84	0.77	0.64	0.750
PCNN+ATT	0.86	0.80	0.68	0.780
APCNN	0.87	0.82	0.72	0.802
APCNN+D	0.87	0.82	0.74	0.813
KBPCNN	**0.89**	0.83	0.74	0.820
KBPCNN+R	0.88	**0.84**	**0.75**	**0.823**

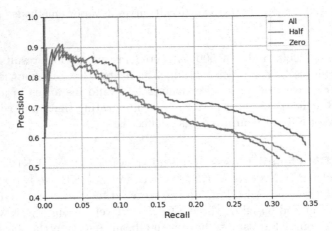

Fig. 4. Effects of the KBs with different completeness on KBPCNN+R.

valid sentences. Both models assign significantly higher weights to S_2 (0.28877 and 0.31279) compared to other valid sentences, because S_2 contains the words "... the prosperous city in ..." which express the label obviously. That is, the knowledge-aware attention mechanism can effectively select valid sentences. In addition, we also find out that the weights assigned by KBPCNN+R are more reasonable than KBPCNN, such as assigning a lower weight (0.07850) to S_4 and assigning a higher (0.31279) one to S_2. It also shows that further knowledge of the KB can be used to further improve the model.

4   Related Work

Relation Extraction. Relational extraction (RE) is a fundamental task in NLP. The methods proposed earlier can be classified as supervised methods generally, and most of them require a large amount of annotated data, which is

Table 3. Examples of generation probability p_{gen} generated by KBPCNN θ_1 and KBPCNN+R θ_2. Entity pairs of the 1st and 2nd cases are unseen in the KB, but entity pairs of the 3rd and 4th cases are in the KB.

EntityPair	r_{id}	InKB	θ_1	θ_2
italy, fiesol	48	False	0.12677	2.92e-10
pixar, edwin_catmull	6	False	0.05664	0.01381
bernard_marcus, atlanta	36	True	0.96074	0.99933
honest_tea, seth_goldman	6	True	0.85938	0.66321

Table 4. Attention weights generated by KBPCNN θ_1 and KBPCNN+R θ_2.

Bag Label: /location/location/contains(**Germany, Stuttgart**)		
Instances	θ_1	θ_2
S_1. Merz and solitude of **Stuttgart, Germany**, printed 800 copies, which are being sold for $20 each at the angola3. [**valid**]	0.22045	0.23464
S_2. Calling it a shotgun introduction to **Stuttgart**, the prosperous city in southwest **Germany** where car worship, like one's taxi driver, seems to know no limits. [**valid**]	0.28877	0.31279
S_3. Mrs.Somary is the managing director of the mendelssohn project, a nonprofit organization founded in Manhattan and **Stuttgart, Germany**, by the bridegroom, its music director. [**valid**]	0.18494	0.18181
S_4. But as **Germany**'s carmakers each try to outpolish the competition with these brand-driven monuments, only **Stuttgart** can claim a high-octane mix of automotive bravado and seminal tradition. [**invalid**]	0.08699	0.07850
S_5. At the same time, **Stuttgart** is in some ways playing catch-up with other auto-crazed cities in **Germany**. [**valid**]	0.21885	0.19226

labor intensive and time consuming. To solve this problem, Mintz et al. [15] proposed the distant supervision method, which uses structured triplets in the KB to align plain text to generate a large amount of training data. But this method will inevitably bring the wrong labelling problem. To alleviate the problem, Bunescu et al. [4] employed multi-instance learning in distantly supervised RE. But these methods use traditional NLP tools to extract the features of instances. Recent methods are based on deep learning. Zeng et al. [25] proposed the piecewise convolutional neural networks (PCNNs) model, which can automatically extract the sentence-level features. Selective attention model proposed by Lin et al. [11] can filter out meaningless sentences by generating inner attentions relying on the sentences' own features. Sentence-level attention model (APCNN) proposed by Ji et al. [8] assumes that vector $(e_1 - e_2)$ represents the relation, and use it to calculate the attention weight distribution, where e_1 and e_2 are word embeddings of the two given entities.

Knowledge Representation Learning. The task of Knowledge Representation Learning (KRL) [9] is to learn the distributed representations of entities and relations in the knowledge base. This paper focuses on the translation model, especially the TransE model [3]. For each triplet (h, r, t), TransE treats the relation vector l_r as the translation from the head entity vector l_h to the tail entity vector l_t, i.e., $l_h + l_r \approx l_t$. Knowledge embeddings trained by translation models contain a wealth of semantic knowledge. The methods of Knowledge Graph Completion (KGC) [20] task related to KRL directly use the vector operation based on knowledge embeddings to find missing relation connections between entities, thus adding new triplets to KBs. However, compared to distantly supervised RE, KGC suffers from the problem that KGC can not find out the triplets containing new entities that are unseen in the KB. To the best of our knowledge, our methods are the first efforts to adopt semantic knowledge of the KB in distantly supervised RE.

5   Conclusions

In this paper, we propose the neural knowledge-aware attention and retrieval-based models for distantly supervised RE. The attention mechanism can leverage semantic knowledge in the KB to generate attention weights, and select valid sentences by assigning them higher weights. The knowledge relation retrieval module can further leverage the semantic knowledge to improve the model. The experimental results show that the proposed methods achieve significant and consistent improvements over the popular benchmark methods.

References

1. Auer, S., Bizer, C., Kobilarov, G., Lehmann, J., Cyganiak, R., Ives, Z.: DBpedia: a nucleus for a web of open data. In: Aberer, K., et al. (eds.) ASWC/ISWC -2007. LNCS, vol. 4825, pp. 722–735. Springer, Heidelberg (2007). https://doi.org/10.1007/978-3-540-76298-0_52
2. Bollacker, E.K., Paritosh, C., Sturge, P., Taylor, T.: Freebase: a collaboratively created graph database for structuring human knowledge. In: SIGMOD Conference, pp. 1247–1250 (2008)
3. Bordes, A., Usunier, N., Weston, J.: Translating embeddings for modeling multi-relational. In: International Conference on Neural Information Processing Systems, pp. 2787–2795 (2013)
4. Bunescu, R., Mooney, R.: Learning to extract relations from the web using minimal supervision. In: ACL, pp. 576–583 (2007)
5. Thomas, G., Richard, H.: Solving the multiple instance problem with axis-parallel rectangles. J. Artif. Intell. **89**(12), 31–71 (1997)
6. Geoffrey, E., Srivastava, N., Krizhevsky, A., Sutskever, I., Ruslan, R.: Improving neural networks by preventing co-adaptation of feature detectors. J. Comput. Sci. **3**(4), 212–223 (2012)
7. Hoffmann, H., Zhang, C., Ling, X., Zettlemoyer, L., Daniel, W.: Knowledge-based weak supervision for information extraction of overlapping relations. In: Meeting of the Association for Computational Linguistics (ACL), pp. 541–550 (2011)
8. Ji, G., Liu, K., He, S., Zhao, J.: Distant supervision for relation extraction with sentence-level attention and entity descriptions. In: AAAI, pp. 3060–3066 (2017)
9. Lin, Y., Liu, Z., Sun, M.: Knowledge representation learning with entities, attributes and relations. J. Ethnicity **1**, 41–52 (2016a)

10. Lin, Y., Liu, Z., Zhu, X.: Learning entity and relation embeddings for knowledge graph completion. In: AAAI, pp. 2181–2187 (2015)
11. Lin, Y., Shen, S., Liu, Z., Luan, H., Sun, M.: Neural relation extraction with selective attention over instances. In: ACL, pp. 2124–2133 (2016)
12. Lukovnikov, D., Fischer, A., Lehmann, J., Auer, S.: Neural network-based question answering over knowledge graphs on word and character level. In: Proceedings of International Conference on World Wide Web (WWW), pp. 1211–1220 (2017)
13. Luong, M., Pham, H., Manning, C.: Effective approaches to attention based neural machine translation. J. Comput. Sci. (2015)
14. Mikolov, T., Sutskever, I., Chen, K., Corrado, G., Dean, J.: Distributed representations of words and phrases and their compositionality. In: NIPS, pp. 3111–3119 (2013)
15. Mintz, M., Bills, S., Jurafsky, D.: Distant supervision for relation extraction without labeled data. In: Joint Conference of the Meeting of the ACL and the International Joint Conference on Natural Language Processing of the AFNLP, pp. 1003–1011 (2009)
16. Mitra, B., Craswell, N.: Neural models for information retrieval. CoRR, abs/1705.01509 (2017)
17. Riedel, S., Yao, L., Mccallum, A.: Modeling relations and their mentions without labeled text. In: European Conference on Machine Learning and Knowledge Discovery in Databases, pp. 148–163 (2010)
18. See, A., Liu, P., Manning, C.: Get to the point: summarization with pointer generator networks. In: ACL, pp. 1073–1083 (2017)
19. Shen, Y., He, X., Gao, J., Deng, L., Mesnil, G.: A latent semantic model with convolutional-pooling structure for information retrieval. In: Proceedings of ACM International Conference on Conference on Information and Knowledge Management, (CIKM), pp. 101–110 (2014)
20. Baoxu, S.B., Weninger, T.: Open-world knowledge graph completion. In: AAAI (2018)
21. Surdeanu, M., Tibshirani, J., Nallapati, R., Manning, C.: Multi-instance multi-label learning for relation extraction. In: Joint Conference on Empirical Methods in Natural Language Processing and Computational Natural Language Learning, pp. 455–465 (2012)
22. Tan, M., Santos, C., Xiang, B., Zhou, B.: Improved representation learning for question answer matching. In: ACL, pp. 464–473 (2016)
23. Yang, B., Mitchell, T.: Leveraging knowledge bases in LSTMs for improving machine reading. In: ACL, pp. 1436–1446 (2017)
24. Yao, X., Durme, B.: Information extraction over structured data: question answering with freebase. In: ACL, pp. 956–966 (2014)
25. Zeng, D., Liu, K., Chen, Y., Zhao, J.: Distant supervision for relation extraction via piecewise convolutional neural networks. In: EMNLP, pp. 1753–1762 (2015)

Two-Stage Entity Alignment: Combining Hybrid Knowledge Graph Embedding with Similarity-Based Relation Alignment

Tingting Jiang[1,2,3], Chenyang Bu[1,2,3(✉)], Yi Zhu[1,2,3], and Xindong Wu[1,3,4]

[1] Key Laboratory of Knowledge Engineering with Big Data
(Hefei University of Technology), Ministry of Education, Hefei, China
jiangtt@mail.hfut.edu.cn,
{chenyangbu,xwu}@hfut.edu.cn, z8d1177@126.com
[2] School of Computer Science and Information Engineering,
Hefei University of Technology, Hefei, China
[3] Institute of Big Knowledge Science, Hefei University of Technology,
Hefei, China
[4] Mininglamp Academy of Sciences, Mininglamp Technology, Beijing, China

Abstract. Entity alignment aims to automatically determine whether an entity pair in different knowledge graphs refers to the same entity in reality. Existing entity alignment methods can be classified into two categories: string-similarity-based methods and embedding-based methods. String-similarity-based methods have higher accuracy, however, they might have difficulty in dealing with literal heterogeneity, i.e., an entity pair in diverse forms. Though embedding-based entity alignment can deal with literal heterogeneity, they also suffer the shortcomings of higher time complexity and lower accuracy. Moreover, there remain limitations and challenges due to only using the structure information of triples for existing embedding methods. Therefore, in this study, we propose a two-stage entity alignment framework, which can combine the advantages of both methods. In addition, to enhance the embedding performance, a hybrid knowledge graph embedding model with both fact triples and logical rules is introduced for entity alignment. Experimental results on two real-world datasets show that the proposed method is significantly better than the state-of-the-art embedding-based entity alignment methods.

Keywords: Entity alignment · Knowledge graph embedding · Relation alignment

1 Introduction

Knowledge graphs (KGs) have been widely used by researchers in many AI-related applications, such as knowledge acquisition [1], answering questions [2], and recommendation systems [3]. Because different knowledge graphs are multi-source heterogeneous, it is particularly important that heterogeneous entities are integrated in multi-

© Springer Nature Switzerland AG 2019
A. C. Nayak and A. Sharma (Eds.): PRICAI 2019, LNAI 11670, pp. 162–175, 2019.
https://doi.org/10.1007/978-3-030-29908-8_13

knowledge graphs to obtain a consistent form (called entity alignment) in the process of building a more complete and richer knowledge graph.

Conventional entity alignment methods have achieved excellent performance in the past decade. Most of these methods are based on similarity calculation [4–6] or propagation [7, 8]. That is, two entities are determined to be equal based on the equality or similarity of character strings, attributes, or neighboring nodes. However, due to the reliance on symbolic characteristics, conventional methods may not work well in the condition of literal heterogeneity (see Sect. 4.1 for details).

In order to ignore the literary form of entities, embedding-based entity alignment was proposed and yielded considerable results [9–11]. Embedding-based entity alignment embeds entities and relations into low-dimensional vector spaces, and a model using structural information between triples is built subsequently. The embedding-based entity alignment determines counterparts for entities through the distance between vectors, regardless of literal characteristics such as characters, words, or even languages. For example, Yang et al. [9] proposed a cross-lingual knowledge alignment model from various transfer matrices; Liu et al. [11] proposed an iterative entity alignment method using joint knowledge embeddings; and Sun et al. proposed an alignment-oriented KG embedding with bootstrapping alignment [10]. However, embedding models also have the low accuracy problem.

Therefore, in this study, we propose a two-stage framework to combine the advantages of the above two methods. Specifically, in Stage I, a string-similarity-based method was adopted to align relation pairs of different KGs, instead of all entity pairs. It is based on the following observations on real-world datasets: (1) the number of relations is far less than the number of entities, which means aligning relation pairs requires less time than aligning entity pairs; (2) a relation corresponds to a number of entities in KGs under normal conditions, that is, it is very helpful for the task of aligning entity pairs if the performance of relation alignment is enhanced; and (3) string-similarity-based methods have high accuracy. Therefore, even if only a part of relation pairs can be aligned, the performance of aligning entity pairs in Stage II is expected to be enhanced.

In Stage II, a hybrid embedding model which simultaneously considers both logical rules and fact triples is introduced. To the best of our knowledge, embedding-based methods for entity alignment has not drawn much attention yet. What is more, most of existing methods on entity alignment only use facts to learn an embedding model and ignore abundant logical rule information contained in the knowledge graphs. Some existing studies indicate that incorporating path information [12], attribute information [13], internal character information [14], and other information into the KG embedding process can effectively improve the performance of the model. Following this idea, logical rules can be used to deduce new triples and mine hidden information in the knowledge graph for enriching the semantic information of entities. We inject logical rules into the embedding model to produce hybrid embedding of rules and triples. Although the use of logical rules proposed by some papers enhance the representation learning, the domain characteristics of entity alignment tasks were not taken into account.

The main contributions of this paper are threefold. (1) A two-stage entity alignment framework is proposed, which combines the advantages of string-similarity-based methods and embedding-based methods. (2) A hybrid embedding model for entity alignment is proposed, which represents triples and logical rules into a uniform space in order to enhance the embedding of individual entities. (3) We evaluate the proposed method with several real-world datasets, and the experimental results show that our model is superior to state-of-the-art methods.

The rest of this paper is organized as follows: Sect. 2 introduce the background, including the concepts of KG embedding, entity alignment, and some learning resources. The preliminary is introduced in Sect. 3, and the proposed framework is detailed in Sect. 4. Experimental results and analysis are presented in Sect. 5, followed by our main conclusion and future work in the final section.

2 Background

In this section, we discuss the background including KG embedding and entity alignment. In addition, the learning resources including rule and grounding rule are introduced.

KG Embedding. Knowledge graphs are usually not easy to manipulate because of the underlying symbolic nature of fact triples. To tackle this issue, KG embedding has been proposed and gained considerable attention. KG embedding refers to embed the entities and relations in the knowledge graphs into continuous vector spaces [15].

Entity Alignment. Entity alignment refers to automatically determining entity pairs in different knowledge graphs that have the same representation in reality. For example, subjects $KG_1{:}e_1$ and $KG_2{:}e_1$ from two different sources refer to the same entity e_1, even though they are in diverse forms. The illustration of entity alignment is shown in Fig. 1.

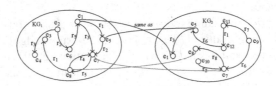

Fig. 1. Illustration of entity alignment (reproduced from [11]). KG1 and KG2 are two different knowledge graphs. Given some aligned entity pairs, called aligned seeds (connected by black lines) and the goal of entity alignment is to find all potential entity pairs (connected by red lines) that can be aligned. (Color figure online)

Rule. We call the triple that the position of the head or tail is a variable an atom, and a rule refers to the complex form expressing certain logical meanings formed by several atoms connected by logical conjunction. The atom to the left of the conjunction is called *condition* and the atom to the right of the conjunction is called *conclusion*. For example, *countryofbirth<?a, ?b>* ⇒ *placeofbirth<?a, ?b>* means that if there is a triple *(?a, countryofbirth, ?b)* in the knowledge graph, then there should be a triple *(?a, placeofbirth, ?b)* in the knowledge graph. In our method, we only account for rules where the number of atoms in *condition* is 1 and the number of atoms in *conclusion* is less than or equal to 2. The details are shown in Table 1.

Table 1. Types of rules.

Types of rules, where *?a*, *?b*, and *?c* denote entities, and r_1, r_2, and r_3 are relations.
$\forall$ *?a*, *?b*: r_1 <*?a*, *?b*> $\Rightarrow$ r_2<*?a*, *?b*>
$\forall$ *?a*, *?b*, *?c*: r_1 <*?a*, *?b*> && r_2<*?b*, *?c*> $\Rightarrow$ r_3<*?a*, *?c*>

Grounding Rule. The process of replacing a variable located in an atom with a concrete entity is called instantiation, and we will obtain grounding rules after this step. For example, *countryofbirth* <*?a*, *?b*> $\Rightarrow$ *placeofbirth*<*?a*, *?b*> can transform to *countryofbirth*<*Lisa, America*> $\Rightarrow$ *placeofbirth*<*Lisa, America*> after instantiation. The types of grounding rules are shown in Table 2.

Table 2. Types of grounding rules.

Types of grounding rules, where a_1, b_1, a_2, b_2, and c_2 denote concrete entities, and r_1, r_2, and r_3 are relations.
R_1: r_1 <a_1, b_1> $\Rightarrow$ r_2<a_1, b_1>
R_2: r_1 <a_2, b_2> && r_2<b_2, c_2> $\Rightarrow$ r_3<a_2, c_2>

3 Preliminary

Sun et al. proposed an alignment-oriented KG embedding called AlignE [10]. Before introducing our method, the details of AlignE are first described.

The entity alignment task assumes a significant difference between the positive and negative samples in order to capture the semantic information of the equivalent entities in different knowledge graphs. Based on this idea, AlignE includes the proposed alignment-oriented loss function for triples:

$$O_T = \sum_{T \in \mathbb{T}^+} [f(T) - \gamma_1]_+ + \mu_1 \sum_{T' \in \mathbb{T}^-} [\gamma_2 - f(T')]_+ \tag{1}$$

$$f(T) = \| \boldsymbol{h} + \boldsymbol{r} - \boldsymbol{t} \|_2^2 \tag{2}$$

Where $(\boldsymbol{h}, \boldsymbol{r}, \boldsymbol{t})$ is the vector embedding of triple (h, r, t); $\| \cdot \|_2^2$ is the L2-norm; $[\cdot]_+ = max(\cdot, 0)$, γ_1, μ_1, and γ_2 are hyper-parameters; T and T' stand for positive and negative sample triples, respectively; $\mathbb{T}^+$ is the set of positive sample triples and $\mathbb{T}^-$ is the set of negative sample triples. As for negative sampling, the *s*-nearest neighbor set of entities are first found and a random element in the set is chosen as the negative sample.

In contrast to the previous embedding-based entity alignment method, AlignE abandons the tradition of direct reuse in the TransE model. The proposed loss function significantly expands the gap between positive and negative sample scores and reduces drift of entity vector embedding. However, representation learning in AlignE is still based on the use of triples. Semantic entity information can be enriched if a large amount of logical rule information in the knowledge graph is used for learning with triples, yielding fuller entity vectors.

4   Method

In this section, the motivation is introduced in Sect. 4.1 and an overview of our proposed model is presented in Sect. 4.2. Next, we detail the two-stages of the model in Sects. 4.3 and 4.4.

4.1   Motivation

(1)   **Similarity-based Relation Alignment with KG Embedding**

Existing study on entity alignment can be classified into two types, i.e., traditional string-similarity-based methods and embedding-based methods [16]. The basic idea of string-similarity-based methods is calculating the string similarity of two entities to judge whether they represent the same entity. Although string-similarity-based methods have higher accuracy, they may have difficulties in dealing with the situations of cross-language [9] and literal heterogeneity [10]. For example, the name "Kobe" (in English) and "科比" (in Chinese) shown in Fig. 2 (a) actually refer to the same person, and this person may also be called as "Black Manba" in some occasions as shown in Fig. 2(b). That is to say, string-similarity-based methods are not able to handle these two situations directly.

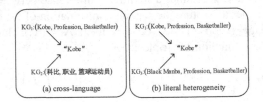

Fig. 2. Illustration of the shortcomings of string-similarity-based methods. The string-similarity-based methods might not work well when encountering the situations of: (a) cross-language; or (b) literal heterogeneity.

The basic idea of embedding-based methods, however, is to embed the semantic information of these entities into low-dimensional vectors. Compared with string-similarity-based methods, this kind of methods can deal with the above two situations efficiently [9, 10]. However, embedding-based methods also have the shortcomings of higher time complexity and lower accuracy.

Therefore, we propose a two-stage framework in this study to combine the advantages of these two methods. The idea is given as follows. In the first stage, a string-similarity-based method was adopted to align relations, rather than align all the entities. The task of relation alignment is regarded as more important than entity alignment, because a relation generally corresponds to multiple entities. That is, an aligned relation pair can be beneficial for aligning multiple entities in the next step. Moreover, relation pairs are regarded as more suitable to be aligned by string-similarity methods, because the cases of literal heterogeneity in relation pairs might be much less than those in entity pairs in general. In the second stage, an embedding-based method was adopted to align entity pairs based on the previously aligned relations and some aligned entity pairs given in advance, which can address the two problems shown in Fig. 2.

(2) **Hybrid Embedding with both Fact Triples and Logical Rules**

In recent years, KG embedding based on translational models (e.g. TransE) has attracted a lot of attention because of its simplicity and efficiency, and has achieved excellent performances in many real-world applications like link pre-diction and question answering [15]. TransE [17] is a representative translation model that states the translational rule, i.e., $h + r - t \approx 0$, must be true if the triples (h, r, t) are contained in a knowledge graph. In other words, vector embedding of the head entity plus vector embedding of the relation is approximately equal to the vector embedding of the tail entity. This translational rule was inspired by the famous word2vec [18], which found the translation invariance in the embedding spaces. For example, the word embedding of "king" minus the embedding of "queen" is approximately equal to the embedding of "man" minus "woman", i.e., $C(king) - C(queen) = C(man) - C(woman)$.

Although translating embedding methods have been demonstrated effective in many other applications, there are not enough studies on the task of entity alignment [9–11]. Moreover, existing studies on entity alignment mainly focus on learning an embedding model based on only the triples facts, ignoring the information of logical rules which can certainly enrich and enhance the expression of the embedding model by deducing new triples [19]. Although several existing methods have already attempt to combine fact triples and logical rules to learn an embedding model for other applications [20, 21], as far as we know, the information of logical rules has not been considered in the task of entity alignment yet.

Therefore, in this study, we propose a hybrid embedding with both fact triples and logical rules to enhance the embedding of entities. And we introduce an alignment-oriented loss function for logical rules that improve the accuracy of entity alignment.

4.2   Overall Architecture

The overall architecture is illustrated in Fig. 3, which contains two components including similarity-based relation alignment with KG embedding (Fig. 3a), and hybrid embedding with both fact triples and logical rules (Fig. 3b).

The relation alignment can be used to enhance the hybrid embedding. We use string-similarity-based method for relation alignment to gain a part of aligned relations and then adjust the hybrid embedding through relation exchange. The relation align-ment is detailed in Sect. 4.3.

The hybrid embedding is intended to facilitate hybrid embedding of hidden logical rule information and triple information in a knowledge graph, and to orient rule embedding towards alignment. This is divided into three parts: triple embedding, rule embedding, and hybrid embedding. The triple embedding adopts the method described in Sect. 3, and the rule embedding and hybrid embedding are described in detail in Sect. 4.4. The details of the proposed model are summarized in Algorithm 1.

Algorithm 1. KG embedding procedure of the proposed model

Require: triples $\mathbb{T}^+$
          grounding rules $\mathbb{R}^+$
          aligned seeds
1. Align relations according to edit distance.
2. Randomly initialize entity and relation embeddings.
3. Exchange the aligned relations to obtain new triples, and then add these new triples to $\mathbb{T}^+$.
4. Calculate the number of s -nearest neighbors for relations ($\triangleright$cf. Eq. (9)).
5. **for** $n = 1: N$ **do**
6.   Generate the dictionary of neighbor nodes for relations;
7.   **for** $i = 1: k$ **do**
8.   Generate negative triples and negative grounding rules;
9.   Calculate the loss function of triples $\triangleright$cf. Eq. (1);
10.   Calculate the loss function of grounding rules $\triangleright$cf. Eq. (8);
11.   Minimize the embedding loss function $\triangleright$cf. Eq. (10).
12.   **end for**
13. **end for**
Ensure: Embeddings of entities $\Theta^{(N)}$

4.3   Stage I: Similarity-Based Relation Alignment with KG Embedding

Relation Alignment. The relations are aligned based on edit distance. Knowledge graphs often use a URL address to describe entities or relations, thus we first find the keywords in the URL. Some techniques are required to finish this step, such as regular expressions and web crawlers. And then the edit distance between two relation strings is calculated. The editing distance between two strings is the minimum operation cost of converting one string to another by adding, deleting, or replacing characters, and each operation will be assigned a certain cost, e.g., 1. The editing distance is generally calculated using dynamic programming with the following recursive formula [22]:

$$d[i,j] = min \begin{cases} d[i,j-1]+1 \\ d[i-1,j]+1 \\ d[i-1,j-1]+c(s_1[i],s_2[j]) \end{cases} \quad (3)$$

$$c(s_1[i],s_2[j]) = \begin{cases} 1, s_1[i] \neq s_2[j] \\ 0, s_1[i] = s_2[j] \end{cases} \quad (4)$$

where $s_1[i]$ is the ith character in the string s_1 and $s_2[j]$ is the jth character in the string s_2.

Relation Exchange. We generate aligned relations with a relation alignment and then apply these aligned relations to coordinate entity embedding in different knowledge graphs based on the relation exchange strategy. The relation exchange strategy unifies the embedding of triples with the same relation in different knowledge graphs. As follows, where a pair of aligned relations (r_s, r_k) exists, then we can obtain the triples as Eq. (5):

$$\mathbb{T}'_{(r_s, r_k)} = \{(h, r_k, t) | (h, r_s, t) \in \mathbb{T}_1^+\} \cup \{(h, r_s, t) | (h, r_k, t) \in \mathbb{T}_2^+\} \tag{5}$$

where $\mathbb{T}_1^+$ and $\mathbb{T}_2^+$ are the set of positive sample triples from KG_1 and KG_2, respectively. The final positive sample set is $\mathbb{T}^+ = \mathbb{T}_1^+ \cup \mathbb{T}_2^+ \cup \mathbb{T}^s_{(x,y)} \cup \mathbb{T}'_{(r_s, r_k)}$, where $\mathbb{T}^s_{(x,y)}$ is the triple set after executing the exchange strategy [10].

4.4 Stage II: Hybrid Embedding with Both Fact Triples and Logical Rules

Rules Embedding. Rule embedding is specific to the grounding rules. Grounding rules are regarded as special complex triples whose scores are determined by the triples and their logical connectives. According to [19], the scores of the two types of grounding rules in Table 2 can be defined as follows:

$$f(R_1) = f(a_1, r_1, b_1) \cdot f(a_1, r_2, b_1) - f(a_1, r_1, b_1) + 1 \tag{6}$$

$$f(R_2) = f(a_2, r_1, b_2) \cdot f(b_2, r_2, c_2) \cdot f(a_2, r_3, c_2) - f(a_2, r_1, b_2) \cdot f(b_2, r_2, c_2) + 1 \tag{7}$$

After obtaining rule scores, we design a loss function for optimizing the vectors of rules. Optimization of the traditional translation model uses margin-based ranking loss and attempts to reduce the score of positive samples below that of negative samples, which improves the effectiveness of the model. However, the difference between the score of positive and negative samples is not obvious in the traditional translation model. Furthermore, similar scores between positive and negative sample scores will reduce the accuracy of entity alignment. Therefore, based on [10], the following objective function for grounding rules is introduced:

$$O_R = \sum_{R_* \in \mathbb{R}^+} [f(R_*) - \gamma_1]_+ + \mu_1 \sum_{R'_* \in \mathbb{R}^-} [\gamma_2 - f(R'_*)]_+ \tag{8}$$

where R_* is a grounding rule (R_1 or R_2) and R'_* is the negative sample of R_*; $\mathbb{R}^+$ and $\mathbb{R}^-$ refer to the positive sample set and negative sample set, respectively. Equation (8) shows that the positive sample will tend to have a lower score and the negative sample will tend to have a higher score in some cases, e.g., when $f(R_*) \leq \gamma_1$ and $f(R'_*) \geq \gamma_2$. At this point, the score difference between positive and negative samples will be more significant than that from TransE.

Negative Sampling. Negative sample triples can help modify the embedding model. For triples, we choose to replace the head or tail entities with negative sample triples. When generating negative samples of rules, the strategy is to first find the triple of the *conclusion* of each grounding rule, and then replace the relation in the triple with another relation. Specifically, the s-nearest neighbor of the relation are calculated, and then a relation is randomly selected from the neighbor to replace the original relation, yielding the negative sample of the rule. The value of s can be determined from the following equation:

$$s = \lceil (1 - \epsilon) * M \rceil \tag{9}$$

where M is the total number of relations and ϵ is a balance parameter.

Hybrid Embedding. The purpose of hybrid embedding is to unify triples and grounding rules such that they reflect the actual information in the knowledge graph. The grounding rules embody the process of logical reasoning, while the reasoning itself is inevitably flawed. Therefore, when building hybrid embedding, we should pay attention to the proportion of triples and grounding rules in the final embedding model. Then, we introduce the balance hyper-parameter w and obtain the following model as Eq. (10):

$$O_e = O_T + wO_R \tag{10}$$

where O_T (defined in Eq. 1) is the loss function of triples, O_R (defined in Eq. 8) is the loss function of grounding rules, and O_e is the overall loss function of triples and grounding rules.

5   Experiments

In this section, the validity of the proposed method for entity alignment is evaluated with different data sets. We use TensorFlow, a popular open source framework for tensor computing, to complete our experiments. Our experiments were performed on a PC with an Intel Xeon E5 2.40 GHz CPU with 128 GB RAM.

5.1   Datasets

The DBP15K [10] dataset and the large-scale DBP-WD [10] dataset were adopted for the experiments. DBP15 K was extracted from DBpedia and consists of three cross-language datasets: DBP_{ZH-EN} (Chinese to English), DBP_{JA-EN} (Japanese to English), and DBP_{FR-EN} (French to English). Each dataset contains approximately 100,000 triples (positive sample triples) and 15,000 aligned entity pairs; 30% of the data were used for reference alignment, and the remaining 70% were used as training data. And DBP-WD [10] is a large-scale dataset containing triples from DBpedia and Wikidata, totaling approximately 900,000 triples and 100,000 aligned entity pairs.

5.2   Experiment Setup

Four state-of-the-art algorithms were used for comparison: MTransE [9], IPTransE [11], JAPE [13], and AlignE [10]. The source codes[1] and the experimental results of these algorithms were taken from [10]. The proposed algorithm is called RTEA-RA. In addition, the proposed algorithm without the relation alignment strategy (called RTEA) was tested to analyze the effectiveness of stage I.

The parameter settings of the proposed algorithms are given as follows: $\gamma_1 = 0.03$, $\gamma_2 = 1.7$, $\mu_1 = 0.2$, and $w = 0.2$. During the process of iterative learning, the learning rate is 0.01 and the maximum iteration number is 500. For the negative sampling strategy, the parameter ϵ is set to 0.98 for the DBP-WD dataset and 0.9 for other datasets; and 10 negative samples were generated for each sample.

The AMIE+ [23] was adopted to extract logical rules for each knowledge graph. In order to obtain rules with higher reliability, we take the PCA (partial completeness assumption) confidence [23] as the filter and then set the threshold value to 1 to filter out rules that may have defects. Furthermore, instantiation is required after a rule is obtained. The grounding rules must be screened again because inferred triples that are not included in the knowledge graph are the only effective logical inferences. Grounding rules that only *condition* triples are contained in the knowledge graph were considered valid. In the process of relation alignment, we require the edit distance of two relation strings to be 0, i.e., they are completely equal. This ensures the effectiveness of the relation exchange strategy.

5.3   Results and Analysis

By convention, Hits@k and MRR were selected as our metrics, where Hits@k represents the proportion of correct alignments in the first k results, and Hits@1 represents the accuracy. MRR represents the average value of the reciprocal ranks of the results. Higher Hits@k and MRR values indicate better alignment.

The results of our method for entity alignment with different datasets are shown in Table 3. We observe that:

- RTEA outperforms MTransE, IPTransE, JAPE and AlignE significantly, due to its hybrid embedding.

---

[1] https://github.com/nju-websoft/BootEA.

Table 3. Result comparison on entity alignment.

Approaches	DBP$_{ZH-EN}$			DBP$_{JA-EN}$		
	Hits@1	Hits@10	MRR	Hits@1	Hits@10	MRR
MTransE [9]	30.83	61.41	0.364	27.86	57.45	0.349
AlignE [10]	47.18	79.19	0.581	44.76	78.89	0.563
IPTransE [11]	40.59	73.47	0.516	36.69	69.26	0.474
JAPE [13]	41.18	74.46	0.490	36.25	68.50	0.476
RTEA	50.89	82.56	0.617	49.04	82.72	0.604
RTEA-RA	**57.30**	**86.44**	**0.674**	**53.39**	**85.73**	**0.644**
Approaches	DBP$_{FR-EN}$			DBP-WD		
	Hits@1	Hits@10	MRR	Hits@1	Hits@10	MRR
MTransE [9]	24.41	55.55	0.335	28.12	51.95	0.363
AlignE [10]	47.36	82.06	0.593	56.55	82.70	0.655
IPTransE [11]	33.30	68.54	0.451	34.85	63.84	0.447
JAPE [13]	32.39	66.68	0.430	31.84	58.88	0.411
RTEA	52.71	86.54	0.643	58.19	84.60	0.673
RTEA-RA	**53.84**	**86.78**	**0.652**	**58.44**	**84.75**	**0.675**

- As expected, RTEA-RA gets better results than RTEA, because it used the relation alignment to help enhancing the hybrid embedding.
- In general, both RTEA and RTEA-RA consistently outperform the baseline methods, which demonstrates that the proposed two-stage method works well for entity alignment.

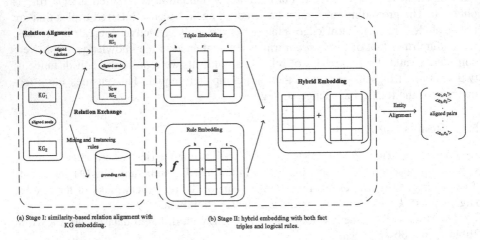

(a) Stage I: similarity-based relation alignment with KG embedding.

(b) Stage II: hybrid embedding with both fact triples and logical rules.

Fig. 3. Overview of our proposed two-stage model. Firstly, in stage I, a set of aligned relations are gained through the string-similarity-based method. Secondly, the vector embedding of entities are learned base on both the logical rules and the triples through stage II. Thus, we can determine whether they are equal entities according to the distance between the vector embedding of entities.

5.4   Parameter Sensitivity

To analyze whether RTEA is sensitive to the proportion of the grounding rules, the changes of Hits@k (Hits@1 and Hits@10) through different values of w are given in Fig. 4. From Fig. 4 we can observe that the results are related to the value of w, and the best configuration on both datasets is $w = 0.2$ for the experiments in this study.

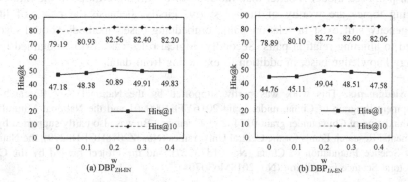

Fig. 4. Hits@k on entity alignment w.r.t. the proportion of the grounding rules.

To evaluate the relation alignment strategy in stage I, we tested the proportion of aligned relations from 0 to 100% with step 25% on DBP$_{ZH-EN}$ and DBP$_{JA-EN}$. Figure 5 depicts the evaluation results with different proportion of aligned relations. As expected, the proposed algorithm performs better with an increase in the proportion of aligned relations on both datasets, which demonstrate that the relation alignment strategy enhanced embedding and worked well for entity alignment.

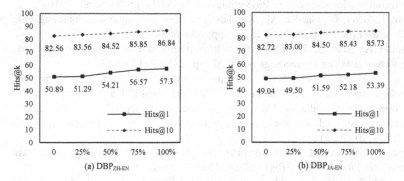

Fig. 5. Hits@k on entity alignment w.r.t. the proportion of the aligned relations.

6   Conclusion and Future Work

In this paper, we proposed a two-stage entity alignment framework combining the advantages of string-similarity-based methods and embedding-based methods. Moreover, the KG embedding model considering both logical rules and fact triples was introduced for entity alignment. The experimental results with several datasets showed that our proposed model is better than the state-of-the-art algorithms. In the future, the following topics are worthy of further study. Firstly, because the cases of literal heterogeneity still exist in relation pairs, embedding-based methods should also be adopted to aligning relation pairs. Secondly, logical rules can also be obtained by an ontology knowledge base, in addition to extracting from data.

Acknowledgments. This work was partly supported by the National Key Research and Development Program of China, under grant 2016YFB1000901 and the National Natural Science Foundation of China under grant 91746209. Chenyang Bu was also partly supported by the Fundamental Research Funds for the Central Universities (No. JZ2018HGBH0279), the National Natural Science Foundation of China (No. 61573327), and the Project funded by the China Postdoctoral Science Foundation (No. 2018M630704).

References

1. Weston, J., Bordes, A., Yakhnenko, O., Usunier, N.: Connecting language and knowledge bases with embedding models for relation extraction. In: Proceedings of EMNLP, pp. 1366–1371 (2013)
2. Yih, W.T., Chang, M.W., He, X., Gao, J.: Semantic parsing via staged query graph generation: question answering with knowledge base. In: Proceedings of ACL, pp. 1321–1331 (2015)
3. Zhang, F., Yuan, N.J., Lian, D., Xie, X., Ma, W.: Collaborative knowledge base embedding for recommender systems. In: Proceedings of SIGKDD, pp. 353–362 (2016)
4. Volz, J., Bizer, C., Gaedke, M., Kobilarov, G.: Discovering and maintaining links on the web of data. In: Proceedings of ISWC, pp. 650–665 (2009)
5. Ngomo, A., Auer, S.: LIMES—a time-efficient approach for large-scale link discovery on the web of data. In: Proceedings of IJCAI, pp. 2312–2317 (2011)
6. Sun, Y., Ma, L., Wang, S.: A comparative evaluation of string similarity metrics for ontology alignment. J. Inf. Comput. Sci. **12**(3), 957–964 (2015)
7. Lacoste Julien, S., Palla, K., Davies, A., Kasneci, G., Graepel, T., Ghahramani, Z.: Sigma: simple greedy matching for aligning large knowledge bases. In: Proceedings of SIGKDD, pp. 572–580 (2013)
8. Pershina, M., Yakout, M., Chakrabarti, K.: Holistic entity matching across knowledge graphs. In: Proceedings of BigData, pp. 1585–1590 (2015)
9. Chen, M., Tian, Y., Yang, M., Zaniolo, C.: Multilingual knowledge graph embeddings for cross-lingual knowledge alignment. In: Proceedings of IJCAI, pp. 1511–1517 (2016)
10. Sun, Z., Hu, W., Zhang, Q., Qu, Y.: Bootstrapping entity alignment with knowledge graph embedding. In: Proceddings of IJCAI, pp. 4396–4402 (2018)
11. Zhu, H., Xie, R., Liu, Z., Sun, M.: Iterative entity alignment via joint knowledge embeddings. In: Proceedings of IJCAI, pp. 4258–4264 (2017)

12. Lin, Y., Liu, Z., Luan, H., Sun, M., Rao, S., Liu, S.: Modeling relation paths for representation learning of knowledge bases. In: Proceedings of EMNLP, pp. 705–714 (2015)
13. Sun, Z., Hu, W., Li, C.: Cross-lingual entity alignment via joint attribute-preserving embedding. In: d'Amato, C., et al. (eds.) ISWC 2017. LNCS, vol. 10587, pp. 628–644. Springer, Cham (2017). https://doi.org/10.1007/978-3-319-68288-4_37
14. Jin, H., et al.: Incorporating chinese characters of words for lexical sememe prediction. In: Proceedings of ACL, pp. 2439–2449 (2018)
15. Wang, Q., Mao, Z., Wang, B., Guo, L.: Knowledge graph embedding: a survey of approaches and applications. IEEE Trans. Knowl. Data Eng. 29(12), 2724–2743 (2017)
16. Trsedya, B.D., Qi, J., Zhang, R.: Entity alignment between knowledge graphs using attribute embeddings. In: Proceedings of AAAI (2019)
17. Bordes, A., Usunier, N., Garcia-Duran, A., Weston, J., Yakhnenko, O.: Translating embeddings for modeling multi-relational data. In: Proceedings of NIPS, pp. 2787–2795 (2013)
18. Mikolov, T., Sutskever, I., Chen, K., Corrado, G.S., Dean, J.: Distributed representations of words and phrases and their compositionality. In: Proceedings of NIPS, pp. 3111–3119 (2013)
19. Guo, S., Wang, Q., Wang, L., Wang, B., Guo, L.: Jointly embedding knowledge graphs and logical rules. In: Proceedings of EMNLP, pp. 192–202 (2016)
20. Wang, Q., Wang, B., Guo, L.: Knowledge base completion using embeddings and rules. In: Proceedings of IJCAI, pp. 1859–1865 (2015)
21. Rocktäschel, T., Bošnjak, M., Singh, S., Riedel, S.: Low-dimensional embeddings of logic. In: Proceedings of ACL Workshop, pp. 45–49 (2014)
22. Klabunde, R.: Daniel jurafsky/james h. martin, speech and language processing. Zeitschrift für Sprachwissenschaft 21(1), 134–135 (2002)
23. Galárraga, L., Teflioudi, C., Hose, K., Suchanek, F.M.: Fast rule mining in ontological knowledge bases with AMIE+. VLDB J. 24(6), 707–730 (2015)

A Neural User Preference Modeling Framework for Recommendation Based on Knowledge Graph

Guiming Zhu[1,2], Chenzhong Bin[2(✉)], Tianlong Gu[2], Liang Chang[2],
Yanpeng Sun[2], Wei Chen[2], and Zhonghao Jia[2]

[1] School of Computer Science and Information Security,
Guilin University of Electronic Technology, Guilin 541004, China
jackming555@gmail.com
[2] Guangxi Key Lab of Trusted Software,
Guilin University of Electronic Technology, Guilin 541004, China
binchenzhong@163.com, w_chen369@163.com,
{cctlgu, changl}@guet.edu.cn, yanpeng_sun@yeah.net,
1090994959@qq.com

Abstract. To address the data sparsity and cold start problems in the traditional recommender systems, lots of researchers aim at incorporating knowledge graphs (KG) into recommender systems to enhance the recommendation performance. However, existing efforts mainly rely on hand-engineered features from KG (e.g., meta paths), which requires domain knowledge. What's more, as relations are usually excluded from meta paths, they hardly specify the holistic semantics of paths. To address the limitations of existing methods, we propose an end-to-end neural user preference modeling framework (UPM) to incorporate features of entity and relation of KG into the representations of users and items, so as to learn user latent interests precisely. Specifically, UPM first propagate user's interests along links between entities in KG iteratively to learn user's potential preferences for the item. Furthermore, these preference features are dynamically during the preference propagation process. That is to say, the importance of these preference features to characterize user is different. Therefore, an attention network is used in UPM to calculate the influence of preference features at different propagating stages, then the final preference vector of the user is calculated from the preference features and the corresponding weights. Lastly, the final prediction probability of user-item interaction is obtained by inner product operation between the embedding of item and user. To evaluate our framework, extensive experiments on two real-world datasets demonstrate significant performance improvements over state-of-the-art methods.

Keywords: Recommender systems · Knowledge graph · User modeling ·
Preference propagation

© Springer Nature Switzerland AG 2019
A. C. Nayak and A. Sharma (Eds.): PRICAI 2019, LNAI 11670, pp. 176–189, 2019.
https://doi.org/10.1007/978-3-030-29908-8_14

1 Introduction

With the rapid development of the Internet, user's personalized needs have been constantly improving. How to help users get the information they need and how to address the information overload are the research hotspots in recommender systems field. Traditional collaborative filtering based recommender systems only use historical interactive information (explicit or implicit feedback) of user and item as input. This brings two problems: First, the interactive information between users and items is usually very sparse. Second, since the systems do not have historical interactive information, it cannot represent user accurately by historical interests and preferences of user, nor can it push personalized information to users. This situation is called cold start problem.

A common way to address the problems of data sparsity and cold start is to introduce some additional auxiliary information as a complementary of the recommendation algorithm. Recently, KG, which is a type of directed heterogeneous graph, has attracted a lot of researcher's attention due to large quantity of entities and concepts and rich semantic relations [1]. KG contains various types of information related to entities in the form of triplet which is expressed as (h, r, t), where h, r and t are head entity, relation and tail entity respectively, e.g. (Saving Private Ryan, directed, Spielberg). The form of triplet can seamlessly integrate user-item interactive data and improve the sparsity of interactive data.

At present, the methods of introducing KG into recommender systems can be divided into two categories: feature-based method and path-based method. The feature-based approaches unify features of users and items as input of recommendation algorithms [2]. However, these methods are not specifically designed for KG, so it cannot utilize all the information of KG effectively. For example, feature-based methods fail to learn multi-hop relational knowledge. To address this weakness, path-based approaches regard KG as a heterogeneous information network, and constructs meta path-based features between items [3]. A meta path is a specific path linking two entities. For example, there is a path (Tom Hanks → The Terminal → Stephen Spielberg → Schindler List) linking Tom Hanks and Schindler List, so this path can be used as a way to mining the potential relation between actors and movies. However, these methods heavily rely on handcrafted features to encode the semantics of path, which further relies on domain knowledge. Furthermore, this approach cannot be applied in where entities do not belong to the same domain (e.g. news recommendation) [4], and the meta paths cannot be predefined.

To address the problems mentioned above, we propose a novel neural user preference modeling framework (abbr. UPM), which takes user-item interaction as input data and predicts the probability of a user interact with a particular item. Specifically, for each user, each item he has interacted with is regarded as a seed item in the KG, and extends the user's interests iteratively along the links in the KG. In this process, the preference features at different stages of the user with respect to the candidate item can be learned, and the influence of the preference features are different to characterize user, thus, we propose get the weights of different preference features through an attention mechanism. After get the weights, UPM takes the sum of different preference features weighted by the corresponding weight, and the final preference vector of a user is generated. Finally, the probability of user-item interaction (e.g. a clicking or browsing action) is calculated by inner product of embedding of user and item. The experimental results on real-world datasets show that the proposed framework outperforms all of the baseline methods in click through rate (CTR) task.

The major contributions of this paper are as follows:

- We propose innovatively combines feature-based methods, path-based methods and attention mechanism in KG-aware recommendation.
- In order to introduce KG into recommender systems, an end-to-end user preference modeling framework (UPM) is proposed to mine the potential preference of user automatically by a user preference propagating process in the KG.
- To distinguish the importance of preference features at different propagating stages to characterize user, we propose calculate a weight for each preference features by an attention network, and make the preference features contribute to the preference vector of user according to the importance weights.
- Compared with the baseline methods, the proposed model performs best on two real-world datasets, indicating the superiority of our model.

2   Related Work

In this section, we mainly introduce the related work of introducing KG into recommender systems, i.e. feature-based and path-based methods. And the attention mechanism used in KG-aware recommendation.

2.1   Introducing KG into Recommender Systems

Feature-Based Methods. In the news recommendation scenario, Wang et al. [4] proposed to fuse the word vectors of news headlines, the entity vectors of KG and the entity context vectors, to generate the vector representation of news. Huang et al. [5] used TransE [6] to generate vector representations of entities and item, and then updates user's vector representations through memory networks based on user preferences for specific entities. Compared with other existing methods, feature-based methods have better performance. However, these methods ignore the semantics of the relations between entities represented by paths, so it cannot fully obtain the rich semantics of KG. On the other hand, since the links between users and items are realized by an implicit way, the regularization term of KG feature learning cannot fully discover the links between users and items.

Path-Based Methods. In the path-based approaches, some previous studies [7, 8] referred to the link patterns between KG entities as meta paths, and used meta paths to improve the performance of recommendations. Meta paths are defined as a sequence of entity types, e.g. a meta path (user $\rightarrow$ movie $\rightarrow$ director $\rightarrow$ movie) obtain user-item related attributes contained in KG. Yu et al. [3] proposed HeteMF to factorize the user-item rating matrix and constrain the distance between latent vectors of similar entities by a graph regularization method.

Meanwhile, there are some other works aim at using meta paths to model user-user or user-item relations. Luo et al. [9] proposed Hete-CF to model user-item, user-user and item-item relations based on the similarity of meta paths. Shi et al. [10] proposed SemRec model and introduced the concept of weighted meta path, which aims at describing the path semantics by distinguishing the nuances between link attribute values. Wang et al. [11] design a matrix factorization method by regularizing the user-user relation using the calculated similarity based on meta paths.

However, the above methods heavily depend on the quality and quantity of meta paths, what's more, the sequence dependencies of entities and relations in meta paths are neglected, which limits the quality of the generated recommendations.

2.2 Attention Mechanism in Recommendation

Attention mechanism shows the effectiveness in various machine learning tasks, such as machine translation [12], text categorization [13] et al. Recently, more and more researchers have applied attention mechanism to recommendation tasks. For example, Pei et al. [14] used the attention network to capture the joint effects of user-item interaction and measure the relevance between users and item. Chen et al. [15] proposed item-level and component-level attention mechanisms to model implicit feedback in multimedia recommendation.

Compared with the simple path-based and feature-based approaches, UPM combines merits of path-based and feature-based approaches to model user's preferences through rich semantic information contained in the KG, and obtain embedding of users by an attention network. Compared with the existing methods, UPM can automatically learning the semantic relations of entities and the sequence dependencies of entities and relations in the path.

3 Neural User Preference Modeling Framework

In this section, we present the proposed UPM framework in detail.

3.1 Notations and Definition

Table 1 summarizes all the notations used in this paper. The user-item interaction matrix $Y = \{y_{uv} | u \in U, v \in V\}$, if the interaction between u and v is observed $y_{uv} = 1$, otherwise $y_{uv} = 0$. A KG G consists of a large number of triplets (h, r, t), where h, r and t are the head entity, relation and tail entity of G respectively.

The relevant definitions are as follows:

Table 1. Notations and descriptions

Notations	Descriptions
$U = \{u_1, u_2, \ldots, u_m\}$	User set
$V = \{v_1, v_2, \ldots, v_n\}$	Item set
$Y \in R^{m \times n}$	User-item interaction matrix
$\epsilon = \{e_1, e_2, \ldots e_e\}$	Entity set
$R = \{r_1, r_2, \ldots, r_r\}$	Relation set
G	KG
ε_u^k	k-hop relevant entities set of user u
S_u^k	k-hop triplets set of user u
$H_u = \{h_1, h_2, \ldots, h_t\}$	Historical interaction record of u
P_i	Relevance probability
O_k^u	k-hop preference features of user u
Att	Attention network
w_k	Weight of k-hop preference features of u
v	Embedding of item v
u	Embedding of user u
y_{uv}	Predicted probability that u interact with v

Definition 1 (KG). Define $\varepsilon = \{e_1, e_2, \ldots, e_e\}$, $R = \{r_1, r_2, \ldots, r_r\}$ denote the sets of entities and relations respectively. $G = (\varepsilon, L)$ is a directed graph with an entity type mapping function $\phi : \varepsilon \rightarrow A$ and a link type mapping function $\psi : L \rightarrow R$. Each entity $e \in \varepsilon$ belongs to an entity type $\phi(e) \in A$, and each link $r \in L$ belongs to a link type (relation) $\psi(r) \in R$ [3].

Definition 2 (Relevant Entity). Given user-item interaction matrix Y and the k-hop relevant entities set of user u is defined as follows:

$$\varepsilon_u^k = \left\{ t | (h, r, t) \in G \& h \in \varepsilon_u^{k-1} \right\}, k = 1, 2, \ldots, K \tag{1}$$

Where $\varepsilon_u^0 = H_u = \{v | y_{uv} = 1\}$, i.e. the historical interaction record of user u [4].

Relevant entities can be regarded as the natural extensions of a user's interest in the KG. Given the definition of the relevant entity, the k-hop triplets set of user u is defined as follows:

Definition 3 (Set of Triplets). The k-hop triplets set of user u is defined as the set of triplets from ε_u^{k-1} [4]:

$$S_u^k = \left\{ (h, r, t) | (h, r, t) \in G \& h \in \varepsilon_u^{k-1} \right\}, k = 1, 2, \ldots, K \tag{2}$$

With the increase of hop number k, the set of triplets may become very large, which will greatly increase the computational overhead. In order to address the problems, we proposes the following restrictions: (1) In a specific recommendation scenario (such as movie recommendation), the relations in the KG can be limited to movie-related attributes. (2) In practice, the total number of hop K is generally not very large, because entities locating far away from user history interaction items may be irrelevant to user latent preference. In this paper, $K = 2$ or 3.

3.2 Architecture of Framework

The framework of UPM is illustrated in Fig. 1. UPM takes a user u and an item v as input of the framework, and outputs the probability that the user u will interact (click, browse, etc.) with the item v. Specifically, for the input user u, his historical interaction record H_u is treated as seeds in the KG, then extended along links to form multiple triplet sets S_u^k ($k = 1, 2,..., K$). A triplet set S_u^k is the set of knowledge triplets that are k-hops away from the seed set H_u. And the user's preference features (the dark blue, olive and yellow blocks) at different hops are obtained through extended interests of user iteratively along the links in triplet sets S_u^k. Then the preference features of user and embedding (the light blue block) of item are input into the attention network simultaneously, and the final preference vector (the pink block) of user is calculated. The probability y_{uv} of user-item interaction can be obtained by inner product between the embedding of item v and user u.

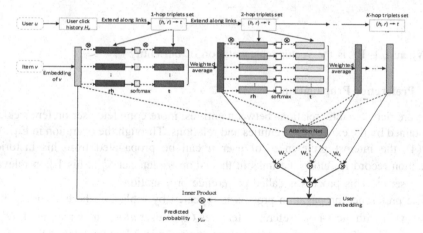

Fig. 1. The framework architecture. The light blue part is the embedding of item v, the dark blue part, the olive part and the yellow part are 1-hop, 2-hop and K-hop preference features of user respectively. (Color figure online)

3.3 1-Hop Preference Feature of User

The traditional collaborative filtering methods firstly learns the latent representation of users and items, then calculates the predicted probability through the inner product. In order to model user-item interaction more accurately, we proposes a neural user preference modeling framework to represent potential preferences of users.

As shown in Fig. 1, each item v has an associated embedding $\mathbf{v} \in R^d$, d is dimension of the embedding. Each item embedding is generated by the attributes of this item. Given $\mathbf{v}$ and 1-hop triplets set S_u^1 of user u, by calculating similarity between item v, head entity h_i and entity relations r_i in S_u^1, each triplet in S_u^1 is assigned a relevance probability P_i:

$$P_i = \mathrm{softmax}\left(\mathbf{v}^T \mathbf{r}_i \mathbf{h}_i\right) = \frac{\exp(\mathbf{v}^T \mathbf{r}_i \mathbf{h}_i)}{\sum_{(h,r,t)\in S_u^1} \exp(\mathbf{v}^T \mathbf{r}\mathbf{h})} \tag{3}$$

Where $\mathbf{r}_i \in R^{d\times d}$, $\mathbf{h}_i \in R^d$ are the vector representation of r_i and h_i respectively, softmax function ensures that the sum of all calculated relevance probabilities is 1. P_i can be regarded as the similarity between item v and head entity h_i on entities relations r_i. It should be noted that the vector representation $\mathbf{r}_i$ of r_i must be taken into account when calculating the above relevance probability P_i, because the similarity between item v and head entity h_i may be different on different entities relations. For example, "Saving Private Ryan" and "Schindler's List" are highly similar when considering director and genre, but they are completely different from the actor attribute.

After obtaining the relevance probability P_i of each triplet (h_i, r_i, t_i) in 1-hop triplets set S_u^1, all tail entity t_i of triplets in S_u^1 are weighted by the corresponding relevance probability P_i, and the 1-hop preference feature O_u^1 of user u is given by:

$$O_u^1 = \sum_{(h_i,r_i,t_i)\in S_u^1} P_i \mathbf{t}_i \tag{4}$$

Where $\mathbf{t}_i \in R^d$ is the vector representation of tail entity t_i.

3.4 Preference Propagation

There are rich semantic relations between entities, more complete user preferences can be obtained by the extension of entities and relations. Through the operation in Eqs. (3) and (4), the interest preferences of user u can be propagated from his historical interaction record H_u along the links in the 1-hop triplets set S_u^1 to his 1-hop relevant entities set ε_u^1. This process is called preference propagation.

The preference propagation process is repeated by replacing the embedding $\mathbf{v}$ of item v in (3) with the 1-hop preference features O_u^1 of user u. As shown in Fig. 1, O_u^1 as u's historical preference is propagated along the links in 2-hop triplets set S_u^1 to his 2-hop relevant entities ε_u^2, repeating the operation in Eqs. (3) and (4) to obtain u's 2-hop preference features O_u^2, which is iteratively performed on user u's k-hop triplets $S_u^k(k = 1, 2, \ldots, K)$. Therefore, a user's preference is propagated from his historical interaction record H_u to K-hop relevant entities ε_u^K. Thus, the preference features of user u at different hops can be obtained: $O_u^1, O_u^2, \ldots O_u^K$. The final preference vector of user u with respect to item v can be obtained by simply combining the preference features of user u at different hops:

$$\mathbf{u} = O_u^1 + O_u^2 + \ldots + O_u^K \tag{5}$$

In theory, with the increase of hop number k, the preference feature O_u^k of user u in the last hop contains all the information of the previous preference features, but they may be weakened in O_u^K, so the preference features at all hops must be superimposed.

3.5   Attention-Based User Preference Extraction

The above method does not take into account that the weights of user preference features $O_u^1, O_u^2, \ldots O_u^K$ at different hops to user's final preference vector are different. As shown in Fig. 1, to model the different effects of user preference features $O_u^1, O_u^2, \ldots O_u^K$ on the final preference vector of user u, we proposes calculate the weight w_k of k-hop preference features of user u by an attention network Att, w_k formulated by:

$$w_k = softmax\left(Att\left(\mathbf{v}, O_u^k\right)\right) = \frac{exp\left(Att\left(\mathbf{v}, O_u^k\right)\right)}{\sum_{k=1}^{K} exp\left(Att\left(\mathbf{v}, O_u^k\right)\right)} \quad k = 1, 2, \ldots, K \tag{6}$$

Attention network Att takes user preference features $O_u^1, O_u^2, \ldots O_u^K$ at different hops and embedding of item v as input, and outputs the corresponding weights w_k of $O_u^1, O_u^2, \ldots O_u^K$. The weight w_k can be regarded as the important scores of user preference features at different hops, w_k adaptively select the informative preference features with different importance, and make the informative preference features contribute more to characterize preference vector of user u. Then we sum up the user preference features $O_u^1, O_u^2, \ldots O_u^K$ at different hops according to the weight w_k provided by Att to get the final preference vector of user u:

$$\mathbf{u} = \sum_{k=1}^{K} w_k O_u^k \tag{7}$$

Finally, given the embedding of user u and item v, the probability of the user interact with the item is calculated by inner product:

$$y_{uv} = \sigma\left(\mathbf{u}^T \mathbf{v}\right) \tag{8}$$

Where $\sigma(x) = \frac{1}{1+exp(-x)}$ is the sigmoid function.

3.6   Model Optimization

Given G and implicit feedback matrix Y, the objective of model optimization is to maximize the posterior probability of model parameter Θ:

$$maxp(\Theta|G, Y) \tag{9}$$

Θ includes the vector representations of all entities, relations. So it's equivalent to maximizing:

$$p(\Theta|G,Y) = \frac{p(\Theta,G,Y)}{p(G,Y)} \propto p(\Theta) \cdot p(G|\Theta) \cdot p(Y|\Theta,G) \tag{10}$$

Taking the negative logarithm of (10) and have the following loss function:

$$
\begin{aligned}
minL &= -\log(p(\Theta) \cdot p(G|\Theta) \cdot p(Y|\Theta,G)) \\
&= \sum_{(u,v)\in Y} -\left(y_{uv} \log \sigma(u^T v) + (1 - y_{uv}) \log\left(1 - \sigma(u^T v)\right)\right) \\
&+ \frac{\lambda_2}{2} \sum_{r\in R} \|I_r - E^T RE\|_2^2 + \frac{\lambda_1}{2}\left(\|V\|_2^2 + \|E\|_2^2 + \sum_{r\in R} \|R\|_2^2\right)
\end{aligned}
\tag{11}
$$

Where V, R and E are the embedding matrices for all items, relation and entities, respectively, I_r is the slice of the indicator tensor I in the KG. The stochastic gradient descent (SGD) algorithm is used to iteratively optimize the loss function. In order to make the calculation more efficient in each training process, positive (negative) records of the smallest batch are sampled randomly from Y and positive (negative) triplets are sampled from G. The gradient of loss L relative to model parameter Θ is calculated, and all parameters are updated by back propagation algorithm.

4   Experiments and Analysis

In this section, the framework is evaluated by compared with the baseline methods on MovieLens-1M and Book-Crossing datasets.

4.1   Datasets and Preprocessing

The proposed framework is evaluated on two real-world datasets from different domains: MovieLens-1M and Book-Crossing. MovieLens-1M contains about 1 million user ratings (ranging from 1 to 5) on movie websites. Book-Crossing contains 1,149,780 explicit ratings (ranging from 0 to10) of books. In this experiment, we use the pre-processed data in [4]. Because MovieLens-1M and Book-Crossing are explicit feedback data, we transform them into implicit feedback data. Similar to [4], the ID embedding of users and items are used as the original input of framework in this experiment. The data statistics are shown in Table 2.

Table 2. The statistics of datasets

Datasets		MovieLens-1 M	Book-Crossing
User-item interaction	#Users	6,036	17,860
	#Items	2,445	14,967
	#Ratings	753,772	139,746
	#Data Density	5.108%	0.0523%
KG	#Entities	182,011	77,903
	#Links	12	25
	#The first 4-hops triplets	1,440,815	241,163

4.2 Baselines

We use the following methods to compare with the framework proposed in this paper:

- CKE [1] unifies collaborative filtering with structured knowledge, text knowledge and pictures information etc. in a framework for recommendation.
- DKN [4] treats word vectors, entity vectors and entity context vectors as multiple channels to fuse in the framework of CNN for click rate prediction.
- SHINE [16] designed a deep self-encoder to combine semantic network, social network and user profile network for celebrity recommendation.
- LibFM [2] is a widely used feature-based factorization framework for click-through rate prediction. In this experiment, user ID, item ID and corresponding entity embedding learned through TransR are used as input of LibFM.
- Wide&Deep [17] is a general deep framework for recommendation, which combines linear and non-linear channels. The embedding of users, items and entities are used as input for Wide&Deep.

4.3 Experiment Setup

In the experiments, $d = 16$ denotes the dimension of the embedding of items and KG, and $\eta = 0.008$ denotes the learning rate. Specific hyper-parameter settings are shown in Table 3. For fairness, all baseline methods have the same dimension settings as Table 3, while other baseline hyper-parameters are based on grid search. The ratio of training, evaluation and test set is 6:2:2. Each experiment was repeated 5 times and the average results is reported. Accuracy and area under curve (AUC) were used to evaluate the performance of click through rate (CTR) prediction.

Table 3. Hyper-parameter settings for the two datasets

Datasets	Hyper-parameter settings
MovieLens-1M	$d = 16$, $T = 2$, $\lambda_1 = 10^{-7}$, $\lambda_2 = 0.01$, $\eta = 0.008$
Book crossing	$d = 4$, $T = 3$, $\lambda_1 = 10^{-5}$, $\lambda_2 = 0.01$, $\eta = 0.001$

4.4 Performance Comparison

The results of all methods in click through rate prediction are shown in Table 4.

The proposed framework UPM achieves the best performance on two datasets with all evaluation metrics. CKE performs poorly than LibFM and Wide&Deep, since there is no text and visual information, and structural knowledge cannot characterize users completely. DKN performs worst among all methods in the two datasets, because film titles and book titles are usually short and contains limited information. SHINE performs better than DKN only, because we have no social and user profile networks. As two general recommendation algorithms, LibFM and Wide&Deep performs satisfactorily, which shows that LibFM and Wide&Deep can make full use of semantic information from KG.

Table 4. The results of AUC and accuracy in click through rate prediction

Framework	MovieLens-1M		Book crossing	
	AUC	ACC	AUC	ACC
CKE	0.796	0.739	0.674	0.635
SHINE	0.778	0.732	0.668	0.631
DKN	0.655	0.589	0.621	0.598
LibFM	0.892	0.812	0.685	0.639
Wide&Deep	0.903	0.822	0.711	0.623
UPM	**0.928**	**0.855**	**0.740**	**0.695**

4.5 The Sensitivity of Hyper-parameters

The effect of dimension of embedding d and training weight of KG term λ_2 on AUC and ACC are shown in Fig. 2, which have similar trends on Book Crossing dataset. d range from 2 to 64, λ_2 range from 0 to 1, while keeping other parameters fixed.

With the increase of d, both AUC and ACC improves and becomes stable, because embedding with larger dimensions can encode more useful information, but when d is greater than 16, both AUC and ACC begin to drops because of possible overfitting. AUC and ACC performed best when $\lambda_2 = 0.01$. This is because when training weight of KG· term is very small, it is not enough to provide effective regularization constraints, while a large training weight may mislead the objective function.

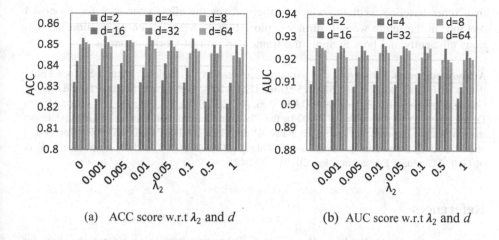

(a)   ACC score w.r.t λ_2 and d (b)   AUC score w.r.t λ_2 and d

Fig. 2. Parameter sensitivity of the proposes framework on MovieLens-1M.

In order to further explore the relationships between the performance of the framework and the maximal hop number K, we vary the maximal hop number K to see how AUC changes in UPM, the results as shown in Table 5.

Table 5. The results of AUC w.r.t. different hop numbers

Hop number K	1	2	3	4
MovieLens-1M	0.927	**0.928**	0.925	0.926
Book crossing	0.739	0.734	**0.740**	0.732

As shown in Table 5, the best performance is achieved when K is 2 or 3. This is because too small of an K can hardly explore inter-entity relatedness and dependency of long distance, while too large of an K brings much more noises than useful signals.

5   Conclusion

To address the challenges of traditional KG-aware recommendation methods, we innovatively combine feature-based methods, path-based methods and attention mechanism in KG-aware recommendation. Specifically, we proposed an end-to-end neural user preference modeling framework (UPM) for recommendation, which introduces KG into recommender systems effectively. UPM mine potential preferences of a user by propagating the user's interests in KG. The attention network is used to adaptively discriminate the importance of the preference features of user at different propagation stages for the final preference vector of user. Experimental results on two real-world datasets shows that the performance of the proposed framework is better than other baseline methods, which further proves the effectiveness of the proposed method. In the future we will further explore how to represent entity-relation interactions efficiently and how to apply the framework to real-world scenarios.

Acknowledgments. This work was partially supported by the National Natural Science Foundation of China (Nos. U1501252, 61572146, U1711263), the Natural Science Foundation of Guangxi Province (No. 2016GXNSFDA380006, AC16380122), the Guangxi Innovation Driven Development Project (No. AA17202024), the Platform Construction Project of Guangxi Information Science Experiment Center (No. PT1601), the Basic Ability Promotion Project for Young and Middle-aged Teachers in Universities of Guangxi (2018KY0203) and the Innovation Project of GUET Graduate Education (Nos. 2019YCXS042).

References

1. Zhang, F., Yuan, N., Lian, D., et al.: Collaborative knowledge base embedding for recommender systems. In: 22th ACM SIGKDD International Conference on Knowledge Discovery and Data Mining, pp. 353–362. ACM, New York (2016)
2. Rendle, S.: Factorization machines with libFM. Trans. Intell. Syst. Technol. 3(3), 57:1–57:22 (2012)
3. Yu, X., Ren, X., Sun, Y., et al.: Personalized entity recommendation: a heterogeneous information network approach. In: 7th International Conference on Web Search and Data Mining, pp. 283–292. ACM, New York (2014)
4. Wang, H., Zhang, F., Xie, X., et al.: DKN: deep knowledge-aware network for news recommendation. In: 27th International Conference on World Wide Web, pp. 1835–1844. ACM, New York (2018)
5. Huang, J., Zhao, W., Dou, H., et al.: Improving sequential recommendation with knowledge-enhanced memory networks. In: 41th International ACM SIGIR Conference on Research & Development in Information Retrieval, pp. 505–514. ACM, New York (2018)
6. Bordes, A., Usunier, N., Garcia-D, A., et al.: Translating embeddings for frameworking multi-relational data. In: 26th International Conference on Neural Information Processing Systems, pp. 2787–2795. MIT Press, Cambridge (2013)
7. Sun, Y., Han, J.: Mining heterogeneous information networks: a structural analysis approach. ACM SIGKDD Expl. Newslett. 14(2), 20–28 (2013)
8. Yu, X., Ren, X., Gu, Q., et al.: Collaborative filtering with entity similarity regularization in heterogeneous information networks. In: 23th International Joint Conference on Artificial Intelligence. Elsevier, Burling (2013)
9. Luo, C., Pang, W., Wang, Z., et al.: Hete-CF: social-based collaborative filtering recommendation using heterogeneous relations. In: 2014 IEEE International Conference on Data Mining. IEEE Computer Society, Washington (2015)
10. Shi, C., Zhang, Z., Luo, P., et al.: Semantic path based personalized recommendation on weighted heterogeneous information networks. In: 24th ACM International on Conference on Information and Knowledge Management, pp. 453–462. ACM, New York (2015)
11. Wang, Y., Xia, Y., Tang, S., et al.: Flickr group recommendation with auxiliary information in heterogeneous information networks. Multimedia Syst. 23(6), 703–712 (2017)
12. Vaswani, A., Shazeer, N., Parmar, N., et al.: Attention is all you need. In: 31th Conference on Neural Information Processing Systems, pp. 6000–6010. MIT Press, Cambridge (2017)
13. Yang, Z., Yang, D., Dyer, C., et al.: Hierarchical attention networks for document classification. In: NAACL-HLT 2016, pp. 1480–1489. ACL, Stroudsburg (2016)

14. Pei, W., Yang, J., Sun, Z., et al.: Interacting attention-gated recurrent networks for recommendation. In: 26th ACM Conference on Information and Knowledge Management, pp. 1459–1468. ACM, New York (2017)
15. Chen, J., Zhang, H., He, X., et al.: Attentive collaborative filtering: multimedia recommendation with item- and component-level attention. In: 40th International ACM SIGIR Conference on Research and Development in Information Retrieval, pp. 335–344. ACM, New York (2017)
16. Wang, H., Zhang, F., Hou, M., et al.: Shine: Signed heterogeneous information network embedding for sentiment link prediction. In: 11th ACM International Conference on Web Search and Data Mining, pp. 592–600. ACM, New York (2018)
17. Cheng, H., Levent, K., Jeremiah, H., et al.: Wide & deep learning for recommender systems. In: 1st Workshop on Deep Learning for Recommender Systems, pp. 7–10. ACM, New York (2016)

Jointing Knowledge Graph and Neural Network for Top-N Recommendation

Wei Chen, Liang Chang, Chenzhong Bin$^{(\boxtimes)}$, Tianlong Gu,
and Zhonghao Jia

Guangxi Key Lab of Trusted Software, Guilin University of Electronic
Technology, Guilin 541004, China
w_chen369@163.com, binchenzhong@163.com,
changl@guet.edu.cn

Abstract. Currently, neutral networks attract much attention and show great potential in recommendation systems. The existing works mainly aim at leveraging neural network to model the nonlinear representations of users and items. However, they only use historical interaction sequence of user-items to learn the latent features of users and items, while ignoring the rich self-attributes of items. Recent methods utilize knowledge graphs as auxiliary information to learn the latent features between users and items, but they fail to represent the relevance and similarity of attributes among items. Based on this observation, we propose a novel model named JKN that incorporates knowledge graph and a neural network for item recommendation. The key point of JKN is to learn accurate latent representations of item attributes through knowledge graph, then to integrate them into a feedforward neural network to model user-item interactions in nonlinear. Empirical results on a real-world dataset demonstrate the superior performance of our model in Top-n recommendation task.

Keywords: Recommendation system · Knowledge graph · Neural network · Implicit feedback

1 Introduction

Recommendation system that effectively alleviates information overload has been widely used in various online services. There are two major approaches (explicit or implicit feedback) to predict items in recommendation system. Implicit feedback generates the ranking list through the historical records of user-item interactions and is more sufficient for personalized recommendation. As a typical recommendation method, Matrix Factorization(MF) [1, 2] based on implicit feedback, projects users and items into a low-dimensional shared vector space as latent features, and produces ranking lists through inner product. It adopts a simple inner product of linear combination, but may not satisfy complex user-item interactions in different semantic environments.

In recent years, researchers have focused on the nonlinear structure of neural network can effectively model the complex interactions of user-item [5]. However, they only use interactive sequences to construct latent features of items and these items features are acquired by randomly initializing the sequence of items during the training of the model. The item features are not actually used, so these models may not be able

© Springer Nature Switzerland AG 2019
A. C. Nayak and A. Sharma (Eds.): PRICAI 2019, LNAI 11670, pp. 190–195, 2019.
https://doi.org/10.1007/978-3-030-29908-8_15

to realize the fine-grained model where users have preferences on the features of each item. To overcome the mentioned disadvantages, some researchers study to introduce knowledge graph (KG) to the recommendations systems [4]. Since KG contains various types of relationships, the correct preferences of the users can be reasonably extracted, improving the accuracy of the recommendation system. These recommendation methods of KG can improve recommendation performance by flexibly using item embedded in recommendation system. However, there is a shortcoming in ignoring the relevance and similarity of attributes among items and they have weakness to model user-item interactions. Here, we put forward a jointing KG and neural network recommendation model (JKN). In this model, we first obtain the latent features of all items through the method of graph embedding, then model user-item interactions by applying multi-layer neural networks. The proposed model not only accurately acquires item embedding, but also maintains the interactive ability of user-item of neural network, and enhances the interactive ability of data by using the attributes of items.

Our contributions are summarized as follows: (1) We show that the method accurately obtain the item vectors with various attributes from KG. (2) We propose a recommendation model that combines the attributes of the items to represent latent features and incorporates them into a neural network to learn the users' preference for the item attributes. (3) We perform experiments on a real-world dataset to show that the model is more effective in recommendation systems.

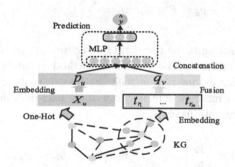

Fig. 1. The framework of JKN. On the bottom of the figure, the KG contains the user (blue), movie (yellow) and other entities (green) and entities are connected by different relationships. (Color figure online)

2 The Proposed Method

2.1 Notations

In our work, the structure of KG consists of triples $G = (E, R, T)$, where a set of entities, a set of relationships and various relational types are denoted as E, R and T, respectively. The set T of relationship types are stored in the knowledge base DBpedia and the item type handled in this paper is $\beta = abo{:}Movie$ [3]. We project the knowledge base to get the entitys of Movie type. As is shown in Fig. 1, we define that the item of attribute-values directly connecting to the item as the subgraph of item in KG.

The subgraph of item contains other attributes of the item except the preference. According to the observation of subgraph of items, we find two different items' subgraph are structurally similar and represent similarities in vector space and obtain the item embedding vector by fusing the attributes of each item so that the user's preference for the item attributes can be accurately trained. Following, a set of users and items are denoted as U and I, respectively. Let a user is denoted as $u \in U \subset E$, an item is denoted as $v \in V \subset E$, and $y_{u,v}$ be the label of the interaction of u and v. If u has preference with v, the value of $y_{u,v}$ be 1, and 0 otherwise. Here, 0 does not mean that u hasn't preference with v, it may be that u does not observe v.

2.2 The Model of Recommendation

Encoding the User and Item. Our model takes in two inputs, namely user and item. We define the one-hot coding representation of user u as $X_u \in R^{|u| \times 1}$ and convert it into a vector P_u by applying a embedding layer. We randomly walk through the subgraph of item v to learn the attributes of item v by node2vec [4] and obtain the item vectors $t_r(v)$ with the corresponding attribute. The ultimate item vector can be represented as q_v by fusing item vector $t_{r_m}(v)$ in the vector space. The preference pairs (u, v) are transformed into the vector of latent features as below:

$$p_u = P^T \cdot X_u, \quad q_v = \sum_m^1 t_{r_m}(v), \tag{1}$$

Predict: In a real-life scenario, the user-item interactions are modeled mainly through linear and nonlinear methods. The nonlinear method is more flexible and more adaptable to characterize the interaction with the fusion attributes of item in various scenarios. In order to increase the effectiveness of the user-item interactions, we apply MLP to our model. Next, we have

$$h_0 = f(p_u, q_v) = \begin{bmatrix} p_u \\ q_v \end{bmatrix}, \ h_L = f^k(W_{l-1}^T h_{l-1} + b_{l-1}), \ \hat{p}_{u,v} = \sigma(W_L^T h_L), \tag{2}$$

Among them, $l = 1, \ldots, L$, $f(*)$ is a function that concatenates two embedding vectors, $f^k(*)$ is Rectifier Linear Unit (*relu*) nonlinear activation function. And W_x and b_x are the weight matrix and the offset vector, respectively. In the prediction layer, σ is the sigmoid function, which is defined as $\delta(x) = 1/1 + \exp(x)$, $\hat{p}_{u,v}$ is the conditional probability after MLP layer and stands for the preference relationship of user for item.

2.3 The Loss Function for Optimization

In order to train my model, we utilize cross-entropy loss as the objective function,

$$\mathcal{L} = - \sum_{<u,v,y_{u,v}>} (y_{u,v} \cdot \log \hat{p}_{u,v} + (1 - y_{u,v}) \cdot log(1 - \hat{p}_{u,v})), \tag{3}$$

The objective function mainly minimizes the loss values of our model and the parameters of our model that are optimized using the SGD. Since there are no explicit negative examples in the data, we uniformly sample the data which user does not observe as a negative example in each iteration. In order to prevent over-fitting caused by excessive dimension, we introduce dropout to reduce the shared neurons randomly during training and improve the robustness of our model. In each training, we randomly select four negative examples that is unobserved corresponding to a positive example. We test the min-batch Adam size of [128, 256, 512, 1024] respectively, and the learning rate is set to 0.001. Since the dimensions of the last layer of JKN directly affect the predicted results, we use the tested dimensions as [8, 16, 32, 64] and the dropout rate is set to 0.3. In the general test, the number of hidden layers that we employ JKN is three.

3 Experiment

3.1 Experimental Settings

Dataset and Evaluation Metrics. The dataset used in the experiment is Movielens 1M which is widely used in recommendation system. We project the movies of Movielens to the corresponding movie entities in DBpedia and obtain the entities and relationships about movie. The processed dataset has seven attributes and 915,100 rating records consisting of 6040 users and 3,125 movies. We evaluate the performance of item recommended by the *leave-one-out*. For each user, we randomly select 100 negative items which have been unobserved. Evaluation indicators include Hit Ratio (HR) and Normalized Discounted Cumulative Gain (NDCG).

Baseline. We will compare the performance of the following baselines. (1) *MF-BPR* [2]: MF-BPR uses pairwise loss to optimize MF to perform implicitly recommended tasks. (2) *GMF* [5]: This method applies *element-wise* above user and item embedding to predict item from data. (3) *MLP* [5]: This is a state-of-the-art model which adopts a non-linear approach to model user-item interactions. (4) *NAIS* [6]: NAIS Based on the similarity between items, distinguishes the importance of items by attention network.

3.2 Experimental Results

Top-N Item Recommendation Results. As is shown in Fig. 2, we can observe the recommendation result by Top-N (1-10) on Movielens. BPR and GMF are similar in performance, MLP performance are better than the former two methods, mainly due to its nonlinear structure. JKN is superior to other models in ranking from 1 to 10. The performance of JKN is much better than that of NAIS and MLP in top-10 recommendation, especially on NDCG. Based on this observation, it is proved that our proposed model can generate high quality data in Top-N recommendation.

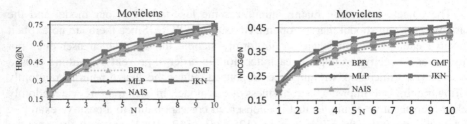

Fig. 2. HR@N and NDCG@N results of Top-N item recommendation (dimension = 64).

Efficacy of Deep. To further verify the impact of different numbers of hidden layers on recommended performance, the experimental results are shown in Table 1. Layer3 means that there are three hidden layers (except the embedding layer). As we have seen, even if the models have the same dimension which is at the last layer, adding more layers is beneficial to the performance of the model. This result is mainly due to the multi-layer nonlinear layer that improves the interactive information of feature semantics.

Table 1. HR@10 and NDCG@10 of JKN with different layers.

Dimensions	HR@10				NDCG@10			
	Layer1	Layer2	Layer3	Layer4	Layer1	Layer2	Layer3	Layer4
8	0.5654	0.6204	0.6896	0.6859	0.3185	0.3561	0.4161	0.4125
16	0.6325	0.6803	0.7204	0.7199	0.3649	0.4059	0.4426	0.4482
32	0.6685	0.7118	0.7382	0.7482	0.3923	0.4368	0.4474	0.4624
64	0.6962	0.7272	0.7411	0.7429	0.425	0.4321	0.4603	0.458

4 Conclusion

This paper presents a novel model named JKN that incorporates KG into neural network. JKN learns the relationship between the item and the item's attribute-values from KG, applies the feature learning to accurately obtain the latent feature representation of the item attributes and finally utilizes the multi-layer perceptron to model the user-item interactions for top-n recommendation. Experiment results on a real-world dataset show the effectiveness of our model.

Acknowledgments. This work was partially supported by the National Natural Science Foundation of China (Nos. U1501252, 61572146, U1711263), the Project of Cultivating Excellent Dissertations for Graduate of GUET (Nos.17YJPYSS16) and the Innovation Project of GUET Graduate Education (Nos. 2019YCXS041).

References

1. Hu, Y., Koren, Y., Volinsky, C.: Collaborative filtering for implicit feedback datasets. In: ICDM IEEE International Conference on Data Mining, 15–19 December 2008, Pisa, Italy, pp. 263–272 (2008)
2. Rendle, S., Freudenthaler, C., Gantner, Z., et al.: BPR: Bayesian personalized ranking from implicit feedback. In: UAI, pp. 452–461 (2009)
3. Rizzo, G., Palumbo, E.: Entity2rec: learning user-item relatedness from knowledge graphs for top-n item recommendation. In: RecSys, pp. 32–36. ACM (2017)
4. Grover, A., Leskovec, J.: node2vec: scalable feature learning for networks. In: ACM SIGKDD International Conference on Knowledge Discovery and Data Mining, pp. 855–864 (2016)
5. He, X., Liao, L., Zhang, H., et al.: Neural collaborative filtering. In: International Conference on World Wide Web, pp. 173–182 (2017)
6. He, X., He, Z., Song, J., et al.: NAIS: neural attentive item similarity model for recommendation. IEEE Trans. Knowl. Data Eng. **30**, 2354–2366 (2018)

A Novel Genetic Programming Algorithm with Knowledge Transfer for Uncertain Capacitated Arc Routing Problem

Mazhar Ansari Ardeh[✉], Yi Mei[✉], and Mengjie Zhang[✉]

Victoria University of Wellington, Wellington, New Zealand
{mazhar.ansariardeh,yi.mei,mengjie.zhang}@ecs.vuw.ac.nz

Abstract. Uncertain Capacitated Arc Routing Problem (UCARP) is a challenging optimization problem. Genetic Programming (GP) has been successfully applied to train routing policies (heuristics to make decisions in real time rather than a fixed solution) to respond to uncertain environments effectively. However, the effectiveness of routing policy is scenario dependent, and it takes time to train a new routing policy for each scenario. In this paper, we investigate GP with knowledge transfer to improve the training efficiency by reusing useful knowledge from previously solved related scenarios. We propose a novel knowledge transfer approach which our experimental results show that it obtained significantly higher training efficiency than the existing GP knowledge transfer methods, and the vanilla training process without knowledge transfer.

Keywords: Uncertain arc routing · Genetic programming · Hyper-heuristics · Transfer learning

1 Introduction

Uncertain Capacitated Arc Routing Problem (UCARP) has many important real-world applications in supply chain and logistics. In UCARP, a graph $G(V, E)$ is given, where V and E are the set of nodes and edges. Each edge $e \in E$ has a positive stochastic deadheading cost $dc(e)$, a non-negative serving cost $sc(e)$, and a non-negative stochastic demand $d(e)$. An edge with positive demand is called a *task*. A number of vehicles with capacity Q are located at the depot $v_0 \in V$. The problem is to find the optimal routes for the vehicles subject to the constraints: (1) each vehicle needs to start and end its route at the depot; (2) between two refills, the total demand served by each vehicle cannot exceed its capacity.

There have been several studies dedicated to solving UCARP (e.g. (Mei et al. 2010)), among which the Genetic Programming (GP) based approaches have achieved great success. GP evolves (trains) routing policies, which are decision-making heuristics, rather than solutions. A routing policy can generate the solution in an online fashion based on the latest information, and thus is effective to handle uncertain environments.

© Springer Nature Switzerland AG 2019
A. C. Nayak and A. Sharma (Eds.): PRICAI 2019, LNAI 11670, pp. 196–200, 2019.
https://doi.org/10.1007/978-3-030-29908-8_16

The effectiveness of routing policies depends on the problem scenario (e.g. the topology of the graph, and the number of vehicles to be used). The performance of a routing policy can dramatically decrease when changing from one scenario to another. Intuitively, one can retrain the routing policy in the new scenario from scratch. However, it can be time consuming and inefficient. In this case, we propose GP with transfer learning to improve the efficiency of the retraining.

Transfer learning can be defined as "the improvement of learning in a new task through the transfer of knowledge from a related task that has already been learned" (Torrey and Shavlik 2010). For transfer learning in GP, a commonly used strategy is to transfer sub-trees from the source domain to the target domain. Intuitively, different subtrees in the source domain should have different levels of importance and should be more likely to be transferred. However, it is challenging to quantitatively measure the re-usability of a subtree. Existing studies mostly select the subtree in promising individuals randomly (e.g. (Dinh et al. 2015)), which is not an optimal strategy. Another measure of defining importance of subtrees is to consider the number of times that they appeared in the source domain (Ansari Ardeh et al. 2019). However, frequency may be misleading because the final GP tree may have many redundant branches and some frequent subtrees can be in the redundant branches. In addition, subtrees can be structurally different but essentially the same.

In this paper we aim to propose a novel GP with subtree transfer to improve the effectiveness of retraining routing policies for UCARP. The research objectives that we follow in this paper are (1) propose a new and more accurate measure for the reusability of subtrees based on their contribution to the individuals; (2) develop a novel GP with knowledge transfer based on the new reusability measure to transfer subtrees from source domains to the target domain of UCARP; (3) verify the efficacy of the proposed algorithm on different transfer scenarios.

2   Novel Subtree Transfer for Genetic Programming Hyper-heuristic

We propose a novel method for filtering good transferable knowledge by evaluating the reusability of subtrees to distinguish their potential for transfer. We choose the final GP population in the source domain as the knowledge source.

When identifying transferable subtrees, it is natural to conjecture that individuals with good fitness value are better sources for knowledge extraction. Therefore, we consider the subtrees of the top 50% individuals in the final population in terms of their test performance in the source domain.

To form the pool of the candidate subtrees, we adopt the following two strategies that are commonly used by existing works: (1) **All**: All the subtrees of all the considered individuals are included in the pool; (2) **Root Subtrees**: Immediate subtrees of the roots of the considered individuals are included in the pool. The subtrees in the pool have different reusability. To select the transferred subtrees more intelligently, we propose a new reusability measure based on the contribution of a subtree to its tree (Mei et al. 2017).

Given a GP tree x, the contribution $\xi(x, \tau)$ of its subtree τ is defined as:

$$\xi(x, \tau) = fit(x|\tau = 1) - fit(x). \tag{1}$$

Then, the weight (importance) of the subtree τ is defined as follows:

$$w(\tau) = \sum_{x \in \Omega} \xi(x, \tau) pow(x), pow(x) = \frac{g(x) - g_{min}}{g_{max} - g_{min}}, \tag{2}$$

in which Ω is the set of all the considered individuals, and $pow(x)$ is the normalised fitness of individual x. Let $fit(x)$ be the fitness of individual x and Ψ the set of all individuals that were evaluated in the source domain,

$$g(x) = \frac{1}{(1 + fit(x))}, g_{min} = min\{g(x)|x \in \Psi\}, g_{max} = max\{g(x)|x \in \Psi\} \tag{3}$$

The motivation behind Eq. (2) is to let good subtrees of good individuals have higher weights. In the target domain, the subtrees in the pool are sorted by their weights and the top subtrees form 50% of the initial population. The corresponding algorithms are named (1) *ContribSub-all* and (2) *ContribSub-subtree*.

3   Experimental Studies

A collection of experimental source and target domain settings are designed to evaluate the proposed methods. In our design, the difference between source and target domain is in terms of the number of vehicles. Several UCARP instances with different sizes are chosen to have a thorough investigation of knowledge transfer in different scenarios. GP settings and datasets in this paper are based on the work in (Mei and Zhang 2018). All algorithms are run 30 times independently. The compared algorithms include FrequentSub-all, FrequentSub-subtree (Ansari Ardeh et al. 2019), SubTree50 (Dinh et al. 2015), ContribSub-all and ContribSub-subtree and GPHH without any knowledge transfer. The reason for including SubTree50 is that it has the same pool of candidate trees as FrequentSub-subtree and ContribSub-subtree.

Figures 1 and 2 show the convergence curves of the test performance in the target domain. We conducted Wilcoxon's rank sum test to compare between the final test performance of the algorithms, and the results showed no significant difference.

Overall, we have the following observations:

- Subtree transfer can improve the efficiency of the retraining process of routing policies in the target UCARP domain.
- The contribution measure is an effective indicator for the reusability of subtrees, and can identify better subtrees to the target domain than the random selection and frequency-based selection.

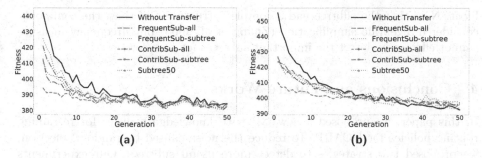

Fig. 1. Convergence curves of the compared algorithms on **gdb9** from 10 to (a) 9 and (b) 11 vehicles.

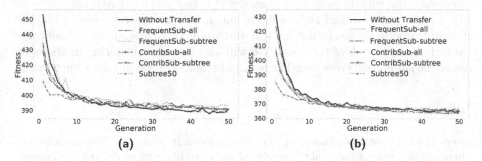

Fig. 2. Convergence curves of the compared algorithms on **val9C** from 5 to (a) 4 and (b) 6 vehicles.

– The frequency measure for both the "All" and "root" pools performed comparable with the random subtree selection for the "root subtrees" pool.

The possible reasons for the above observations are that the subtrees of the root are large and may not appear more than once. Thus, frequency-based method is very similar to random selection. If considering all the subtrees, then the small subtrees are more likely to have higher frequency, and tend to be selected. However, the frequency can be misleading as the occurrences can be in redundant branches but the contribution-based measure can handle this, and identify the truly important subtrees regardless of their frequency. Therefore, the contribution-based transfer methods can work better.

In our experiments, we noticed subtrees could receive different opinions from the frequency and contribution measures. For example, the subtree $max(max(min(FUT, FRT), CFH/FULL), CFH/FULL)$ appeared 47 times in the source domain and was transferred by the FrequentSub-all method. However, its contribution to one of its trees $min(CTT1, max(min(FUT, (max(DEM, DC)/max (CFR1/CTT1))/((FRT * CR)/DEM1)), min(FUT, FRT))) + (((CR + FULL) * (FUT/CTD) * (FULL * max(DC, FUT))) * ((max(max(min(FUT, FRT), CFH/FULL), CFH/FULL))/(max(min(CFD, RQ), RQ/DEM))/ RQ)$ was −47.57. Thus, ContribSub-all considered it as useless and did not transfer it. Note that the subtree had a complex structure and thus, the commonly considered algebraic simplification (Zhang et al. 2005) plus frequency measure cannot effectively detect the important subtrees for transfer.

4 Conclusions and Future Works

In this paper, we proposed a new GP with knowledge transfer for retraining routing policies for UCARP. To reduce the noise caused by random selection, we proposed two strategies to detect more useful subtrees. Our experiments showed that subtree transfer can effectively improve the efficiency of the retraining process, making GP achieve the desired performance in a much shorter time. Specifically, the contribution measure showed better efficiency and effectiveness than random selection and thus achieved much better convergence speed in the retraining process. In the future, we will develop more advanced tree transformation techniques to reduce the noise. We will also consider clustering methods to cluster the similar subtrees together to avoid transferring redundant knowledge.

References

Ansari Ardeh, M., Mei, Y., Zhang, M.: Transfer learning in genetic programming hyper-heuristic for solving uncertain capacitated arc routing problem. In: IEEE Congress on Evolutionary Computation (2019)

Dinh, T.T.H., Chu, T.H., Nguyen, Q.U.: Transfer learning in genetic programming. In: IEEE Congress on Evolutionary Computation (2015)

Mei, Y., Zhang, M.: Genetic programming hyper-heuristic for multi-vehicle uncertain capacitated arc routing problem. In: Proceedings of the Genetic and Evolutionary Computation Conference Companion, pp. 141–142 (2018)

Mei, Y., Tang, K., Yao, X.: Capacitated arc routing problem in uncertain environments. In: IEEE Congress on Evolutionary Computation (2010)

Mei, Y., Nguyen, S., Xue, B., Zhang, M.: An efficient feature selection algorithm for evolving job shop scheduling rules with genetic programming. IEEE Trans. Emerg. Top. Comput. Intell. **1**(5), 339–353 (2017)

Torrey, L., Shavlik, J.: Transfer learning. In: Handbook of Research on Machine Learning Applications and Trends: Algorithms, Methods, and Techniques, pp. 242–264. IGI Global (2010)

Zhang, M., Zhang, Y., Smart, W.: Program simplification in genetic programming for object classification. In: Khosla, R., Howlett, R.J., Jain, L.C. (eds.) KES 2005. LNCS (LNAI), vol. 3683, pp. 988–996. Springer, Heidelberg (2005). https://doi.org/10.1007/11553939_139

Image Recognition and Manipulation

Multi-label Recognition of Paintings with Cascaded Attention Network

Yue Li[✉], Tingting Wang, Guangwei Huang, and Xiaojun Tang

BOE Technology Group Co., Ltd., Beijing, China
{liyue111,wangtingt,huangguangwei,tangxiaojun}@boe.com.cn

Abstract. Convolutional neural networks (CNNs) have demonstrated advanced performance on image multi-label classification. However, recognizing labels of paintings is still a challenging problem due to the huge collection and labeling cost on painting training set. Inspired by the similarity between natural image and painting image, we propose an approach based on progressive learning to solve this issue by use of a few labeled paintings. In addition, we set up an effective framework built upon visual cascaded attention for multi-label image classification. Different from the existing approaches, the proposed model extracts and integrates multi-scale features to learn discriminative feature representations, which are then fed to the class-wise attention module with a simple scheme. Experimental results on the challenging benchmark MS-COCO dataset show that our proposed model achieves the best performance compared to the state-of-the-art models. We also demonstrate the effectiveness of the model on our constructed painting testing datasets (Datasets will be made publicly available soon.).

Keywords: Deep neural network · Multi-label recognition · Cascaded attention

1 Introduction

Multi-label classification of natural image has recently witnessed a rapid progress due to the large scale labeled datasets and the fast development of convolutional neural networks (CNNs) [1–5]. However, little research has been made in the label recognition of paintings as described in Fig. 1. The challenges of the task are mainly two-fold: one is from the variations and differences of the same object category among different paintings, and the other is the construction of a large painting dataset because it is much more difficult to collect painting images than natural images. To tackle these problems, we introduce a progressive learning scheme considering the similarity and difference between natural image and painting. The artificially generated painting based on image-to-image translation algorithm [6] is utilized to bridge the gap. To the best of our knowledge, this work represents the first attempt to achieve multi-label recognition of paintings by use of a few labeled paintings.

Y. Li and T. Wang—Equal contribution.

© Springer Nature Switzerland AG 2019
A. C. Nayak and A. Sharma (Eds.): PRICAI 2019, LNAI 11670, pp. 203–216, 2019.
https://doi.org/10.1007/978-3-030-29908-8_17

Fig. 1. Multi-label paintings. The variations and differences of the same object category among different paintings make the task of multi-label recognition of painting more challenging.

Recently, there have been attempts to apply the visual attention mechanism in multi-label classification [5,7–10]. Most related methods do not take full advantage of high resolution information of low-level features and semantic information of high-level features simultaneously. Some work [11] aggregates the predictions from multiple attention masks on different representation scales, however, multi-scale features are not fused and redundant computational cost is taken with additional learning parameters.

Inspired by the effective attention mechanism, we propose a cascaded attention framework built on multi-scale features extraction. Specially, the proposed model consists of the main net, multi-scale feature extraction module and attention branch, as shown in Fig. 3. Our attention mechanism includes two cascaded modules to learn attention maps. One is used to improve the discriminability of feature representations, and the other is class-wise feature-based attention module with a more lightweight (with less parameters) architecture.

Extensive comparative evaluations demonstrate the superiority of the proposed network over a wide range of state-of-the-art methods. Besides, our experimental results on the constructed painting dataset show that our method can make the multi-label recognition task for painting achievable and practical, without involving a large scale of annotated painting dataset.

To sum up, our main contributions are as follows: (I) We develop a multi-label painting classification framework with no need for a large number of annotated paintings as training set. (II) We further formulate a cascaded attention neural network by use of more than one scale of image features. The lightweight network is designed to learn discriminative class-specific features efficiently. (III) The experimental results show the superiority of the proposed model. The constructed painting dataset is used to evaluate the performance of the proposed multi-label classification scheme for paintings.

2   Methodology

Inspired by the similarity between natural images and paintings, we propose a transfer learning method to realize multi-label recognition of paintings. Here, most training images are from natural images and artificially generated images. Only a few real painting images are used.

2.1   Workflow Overview

Our goal is to generate labels of paintings through progressive learning from natural images to real painting images, as shown in Fig. 2(left). Assuming that the training dataset consists of N images and has C categories, the labeled training set is denoted as $\mathbb{1} = \{(I^1, y^1), \dots, (I^N, y^N)\}$, where I is the image data and corresponding C-dimensional label vector is $y = [y_1, \dots, y_i, \dots, y_C]^T$, $i \in \{0, \dots, C\}$. y_i is a binary value indicating whether the specific object is present in the image. The natural image training set is denoted as $\mathbb{1}_S$, and the real painting training set is collected as $\mathbb{1}_T$. The categories to be detected in $\mathbb{1}_T$ are a subset of the categories in $\mathbb{1}_S$.

The challenge mainly arises from the large gap between natural image and painting. We propose to bridge the gap with three main steps as depicted in Fig. 2(right): (1) content learning, (2) texture adaptation and (3) fine-tuning both content and texture. In content learning step, we train the multi-label classification network illustrated below using the natural images $\mathbb{1}_S$. In texture adaptation, the Image-to-Image Translation method [6] is employed to obtain artificially generated paintings. Afterwards, the model is fine-tuned on the generated paintings to learn the discriminative representations specific to painting. In the last adaptation step, we further fine-tune the network with real paintings $\mathbb{1}_T$ to increase the specificity of painting feature representation.

Remarks. The natural image represents the common content characteristics of different types of paintings. But for the generated image, it has the texture features of paintings, but loses content information details. So only using the synthetic and real paintings for the model training without natural images will result in worse performance.

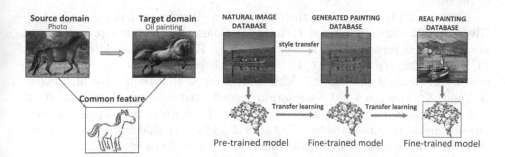

Fig. 2. Left: content feature similarity exists between the natural image and painting of the same object category, so the label recognition of paintings can be realized through progressive learning. Right: the progressive representation adaptation workflow of our framework.

2.2   Proposed Multi-label Classification Network

Framework Overview. The proposed multi-label recognition model is a multi-branch scheme composed of three components: a main net, a multi-scale feature extraction module and a cascaded attention learning sub-network. The feature extraction module employs a stack of features at different scales, which are then fed into the cascaded attention module to produce class-specific feature maps. Figure 3 shows the entire network of our proposed model.

More specifically, we select the ResNet-101 [14] for building the main net modeled as $f_{cnn}(\cdot)$. It extracts the visual features from input image I, then ends with a global average pooling layer and a fully-connected (FC) layer with C-dimensional output $\hat{y}_1$, where C is the number of labels of interest.

For the local branch, the feature extraction module is first applied to combine local context information with semantically strong features. Then the cascaded attention network is attached to explore features with respect to each category. As illustrated in the bottom of Fig. 3, it consists of two attention modules: (1) the module to fuse and strengthen the feature representations, and (2) the class-wise attention feature extraction model. The local branch ends up with a global sum pooling layer to get the C-dimensional output. At last, the predictions from the attention branch and the main net classifier are aggregated at a score level.

Multi-scale Features Extraction. For the main net, the activations of lower layers are more accurately localized and higher layers extract the better semantic information, which motivates us to incorporate multi-layer information to infer the visual attention features. Using the feature maps from the second or third stage with shallow neural networks will result in bad performance. In consideration of both resolution and expense of computing resources, we opt for the feature maps 14×14 and 7×7 from the outputs of the fourth and fifth stage respectively.

The top of Fig. 3 illustrates the feature extraction block denoted as $f_{ex}(\cdot)$. Given the 224×224 input image I, the three-dimensional feature representations are denoted as tensors $X^1 \in \mathbb{R}^{H_1 \times W_1 \times C_1}$, $X^2 \in \mathbb{R}^{H_2 \times W_2 \times C_2}$, and $H_1 = W_1 = 14$, $C_1 = 1024$, $H_2 = W_2 = 7$, $C_2 = 2048$, where H_i, W_i, and C_i denote the number of pixel in the height, width and channel dimensions. We first make X^1 and X^2 undergo 1×1 convolution layer to reduce dimensions from 1024 and 2048 channels to 256 channels separately. Then we directly up-sample the low-dimension feature maps by a factor of 2 and the spatial size is the same as the finer feature maps. Finally, the different levels of features are concatenated to achieve the different scales aggregation. The concatenated feature maps are denoted as $X^C \in \mathbb{R}^{14 \times 14 \times 512}$, which fuse these features under two different scales to obtain both semantically strong and high-resolution feature representations.

Cascaded Attention Network. In order to refine more discriminative features and accelerate the model convergence, we embed a feature enhancement module before the class-wise attention learning subnet. As depicted in the bottom half of Fig. 3, such stacking mechanism can naturally be regarded as a sequential refinement process of generating class specific attention maps.

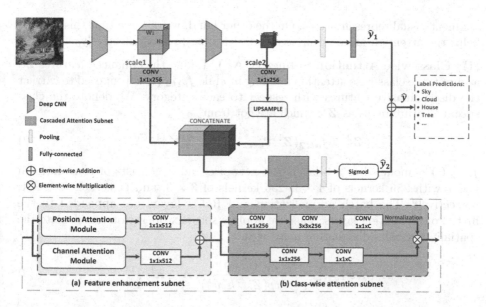

Fig. 3. An overview of our network architecture. Given an input image, deep CNN (Top) is used to get feature maps and get predictions from fully-connected layer. Then a multi-scale feature extraction module is applied to form the feature representations which carry both local context and semantically strong information. After that, the concatenating features are fed into the cascaded attention network (Bottom) which consists of feature enhancement subnet (a) and class-wise attention subnet (b). Finally, the combination of outputs of main net and attention branch is taken as the final label prediction. The ReLU and Batch Normalisation (applied to each conv layer) are not shown for brevity.

(I) Feature Enhancement Subnet (FES). For the feature enhancement subnet modeled as $f_{att1}(\cdot)$, we first generate the strengthened feature representations using the position attention module (PAM) and channel attention module (CAM) proposed in the works as [13]. It is convenient to embed the two blocks in the network pipeline with only a small number of additional parameters. PAM generates new features of spatial long-range dependencies through modeling the spatial relationship between any two pixels of the features. CAM captures the long-range dependencies through calculating the correlation matrix between channels. In $f_{att1}(\cdot)$, we apply a convolution layer with 512 kernels of 1×1, batch normalization and ReLU layers to follow PAM and CAM respectively, then perform an element-wise sum to accomplish feature fusion. Given the input visual features $\boldsymbol{X}^C \in \mathbb{R}^{14 \times 14 \times 512}$, our attention model extracts the strengthened representations $\boldsymbol{Z}^1$ by applying the function expressed as

$$\boldsymbol{Z}^1 = \boldsymbol{f}_{att1}(\boldsymbol{X}^C; \theta_{att1}), \boldsymbol{Z}^1 \in \mathbb{R}^{14 \times 14 \times 512}. \tag{1}$$

FES module plays dual roles in our model. On one hand, the multi-scale features are fully fusion in the spatial (inter-pixel) and channel (inter-scale) dimensions to augment visual representations. On the other hand, we observe that this module helps the attention mechanism accelerate learning process effectively.

(II) Class-wise Attention Subnet (CAS). Taking the enhanced features $\boldsymbol{Z}^1$ as input, the class-wise attention learning module $f_{att2}(\cdot)$ is proposed to extract the discriminative features with respect to each category. We denote the class specific feature maps as $\boldsymbol{Z}^2$, which are got from

$$\boldsymbol{Z}^2 = f_{att2}(\boldsymbol{Z}^1; \theta_{att2}), \boldsymbol{Z}^2 \in \mathbb{R}^{14 \times 14 \times C}. \tag{2}$$

$f_{att2}(\cdot)$ is modeled as a 3-layers sub-network, and it consists of convolutional layers with 256 kernels of 1×1, 256 kernels of 3×3, and C kernels of 1×1, respectively. The batch normalization and ReLU layers are appended to the first two convolution layers. Then we apply a softmax operation to calculate the spatially normalized attention maps with

$$\boldsymbol{A}_c(h, w) = \frac{exp(\boldsymbol{Z_c}^2(h, w))}{\sum_{h,w} exp(\boldsymbol{Z_c}^2(h, w))}, \boldsymbol{A} \in \mathbb{R}^{14 \times 14 \times C}, \tag{3}$$

where $\boldsymbol{Z_c}^2$ represents the unnormalized value at (h, w), and c corresponds to the class label. For each label c, the corresponding attention feature map has the property of $\sum_{h,w} \boldsymbol{A}_c(h, w) = 1$ and the most relevant region is salient.

Following previous attention mechanism idea in SRN [5], a mask branch also with the features $\boldsymbol{Z}^1$ as input, is designed to assign the label confidence weights to the class-wise attention features. The weighted attention learning aims at discriminating the label confidence information at different spatial positions. The mask branch comprises of two consecutive 1×1 convolution layers and aims to produce salient confidence maps $\boldsymbol{S} \in \mathbb{R}^{14 \times 14 \times C}$ with the same size as attention maps $\boldsymbol{A}$. Different from main net classifier, the classification output $\widehat{y}_2$ of the attention subnet is computed through element-wise multiplying attention maps $\boldsymbol{A}$ and label confidence maps $\boldsymbol{S}$, and then followed by a global pooling.

Finally, we make the combination of the outputs of main net and attention branch as $\widehat{y} = \alpha \widehat{y}_1 + (1 - \alpha) \widehat{y}_2$, where α is a weighting factor. Then $\widehat{y}$ is taken as the final label prediction of an image.

Remarks. The proposed attention model is conceptually similar to SRN [5] because both are designed as multi-branch scheme with aggregating the outputs of two branches. However, they differ significantly in design: (1) SRN only uses one scale of feature map, and applies some extra layers to refine features as the input of attention branch, while the model we designed directly extracts the multi-scale features from main net. The latter has smaller parameters but more information is utilized. (2) SRN explores semantic and spatial relations of labels through learning spatial regularizations, which makes the network deeper. Our model embeds a module with fewer parameters to fuse features in spatial and channel dimensions to strengthen visual representation. The latter model is trained easier and faster. (3) Compared to SRN, our attention scheme is designed to get outputs without using fully-connected layer.

2.3   Training Scheme

We organize our network training in a progressive learning mode with three steps via Stochastic Gradient Descent method (SGD). The loss term is defined as sigmoid cross-entropy loss function for multi-label image classification.

$$L_s(y, \widehat{y}) = \sum_{c=1}^{C} y^c log\sigma(\widehat{y}_c) + (1 - y^c)log(1 - \sigma(\widehat{y}_c)) \tag{4}$$

where y^c and $\widehat{y}^c$ denote the ground truth and prediction of the c-th label, and $\sigma(\cdot)$ is the sigmoid activation function.

(1) Train the network on natural image set. First, we initialize main net except FC layer with pre-trained parameters of ResNet-101 network on ImageNet [15], and randomly initialize the weights of the modified classification layers and added parameters. Only main net $f_{cnn}(\cdot)$ is trained with cross-entropy loss $L_1(y, \widehat{y}_1)$. Second, main net is frozen and we only train all other sub-networks as $f_{ex}(\cdot)$, $f_{att1}(\cdot)$ and $f_{att2}(\cdot)$ with cross-entropy loss $L_2(y, \widehat{y}_2)$. Finally, the whole network is fine-tuned to get model M_1 with loss term

$$L(y, \widehat{y}) = L_1(y, \widehat{y}_1) + L_2(y, \widehat{y}_2), \tag{5}$$

(2) Fine tune the model on artificially generated paintings. Taking M_1 as pre-trained model, we train end-to-end to fine tune the whole network to obtain model M_2.

(3) Fine tune the model on the real paintings. Based on model M_2, the whole network is fine-tuned to obtain the final learned model M_3.

3   Experiments

Our experiment consists of two parts. (I) To evaluate the proposed network architecture, we carry out experiments on benchmark dataset MS-COCO [16]. (II) To verify the effectiveness of transfer learning approach for painting label recognition, we evaluate the method on our collected datasets including natural images, watercolors and oil paintings.

3.1   Dataset, Evaluation Metrics and Implementation

Dataset Description. MS-COCO dataset is primarily built for object recognition task in the context of scene understanding, also used for multi-label recognition without bounding box information. The training set is composed of 82,783 images and the validation set is composed of 40504 images. The dataset covers 80 classes of common objects in the scenes.

As there is no suitable dataset for transfer learning from natural image to painting, we construct three datasets: natural image dataset, generated painting dataset and a small dataset of real paintings. We collect and annotate 10,000

Fig. 4. Example paintings generated by CycleGAN which makes natural image with painting style (oil painting and watercolor).

photography images from the search engines by crawlers. These images contain 12 classes that usually appear in art paintings. We collect and annotate 1200 oil paintings and 1200 watercolors as our real painting dataset which has the same categories as natural images.

Besides, the generated oil paintings and watercolors from natural images are needed. We employ an image-to-image translation method CycleGAN [6] to train a model to achieve this, as shown in Fig. 4. The color and texture of generated paintings almost imitate the real paintings, and most of the edges and semantics of the natural images are kept. Then 10,000 generated oil paintings and 10,000 generated watercolors are respectively got based on the image-to-image translation model and 10,000 photography images above.

Implementation Details. We train our proposed model with Pytorch. All the experiments in this paper are conducted on a NVIDIA Tesla P100 GPU, 16 GB. For MS-COCO and the natural image dataset, we employ SGD algorithm with multi-step learning strategy, with a batch size of 24, a momentum of 0.9 and weight decay of 0.0005. The initial learning rate is set as 0.001, 0.01 and 0.0005 respectively during training steps, and the epoch is set as 7, 15 and 15 respectively. For the generated painting dataset, the fine tuning process is with initial learning rate of 0.0005 and 10 epochs. For the real painting dataset, the fine tuning process is with initial learning rate of 0.00005 and 10 epochs.

3.2   Results on MS-COCO

Baselines. We compare our results with the following state-of-the-art methods: RIA [4], CNN-RNN [2], LESP [12], RLSD [3], Resnet-101 and SRN [5] on the MS-COCO dataset in Table 1. We select the top-3 labels for each image to compute

Table 1. Evaluation results with top-3 label predictions of our proposed approach and other methods on MS-COCO validation set. The method with asterisk indicates that it is our re-implementation for fair. "mAP", "F1-C", "P-C" and "R-C" are evaluated for each class before averaging. "F1-O", "P-O" and "R-O" are averaged over all sample-label pairs.

Method	mAP	F1-C	P-C	R-C	F1-O	P-O	R-O
LESP [12]	-	63.8	73.5	56.4	68.3	76.3	61.8
RIA [4]	-	58.7	64.3	54.1	69.1	74.2	64.5
CNN-RNN [2]	-	60.4	66.0	55.6	67.8	69.2	**66.4**
RLSD [3]	67.4	62.0	67.6	**57.2**	66.5	70.1	63.4
ResNet-101*	73	63.6	83.4	54.6	70.1	86.1	59.1
SRN*	74.2	64.6	83.2	55.7	70.7	85.7	60.1
Ours	**74.7**	**65.1**	**84.6**	55.9	**71.1**	**86.7**	60.3

Table 2. Ablation experiments with threshold predictions on MS-COCO validation set to analyze the impact of each submodule to the final result. "mAP" and "F1-C" are evaluated for each class before averaging. "F1-O" is averaged over all sample-label pairs. FC: Fully-connected layer. CAS: Class-wise Attention Subnet. FES: Feature Enhancement Subnet.

Main Net	FC	CAS	FES	F1-C	F1-O	mAP
ResNet-101	✓			67.0	72.6	73.0
ResNet-101		✓		67.5	73.1	73.7
ResNet-101		✓	✓	67.9	73.4	73.9
ResNet-101	✓	✓	✓	**68.6**	**73.8**	**74.7**

metrics following [2]. We can see that the baseline models ResNet-101 and SRN show a superior performance over many other recent models. In order to allow for a fair comparison with the rest of methods, we re-implement the both methods ResNet-101 and SRN (there is an asterisk marked in Table 1) on the MS-COCO dataset.

Evaluation Results. As shown in Table 1, our proposed approach achieves F1-C score of 65.1%, F1-O score of 71.1% and boots mAP to 74.7%. The improvement of mAP metric reaches almost 1.7% and 0.5% compared with ResNet-101 and SRN method respectively. From the results above, our method performs better than state-of-the-art approaches, and shows an improvement over SRN with a smaller model size and a simpler training scheme.

Ablation Experiment. In order to investigate the contributions of different components of our work, we further perform ablation experiments with threshold criterion as described in Table 2.

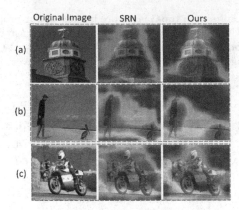

Fig. 5. Attention map visualizations for SRN model and our model in COCO. Red region corresponds to high activation for class. Figure best viewed in color.

- Main Net + FC: only using main net and fully connected layer, which is taken as baseline.
- Main Net + CAS: using main net and attention mechanism only with CAS.
- Main Net + CAS + FES: using main net and cascade attention mechanism.
- Main Net + FC + CAS + FES: using main net, FC and cascaded attention mechanism.

Compared with the baseline, the other three architectures enhance the mAP metrics gradually. The attention branch based on multi-scale features fusion (the second row in Table 2) boots 0.7% in terms of mAP, whereas the cascaded attention mechanism (the third row in Table 2) boots 0.9% which verifies the efficiency of FES module. Ensemble mechanism of main net and cascaded attention mechanism (last row in Table 2) is proved to be with the best performance and boots mAP to 74.7%.

Visualization and Analysis. We visualize and analyze the output of the proposed cascaded attention modules compared with SRN attention maps. Some examples on COCO are shown in Fig. 5. The attention maps highlight discriminative areas for different categories and exhibit almost no activations with respect to absent classes. It is observed that our model is highly class-discriminative and more effective for co-existing labels in the same image. For example in Fig. 5(b), the region between "person" and "bird" is also highlighted in SRN attention map, whereas the proposed class attention learning is capable of tracking the object localization precisely. The final predicted result is highly related with the localization precision of attention maps.

3.3 Results on Painting Dataset

Oil Painting Dataset. To evaluate the effectiveness and contribution of each step of progressive learning approach, we train three models corresponding to

Table 3. Progressive learning results: mAP for Oil-painting1K testing set.

	Model	Datasets	mAP
Training Process	M_1	Natural images	84.3
	M_2	Generated paintings	85.2
	M_3	Real oil paintings	87.2
	M_2'	Only generated paintings	83.0

Table 4. Progressive learning results: mAP for Watercolor1K testing set.

	Model	Datasets	mAP
Training Process	M_1	Natural images	82.8
	M_2	Generated paintings	84.0
	M_3	Real watercolors	85.7
	M_2'	Only generated paintings	82.4

three steps respectively: the model M_1 trained only using 10,000 natural images, the model M_2 fine-tuned on M_1 using 10,000 generated oil paintings, and the model M_3 fine-tuned on M_2 using 200 real oil paintings. The performance of each model is examined based on testing set composed of 1,000 real oil paintings.

The progressive training results are displayed in Table 3. It can be seen that fine-tuning the model on both generated paintings and real oil paintings could provide an obvious improvement in terms of mAP. The last fine-tuning step boots mAP metric from 84.3% to 87.2%. Note that we also train the model M_2' only on generated paintings and the mAP metric is 83.0% which is lower than M_2. This result proves that the pre-trained process on natural images is quite necessary.

Figure 6(left) shows AP metric of each category in progressive learning process. It can be seen that AP metrics of most classes are gradually improved. Especially for the classes of "person", "hill", "house", "boat", "bridge", "horse", "flower" and "table", the AP metric of model M_3 rises more than 3% compared with M_1. The AP metrics of a few classes decline slightly after fine-tuning the model with generated paintings, which is caused by the large differences of some object representations between style-transfer and real painting.

Watercolor Dataset. The evaluation of our method on watercolor dataset adopts the same process as oil paintings. Table 4 displays that the last fine-tuning step boots mAP metric from 82.8% to 85.7%. We also train the model M_2' only on generated paintings and the mAP metric is 82.4% which is lower than M_2. From Fig. 6(right), we can see that AP metrics of all classes rise after transfer learning except for the classes of "water" and "bridge", and we believe that is because of the big difference between watercolor and natural image on the color and texture for the objects of "water" and "bridge". To solve this problem, more training images of real paintings are needed for these

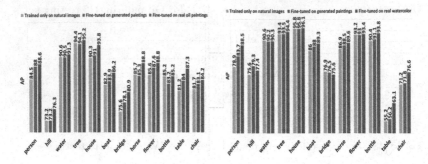

Fig. 6. Progressive learning results. Left: each category AP metric for Oil-painting1K testing set. Right: each category AP metric for Watercolor1K testing set.

kinds of objects. In brief, the results above demonstrate the effectiveness of our multi-label painting recognition method, which does not need a large dataset of annotated real paintings.

Visualization and Analysis. We further compute the class-wise heat maps on all 12 oil-painting categories visualizations learned from the proposed attention layers. The class attention maps are shown in Fig. 7, where the left is original oil painting and the predicted label score is above each feature map. As we can see, the attention maps highly correlate the scores of predicted labels and exhibit excellent spatially localization characteristics. The maps highlight discriminative regions for positive classes, for example, the first row leads to positive predictions. From the visualizations, we can see that the class "bridge" at the second row is a fake label but activated because of the similarity between "bridge" and "housetop" in the painting. Besides the classification, the proposed attention modules can also be utilized in the field of weakly supervised object detection.

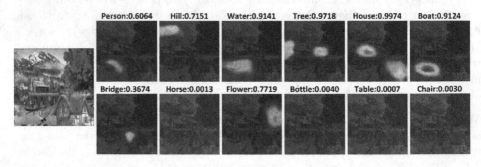

Fig. 7. Class attention maps of an oil painting example with respect to different classes, with predicted label and score above the attention map. Figure best viewed in color.

4 Conclusion

In this paper, we aim to achieve the multi-label painting recognition without a large scale painting training set. It is solved by transfer learning and an effective classification framework which is based on multi-scale features extraction and cascaded attention scheme. The experimental results on the MS-COCO dataset and our constructed painting datasets demonstrate that the proposed approach achieves superior performance to the state-of-the-art methods.

References

1. Krizhevsky, A., Sutskever, I., Hinton, G.E.: Imagenet classification with deep convolutional neural networks. In: Advances in Neural Information Processing Systems, pp. 1097–1105 (2012)
2. Wang, J., Yang, Y., Mao, J., Huang, Z., Huang, C., Xu, W.: CNN-RNN: a unified framework for multi-label image classification. In: Proceedings of the IEEE Conference on Computer Vision and Pattern Recognition, pp. 2285–2294 (2016)
3. Zhang, J., Wu, Q., Shen, C., Zhang, J., Lu, J.: Multi-label image classification with regional latent semantic dependencies. IEEE Trans. Multimedia 20, 2801–2813 (2018)
4. Jin, J., Nakayama, H.: Annotation order matters: recurrent image annotator for arbitrary length image tagging. In: 2016 23rd International Conference on Pattern Recognition (ICPR), pp. 2452–2457. IEEE (2016)
5. Zhu, F., Li, H., Ouyang, W., Yu, N., Wang, X.: Learning spatial regularization with image-level supervisions for multi-label image classification. arXiv preprint arXiv:1702.05891 (2017)
6. Zhu, J.-Y., Park, T., Isola, P., Efros, A.A.: Unpaired image-to-image translation using cycle-consistent adversarial networks. arXiv preprint (2017)
7. Chu, X., Yang, W., Ouyang, W., Ma, C., Yuille, A.L., Wang, X.: Multi-context attention for human pose estimation. arXiv preprint arXiv:1702.07432, 1(2) (2017)
8. Chen, S.-F., Chen, Y.-C., Yeh, C.-K., Wang, Y.-C.F.: Order-free RNN with visual attention for multi-label classification. arXiv preprint arXiv:1707.05495 (2017)
9. Simonyan, K., Zisserman, A.: Very deep convolutional networks for large-scale image recognition. arXiv preprint arXiv:1409.1556 (2014)
10. Jaderberg, M., Simonyan, K., Zisserman, A., et al.: Spatial transformer networks. In: Advances in Neural Information Processing Systems, pp. 2017–2025 (2015)
11. Sarafianos, N., Xu, X., Kakadiaris, I.A.: Deep imbalanced attribute classification using visual attention aggregation. arXiv preprint arXiv:1807.03903 (2018)
12. Li, Y., Song, Y., Luo, J.: Improving pairwise ranking for multi-label image classification, pp. 1837–1845 (2017)
13. Fu, J., Liu, J., Tian, H., Fang, Z., Lu, H.: Dual attention network for scene segmentation. arXiv preprint arXiv:1809.02983 (2018)
14. He, K., Zhang, X., Ren, S., Sun, J.: Deep residual learning for image recognition. In: Proceedings of the IEEE Conference on Computer Vision and Pattern Recognition, pp. 770–778 (2016)

15. Deng, J., Dong, W., Socher, R., Li, L.-J., Li, K., Fei-Fei, L.: Imagenet: a large-scale hierarchical image database. In: IEEE Conference on Computer Vision and Pattern Recognition, CVPR 2009, pp. 248–255. IEEE (2009)
16. Lin, T.-Y., et al.: Microsoft COCO: common objects in context. In: Fleet, D., Pajdla, T., Schiele, B., Tuytelaars, T. (eds.) ECCV 2014. LNCS, vol. 8693, pp. 740–755. Springer, Cham (2014). https://doi.org/10.1007/978-3-319-10602-1_48

Stacked Mixed-Scale Networks
for Human Pose Estimation

Xuan Wang[1], Zhi Li[2(✉)], Yanan Chen[3], Peilin Jiang[2], and Fei Wang[1]

[1] School of Electronic and Information Engineering,
Xi'an Jiaotong University, Xi'an, China
xwang.cv@gmail.com, wfx@mail.xjtu.edu.cn
[2] School of Software Engineering, Xi'an Jiaotong University, Xi'an, China
maleficentlee@gmail.com, pljiang@xjtu.edu.cn
[3] Megvii Inc. (Face++), Beijing, China
chenyanan@megvii.com

Abstract. Human pose estimation is an important problem in computer vision, which has been dominated by deep learning techniques in recent years. In this paper, we propose a novel model, named Mixed-Scale Dense Block, that exploits dilation convolution layers and dense concatenation connections to maximise the information flow through the block. Consequently, it captures the feature representation in different scales more effectively and efficiently. Comparing with the baseline method, Hourglass models, our model employs fewer learning parameters. Nevertheless, experiments demonstrate that the proposed model produces more accurate predictions. Meanwhile, our method achieves the comparable accuracy to state-of-the-art techniques. Especially in some indicators, our approach has better performance. In addition, this model is easy to implement and could be improved by most existing techniques that are adopted to promote the hourglass models.

1 Introduction

Estimating the location of keypoints of human body given a single RGB image, which is defined as human pose estimation, is an important problem attracting lots of attention in computer vision community. Obtaining the accurate human pose often serves as the crucial part in many research fields such as entertainment interaction, animation and action recognition.

Many early approaches adopt tree-structured graphical models [1,7], and formulate the pose estimation as the inference problem on the graph. The spatial dependencies between adjacent joints are encoded. In the recent years, the research of pose estimation has shifted to Deep Neural Network (DNN) techniques, especially the Convolutional Neural Network (CNN), which have already made the great progress on most computer vision tasks. One of the most popular methods, in pose estimation, is Stacked Hourglass Network [16] with encoder-decoder structure and skip connections. In conjunction with the use of intermediate supervision, the stacked multiple hourglass networks are trained to capture

© Springer Nature Switzerland AG 2019
A. C. Nayak and A. Sharma (Eds.): PRICAI 2019, LNAI 11670, pp. 217–229, 2019.
https://doi.org/10.1007/978-3-030-29908-8_18

and consolidate the information across different scales. The design of hourglass networks are adopted as the basic unit or extended by many other approaches.

In this work, we propose a novel architecture, called Mixed-Scale Dense (MSD) network (or block), that can effectively and efficiently learn the feature representation across all scales more. Figure 1 shows a demonstration of our MSD block. The block is formed with a few groups of dilation convolution layer(s). Without using the statistical down-sampling operation, adopted in classical encoder-decoder architecture, the detailed spatial information is naturally preserved. It mixes scales within each group via densely concatenates outputs of all the groups together. Hence in the block, the information flow between groups is maximised. Specifically, the information passing about input and gradient are more effective and efficient. Therefore, less learning parameters are exploited in the block, comparing with the standard hourglass network. Given the designed MSD block, we construct the entire network by stacking multiple blocks together end-to-end. Relying on the intermediate supervision, the stacked MSD blocks repeatedly infer the human pose. Finally, the accurate estimation is yielded by the last block.

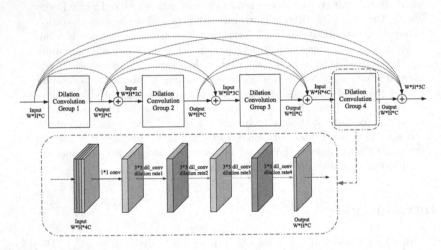

Fig. 1. Demonstration of an MSD block.

We demonstrate that our method is more effective and efficiency by evaluate it on standard benchmark. The experimental results show that our network outperforms the state-of-the-art approaches. Comparing against the hourglass network, only 1/2 parameters are employed though, the presented method still yields more accurate results on standard benchmark [2].

2   Related Works

Coming with the popularities of deep learning techniques in computer vision, the CNN-based methods add a huge boost to the field of human pose estimation.

A variety of architectures are introduced to tackling this challenging task. We follow the same pipeline of the CNN-based methods. A novel architecture is presented to tackle this challenging problem by effectively and efficiently capturing the feature representation in all the scales.

In Yu and Koltun [22], Dilation Convolution (also called atrous convolution) is proposed to solve the problem of semantic image segmentation. The main advantages of dilation convolution is two-fold. On the one hand, by removing the down-sampling operations in network, the detailed spatial information is preserved. On the other hand, the resolution of intermediate image can be controlled, when enlarging the receptive field.

To capture more information across multiple scales, the deeper networks are desired. However, this often leads to the vanishing-gradient problem. ResiNets [8], Highway Networks [18], Stochastic depth [10] and FractalNets [13] tackle this problem by creating short paths between layers. A more efficient approach, DenseNets [11] is presented. The featuremaps are concatenated with all the subsequent layers, such that it can exploit less parameters to reach the better performance on public datasets.

Recently, several deep learning techniques [3–6, 12, 14, 16, 17, 20, 21] have been proposed to tackle the problem of human pose estimation, the most adopted one of which is stacked hourglass models in Newell *et al.* [16]. It resembles several hourglass networks, in which successive steps of pooling and up-sampling are employed to produce a set of predictions. The encoder-decoder model captures multi-scale features, and the skip connections have to be exploited to preserve the detailed information. Based on the hourglass models, several adaptations [6, 17, 21] have been made. Chu *et al.* [6] rely on CRF (Conditional Random Field) based attention map and increase the network complexity to extend the standard hourglass network. In Ning *et al.* [17], the inception-resnet is employed as the building block in hourglass design. The Pyramid Residual module [21] is added to the network to promote the performance. In addition to the hourglass-based approaches, in Luvizon *et al.* [14], the residual separable convolution is proposed to replace the residual block in the stacked hourglass model. Moreover, some approaches relying adversarial networks are proposed to tackle the problem of human pose estimation and reach a good performance.

3 Design of Mixed-Scale Dense Networks

Most existing human pose estimation approaches fall into two categories: detection based methods and regression based methods. In this paper, we aim at the detection based method that produces the likelihood heatmap for each joint of human body. In hourglass networks, the encoder-decoder architecture is exploited to capture multi-scale features. However, using the statistical sampling operation, E.g.max pooling, the details of spatial information could not be preserved. Relying on the skip connection of matching featuremaps, the information flow in hourglass networks nonetheless can be promoted. To this end, we present a novel architecture, called mixed-scale dense block. The block consists of several groups of the dilation convolution layer, which are exploited to capture multi-scale features. By concatenating the output of all the groups, the feature-maps across all the scales are densely mixed. Figure 2 briefly demonstrates the full configuration of our network.

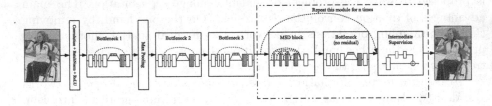

Fig. 2. Demonstration of our full network.

Dense Connection. Similar to previous methods [11, 16], our network is formed with multiple stacked basic blocks. Inspired by ResNets [8], hourglass networks employs skip connections to avoid vanishing gradient problem. The skip connection bypasses the non-linear transformations by connecting the early layer and the later layer. Relying on the skip connection, the network can be designed extremely deep. This brings great capacity of the feature representation, however the number of learning parameters explosively increases. Relying on dense connections [11] is an efficient method of drawing the representational power. Instead of using deep or wide architectures, the information flow is improved by densely connecting the featuremaps. Motivated by DenseNets, in MSD block, the outputs of all the groups of dilation convolutional layers are connected. Such that, the comparable performance can be achieved by using significantly fewer parameters.

Dilation Convolution. In DenseNets, the existence of pooling layers makes the use of concatenation operation unavailable, because the size of feature-maps changes. To tackle this problem, the DenseNets is divided into multiple dense blocks. The pooling layers are only placed between the adjacent dense blocks, hence the feature-maps in a single block have the same size and can be concatenated. However, these connections can only maximize the in-block information flow. In these blocks, the scale of feature-maps changes slightly. Instead, we make use of the dilation convolution layer and remove the pooling layer from our network. The adopted dilation convolution layers can enlarge the receptive field and preserve the resolution of feature-maps simultaneously. As a result, we can concatenate the feature-maps produced by all the groups of dilation convolution layers together. In addition, the feature-maps in each block of our network are captured in all the scales. By another words, the information flow between different scales is improved.

Intermediate Supervision. Our method adopts the intermediate supervision similarly to hourglass networks. Nevertheless, the more effective and efficient

MSD blocks are employed as the building blocks in such framework. Following the standard pipeline, the intermediate inference yielded by each block is supervised by applying a loss to it. Given the use of intermediate supervision, the high level features and high order spatial relationships are repeatedly processed through the consecutive blocks. Incorporating MSD blocks, the presented network produces more accurate estimation than the state-of-the-art approaches.

Layer Implementation. In our experiments, several mixed-scale dense (MSD) blocks are stacked to fully extract and utilise the feature information of different scales contained in the images. Before entering the MSD blocks, the input images are firstly convoluted, batch normalised and ReLU activated to convert the 3 image channels to 64 feature channels. Then, the features go through a sequence of convolutional layers to bring their number of channels from 64 up to 256. This sequence contains three bottleneck layers (whose configurations are shown in Figure 3 bottleneck), each with its own residual connection, and between the first and second bottleneck layer there is a size 2, stride 2 max-pooling layer. Here the bottleneck layer indicates a sequence of layers which contains three convolutional layers, respectively of size 1×1, 3×3 and 1×1, together with their batchnorm and ReLU layers. Behind the third convolutional layer and its batchnorm layer, we specially insert an SE layer [9], then the ReLU layer is appended. After that, our MSD blocks with bottleneck layers and residual connections officially show up. For MSD blocks, there are several groups of dilated convolutional layers (each dilated convolutional layer is followed by a batchnorm and a ReLU layer), each with its own dilation rate set defined beforehand (e.g. [1], [1,2], [1,2,4], [1,2,4,8], etc.). Each time the features produced by the last group of dilated convolutional layers are entering the next group, they are firstly concatenated with every set of features produced by all of the previous dilated convolutional layer groups. That is to say, every group of dilated convolutional layers is utilising the features produced by all of the previous groups, thus the feature information with different scales is shared between every two groups. Next, the features go through a bottleneck layer. Now at this point, the output produced by the bottleneck layer and the input before entering the MSD blocks are added together to create a residual connection. Then, the output is remapped by a 1×1 convolution and added back in order to implement the intermediate supervision. To achieve better performance, this MSD blocks + bottleneck layers + intermediate supervision setting is repeated for several

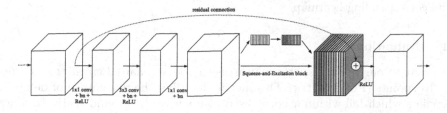

Fig. 3. Demonstration of a bottleneck layer (with residual connection).

times, which resembles the configuration in the stacked hourglass network, in which several "hourglasses" are "stacked" (Fig. 3).

4   Experiments

In this section, several experiments are conducted to evaluate our approach by comparing with the state-of-the arts and doing ablation studies.

Datasets. Our network is trained on the well-adopted MPII human pose dataset [2] which consists of about 25k images of multiple people with annotations, providing altogether 40k annotated samples (about 28k for training and 11k for testing). Since MPII doesn't provide annotations for the test set, we validate our results on a subset of the training set containing about 3,000 images. Each provided annotation consists of 16 landmarks on the whole human body, which marks the 16 body joints of a certain person.

Experimental Settings. The network is implemented by PyTorch, and the code for handling the data follows closely to the PyTorch implementation of the stacked hourglass network [16]. The images are cropped around the target person using the scale and centre annotations provided by the dataset, and then resized to 256×256 pixels. We also augmented the data by 0.75–1.25 random scaling, +/-30° random rotation and left-right flipping.

In our experiments there are two settings inside a single MSD block. Setting I is that one MSD block contains 10 dilated convolutional layer groups, and the groups iteratively pick the dilation rate arrays in the set [[1], [1, 2], [1, 2, 4], [1, 2, 4, 8], [1, 2, 4, 8, 16]]. In setting II, there are 18 dilated convolutional layer groups in a single MSD block, and the groups iteratively follow the dilation rate arrays of [[1], [1, 2], [1, 2, 4], [1, 2, 4, 8], [1, 2, 4, 8, 12], [1, 2, 4, 8, 16], [1, 2, 4, 8, 16, 24]].

Training Details. We train our network for 90 epochs with a learning rate of 2.5e−4 and rmsprop optimization. For the loss function we adopt the Mean Squared Error (MSE) loss between the predicted heatmap and the groundtruth heatmap of each joint. A 8-stack network with setting I takes about 2 days on two 11 GB NVIDIA 1080ti GPUs, and setting II with 8 stacks takes about 4 days. For evaluation we adopt the same trick as is in [16] that the heatmaps are generated on both original and flipped versions of the input images and averaged together, and the final prediction for a given joint is the max activating location of its corresponding heatmap.

4.1   Evaluation

We evaluate our network performance using the standard Percentage of Correct Keypoints (PCK) metric. This metric describes the percentage of detected keypoints which fall within a normalised distance of the ground truth. Here we adopt the PCKh configuration, in which the distance threshold is set to 0.5, and the fraction of head size is used to normalise the distance.

Table 1. PCKh@0.5 of an 8-stack hourglass network (hg-s8) and our 8-stack mixed-scale dense network with setting II (msd2-s8). Δ indicates how much better our approach is to the hg-s8 network. The results are obtained on validation set.

Method	Head	Sho.	Elb.	Wri.	Hip	Knee	Ank.	**Mean**
hg-s8	96.59	94.97	89.02	83.57	87.45	83.48	78.53	**87.72**
msd2-s8	96.86	95.84	90.80	86.33	89.30	86.34	82.40	**89.78**
Δ	+0.27	+0.87	+1.60	+2.76	+1.85	+2.86	+3.87	**+2.06**

Fig. 4. Example predictions of our MSD network.

Table 1 shows a comparison on the validation set between our 8-stack MSD network with setting II and an 8-stack hourglass network. From this we can see that our network achieves a competitive 2% increase in PCKh scores on the validation set. Example predictions made by our network are shown in Fig. 4.

We then generate predictions on the MPII test set using our 8-stack mixed-scale dense network with setting I, and the test results are shown in Table 2, together with some state-of-the-art results from recent years' works. Just as expected, our MSD network outperforms the stacked hourglass network [16] and the ones [3,12,20] that come before and several [14,17] which go after. What is more, our network produces comparable results to that of [6], which is yet another modification of the stacked hourglass network. There are several works that perform slightly better than this version of our network, in which [5] and [4] utilise the adversarial architecture, which may not be very stable to train.

In fact, msd1-s8 is a relatively simple setting of our network. With this setting we can acquire the performance comparable to the state-of-the-arts, in the
meantime maintain our advantage of training with a small number of hyper-
parameters. An 8-stack MSD network with setting 2 has already gotten higher
accuracy on the validation set, which indicates that our test results may still be
tuneable with proper settings of the dilation rate groups and some increase on
the scale of the network.

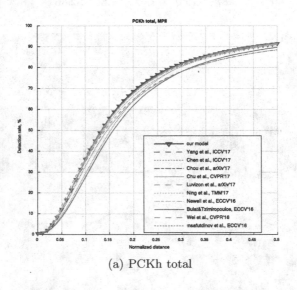

(a) PCKh total

Fig. 5. Visualisation of test accuracies for total joints, PCKh@0-0.5.

Figure 5 visualises the test accuracy of the average PCKh scores of total
joints. Meanwhile, the accuracies of some specific joints (i.e. ankle, wrist, shoulder, knee, elbow and hip) are illustrated in Fig. 6. For both cases, the scope
of PCKh scores varies from 0 to 0.5. From the graph we can observe that our
network produces satisfying results across the scope, and outperforms all of the
state-of-the-arts in around PCKh@0.15-0.30, indicating that our network excels
in predicting at micro-scales because of our well-preserved multi-scale features.

4.2   Component Analysis

In this section we conduct some ablation studies to look into the properties of
our network. Two main design configurations in our network are explored: the
number MSD blocks stacked in the network, and the design of dilation groups
inside a single MSD block.

Stacks of MSD Blocks. In [16], the effect of stacking several hourglass modules
is studied. Here, we compare 2- and 8-stack hourglasses with our 2- and 8-stack
MSD blocks with setting I on the validation set, and the results of which are

Table 2. Comparison with the state-of-the-arts on test set, PCKh@0.5.

Method	Head	Sho.	Elb.	Wri.	Hip	Knee	Ank.	Total	AUC
Insafutdinov et al. [12]	96.8	95.2	89.3	84.4	88.4	83.4	78.0	88.5	60.8
Wei et al. [20]	97.8	95.0	88.7	84.0	88.4	82.8	79.4	88.5	61.4
Bulat and Tzimiropoulos [3]	97.9	95.1	89.9	85.3	89.4	85.7	81.7	89.7	59.6
Newell et al. [16]	98.2	96.3	91.2	87.1	90.1	87.4	83.6	90.9	62.9
Ning et al. [17]	98.1	96.3	92.2	87.8	90.6	87.6	82.7	91.2	63.6
Luvizon et al. [14]	98.1	96.6	92.0	87.5	90.6	88.0	82.7	91.2	63.9
Chu et al. [6]	98.5	96.3	91.9	88.1	90.6	88.0	85.0	91.5	63.8
Chou et al. [5]	98.2	96.8	92.2	88.0	91.3	89.1	84.9	91.8	63.9
Chen et al. [4]	98.1	96.5	92.5	88.5	90.2	89.6	86.0	91.9	61.6
Yang et al. [21]	98.5	96.7	92.5	88.7	91.1	88.6	86.0	92.0	64.2
msd1-s8	98.1	96.4	91.8	87.8	90.7	87.9	84.4	91.4	64.1

shown in Table 3. From this we can observe that stacking more blocks increases the detection accuracies, and in both cases our MSD network performs better than HG network. Our 2-stack MSD network even yields slightly better results than the 8-stack hourglass network. What is more, by achieving higher accuracies our models have less parameters to train, which indicates that our MSD block design is more expressive and require potentially less training time.

Table 3. PCKh@0.5 of different stacks of hourglass networks and mixed scale dense networks with setting I and II. Pars(M) indicates the number of training parameters (in million) of each network.

Method	Pars(M)	Head	Sho.	Elb.	Wri.	Hip	Knee	Ank.	Mean
hg-s2	6.73	95.80	94.57	88.12	83.31	86.24	80.88	77.44	86.76
hg-s8	25.59	96.59	94.97	89.02	83.57	87.45	83.48	78.53	87.72
msd1-s2	**4.54**	96.28	95.26	88.89	83.07	88.71	82.99	79.15	87.88
msd1-s8	16.58	**96.96**	95.79	90.52	85.76	**89.80**	86.00	81.93	89.61
msd2-s8	28.23	96.86	**95.84**	**90.80**	**86.33**	89.30	**86.34**	**82.40**	**89.78**

Groups of Dilated Convolutional Layers. In Table 3 we can also see the comparison between the performance of settings I and II of our MSD network when 8 MSD blocks are stacked. Setting II slightly outperforms setting I, indicating that the increase of dilated convolutional groups in a single MSD block does help with the network performance. In setting II, however, the network is much larger than that in setting I, and has much more parameters to train. Therefore, in practical design of the MSD blocks, a tradeoff need to be considered between the number of groups and the network performance (Fig. 7).

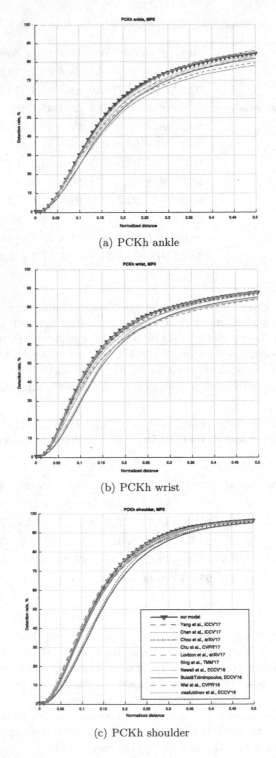

(a) PCKh ankle

(b) PCKh wrist

(c) PCKh shoulder

Fig. 6. Visualisation of test accuracies for single joints, PCKh@0-0.5.

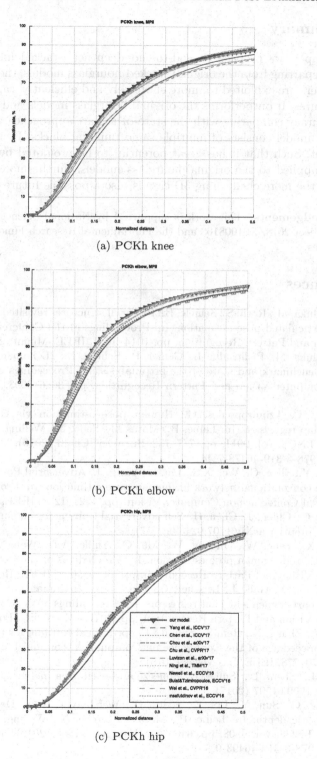

(a) PCKh knee

(b) PCKh elbow

(c) PCKh hip

Fig. 7. Visualisation of test accuracies for single joints, PCKh@0-0.5.

5   Summary

In this paper, we propose an end-to-end deep net, stacked mixed-scale dense blocks. Departing from the original stacked hourglass models, the dilation convolution layers are exploited to, more effectively and efficiently, capture the multiscale features. It outperforms the original hourglass models and achieve comparable accuracy comparing with the state-of-the-art approaches. In addition, the presented model consists of multiple same building blocks. Hence, it is easy to implement. Such that it has great potential to be promoted by the techniques that are applied to the original hourglass models. Furthermore, to extend our model to the more challenging 3D case is also a possible future works.

Acknowledgements. This work is supported by National Science and Technology Major Project 2018ZX01008103 and the Fundamental Research Funds for the Central Universities.

References

1. Andriluka, M., Roth, S., Schiele, B.: Pictorial structures revisited: people detection and articulated pose estimation. In: Proceedings of the Conference on Computer Vision and Pattern Recognition, pp. 1014–1021. IEEE, Miami (2009)
2. Andriluka, M., Pishchulin, L., Gehler, P., Schiele, B.: 2D human pose estimation: new benchmark and state of the art analysis. In: Proceedings of the Conference on Computer Vision and Pattern Recognition, pp. 3686–3693. IEEE, Columbus (2014)
3. Bulat, A., Tzimiropoulos, G.: Human pose estimation via convolutional part heatmap regression. In: Leibe, B., Matas, J., Sebe, N., Welling, M. (eds.) ECCV 2016. LNCS, vol. 9911, pp. 717–732. Springer, Cham (2016). https://doi.org/10.1007/978-3-319-46478-7_44
4. Chen, Y., Shen, C., Wei, X.S., Liu, L., Yang, J.: Adversarial PoseNet: a structure-aware convolutional network for human pose estimation. In: Proceedings of International Conference on Computer Vision, pp. 1221–1230. IEEE, Venice (2017)
5. Chou, C., Chien, J., Chen, H.: Self adversarial training for human pose estimation. arXiv preprint arXiv:1707.02439 (2017)
6. Chu, X., Yang, W., Ouyang, W., Ma, C., Yuille, A.L., Wang X.: Multi-context attention for human pose estimation. In: Proceedings of the Conference on Computer Vision and Pattern Recognition, pp. 5669–5678. IEEE, Honolulu (2017)
7. Dantone, M., Gall, J., Leistner, C., van Gool, L.: Human pose estimation using body parts dependent joint regressors. In: Proceedings of the Conference on Computer Vision and Pattern Recognition, pp. 3041–3048. IEEE, Portland (2013)
8. He, K., Zhang, X., Ren, S., Sun, J.: Deep residual learning for image recognition. In: Proceedings of the Conference on Computer Vision and Pattern Recognition, pp. 770–778. IEEE, Las Vegas (2016)
9. Hu, J., Shen, L., Sun, G.: Squeeze-and-excitation networks. arXiv preprint arXiv:1709.01507 (2017)
10. Huang, G., Sun, Y., Liu, Z., Sedra, D., Weinberger, K.Q.: Deep networks with stochastic depth. In: Leibe, B., Matas, J., Sebe, N., Welling, M. (eds.) ECCV 2016. LNCS, vol. 9908, pp. 646–661. Springer, Cham (2016). https://doi.org/10.1007/978-3-319-46493-0_39

11. Huang, G., Liu, Z., Maaten, L.V.D., Weinberger, K.Q.: Densely connected convolutional networks. In: Proceedings of the Conference on Computer Vision and Pattern Recognition, pp. 2261–2269. IEEE, Honolulu (2017)

12. Insafutdinov, E., Pishchulin, L., Andres, B., Andriluka, M., Schiele, B.: DeeperCut: a deeper, stronger, and faster multi-person pose estimation model. In: Leibe, B., Matas, J., Sebe, N., Welling, M. (eds.) ECCV 2016. LNCS, vol. 9910, pp. 34–50. Springer, Cham (2016). https://doi.org/10.1007/978-3-319-46466-4_3

13. Larsson, G., Maire, M., Shakhnarovich, G.: FractalNet: ultra-deep neural networks without residuals. In: Proceedings of International Conference on Learning Representations, Toulon (2017)

14. Luvizon, D.C., Tabia, H., Picard, D.: Human pose regression by combining indirect part detection and contextual information. arXiv preprint arXiv:1710.02322 (2017)

15. Mehta, S., Mercan, E., Bartlett, J., Weaver, D.L., Elmore, J.G., Shapiro, L.G.: Learning to segment breast biopsy whole slide images. arXiv preprint arXiv:1709.02554 (2017)

16. Newell, A., Yang, K., Deng, J.: Stacked Hourglass networks for human pose estimation. In: Leibe, B., Matas, J., Sebe, N., Welling, M. (eds.) ECCV 2016. LNCS, vol. 9912, pp. 483–499. Springer, Cham (2016). https://doi.org/10.1007/978-3-319-46484-8_29

17. Ning, G., Zhang, Z., He, Z.: Knowledge-guided deep fractal neural networks for human pose estimation. IEEE Trans. Multimedia **20**(5), 1246–1259 (2018)

18. Srivastava, R.K., Greff, K., Schmidhuber, J.: Training very deep networks. In: Proceedings of Advances in Neural Information Processing Systems, pp. 2377–2385. Curran Associates, Montreal (2015)

19. Sun, D., Yang, X., Liu, M., Kautz, J.: PWC-net: CNNs for optical flow using pyramid, warping, and cost volume. arXiv preprint arXiv:1709.02371 (2017)

20. Wei, S., Ramakrishna, V., Kanade, T., Sheikh, Y.: Convolutional pose machines. In: Proceedings of the Conference on Computer Vision and Pattern Recognition, pp. 4724–4732. IEEE, Las Vegas (2016)

21. Yang, W., Li, S., Ouyang, W., Li, H., Wang, X.: Learning feature pyramids for human pose estimation. In: Proceedings of International Conference on Computer Vision, pp. 1290–1299. IEEE, Venice (2017)

22. Yu, F., Koltun, V.: Multi-scale context aggregation by dilated convolutions. In: Proceedings of International Conference on Learning Representations, San Juan (2016)

MIDCN: A Multiple Instance Deep Convolutional Network for Image Classification

Kelei He[1], Jing Huo[1], Yinghuan Shi[1], Yang Gao[1(✉)], and Dinggang Shen[2]

[1] State Key Laboratory for Novel Software Technology, Nanjing University, Nanjing, People's Republic of China
gaoy@nju.edu.cn
[2] Biomedical Research Imaging Center, University of North Carolina, Chapel Hill, NC, USA

Abstract. For the image classification task, usually, the image collected in the wild contains multiple objects instead of a single dominant one. Besides, the image label is not explicitly associated with the object region, i.e., it is weakly annotated. In this paper, we propose a novel deep convolutional network for image classification under a weakly supervised condition. The proposed method, namely MIDCN, formulate the problem into Multiple Instance Learning (MIL), where each image is a bag which contains multiple instances (objects). Different with previous deep MIL methods which predict the label of each bag (i.e., image) by simply performing pooling/voting strategy over their instance (i.e., region) predictions, MIDCN directly predicts the label of a bag via bag features learned by measuring the similarities between instance features and a set of learned informative prototypes. Specifically, the prototypes are obtained by a newly proposed Global Contrast Pooling (GCP) layer which leverages instances not only coming from the current bag but also the other bags. Thus the learned bag features also contain global information of all the training bags, which is more robust and noise free. We did extensive experiments on two real-world image datasets, including both natural image dataset (PASCAL VOC 07) and pathological lung cancer image dataset, and show the results of the proposed MIDCN consistently outperforms the state-of-the-art methods.

Keywords: Multiple instance learning ·
Convolutional neural network · Lung cancer · Image classification

1 Introduction

Convolutional Neural Networks (CNNs) have demonstrated their efficacy in various computer vision tasks, including image classification [15], object detection [9], and image captioning [14], etc. CNN and its variations have nearly reached the human-level performance in many tasks, such as, face recognition [28]. However, for common tasks like image classification, there are still problems to solve

© Springer Nature Switzerland AG 2019
A. C. Nayak and A. Sharma (Eds.): PRICAI 2019, LNAI 11670, pp. 230–243, 2019.
https://doi.org/10.1007/978-3-030-29908-8_19

and it remains to achieve higher performance. One problem of image classification is that it is usually weakly supervised, i.e., global image-level label is given without been associated with specific image regions. In many real applications, there are usually multiple objects on an image. However, only a small region of the image (contains a label related object) is related to the image label. The basic CNN treats every pixel of the image equally for classification which leads to deteriorating performance. Another problem is that even a single object can have various appearance under different situations, making robust feature learning very difficult.

To solve the weakly supervised problem, Oquab *et al.* [20] proposed to train a CNN with multiple image patches of multiple scales as input, and used a max-pooling operation to aggregate the prediction results of multiple inputs. However, they use sliding window to sample input image patches which makes it sensitive to object size and this operation may also lead to many image patches that are meaningless. Besides, Xu *et al.* [31] presented a method that directly fed the deep features to a traditional Multiple Instance Learning (MIL) classifier, where the deep feature extractor and MIL classifier are learned separately. Another work of Wu *et al.* [30] proposed to address the problem by formulating image classification as a MIL problem.

Moreover, a few other works [8,27] also formulate image classification as a MIL problem. These previous works all demonstrated that learn in a weakly-supervised setting for image classification can further improve the performance of CNNs that are originally trained in a fully-supervised way. However, one drawback of these methods is that they predict the label or feature of a bag by simply performing pooling/voting strategy over their instance predictions or instance features. This scheme may weaken the contribution of instances that are most correlated with the image label.

In this paper, we also formulate image classification as a MIL problem. Formally, in MIL setting, both training and testing data are formed of a number of bags, where each bag has an arbitrary number of feature vectors (called instances). A bag is labeled as positive if at least one instance is labeled as positive. Otherwise, the bag is labeled as negative. The goal of MIL is to predict the bag label without given instance labels. For image classification tasks, according to the previous studies [1,11], the aforementioned assumption in MIL is usually too strict to follow in practice. Thus, an extended assumption of MIL has been widely applied: the images to be classified are regarded as bags, and the regions of these images are regarded as instances, where instances are labeled as positive if they are highly relevant to the positive labels. We also adopted this extended assumption in this work.

Based on the assumption, we proposed a Multiple Instance Deep Convolutional Network (MIDCN) for image classification. The major difference of MIDCN compared with the previous works is that it is based on learning prototypes to learn more representative bag features. MIDCN can learn a set of representative concepts from the training images to quantize the similarity of instance features and prototypes. Specifically, the learning is done by a newly proposed Global Contrast Pooling (GCP) layer from all the training instances. The network is learned in an end-to-end manner by minimizing the classification loss of training bags. With the learned prototypes, the bag feature is defined as the similarities of the instances and the prototypes. In this way, prototypes are correlated to a set of informative concepts. A concept is related to a specific object class. If multiple prototypes are used to represent a class, it relates to multiple appearance of the class is learned. Therefore, an instance of a specific class can always find a best matched prototype with a high similarity score, making the extracted bag feature more robust and noise free.

The pipeline of MIDCN is: Firstly, some overlapped region proposals are generated from images. Secondly, a shared-weight pre-trained Caffenet [13] (or some other pre-trained networks) is used to extract fixed size features for these regions. In this case, the output of the fully-connected layer 7 ($fc7$) with 4096-dimension is used as instance feature. Thirdly, the novel GCP layer is proposed to calculate the inner product of all the obtained instance features from one bag with the prototype weights to obtain a bag feature.

Typically, each dimension of a bag feature represents the maximum similarity of its instances to one specific prototype. The GCP layer is named because all the instances are employed to generate a pooling operation in the proposed prototype learning, which can be regarded as pooling in the global perspective. Finally, a softmax classifier is utilized to predict a bag. The whole network is trained in a standard Back-Propagation (BP) procedure.

Our implementation is based on the open source framework *pytorch* [21]. The implementation details are introduced in the experimental results section. The experiments are conducted on a popular benchmark dataset PASCAL VOC 07 [7] and a pathological lung cancer image dataset [26]. These two datasets are challenging with multiple objects contained in their images. The objects have complex structures and large appearance variations. Thus they naturally fit the MIL assumption and the problems we aim to address.

The contributions of MIDCN are three-fold:

- A novel multiple instances deep convolutional network is proposed to learn bag-level features based on the learned prototypes. Results show learning in a weakly supervised way can further improve the performance of existing CNNs.
- A novel Global Contrast Pooling (GCP) layer is proposed which offers an efficient prototype learning scheme for neural networks and can be easily embedded into various pre-trained networks for performance improvement.
- Extensive experiments on two real-world datasets (including both the natural image dataset PASCAL VOC 07 and a pathological lung cancer image dataset) show promising results are achieved by MIDCN.

2   Related Work

We review the methods of multiple instance learning and weakly supervised deep networks.

2.1   Multiple Instance Learning

Basically, previous MIL methods for image classification can be roughly divided into two categories, instance-level methods and bag-level methods. The instance-level methods follow the assumption that all instances in a bag contribute equally to the bag label [1]. Under the above assumption, the prediction of the label of a bag is conducted by aggregating (e.g., voting or pooling) the predictions of instances. This kind of learning approaches (e.g., [23,30]) predict the labels of instances separately. Besides, while aggregating instance labels to make bag label predictions, the relation of instances is not considered. However, information from other instances in the same bag is usually considered useful. As proved by [1], compared with the instance-level methods, predicting the labels in bag-level usually achieves higher accuracy as well as better time efficiency. As MIL naturally fits the weakly supervised image classification problem, more and more works are working on using MIL for image classification [3,27,30]. Specially, with the development of deep features, MIL is consistently combined with deep neural networks for solving the problem.

2.2   Weakly Supervised Deep Neural Networks

A few studies for developing deep learning methods under MIL assumption have been proposed in recent years [30,31]. Besides, the methods developed in weakly supervised learning setting can also be adapted into MIL assumption. [20] developed a method using CNN for object recognition in a weakly supervised condition. The goal of this work is achieved by first constructing the network with an extended score mapping for the region of the image. Then average pool the score map under a weakly supervised assumption to get the final prediction. [18] proposed a weakly supervised approach to learn features for object recognition. [22] presented a network inferring in a weakly supervised condition, for which the output was encouraged to follow a latent probability distribution which lies in a constrained manifold. It is obvious that these methods developed in weakly supervised setting can be equivalently transformed to that in MIL setting, by regarding the mapped scores/mid-level features as instance predictions and final scores/features as bag predictions. Therefore, all of these methods can be categorized into instance-level methods. It inspired us to explore the bag-level methods for deep networks.

3   Multiple Instance Deep Convolutional Network

An overview of the MIDCN framework is shown in Fig. 1, which has three major components. Firstly, a region proposal extractor is used to detect class relevant saliency regions from all the images. Secondly, the detected regions are resized to a fixed size and fed into a shared-weight pre-trained Caffenet [13] in order to get the instance features for all the image regions. Then, the proposed GCP layer combines the instance features to generate a bag feature. Finally, the bag

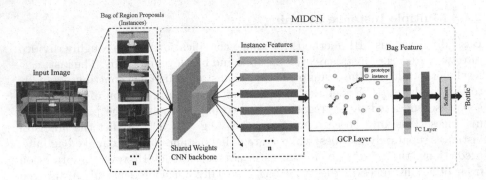

Fig. 1. Overall framework of MIDCN. It predicts image-level labels without region-level labels. With the region proposals generated, it is an n to 1 system, meaning that 1 image label supervises the learning of a whole network with n inputs. Here n is the number of region proposals.

feature is mapped by a fully connected (FC) layer, followed by a softmax classifier to decide the label. In this following, we present the whole architecture of the network, the design of the GCP learning layer and the detailed learning procedure.

3.1   Generate Instance Features

Based on the MIL assumption mentioned in the first section, the images to be classified are treated as bags. The region proposals of images are treated as instances. Therefore, one important step of the proposed method is to detect image regions that may contain objects. In our method, most of the popular region proposal methods can be adopted, for example, BING [6], R-CNN [9] and Region Proposal Network (RPN) [24]. In the PASCAL VOC 07 experiment, BING is chosen to efficiently detect salient regions. For the pathological lung cancer dataset, a specifically designed detection method is used to detect cells. Details can be found in the experimental section. The detected regions are then resized to fixed size required by the instance feature extractor.

An intuitive way to extract all the instance features is to use multiple CNNs with respect to all the region proposals. However, it is very computationally expensive and the number of region proposals of each image is not fixed. Therefore one simple shared-weights CNN is adopted to extract all the instance features.

3.2   Global Contrast Pooling Layer

Preliminaries and Bag Feature Encoding. Given the i-th and j-th bags as $\mathcal{B}_i = \{\boldsymbol{x}_{i1}, \boldsymbol{x}_{i2}, ..., \boldsymbol{x}_{in_i}\}$, and $\mathcal{B}_j = \{\boldsymbol{x}_{j1}, \boldsymbol{x}_{j2}, ..., \boldsymbol{x}_{jn_j}\}$, respectively. n_i and n_j are the number of instances in bag $\mathcal{B}_i$ and $\mathcal{B}_j$. $\boldsymbol{x}_{ik} \in \mathbb{R}^d$ $(k = 1, 2, ..., n_i)$ is the

k-th instance contained in $\mathcal{B}_i$, where d is the feature dimensionality. In our case, instance feature is the output of the CNN based feature extraction module.

As can be seen, under the MIL setting, bags (images) are sets of instances (regions). Previous works predict the label of a bag use max-pooling or voting over the predicted label of instances. This kind of method predicts the label of instances separately and does not take the relationship of instances into consideration. Although there are also works using max-pooling over features of instances to extract bag features, such scheme is relatively simple and may weaken the importance of key instance's feature. The aim of the GCP layer is also to extract bag features while take the relationship of instances into consideration.

To this end, prototypes are introduced to quantize bag features. We define the bag feature as a vector of similarity scores. Each element of the vector is a similarity score of the most similar instance and a specific prototype. As prototypes correlate to class of concepts, such as concepts of 'aeroplane', 'car' and so on, a high similarity score indicates the existence of an instance that has the concept of the prototype. In this way, the bag feature encodes the possibility of existence of a set of basic concepts. It is this bag feature that is used for further classification. Therefore, co-existence of some concepts can help distinguish the label. For example, coexistence aerofoil and wheels indict the label of 'aeroplane'.

Formally, the bag feature is defined as follows. Denote the m-th prototype as $z_m \in \mathbb{R}^d$, a bag feature is defined as the similarity score of this prototype with the most similar instance in this bag. Typically, the similarity score can be defined as the negative distance (usually the Euclidean distance). However, the neural network may suffers during the back propagation process when an Euclidean distance calculation module is embedded in it.

To address this problem, [16] advised to use a sigmoid function instead of directly using the Euclidean distance. However, this solution will lead to heavy calculation burden because of the non-linear sigmoid function. Besides, it will binarize the continuous similarity values. We here propose a new solution that is equivalent to transform the distance between the i-th bag and m-th prototype to a similarity value s_{im} with parameter w_m. The equivalence relationship is as follows:

$$d_{im} = \min_{k=1}^{n_i} \|x_{ik} - z_m\| \sim \max_{k=1}^{n_i} w_m^\top x_{ik} = s_{im}, \tag{1}$$

where d_{im} is the distance of the i-th bag and m-th prototype. It is defined as the minimum distance of all the instances in the ith bag and m-th prototype. $\|\cdot\|$ is the l_2 norm. Therefore, $\|x_{ik} - z_m\|$ is the Euclidean distance of the kth instance and the m-th prototype. s_{im} is defined as the maximum similarity score of all the instances with the m-th parameter w_m. Suppose x_{ik} and z_m are normalized to unit vectors with their l_2 norm equal to 1 and set w_m with the same value as z_m, the two equations are equivalent. Therefore, instead of embedding the Euclidean distance module into the neural network, we used the similarity score based module for prototype learning and bag feature encoding. The learning of prototype z_m is now changed to learning w_m. In the following section, w_m is also referred as prototype indistinctively.

The transformation introduced above has two additional advantages: (1) Parameters of prototype can be learned according to image labels. A loss defined on the classification score of the bag label can be used to tune the network. (2) It prevents heavy computational cost of prototype initialization. If the distance module is used, clustering methods are usually needed to initialize the prototype parameter to achieve good results. However, in the proposed method, the weights $\boldsymbol{w}_m$ can be initialized randomly.

More prototypes will usually strengthen the learned network and make the extracted bag feature more robust. However, too many prototypes will lead to heavy computational burden. Thus there should be a balance between the number of prototypes and computational cost. With all the prototypes learned, suppose p weights of prototypes are learned and is defined as $\mathcal{W} = \{\boldsymbol{w}_1, ..., \boldsymbol{w}_p\}$. Maximum similarity values of the instances in a bag and the learned prototypes are used to represent a bag. Using the aforementioned prototype weights, the feature of the i-th bag can be formulated as,

$$\boldsymbol{b}_i = [s_{i1}, s_{i2}, ..., s_{im}, ..., s_{ip}], \tag{2}$$

$$s_{im} = \max_{k=1}^{n_i} \boldsymbol{w}_m^\top \boldsymbol{x}_{ik} + \boldsymbol{R}(\boldsymbol{w}_m), \tag{3}$$

$\boldsymbol{b}_i$ is the feature of the ith bag. The set of weights of prototypes $\mathcal{W}$ is the parameter used to control the bag feature encoding. Also, a L2 norm regularization term which is usually used in normal convolutional neural networks is also adopted here to constraint the weight parameter $\boldsymbol{w}_m$, denote as $\boldsymbol{R}(\boldsymbol{w}_m)$. It can be used to constraint the magnitude of the values in the weight parameter.

Principle for Prototype Learning. We now give the overall objective function to learn the multiple instance based neural network. Firstly, we give a brief principle of prototype learning. As introduced before, a bag feature is encoded by the GCP layer with the similarity scores as its feature values. The prototypes used by the GCP layer can be coarsely divided into two categories, prototypes that are highly relevant to the ground truth label and others that are not. In the meanwhile, instances can also be divided into these two categories. Therefore the learning principle should be prototypes and instances that are relevant to the same label should be close with high similarity score. Therefore, they should be pulled together during training. Otherwise they will be pushed apart. The learning principle of prototypes is shown in Fig. 2.

However, the relationship between the learned prototypes and the labels is unknown. The prototypes are latent values. Instead of using a separate learning objective function to learn prototypes, we used an overall objective function to learn the whole network. The final objective function is parameterized with a set of classification parameter to classify the bag feature. Thus, the final objective is somewhat like a logsoftmax loss calculated on the bag feature. The final prediction w.r.t the network parameter θ is formulated as,

$$p(Y^i = j|\boldsymbol{b}_i; \theta) = \frac{e^{\theta_j^\top(\boldsymbol{b}_i)}}{\sum_{l=1}^c e^{\theta_l^\top(\boldsymbol{b}_i)}}. \tag{4}$$

Then, the final loss is:

$$J(\theta) = -\frac{1}{b}\left[\sum_{i=1}^{b}\sum_{j=1}^{c}1\{Y^i = j\}\log p(Y^i = j|\boldsymbol{b}_i;\theta)\right],\tag{5}$$

where b is the number of bags in one batch, c is the number of total classes, Y^i is the probability of the i-th class, and $1\{\cdot\}$ is an indicator function, with a value 1 if the expression is true and with value 0 otherwise. One difference of the above loss function of the proposed multiple instance based network compared with traditional CNN network is that the feature $\boldsymbol{b}_i$ in Eq. (5) is the output of the GCP layer, which are similarity values of the instances and prototypes. During learning, the loss function will automatically correlates the similarity score in the bag feature to the bags' label. Therefore, losses on the bag feature can be used to guide the learning of prototypes, making prototypes learned implicitly correlate to the bag labels (concepts). This learning method has two advantages:

- Prototypes are learned from the whole training set which encodes global information, making bag feature also contains global information and is therefore more robust.
- The method is trained with an overall loss to classify bags, which avoids inferring from the unknown instance and prototype relation.

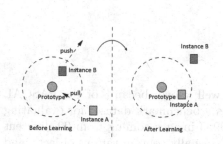

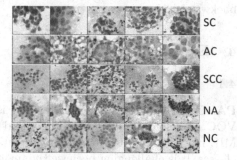

Fig. 2. Brief learning principle of prototypes. Correlated prototype and instance are in same color. (Color figure online)

Fig. 3. Images of pathological lung cancer image set in five categories: Squamous Carcinoma (SC), Adenocarcinoma (AC), Small Cell Cancer (SCC), Nuclear Atypia (NA) and Normal type (NC).

3.3   Optimization

Stochastic Gradient Descent (SGD) algorithm with mini-batches is used for weight optimization. The whole network is trained under standard back propagation. Prototype parameters $\mathcal{W}$ in the GCP layer can be treated as normal parameters of the neural network and updated as usual. This works well in the experiments. Given $\mathcal{L}$ the final loss calculated for the whole network, b the number of bags contained in one batch, and the partial derivative with respect to the GCP layer output $\frac{\partial \mathcal{L}}{\partial y}$, thus the gradient of the GCP layer parameters can be easily calculated. The partial derivatives with respect to weights $\boldsymbol{w}_m$ are,

$$\frac{\partial \mathcal{L}}{\partial \boldsymbol{w}_m} = \frac{\partial \mathcal{L}}{\partial y} \times \sum_{i=1}^{b} \boldsymbol{x}_{im*}^{\top} + \frac{\partial \boldsymbol{R}(\boldsymbol{w}_m)}{\partial \boldsymbol{w}_m}, \tag{6}$$

where "$*$" denotes the instance feature which has the maximum similarity to a prototype among all the instances of a bag. For example, $\boldsymbol{x}_{im*}$ denotes the instance feature in bag $\mathcal{B}_i$ which has the maximum similarity to the m-th prototype. Besides, the partial derivatives with respect to weights $\boldsymbol{x}_{ik*}$ are,

$$\frac{\partial \mathcal{L}}{\partial \boldsymbol{x}_{ik*}} = \frac{\partial \mathcal{L}}{\partial y} \times \frac{1}{r} \sum_{m \in S_k} \boldsymbol{w}_m, \tag{7}$$

Notice only those instances that have the maximum similarities to prototypes will get updated, since only these instances contribute to the final prediction. This is constrained by the weakly supervised condition. r denotes the number of times $\boldsymbol{x}_{ik*}$ been the most similar instance feature to all the prototypes in bag $\mathcal{B}_i$. S_k is the indices of prototypes for which $\boldsymbol{x}_{ik*}$ has maximum similarity values. Finally, by using Eq. (2)–(7), learning can be conducted in a simple forward and backward pass.

4   Experimental Results

4.1   Datasets

PASCAL VOC 07 [7]: This dataset is well known because of the PASCAL VOC challenge. It has been widely used as a benchmark dataset for evaluating MIL methods [22,30]. It contains about 9963 images coming from 20 different classes. It is challenging because the images usually contain various objects, and these objects are not well centralized. We used both the commonly suggested training and validation set division for training. Results are reported on the test set.

Pathological Lung Cancer Image Set [26]: This dataset has about 1200 histopathological lung cancer images of five classes: four types of lung cancer (NA, SC, AC, SCC), and normal (NC), with 200–400 images per class. The images have a higher resolution than PASCAL VOC 07, which is 576×768. The major challenges include large intra-class appearance variations, obscure boundaries between cellula and background, and the variations of the cellula shapes. Actually, this pathological image dataset is suitable to evaluate MIL methods, as the intra-class images have both private structures (cancer cells) as well as public structures (common tissues) (see Fig. 3).

Table 1. Quantitative comparison of MIDCN and the state-of-the-art methods in AP (%) on PASCAL VOC 07 (best are in **bold**). We only display selected classes as limited by spaces for better visualization. The mAP is calculated on all 20 classes.

Methods	aero	btl	bus	car	cat	chair	table	dog	mbk	per	plant	sofa	tv	mAP
OverFeat	91.2	51.6	81.6	84.4	83.9	54.5	53.8	72.3	75.6	83.7	47.4	60.0	79.4	73.0
DMIL-R	92.9	53.9	81.8	86.8	83.4	53.7	51.8	72.3	77.3	86.1	50.1	61.7	80.1	74.7
DMIL-J	93.5	54.2	81.6	86.6	85.2	54.5	53.8	73.2	79.0	86.6	51.2	63.7	80.4	75.5
[19]	88.5	47.5	75.5	90.1	87.2	61.6	67.3	85.5	80.0	**95.6**	60.8	58.0	77.9	77.7
[29]	95.1	51.5	80.0	91.7	91.6	57.7	70.9	89.3	85.2	93.0	64.0	62.7	78.3	81.5
VGG-F	88.7	46.9	77.5	86.3	85.4	58.6	72.6	82.0	80.7	91.8	58.5	66.3	71.3	77.4
VGG-S	**95.3**	54.4	81.9	91.5	**91.9**	64.1	74.9	**89.7**	**86.9**	95.2	60.7	68.0	74.4	82.4
MIDCN-C	90.8	76.0	82.0	91.1	86.9	71.5	74.2	84.3	81.5	94.5	**83.3**	67.0	86.3	82.6
MIDCN-F	86.5	78.8	78.3	91.3	85.9	**74.6**	70.9	83.5	83.9	94.6	80.9	**69.9**	85.5	83.3
MIDCN-S	91.7	**81.1**	**81.9**	**92.4**	89.1	72.5	**75.3**	88.2	86.4	94.8	81.2	68.0	**87.6**	**84.2**

4.2 Implementation Details

We construct our model with pre-trained Caffenet [13] as instance feature-extractor (named as "MIDCN-C") to keep the consistency and conduct the fair comparison with the previous work [30]. However, for PASCAL VOC 07 experiments, we also employ VGG-F and VGG-S net [5] which are reported to have a better performance of top5 error rate in ILSVRC compared with Caffenet, named "MIDCN-F" and "MIDCN-S". We do not use more deeper ResNet [10] or DenseNet [12] as we only want to demonstrate our assumption, not to achieve the optimal performance. For PASCAL VOC 07, we use BING [6] trained with PASCAL VOC 07 as the region proposal extractor. In addition, for the pathological lung cancer image dataset, a simple ROI detection method is performed to crop patches of cancer cells in images. According to the settings of BING, we adopt the combined along with the high confidence scored frames in order to get 20 instances for each image. It is easy to observe that region proposals are important for the final classification performance. Here, we did not illustrate and compare all the possible region proposal methods for MIDCN, as our goal is to validate the efficacy of the proposed MIL-based method. The batch size we used is set to 20 for bags, as each bag has 20 image regions. Therefore, the total batch size is 400 of instance features for the neural network. Prototype number is also a data-relevant parameter, where we chose 200 to make a tradeoff between time efficiency and performance efficacy. The prototype weights are initialized by sampling from a normal distribution with a variance of 0.01.

4.3 Image Classification on PASCAL VOC 07 Dataset

We followed the previous works to pre-train the feature extractor network with extra training data ILSVRC in our method [25,30], and adopted Average Precision (AP) as the metric to evaluate the performance.

The quantitative comparison with the state-of-the-art methods in AP are reported in Table 1. In the first bar, we compared with the methods that either under purely fully-supervised condition (i.e. OverFeat [25]) or instance-based MIL assumption (i.e. DMIL-R/J [30]). For a fair comparison, MIDCN-C is constructed on Caffenet. The performance of this network was reported to have similar performance to the Alexnet [13] which is used in OverFeat method. OverFeat (CNN-SVM) is a widely used deep learning based baseline method. It is worth noting that we also compared with DMIL-R and DMIL-J, which are deep multiple instance learning frameworks that performs predictions at the instance-level. Specifically, the method DMIL-J is DMIL-R joint learned with text annotations, which leveraged information of an additional modality. According to Table 1, the first observation is, the three proposed methods achieved the highest performance in mAP. This shows the general performance of the proposed method outperforms the state-of-the-art methods. Besides, compared with OverFeat which is trained without weakly supervised condition, the methods such as DMIL-R, DMIL-J and MIDCN-C can obtain superior performance. Also, compared with DMIL R/J which adopts the deep features purely trained on instances, MIDCN shows better performance because of the more reasonable inference on bag-level. The advantage of using prototype learning is construction of the bag features by incorporating global information, hence previous challenging classes (e.g., dinning table, chair, potted plant) can be well classified by MIDCN.

In the last two bars, we investigated the improvement of performance given specific instance feature extractors. Compared with the original networks, i.e., OverFeat, VGG-F and VGG-S, MIDCNs constructed with these instance feature extractors boost the performance of mAP by 9.6%, 5.9%, and 1.8%.

4.4   Five Class Classification on Pathological Lung Cancer Image Set

Table 2 presents the classification results on the pathological lung cancer image dataset. Accuracy, Precision, Recall, F1 and TNR of MIDCN and competitive methods are reported on this dataset. The first four methods mcSVM [4], ESRC [17], KSRC [32], mSRC [26], belong to non-deep methods. These methods adopted hand crafted features to extracted features for lung cellula, including shape, color, texture, etc. The last four rows in Table 2 provide results of deep learning based methods. CNN-SVM is an OverFeat like method which is regarded as the baseline of deep models. An general observation is that deep methods are significantly better compared with non-deep methods. This indicates the quality of deep features are much better than traditional hand-crafted features. Besides, results of weakly supervised mi-SVM and MI-SVM [2] trained over the deep features are even better than CNN-SVM which indicates weakly-supervised learning is more suitable than fully-supervised learning in this application. The best results are achieved by the proposed MIDCN-C indicates the effectiveness of the proposed method. Also, we found that the feature extractor trained on ILSVRC still works well in pathological images, which expanded the limits of using the pre-trained feature extractor across different datasets while only for natural images in previous works [25, 30].

Table 2. Comparison of performance on pathological lung cancer image set.

Methods	Accuracy	Precision	Recall	F1	TNR
mcSVM	0.674	0.598	0.577	0.576	0.921
ESRC	0.800	0.730	0.884	0.777	0.940
KSRC	0.830	0.782	0.843	0.804	0.953
mSRC	0.867	0.834	0.913	0.862	0.962
CNN-SVM	0.968	0.968	0.968	0.968	0.992
CNN-miSVM	0.987	0.988	0.987	0.987	0.997
CNN-MISVM	0.981	0.981	0.981	0.981	0.995
MIDCN-C	**0.996**	**0.996**	**0.996**	**0.996**	**0.998**

5   Conclusion

In this paper, we propose a multiple instance deep convolutional network (MIDCN). In MIDCN, we introduce a global contrast pooling (GCP) layer for prototype learning above instance features, which is able to create more informative bag-level representations. The proposed GCP layer can be easily embedded into other pre-trained networks to further improve the performance. Experimental results evaluated on PASCAL VOC 07 and pathological lung cancer image set, show promising results by MIDCN.

Acknowledgment. This work was supported in part by the National Key Research and Development Program of China (2017YFB0702601), the National Natural Science Foundation of China (Grant Nos. 61673203, 61806092), Jiangsu Natural Science Foundation (BK20180326), and the Fundamental Research Funds for the Central Universities (14380056).

References

1. Amores, J.: Multiple instance classification: review, taxonomy and comparative study. Artif. Intell. **201**, 81–105 (2013). https://doi.org/10.1016/j.artint.2013.06.003

2. Andrews, S., Tsochantaridis, I., Hofmann, T.: Support vector machines for multiple-instance learning. In: NIPS, Vancouver, BC, Canada, 9–14 December 2002, pp. 561–568 (2002)

3. Babenko, B., Verma, N., Dollár, P., Belongie, S.J.: Multiple instance learning with manifold bags. In: ICML 2011, Bellevue, WA, USA, 28 June–2 July 2011, pp. 81–88 (2011)

4. Chang, C., Lin, C.: LIBSVM: a library for support vector machines. ACM TIST **2**(3), 27 (2011). https://doi.org/10.1145/1961189.1961199

5. Chatfield, K., Simonyan, K., Vedaldi, A., Zisserman, A.: Return of the devil in the details: delving deep into convolutional nets. In: British Machine Vision Conference (2014)

6. Cheng, M., Zhang, Z., Lin, W., Torr, P.H.S.: BING: binarized normed gradients for objectness estimation at 300 fps. In: CVPR 2014, Columbus, OH, USA, 23–28 June 2014, pp. 3286–3293 (2014). https://doi.org/10.1109/CVPR.2014.414

7. Everingham, M., Eslami, S.M.A., Gool, L.J.V., Williams, C.K.I., Winn, J.M., Zisserman, A.: The Pascal visual object classes challenge: a retrospective. Int. J. Comput. Vis. 111(1), 98–136 (2015). https://doi.org/10.1007/s11263-014-0733-5

8. Feng, J., Zhou, Z.H.: Deep MIML network. In: AAAI, pp. 1884–1890 (2017)

9. Girshick, R.B., Donahue, J., Darrell, T., Malik, J.: Rich feature hierarchies for accurate object detection and semantic segmentation. In: CVPR 2014, Columbus, OH, USA, 23–28 June 2014, pp. 580–587 (2014). https://doi.org/10.1109/CVPR.2014.81

10. He, K., Zhang, X., Ren, S., Sun, J.: Deep residual learning for image recognition. In: Proceedings of the IEEE Conference on Computer Vision and Pattern Recognition, pp. 770–778 (2016)

11. Hoffman, J., Pathak, D., Darrell, T., Saenko, K.: Detector discovery in the wild: joint multiple instance and representation learning. In: CVPR 2015, Boston, MA, USA, 7–12 June 2015, pp. 2883–2891 (2015). https://doi.org/10.1109/CVPR.2015.7298906

12. Huang, G., Liu, Z., Van Der Maaten, L., Weinberger, K.Q.: Densely connected convolutional networks. In: Proceedings of the IEEE Conference on Computer Vision and Pattern Recognition, pp. 4700–4708 (2017)

13. Jia, Y., et al.: Caffe: convolutional architecture for fast feature embedding. In: MM 2014, Orlando, FL, USA, 03–07 November 2014, pp. 675–678 (2014). https://doi.org/10.1145/2647868.2654889

14. Karpathy, A., Li, F.: Deep visual-semantic alignments for generating image descriptions. In: CVPR 2015, Boston, MA, USA, 7–12 June 2015, pp. 3128–3137 (2015). https://doi.org/10.1109/CVPR.2015.7298932

15. Krizhevsky, A., Sutskever, I., Hinton, G.E.: Imagenet classification with deep convolutional neural networks. In: NIPS, Lake Tahoe, NV, USA, 3–6 December 2012, pp. 1106–1114 (2012)

16. LeCun, Y., et al.: Backpropagation applied to handwritten zip code recognition. Neural Comput. 1(4), 541–551 (1989). https://doi.org/10.1162/neco.1989.1.4.541

17. Liu, M., Zhang, D., Shen, D.: Ensemble sparse classification of Alzheimer's disease. NeuroImage 60(2), 1106–1116 (2012). https://doi.org/10.1016/j.neuroimage.2012.01.055

18. Mittelman, R., Lee, H., Kuipers, B., Savarese, S.: Weakly supervised learning of mid-level features with Beta-Bernoulli process restricted Boltzmann machines. In: CVPR, Portland, OR, USA, 23–28 June 2013, pp. 476–483 (2013). https://doi.org/10.1109/CVPR.2013.68

19. Oquab, M., Bottou, L., Laptev, I., Sivic, J.: Learning and transferring mid-level image representations using convolutional neural networks. In: CVPR 2014, Columbus, OH, USA, 23–28 June 2014, pp. 1717–1724 (2014). https://doi.org/10.1109/CVPR.2014.222

20. Oquab, M., Bottou, L., Laptev, I., Sivic, J.: Is object localization for free? Weakly-supervised learning with convolutional neural networks. In: CVPR, Boston, USA, June 2015

21. Paszke, A., et al.: Automatic differentiation in PyTorch. In: NIPS-W (2017)

22. Pathak, D., Krähenbühl, P., Darrell, T.: Constrained convolutional neural networks for weakly supervised segmentation. In: ICCV 2015, Santiago, Chile, 7–13 December 2015, pp. 1796–1804 (2015). https://doi.org/10.1109/ICCV.2015.209

23. Pinheiro, P.H.O., Collobert, R.: From image-level to pixel-level labeling with convolutional networks. In: CVPR 2015, Boston, MA, USA, 7–12 June 2015, pp. 1713–1721 (2015). https://doi.org/10.1109/CVPR.2015.7298780

24. Ren, S., He, K., Girshick, R., Sun, J.: Faster R-CNN: towards real-time object detection with region proposal networks. In: Advances in Neural Information Processing Systems, pp. 91–99 (2015)

25. Sermanet, P., Eigen, D., Zhang, X., Mathieu, M., Fergus, R., LeCun, Y.: Overfeat: integrated recognition, localization and detection using convolutional networks. CoRR abs/1312.6229 (2013)

26. Shi, Y., Gao, Y., Yang, Y., Zhang, Y., Wang, D.: Multimodal sparse representation-based classification for lung needle biopsy images. IEEE Trans. Biomed. Eng. 60(10), 2675–2685 (2013). https://doi.org/10.1109/TBME.2013.2262099

27. Sun, M., Han, T.X., Liu, M.C., Khodayari-Rostamabad, A.: Multiple instance learning convolutional neural networks for object recognition. In: 2016 International Conference on Pattern Recognition, pp. 3270–3275. IEEE (2016)

28. Taigman, Y., Yang, M., Ranzato, M., Wolf, L.: Deepface: closing the gap to human-level performance in face verification. In: CVPR 2014, Columbus, OH, USA, 23–28 June 2014, pp. 1701–1708 (2014). https://doi.org/10.1109/CVPR.2014.220

29. Wei, Y., et al.: CNN: single-label to multi-label. CoRR abs/1406.5726 (2014)

30. Wu, J., Yu, Y., Huang, C., Yu, K.: Deep multiple instance learning for image classification and auto-annotation. In: CVPR 2015, Boston, MA, USA, 7–12 June 2015, pp. 3460–3469 (2015). https://doi.org/10.1109/CVPR.2015.7298968

31. Xu, Y., Mo, T., Feng, Q., Zhong, P., Lai, M., Chang, E.I.: Deep learning of feature representation with multiple instance learning for medical image analysis. In: ICASSP 2014, Florence, Italy, 4–9 May 2014, pp. 1626–1630 (2014). https://doi.org/10.1109/ICASSP.2014.6853873

32. Zhang, L., et al.: Kernel sparse representation-based classifier. IEEE Trans. Signal Process. 60(4), 1684–1695 (2012). https://doi.org/10.1109/TSP.2011.2179539

Discriminative Deep Attention-Aware Hashing for Face Image Retrieval

Zhi Xiong[1,2], Bo Li[1(✉)], Xiaoyan Gu[1], Wen Gu[1,2], and Weiping Wang[1]

[1] Institute of Information Engineering, Chinese Academy of Sciences, Beijing, China
{xiongzhi,libo,guxiaoyan,guwen,wangweiping}@iie.ac.cn
[2] School of Cyber Security, University of Chinese Academy of Sciences, Beijing, China

Abstract. Although the power of hashing methods has been proved in image retrieval, they cannot effectively extract discriminative features for face image retrieval as the discriminative differences in face regions are subtle and the background information interferes with the feature expression. To solve this problem, we propose an end-to-end deep hashing method with attention mechanisms to learn discriminative hash codes. Specifically, a face spatial network is designed to enhance the discrimination of face features from the spatial aspect. With a specially designed face spatial loss, it can automatically mine differentiated facial regions, and reduce the interference of background information. Furthermore, an attention-aware hash network, in which facial features could be enhanced by fusing strategy and channel attention module, is designed to learn compact and discriminative hash codes. Experimental results on two widely used datasets demonstrate the inspiring performance over several state-of-the-art hashing methods.

Keywords: Deep hashing · Face image retrieval · Deep learning

1 Introduction

With the growth of social media users, a large number of face-containing images have been uploaded to the Internet [9]. Face image retrieval, which aims to return images containing the specific query face, has attracted increasing interest. As a popular image retrieval solution, hashing shows its power in image retrieval. Owing to hashing methods, the similarity search of images could be done in Hamming space, which is efficient in terms of time and storage costs. In this work, we are focusing on learning effective hashing functions for face image retrieval.

Existing hashing methods can be divided into non-deep methods and deep methods. Representative non-deep methods include [2–5,15,18]. These methods usually try to generate hash code based on handcraft features. However, the quality of handcraft features affects the performance of the generated hash codes.

© Springer Nature Switzerland AG 2019
A. C. Nayak and A. Sharma (Eds.): PRICAI 2019, LNAI 11670, pp. 244–256, 2019.
https://doi.org/10.1007/978-3-030-29908-8_20

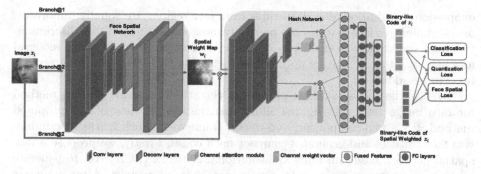

Fig. 1. The overall architecture of the proposed method. Our proposed method is comprised of three components: (1) a face spatial network based on FCN for the learning of spatial weight map. (2) a hashing network based on CNN for the learning hashing codes. (3) a set of loss functions including classification loss, quantization loss and face spatial loss for the optimization. In the training stage, the face spatial network and the hashing network are trained alternately. Classification loss and quantization loss are used to supervise the learning of hash network while face spatial loss guides the learning of face spatial network. When training, the original face image (Branch@1) and the spatial weighed face images (Branch@2) are fed into hash network respectively to get their binary-like codes. When testing, only spatial weighed face images (Branch@2) are used to generate their hash codes.

On the contrary, deep hashing methods adopt deep learning techniques to perform feature learning and hash code generation simultaneously [6,7,11,13,20,21], reducing the information loss between the feature learning stage and hash code generation stage. For example, Deep Supervised Hashing (DSH) [11] learns hash codes for input raw images by optimizing the binary-like output to minimize the Hamming distance of relevant images. Nevertheless, most of the hashing methods are designed for general image retrieval instead of face image retrieval.

Recently, some works try to apply deep hashing methods to face image retrieval tasks, such as Discriminative Deep Hash (DDH) [10] and Discriminative Deep Quantization Hash (DDQH) [17]. DDH proposes a deep network to extract multi-scale face image features and uses a divide-and-encode module to generate compact binary codes. Further, DDQH proposes a batch normalization quantization module to improve the performance. These methods mainly focus on designing network structures to generate hash codes with features directly extracted from the image. However, as face features are critical to generating compact hash codes, it is important to optimize the face feature extraction process for face image retrieval task.

Unlike general image retrieval, where shapes and appearances are obviously different from class to class, images in face image retrieval often have an overall facial appearance but with subtle facial differences. Therefore, we need to focus on features from the face regions that represent the facial differences. Extracting face features directly from images may result in an insufficient expression of the subtle facial differences, as background information may interfere with the expression. In addition, not all features from face regions are equally important, as discriminative face features are more advantageous for generating discriminative hash codes. How to effectively optimize the features for face image retrieval and how to integrate feature optimization with the hash codes learning process need to be addressed.

To solve the above issues, we propose a deep attention-aware hashing method for face image retrieval. Figure 1 illustrates the overall view of our proposed method. It's a two-component, end-to-end framework which learns discriminative face features and generates compact hash codes. Firstly, we propose a face spatial network based on Fully Convolutional Network (FCN) [14] to generate spatial weight maps to indicate discriminative facial regions. After a matrix dot production between the original data points and the corresponding spatial weights, data points are fed into a Convolutional Neural Network (CNN) to extract multi-scale face features and generate compact binary codes. Furthermore, we use channel attention modules to generate channel weight vectors for face feature maps, thus, discriminative face features are enhanced from the channel aspect. Finally, an alternate training strategy is introduced to the network training process with three loss functions. Classification loss and quantization loss are used to supervise the hash network, while a specially designed face spatial loss to supervise the training of face spatial network.

Our contributions can be summarized as follows:

- We propose a new deep hashing method for face image retrieval, in which face features extracted from images are enhanced by both spatial-wise and channel-wise attention mechanisms.
- We propose a face spatial loss, combined with an alternating training strategy, to guide the face spatial network to mine discriminative facial regions and alleviate the interference of background information of face images.
- Extensive experiments conducted on two widely-used face image datasets demonstrate that our proposed method achieves inspiring improvements compared with several state-of-the-art methods.

2 The Proposed Approach

Given N face images $X = \{x_i\}_1^N$ associated with label information $Y \in \mathbb{R}^{N \times M}$, the goal of the proposed method is to learn hashing function $\mathcal{H}(\cdot)$ that projects X into binary hash codes $B = \{b_i\}_1^N$, where M denotes the class number and $y_{ij} = 1$ if x_i belongs to the j-th class and 0 otherwise. $b_i \in \{-1, 1\}^K$ denotes the corresponding K-bit hash code of the i-th image $x_i \in \mathbb{R}^{W \times H \times C}$, with W, H, C stand for the image width, height and channel number respectively.

As shown in Fig. 1, our framework consists of two networks. (1) Face spatial network: in order to effectively utilize features from facial regions and reduce the interference of background information on images, we use the FCN-based network as a spatial attention mechanism to capture spatial information of images. (2) Hash network: a CNN which maps images to hash codes is deployed in which two feature optimization strategies are used to enhance face features. The two networks are trained alternately. Classification loss requires the learned hash codes to preserve class information while quantization loss forces the output of hash network to be more binary-like codes. Face spatial loss guides the face spatial network to exploit discriminative facial regions.

2.1   Face Spatial Network

In face image retrieval, only features from facial regions are helpful. Background information may interfere with the expression of face features. Therefore, we introduce a FCN-based network called face spatial network in our architecture to mine discriminative facial regions. As shown in Fig. 1, it is composed of several convolutional layers and deconvolutional layers. Instead of generating the label vector for each pixel as it is in semantic segmentation, we generate a spatial weight map $w_i \in \mathbb{R}^{W \times H}$ for each input face image x_i, so that each pixel of the image is assigned a value in the spatial weight map which indicates its importance:

$$w_i = f_{spatial}(x_i; \theta_{spatial}), \tag{1}$$

where $f_{spatial}$ denotes the face spatial network and $\theta_{spatial}$ represents its parameters. The following normalization is used to limit spatial weight maps in the range of 0 to 1:

$$w_i(p, q) = \frac{w_i(p, q) - \min(w_i)}{\max(w_i) - \min(w_i)}, \tag{2}$$

where $p \in [1, W]$ and $q \in [1, H]$. After normalization, the spatial weighted face image $x_i^{spatial}$ can be obtained through a matrix dot product operation is conducted:

$$x_i^{spatial} = x_i * w_i, \tag{3}$$

where $*$ denotes matrix dot production. With the face spatial loss defined in Eq. 9, the face spatial network learns to mine discriminative facial regions automatically.

2.2   Hash Network

The aim of the proposed hash network is to learn a compact and discriminative hash code for the input data. It consists of several convolutional layers, max-pooling layers, channel attention modules, and fully connected layers. Since effective face features could improve the quality of the generated hash codes, we introduce the following two strategies in the hash network:

(1) **Feature fusing.** As shallow layers of CNN often extract shallow-level features like edges and textures and deep layers extract high-level features [16], fusing features from different layers could generate multi-scale features. Therefore, features from the last two convolutional layers are passed to the channel attention modules respectively, and then, fused together to get a weighted fused feature.

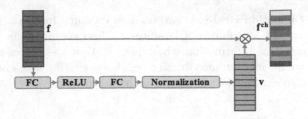

Fig. 2. Demonstration of the channel attention module.

(2) **Channel attention module.** Considering that not all features from facial regions are equally important, we use channel attention module to highlight features from discriminative channels. The architecture of the channel attention module is shown in Fig. 2. Input feature f is unfolded as $f = [f_1, f_2, ..., f_c]$, where c denotes its channel number and $f_i \in \mathbb{R}^{w \times h}$ denotes the i-th channel slice of f and w and h denotes its width and height respectively. Channel weight vector $v \in \mathbb{R}^c$, which indicates the importance of different channels, could be obtained through channel attention modules. Following normalization is applied to limit the weights in v to be in the range of 0 to 1:

$$v(i) = \frac{v(i) - \min(v)}{\max(v) - \max(v)}, \tag{4}$$

where $i \in [1, c]$. Then, channel-weighted face feature f^{ch} is generated through a channel-wise multiplication between f and v.

Channel-weighted face features are concatenated and passed to fully connected layers. The last fully connected layer is named as hash layer, in which the number of features is the same as the length of hash codes. In the training process, the spatial-weighted image $x_i^{spatial}$ and the original image x_i are encoded by the hash network respectively to get their corresponding binary-like codes. Parameters in hash network are updated with the classification loss and quantization loss defined in Sect. 2.3. Once the training process is finished, only spatial-weighted images are encoded through hash network. Then, binary hash code can be obtained by applying $sign(\cdot)$ function on the output of hash layer, where $sign(x) = 1$ when $x > 0$ and $sign(x) = -1$ otherwise.

2.3   Loss Functions

There are three loss functions in our method, classification loss and quantization loss are used to supervise the learning of hash network while face spatial loss guides the learning of face spatial network. Since discrete optimization is difficult to be solved by deep networks, we relax the binary constraint to binary-like code for the training process. We use h_i to denote the binary-like code of original image x_i, and h_i' to denote the binary-like code of the spatial weighted image $x_i^{spatial}$ in this sections.

Classification Loss. The generated hash codes are expect to preserve their class information, thus, we assume that class labels can be well predicted by the generated hash codes. We deploy a softmax function for the classification loss, which can be formulated as follows:

$$L_c^{origin} = \sum_{i=1}^{N} \sum_{j=1}^{M} -y_{ij} \log \frac{e^{\sigma_j^T h_i}}{\sum_{k=1}^{M} e^{\sigma_k^T h_i}} = \sum_{i=1}^{N} l_{c,i}, \tag{5}$$

$$L_c^{spatial} = \sum_{i=1}^{N} \sum_{j=1}^{M} -y_{ij} \log \frac{e^{\sigma_j^T h_i'}}{\sum_{k=1}^{M} e^{\sigma_k^T h_i'}} = \sum_{i=1}^{N} l_{c,i}', \tag{6}$$

where L_c^{origin} and $L_c^{spatial}$ respectively represent the classification loss of the original face images and the classification loss of spatial weighted face images. σ_j is a prediction function for the j-th class. $l_{c,i}$ and $l_{c,i}'$ denote the classification loss of x_i and the classification loss of $x_i^{spatial}$ respectively.

Quantization Loss. To minimize the information loss caused by $sign(\cdot)$ function, we want the output of hash layer to be more binary-liked, which means values in h_i and h_i' are close to $+1/-1$. Therefore, the quantization loss can be formulated as follows:

$$L_q^{origin} = \sum_{i=1}^{N} |||h_i| - 1||_1 = \sum_{i=1}^{N} l_{q,i}, \tag{7}$$

$$L_q^{spatial} = \sum_{i=1}^{N} |||h_i'| - 1||_1 = \sum_{i=1}^{N} l_{q,i}', \tag{8}$$

where L_q^{origin} and $L_q^{spatial}$ respectively stand for the quantization loss of original face images and the quantization loss of spatial weighted face images. **1** denotes the vector of ones, $|\cdot|$ is an element-wise absolute value operation, and $\|\cdot\|_1$ is l_1-norm. $l_{q,i}$ and $l_{q,i}'$ denote the quantization loss of x_i and the quantization loss of $x_i^{spatial}$ respectively.

Face Spatial Loss. Face spatial loss is designed to guide the face spatial network to mine discriminative facial regions so that effective face spatial maps could be generated. The key idea is that the hash code of the spatial weighed images are more discriminative, thus, the binary-liked code of spatial weighed facial images will have a smaller classification loss and quantization loss compared to the ones the original images. Considering that, we define face spatial loss as follows:

$$L_{spatial} = \sum_{i=1}^{N} \max(l_{c,i}' + \alpha l_{q,i}' - (1-m)(l_{c,i} + \alpha l_{q,i}), 0), \tag{9}$$

Algorithm 1. Training Strategy.
Input:
    Training face image set X with label Y;
    Max training epoch T;
Output:
    Face spatial network $f_{spatial}(x; \theta_{spatial})$;
    Hash network $f_{hash}(x; \theta_{hash})$;
1: **for** t = 1: T **do**
2:    Compute $X^{spatial}$ according to Eq. 3;
3:    Fixing $\theta_{spatial}$, update θ_{hash} according to Eq. 10 through Back Propagation;
4:    Fixing θ_{hash}, update $\theta_{spatial}$ according to Eq. 9 through Back Propagation;
5: **end for**
6: **return** $f_{spatial}(x; \theta_{spatial})$, $f_{hash}(x; \theta_{hash})$;

---

where α is a trade-off parameter to balance classification loss and quantization loss. $m \in [0, 1]$ is the margin parameter. When m is set to 0, the face spatial loss will push the face spatial network to mining a better face spatial map every time the weighted sum of classification loss and quantization loss of the spatial weighted face images ($l'_{c,i} + \alpha l'_{q,i}$) is larger than the ones of the original face images ($l_{c,i} + \alpha l_{q,i}$).

2.4   Training Strategy

Our proposed method is a two-stage, end-to-end deep model which contains a face spatial network and a hash network. Face spatial network is used to generate spatial weight maps for face images and the hash network is used to generate hash codes. As shown in Algorithm 1, we train the two networks alternately.

For the training process, the hash network is expected to generate effective hash codes for both original images and spatial weighted images, thus, both classification loss defined in Eqs. 5 and 6 and quantization loss defined in Eqs. 7 and 8 are used to form the overall hash loss:

$$L_{hash} = \sum_{i=0}^{N} l_{c,i} + l'_{c,i} + \alpha(l_{q,i} + l'_{q,i}), \tag{10}$$

where α is the same trade-off parameter defined in Eq. 9. By minimizing this term, the hash network is trained to generate class-preserving hash codes for both the original face images and the spatial weighed face images.

For the training of face spatial network, the face spatial loss defined in Eq. 9 is used. By minimizing this term, the face spatial network is trained to mining discriminative face regions of the input face images. Thus, after the matrix dot production between the original images and the spatial weight map produced by face spatial network, the background information is reduced, and the face regions are highlighted, leading to more discriminative hash codes.

2.5 Out-of-Sample Extension

When the training process is finished, we can use this model to generate K-bit length hash codes for face images.

Since our proposed model consists of two networks, firstly, as discussed in Sect. 2.1, the face image x_i is mapped to spatial weighted face image $x_i^{spatial}$ by face spatial network.

Then, $x_i^{spatial}$ is fed into hash network to generate binary-like code h_i'.

Finally, as discussed in Sect. 2.2, the binary hash code can be obtained by applying $sign(\cdot)$ function to h_i'.

3 Experiments

In order to validate the performance of our proposed method, we conduct experiments with several state-of-the-art hashing methods on two widely-used face image datasets, YouTube Faces [19] and FaceScrub [12]. Experiments are implemented with PyTorch on NVIDIA Tesla M40 with CUDA9.0 and cuDNN v7.1.2. Our source codes are released at https://github.com/deephashface/DDAH.

3.1 Datasets and Evaluation Metric

YouTube Faces is a video face dataset which contains 3,425 videos of 1,595 different people. We randomly select 40 face images for every person as the training set and 5 face images per person as the testing set. Therefore, we get 63,800 training face images and 7,975 testing face images. All face images are resized to 32×32.

FaceScrub comprises a total of 106,863 face images of 530 celebrities, with about 200 images per person. In our experiments, 5 face images for every person are randomly selected as the testing set and the remaining face images as the training set. All face images are resized to 32×32.

To evaluate our proposed method, following four evaluation metrics are employed: mean average precision (MAP), precision recall curves, precision with Hamming distance 2, and precisions w.r.t different top returned samples. When evaluation, images from the test sets are used as queries and the training sets are regarded as the galleries.

3.2 Experimental Settings

We compare our proposed method with several non-deep methods and deep methods. Non-deep methods include ITQ [4], SH [18], LSH [2], KSH [3], SDH [15], SpH [5], and deep methods include DSH [11], DDH [10] and DDQH [17]. For non-deep methods, 256-D local binary pattern (LBP) [1] features are extracted to represent the face images. For fair comparisons, deep features from CNN are also extracted for the non-deep methods which are denoted as "+CNN". We reimplement DDH and DDQH with PyTorch and results of other methods are obtained with public available source codes.

Table 1. Mean Average Precision (MAP) results for different number of bits on Face-Scrub and YouTube Faces.

Method	YouTube Faces				FaceScrub			
	12 bits	24 bits	36 bits	48 bits	12 bits	24 bits	36 bits	48 bits
ITQ [4]	0.0089	0.0138	0.0209	0.0198	0.0024	0.0026	0.0027	0.0028
LSH [2]	0.0039	0.0042	0.0059	0.0120	0.0022	0.0024	0.0027	0.0025
SH [18]	0.0104	0.0170	0.0289	0.0366	0.0026	0.0031	0.0033	0.0034
SpH [5]	0.0145	0.0373	0.0541	0.0656	0.0030	0.0035	0.0038	0.0039
KSH [3]	0.0176	0.0466	0.0740	0.1000	0.0037	0.0038	0.0039	0.0039
SDH [15]	0.0160	0.0410	0.0667	0.0844	0.0027	0.0033	0.0036	0.0039
ITQ+CNN	0.0248	0.1900	0.3420	0.4394	0.0186	0.0352	0.0504	0.0667
LSH+CNN	0.0391	0.1926	0.3354	0.4439	0.0064	0.0132	0.0238	0.0366
SH+CNN	0.0154	0.0851	0.1603	0.2421	0.0036	0.0081	0.0114	0.0145
SpH+CNN	0.0524	0.2006	0.3245	0.4080	0.0093	0.0165	0.0230	0.0295
KSH+CNN	0.0481	0.2663	0.4167	0.5047	0.0230	0.0348	0.0767	0.1026
SDH+CNN	0.5474	0.7676	0.8100	0.8331	0.1281	0.2388	0.2934	0.3291
DSH [11]	0.1538	0.4274	0.5341	0.5718	0.0122	0.0186	0.0246	0.0247
DDH [10]	0.5681	0.8526	0.9208	0.9437	0.0798	0.1109	0.1284	0.1419
DDQH [17]	0.7334	0.9721	0.9789	0.9878	0.2103	0.3431	0.4748	0.5532
Our method	**0.8987**	**0.9831**	**0.9838**	**0.9911**	**0.3574**	**0.4842**	**0.6247**	**0.6475**

For FCN, there are four convolutional layers and three deconvolutional layers. The filter size of the first convolutional layer is 3×3 while the next two are 2×2. The fourth convolutional layer has a filter size 1×1 with stride 1×1. For convolutional layers, the number of feature maps is set to 32, 64, 128 and 128 respectively. For deconvolutional layers, filter size is set to 3×3 with stride 2×2, padding 1, dilation 1, and their feature map numbers are set to 64, 32 and 16 respectively.

For CNN, the first three convolutional layers are the same as FCN except for additional max-pooling layers with filter size 2×2, stride 2×2. The number of feature maps of the fourth convolutional layer is set to 256, with a filter size 2, and stride 2×2. The number of features of the first fully connected layer is set to 1024. For channel attention module, feature number of the first fully connected layer is set to 512 while the second fully connected layer is set according to its input feature number, in our experiments, 128 and 256 respectively.

In the training process, we use Adam algorithm [8] to optimize the network with batch size fixed as 256 and weight decay parameter as 0.0001. We select the hyper-parameters m and α, by cross-validation.

3.3   Results and Discussions

Results on Mean Average Precision. The MAP results for different bits on FaceScrub and YouTube Faces are shown as Table 1. In our experiments, the

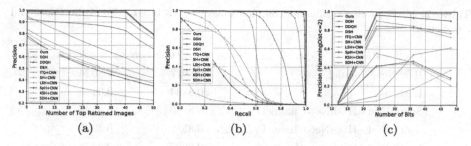

Fig. 3. Comparison of different methods on YouTube Faces. (a) Precision curves of different top returned images with 48 bits. (b) Precision recall curves of Hamming ranking with 48 bits. (c) Precision curves with Hamming distance 2.

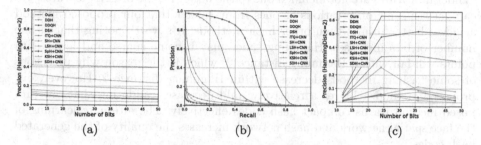

Fig. 4. Comparison of different methods on Facescrub. (a) Precision curves of different top returned images with 48 bits. (b) Precision recall curves of Hamming ranking with 48 bits. (c) Precision curves with Hamming distance 2.

bit length of the hash codes is set to 12, 24, 36 and 48, respectively. It can be observed that, in general, non-deep methods using deep features achieve better results compared with the ones using LBP features. This is because deep features are more robust and selective. Second, the performances of deep methods are better than non-deep methods. In deep methods, face feature extracting and hash code learning are conducted in a unified framework, reducing the information loss between the two stages. Third, compared to DDH and DDQH, our proposed method achieves better performance. This is because DDH and DDQH directly extract features from images, but our proposed method is specially designed for face images. In our proposed method, the face spatial network learns to exploit discriminative facial regions to reduce the interference of the background information and highlight the facial regions, the channel attention module enhances discriminative face features from channel aspect, leading to a more discriminative hash code.

Results on Precision Recall Curves, Precision Curves. Experimental results of precision recall curves, precision curves w.r.t. different top returned images and precision curves within Hamming distance 2 for FaceScrub and YouTube Faces are shown in Figs. 3 and 4 respectively. Non-deep methods with

Table 2. MAP result of different combinations with 12 bits on YouTube Faces and FaceScrub.

Combinations	YouTube Faces	FaceScrub
HashNet-Channel	0.8017	0.3024
HashNet	0.8184	0.3166
SpatialNet+HashNet-Channel	0.8130	0.3127
SpatialNet+HashNet-Channel+$L_{spatial}$	0.8586	0.3292
SpatialNet+HashNet+$L_{spatial}$	0.8987	0.3574

handcraft features are not included due to the better results of CNN features. We can get the same observations as ones from Table 1 that our proposed method achieves better results than other hash methods. It's worth noting that as the number of bits increases, Hamming space may become sparse and fewer data points will fall within Hamming distance 2, which is the reason why precisions of some hashing methods decrease within Hamming distance 2. However, our proposed method achieves a relatively mild decrease, validating that our proposed method learns compact hash codes effectively. The alternate learning of the face spatial network and hash network increases the quality of the generated hash codes.

Ablation Study. Since our proposed method contains two attention mechanisms: face spatial network and channel attention module, to evaluate their contribution, we further study the retrieval performance of their different combinations. The results are shown in Table 2. We use "HashNet", "Channel" and "SpatialNet" to denote the hash network trained with hash loss, channel attention module and the face spatial network respectively. "$L_{spatial}$" denotes the alternate learning with face spatial loss. For example, "SpatialNet + HashNet-Channel" denotes that both face spatial network and hash network are used but there's no channel attention module and face spatial loss. It can be observed that both these two mechanisms improve the results. With the guidance of face spatial loss, the face spatial network learns to enhance features from discriminative facial regions, which improves the performance. Furthermore, combining these two mechanisms together achieves a better performance, as the two attention mechanisms improve face features from two different aspects, spatial aspect and channel aspects respectively.

Sensitivity Study. We further study the influence of parameters α and m. We set α within $\{10^{-4}, 10^{-3}, 10^{-2}, 10^{-1}\}$ and m within $\{0, 2^{-4}, 2^{-3}, 2^{-2}, 2^{-1}\}$. The results on FaceScrub dataset are shown in Fig. 5. The best result is obtained with $\alpha = 10^{-3}$ and $m = 2^{-4}$. It can be observed that α affects the performance more than m. As α balances the classification loss and quantization loss, an inappropriate value will affect the quality of hash codes generated by hash network. m

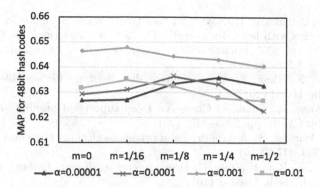

Fig. 5. Sensitivity to hyper-parameters on FaceScrub dataset.

controls the learning of face spatial network, a too large or too small value may lead to the face spatial network not learned well.

4   Conclusion

This paper presents a deep hashing approach specially designed for face image retrieval. With the integration of face spatial network and hash network enhanced by channel attention module, discriminative face features can be exploited and effective binary hash codes can be generated. Experiments on two widely-used datasets compared with some state-of-the-art methods show that our proposed method achieves inspiring performance.

Acknowledgements. This work was supported by the Strategic Priority Research Program of the Chinese Academy of Sciences (XDC02050200).

References

1. Ahonen, T., Hadid, A., Pietikainen, M.: Face description with local binary patterns: application to face recognition. IEEE TPAMI **28**(12), 2037–2041 (2006)
2. Andoni, A., Indyk, P.: Near-optimal hashing algorithms for approximate nearest neighbor in high dimensions. Commun. ACM **51**, 117–122 (2008)
3. Chang, S.F., Jiang, Y.G., Ji, R., Wang, J., Liu, W.: Supervised hashing with kernels. In: CVPR, pp. 2074–2081 (2012)
4. Gong, Y., Lazebnik, S.: Iterative quantization: a procrustean approach to learning binary codes. In: CVPR, pp. 817–824 (2011)
5. Heo, J.P., Lee, Y., He, J., Chang, S.F., Yoon, S.E.: Spherical hashing: binary code embedding with hyperspheres. IEEE TPAMI **37**(11), 2304–2316 (2015)
6. Jiang, Q., Li, W.: Asymmetric deep supervised hashing. In: AAAI, pp. 3342–3349 (2018)
7. Jin, S.: Deep saliency hashing. CoRR abs/1807.01459 (2018)
8. Kingma, D.P., Ba, J.: Adam: a method for stochastic optimization. CoRR abs/1412.6980 (2014)

9. Kumar, N., Belhumeur, P., Nayar, S.: FaceTracer: a search engine for large collections of images with faces. In: Forsyth, D., Torr, P., Zisserman, A. (eds.) ECCV 2008. LNCS, vol. 5305, pp. 340–353. Springer, Heidelberg (2008). https://doi.org/10.1007/978-3-540-88693-8_25

10. Lin, J., Li, Z., Tang, J.: Discriminative deep hashing for scalable face image retrieval. In: IJCAI (2017)

11. Liu, H., Wang, R., Shan, S., Chen, X.: Deep supervised hashing for fast image retrieval. In: CVPR, pp. 2064–2072 (2016)

12. Ng, H.W., Winkler, S.: A data-driven approach to cleaning large face datasets. In: ICIP, pp. 343–347 (2015)

13. Qi, L., Sun, Z., Ran, H., Tan, T.: Deep supervised discrete hashing. IEEE Trans. Pattern Anal. Mach. Intell. 1 (2017)

14. Shelhamer, E., Long, J., Darrell, T.: Fully convolutional networks for semantic segmentation. IEEE TPAMI **39**(4), 640–651 (2014)

15. Shen, F., Shen, C., Liu, W., Shen, H.T.: Supervised discrete hashing. In: CVPR (2015)

16. Taigman, Y., Yang, M., Ranzato, M., Wolf, L.: DeepFace: closing the gap to human-level performance in face verification. In: IEEE CVPR, pp. 1701–1708 (2014)

17. Tang, J., Jie, L., Li, Z., Jian, Y.: Discriminative deep quantization hashing for face image retrieval. IEEE TNNLS **29**, 6154–6162 (2018)

18. Weiss, Y., Torralba, A., Fergus, R.: Spectral hashing. In: NIPS, pp. 1753–1760 (2008)

19. Wolf, L., Hassner, T., Maoz, I.: Face recognition in unconstrained videos with matched background similarity. In: CVPR, pp. 529–534 (2011)

20. Wu, D., Lin, Z., Li, B., Ye, M., Wang, W.: Deep supervised hashing for multi-label and large-scale image retrieval. In: ICMR, pp. 150–158 (2017)

21. Zhu, H., Long, M., Wang, J., Cao, Y.: Deep hashing network for efficient similarity retrieval. In: AAAI, pp. 2415–2421 (2016)

Texture Deformation Based Generative Adversarial Networks for Multi-domain Face Editing

Wenting Chen[1,2], Xinpeng Xie[1,2], Xi Jia[1,2], and Linlin Shen[1,2(✉)]

[1] Computer Vision Institute, School of Computer Science and Software
Engineering, Shenzhen University, Shenzhen 518060,
People's Republic of China
{chenwenting2017,xiexinpeng2017,
jiaxi}@email.szu.edu.cn, llshen@szu.edu.cn
[2] Guangdong Key Laboratory of Intelligent Information Processing,
Shenzhen University, Shenzhen 518060, People's Republic of China

Abstract. Despite the significant success in image-to-image translation and latent representation based facial attribute editing and expression synthesis, the existing approaches still have limitations of preserving the identity and sharpness of details, and generating distinct image translations. To address these issues, we propose a Texture Deformation Based GAN, namely TDB-GAN, to disentangle texture from original image. The disentangled texture is used to transfer facial attributes and expressions before the deformation to target shape and poses. Sharper details and more distinct visual effects are observed in the synthesized faces. In addition, it brings faster convergence during training. In the extensive ablation studies, we also evaluate our method qualitatively and quantitatively on facial attribute and expression synthesis. The results on both the CelebA and RaFD datasets suggest that TDB-GAN achieves better performance.

Keywords: Texture · Deformation · Generative Adversarial Networks · Multi-domain face editing

1 Introduction

Face editing aims to change or enhance facial attributes such as hair color, expression, gender and age, and add virtual makeup to human faces etc. In recent years, face editing has attracted great interests in computer vision fields [1, 16, 22]. Several methods [8, 24, 27] have achieved facial attributes and expressions manipulation on single or multiple domains. Most of them are based on the generative adversarial networks (GANs) [3] like Cycle GAN [27], IcGAN [14], StarGAN [2], etc. These

The work is supported by National Natural Science Foundation of China (Grant No. 61672357 and U1713214), and the Science and Technology Project of Guangdong Province (Grant No. 2018A050501014).

© Springer Nature Switzerland AG 2019
A. C. Nayak and A. Sharma (Eds.): PRICAI 2019, LNAI 11670, pp. 257–269, 2019.
https://doi.org/10.1007/978-3-030-29908-8_21

image-to-image translation approaches feed the generator with the input image directly, which might lead to indistinctive modification of the input.

Input	Texture	Smiling	Pale skin	Eyeglasses	Gender	Age

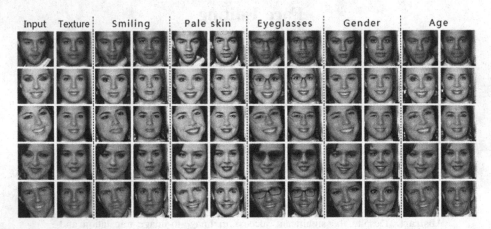

Fig. 1. TDB-GAN transfer attributes by texture based deformation. The first and second columns are the input images and the texture generated by DAE, respectively. In the remaining columns, while the even columns show the textures produced by generator for the target attribute, the odd ones show the synthesized images by warping the texture.

The recent works [21, 26] demonstrate that intuitively decomposing a single image into shading and albedo components can benefit the perception of images. The intrinsic image model assumes that color image I can be reconstructed by the point-wise product of shading S and albedo $A : I = A \cdot S$. Here, albedo is the reflectance of surfaces in the scene. Compared to standard autoencoder architectures, intrinsic image decomposition can lead to better synthesis results for tasks such as facial attributes and expression manipulation. In [15], the author extended the intrinsic image decomposition by decompose the image into shading, albedo and deformation components. The deformation component can dispel the variation of rotations, translation, or scaling, which makes it more feasible to control and understand deep networks. Note that, in [15] I is termed texture by the author. By dispelling the deformation, the face textures extracted by shading and albedo are all well-aligned and have more strong representation of intrinsic images.

Inspired by these works, we first adopt the DAE [15] model to extract a well-aligned texture image from the input image. Then, we feed both the generated texture and target domain labels into a GAN model to synthesize a new texture image with target attributes. Finally, we warp the generated texture with the spatial deformation and employ an identity loss to preserve the identity. Overall, our main contributes are summarized as follows: (1) We propose the Texture Deformation Based GAN, a novel framework that learns the mappings among multiple domains based on disentangled texture and warps the generated texture spatially to generate the face image in target domain. (2) We empirically demonstrate the effectiveness of our TDB-GAN through the ablation studies on facial attribute editing and expression synthesis. The results justify the superiority of texture-to-image translation over the image-to-image translation.

2   Related Work

The popularity of generative models has a great effect on face editing. The Encoder-Decoder architecture and Generative Adversarial Network (GAN) [3] are the two major categories of methods for this task.

Generative Adversarial Networks (GANs) [3] is a promising generative model and can be used to solve various computer vision tasks such as image generation [6, 20, 23], image translation [8, 11, 24, 27], and face image editing [2, 13, 22]. The GAN model is mainly designed to learn a generator to generate fake samples and a discriminator to distinguish between real and fake samples. Besides leveraging the typical adversarial loss, a reconstruction loss is often employed [2, 4] to generate the faces as realistic as possible. Additionally, an identity loss is proposed to assure that the generated faces preserve the original identity in our approach.

Image-to-image translation based methods, e.g. MUNIT [7], CycleGAN [27], and IcGAN [14], are commonly used to transfer style, as they can learn the mapping between input and output domains. As for the multi-domain image translation, Star-GAN [2] and AttGAN [4] are proposed recently. StarGAN employs only one generator to achieve this translation across different datasets, and its generator is allowed to reconstruct the original image from the fake image given the original domain label. However, StarGAN does not involve any latent representation, so its capability of changing facial attributes is limited. As for AttGAN, it contains three components at training: the attribute classification constraint, the reconstruction learning and the adversarial learning. AttGAN tries to generate fake images from the attribute-independent latent representation, while our approach encodes the input image to texture and employs an image-to-image translation to achieve face editing. TextureGAN [17] using the same terminology can synthesize objects consistent with the given texture suggestions, which is completely in a different context.

Intrinsic Image Decomposition decomposes a single image into shading, albedo and deformation components. DAE [15] is a novel intrinsic image decomposition model which decomposes the input image into texture and deformation. DAE follows the deformable template paradigm and models image generation through texture synthesis and spatial deformation. DAE can obtain the prototypical object by removing the deformation. Discarding variability due to deformations, the texture encoded from the original image is a disentangled representation. Moreover, by modeling the face image with a low-dimensional latent code, we can more easily control the facial attributes and expression over the generation process. However, DAE is only proposed for reconstructing the intrinsic image, which is not applicable for multi-domain face editing. Thus, we integrate disentangled representation capability of DAE to the popular GAN-based framework in this paper, for multi-domain translation. Combing these two architectures, we can more easily control the process of face attributes editing on textures separated from the shape variability.

3   Texture Deformation Based GAN

StarGAN takes an image as input, which cannot transfer attributes specifically and disentangle the attribute-related information from the attribute-independent information. Inspired by AttGAN, we utilize DAE to synthesize an attribute-independent texture. Furthermore, we adopt an identity loss to strength the identity preservation between the input and the generated texture.

3.1   Intrinsic Deforming Autoencoder

While editing facial attributes and synthesizing expression, an ideal algorithm shall be able to disentangle the pose and shape of face from such process. Thus, we utilize the Intrinsic DAE [15] to separate a face image into texture and deformation to disentangle the pose and shape. Without the geometric information, the identity, illumination and face attributes etc. contained in the texture can be deformed by the spatial gradient of the warping field (spatial transformation).

As visualized in Fig. 2, the encoder $E_{\theta_{enc}}$, a densely connected convolutional network, takes an input image I_{Input} as input and generates a latent representation $Z = [Z_S, Z_A, Z_D]$, where Z_S, Z_A and Z_D are shading-related, albedo-related and deformation-related representations, respectively. Then, three separate decoders for shading, albedo and deformation, including D_S, D_A and D_D, are fed with the latent representations Z_S, Z_A and Z_D respectively. The decoders can provide us with a clear separation of shading, albedo and deformation. Next, the texture T of the input image can be computed by the Hadamard product of the shading S and albedo A. Finally, we spatially warp the generated texture with the deformation De to synthesize the ultimate image I_{Output}.

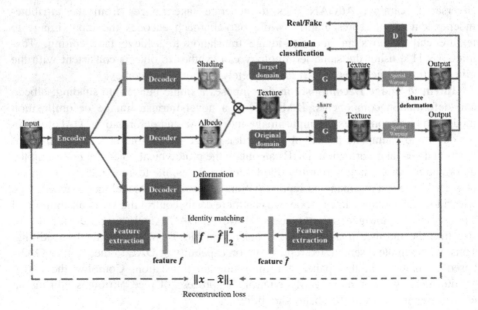

Fig. 2. Overview of texture deformation based GAN.

3.2 Multi-domain Texture-to-Image Translation

To achieve multi-domain texture-to-image translation, we first feed the generator with texture t and target domain label c randomly sampled from training data. Then, we warp the generated texture $\hat{t}$ with deformation De of input image to synthesize the fake face image. We also introduce texture reconstruction loss and identity loss to synthesize more realistic texture and identity-preserved face images, respectively.

Adversarial Loss. We utilize the adversarial loss to enable the generated images as genuine as the real samples. It can be written as:

$$L_{adv} = E_x[logD_{src}(x)] + E_{t,c}[log(1 - D_{src}(W(G(t,c),De)))]. \qquad (1)$$

In adversarial loss, G generates a new texture $G(t,c)$ conditioned on both the face texture t and target domain label c, while D strives to differentiate the real face texture from the generated face texture. In Eq. (1), $D_{src}(x)$ denotes a probability distribution over sources given by D. The discriminator tries to maximize this objective, whereas the generator tries to minimize it.

Domain Classification Loss. To enable the generator to generate the fake image with the target domain, we add a domain classifier on the top of D. For the optimization of D and G, we define the domain classification of the real image as follow:

$$L_{cls}^r = E_{t,c'}\left[-logD_{cls}\left(c'|x\right)\right], \qquad (2)$$

where c' stands for the original domain label for the real face image. The term $D_{cls}(c'|x)$ represents a probability distribution over domain labels produced by D. In addition, the domain classification loss of the fake face texture is defined as:

$$L_{cls}^f = E_{t,c}[-logD_{cls}(c|W(G(t,c),De))], \qquad (3)$$

where W denotes spatially warping.

Reconstruction Loss. By optimizing the adversarial and classification loss, G is able to generate the realistic face texture with proper attributes. Nonetheless, we cannot guarantee that the generated face texture preserves the content of the input face texture while changing the domain-related parts of the input face texture. Therefore, the reconstruction loss is imposed to the reconstructed texture and image, respectively. For the texture image, we apply a cycle consistency loss proposed by Zhu et al. [27] to our generator, which is defined as:

$$L_{rec}^t = E_{t,c,c'}\left[\left\|t - G\left(G(t,c),c'\right)\right\|_1\right], \qquad (4)$$

where G takes the generated face texture $G(t, c)$, the original domain label c' as input and tries to reconstruct the original face texture and $\|\cdot\|_1$ denotes the L1 norm.

For the reconstructed image, L1 norm of the difference between the input and the generated image is defined as below:

$$L_{rec}^i = E_{t,c'} \left[\left\| x - W\left(G\left(t,c'\right),De\right) \right\|_1 \right] \tag{5}$$

$$L_{rec} = L_{rec}^t + L_{rec}^i \tag{6}$$

Identity Loss. To preserve the identity while transferring attributes, we exploit an identity preserving network F_{ip} to retain the identity discrimination of the synthesized face texture, and an identity loss L_{ip} to preserve personal facial features, which is derived from the work proposed by Huang [5]. F_{ip} denotes a feature extractor to extract the feature of the synthesized face texture $\hat{t}$ and the real face texture t. We select the LightCNN [19] as our feature extractor and apply the output of the second to last fully connected layer of F_{ip} to the identity loss L_{ip}:

$$L_{ip} = \left\| F_{ip}(t) - F_{ip}(\hat{t}) \right\|_2^2, \tag{7}$$

where $\| \cdot \|_1$ denotes the L2-norm.

GAN-Related Objective Function. Overall, the final objective functions to optimize G and D are illustrated as:

$$L_D = -L_{adv} + \lambda_{cls} L_{cls}^r, \tag{8}$$

$$L_G = L_{adv} + \lambda_{cls} L_{cls}^f + \lambda_{rec} L_{rec} + \lambda_{ip} L_{ip}, \tag{9}$$

where λ_{cls}, λ_{rec} and λ_{ip} are hyper-parameters to control the weight of domain classification, reconstruction and identity loss.

4 Implementation

The proposed TDB-GAN is implemented using Pytorch toolbox with reference to the source code of DAE [15] and StarGAN [2]. We use DAE [15] as our backbone and adopt a multi-stage training strategy to stabilize and accelerate the training procedure of TDB-GAN. In the first stage, we only optimize the DAE model, namely the L_R, L_{smooth}, L_B and $L_{Shading}$. Then, we fix the pretrained weights of DAE model. Simultaneously, the generator G and discriminator D are trained with the L_G (with $\lambda_{ip} = 0$) and L_D loss, respectively. Finally, we jointly train L_{DAE}, L_G and L_D. Note that, we impose the identity loss L_{ip} in the final training stage to ensure that the generated image preserves the identity.

5 Experiments

5.1 Datasets

The CelebFaces Attributes (CelebA) dataset [12] contains 202,599 face images of 10,177 celebrities, each annotated with 40 binary attributes. We resize all aligned images from the 178×218 into 64×64. We randomly select 2,000 images as test set and use the remaining images for training. We mainly test ten domains with following attributes: expression (smiling/not smiling), skin color (pale skin/normal skin), accessory (eyeglasses/no eyeglasses), gender (male/female) and age (young/old).

The Radboud Faces Database (RaFD) [10] consists of 4,824 images collected from 67 subjects. Each subject has eight facial expressions in three different gaze directions, which are captured from three different angles. We first detect all face images with MTCNN [25] and crop out the images with size 384×384, where the faces are centered, and resized to 64×64. In all the experiment, we fix the input domain as the 'neutral' expression and set the target domain to the seven remaining expressions. Thus, the proposed task aims to impose a particular expression to a neutral face. Then, we randomly split the RaFD dataset into training and testing sets with a 90%:10% ratio, namely 4,320 training images and 504 testing images including 63 neutral faces.

5.2 Training

All the models are optimized with Adam [9], where $\beta_1 = 0.5$ and $\beta_2 = 0.999$. We flip the images horizontally with a probability of 0.5. We perform one generator update after five discriminator updates as described in [2]. The batch size is 100. We first train the DAE module for 5 epochs with a learning rate of 0.0002. Then, we train the GAN module for 200 epochs. Next, we impose the identity loss to the GAN module and train the GAN-related part for 29 epochs with a learning rate of 0.0001 and apply the aforementioned decaying strategy over the next 29 epochs. The learning rate starts with 0.0001 and is decreased linearly to 0 after 100 epochs. We set $\lambda_{cls} = 1$, $\lambda_{rec} = 10$ and $\lambda_{ip} = 0.001$.

5.3 Qualitative and Quantitative Evaluation on CelebA

Qualitative Evaluation. Figure 1 shows the images and textures transferred to different attributes, which are synthesized by our method. Figure 3 shows the face images generated by IcGAN [14], CycleGAN [27], StarGAN [2] and our TDB-GAN for attribute transfer in smiling, pale, eyeglasses, gender and age. Note that, we used the hyper-parameters mentioned in their papers without any further parameters tuning. As visualized in the Fig. 3, the images generated by image-to-image translation approaches are better than that generated by IcGAN. Our approach contains more information than the low-dimension latent representation and also preserves the attribute-independent information, like hairstyle. The faces generated by TDB-GAN for gender and age transfer are better than that generated by StarGAN, and the eyeglasses added by TDB-GAN are more natural than that added by CycleGAN. Furthermore, our method not only achieves higher visual quality but also preserves the identity related to the input image due to the proposed identity loss.

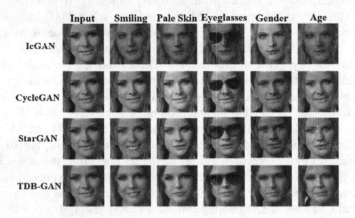

Fig. 3. Facial expression synthesis results on CelebA dataset.

Quantitative Evaluation. For quantitative evaluation, we perform a user study on the visual effect of transferred facial attributes to evaluate IcGAN, CycleGAN, StarGAN and TDB-GAN. Each of the four approaches were applied to transfer smile, pale skin, eyeglasses, gender, and age of faces from randomly selected twenty subjects of the CelebA dataset. For each of the five attributes, four images synthesized by different models were shuffled and randomly shown to volunteers. As a number of 15 volunteers participated the questionnaire, a maximum of $20 \times 5 = 300$ votes can be received for each approach and attribute. They were asked to select the best one, in terms of the realism, preservation of identity and quality of the facial attribute synthesis. Table 1 lists the ratio of votes received for each model and attribute. While StarGAN received the highest votes for pale skin transfer, our TDB-GAN received the highest votes for four of the five attributes.

Table 1. The perceptual evaluation of different models. Note that, the sum of probability of each row is not strictly equal to 100% due to numerical precision loss.

Models	Smile	Pale skin	Eyeglasses	Gender	Age
IcGAN	2.33%	2.00%	0	1.33%	0.33%
CycleGAN	21.33%	**37.00%**	28.00%	35.00%	20.00%
StarGAN	19.00%	36.67%	30.33%	9.67%	17.67%
TDB-GAN	**57.33%**	24.33%	**41.67%**	**54.00%**	**62.00%**

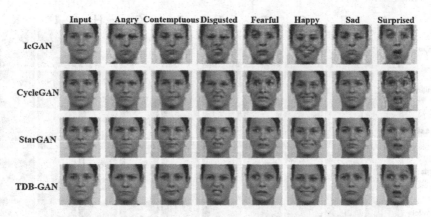

Fig. 4. Facial expression synthesis results on RaFD dataset.

5.4   Qualitative and Quantitative Evaluation on RaFD

Qualitative Evaluation. Figure 4 shows an example of seven facial expressions synthesized by IcGAN, CycleGAN and StarGAN and our TDB-GAN. In Fig. 4, the images generated by StarGAN and our TDB-GAN have better visual quality than that generated by IcGAN and CycleGAN. The images generated by IcGAN have the lowest quality. We believe that the latent vector extracted from IcGAN lacks effective representability. While the performance of CycleGAN is considerably better than that of IcGAN, the fake images generated by CycleGAN are still ambiguous. The fake faces synthesized by StarGAN have much more natural and distinct expressions. Nonetheless, TDB-GAN is superior to StarGAN for the sharper details and the more distinguishable expressions. For example, the faces generated by our TDB-GAN for angry, fearful and surprised are much more representative than that of StarGAN in the eye regions.

Quantitative Evaluation. For a quantitative evaluation, we compute the classification error of facial expression recognition on the generated images. We first train a facial expression classifier with the 4,320 training images. And then we train all the GAN models using the same training set. For testing, we first use the trained GANs to transfer all the neutral expression of the testing images to seven different expressions. Then we use the aforementioned classifier to classify these synthesized expressions. Table 2 lists the accuracies of the facial expression classifier on the images synthesized by different GAN models. As shown in Table 2, the images synthesized by TDB-GAN model achieves the highest accuracy, which suggests that it synthesizes the most realistic facial expressions compared with the other methods.

Table 2. The expression classification accuracies of images synthesized by different GAN models.

Models	IcGAN	CycleGAN	StarGAN	TDB-GAN
Accuracy (%)	91.61	88.44	92.06	**97.28**

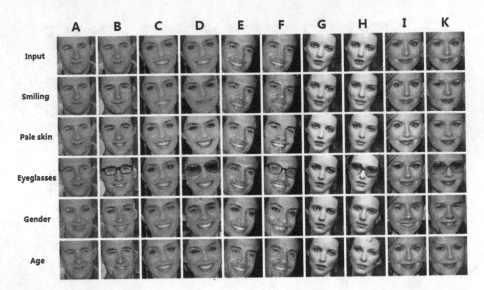

Fig. 5. Facial attribute transfer results on the CelebA dataset. The first row demonstrates the input image, next five rows show the single attribute transfer results. The odd columns display the results generated by the TDB-GAN without DAE module, while the even columns show the results produced with DAE.

5.5 Ablation Studies

In this section, we conduct an experiment on TDB-GAN with/without DAE module and prove that the proposed identity loss helps to preserve more identity information through a verification result.

Results with/without DAE. Note that, TDB-GAN without DAE is a typical StarGAN. We argue that deformation would significantly affect the quality of face editing and the convergence of the domain classification loss of fake face texture during training. As illustrated in Fig. 5, the eyeglasses generated by TDB-GAN with DAE are more obvious. For example, no glasses can be observed for the faces in column C, E, G and I generated by TDB-GAN without DAE. The images generated by TDB-GAN without DAE (A, C) do not show the pale skin as realistic as those generated by TDB-GAN. While the faces of C and E generated by TDB-GAN without DAE are still smiling, TDB-GAN with DAE correctly and naturally transfers the face image to smile or not. Lastly, our method generates more genuine transfer of feminization, masculinity, aging and rejuvenation than the TDB-GAN without DAE module.

As depicted in Fig. 6 (a), TDB-GAN with DAE achieves a lower domain classification loss of fake face textures than the TDB-GAN without DAE. There is a clear margin between the curves in the chart. The lower domain classification loss of fake face textures indicates the better attributes transferring.

Results for Identity Loss. To verify the effectiveness of the identity loss, we evaluate the performance of the domain transferring in terms of face recognition accuracy generated by the identity loss.

For each of the neutral face, we apply our network to generate seven facial expression images, i.e. in total $63 \times 7 = 441$ fake facial expression images were generated. Based on the 441 generated faces and 504 test images, we randomly generate 3,000 client and 3,000 impostor accesses. The network proposed by Wen and Zhang [18] is employed to extract 512-dimension identity features from the face images. The cosine distance is adopted to measure the similarity of two faces. The similarity was compared with a threshold (e.g. 0.5) to decide whether they are from the same person, or not. In this work, TPR (True Positive Rate), FPR (False Positive Rate), EER (Equal Error Rate), AP (Average Precision) and AUC (Area under curve) are used to evaluate the performance of face verification. The higher scores of these metrics, except EER, the better results.

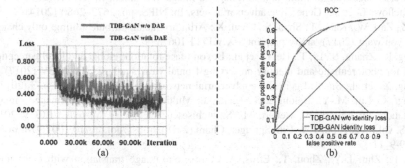

Fig. 6. (a) The domain classification loss of the fake face textures generated by the TDB-GAN with/without DAE module. (b) ROC curves on the test set of RaFD dataset.

Table 3. Verification performance on RaFD dataset.

Method	TDB-GAN with identity loss	TDB-GAN w/o identity loss
TPR@FPR = 1%	11.07	8.7
TPR@FPR = 0.1%	1.6	0.6
EER (%)	23.6	24.5
AP (%)	81.89	80.29
AUC (%)	83.73	82.82

Figure 6 (b) and Table 3 show the ROC curves and the verification results of the TDB-GAN with/without identity loss. From Table 3, while the TPR@FPR = 1% for TDB-GAN without identity loss is 8.70, the identity loss significantly increases the TPR@FRP = 1% to as high as 11.07. Identity loss almost doubles the TPR@FPR = 0.1% of the TDB-GAN. Table 3 also suggests that the TDB-GAN with identity loss achieves the lower EER and higher AP and AUC than the TDB-GAN without identity loss.

6 Conclusion

In this paper, we proposed Texture Deformation Based GAN to perform texture-to-image translation among multiple domains. The proposed TDB-GAN can generate images with higher quality and preserved identity compared to the existing methods, due to the disentangled texture and deformation, and the identity loss.

References

1. Chen, Y.-C., et al.: Facelet-bank for fast portrait manipulation. In: CVPR (2018)
2. Choi, Y., Choi, M., Kim, M., Ha, J.-W., Kim, S., Choo, J.: StarGAN: unified generative adversarial networks for multi-domain image-to-image translation. In: CVPR (2018)
3. Goodfellow, I., et al.: Generative adversarial nets. In: NIPS, pp. 2672–2680 (2014)
4. He, Z., Zuo, W., Kan, M., Shan, S., Chen, X.: Arbitrary facial attribute editing: only change what you want (2017). arXiv preprint arXiv:1711.10678
5. Huang, R., Zhang, S., Li, T., He, R., et al.: Beyond face rotation: global and local perception GAN for photorealistic and identity preserving frontal view synthesis. In: ICCV (2017)
6. Huang, X., et al.: Stacked generative adversarial networks. In: CVPR, vol. 2, p. 3 (2017)
7. Huang, X., Liu, M.-Y., Belongie, S., Kautz, J.: Multimodal unsupervised image-to-image translation. In: Ferrari, V., Hebert, M., Sminchisescu, C., Weiss, Y. (eds.) ECCV 2018. LNCS, vol. 11207, pp. 179–196. Springer, Cham (2018). https://doi.org/10.1007/978-3-030-01219-9_11
8. Isola, P., Zhu, J.-Y., Zhou, T., Efros, A.A.: Image-to-image translation with conditional adversarial networks (2017). arXiv preprint
9. Kingma, D.P., Ba, J.: Adam: a method for stochastic optimization (2014). arXiv preprint arXiv:1412.6980
10. Langner, O., et al.: Presentation and validation of the Radboud Faces Database. Cogn. Emot. **24**(8), 1377–1388 (2010)
11. Liu, M.-Y., Breuel, T., Kautz, J.: Unsupervised image-to-image translation networks. In: NIPS, pp. 700–708 (2017)
12. Liu, Z., et al.: Deep learning face attributes in the wild. In: ICCV, pp. 3730–3738 (2015)
13. Natsume, R., Yatagawa, T., Morishima, S.: RSGAN: face swapping and editing using face and hair representation in latent spaces (2018). arXiv preprint arXiv:1804.03447
14. Perarnau, G., van de Weijer, J., Raducanu, B., Álvarez, J.M.: Invertible conditional GANs for image editing. In: NIPS Workshop on Adversarial Training (2016)
15. Shu, Z., Sahasrabudhe, M., Alp Güler, R., Samaras, D., Paragios, N., Kokkinos, I.: Deforming autoencoders: unsupervised disentangling of shape and appearance. In: Ferrari, V., Hebert, M., Sminchisescu, C., Weiss, Y. (eds.) ECCV 2018. LNCS, vol. 11214, pp. 664–680. Springer, Cham (2018). https://doi.org/10.1007/978-3-030-01249-6_40
16. Shu, Z., Yumer, E., Hadap, S., Sunkavalli, K., Shechtman, E., Samaras, D.: Neural face editing with intrinsic image disentangling. In: CVPR, pp. 5444–5453. IEEE (2017)
17. Xian, W., Sangkloy, P., Agrawal, V., et al.: TextureGAN: controlling deep image synthesis with texture patches. In: CVPR (2018)
18. Wen, Y., Zhang, K., Li, Z., Qiao, Yu.: A discriminative feature learning approach for deep face recognition. In: Leibe, B., Matas, J., Sebe, N., Welling, M. (eds.) ECCV 2016. LNCS, vol. 9911, pp. 499–515. Springer, Cham (2016). https://doi.org/10.1007/978-3-319-46478-7_31

19. Wu, X., et al.: A light cnn for deep face representation with noisy labels. IEEE Trans. Inf. Forensics Secur. **13**(11), 2884–2896 (2018)
20. Xian, W., Sangkloy, P., Lu, J., Fang, C., Yu, F., Hays, J.: TextureGAN: controlling deep image synthesis with texture patches (2017). arXiv preprint
21. Narihira, T., Maire, M., Yu, S.X.: Direct intrinsics: learning albedo-shading decomposition by convolutional regression. In: ICCV (2015)
22. Xiao, T., Hong, J., Ma, J.: ELEGANT: exchanging latent encodings with GAN for transferring multiple face attributes. In: Ferrari, V., Hebert, M., Sminchisescu, C., Weiss, Y. (eds.) ECCV 2018. LNCS, vol. 11214, pp. 172–187. Springer, Cham (2018). https://doi.org/10.1007/978-3-030-01249-6_11
23. Yan, X., Yang, J., Sohn, K., Lee, H.: Attribute2Image: conditional image generation from visual attributes. In: Leibe, B., Matas, J., Sebe, N., Welling, M. (eds.) ECCV 2016. LNCS, vol. 9908, pp. 776–791. Springer, Cham (2016). https://doi.org/10.1007/978-3-319-46493-0_47
24. Yi, Z., Zhang, H.R., Tan, P., Gong, M.: DualGAN: unsupervised dual learning for image-to-image translation. In: ICCV, pp. 2868–2876 (2017)
25. Zhang, K., et al.: Joint face detection and alignment using multitask cascaded convolutional networks. IEEE Signal Process. Lett. **23**(10), 1499–1503 (2016)
26. Tappen, M.F., et al.: Recovering intrinsic images from a single image. In: NIPS (2003)
27. Zhu, J.-Y., Park, T., Isola, P., Efros, A.A.: Unpaired image-to-image translation using cycle-consistent adversarial networks. In: ICCV (2017)

Towards Generating Stylized Image Captions via Adversarial Training

Omid Mohamad Nezami[1,2]($\boxtimes$), Mark Dras[1], Stephen Wan[2],
Cécile Paris[1,2], and Len Hamey[1]

[1] Macquarie University, Sydney, NSW, Australia
omid.mohamad-nezami@hdr.mq.edu.au
{mark.dras,len.hamey}@mq.edu.au
[2] CSIRO's Data61, Sydney, NSW, Australia
{stephen.wan,cecile.paris}@data61.csiro.au

Abstract. While most image captioning aims to generate objective descriptions of images, the last few years have seen work on generating visually grounded image captions which have a specific style (e.g., incorporating positive or negative sentiment). However, because the stylistic component is typically the last part of training, current models usually pay more attention to the style at the expense of accurate content description. In addition, there is a lack of variability in terms of the stylistic aspects. To address these issues, we propose an image captioning model called ATTEND-GAN which has two core components: first, an attention-based caption generator to strongly correlate different parts of an image with different parts of a caption; and second, an adversarial training mechanism to assist the caption generator to add diverse stylistic components to the generated captions. Because of these components, ATTEND-GAN can generate correlated captions as well as more human-like variability of stylistic patterns. Our system outperforms the state-of-the-art as well as a collection of our baseline models. A linguistic analysis of the generated captions demonstrates that captions generated using ATTEND-GAN have a wider range of stylistic adjectives and adjective-noun pairs.

Keywords: Image captioning · Attention mechanism ·
Adversarial training

1 Introduction

Deep learning has facilitated the task of supplying images with captions. Current image captioning models [2,27,29] have gained considerable success due to powerful deep learning architectures and large image-caption datasets including the MSCOCO dataset [17]. These models mostly aim to describe an image in a factual way. Humans, however, describe an image in a way that combines subjective and stylistic properties, such as positive and negative sentiment, as in

© Springer Nature Switzerland AG 2019
A. C. Nayak and A. Sharma (Eds.): PRICAI 2019, LNAI 11670, pp. 270–284, 2019.
https://doi.org/10.1007/978-3-030-29908-8_22

1. the gorgeous sky really makes the man on the board stand out!
2. a great man flying through the air while riding a kite board.

1. a group of horses have a tough race around the track.
2. small number of horses with jockeys in a race on a track.

Fig. 1. Examples of positive (green) and negative (red) captions. (Color figure online)

the captions of Fig. 1. Users often find such captions more expressive and more attractive [8]; they have the practical purpose of enhancing the engagement level of users in social applications (e.g., chatbots) [14], and can assist people to make interesting image captions in social media content [8]. Moreover, Mathews *et al.* [19] found that they are more common in the descriptions of online images, and can have a role in transferring visual content clearly [18].

In stylistically enhanced descriptions, the content of images should still be reflected correctly. Moreover, the descriptions should fluently include stylistic words or phrases. To meet these criteria, previous models have used two-stage training: first, training on a large factual dataset to describe the content of an image; and then training on a small stylistic dataset to apply stylistic properties to a caption. The models have different strategies for integrating the learned information from the datasets. SentiCap has two Long Short-Term Memory (LSTM) networks: one learns from a factual dataset and the other one learns from a stylistic dataset [19]. In comparison, Gan *et al.* [8] proposed a new type of LSTM network, factored LSTM, to learn both factual and stylistic information. The factored LSTM has three matrices instead of one multiplied to the input caption: two matrices are learned to preserve the factual aspect of the input caption and one is learned to transfer the style aspect of the input caption. Chen *et al.* [5] applied an attention-based model which is similar to the factored LSTM, but it has an attention mechanism to differentiate attending to the factual and sentiment information of the input caption.

However, since the stylistic dataset is usually small, preserving the correlations between images and captions as well as generating a wide variety of stylistic patterns is very difficult. An imperfect caption from the system of Mathews *et al.* [19]—"a dead man doing a clever trick on a skateboard at a skate park"— illustrates the problem: the man is not actually dead; this is just a frequently used negative adjective.

Recently, Mathews *et al.* [18] dealt with this by applying a large stylistic dataset to separate the semantic and stylistic aspects of the generated captions. However, evaluation in this work was more difficult because the dataset includes stylistic captions which are not aligned to images. To address this challenge without any large stylistic dataset, we propose ATTEND-GAN, an image captioning model using an attention mechanism and a Generative Adversarial Network (GAN); our particular goal is to better apply stylistic information in the sort of two-stage architecture in previous work. Similar to this previous work, we first train a caption generator on a large factual dataset, although ATTEND-GAN uses an attention-based version attending to different image regions in the caption generation process [2]. Because of this, each word of a generated caption is conditioned upon a relevant fine-grained region of the corresponding image, ensuring a direct correlation between the caption and the image. Then we train a caption discriminator to distinguish between captions generated by our caption generator, and real captions, generated by humans. In the next step, on a small stylistic dataset, we implement an adversarial training mechanism to guide the generator to generate sentiment-bearing captions. To do so, the generator is trained to fool the discriminator by generating correlated and highly diversified captions similar to human-generated ones. The discriminator also periodically improves itself to further challenge the generator. Because GANs are originally designed to face continuous data distributions not discrete ones like texts [9], we use a gradient policy [31] to guide our caption generator using the rewards received from our caption discriminator for the next generated word, as in reinforcement learning [23]. The contributions of this paper are[1]:

- To generate human-like stylistic captions in a two-stage architecture, we propose ATTEND-GAN (Sect. 3) using both the designed attention-based caption generator and the adversarial training mechanism [9].
- ATTEND-GAN achieves results which are significantly better than the state-of-the-art (Sect. 4.5) and a comprehensive range of our baseline models (Sect. 4.6) for generating image captions with styles.
- On the SentiCap dataset [19], we show how ATTEND-GAN can result in stylistic captions which are strongly correlated with visual content (Sect. 4.8). ATTEND-GAN also exhibits significant variety in generating adjectives and adjective-noun pairs (Sect. 4.7).

2   Related Work

2.1   Image Captioning

The encoder-decoder framework of Vinyals et al. [27] where the encoder learns to encode visual content, using a Convolutional Neural Network (CNN), and the decoder learns to describe the visual content, using a long-short term memory (LSTM) network, is the basis of modern image captioning systems. Having an attention-based component has resulted in the most successful image captioning models [2,22,29,30]. These models use attention in either the image side or the caption side. For instance, Xu et al. [29] and Rennie et al. [22] attended to

---

[1] Our code and trained model are publicly available from https://github.com/omidmnezami/ATTEND-GAN.

the spatial visual features of an image. In comparison, You *et al.* [30] applied semantic attention attending to visual concepts detected in an image. Anderson *et al.* [2] applied an attention mechanism to attend to spatial visual features and discriminate not only the visual regions but also the detected concepts in the regions [2]. In addition to factual image captioning, the ability to generate stylistic image captions has recently become popular. The key published work [5,8,18,19] uses a two-stage architecture, although end-to-end is possible. None of the existing work uses an adversarial training mechanism; we show this, combined with attention, significantly outperforms the previous work.

2.2 Generative Adversarial Network

Goodfellow *et al.* [9] introduced Generative Adversarial Networks (GANs), whose training mechanism consists of a generator and a discriminator; they have been applied with great success in different applications [12,15,21,28,31]. The discriminator is trained to recognize real and synthesized samples generated by the generator. In contrast, the generator wants to generate realistic data to mislead the discriminator in distinguishing the source of data.

GANs were originally established for a continuous data space [9,31] rather than a discrete data distribution as in our work. To handle this, a form of reinforcement learning is usually applied, where the sentence generation process is formulated as a reinforcement learning problem [23]; the discriminator provides a reward for the next action (in our context the next generated word), and the generator uses the reward to calculate gradients and update its parameters, as proposed in Yu *et al.* [31]. Wang and Wan [28] applied this to generating sentiment-bearing text (although not conditioned on any input, such as the images in our captioning task).

3   ATTEND-GAN Model

The purpose of our image captioning model is to generate sentiment-bearing captions. Our caption generator employs an attention mechanism, described in

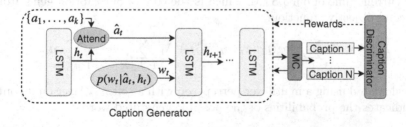

Fig. 2. The architecture of the ATTEND-GAN model. $\{a_1, ..., a_K\}$ are spatial visual features generated by ResNet-152 network. Attend and MC modules are our attention mechanism and Monte Carlo search, respectively.

Sect. 3.1, to attend to fine-grained image regions $a = \{a_1, ..., a_K\}, a_i \in \mathbb{R}^D$, where the number of regions is K with D dimensions, in different time steps so as to generate an image caption $x = \{x_1, ..., x_T\}, x_i \in \mathbb{R}^N$, where the size of our vocabulary is N and the length of the generated caption is T. We also propose a caption discriminator, explained in Sect. 3.2, to distinguish between the generated captions and human-produced ones. We describe our training in Sect. 3.3. Our proposed model is called ATTEND-GAN (Fig. 2).

3.1   Caption Generator

The goal of our caption generator $G_\theta(x_t|x_{1:t-1}, \hat{a}_t)$ is to generate an image caption to achieve a maximum reward value from our caption discriminator $D_\phi(x_{1:T})$, where θ and ϕ are the parameters of the generator and the discriminator, respectively. The objective function of the generator, which is dependent on the discriminator, is to minimize:

$$L_1(\theta) = \sum_{1 \leq t \leq T} G_\theta(x_t|x_{1:t-1}, \hat{a}_t).Z_{D_\phi}^{G_\theta}(x_{1:t}) \tag{1}$$

where $Z_{D_\phi}^{G_\theta}(x_{1:t})$ is the reward value of the partially generated sequence, $x_{1:t}$, and is estimated using the discriminator. The reward value can be interpreted as a score value that $x_{1:t}$ is real. Since the discriminator can only generate a reward value for a complete sequence, Monte Carlo (MC) search is applied, which uses the generator to roll out the remaining part of the sequence at each time step. We apply MC search N times, and calculate the average reward (to decrease the variance of the next generated words):

$$Z_{D_\phi}^{G_\theta}(x_{1:t}) = \begin{cases} \frac{1}{N} \sum_{n=1}^{N} D_\phi(x_{1:T}^n), \ x_{1:T}^n \in MC_{G_\theta}(x_{1:t;N}) & \text{if } t < T \\ D_\phi(x_{1:t}) & \text{if } t = T \end{cases} \tag{2}$$

$x_{1:T}^n$ is the n-th MC-completed sequence at current time step t. In addition to Eq. (1), we calculate the maximum likelihood estimation (MLE) of the generated word with respect to the attention-based content ($\hat{a}_t$) and the hidden state (h_t) at the current time of our LSTM, which is the core of our caption generator, as the second objective function:

$$L_2(\theta) = - \sum_{1 \leq t \leq T} \log(p_w(x_t \mid \hat{a}_t, h_t)) + \lambda_1 \sum_{1 \leq k \leq K} (1 - \sum_{1 \leq t \leq T} a_{tk})^2 \tag{3}$$

p_w is calculated using a multilayer perceptron with a softmax layer on its output and indicates the probabilities of the possible generated words:

$$p_w(x_t \mid \hat{a}_t, h_t) = \text{softmax}(\hat{a}_t W_a + h_t W_h + b_w) \tag{4}$$

W_x and b_w are the learned weights and biases. The last term in Eq. (3) is to encourage our caption generator to equally consider diverse regions of the given

Algorithm 1. ATTEND-GAN Training Mechanism.

1: Pre-train the caption generator (G_θ) using Eq. (9).
2: Use G_θ to generate sample captions $\mathbb{P}_G$ and select ground-truth captions $\mathbb{P}_H$.
3: Pre-train the caption discriminator (D_ϕ) using Eq. (10) and the combination of $\mathbb{P}_G$ and $\mathbb{P}_H$.
4: **repeat**
5:     **for** g steps **do**
6:         Apply G_θ to generate image captions.
7:         Calculate $Z_{D_\phi}^{G_\theta}$ using Eq. (2).
8:         Update θ, the parameters of G_θ, using Eq. (8).
9:     **end for**
10:     **for** d steps **do**
11:         Generate sample captions $\mathbb{P}_G$ by G_θ and select human-generated captions $\mathbb{P}_H$.

12:         Update ϕ, the parameters of D_ϕ, using Eq. (10).
13:     **end for**
14: **until** ATTEND-GAN converges

image at the end of the caption generation process. λ_1 is a regularization parameter. h_t is calculated using our LSTM:

$$
\begin{aligned}
i_t &= \sigma(H_i h_{t-1} + W_i w_{t-1} + A_i \hat{a}_t + b_i) \\
f_t &= \sigma(H_f h_{t-1} + W_f w_{t-1} + A_f \hat{a}_t + b_f) \\
g_t &= \tanh(H_g h_{t-1} + W_g w_{t-1} + A_g \hat{a}_t + b_g) \\
o_t &= \sigma(H_o h_{t-1} + W_o w_{t-1} + A_o \hat{a}_t + b_o) \\
c_t &= f_t c_{t-1} + i_t g_t \\
h_t &= o_t \tanh(c_t)
\end{aligned}
\tag{5}
$$

Here, i_t, f_t, g_t, o_t, and c_t are the LSTM's gates and represent input, forget, modulation, output, and memory gates, respectively. w_{t-1} is the embedded previous word in M dimensions, $w_x \in \mathbb{R}^M$. H_x, W_x, A_x, and b_x are learned weights and biases; and σ is the Sigmoid function. Using h_t, our soft attention module generates unnormalized weights $e_{j,t}$ for each image region a_j. Then, the weights are normalized using a softmax layer, e'_t:

$$
e_{j,t} = W_e^T \tanh(W_a' a_j + W_h' h_t), e_t' = \mathrm{softmax}(e_t)
\tag{6}
$$

W_e^T and W_x' are our trained weights. Finally, $\hat{a}_t$, our attention-based content, is calculated using Eq. (7):

$$
\hat{a}_t = \sum_{1 \le j \le K} e'_{j,t} a_j
\tag{7}
$$

During the adversarial training, the objective function of the caption generator is a combination of Eqs. (1) and (3):

$$
L_G(\theta) = \lambda_2 L_1(\theta) + L_2(\theta)
\tag{8}
$$

λ_2 is a balance parameter. The discriminator cannot be learned effectively from a random initialization of the generator; we therefore pretrain the generator with the MLE objective function:

$$L_G(\theta) = L_2(\theta) \tag{9}$$

3.2   Caption Discriminator

Our caption discriminator is inspired by the Wasserstein GAN (WGAN) [3] which is an improved version of the GAN [9]. The WGAN generates continuous values and solves the problem of the GAN generating non-continuous outputs leading to some training difficulties (e.g. vanishing gradients). The objective function of our WGAN is:

$$L_D(\phi) = \mathbb{E}_{x \sim \mathbb{P}_H}[D_\phi(x)] - \mathbb{E}_{\bar{x} \sim \mathbb{P}_G}[D_\phi(\bar{x})] \tag{10}$$

where ϕ are the parameters of the discriminator (D_ϕ); $\mathbb{P}_H$ is the set of the generated captions by humans; and $\mathbb{P}_G$ is the set of the generated captions by the generator. D_ϕ is implemented via a Convolutional Neural Network (CNN) that calculates the score value of the input caption. To feed a caption to our CNN model, we first embed all words in the caption into M embedding dimensions, $\{w'_1, \ldots, w'_T\}, w'_i \in \mathbb{R}^M$, and build a 2-dimensional matrix for the caption, $S \in \mathbb{R}^{T \times M}$ [31]. Our CNN model includes Convolutional (Conv.) layers with P different kernel sizes $\{k_1, \ldots, k_P\}, k_i \in \mathbb{R}^{C \times M}$, where C indicates the number of the words ($C \in [1, T]$). Applying each Conv. layer to S results a number of feature maps, $v_{ij} = k_i \otimes S_{j:j+C-1} + b_j$, where $\otimes$ is a convolution operation and b_j is a bias vector. We apply a batch normalization layer [11], and a nonlinearity, a rectified linear unit (ReLU), respectively. Then, we apply a max-pooling layer, $v_i^* = \max v_{ij}$. Finally, a fully connected layer is applied to output the score value of the caption. The weights of our CNN model are clipped to be in a compact space.

3.3   ATTEND-GAN Training

As shown in Algorithm 1, we first pre-train our caption generator for a specific number of epochs. Then, we apply the best generator model to generate sample captions. The real captions are selected from the ground truth. In Step 3, our caption discriminator is pre-trained using a combination of the generated and real captions for a specific number of epochs. Here, both the caption generator and discriminator are pre-trained on a factual dataset. In Step 4, we start our adversarial training on a sentiment-bearing dataset with positive or negative sentiment. We continue the training of the caption generator and discriminator for g-steps and d-steps, respectively. Using this mechanism, we improve both the caption generator and discriminator. Here, the caption generator applies the received rewards from the caption discriminator to update its parameters using Eq. (8).

4   Experiments

4.1   Datasets

Microsoft COCO Dataset. We use the MSCOCO image-caption dataset [17] to train our models. Specifically, we use the training set of the dataset including 82K+ images and 413K+ captions.

SentiCap Dataset. To add sentiment to the generated captions, our models are trained on the SentiCap dataset [19] including sentiment-bearing image captions. The dataset has two separate sections of sentiments: *positive* and *negative*. 2,873 captions paired with 998 images (409 captions with 174 images are for validation) are for training and 2019 captions paired with 673 images are for testing in the positive section. 2,468 captions paired with 997 images (429 captions with 174 images are for validation) are for training and 1,509 captions paired with 503 images are for testing in the negative section. We use the same training/test folds as in the previous work [5,19].

4.2   Evaluation Metrics

ATTEND-GAN is evaluated using standard image captioning metrics: METEOR [7], BLEU [20], CIDEr [26] and ROUGE-L [16]. SPICE has not previously been used in the literature; however, it is reported for future comparisons because it has shown a close correlation with human-based evaluations [1]. Larger values of these metrics indicated better results.

4.3   Models for Comparison

We first trained our models on the MSCOCO dataset to generate factual captions. Then, we trained our models on the SentiCap dataset to add sentiment properties to the generated captions. This two-stage training mechanism is similar to the training methods of [19] and [8]. The work of [5], the newest one in this domain, was also implemented in a similar way. Following this training approach makes our results directly comparable to the previous ones. Our models are compared with a range of baseline models from Mathews *et al.* [19]: CNN+RNN, which is only trained using the MSCOCO dataset; ANP-REPLACE, which adds the most common adjectives to a randomly chosen noun; ANP-SCORING, which applies multi-class logistic regression to select an adjective for the chosen noun; RNN-TRANSFER, which is CNN+RNN fine-tuned on the SentiCap dataset; and their key system SENTICAP, which uses two LSTM modules to learn from factual and sentiment-bearing caption. We also compare with SF-LSTM+ADAP, which applies an attention mechanism to weight factual and sentiment-based information [5]. The results of all these models in Table 1 are obtained from the corresponding references. Moreover, we first train our attention-based model only on the factual dataset MSCOCO (we name this model ATTEND-GAN$_{-SA}$). Second, we train our model additionally on the SentiCap dataset but without our caption discriminator (ATTEND-GAN$_{-A}$). Finally, we train our full model using the caption discriminator (ATTEND-GAN).

4.4   Implementation Details

Encoder. In this work, we apply ResNet-152 [10] as our visual encoder model pre-trained using the ImageNet dataset [6]. In comparison with other CNN models, ResNet-152 has shown more effective results on different image-caption datasets [4]. We specifically use its Res5c layer to extract the spatial features of an image. The layer gives us $7 \times 7 \times 2048$ feature map converted to 49×2048 representing 49 semantic-based regions with 2048 dimensions.

Vocabulary. Our vocabulary has 9703 words, coming form both the MSCOCO and SentiCap datasets, for all our models. Each word is embedded into a 300 dimensional vector.

Generator and Discriminator. The size of the hidden state and the memory cell of our LSTM is set to 512. For the caption generator, we use the Adam function [13] for optimization and set the learning rate to 0.0001. We set the size of our mini-batches to 64. To optimize the caption discriminator, we use the RMSprop solver [24] and clip the weights to $[-0.01, 0.01]$. The mini-batches are fixed to 80 for the discriminator. We apply Monte Carlo search 5 times (Eq. (2)). We set λ_1 and λ_2 to 1.0 and 0.1 in Eqs. (3) and (8), respectively. During the adversarial training, we alternate between Eqs. (8) and (10) to optimize the generator and the discriminator, respectively. We particularly operate a single gradient decent phase on the generator (g steps) and 3 gradient phases (d steps) on the discriminator every time. The models are trained for 20 epochs to converge. The METEOR metric is used to select the model with the best performance on the validation sets of positive and negative datasets of SentiCap because it has a close correlation with human judgments and is less computationally expensive than SPICE which requires dependency parsing [1].

4.5   Results: Comparison with the State-of-the-Art

All models in Table 1 used the same training/test folds of the SentiCap dataset to make them comparable. In comparison with the state-of-the-art, our full model (ATTEND-GAN) achieves the best results for all image captioning metrics in both positive and negative parts of the SentiCap dataset. We report the average results to show the average improvements of our models over the state-of-the-art model. ATTEND-GAN achieved large gains of 6.15, 6.45, 3.00, and 2.95 points with respect to the best previous model using BLEU-1, ROUGE-L, CIDEr and BLEU-2 metrics, respectively. Other metrics show smaller but still positive improvements.

4.6   Results: Comparison with Our Baseline Models

Our models are compared in Table 1 in terms of image captioning metrics. ATTEND-GAN outperforms ATTEND-GAN$_{-A}$ over all metrics across both positive and negative parts of the SentiCap dataset; the discriminator is thus

Table 1. The compared performances on different sections of SentiCap and their average. BLEU-N metric is shown by B-N. (The best results are bold.)

Senti	Model	B-1	B-2	B-3	B-4	ROUGE-L	METEOR	CIDEr	SPICE
Pos	CNN+RNN	48.7	28.1	17.0	10.7	36.6	15.3	55.6	-
	ANP-Replace	48.2	27.8	16.4	10.1	36.6	16.5	55.2	-
	ANP-Scoring	48.3	27.9	16.6	10.1	36.5	16.6	55.4	-
	RNN-Transfer	49.3	29.5	17.9	10.9	37.2	17.0	54.1	-
	SentiCap	49.1	29.1	17.5	10.8	36.5	16.8	54.4	-
	SF-LSTM + Adap	50.5	30.8	19.1	12.1	38.0	16.6	60.0	-
	Ours: ATTEND-GAN$_{-SA}$	56.1	32.5	19.4	11.8	44.8	17.1	63.0	15.9
	Ours: ATTEND-GAN$_{-A}$	55.8	33.4	20.1	12.4	44.2	18.6	61.1	15.7
	Ours: ATTEND-GAN	56.9	33.6	20.3	12.5	44.3	18.8	61.6	15.9
Neg	CNN+RNN	47.6	27.5	16.3	9.8	36.1	15.0	54.6	-
	ANP-Replace	48.1	28.8	17.7	10.9	36.3	16.0	56.5	-
	ANP-Scoring	47.9	28.7	17.7	11.1	36.2	16.0	57.1	-
	RNN-Transfer	47.8	29.0	18.7	12.1	36.7	16.2	55.9	-
	SentiCap	50.0	31.2	20.3	13.1	37.9	16.8	61.8	-
	SF-LSTM + Adap	50.3	31.0	20.1	13.3	38.0	16.2	59.7	-
	Ours: ATTEND-GAN$_{-SA}$	55.4	32.4	19.4	11.9	44.4	17.0	63.4	15.6
	Ours: ATTEND-GAN$_{-A}$	54.7	32.6	20.4	12.9	43.2	17.7	60.4	16.1
	Ours: ATTEND-GAN	56.2	34.1	21.3	13.6	44.6	17.9	64.1	16.2
Avg	CNN+RNN	48.15	27.80	16.65	10.25	36.35	15.15	55.10	-
	ANP-Replace	48.15	28.30	17.05	10.50	36.45	16.25	55.85	-
	ANP-Scoring	48.10	28.30	17.15	10.60	36.35	16.30	56.25	-
	RNN-Transfer	48.55	29.25	18.30	11.50	36.95	16.60	55.00	-
	SentiCap	49.55	30.15	18.90	11.95	37.20	16.80	58.10	-
	SF-LSTM + Adap	50.40	30.90	19.60	12.70	38.00	16.40	59.85	-
	Ours: ATTEND-GAN$_{-SA}$	55.75	32.45	19.40	11.85	**44.60**	17.05	**63.20**	15.75
	Ours: ATTEND-GAN$_{-A}$	55.25	33.00	20.25	12.65	43.70	18.15	60.75	15.90
	Ours: ATTEND-GAN	**56.55**	**33.85**	**20.80**	**13.05**	44.45	**18.35**	62.85	**16.05**

an important part of the architecture. ATTEND-GAN outperforms ATTEND-GAN$_{-SA}$ for all metrics except, by a small margin, CIDEr and ROUGE-L. Recall that ATTEND-GAN$_{-SA}$ is trained only on the large MSCOCO (with many captions), and so is in a sense encouraged to have diverse captions; second-stage training for ATTEND-GAN$_{-A}$ and ATTEND-GAN leads to more focussed captions relevant to SentiCap. As CIDEr and ROUGE-L are the two recall-oriented metrics, they suffer in this two-stage process, illustrating the issue we noted in Sect. 1. The discriminator, however, removes almost all of this penalty, as well as boosting the other metrics beyond ATTEND-GAN$_{-SA}$. Furthermore, Sect. 4.7 illustrates how ATTEND-GAN$_{-SA}$ produces unsatisfactory captions in terms of sentiment.

Table 2. Entropy and Top_4 of the generated adjectives using different models.

Senti	Model	Entropy	Top_4
Pos	ATTEND-GAN$_{-SA}$	2.2457	93.33%
	ATTEND-GAN$_{-A}$	3.0324	72.11%
	ATTEND-GAN	3.5671	62.33%
Neg	ATTEND-GAN$_{-SA}$	2.2448	91.67%
	ATTEND-GAN$_{-A}$	4.1040	48.44%
	ATTEND-GAN	3.9562	50.51%
Avg	ATTEND-GAN$_{-SA}$	2.2453	92.50%
	ATTEND-GAN$_{-A}$	3.5682	60.28%
	ATTEND-GAN	**3.7617**	**56.42%**

Table 3. The top-10 adjectives that are generated by our models and are in the adjective-noun pairs of the SentiCap dataset.

Senti	Model	Top 10 adjectives
Pos	ATTEND-GAN$_{-SA}$	white, black, small, blue, different, little, busy, _, _, _
	ATTEND-GAN$_{-A}$	nice, beautiful, happy, busy, great, sunny, good, cute, pretty, white
	ATTEND-GAN	nice, beautiful, happy, great, good, sunny, busy, white, pretty, delicious
Neg	ATTEND-GAN$_{-SA}$	black, white, small, blue, different, tall, little, _, _, _
	ATTEND-GAN$_{-A}$	lonely, dead, broken, stupid, dirty, bad, cold, little, crazy, lazy
	ATTEND-GAN	lonely, stupid, broken, dirty, dead, cold, bad, white, crazy, little

4.7 Qualitative Results

To analyze the quality of language generated by our models, we extract all generated adjectives using the Stanford part-of-speech tagger software [25], and select the adjectives found in the adjective-noun pairs (ANPs) of the SentiCap dataset. Then, we calculate Entropy of the distribution of these adjectives as a measure of variety in lexical selection (higher scores mean more variety) using Eq. (11).

$$\text{Entropy} = -\sum_{1 \leq j \leq U} \log_2[p(A_j)] \times p(A_j) \tag{11}$$

where $p(A_j)$ is the probability of the adjective (A_j) and U indicates the number of all unique adjectives. Moreover, we calculate the total probability mass of the four most frequent adjectives (Top_4) generated by our models. Here, lower values mean that the model allocates more probability to other generated adjectives, also indicating greater variety.

Table 2 shows that ATTEND-GAN achieves the best results on average for Entropy (highest score) and Top$_4$ (lowest) compared to other models, by a large margin with respect to ATTEND-GAN$_{-SA}$. It is not surprising that ATTEND-GAN$_{-SA}$ has the lowest variability of use of sentiment-bearing adjectives because it does not use the stylistic dataset. As demonstrated by the improvement of ATTEND-GAN over ATTEND-GAN$_{-A}$, the discriminator helps in generating a greater diversity of adjectives.

The top-10 adjectives generated by our models are shown in Table 3. "white" is generated for both negative and positive sections because they are common in both sections. ATTEND-GAN and ATTEND-GAN$_{-A}$ produce a natural ranking of sentiment-bearing adjectives for both sections. For example, these models rank "nice" as the most positive adjective, and "lonely" as the most negative. As ATTEND-GAN$_{-SA}$ does not use the stylistic dataset, it generates a similar and limited (<10) range of adjectives for both.

Fig. 3. Examples on the positive (first 3) and negative (last 3) datasets (AS for ATTEND-GAN$_{-SA}$, A for ATTEND-GAN$_{-A}$ and AG for ATTEND-GAN). Green and red colors indicate the generated positive and negative adjective-noun pairs in SentiCap, respectively. (Color figure online)

4.8   Generated Captions

Figure 3 shows sample sentiment-bearing captions generated by our models for the positive and negative sections of the SentiCap dataset.[2] For instance, for the first two images, ATTEND-GAN correctly applies positive sentiments to describe the corresponding images (e.g., "nice street", "tasty food"). Here, ATTEND-GAN$_{-A}$ also succeeds in generating captions with positive sentiments, but less well. In the third image, ATTEND-GAN uses "pretty woman" to describe the image which is better than the "beautiful court" of ATTEND-GAN$_{-A}$: for this image, all ground-truth captions have positive sentiment for the noun "girl" (e.g. "a beautiful girl is running and swinging a tennis racket"); none of them describes the noun "court" with a sentiment-bearing adjective as

---

[2] See a link to supplementary materials for additional samples: https://github.com/omidmnezami/ATTEND-GAN/blob/master/st.pdf.

ATTEND-GAN$_{-A}$ does. For all images, since ATTEND-GAN$_{-SA}$ is not trained using the SentiCap dataset, it does not generate any caption with sentiment. For the fourth image, ATTEND-GAN generates "a group of stupid people are playing frisbee on a field", applying "stupid people" to describe the image negatively. Here, one of the ground-truth captions exactly includes "stupid people" ("two stupid people in open field watching yellow tent blown away"). ATTEND-GAN$_{-A}$, like our flawed example from Sect. 1, refers instead inaccurately to a dead man. For the fifth image (as for the first image), ATTEND-GAN has incorporates more (appropriate) sentiment in comparison to ATTEND-GAN$_{-A}$. It generates "rough hill" and "cold day", while ATTEND-GAN$_{-A}$ only generates the former. It also uses "skier" which is more appropriate than "person". In the last image, ATTEND-GAN adds "bad picture" and ATTEND-GAN$_{-A}$ generates "bad food". One of the ground-truth captions exactly includes "bad picture".

5　Conclusion

In this paper, we proposed ATTEND-GAN, an attention-based image captioning model using an adversarial training mechanism. Our model is capable of generating stylistic captions which are strongly correlated with images and contain diverse stylistic components. ATTEND-GAN achieves the state-of-the-art performance on the SentiCap dataset. It also outperforms our baseline models and generates stylistic captions with a high level of variety. Future work includes developing ATTEND-GAN to generate a wider range of captions and developing further mechanisms to ensure compatibility with the visual content.

References

1. Anderson, P., Fernando, B., Johnson, M., Gould, S.: SPICE: semantic propositional image caption evaluation. In: Leibe, B., Matas, J., Sebe, N., Welling, M. (eds.) ECCV 2016. LNCS, vol. 9909, pp. 382–398. Springer, Cham (2016). https://doi.org/10.1007/978-3-319-46454-1_24
2. Anderson, P., et al.: Bottom-up and top-down attention for image captioning and visual question answering. In: CVPR, vol. 3, p. 6 (2018)
3. Arjovsky, M., Chintala, S., Bottou, L.: Wasserstein GAN (2017). arXiv preprint arXiv:1701.07875
4. Chen, L., et al.: SCA-CNN: spatial and channel-wise attention in convolutional networks for image captioning. In: 2017 IEEE Conference on Computer Vision and Pattern Recognition (CVPR), pp. 6298–6306. IEEE (2017)
5. Chen, T., et al.: "Factual" or "emotional": stylized image captioning with adaptive learning and attention (2018). arXiv preprint arXiv:1807.03871
6. Deng, J., Dong, W., Socher, R., Li, L.J., Li, K., Fei-Fei, L.: ImageNet: a large-scale hierarchical image database (2009)
7. Denkowski, M., Lavie, A.: Meteor universal: language specific translation evaluation for any target language. In: WMT, pp. 376–380 (2014)
8. Gan, C., Gan, Z., He, X., Gao, J., Deng, L.: StyleNet: generating attractive visual captions with styles. In: CVPR. IEEE (2017)

9. Goodfellow, I., et al.: Generative adversarial nets. In: Advances in Neural Information Processing Systems, pp. 2672–2680 (2014)
10. He, K., Zhang, X., Ren, S., Sun, J.: Deep residual learning for image recognition. In: CVPR, pp. 770–778 (2016)
11. Ioffe, S., Szegedy, C.: Batch normalization: accelerating deep network training by reducing internal covariate shift (2015). arXiv preprint arXiv:1502.03167
12. Isola, P., Zhu, J.Y., Zhou, T., Efros, A.A.: Image-to-image translation with conditional adversarial networks. In: Proceedings of the IEEE Conference on Computer Vision and Pattern Recognition, pp. 1125–1134 (2017)
13. Kingma, D.P., Ba, J.: Adam: a method for stochastic optimization (2014). arXiv preprint arXiv:1412.6980
14. Li, Y., Yao, T., Mei, T., Chao, H., Rui, Y.: Share-and-chat: achieving human-level video commenting by search and multi-view embedding. In: Proceedings of the 24th ACM International Conference on Multimedia, pp. 928–937. ACM (2016)
15. Liang, X., Hu, Z., Zhang, H., Gan, C., Xing, E.P.: Recurrent topic-transition GAN for visual paragraph generation (2017). arXiv preprint arXiv:1703.07022
16. Lin, C.Y.: Rouge: a package for automatic evaluation of summaries. In: Text Summarization Branches Out (2004)
17. Lin, T.-Y., et al.: Microsoft COCO: common objects in context. In: Fleet, D., Pajdla, T., Schiele, B., Tuytelaars, T. (eds.) ECCV 2014. LNCS, vol. 8693, pp. 740–755. Springer, Cham (2014). https://doi.org/10.1007/978-3-319-10602-1_48
18. Mathews, A., Xie, L., He, X.: SemStyle: learning to generate stylised image captions using unaligned text. In: Proceedings of the IEEE Conference on Computer Vision and Pattern Recognition, pp. 8591–8600 (2018)
19. Mathews, A.P., Xie, L., He, X.: SentiCap: generating image descriptions with sentiments. In: AAAI, pp. 3574–3580 (2016)
20. Papineni, K., Roukos, S., Ward, T., Zhu, W.J.: BLEU: a method for automatic evaluation of machine translation. In: ACL, pp. 311–318. Association for Computational Linguistics (2002)
21. Radford, A., Metz, L., Chintala, S.: Unsupervised representation learning with deep convolutional generative adversarial networks (2015). arXiv preprint arXiv:1511.06434
22. Rennie, S.J., Marcheret, E., Mroueh, Y., Ross, J., Goel, V.: Self-critical sequence training for image captioning. In: CVPR, vol. 1, p. 3 (2017)
23. Silver, D., et al.: Mastering the game of go with deep neural networks and tree search. Nature 529(7587), 484 (2016)
24. Tieleman, T., Hinton, G.: Lecture 6.5-rmsprop: divide the gradient by a running average of its recent magnitude. COURSERA: Neural Netw. Mach. Learn. 4(2), 26–31 (2012)
25. Toutanova, K., Klein, D., Manning, C.D., Singer, Y.: Feature-rich part-of-speech tagging with a cyclic dependency network. In: NAACL HLT, pp. 173–180. Association for Computational Linguistics (2003)
26. Vedantam, R., Lawrence Zitnick, C., Parikh, D.: Cider: consensus-based image description evaluation. In: CVPR, pp. 4566–4575. IEEE (2015)
27. Vinyals, O., Toshev, A., Bengio, S., Erhan, D.: Show and tell: a neural image caption generator. In: CVPR, pp. 3156–3164. IEEE (2015)
28. Wang, K., Wan, X.: SentiGAN: generating sentimental texts via mixture adversarial networks. In: IJCAI, pp. 4446–4452 (2018)
29. Xu, K., et al.: Show, attend and tell: neural image caption generation with visual attention. In: ICML, pp. 2048–2057 (2015)

30. You, Q., Jin, H., Wang, Z., Fang, C., Luo, J.: Image captioning with semantic attention. In: CVPR, pp. 4651–4659. IEEE (2016)
31. Yu, L., Zhang, W., Wang, J., Yu, Y.: SeqGAN: sequence generative adversarial nets with policy gradient. In: AAAI, pp. 2852–2858 (2017)

Language and Speech

Keywords-Based Auxiliary Information Network for Abstractive Summarization

Haihan Wang, Jinlong Li(✉), and Xuewen Chen

School of Computer Science, University of Science and Technology of China,
Hefei 230027, China
jlli@ustc.edu.cn
{whh,shirveon}@mail.ustc.edu.cn

Abstract. Automatic text summarization is an important research task in the field of natural language processing (NLP). The abstractive approach to automatic text summarization produces the condensed version of the source text by generating new words and phrases. Recently, the attentional sequence-to-sequence models have shown good ability in abstractive text summarization. Nevertheless, these neural network models are still hard to cover most key points of the source text and may produce unfactual details. To address these issues, we proposed a keywords-based auxiliary information model to guide the process of encoding and decoding. Firstly, we proposed an auxiliary information network based on the keywords of the document, which aims to generate the modified encoded representation. In addition, we designed a novel selective beam search mechanism to keep more keywords and reduce redundancy in the decoded summaries. We evaluated our model on different datasets including the benchmark *CNN/Daily Mail* dataset. The experimental results show that our model leads to significant improvements compared with abstractive baseline models.

Keywords: Abstractive text summarization ·
Keywords-based auxiliary information network ·
Selective beam search mechanism

1 Introduction

Automatic text summarization has become an important and effective method for processing and interpreting text information [6] and its task is to generate a brief and informative summary text from an input text. The approaches to automatic text summarization can roughly fall into two categories: extractive and abstractive [19]. Extractive models usually assemble summaries by extracting a set of sentences or keywords directly from the source text [12,24], thus extractive models unavoidably suffer from redundancy and incoherence. On the other hand, abstractive models can generate novel phrases and restructure sentences to form concise summary [5,23].

© Springer Nature Switzerland AG 2019
A. C. Nayak and A. Sharma (Eds.): PRICAI 2019, LNAI 11670, pp. 287–299, 2019.
https://doi.org/10.1007/978-3-030-29908-8_23

Based on the attentional sequence-to-sequence (seq2seq) framework [1, 26], neural network models are capable to generate the summaries with good performance [2,21]. However, these models still have weaknesses such as producing unfactual details and redundant sequences in summary text [4]. To solve these issues, the model need to capture the information accurately and control the generation process, which requires improving the quality of encoded representation and optimizing the decoding strategy [7,29]. Inspired by the process of human writing summaries—the key information, such as key concepts and key entities of the source text, is first identified and then summaries are generated based on these key information, we propose a keywords-based auxiliary information model, which uses the keywords of the source text as a guidance to control the procedures for both encoding and decoding. Specifically, We construct an auxiliary information network that uses the key vector derived from the keywords to guide the generation process of the encoded word representation. Moreover, we employ a selective beam search mechanism to control the selection of the candidate summaries by applying a hypotheses scoring function. Compared with existing work on abstractive text summarization, the proposed model utilizes the auxiliary information to guide the generation and has better ability of grasping the key points of the text.

To summarize, we make the following contributions:

- We propose a keywords-based auxiliary information network that injects the auxiliary information into the encoding process to generate the modified encoded representation.
- A novel selective beam search mechanism is introduced to rerank and select the candidate summary with more keywords and less redundancy in the decoding process.
- We evaluate our model on different datasets compared with the state-of-the-art abstractive model with pointer mechanism. Experimental results indicate that our model achieves significant improvements and obtains higher ROUGE scores on benchmark *CNN/Daily Mail* dataset.

The rest of the paper is organized as follows. Section 2 reviews the related research. Section 3 introduces the concrete implementation of our model. Section 4 presents the experiment settings and performs a qualitative analysis of the results. Section 5 concludes the paper and discusses the future direction.

2   Related Work

Sequence-to-sequence model was first proposed for the natural language process task of machine translation [18,26]. The seq2seq model belongs to a family of encoder-decoders, which first encodes the input text to the abstract representation and then decodes the summary based on the encoded representation [9]. Motivated by its success on machine translation, this model was then applied to the task of abstractive text summarization [21,23].

Many abstractive models attempt to improve the encoding process in different ways [3,15]. Nallapati et al. [21] used the lexical and statistic features to enrich the encoder by concatenating them to embedding vector, which boosts performance incrementally. Zhou et al. [29] proposed a selective encoding model to improve encoding efficiency. This model is capable of selecting the encoded information to construct the second level representation by applying a selective gate network, which leads to better performance. Lin et al. [17] designed a global encoding framework, which implements a convolutional gated unit to improve the influence of the global context on word representation. These models employed diverse ways to improve the representation of the source-side information, which leads to certain progress on the task of short text summarization. However, on the task of multi-sentence summarization, obtaining a high-quality encoded representation remains a challenge because the lengthy source text have too much redundancy information [25]. Focusing on refining the encoded representation on the task of multi-sentence summarization, in our work, we propose an keywords-based auxiliary information network to generate the tailored representation by incorporating the word representation with the key information representation derived from the keywords.

Besides the work on improving the encoded representation, some research also consider improving the decoding strategy [13,14]. In the decoding process, there are some notable issues such as out-of-vocabulary (OOV) words problem, the repetition and the suboptimization problem of beam search method [25]. To handle these issues, great efforts have been made recently. Based on the pointer network [28], copy mechanism and pointer-generator network are proposed to alleviate the OOV words problem [11,25]. Coverage mechanism and intra-attention are utilized to solve the repetition problem [22,25]. Diverse beam search method is leveraged to improve the diversity of the hypotheses [27]. These studies demonstrate that optimizations in the decoding portion can improve the quality of the summary text effectively. In our work, we introduce a novel selective beam search mechanism, which applies a hypotheses scorer that considers not only the conditional probability but also the keywords and the redundancy during decoding summary.

3 Our Model

An overview of our approach is illustrated in Fig. 1. Given a source text, we first apply a sentence encoder to read the input words $x = (x_1, x_2, ..., x_n)$ and build its representation $(h_1, h_2, ..., h_n)$. Then the keywords-based auxiliary information network incorporate the auxiliary information with the word representation to produce the modified encoded representation $(h_1^*, h_2^*, ..., h_n^*)$, which is used as input of the attentional-equipped decoder. At last, we use the selective beam search mechanism to score the candidate summaries and select the highest-ranking hypothesis as final summary. In the following, we will introduce our sentence encoder, the keywords-based auxiliary information network, attentional summary decoder and the selective beam search mechanism respectively.

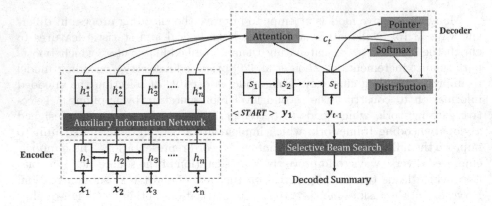

Fig. 1. Overview of the keywords-based auxiliary information model.

3.1 Sentence Encoder

We employ a single-layer bidirectional LSTM as encoder, which reads the tokens w_i of the input text one-by-one to generate the bidirectional hidden states.

The BiLSTM consists of a forward LSTM and a backward LSTM. The forward LSTM reads the word embeddings of the input text from left to right and produces a sequence of hidden states ($\overrightarrow{h}_1, \overrightarrow{h}_2, ..., \overrightarrow{h}_n$). The backward LSTM reads the input sequence in reverse order and produces another sequence of hidden states ($\overleftarrow{h}_1, \overleftarrow{h}_2, ..., \overleftarrow{h}_n$). The process can be defined by the Eqs. (1) and (2):

$$\overrightarrow{h}_i = \text{LSTM}(x_i, \overrightarrow{h}_{i-1}) \tag{1}$$

$$\overleftarrow{h}_i = \text{LSTM}(x_i, \overleftarrow{h}_{i+1}) \tag{2}$$

After that, the forward and backward hidden states are concatenated to produce a sequence of encoded hidden states, i.e., $h_i = [\overrightarrow{h}_i; \overleftarrow{h}_i]$. These basic encoded hidden states are then used as the input of keywords-based auxiliary information network to generate the modified encoded word presentation.

3.2 Keywords-Based Auxiliary Information Network

In multi-sentence summarization, covering the key points of the source text is still a challenge. In order to accomplish this goal, we propose a keywords-based auxiliary information network (KAIN) as shown in Fig. 2.

We use the extractive method to obtain keywords of the source text by TextRank algorithm [20] and then concatenate the keywords to obtain the key information representation z. For each word x_i, the auxiliary information network generates a key vector k_i by feeding the initial encoded state h_i and key information representation z through a linear layer with activation function. After that, we incorporate each initial hidden state h_i with the key vector k_i to construct the sequence of tailored representation $(h_1^*, h_2^*, \ldots, h_n^*)$.

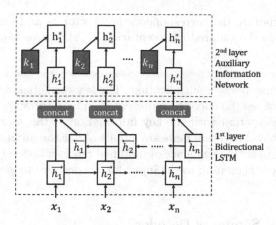

Fig. 2. Structure of encoder with the auxiliary information network

In detail, we apply the extractive method to get m keywords. To obtain the key information representation, we firstly make a keywords mask, shown as Eq. (3), to get the corresponding keywords' hidden state h_i by determining if the word x_i is a keyword.

$$Mask_i = \begin{cases} 1 & x_i \in keywords \\ 0 & otherwise \end{cases} \tag{3}$$

After that, we concatenate the extracted keywords' hidden states $h_i, ..., h_m$ to form a key information representation z as shown by Eq. (4), which can be viewed as representing the key points of the text.

$$z = \begin{bmatrix} h_1 \\ ... \\ h_m \end{bmatrix} \tag{4}$$

To improve the representation ability of encoder, we construct a keywords-based auxiliary information network, which mainly falls into two steps: firstly, generating the key vector k_i, and then building the modified encoded representation $(h_1^*, h_2^*, ..., h_n^*)$.

Concretely, this auxiliary network in our model takes two inputs, the encoder hidden state h_i and the key information representation z. The encoder hidden state h_i represents the semantic and context information of word x_i. The representation z can be regarded as containing the core information of source text. For each timestep i, the network feed the key information representation z and BiLSTM hidden state h_i through a linear layer to compute the key vector k_i, shown as Eq. (5):

$$k_i = ReLU(\mathbf{W}_h h_i + \mathbf{W}_z z + b) \tag{5}$$

where $\mathbf{W}_h$ and $\mathbf{W}_z$ are weight matrices, b is the bias vector, we use ReLU as activation function. The key vector k_i can be seen as a feature-wise weight vector, which can select the highlights and filter out the unnecessary information.

Then we incorporate the corresponding key vector k_i and the initial hidden state h_i to generate the tailored representation h_i^*, shown as Eq. (6):

$$h_i^* = h_i \odot k_i \tag{6}$$

where $\odot$ is the element-wise multiplication. The keywords-based auxiliary information network refines the encoded representation by improving the connection of the word representation with the key information of the source text so that the model is capable of covering more key points. After the auxiliary information network, we obtain another sequence of modified representation $(h_1^*, h_2^*, \ldots, h_n^*)$. This new sequence is then used as the input for the decoder to generate the summary.

3.3   Attentional Summary Decoder

We employ a single-layer unidirectional LSTM as decoder. On each timestep t, the decoder receives the previous word embedding y_{t-1} and the context vector c_{t-1} to produce the decoder hidden state s_t. The context vector c_t can be seen as a fixed-size representation of the input content for this step, which can be derived by the attention mechanism shown as Eqs. (7)–(9):

$$e_i^{\,t} = v^T \tanh(\mathbf{W}_{h^*} h_i^* + \mathbf{W}_s s_t + \mathbf{W}_v v_i^{\,t} + b_{attn}) \tag{7}$$

$$a^t = softmax(e^t) \tag{8}$$

$$c_t = \sum_i a_i^t h_i^* \tag{9}$$

where $\mathbf{W}_{h^*}$, $\mathbf{W}_s$, $\mathbf{W}_v$ and b_{attn} are learnable parameters, the attention distribution a^t can be viewed as a probability distribution over the input words, and the coverage vector v^t is the sum of attention distributions over all previous decoder timesteps.

The vocabulary distribution P_w, shown as Eq. (10), is then calculated by concatenating the context vector c_t with the decoder state s_t and then passing them through two linear layers and a softmax layer.

$$P_w(y_t|y_1, \ldots, y_{t-1}) = softmax(f'(f[s_t, c_t])) \tag{10}$$

The pointer mechanism in our model is used to copy words from the input text, which can deal with the out-of-vocabulary words. The calculation of p_{gen} is presented by Eq. (11):

$$P_{gen} = \sigma(\mathbf{W}_{c'}^T c_t + \mathbf{W}_{s'}^T s_t + \mathbf{W}_{y'}^T y_t + b_{gen}) \tag{11}$$

where $\mathbf{W}_{c'}^T$, $\mathbf{W}_{s'}^T$, $\mathbf{W}_{y'}^T$ and b_{gen} are trainable variables, y_t is the decoder input, p_{gen} denotes a soft switch to choose between generating a word from the vocabulary or copying a word from the input sequence. Hence, the final probability distribution $P(w)$, shown as Eq. (12), is computed as a convex combination.

$$P(w) = p_{gen} P_w + (1 - p_{gen}) \sum_{i:w_i=w} a_i^t \tag{12}$$

During training, the loss for every timestep t is the negative log likelihood of the target word y_t^* and the overall loss for the input text is described as Eq. (13):

$$L = -\frac{1}{T} \sum_{t=0}^{T} \log P(y_t^* | y_1^*, ..., y_{t-1}^*, x, \theta) \tag{13}$$

where θ denotes the parameters of the model, x represents the source sequence. At the generation process in our model, the selective beam search mechanism is employed to decode the output summary, which considers both the conditional probability and some additional factors.

3.4 Selective Beam Search Mechanism

Regular beam search method selects the candidate summaries for the next timestep only based on conditional probability. At the start of timestep t, the beam search mechanism held B hypotheses $Y_{t-1} = \{y_{1,t-1}, ..., y_{B,t-1}\}$, where B is the beam width. Then beam search mechanism considers all possible single token extensions of held hypotheses and selects the B most likely extensions, described as Eq. (14):

$$Y_t = \underset{y_{1,t},...,y_{B,t} \in V_t}{\arg\max} \sum_{b \in \{1,2,...,B\}} \Theta(y_{b,t}) \quad s.t. \ y_{i,t} \neq y_{j,t} \tag{14}$$

where V is the vocabulary, $V_t = Y_t \times V$ is the set of all possible token extensions and $\Theta(y_{i,t})$ is the log probability of a partial solution.

The beam search method has been observed to produce suboptimal results in neural sequence generation [27] and whether or not the hypotheses contain the key information of source text does not appear its importance [8]. To solve this issue, we propose a novel selective beam search mechanism (SeBS).

Concretely, in selective beam search mechanism, we design a hypotheses scoring function for beam search method to rerank and select the candidate summaries based on auxiliary information including the number of keywords and the redundant words. Specifically, this selective beam search mechanism is performed every K timesteps during the summary decoding process, at other times the selection is only based on the conditional probability. The hypotheses scoring function J_{score} for selective beam search is defined by Eq. (15):

$$J_{score}(y_{i,t}) = \Theta(y_{i,t}) + \alpha \cdot \log(1 - \frac{count_r(y_{i,t})}{|y_{i,t}|}) + \beta \cdot \frac{count_k(y_{i,t})}{|y_{i,t}|} \tag{15}$$

where α and β are the hyperparameters, $y_{i,t}$ is the i-th candidate summary text for timestep t, $count_r$ is employed to calculate the number of the redundant word that appears more than four times in the summary sequence, $count_k$ calculates the number of keywords of the hypothesis. By employing this novel selective beam search mechanism at decoding process, the hypothesis with less repetition and covering more keywords from the source text are ranked higher. Finally, the top-ranked hypothesis is selected as the generated summary.

4   Experiments

4.1   Datasets and Settings

Datasets. We use the benchmark *CNN/Daily Mail* dataset, which contains online news articles (781 tokens on average) paired with multi-sentence summaries (3.75 sentences or 56 tokens on average). We use scripts to obtain the version of data, which has 287,226 training pairs, 13,368 validation pairs and 11,490 test pairs [25]. We also test our model on the small-scale *BBC* dataset [10], which consists of 2225 medium-length documents (258 tokens on average) paired with the short-length abstracts (10 tokens on average).

Table 1. ROUGE F1 scores on the *CNN/Daily Mail* test set.

Model	ROUGE-1	ROUGE-2	ROUGE-L
seq2seq+attn (50k vocab)	31.33	11.81	28.83
words-lvt2k-temp-att	35.46	13.30	32.65
pointer-generator	36.44	15.66	33.42
pointer-generator+coverage	39.53	17.28	36.38
KAIN, no coverage	37.45	16.06	34.01
KAIN+SeBS, no coverage	38.15	17.03	34.72
KAIN+SeBS	**39.84**	**17.40**	**36.58**

Experiment Settings. For our experiments, we used bidirectional LSTM for encoder and uni-directional LSTM for decoder both with 256-dimensional hidden states. We set the vocabulary with a size of 50k for both source and target. Moreover, we used 128-dimensional word embedding without pre-training – they are learned from scratch during training and set the batch size as 16. We trained our model using Adagrad with learning rate 0.15. Besides, early stopping was implemented using the loss on the validation set. In our keywords-based auxiliary information fusion network, we choosed the top 10 keywords from the source text using TextRank method [20]. At test time, we set the beam size of 4. In our selective beam search mechanism, the hyperparameter α and β were in the range of [0.5, 2.5] and we found it was an optimal choice to set the α as 1 and the β as 1.5. Moreover, for the hyperparameter K, We set it of 8 for reranking every K timesteps.

We implemented our experiments in TensorFlow on a single NVIDIA 1080Ti GPU. We truncated the input tokens to 400 and we limit the length of the decoded summary to 100 tokens at test time on *CNN/Daily Mail* dataset and 20 tokens on *BBC* dataset. The training took 3 days and 16 h requiring less than 250000 iterations (about 13 epochs). Following the previous studies, we use ROUGE metric [16] for evaluation. All our ROUGE scores have a 95% confidence interval of at most 0.25 as reported by the official ROUGE script.

4.2   Results and Discussions

We evaluated the effectiveness of our approach on *CNN/Daily Mail* dataset. Apart from our own model, we also reported some compared abstractive models carried out on the same dataset, including seq2seq+attn (50k vocab) [23], words-lvt2k-temp-att [21], pointer-generator (PGNet) and the state-of-the-art model, pointer-generator+coverage (PGNet+coverage) [25]. We exploited the *pyrouge*[1] package to obtain the F1 scores for ROUGE-1, ROUGE-2, and ROUGE-L respectively. The experimental results of these compared models are obtained from their papers.

The results are presented on Table 1. Specifically, under the condition that models does not use coverage mechanism, the scores of the model equipped with KAIN exceeded the pointer-generator baseline model by (+1.0 ROUGE-1, +0.4 ROUGE-2, +0.6 ROUGE-L). The scores of our KAIN+SeBS model substantially exceeded the pointer-generator baseline model by (+1.7 ROUGE-1, +1.4 ROUGE-2, +1.3 ROUGE-L). Results in these two cases demonstrated that our model has better capability to generate high-quality summaries than PGNet without using coverage as well.

Moreover, we compared our complete model with the PGNet+coverage. The scores of our model exceeded the state-of-the-art abstractive model by (+0.3 ROUGE-1, +0.1 ROUGE-2, +0.2 ROUGE-L). The experimental results indicated that both the auxiliary information network and the selective beam search mechanism bring benefits to the abstractive model, which improve the performance of generated summary with higher ROUGE scores. Moreover, our proposed structures only increased about 10000 additional iterations during model training.

Table 2. ROUGE F1 scores on the *BBC* test set.

Model	R-1	R-2	R-L
seq2seq+attn	12.02	1.02	11.32
PGNet	14.02	1.21	13.53
PGNet+coverage	18.54	1.78	17.68
KAIN+SeBS	**22.04**	**2.14**	**21.21**

Except for the experiment on *CNN/Daily Mail* dataset, we tested our model in the task of short text summarization with small-scale dataset. Table 2 shows that the scores of our model exceed the baseline model by (+3.5 ROUGE-1, +0.4 ROUGE-2, +3.5 ROUGE-L) on *BBC* dataset. The results revealed that our model outperformed the baseline models and can generate summaries that contain more key points of the source text even on small dataset.

Deeply, we explored the ROUGE-1 score of our model with different input length on *CNN/Daily Mail* dataset. We selected input text of different length

---

[1] https://github.com/bheinzerling/pyrouge.

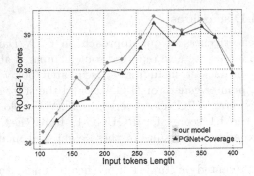

Fig. 3. ROUGE-1 F1 score of our model and PGNet+coverage model on *CNN/Daily Mail* dataset.

(the input length is less than 400 due to the truncation) from the dataset randomly and use two models to generate summaries respectively. As seen in Fig. 3, for the source text with different length, the ROUGE-1 scores of our model had consistently been higher than PGNet+coverage with a certain range, which shows that our model have the stable ability to capture the key information accurately.

Case Study. Table 3 lists the summaries generated by PGNet+coverage model and our model. We compared the generated summaries of two models with the

Table 3. An example of summaries generated by our model and baseline model. Bold words are the key information of the source text.

Source (truncated): prince harry will tonight **fly out of the uk to australia** without seeing his new niece or nephew. A disappointed prince harry will tonight fly out of the uk to australia without seeing his new niece or nephew. With the duchess of cambridge now **overdue for the birth of her second child**, the prince will not be able to meet the new **royal baby** until he returns to this country **in mid may**. Harry, who will be bumped down to fifth in the line of succession by the new arrival, had returned briefly at the weekend to hand out prizes at the london marathon after undertaking several engagements in turkey to mark the centenary of the gallipoli campaign
Reference: prince harry will fly out of uk tonight to continue placement in australia. This means he will not be able to meet royal baby until return in mid-may
PGNet+coverage: prince harry will tonight fly out of the uk to australia. A disappointed prince harry will tonight fly out of the uk to australia without seeing his new niece or nephew, the prince will not be able to meet the new royal baby
KAIN+SeBS: prince harry will fly out of the uk to australia without seeing his new niece or nephew. The prince will not be able to meet the new royal baby until he returns to this country in mid may

reference summary and we observed that the baseline model captures a part of key information, however, the summary still misses some keywords and has a certain degree of redundancy. On the other hand, our model covers the almost all key information of the source text and there is no redundant sequence owing to our keywords-based auxiliary information network and the selective beam search mechanism. Obviously, the summary generated by our model condense more important content and has better readability compared with the summary generated by PGNet+coverage.

5 Conclusion

In this paper we propose an keywords-based auxiliary information model that guides the summary generation. For encoding, the keywords-based auxiliary information network is proposed to refine the encoded representation by incorporating the word representation with the key information of the source text. In addition, the selective beam search mechanism is employed to control the decoding process by applying a novel hypotheses scorer to rerank and select the candidate summary. Experiments on benchmark dataset showed that our model leads to significant improvements compared with the state-of-the-art abstractive model equipped with pointer mechanism. In the future work, we attempt to apply this approach to other related sequence generation tasks and focus on improving the attention mechanism for language processing.

References

1. Bahdanau, D., Cho, K., Bengio, Y.: Neural machine translation by jointly learning to align and translate. In: Proceedings of the International Conference on Learning Representations (ICLR) (2014)
2. Cao, Z., Wei, F., Li, W., Li, S.: Faithful to the original: fact aware neural abstractive summarization. In: Proceedings of the Association for the Advancement of Artificial Intelligence (AAAI) (2018)
3. Celikyilmaz, A., Bosselut, A., He, X., Choi, Y.: Deep communicating agents for abstractive summarization. In: Proceedings of the 2018 Conference of the North American Chapter of the Association for Computational Linguistics: Human Language Technologies (Volume 1: Long Papers), pp. 1662–1675 (2018)
4. Chen, Y.C., Bansal, M.: Fast abstractive summarization with reinforce-selected sentence rewriting. In: Proceedings of the 56th Annual Meeting of the Association for Computational Linguistics (Volume 1: Long Papers), pp. 675–686 (2018)
5. Chopra, S., Auli, M., Rush, A.M.: Abstractive sentence summarization with attentive recurrent neural networks. In: Proceedings of the 2016 Conference of the North American Chapter of the Association for Computational Linguistics: Human Language Technologies, pp. 93–98 (2016)
6. Das, D., Martins, A.F.: A survey on automatic text summarization. Lit. Surv. Lang. Stat. II Course CMU 4, 192–195 (2007)
7. Fan, L., Yu, D., Wang, L.: Robust neural abstractive summarization systems and evaluation against adversarial information (2018). arXiv preprint arXiv:1810.06065

8. Freitag, M., Al-Onaizan, Y.: Beam search strategies for neural machine translation. In: Proceedings of the First Workshop on Neural Machine Translation, pp. 56–60 (2017)
9. Gehrmann, S., Deng, Y., Rush, A.M.: Bottom-up abstractive summarization. In: Proceedings of the 2018 Conference on Empirical Methods in Natural Language Processing, pp. 4098–4109 (2018)
10. Greene, D., Cunningham, P.: Practical solutions to the problem of diagonal dominance in kernel document clustering. In: Proceedings of the 23rd International Conference on Machine Learning, pp. 377–384. ACM (2006)
11. Gu, J., Lu, Z., Li, H., Li, V.O.: Incorporating copying mechanism in sequence-to-sequence learning. In: Proceedings of the 54th Annual Meeting of the Association for Computational Linguistics (Volume 1: Long Papers), pp. 1631–1640 (2016)
12. Isonuma, M., Fujino, T., Mori, J., Matsuo, Y., Sakata, I.: Extractive summarization using multi-task learning with document classification. In: Proceedings of the 2017 Conference on Empirical Methods in Natural Language Processing, pp. 2101–2110 (2017)
13. Lebanoff, L., Song, K., Liu, F.: Adapting the neural encoder-decoder framework from single to multi-document summarization. In: Proceedings of the 2018 Conference on Empirical Methods in Natural Language Processing, pp. 4131–4141 (2018)
14. Li, C., Xu, W., Li, S., Gao, S.: Guiding generation for abstractive text summarization based on key information guide network. In: Proceedings of the 2018 Conference of the North American Chapter of the Association for Computational Linguistics: Human Language Technologies (Volume 2: Short Papers), pp. 55–60 (2018)
15. Li, W., Xiao, X., Lyu, Y., Wang, Y.: Improving neural abstractive document summarization with explicit information selection modeling. In: Proceedings of the 2018 Conference on Empirical Methods in Natural Language Processing, pp. 1787–1796 (2018)
16. Lin, C.Y., Hovy, E.: Automatic evaluation of summaries using n-gram co-occurrence statistics. In: Proceedings of the 2003 Human Language Technology Conference of the North American Chapter of the Association for Computational Linguistics (2003)
17. Lin, J., Sun, X., Ma, S., Su, Q.: Global encoding for abstractive summarization. In: Proceedings of the 56th Annual Meeting of the Association for Computational Linguistics (Volume 2: Short Papers), pp. 163–169 (2018)
18. Luong, M.T., Pham, H., Manning, C.D.: Effective approaches to attention-based neural machine translation. In: Proceedings of the 2015 Conference on Empirical Methods in Natural Language Processing, pp. 1412–1421 (2015)
19. Mehta, P.: From extractive to abstractive summarization: a journey. In: Proceedings of the ACL 2016 Student Research Workshop, pp. 100–106 (2016)
20. Mihalcea, R., Tarau, P.: Textrank: bringing order into text. In: Proceedings of the 2004 Conference on Empirical Methods in Natural Language Processing (2004)
21. Nallapati, R., Zhou, B., Gulcehre, C., Xiang, B., et al.: Abstractive text summarization using sequence-to-sequence RNNs and beyond. In: Proceedings of the 20th SIGNLL Conference on Computational Natural Language Learning, pp. 280–290 (2016)
22. Paulus, R., Xiong, C., Socher, R.: A deep reinforced model for abstractive summarization (2017). arXiv preprint arXiv:1705.04304
23. Rush, A.M., Chopra, S., Weston, J.: A neural attention model for abstractive sentence summarization. In: Proceedings of the 2015 Conference on Empirical Methods in Natural Language Processing, pp. 379–389 (2015)

24. Saggion, H., Poibeau, T.: Automatic text summarization: past, present and future. In: Poibeau, T., Saggion, H., Piskorski, J., Yangarber, R. (eds.) Multi-source, Multilingual Information Extraction and Summarization. Theory and Applications of Natural Language Processing, pp. 3–21. Springer, Heidelberg (2013). https://doi.org/10.1007/978-3-642-28569-1_1

25. See, A., Liu, P.J., Manning, C.D.: Get to the point: summarization with pointer-generator networks. In: Proceedings of the 55th Annual Meeting of the Association for Computational Linguistics (Volume 1: Long Papers), pp. 1073–1083 (2017)

26. Sutskever, I., Vinyals, O., Le, Q.V.: Sequence to sequence learning with neural networks. In: Advances in Neural Information Processing Systems, pp. 3104–3112 (2014)

27. Vijayakumar, A.K., et al.: Diverse beam search: decoding diverse solutions from neural sequence models. In: Proceedings of the Association for the Advancement of Artificial Intelligence (AAAI) (2016)

28. Vinyals, O., Fortunato, M., Jaitly, N.: Pointer networks. In: Advances in Neural Information Processing Systems, pp. 2692–2700 (2015)

29. Zhou, Q., Yang, N., Wei, F., Zhou, M.: Selective encoding for abstractive sentence summarization. In: Proceedings of the 55th Annual Meeting of the Association for Computational Linguistics (Volume 1: Long Papers), pp. 1095–1104 (2017)

Training with Additional Semantic Constraints for Enhancing Neural Machine Translation

Yatu Ji, Hongxu Hou$^{(\boxtimes)}$, Junjie Chen, and Nier Wu

Computer Science Department, Inner Mongolia University, Hohhot, China
jiyatu0@126.com, cshhx@imu.edu.cn

Abstract. Replacing the traditional cross-entropy loss with BLEU as the optimization objective is a successful application of reinforcement learning (RL) in neural machine translation (NMT). However, a considerable weakness of the approach is that the monotonic optimization of BLEU's training algorithm ignores the semantic fluency of the translation. One phenomenon is an incomprehensible translation accompanied by an ideal BLEU. In addition, sampling inefficiency as a common shortcoming of RL is more prominent in NMT. In this study, we address these issues in two ways. (1) We use the annealing schedule algorithm to add semantic evaluation for reinforcement training as part of the training objective. (2) We further attach a value iteration network to RL to transform the reward into a decision value, thereby making model training highly targeted and efficient. We use our approach on three representative language machine translation tasks, including low resource Mongolian-Chinese, agglutinative Japanese-English, and common task English-Chinese. Experiments show that our approach achieves significant improvements over the strong baselines, besides, it also saves nearly one-third of training time on different tasks.

Keywords: Neural machine translation · Reinforcement learning · Semantic fluency · Value iteration

1 Introduction

Neural machine translation (NMT) [5,17] is the process of end-to-end encoding and decoding of parallel corpus by neural networks. These models which trained by maximum likelihood estimation (MLE) algorithm, including recurrent neural network (RNN), convolutional neural network (CNN), and Transformer [2,7,18], are used to generate text suffer from two major drawbacks. First, the models are trained to predict the next word given the ground truth words as input. However, at inference, the resulting models are used to generate an entire sequence by predicting one word at a time. The optimization strategy for these models is training with cross-entropy loss (i.e., XENT), which is used to maximize the probability of the next correct word [4,8]. Second, the loss function used to train

© Springer Nature Switzerland AG 2019

A. C. Nayak and A. Sharma (Eds.): PRICAI 2019, LNAI 11670, pp. 300–313, 2019.
https://doi.org/10.1007/978-3-030-29908-8_24

the models is based on the predicted words, namely, word level training. This result is inconsistent with the final evaluation criteria.

Two novel methods can be used to solve the above problems. One is the minimum risk training (MRT) [14] which leads to significant improvements over MLE on a state-of-the-art NMT system. MRT introduces evaluation metrics as loss functions and aims to minimize expected loss on the training data. The approach allows arbitrary sentence-level loss functions, which are not necessarily differentiable. The other is to directly optimize metrics such as BLEU, which is an effective attempt for the problems. However, the non-differentiable nature prevents the model from directly applying BLEU to the training process. An effective solution is sequence level training with an RNN (i.e., MIXER) [10]. The algorithm addresses the issues by training with reinforcement learning (RL) [12] to optimize the BLEU. This approach is a successful attempt to apply RL to machine translation with policy optimization on RNN. Reference [21] proposed a new method to leverage RL to boost further the performance of the Transformer that is trained with source/target monolingual data. We refer to this approach as (Transformer+RL, TR).

These methods of directly optimizing the evaluation objective can improve the final evaluation result to a certain extent. However, in the manual evaluation of the aforementioned typical systems, the translation of the enhanced BLEU corresponds to an incomprehensible sentence, and the semantic fluency is seriously deficient (results of the manual evaluation of the translation in Table 4 will be explained). In addition, in some vocabulary-sparse translation tasks, the optimal parameters updated by RL training can only be effective in short epochs. Then, they return to the initial state of XENT, which will be explained in Sect. 4.2. We attribute the reason to the inefficient training sampling methods, which is a ubiquitous problem in RL. For decision-making tasks, progressive improvements to the original RL, such as the Work proposed in [3], allow the model to ignore some of the empirical parameters selectively for excellent results. However, for NMT tasks aimed at obtaining empirical parameters, such an approach will greatly diminish the advantages of RL. Related studies on the training efficiency of RL in NMT are few, thereby motivating us to add value iteration network (VIN) to RL.

In this study, we conduct the investigation on two aspects. First, we add semantic evaluation as part of the training objective. Second, we use VIN to solve the inefficiency of parameter updating and model training in RL. The problem of excellence is to propose a training strategy based on VIN, which transforms semantic evaluation and BLEU into a "decision value" in each batch to make the model highly targeted and efficient.

We construct the model on long short-term memory (LSTM) because the reward with temporal attributes is suitable for VIN input. By contrast, RNN accurately captures limited semantics for some low-resource translation tasks with sparse corpus. Experimental results demonstrate that the proposed approach produces more accurate and lower consumption than the original LSTM, RNN with MRT, Transformer, MIXER, and TR on three representative machine translation tasks.

2   Background

2.1   Training with Cross-Entropy

We review the methodology used for training translation models that optimize the prediction of only one word or stem ahead of time. The method is the simplest and the most well-known in RNN to optimize the cross-entropy loss at every timestep. The usual practice [1,11,17] is to use beam search to explore multiple alternative paths. The change in gradient depends on the calculation of the loss between the model prediction and ground truth. This process can be summarized through its structure (Fig. 1).

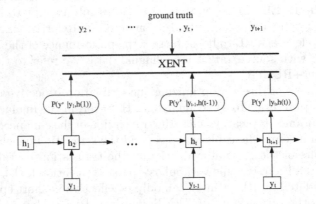

Fig. 1. RNN unfolded based on time series t, which is used to illustrate that each step of the XENT requires the participation of ground truth. This process differs from the inference; prediction y' only depends on the decoding output of the model one step ahead

Here, the calculation of the predicted output $p(y'|y_t, h_t)$ depends on the ground truth y_t and hidden layer state output h_t, and the loss function can be defined as minimizing

$$Loss = -\sum_{t=1} p(y_t|y_1, ..., y_{t-1}) \tag{1}$$

2.2   Reinforce Algorithm

A reinforce algorithm is simply understood as a text-generating task. An agent acts to change its state to gain rewards and interact with the environment in a cyclical process [16,20]. This strategy depends on the current status (only present matters) entirely, which is also a manifestation of its Markov nature. The algorithm can be simply expressed as $M = <S, A, P\{s, a\}, R>$.

Specifically, $s \epsilon S$ is a finite state set, where s represents a specific state; and $a \epsilon A$ is a limited action set, where a denotes a specific action. The transition model predicts the next state s' based on the current state s and action a, denoted as $P\left\{s'|s,a\right\}$. Reward $R = R(s,a)$ represents an instant reward after an agent takes an action. However, this process will cause unbiasedness and an endless loop of states because the rewards are accumulated in an infinite time series. Therefore, the variable of discount rate is used for reward calculation. The reward that is fed back by the state of the subsequent sequence is multiplied by this coefficient; thus, the current reward is more important than the future feedback.

A policy generates an action based on the current state, which indicates the probability of performing in a certain state.

2.3 Value Iteration (VI) Algorithm

We use the VI algorithm because its Markov characteristics fit perfectly into RL. A standard model for sequential decision making [15] and planning can be described as the process of encoding the probability of the next state s' given the current state and action. Formally, the result value $V(s)$ of VI is the expectation of the sum of n sequence rewards when starting from the state s and executing policy. The states evolve following the $P\left\{s'|s,a\right\}$. The optimal value $V^*(s) \approx Max\ Q(s,a)$ is the maximal long-term return possible from a state. We can formulate VI as

$$V_{n+1}(s) = max\ Q(s,a), \tag{2}$$

$$Q(s,a) = R(s,a) + \sum p(s'|s;a)V_n(s'). \tag{3}$$

The value of V_n in VI converges as $V^*(n \to \infty)$, from which an optimal policy may be derived as argmax $Q_\infty(s,a)$.

3 Model

3.1 RL in Machine Translation

We case our problem in the reinforcement framework [9,19]. Specifically, the training model can be viewed as an agent that interacts with the external environment (the words and the context vector that are regarded as input at every timestep). The parameters of this agent define a policy whose execution results in the agent is selecting an action. In machine translation setting, an action refers to predicting the next word in the sequence at each timestep. After taking an action, the agent updates its internal state (i.e., the hidden units).

3.2 Model Description

We consider a standard training process of our model composed of four basic components (Fig. 2).

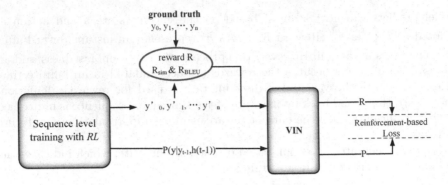

Fig. 2. Sequence-level training model based on VIN

(1) *Sequence level training:* (a) Optimal initial state. We train the model with the cross-entropy loss to ensure that an optimal model result is directly involved in training and take it as an initial value. This process also provides a good search space for greedy search. (b) RL algorithm. ⋆ Sample: The input is a sample from the distribution over words produced at the previous timestep. Different from XENT, the input here completely comes from the sampling based on the *model's* predictions without the ground truth involved. ⋆ Prediction calculation: The sampled prediction is used as the input of the hidden layer for the next timestep. Then, the hidden layer output and prediction for the current timestep are produced, followed by a round of sampling.

(2) *Reward observation:* Process (1-b) is looped once the end of the sentence (or the maximum sequence length) is reached. The reward is then computed (the full interpretation is presented in Sect. 3.3).

(3) *VIN (Fig. 3):* The reward is fed into a convolutional layer and a linear activation function Q. This layer corresponds to a particular action. The next iteration value function layer is then stacked with the reward and fed back into the convolutional layer N times, where N depends on the length of the sequence. Then, a long-term value is obtained by training a batch. The model improves the training efficiency of the model based on this value-based sampling strategy.

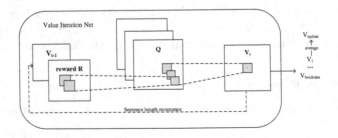

Fig. 3. VI algorithm represented by CNN

(4) *Reinforcement-based loss:* The value of a batch is observed. The model gradient propagation depends on its comparison with the optimal value. When the value obtained from this comparison determines that the network must be updated, the reward involves the calculation of loss function. Subsequently, reinforcement algorithm is used to backpropagate the gradients by the sequence of samplers.

3.3   Reward Calculation

Once the agent has reached the end of a sequence, it observes a reward from BLEU and semantic similarity calculation.

Here, the semantic similarity (denoted as *sim*) is applied to the calculation of reward in each iteration, which compares the ground truth and prediction. The reason for this approach is that the optimization of BLEU will ignore the structure of the sentence pattern to a certain extent. As part of the reward, we calculate the cosine angle between the two sentence vectors (y' and y) on the basis of (4) to use word embedding well. This process can effectively play the role of reinforce algorithm.

$$R_{sim} = sim(y', y) = \frac{\sum_{i=1}^{n}(y_i' \times y_i)}{\sqrt{\sum_{i=1}^{n}(y_i')^2} \times \sqrt{\sum_{i=1}^{n}(y_i)^2}} \tag{4}$$

where i denotes a word vector in a sentence with length of n. The reward associated with BLEU can be expressed as

$$R_{BLEU} = bp \times exp(\sum_{n=1}^{n} w_n log p_n) \tag{5}$$

$$bp = \begin{cases} 1 & if\ c > l \\ e^{1-\frac{l}{c}} & if\ c \le l \end{cases} \tag{6}$$

where p_n represents the precision calculation term of the N-gram; w_n is the corresponding weight; bp is a standard-length penalty term; c and l represent the length of the translation to be evaluated and the reference translation, respectively. Thus, the final reward can be described as

$$Reward = \lambda R_{BLEU} + (1 - \lambda)R_{sim}. \tag{7}$$

The next idea is to introduce model predictions during training with an annealing schedule in order to gradually teach the model to produce stable sequences. For every batch sequences we use the R_{BLEU} reward for the first ($batchsize$-$\triangle$) sequences, and ($R_{BLEU} + R_{sim}$) reward for the remain $\triangle$ sequences. $\triangle$ is set to $\lfloor (5\% \sim 10\%) \times batchsize \rfloor$. Next we anneal the number of sequences for which we use the XENT loss for every batch to ($batchsize$-$2\triangle$) and repeat the training for another epochs.

We use an annealing schedule on β to weigh the two part calculation of the final reward, that is, rewards for BLEU R_{BLEU} and for semantic similarity R_{sim}, starting with β equal to the $batchsize$ and finishing with $\beta = 1$. λ is a hyperparameter.

3.4   VIN

We use a novel interpretation of an approximate VI algorithm as a particular form of a CNN. This approach allows us to treat the planning module conveniently as another NN. We can train the entire policy end-to-end on the basis of its simplification by backpropagation. The VI module is simply an NN architecture with the capability of performing an approximate VI computation. Nevertheless, VI in this form, which made learning the MDP parameters and reward function natural by backpropagation through the network, which similar to a standard CNN. Once a VIN design is selected, implementing the VIN is straightforward because it is simply a form of a CNN. The networks in the experiments all require only several lines of Tensor code.

Selecting a kernel width of not less than 3 is reasonable and valid for three vocabularies. The value of the output continues to use sparse storage to participate in the next round of convolution operations.

Each iteration of VI algorithm may be seen as passing the previous value of V_n and reward R by a convolution layer and mean pooling layer. In this analogy, the active function in the convolution layer corresponds to the Q function. Convolution kernel weights correspond to the discounted transition probabilities that are generated by the decoder model in machine translation. Thus, the value of the sequence is produced by applying the convolution layer recurrently for N times.

The value of the current iteration represents the cost of the current state model decoding. The total value, which is obtained after batch training, represents the performance of the RL model. Therefore, on the basis of the total value of each batch training, we infer whether the parameters of the model are optimal; thus, we can decide the necessity of performing gradient propagation. This process ensures that the optimal parameters participate in the decoding process and enables the model to be highly targeted for samples with poor semantic information and low BLEU.

3.5   Agent Update

The observed reward determines the cost of the prediction, which has the same meaning as the loss function in the usual sense. Therefore, the loss function based on the reinforce algorithm can be defined as

$$Loss = -\sum_{i=1}^{n} p(y_i')r(y_i'), \tag{8}$$

where $p(y_i')$ represents the predicted probability of y_i', and y_i' is the word selected by the model at the i-th timestep. r is the reward associated with the generated sequence. The goal of training is to find the parameters of the agent that can maximize the expected reward. Thus, we quote the conclusion of [12, 22], which can be equated with the expectation of r.

$$Loss = -\sum_{y_i \sim p} r(y_i'), \ i\epsilon(1, 2, ..., n) \tag{9}$$

In other words, the expectation that can obtain a better reward in each time decoding is expected. The partial derivatives and interpretation of the gradients are:

$$\frac{\partial Loss}{\partial o_t} = \frac{\partial Loss(XENT)}{\partial o_t}(r(y_i') - \bar{r}_{t+1})$$
$$= (p(y_{t+1}'|y_t', h_{t+1}, c_t) - y_{t+1}')(r(y_i') - \bar{r}_{t+1}), \quad i\epsilon(1, 2, ..., n), \tag{10}$$

where $\bar{r}_{t+1}$ refers to the average reward of the result at timestep $(t+1)$, which is estimated by a linear regression. The regression uses the output of hidden states at $(t + 1)$ time as its input. Although high variance is not prominent in the NMT, we continue to use $\bar{r}_{t+1}$ to decrease the variance of the gradient estimator considering some sparse language. o_t is inputted to Softmax. When the predicted average result $\bar{r}_{t+1}$ is less than the actual reward r, it is updated in the positive direction; otherwise, it is reversed in the negative direction. The parameters of the regressor are trained by minimizing the mean square loss, $\|\bar{r}_t - r\|^2$. In our implementation, this error is ignored when backpropagating because it will result in feedback loops.

4   Experiments

4.1   Dataset and Model Configuration

We validate the effectiveness of our approach on three typical NMT tasks, namely, low-resource Mongolian-Chinese (M-C), agglutinative language Japanese-English (J-E), and common Chinese-English (C-E) tasks; we use data from CWMT2017, Wikipedia Kyoto Articles, and WMT2017, for the three tasks, respectively (Table 1). To avoid allocating excessive training time on long sentences, all sentence pairs longer than 50 words either on the source or target side are discarded.

Table 1. Statistical analysis of experimental corpus

	Training	Dev	Test
M-C	201643	1001	1000
J-E	500580	1001	1000
E-C	923471	1001	1000

To solve the problem of excessive low-frequency words in the vocabulary, we perform BPE processing on the three training sets to compare the proposed approach's adaptability of the corpus granularity. Furthermore, BPE can alleviate the vocabulary sparse problem of low-resource language to a certain extent. We reference the best empirical operands of in previous studies [6,13] and attempt several num-operations of BPE on three datasets. The final selected BPE operations number are M-C (Mongolian: 35000, Chinese: 15000), J-E (Japanese: 60000, English: 30000), and E-C (English: 55000, Chinese: 25000).

The experiment mainly compares the performance of different language tasks in each system. We use the seq2seq LSTM model (training by XENT), RNN with MRT, MIXER (training by RL), Transformer, and TR as the baseline system. To verify the influence of model innovation on experimental results, we have not compared the monolingual experiments in TR with only RL-based Transformer as the baseline.

For the RNN with MRT, LSTM and MIXER, following the base model of [10] and [14], we set the dimension of word embedding as 512 and dropout rate as 0.1/0.1/0.3. We use a beam search with a beam size of 4 and length penalty of 0.6. We also attempt a standard RNN as our generative model for this task. However, this process does not improve the performance because the generated sentences in M-C and J-E are relatively long.

For the Transformer [18] and TR [21], the original Transformer_base configuration is an effective experience setting for our experiments.

For our approach, we set the hidden units for attentive encoders and decoders as 256 on the basis of the study of [10]. The dimension of the word embedding is also set as 512. During testing, we use a beam search with a beam size of 8; the length penalty is not applied for all the three tasks.

4.2   Main Results and Analysis

We stop training when the model achieves no improvement (the accuracy is not less than 0.8) for the evaluation on the development set. All models are trained on up to single Titan-X GPU, and we count the convergence hours of each model in the three tasks (Table 2).

Table 2. Convergence time when accuracy achieves 0.8 in three languages

	M-C	J-E	E-C
Transformer	25	34	51
TR	32	40	47
RNN with MRT	37	55	82
LSTM	28	41	55
MIXER (LSTM+RL)	33	45	62
Our model	19	25	33

The additional VIN's RL approach can effectively reduce the convergence time of the model while achieving the higher BLEU score, and approximately one-third of training time is saved.

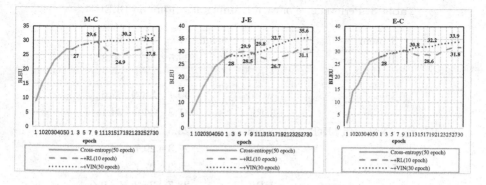

Fig. 4. Influences of training algorithms on different tasks. (Color figure online)

We observe the BLEU score for the three tasks in 50 epochs of initial training and 40 epochs of reinforce training (including 10 epochs of RL training and 30 epochs of RL+VIN training) to illustrate the effect of our approach on the model (Fig. 4). The orange and blue short lines in Fig. 4 indicate the epochs when RL algorithm and VIN begin to affect the model, respectively. When the model obtains optimal results through cross-entropy training, the RL algorithm can be further converged on its basis and the accuracy of prediction can be improved. However, this state only maintains for few iterations with the decline of the model's predictive ability. Multiple iterations are needed to train the RL model and readapt to the input. Notably, the VIN based model can maintain the training process in an optimal state and continuously provide accurate predictions. The optimal parameters make the model stay on the optimal parameters and have a steady growth. In comparison with the baseline models, our approach has achieved greater improvement in BLEU score and can converge to optimal values faster. Table 3 shows the BLEU evaluation results.

All monotonous BLEU enhancements have no remarkable improvement on M-C and J-E, even with a considerable decline in Transformer. Over-translation occurs in the translation of Transformer because the self-attention mechanism greatly improves the accuracy of prediction, and the simple reinforcement will result in the over-fitting of the model. RL+VIN has an average of 3–4 BLEU score improvement on all the three datasets whether at the word or sub-word level. The approach has considerably improved on the M-C and J-E test set.

Furthermore, a single BLEU evaluation does not perfectly measure the semantic fluency of a sentence. Therefore, we use the specification of manual evaluation in the "Machine Translation Evaluation Outline"[1] as a standard to evaluate manually the translation results. Following the intelligibility, the sentence scores range from 0 to 10 points, including two decimal places, and the final score is the arithmetic mean of all scores.

---

[1] http://www.liip.cn/wmt2013/.

Table 3. Performance of each training algorithm on BLEU-4

System	Segmentation	M-C	J-E	E-C
Transformer	word	28.5	31	30.8
	BPE	29.4	31.4	33
TR	word	30.1	**33.9**	32.6
	BPE	31.3	34.8	34.1
RNN with MRT	word	25.5	29.4	31.8
	BPE	27.3	29.7	34
LSTM	word	24.2	27.7	28.1
	BPE	28.7	28.2	28.9
MIXER (LSTM+RL)	word	29.7	28.9	31.7
	BPE	30.9	31.4	32.3
Our Model	word	**32.5**	33.7	**34.1**
	BPE	**34.1**	**36.8**	**36.2**

Table 4. Human evaluation of translations

	M-C	J-E	E-C
Transformer	6.30	6.55	7.15
TR	5.85	6.50	7.40
RNN with MRT	5.30	6.06	7.17
LSTM	6.05	6.20	7.00
MIXER (LSTM+RL)	5.75	6.10	7.35
Our Model	7.65	7.15	7.7

Table 4 clearly illustrates the necessity of increasing semantic similarity as a training objective. The BLEU enhancement for LSTM and Transformer is worse than the original system semantics on M-C and J-E tasks.

We also extract a representational sentence in different languages to illustrate the proposed approach on the translation, as shown in Fig. 5. The Italics indicate mistranslation and paired quotation marks indicate omission. The bold words in Target sentence are the key observation words which are prone to semantic ambiguity.

Sentence		M–C	J–E	E–C
	Source	ᠮᠣᠩ᠁ (Mongolian script text)	結婚後に太ると考えている人が多く。	There are both a scientific and a moral attitude toward the universe.
	Target	在英国，来往车辆都靠左行驶，对吗？	Many people think **that** they **will** **gain weight** after marriage.	人类对宇宙有一种科学的态度，也有一种道德的**态度**。
got	Transformer	在英国，车来回在左边⟨unk⟩，*应该*吗？	Many people think *" "*they will *grow* *fat* after marriage.	宇宙有一个科学的*认为*，和一种道德*" "*。
	TR	在英国，车来回在左边⟨unk⟩，*应该*吗？	Many people think that *that their* *" "* *weight grow* after marriage.	*" "*宇宙有一个科学和一个道德的*脾气*。
	RNN with MRT	在英国中，车辆⟨unk⟩左，可以吗？	Many people think *" "* they will *be* *" "* *fat* after marriage.	*" "*宇宙是*" "*科学*" "*，也是道德的态度。
	LSTM	英国，*都跟刚左边*⟨unk⟩，*应该跟随*吗？	*There are* many people think that they *become* weight after marriage.	*" "*在宇宙方面有一种科学的态度和*" "*道德*" "*。
	MIXER (LSTM+RL)	在英国，*车辆来往都在左*，对吗？	Many people think *" "* they will ⟨unk⟩ weight *when* marriage.	*" "*宇宙有一种科学*" "*和*" "*道德*" "*。
	Our Model	在英国，来回车辆都*在左边*行驶，对吗？	Many people think that they *" "* gain weight after marriage.	*" "*宇宙既是一种科学的态度，*也是*一种道德*" "*。

Fig. 5. Performance of translation under different model

5   Conclusion

In this study, we introduce an approach of attaching VIN to RL to solve the problem of monotonous optimization of BLEU and a lack of semantic information. We use semantic similarity evaluation with BLEU as the optimization objective of training to obtain joint reward information and regard such reward information as the cost of model decoding. Our other contribution is the conversion of such decoding costs into a simple value by the VIN, thereby determining whether the current parameters need to be adjusted and making the model training highly targeted. This approach outperforms several strong baseline systems in all three typical machine translation tasks and saves an average of nearly one-third of training time. We have found that similar problems exist in other NLP tasks that apply RL, and in subsequent studies, we aim to apply our algorithms to additional NLP tasks.

References

1. Bahdanau, D., Cho, K., Bengio, Y.: Neural machine translation by jointly learning to align and translate. In: International Conference on Learning Representations (ICLR), pp. 473–484 (2014)

2. Gehring, J., Auli, M., Grangier, D., et al.: Convolutional sequence to sequence learning. In: International Conference on Machine Learning (ICML), pp. 1243–1252 (2017)
3. Hessel, M., Modayil, J., Van Hasselt, H., et al.: Rainbow: combining improvements in deep reinforcement learning. In: National Conference on Artificial Intelligence (NCAI), pp. 3215–3222 (2018)
4. Hui, K.P., Bean, N., Kraetzl, M., et al.: The cross-entropy method for network reliability estimation. Ann. Oper. Res. **134**(1), 101 (2005)
5. Kalchbrenner, N., Blunsom, P.: Recurrent continuous translation models. In: Conference on Empirical Methods in Natural Language Processing (EMNLP), pp. 1700–1709 (2013)
6. Kunchukuttan, A., Bhattacharyya, P.: Learning variable length units for SMT between related languages via Byte Pair Encoding. In: Conference on Empirical Methods in Natural Language Processing (EMNLP), pp. 14–24 (2017)
7. Mikolov, T., Kombrink, S., Burget, L., et al.: Extensions of recurrent neural network language model. In: IEEE International Conference on Acoustics, Speech, and Signal Processing, pp. 5528–5531 (2011)
8. Morin, F., Bengio, Y.: Hierarchical probabilistic neural network language model. In: International Conference on Artificial Intelligence and Statistics (AISTATS), pp. 246–252 (2005)
9. Nogueira, R., Cho, K.: WebNav: a new large-scale task for natural language based sequential decision making. In: Conference and Workshop on Neural Information Processing Systems (NIPS), pp. 177–186 (2016)
10. Ranzato, M., Chopra, S., Auli, M., Zaremba, W.: Sequence level training with recurrent neural networks (2015). arXiv:1511.06732
11. Rush, A.M., Chopra, S., Weston, J.: A neural attention model for abstractive sentence summarization. In: Conference on Empirical Methods in Natural Language Processing (EMNLP), pp. 379–389 (2015)
12. Schwenker, F., Palm, G.: Artificial development by reinforcement learning can benefit from multiple motivations. Front. Robot. AI **6**(6) (2019)
13. Sennrich, R., Haddow, B., Birch, A.: Neural machine translation of rare words with subword units. In: Association for Computational Linguistics (ACL), pp. 1715–1725 (2015)
14. Shen, S., Cheng, Y., He, Z., et al.: Minimum risk training for neural machine translation. In: Association for Computational Linguistics (ACL), pp. 1683–1692 (2016)
15. Smith, D.K.: Dynamic programming and optimal control. J. Oper. Res. Soc. **47**(06), 833–834 (1996)
16. Sunmola, F.T., Wyatt, J.L.: Model transfer for Markov decision tasks via parameter matching. In: Proceedings of the 25th Workshop of the UK Planning and Scheduling Special Interest Group (PlanSIG 2006), pp. 77–86 (2006)
17. Sutskever, I., et al.: Sequence to sequence learning with neural networks. In: Conference and Workshop on Neural Information Processing Systems (NIPS), pp. 3104–3112 (2014)
18. Vaswani, A., Shazeer, N., Parmar, N., et al.: Attention is all you need. In: Conference and Workshop on Neural Information Processing Systems (NIPS), pp. 5998–6008 (2017)
19. Volodymyr, M., Koray, K., et al.: Human-level control through deep reinforcement learning. Nature **518**(7540), 529 (2015)
20. From Wikipedia Dynamic Programming and Markov Processes: Markov decision process. J. Oper. Res. Soc. **112**(4), 217–243 (2010)

21. Wu, L., Tian, F., Qin, T., et al.: A study of reinforcement learning for neural machine translation. In: Conference on Empirical Methods in Natural Language Processing (EMNLP), pp. 3215–3222 (2018)
22. Zaremba, W., Sutskever, I.: Reinforcement learning neural turing machines-revised (2015). arXiv:1505.00521

Automatic Acrostic Couplet Generation with Three-Stage Neural Network Pipelines

Haoshen Fan[1,2], Jie Wang[2(✉)], Bojin Zhuang[2], Shaojun Wang[2], and Jing Xiao[2]

[1] University of Science and Technology of China, Hefei, China
[2] Ping An Technology (Shenzhen) Co., Ltd., Shenzhen, China
photonicsjay@163.com

Abstract. As one of the quintessence of Chinese traditional culture, couplet compromises two syntactically symmetric clauses equal in length, namely, an antecedent and subsequent clause. Moreover, corresponding characters and phrases at the same position of the two clauses are paired with each other under certain constraints of semantic and/or syntactic relatedness. Automatic couplet generation is recognized as a challenging problem even in the Artificial Intelligence field. In this paper, we comprehensively study on automatic generation of acrostic couplet with the first characters defined by users. The complete couplet generation is mainly divided into three stages, that is, antecedent clause generation pipeline, subsequent clause generation pipeline and clause re-ranker. To realize semantic and/or syntactic relatedness between two clauses, attention-based Sequence-to-Sequence (S2S) neural network is employed. Moreover, to provide diverse couplet candidates for re-ranking, a cluster-based beam search approach is incorporated into the S2S network. Both BLEU metrics and human judgments have demonstrated the effectiveness of our proposed method. Eventually, a mini-program based on this generation system is developed and deployed on Wechat for real users.

Keywords: Natural language generation · Couplet generation · Sequence-to-Sequence · Language model · Attention

1 Introduction

Chinese antithetical couplet, (namely "对联"), which consists of two clauses, is an important part of Chinese cultural heritage. As a part of Chinese people's cultural life, couplets have become a popular way to expressing personal emotion, political views, or communicating blessing messages at festive occasions. As an important traditional cultural game, given one antecedent clause, people are challenged to write the subsequent clause. Additionally, couplets expressing blessing and happiness are written on red banners on special days, such as the Chinese New Year, birthday and wedding ceremonies. Literally, Chinese couplet must satisfy certain constraints on syntactic and/or semantic relevance. For example, corresponding characters or phrases from the same position in the two clauses must be paired with each other. For instance, as shown in Fig. 1, the character 'hundred' is paired with 'thousand', 'flower' is antithetical to

© Springer Nature Switzerland AG 2019
A. C. Nayak and A. Sharma (Eds.): PRICAI 2019, LNAI 11670, pp. 314–324, 2019.
https://doi.org/10.1007/978-3-030-29908-8_25

'tree', 'good' correspond to 'new' and 'bloom' is coupled with 'boom'. Compared to common proses such as news and fictions, couplet also exhibits poetic aestheticism, e.g., rhyming and conciseness etc.

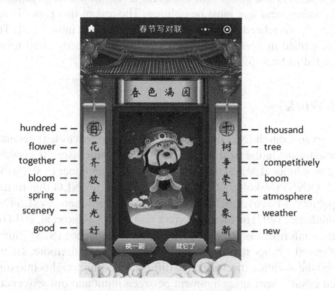

Fig. 1. An acrostic couplet generated by our developed mini-program of couplet generation. Each Chinese character is translated into English for reference. The abstract meaning of this couplet is that many flowers blossom in spring making the spring scenery very beautiful (antecedent clause, left); many trees competitively boom in spring which refresh the atmosphere (subsequent clause, right).

In this paper, we focus on automatic acrostic couplet generation with deep learning methods. Especially at Spring Festival, couplet have been popularly used for expressing blessing for the coming new year. Under the analysis of potential user demand, we had planned to develop an online mini-program of automatic couplet generation on Wechat. Contrary to the couplet game that people are challenged to write subsequent clause given the antecedent one, our automatic acrostic couplet generation can compose a complete clause pair with users' intent defined in both clause heads, which facilitates fluent user interaction. Obviously, as opposed to completing subsequent clause, automatic acrostic couplet generation is more challenging. Herein, we formulate the acrostic couplet generation as a three-stage natural language generation problem. In the first stage, the antecedent clause is generated by a pipeline of recurrent neural network based language model (RNN-LM) given user' intent as head characters. Afterwards, the subsequent clause is generated by an attention-based S2S network by taking in the antecedent one from the previous stage. Moreover, a cluster-based beam search (CBS) method is incorporated to generate a candidate pool of diverse couplets. Eventually, best couplet is selected from the candidate pool with a re-rank pipeline.

In order to creating interesting and excellent Spring Festival acrostic couplets for online users, the re-ranking pipeline is based on the following criterions. For example, the length of single clause is in the range from 5 to 12 characters. In addition, corresponding characters at the same positon of two clauses should have the same part of speech (POS). Ending tone of both clauses must be opposed. For instance, if pronounce tone of last character in the antecedent clause is level, the corresponding tone of last character in the subsequent one must be oblique. The rest of this paper is organized as follows. In Sect. 2, the related work of couplet generation is introduced. The detail of our model is described in Sect. 3. Section 4 summarizes experimental results and our study is concluded in Sect. 5.

2 Related Work

Natural language generation (NLG) (Mann 1982), also known as text generation, is one of most important tasks in the field of natural language processing (Chowdhury 2003). Compared to convolutional neural network (CNN) (Kalchbrenner et al. 2014), recurrent neural network (RNN) (Mikolov 2010) is more suitable for NLG due to its sequential prediction capability. Moreover, RNN with long-short term memory (LSTM) (Hochreither and Schmidhuber 1997) or gated recurrent unit (GRU) (Cho et al. 2014) can capture longer contextual information. Recently, Sequence-to-Sequence (S2S) (Sutskever et al. 2014) was proposed for heterogeneous data translation. Furthermore, Bahdanau et al. (2015) proposed the attention mechanism to diffuse decoding weights into different parts of input, which ensures semantic alignment between input and out sequences.

To some extent, couplet generation can be considered as a similar case of statistical machine translation or poetry generation. There are two main methods for machine translation: Statistical Machine Translation (SMT) (Koehn 2010) and Neural Machine Translation (NLT) (Koehn 2017). For example, Koehn et al. (2003) proposed an approach of Statistical phrase-based translation and Devlin et al. (2014) proposed Neural Network Joint Model (NNJM) which was constructed using the context of both source and target language. Recently, Ahmed et al. (2018) applied the state-of-the-art transformer structure for machine translation. On the other hand, some researchers proposed the methods based on rules or templates, e.g., phrase search approach (Wu et al. 2009), template search approach (Oliveira 2012) and summarization approach (Yan et al. 2013) for poetry generation. Furthermore, Zhang and Lapata (2014) proposed a poetry generation model based on RNN which generates each line character by character. In order to achieve semantic coherence, a novel two-stage poetry generating method (Wang et al. 2016) was presented. In order to create flexible and creative Chinese poetry, Zhang et al. (2017) extended the neural model with memory augment, which balanced the requirements of linguistic accordance and aesthetic innovation.

Under most circumstance, the results generated by end-to-end (E2E) system are not guaranteed to be always satisfied. To address this problem, few researchers re-rank generated texts to select desired results. Jiang and Zhou (2008) used multiple features including Mutual information score and MI-based structural similarity score to train a SVM model for candidates re-ranking. Sordoni et al. (2015) applied the ranking algorithms LambdaMART as supervised ranker.

To the best of our knowledge, nevertheless, few research work focused on the task of couplet generation. Zhang and Sun (2009) proposed a couplet generation model based on statistics and rules. Jiang and Zhou (2008) regarded this task as a kind of machine translation and reported a phrase-based statistical machine translation (SMT) approach. Furthermore, Yan et al. (2016) proposed a novel polishing schema to refine the generated couplets using additional information. However, our study is different from all above methods. Most previous work tended to treat couplet generation as a special translation task and tried to generate subsequent clauses given antecedent ones. However, a complete acrostic couplet can be generated by our method only given few head characters defined by users. Moreover, contrary to previously reported E2E approaches, three pipelines consisting of a RNN-LM, an attention-based S2S and a re-ranker are combined for better couplet generation for real-time online Wechat users. To provide a candidate pool of diverse couplets for selection, a CBS method is incorporated into the S2S network. Furthermore, we believe that our propose three-pipeline generation method can also be extended to other language generation tasks, such as poetry and stories.

3 Model

In this paper, our proposed acrostic couplet generation method mainly consists of three stages, namely, antecedent clause generation pipeline, subsequent clause generation pipeline and the re-rank pipeline as shown in Fig. 2. Among an acrostic couplet, head characters of both clauses are defined by users, denoting as K_1 and K_2. Thus, the two clauses are denoted as: $S_1 = \{K_1, C_{1,2}, C_{1,3}, \ldots, C_{1,m}\}$ and $S_2 = \{K_2, C_{2,2}, C_{2,3}, \ldots, C_{2,m}\}$, where m represents the clause length minus 1. The antecedent clause S_1 can be automatically generated by the RNN-LM pipeline given the head K_1. Based on the generated clause S_1, the subsequent clause S_2 can be generated with an attention-based S2S. Moreover, a CBS method is incorporated to realize a candidate pool of diverse clauses. Eventually, a re-ranking pipeline is used to select the best couplet from the candidate pool.

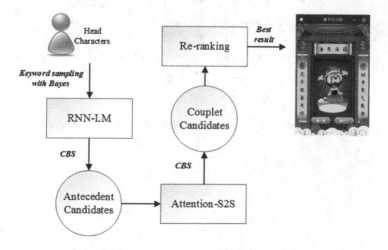

Fig. 2. The main framework of our model.

3.1   Antecedent Clause Generator

Given the head characters, an RNN-LM model is used to generate antecedent clauses. Neural language model (Bengio et al. 2003) was first proposed in 2003, then Mikolov (2010) extended it with RNN. We use the vanilla RNN cell to calculate and store the information of each character $C_{1,2}, C_{1,3}, \ldots, C_{1,m}$ in the antecedent clause. Taking in the word embedding of a character $C_{1,i}$, and the previous state s_{i-1}, the RNN cell can calculate a current hidden state s_i as follows:

$$s_i = f\left(w_s s_{i-1} + w_c C_{1,i} + b\right) \tag{1}$$

where w and b are trainable parameters as weights and bias, and the parametrized non-linear function f is based on hyperbolic tangent. As shown in Fig. 3, next character $C_{1,i+1}$ is predicted by the hidden state s_i. To generate antecedent sentence diversely and effectively, a CBS method was applied in the decoder, which was proposed by Tam et al. (2019) to overcome the shortcoming that beam search tends to output several sentences with slight difference. As shown in Algorithm 1, CBS combines K-means cluster and beam search to generate more meaningful response. In each decoding step of beam search, CBS perform K-means cluster according to the average embedding of candidates and remove half of candidates in each cluster.

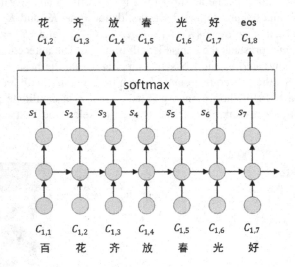

Fig. 3. Schematic diagram of the antecedent clause generator.

Algorithm 1: Cluster-based beam search

Input: Beam width BW, Candidates C initialized with start symbol, maximum decoding step t_{max}, cluster number K
Output: Final result res
While *Number of completed hypothesis does not reach BW or decoding step not reach* t_{max}
do

    for *i in BW* **do**
        tmpHyps = Top-N(Extend($C[i]$, $BS \times 2$));
        Remove hyp in tmpHyps with repeated N-grams or UNK;
        Save tmpHyps to extended candidates C_e;
    end
    Perform K-means over extended candidates C_e;
    for *candidates in each cluster* **do**
        Sort candidates by partial log-prob scores;
        Choose top BW/K candidates;
        Put candidates with end symbol in R;
        Put incomplete candidates in C;
    end
end
$res \leftarrow$ sort R according to log-prob scores;

3.2 Subsequent Clause Generator

The basic idea of subsequent clause generator is to map the antecedent clause into a fix dense vector and then decode the subsequent clause iteratively and sequentially. Sequence-to-sequence (S2S) (Sutskever et al. 2014) is a popular framework for this task. To enhance syntactic and/or semantic relatedness between antecedent and subsequent clauses, attention mechanism is incorporated into the S2S generation model. The generation model iteratively encodes the antecedent clause into a fix dense contextual vector s_i. Contrast to the conventional S2S which rely context vector on the last input hidden state s_m, attention mechanism considers contribution of each input character into a new context vector as follows:

$$v_t = \sum\nolimits_{j=1}^{m} a_{tj}s_j \tag{2}$$

The a_{tj} is determined by the previous hidden state h_{t-1} and each hidden state of encoder, i.e., $\{s_1, s_2, \ldots, s_m\}$. Therefore, the new context vector is a weighted sum of hidden states of the encoder, which can adaptively pay attention to the corresponding input character during decoding. As depicted in Fig. 4, the decoding of subsequent clause with attention mechanism can be expressed as follows:

$$h_t = f\left(w_s h_{t-1} + w_c C_{2,t-1} + w_v v_t\right) \tag{3}$$

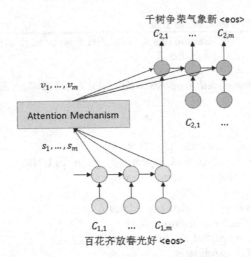

Fig. 4. Illustration of the attention-based S2S model.

3.3   Processing of Head Character

Based on our proposed three-stage model, the quality of generated complete acrostic couplet heavily depends on the first antecedent clause. Due to its sequential and iterative generation procedure, vanilla RNN based antecedent clause generator requires a reasonable head character (i.e. start token) for high-quality generation. As a part of our design of user interaction (UI), user input is constrained in 4 characters. As an optimal method of improving user experience, Naive Bayes is employed for sampling two characters from user input, which is expressed as:

$$P(B|A) = \frac{P(B)P(A|B)}{P(A)} \tag{4}$$

Where event A represents the appearance frequency of character C while event B represents the frequency of character C appearing in antecedent sentence as head character. Based on train corpus, this Naïve Bayes model can be trained with maximum likelihood loss.

3.4   Re-ranking

According to the restrictions of Chinese acrostic couplets, a ranking score is calculated for re-ranking, including length score s_l, repeated score s_r, tone score s_t and sentiment score s_s, which can be denoted as:

$$s = w_l s_l + w_r s_r + w_t s_t + w_s s_s \tag{5}$$

Where weight parameters w_l, w_r, w_t and w_s are empirically set and optimized. Length score means whether the two clauses of couplet have the same length. Repeated score checks whether repeated characters exist in couplets. Tone score determines whether the ending characters of two clauses exhibit opposed tone. The sentiment score is calculated based on a SWM model, which is higher for positive couplets.

4 Experiments and Evaluations

In this section, dataset processing and experimental settings are described at detail. Moreover, re-ranking method for clause candidate selection is introduced. In addition, couplet generation based on LM as baseline models are used.

4.1 Dataset

Firstly, a large couplet corpus is collected to efficiently train the generation model, which consists of approximately 602858 couplets. As a result, a primitive vocabulary of 7318 characters was achieved. After omitting low frequency characters less than 10 times, the vocabulary size is decreased to 5647. Additionally, specific symbols are added into the trained vocabulary, including '<unk>' representing low frequency characters and '<eos>' donating the end of sentence. Moreover, 1000 and 2000 couplets are randomly sampled for validation and testing.

4.2 Parameter Setting and Training

In this paper, word2vec is used for distributed representation of Chinese characters, which is initially pre-trained with the random initialization (Mikolov et al. 2013). Herein, each character are mapped into a low and dense dimensional vectors, where 256-dimensional word embedding is used. The LSTM cell in antecedent and subsequent sentence generator both have 1000 hidden units. The cell layer number of LM and seq 2seq models is 2 and 4, respectively. To ensure generation diversity, group size and width of beam search is set to 4 and 2, respectively. As a result, 16 candidates are achieved according to the constraints of Spring Festival couplet.

All of the trainable parameters are randomly initialized within the range $[-0.5, 0.5]$. They are trained by stochastic gradient descent to minimize the cross-entropy loss with the Adam optimizer (Kingma and Ba 2015). The mini-batch size of 128 is chose for training. Moreover, to prevent gradient explosion, the gradient is clipped to the maximum of 5. The learning rate is initially set to 0.001 and adaptively decreased along with training.

4.3 Evaluation Metrics

Automatic Evaluation
The Bilingual Evaluation Understudy (BLEU) (Papineni et al. 2002) score is widely used for evaluation of machine translation. In this paper, BLEU metric is chosen as an automatic evaluation approach for our couplet generation, where original couplets are used as reference ground truth. Note that BLEU score can represent the similarity between generated couplets and human-written ones. Moreover, some specifications

related to Chinese couplets are taken into consideration, such as Length Matching, Character Structure and Tone Pairing. Among them, Length Matching means that both couplet sentence must have the same length. Character Structure ensures that both sentences shouldn't contain the same characters and/or phrases in the same positon. Tone Pairing requires the last characters in the two sentences exhibit opposed tone.

Human Evaluation

Different from machine translation, BLEU score is not enough for couplet generation evaluation due to its high diversity. Therefore, eight graduate students majored in traditional Chinese are asked to review generated couplets. They are asked to score our generated couplets with 1–5 scores in three aspects, including Structural Symmetry, Semantic Coherence and Topic Relevance. In terms of Structural Symmetry, correspondence of each character in the same position from two sentences in the aspect of part of speech (POS) and semantics. Semantic Coherence means that antecedent and subsequent clauses are semantically coherent but not repetitive. Fluency examines whether both clauses of generated couplets are expressed fluently.

Table 1. Automatic evaluation results.

Method	Length matching	Character structure	Tone pairing	BLEU
LM	1.0	0.60	0.86	0.2830
SPC generator	1.0	0.73	0.99	0.2831

4.4   Experimental Results

Table 1 depicts automatic evaluation results of our proposed couplet generator compared with the baseline. Apparently, both models can easily generate two same length clauses. However, our proposed model performs better in Character Structure and Tone pairing, which is owing to the encoder-decoder network structure and attention mechanism. Note that both models perform comparatively in terms of BLEU score, verifying the ineffectiveness of BLEU metric for literature creation. Human evaluation results are shown in Table 2. Our proposed model performs better in all three aspects than baseline. Other than attention mechanism, both pipelines including selection of head character and re-ranking of candidate clauses contribute to improving the topic relevance of the generated couplets.

Table 2. Human evaluation results.

Method	Character correspondence	Semantic coherence	Topic relevance	Average
LM	3.50	3.46	2.69	3.22
SPC generator	3.82	3.62	3.38	3.61

5 Conclusions

In this paper, we propose a novel three-stage neural network pipeline method for Chinese couplet generation. The antecedent generator consists of a LM model equipped with a CBS method after statistical selection of head character using the Naïve Bayes. Afterwards, an attention-based S2S model is trained to generate diverse subsequent clauses which are submitted to the re-ranking pipeline for selection of better results. Both automatic and human evaluations demonstrate better performance of our proposed generation system. Moreover, a mini-program of acrostic couplet generation based on our model has also been developed and deployed on Wechat for real users.

Acknowledgement. This work was supported by Ping An Technology (Shenzhen) Co., Ltd, China.

References

Mann, W.: Text generation. Comput. Linguist. **8**, 62–69 (1982)

Zhang, K., Sun, M.: An Chinese couplet generation model based on statistics and rules. J. Chin. Inf. Process. **23**(1), 100–105 (2009)

Jiang, L., Zhou, S.: Generating Chinese couplets using a statistical MT approach. In: International Conference on Computational Linguistics, vol. 21, no. 3, pp. 427–437 (2008)

Yan, R., Li, C., Hu, X.: Chinese couplet generation with neural network structures. In: Association for Computer Linguistics, Berlin, pp. 2347–2357 (2016)

Bahdanau, D., Cho, K., Bengio, Y.: Neural machine translation by jointly learning to align and translate. In: International Conference on Learning Representations (2015)

Wang, Z., He, W., Wu, H.: Chinese poetry generation with planning based neural network. In: International Conference on Computational Linguistics, pp. 1051–1060 (2016)

Zhang, J., Feng, Y., Wang, D.: Flexible and creative Chinese poetry generation using neural memory. In: Computing Research Repository (2017)

Bengio, Y., Ducharme, R., Vincent, P.: A neural probabilistic language model. J. Mach. Learn. Res. **3**, 1137–1155 (2003)

Mikolov, T.: Recurrent neural network based language model. Interspeech **2**, 3 (2010)

Sutskever, I., Vinyals, O., Le, Q.: Sequence to sequence learning with neural networks. In: Advances in Neural Information Processing Systems, vol. 4, pp. 3104–3112 (2014)

Mikolov, T., Sutskever, I., Chen, K.: Distributed representations of words and phrases and their compositionality. In: Advances in Neural Information Processing Systems, vol. 26, pp. 3111–3119 (2013)

Kingma, D., Ba, J.: Adam: a method for stochastic optimization. In: International Conference on Learning Representations, San Diego (2015)

Chowdhury, G.G.: Natural language processing. In: Annual Review of Information Science and Technology, pp. 51–89 (2003)

Kalchbrenner, N., Grefenstette, E., Blunsom, P.: A convolutional neural network for modelling sentences (2014). arXiv preprint arXiv:1404.2188

Hochreither, S., Schmidhuber, J.: Long short-term memory. Neural Comput. **9**, 1735–1780 (1997)

Cho, K., et al.: Learning phrase representations using RNN encoder-decoder for statistical machine translation. In: Computer Science (2014)

Koehn, P.: Statistical Machine Translation. Cambridge University Press, Cambridge (2010)

Koehn, P.: Neural machine translation. Comput. Lang. (2017). arXiv:1709.07809

Koehn, P., Och, F.J., Marcu, D., et al.: Statistical phrase-based translation. In: North American Chapter of the Association for Computational Linguistics, pp. 48–54 (2003)

Devlin, J., Zbib, R., Huang, Z., et al.: Fast and robust neural network joint models for statistical machine translation. In: Meeting of the Association for Computational Linguistics, pp. 1370–1380 (2014)

Ahmed, K., Keskar, N.S., Socher, R., et al.: Weighted transformer network for machine translation. Artif. Intell. (2018). arXiv:1711.02132

Wu, X., Tosa, N., Nakatsu, R.: New Hitch Haiku: an interactive Renku poem composition supporting tool applied for sightseeing navigation system. In: Natkin, S., Dupire, J. (eds.) ICEC 2009. LNCS, vol. 5709, pp. 191–196. Springer, Heidelberg (2009). https://doi.org/10.1007/978-3-642-04052-8_19

Oliveira, H.G.: PoeTryMe: a versatile platform for poetry generation. In: Computational Creativity, Concept Invention, and General Intelligence, vol. 1, p. 21 (2012)

Yan, R., Jiang, H., Lapata, M., Lin, S.D., Lv, X., Li, X.: i, Poet: automatic Chinese poetry composition through a generative summarization framework under constrained optimization. In: International Joint Conference on Artificial Intelligence, pp. 2197–2203 (2013)

Zhang, X., Lapata, M.: Chinese poetry generation with recurrent neural networks. In: Empirical Methods in Natural Language Processing, pp. 670–680 (2014)

Sordoni, A., Bengio, Y., Vahabi, H., Lioma, C., Simonsen, J.G., Nie, J.Y.: A hierarchical recurrent encoder-decoder for generative context-aware query suggestion. In: Computing Research Repository, pp. 553–562 (2015)

Tam, Y., Ding, J., Niu, C., Zhou, J.: Cluster-based beam search for pointer-generator chatbot grounded by knowledge. In: Proceedings of the Thirty-Three AAAI Conference on Artificial Intelligence (AAAI 2019) (2019)

Papineni, K.: BLEU: a method for automatic evaluation of machine translation. In: Proceedings of the 40th Annual Meeting of the Association for Computational Linguistics (ACL) (2002)

Named Entity Recognition
with Homophones-Noisy Data

Zhicheng Liu and Gang Wu^(✉)

Shanghai Jiaotong University, Shanghai, China
{zhichengliu,dr.wugang}@sjtu.edu.cn

Abstract. General named entity recognition systems exclusively focus on higher accuracy regardless of dirty data. However, raw source data face serious challenges specially that are originated from automated speech recognition systems' results. In this paper, we propose *Pinyin* (Pinyin is the official romanization system for Standard Chinese, each Chinese character has its own pinyin sequence which is composed of Latin alphabet) *Hierarchical Attention Encoder-Decoder network* and *Character Alternate Network* to overcome Chinese homophones' problems which frequently frustrate researchers in consecutive *Natural Language Understanding* (NLU). Our models present a none word segmentation structure to effectively avoid secondary data corruption and adequately extract words' internal features. Besides, corrupted sequences can be revised by character-level network. Evaluation demonstrates that our proposed method achieves 93.73% F1 scores which are higher than 90.97% F1 scores using baseline models in homophone-noisy dataset. Additional experiments are conducted to show equivalent results in the universal dataset.

Keywords: Named entity recognition · Homophone · Encoder-Decoder network · Autoencoder

1 Introduction

Automated speech recognition (ASR) is playing an essential part in information extraction, manipulation and generation. Speeches are commonly used as sole source data in ASR to improve user experience. Unfortunately, ASR system deviates dramatically from the ground truth when homophones are involved in source data. However, even in the state-of-art ASR system, homophones problems are still ignored due to noisy data. In addition, words segmentation in several languages (e.g., Chinese) will severely affect the precision of ASR system due to it is highly relies on understanding of source data. Recent research efforts, [18] and [29], have developed the speech corrector technique, but the benefits of afterward semantic extraction are limited. The homophones problem is hindering ASR system and the latter semantic extraction after ASR. This paper focuses on solving these problems.

© Springer Nature Switzerland AG 2019
A. C. Nayak and A. Sharma (Eds.): PRICAI 2019, LNAI 11670, pp. 325–337, 2019.
https://doi.org/10.1007/978-3-030-29908-8_26

Sequence tagging, which is the process following ASR in the whole NLP process has been widely applied in text semantic extraction for a few decades. Recent breakthrough in sequence tagging is eminently enhancing performance. [19] provides CharWNN deep neural network to construct a language-independent *Name Entity Recognition* (NER) systems. [10] proposes Bidirectional LSTM-CRF based models for sequence tagging. Based on the rapid development of pre-trained word embeddings, works [16] and [11,17] demonstrate a semi-supervised method by adding pre-trained context embeddings to Bi-LSTM model. [2] distinguishes character-level and word-level embeddings, combining hierarchical Bi-LSTM model, which achieves high performance. However, these efforts have been devoted in sequence tagging regardless of noisy data arising from homophones.

In this paper, we focus on mitigating the effects of *homophones* from ASR in sequence tagging. We argue that Pinyin, which is ignored in other work, has capabilities to retain features of Chinese homophones which inspires us after comprehensive investigations. Based on this idea, a homophones-insensitive Chinese sequence tagging model by combining a variety of *recurrent neural network* (RNN) based models is developed. These models include hierarchical Bi-LSTM networks [2], attention mechanism [1,15], encoder-decoder framework [5], autoencoder network [7]. To further improve accuracy of sequence tagging while encountering homophones, we propose a character-level tagging method to avoid word segmentation. Moreover, an auto-encoder based character alternate network is clarified to search advisable characters. The contribution of this paper can be summarized as follows:

1. To the best of our knowledge, this is the first work which employs Pinyin in sequence tagging tasks. The character-level Chinese tagging network, named *Pinyin Hierarchical Attention Encoder-Decoder network* (PHAED), is proposed as the sequence tagging model.
2. Based on the proposed techniques, there is no need for word segmentation, which greatly decreases error propagation from homophones.
3. Extensive *character alternate network* (CAN) is developed to search advisable characters. CAN eliminates noise to enhance latter data analysis.

Evaluation shows that PHAED obtains equivalent F1 scores needless of word segmentation in no-homophone dataset compared to the state-of-the-art sequence tagging methods. PHAED also achieves 93.73% F1 scores, which is better than 90.97% F1 scores in baseline models in homophone dataset.

The rest of this paper is organized as follows. Section 2 describes the based sequence tagging models. In Sect. 3, two models PHAED and CAN are proposed. The environment and training procedures of all experiments are given in Sect. 4. Evaluation and analysis on the characteristics of proposed algorithms are presented in Sect. 4. Section 5 concludes this paper.

2 Base Model

2.1 Attention-Based Encoder-Decoder Network

Encoder-Decoder framework [5] with attention mechanism [1,15] is particularly suitable for modeling sequential phenomena. In Fig. 1, at step i, an RNN cell takes the input x_i and the hidden state h_{i-1}, and the next hidden state h_i can be obtained as follow:

$$\begin{aligned}
\overrightarrow{\mathbf{h}}_i &= \overrightarrow{R}(\mathbf{x}_i, \overrightarrow{\mathbf{h}}_{i-1}; \theta_{\overrightarrow{R}}) \\
\overleftarrow{\mathbf{h}}_i &= \overleftarrow{R}(\mathbf{x}_i, \overleftarrow{\mathbf{h}}_{i-1}; \theta_{\overleftarrow{R}}) \\
\mathbf{h}_i &= [\overrightarrow{\mathbf{h}}_i; \overleftarrow{\mathbf{h}}_i]
\end{aligned} \tag{1}$$

Here, notations $\overrightarrow{\mathbf{h}}_i$ and $\overleftarrow{\mathbf{h}}_i$ refer to forward and backward RNN network. $\overrightarrow{R}$ and $\overleftarrow{R}$ indicate either GRU cell [6] or LSTM cell [9].

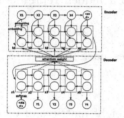

Fig. 1. Encoder-Decoder architecture with attention module.

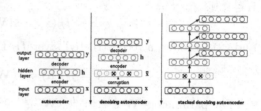

Fig. 2. Differences among autoencoders, denoising autoencoders, stacked denoising autoencoders.

Multiple layers of LSTMs [16] achieve competitive performance in an encoder-decoder framework. Note that, we adopt 2 layers of LSTM cells in all experiments. Attention mechanism is proved to be efficient in [15]. The attention module calculates associated weights at step i can be presented as follows:

$$s_{ij} = b^T \tanh(W_a \mathbf{z}_{i-1} + U_a \cdot \mathbf{h}_j) \tag{2}$$

Then we get attention vectors:

$$\alpha_{ij} = \frac{\exp(s_{ij})}{\sum_{k=0}^{i-1} \exp(s_{ik})} \tag{3}$$

Context vectors are finally calculated as follows,

$$c_i = \sum_{j=0}^{i-1} \alpha_{ij} \mathbf{z}_j \tag{4}$$

where $\mathbf{z}_{i-1}$ is the hidden state when decoding, while $\mathbf{h}_j$ is the hidden state when encoding. W_a, U_a are parameters that learned during training time. The final layer uses a softmax function to predict a score for token from the tag vocabulary V.

$$g(\mathbf{h}_i, v_j, c_i) = O_{v_j}^T (W_h \mathbf{h}_i + W_c c_i) \tag{5}$$

$$P(\omega_i = v_j | \omega_1, \omega_2, \omega_3, ...) = \frac{\exp(g(\mathbf{h}_i, v_j, c_i))}{\sum\limits_{v_k \in V} \exp(g(\mathbf{h}_i, v_j, c_i))} \tag{6}$$

where $O \in \mathbb{R}^{d \times V}$ is a dense layer and O_{v_j} corresponding to tag token v_j, ω_i is the character at position i in output sentence.

2.2 Stacked Denoising Autoencoder

An autoencoder [25] is a special neural network which is similar to encoder-decoder framework. It is composed of two parts: (1) Encoder, a deterministic mapping f_E that transforms an input $\mathbf{x} \in \mathbb{R}^{d_x}$ into a hidden representation $\mathbf{h} \in \mathbb{R}^{d_h}$:

$$\mathbf{h} = f_E(\mathbf{x}) = s(\mathbf{W}\mathbf{x} + \mathbf{b}) \tag{7}$$

and (2) Decoder, a reconstructed layer f_D that mapped back to $\mathbf{y} \in \mathbb{R}^{d_y}$:

$$\mathbf{y} = f_D(\mathbf{h}) = s(\mathbf{W}'\mathbf{h} + \mathbf{b}') \tag{8}$$

The parameters $\mathbf{W} \in \mathbb{R}^{d_x \times d_h}$, $\mathbf{W}' \in \mathbb{R}^{d_h \times d_y}$ are weight matrices of encoder and decoder; $\mathbf{b} \in \mathbb{R}^{d_h}$ and $\mathbf{b}' \in \mathbb{R}^{d_y}$ are called encoder and decoder bias vectors. In practice, $\mathbf{W} = \mathbf{W}'$ may often be used. Those parameters are learned simultaneously on the task of minimizing an associated reconstruction error:

$$J(\mathbf{W}, \mathbf{b}, \mathbf{b}') = L(\mathbf{x}, \mathbf{y}) \tag{9}$$

Where L can be the squared Euclidean distance $L(\mathbf{x}, \mathbf{y}) = ||\mathbf{x} - \mathbf{y}||^2$ or $L(\mathbf{x}, \mathbf{y}) = -\sum\limits_{i=1}^{d_x} x_i \log y_i + (1 - x_i) \log(1 - y_i)$ which refers to cross entropy in case that s is sigmod function and inputs are in $[0, 1]^{d_x}$.

The *denoising autoencoder* (DAE) is neural network aimed at reconstructing a clean input from a corrupted version of it. The raw input vectors x are first corrupted by means of a stochastic mapping $\tilde{\mathbf{x}} = q_D(\tilde{\mathbf{x}} | \mathbf{x})$. After completing whole data corruption, the noisy data $\tilde{\mathbf{x}}$ is mapped, as the same with autoencoder, to hidden representation h from which we reconstruct y, as illustrated in Fig. 2. Parameters $(\mathbf{W}, \mathbf{b}, \mathbf{b}')$ are trained to minimize the reconstruction error $L(\mathbf{x}, \mathbf{y})$ rather than $L(\tilde{\mathbf{x}}, \mathbf{y})$. Multiple layers of DAE compose the stacked denoising autoencoder. There are h layers of DAE that are trained by bottom-up and layer-wise methods. Input corruption is only used for the initial input vector. The hidden layer of the top autoencoder is the output of the stacked denoising autoencoder, which can be further applied into other applications, such as SVM for classification [25].

3   Improved Model and Algorithms

The main architecture of our proposed model is illustrated in Fig. 3. In this example, we split an input Chinese sentence into Chinese characters without word segmentation. Then, we extract the Pinyin sequence into every character from the given sequence. Utilizing the concentration of pre-trained character embedding and Pinyin embedding, internal features which refers to named entities are discerned by RNN tagging model while revised sentences are generated by CAN model during training. But for inference, input sequences are primarily processed by CAN model and subsequently deposited into whole system.

3.1   Hierarchical Character Embedding

The employed hierarchical character embedding structure in PHAED follows several studies [2,4,17,28]. Given a sentence, we consider each character c_i , its tokens t_i and the Pinyin sequence of c_i that comprise Pinyin: $c_i^1, c_i^2, c_i^3, ...$ A character embedding $\mathbf{x}_i$ for each character comprises three parts as shown in Fig. 4: (1) a Pinyin based representation $\mathbf{b}_i$ (2) a pre-trained character embedding $\mathbf{e}_i$ (3) a character token embedding $\mathbf{w}_i$.

$$\begin{aligned}
\mathbf{b}_i &= \phi(c_i^1, c_i^2, c_i^3, ...; \theta_b) \\
\mathbf{e}_i &= E_{pre}(c_i) \\
\mathbf{w}_i &= E(t_i; \theta_w) \\
\mathbf{x}_i &= [\mathbf{b}_i; \mathbf{e}_i; \mathbf{w}_i]
\end{aligned} \tag{10}$$

where θ_b, θ_w are parameters to be computed during training phase, $\phi(\mathbf{x})$ denotes the character embedding which is either RNN [8] or CNN [12,20], $E_{pre}(\mathbf{x})$ denotes pre-trained character embedding model [11,16] and $E(\mathbf{x})$ is a token lookup table which is initialized with random token embedding.

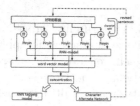

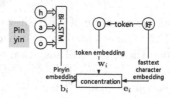

Fig. 3. Overview of the whole NER model.

Fig. 4. Structures of Hierarchical character embedding, 'h', 'a', 'o' is the Pinyin sequence of Chinese character '好' .

Unlike English sentence, there is no space between words in Chinese, which indicates that word segmentation is essential for Chinese sentence. In the homophone condition, word segmentation is more likely to generate mistaken word embedding which introduces noisy data in pre-process procedure. To solve this, we employ hierarchical character embedding which can capture the pronunciation feature (*homophones have similar pronunciation*) by Pinyin representation b_i. Inspired by the idea that each Chinese character shows individual implication, we adopt character embedding instead of word embedding as inputs of the tagging model. Similar approaches of character embedding, which has been elaborated above, are also utilized in Pinyin representation.

3.2 PHAED

We combine a hierarchical character embedding and an attention-based Encoder-Decoder network to form a PHAED model, which is shown in Algorithm 1. Hierarchical character embedding can efficiently extract features of homophones and characters. More meaningfully, our proposal skips the word segmentation step so that precise character embedding is taken into the Encoder-Decoder network. Attention-based Encoder-Decoder network has an attention vector as weight parameters of characters in various position. With such mechanism, we can skillfully predict the current tag using past and future tags. We consider the matrix of scores $f_\theta(x)$ are output by attention-based encoder-decoder network. The element $[f_\theta(x)]_{i,j}$ of the matrix is the score output by previous network with parameters θ, representing the probability that the i-th word in sentence x belongs to the j-th tag in vocabulary V:

$$[f_\theta(x)]_{i,j} = P(\omega_i = v_j | \omega_1, \omega_2, \omega_3, ...) \tag{11}$$

where $P(\omega_i = v_j | \omega_1, \omega_2, \omega_3, ...)$ is the output by attention-based encoder-decoder network according to Eq. (6).

The cost function of a sentence x along with a path of true tags y is given by the cross entropy between scores matrix $f_\theta(x)$ and one-hot encoding matrix y:

$$g(x, y|\theta) = -\sum_{i=1}^{T} \sum_{j=1}^{|V|} y_{i.j} \log [f_\theta(x)]_{i,j} \tag{12}$$

An simple example of NER shown in Fig. 5 presents the running procedures of PHAED. The vocabulary which contains all Chinese characters is applied to transfer them into tokens. Combining tokens which generated by characters sequences '好','听','的','歌','曲' via above vocabulary and Corresponding Pinyin sequences "hao", "ting", "de", "ge", "qu", entity types results are labeled as "quality", "quality", "O", "channel", "channel".

3.3 CAN

Similar to stacked denoising autoencoders, CAN adopts stacked encoders-decoders framework (Fig. 2), but replaces encoders and decoders with Bi-LSTM networks. The motivation of replacement is that standard DAE can't handle sequential data and one linear transformation layer lacks abilities to entirely

Algorithm 1. PHAED($\mathbf{c}, \mathbf{t}, \mathbf{b}, y, l$)

Input: $\mathbf{c}$, Pinyin sequences of input sentences; $\mathbf{t}$, token sequences of input sentences; y, tag sequences of output sentences; l, train steps; $\mathbf{b}$, Pinyin based representation;
Output: $\theta_{Encoder}, \theta_{Decoder}$
1: $\theta_{Encoder}, \theta_{Decoder}$ are randomly initialized
2: **for** each $i \in [1:l]$ **do**
3:      $\mathbf{x} \leftarrow \text{HCE}(\mathbf{c}, \mathbf{t}, \mathbf{b})$ // HCE(), function to construct $\mathbf{x}$ in Eq (10)
4:      $\mathbf{x} \leftarrow encoder(\mathbf{x}, \theta_{Encoder})$ // Apply encoder following Eq (1)
5:      $\mathbf{x} \leftarrow decoder(\mathbf{x}, \theta_{Decoder})$ // Apply attention decoder following Eq (4), (5)
6:      $y' \leftarrow full\_connected(\mathbf{x})$ // Calculate tokens using Eq (11)
7:      $cost \leftarrow cross\_entry(y, y')$ // Calculate entry loss via Eq (12)
8:      **if** $cost$ is lower enough **then**
9:          **return** $\theta_{Encoder}, \theta_{Decoder}$
10:     **else**
11:         $\theta_{Encoder}, \theta_{Decoder} \leftarrow \text{SGD}(\theta_{Encoder}, \theta_{Decoder})$ // minimize $cost$ by stochastic gradient descent
12:     **end if**
13: **end for**
14: **return** $\theta_{Encoder}, \theta_{Decoder}$

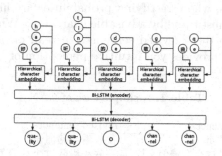

Fig. 5. Architecture of PHAED, "好听的歌曲(satisfying song)" is split as five Chinese characters inputs. "hao", "ting", "de", "ge", "qu" are correspond Pinyin of these characters. "quality", "O", "channel" are entity types that some are beyond official entity types.

express the whole feature set. Training procedures are identical in which multiple layers of autoencoders are successively trained and no upper layers can be trained unless all lower autoencoders' parameters are frozen.

Under corruption steps, input Chinese sentences are initially polluted by two corrupting operations which contain deletion and substitution with respective probability p_1, p_2. Substituted words strictly originate in Chinese Homophones Dictionary [22]. Finishing corrupting, aforementioned hierarchical character embedding is employed to generate corrupted input vectors $\tilde{x}$ to extract characters' features. The detailed procedures of training a CAN model are listed in Algorithm 2. In our implementation, we use cross-entropy loss function which is identical to general NLP tasks.

Algorithm 2. CAN($\mathbf{x}, p_1, p_2, h$)

Input: h, number of layers of autoencoder; $\mathbf{x}$, training data; p_1, deleting probability; p_2, substituting probability

Output: θ_E, θ_D

1: **for** each $i \in [1:h]$ **do**
2:      $\mathbf{x} = substitution(\mathbf{x}, p_1)$ //randomly substitute words in $\mathbf{x}$ with p_1 probability
3:      $\mathbf{x} = deletion(\mathbf{x}, p_2)$ //randomly delete words in $\mathbf{x}$ with p_2 probability
4:      $\mathbf{x} \leftarrow$ HCE($\mathbf{x}$) //the inputs of HCE is the same as Algotithm 1, we use ($\mathbf{x}$) for simplicity
5:      randomly initialize θ_E^i, θ_D^i
6:      **repeat:**
7:          $\mathbf{x} \leftarrow encoder(\mathbf{x}, \theta_E)$ $\mathbf{x} \leftarrow decoder(\mathbf{x}, \theta_D)$ $\theta_E^i, \theta_D^i \leftarrow$ SGD(θ_E^i, θ_D^i)
8:      **until** convergence
9: **end for**
10: **return** θ_E, θ_D

After the above-mentioned training algorithm is executed, CAN is employed to check the probabilities of rectifying original characters into more suitable characters. The characters, whose rectifying probabilities are higher than a certain threshold T, will be substituted by advisable characters. While inference, inputs of PHAED model are these revised sentences instead of original sentences. A simple case in Table 1 provides you intuitive perception of CAN's results.

Table 1. A simple case of CAN

original sentences	revised sentences
我想看猪佩奇	我想看小猪佩奇
海底小中队这个东漫怎么样	海底小纵队这个动漫怎么样
这这部美国队长的导演演是谁	这部叫美国队长的导演是谁
列举一些邓丽君成名歌曲	列举一些邓丽君成名歌曲

4   Experiments

In this section, extensive simulations are conducted to show the advantages of the proposed scheme. Although our research focuses on the homophones task, we still evaluate it in the no-homophone task to verify universality. In both cases, we employ the general F1 score to compare our scheme with baseline sequence tagging model [17,28] and other previous tagging models [14]. In homophones task, we employ character accuracy rate[1] to measure the performance of searching advisable characters.

---

[1] Character accuracy rate = numbers of correct rectifying characters/numbers of wrong characters.

4.1 Datasets Description

The CoNLL-2003 NER task [24] and the CoNLL-2000 Chunking task [23] are benchmarked sequence tagging task. Unfortunately, these datasets are strongly cleaned with rare homophones and lack Chinese corpus. We crawled data from *Weibo* (Chinese micro-blogging) with 100,000 sentences, named Weibo NER Corpus. The whole dataset involves 35 entity types which are labeled manually. According to the state-of-the-art speech recognition technique reported by [3], *word error rate* (WER) reduces to 5.6%. We investigated that 95% of the error samples are provoked by homophones. In Weibo NER Corpus, 5% (5.6% × 95% ≈ 5%) words in Weibo NER Corpus are replaced by corresponding homophones under Chinese Homophone Dictionary [22] to form Noising Weibo NER Corpus which is applied to train PHAED and CAN. Considering Noising Weibo Corpus is just one synthetic dataset, we employ the state-of-the-art ASR system [3] to generate Speech Text Corpus based on open source data set THCHS30 [26] and ST-CMDS-20170001_1. Using real speech recognition text, we confirm the ability of our scheme to overcome Homophone problems.

4.2 Pre-trained Character Embedding

The character embeddings we used in this work are trained on Chinese Wiki Corpus with the duplicate parameters following [11]. Considering unnecessary word segmentation, sentences in Chinese Wiki Corpus[2] are split as independent Chinese characters. Hence, the output word embedding is treated as Chinese character embedding which extensively employed in hierarchical character embedding.

4.3 Setup

All baseline models and PHAED use 30-dimensional character embeddings. [14] uses two layers of 100-dimensional hidden states. 275-dimensional hidden states with only 1 layer LSTM is adopted in [4]. [28] employs a CNN with 30 filters of width 3 characters which is same as [4] and 2 layers of Bi-LSTMs with 200 hidden units and 50% dropout rate [21] while [17] employs two stacked LSTMs with 8192 hidden units and 50% dropout rate which is considered as optimal solution.

We use two layers of Bi-LSTMs in encoder and Pinyin embedding with 300 hidden units and 25-dimensional character embedding. Replicate Bi-LSTM parameters are trained in CAN whose initial parameters could be shared by trained PHAED model. Following [21], we add 50% dropout rate to the recurrent connections in both CAN and PHAED for regularization.

---

[2] https://dumps.wikimedia.org/zhwiki/latest/zhwiki-latest-pages-articles.xml.bz2.

Table 2. F1 results on Speech Text Corpus

Module	Setting	Accuracy
CAN + PHAED	-/-	**93.73%**
CAN + PHAED	separated HCE	92.49%
PHAED	No CAN	91.27%
CAN + PAED[a]	No HCE	87.46%
PAED[a]	No CAN and HCE	86.46%

[a]PAED: PHEAD without HCE

Table 3. F1 results on Weibo NER Corpus

Model	F1±std
Chiu et al. (2015)	90.92%±0.33
Lample et al. (2016)	91.95%±0.27
Yang et al. (2017)	92.57%±0.18
Peters et al. (2017)	94.82%±0.15
CAN + PHAED	**94.85%±0.18**

4.4 Training

Following [17], all experiments are trained by the Adam optimizer [13] with gradient norms clipped at 5.0. As length of Pinyin in Chinese is inevitable no more than 6, maximal Pinyin length is limited as 6. Moreover, the length of maximal character in Hierarchical character embedding is decided as 30 to increase batch size, since [27] proves that larger batch size can achieve higher performance. In all experiments, we freeze all the pre-trained fasttext word embedding and initialize the token embedding by standard Gaussian distribution. For the purpose of explicit dropout regularization, early stopping is adopted to prevent over-fitting, while adaptive learning rate is used to restrict high learning rate during the ultimate training stage. We train with a time-based decay learning rate a = 0.002 on the training data and decrease 1% at every 200 epochs. Then, we monitor the development set performance at each step and stop whole train process at the epoch with the highest development performance.

Following [17] and [28], we train the final model with the same configurations ten times in all experiments using different random seed; therefore the mean and standard deviation of F1 can be checked. Estimating the variance of sequence tagging performance is significant since variance of samples confirms truth but a simple sample not.

4.5 Evaluation

In order to estimate CAN and HCE performance to tagging sequence, we adopt several ablation experiments to analyze performance of PHAED and CAN on Speech Text Corpus. The experiments' results are shown in Table 2. We notice that both PHAED and CAN employ hierarchical character embedding. Therefore generating initial point of PHAED with CAN-shared trained hierarchical character embedding makes it converge faster and improved performance. The reason is that denoising features are simultaneously extracted in PHAED, thus PHAED pays more attention to decode denoised data in training periods.

Table 3 demonstrates the final results on the test dataset of Weibo NER Corpus. Our scheme performs equivalently compared to the state-of-the-art tagging model on normal dataset. The results denote that word segmentation doesn't effectively contribute to deriving named entities from original dataset. PHAED,

Table 4. F1 results on Speech Text Corpus

Model	F1±std
Chiu et al. (2015)	86.76%±0.46
Lample et al. (2016)	87.38%±0.47
Yang et al. (2017)	87.84%±0.27
Peters et al. (2017)	90.97%±0.25
CAN + PHAED	**93.73%±0.21**

Table 5. F1 results on Nosing Speech Text Corpus

Model	F1±std
Chiu et al. (2015)	37.22%±0.61
Lample et al. (2016)	38.01%±0.57
Yang et al. (2017)	39.10%±0.40
Peters et al. (2017)	55.72%±0.36
CAN + PHAED	**79.35%±0.25**

which employs original character sequences without noise introduced by previous steps, captures implicit structures of words, which decreases error rates.

Next, we analyze the performance from the perspective of real speech recognition text. When one word in sentences is mistakenly spilt, one of two adjacent words are remarkably affected which may engender continuous inaccurate word segmentation. As shown in Table 4, with the homophones words generated by ASR system, PHAED presents surprising performance. Besides, our proposal exhibits low degree of increase in experiments' variance, while previous achievements lack adequate robustness which leads to high variance.

To further figure out the reason why PHAED and CAN outperforms on noisy dataset. We ran additional experiments on Noisy Speech Text Corpus (*all selected sentences' confidences generated by prementioned* [3] *are below a certain threshold*). Results presented in Table 5 illustrate that PHAED + CAN improves noise immunity in tagging model. [28] and [17] encouter severe decline in extracting appropriate information when reliable data doesn't occupy overwhelming majority.

5 Conclusions

In this paper, we developed PHAED based on Encoder-Decoder network for sequence tagging. The model integrates pronunciation, attention mechanism and Encoder-Decoder network into a single neural network, and it does not require word segmentation. To improve performances, we investigate various methods and optimization mechanisms. The proposed techniques outperform significantly in homophone NER task, while obtains equivalent results on non-homophone NER task. In experiments, we observe that word segmentation is crucial for errors of sequence tagging under polluted data. Applying pronunciation features, bad cases caused by homophones are highly reduced. Furthermore, we have established a new CAN network to recover original sentences from noisy corpus. This may be essential for NLU in which clear data is urgently needed. The proposed method will be effective to be adopted in latter procedures in automated speech recognition system.

References

1. Bahdanau, D., Cho, K., Bengio, Y.: Neural machine translation by jointly learning to align and translate. In: International Conference on Learning Representations (ICLR) (2015)
2. Boag, W., Sergeeva, E., Kulshreshtha, S., Szolovits, P., Rumshisky, A., Naumann, T.: Cliner 2.0: Accessible and accurate clinical concept extraction. arXiv preprint arXiv:1803.02245 (2018)
3. Chiu, C.C., et al.: State-of-the-art speech recognition with sequence-to-sequence models. In: ICASSP 2018 (2018, submitted)
4. Chiu, J., Nichols, E.: Named entity recognition with bidirectional LSTM-CNNs. Trans. Assoc. Comput. Linguist. 4(1), 357–370 (2016)
5. Cho, K., van Merriënboer, B., Bahdanau, D., Bengio, Y.: On the properties of neural machine translation: encoder-decoder approaches. In: Syntax, Semantics and Structure in Statistical Translation, p. 103 (2014)
6. Cho, K., et al.: Learning phrase representations using RNN encoder-decoder for statistical machine translation. In: Proceedings of the 2014 Conference on Empirical Methods in Natural Language Processing (EMNLP), pp. 1724–1734 (2014)
7. Deng, L., Seltzer, M.L., Yu, D., Acero, A., Mohamed, A.R., Hinton, G.: Binary coding of speech spectrograms using a deep auto-encoder. In: Eleventh Annual Conference of the International Speech Communication Association (2010)
8. Ding, C., Xie, L., Yan, J., Zhang, W., Liu, Y.: Automatic prosody prediction for Chinese speech synthesis using BLSTM-RNN and embedding features. In: 2015 IEEE Workshop on Automatic Speech Recognition and Understanding (ASRU), pp. 98–102. IEEE (2015)
9. Hochreiter, S., Schmidhuber, J.: Long short-term memory. Neural Comput. 9(8), 1735–1780 (1997)
10. Huang, Z., Xu, W., Yu, K.: Bidirectional LSTM-CRF models for sequence tagging. arXiv preprint arXiv:1508.01991 (2015)
11. Joulin, A., Grave, E., Bojanowski, P., Mikolov, T.: Bag of tricks for efficient text classification. In: Proceedings of the 15th Conference of the European Chapter of the Association for Computational Linguistics: Volume 2, Short Papers, vol. 2, pp. 427–431 (2017)
12. Kim, Y., Jernite, Y., Sontag, D., Rush, A.M.: Character-aware neural language models. In: AAAI, pp. 2741–2749 (2016)
13. Kinga, D., Adam, J.B.: A method for stochastic optimization. In: International Conference on Learning Representations (ICLR), vol. 5 (2015)
14. Lample, G., Ballesteros, M., Subramanian, S., Kawakami, K., Dyer, C.: Neural architectures for named entity recognition. In: Proceedings of NAACL-HLT, pp. 260–270 (2016)
15. Luong, T., Pham, H., Manning, C.D.: Effective approaches to attention-based neural machine translation. In: Proceedings of the 2015 Conference on Empirical Methods in Natural Language Processing, pp. 1412–1421 (2015)
16. Mikolov, T., Sutskever, I., Chen, K., Corrado, G.S., Dean, J.: Distributed representations of words and phrases and their compositionality. In: Advances in Neural Information Processing Systems, pp. 3111–3119 (2013)
17. Peters, M., Ammar, W., Bhagavatula, C., Power, R.: Semi-supervised sequence tagging with bidirectional language models. In: Proceedings of the 55th Annual Meeting of the Association for Computational Linguistics (Volume 1: Long Papers), vol. 1, pp. 1756–1765 (2017)

18. Ratinov, P.B., Kanevsky, D.: Multilanguage machine translation speech corrector (2015)
19. dos Santos, C., Guimaraes, V., Niterói, R., de Janeiro, R.: Boosting named entity recognition with neural character embeddings. In: Proceedings of NEWS 2015 the Fifth Named Entities Workshop, p. 25 (2015)
20. Seo, M., Kembhavi, A., Farhadi, A., Hajishirzi, H.: Bidirectional attention flow for machine comprehension. arXiv preprint arXiv:1611.01603 (2016)
21. Srivastava, N., Hinton, G., Krizhevsky, A., Sutskever, I., Salakhutdinov, R.: Dropout: a simple way to prevent neural networks from overfitting. J. Mach. Learn. Res. **15**(1), 1929–1958 (2014)
22. The-Commercial-Press: 中国大辞典编纂处同音字典2版 (1957)
23. Tjong Kim Sang, E.F., Buchholz, S.: Introduction to the CONLL-2000 shared task: chunking. In: Proceedings of the 2nd Workshop on Learning Language in Logic and the 4th Conference on Computational Natural Language Learning-vol. 7, pp. 127–132. Association for Computational Linguistics (2000)
24. Tjong Kim Sang, E.F., De Meulder, F.: Introduction to the CONLL-2003 shared task: language-independent named entity recognition. In: Proceedings of the Seventh Conference on Natural Language Learning at HLT-NAACL 2003-vol. 4, pp. 142–147. Association for Computational Linguistics (2003)
25. Vincent, P., Larochelle, H., Lajoie, I., Bengio, Y., Manzagol, P.A.: Stacked denoising autoencoders: learning useful representations in a deep network with a local denoising criterion. J. Mach. Learn. Res. **11**(Dec), 3371–3408 (2010)
26. Wang, D., Zhang, X.: Thchs-30: A free Chinese speech corpus (2015)
27. Wilson, A.C., Roelofs, R., Stern, M., Srebro, N., Recht, B.: The marginal value of adaptive gradient methods in machine learning. In: Advances in Neural Information Processing Systems, pp. 4148–4158 (2017)
28. Yang, Z., Salakhutdinov, R., Cohen, W.W.: Transfer learning for sequence tagging with hierarchical recurrent networks. In: International Conference on Learning Representations (ICLR) (2017)
29. Zweig, G., Ju, Y.C.: Automated data cleanup by substitution of words of the same pronunciation and different spelling in speech recognition, 4 October 2016. US Patent 9,460,708

Robust Sentence Classification by Solving Out-of-Vocabulary Problem with Auxiliary Word Predictor

Sang-Seok Park[1], Yunseok Noh[1], Seyoung Park[1(✉)], and Seong-Bae Park[2]

[1] School of Computer Science and Engineering, Kyungpook National University,
Daegu 41566, Korea
{sspark,ysnoh}@sejong.knu.ac.kr, seyoung@knu.ac.kr
[2] Department of Computer Engineering, Kyung Hee University,
Yongin 17104, Korea
sbpark71@khu.ac.kr

Abstract. In recent years, deep learning methods have achieved outstanding performances in sentence classification. However, many sentence classification models do not consider the out-of-vocabulary (OOV) problem, which generally appears in sentence classification tasks. Input units smaller than words, such as characters or subword units, have been considered the basic unit for sentence classification to cope with the OOV problem. Although this approach naturally solves the OOV problem, it has obvious performance limitations because a character by itself has no meaning, whereas a word has a definite meaning. In this paper, we propose a neural sentence classification model that is robust to the OOV problem, even though the proposed model utilizes words as the basic unit. To this end, we introduce the *unknown word prediction* (UWP) task as an auxiliary task to train the proposed model. Owing to joint training of the proposed model with the objectives of classification and UWP, the proposed model can represent the meanings of entire sentences robustly even if a sentence includes a number of unseen words. To demonstrate the effectiveness of the proposed model, a number of experiments are conducted using several sentence classification benchmarks. The proposed model consistently outperforms two baselines over all four benchmark datasets in terms of the classification accuracy.

Keywords: Sentence classification · Out-of-vocabulary problem · Neural network · Multi-task learning

1 Introduction

Sentence classification is a fundamental task in natural language processing (NLP), which is being studied extensively for sentiment analysis in social media

This work was supported by Institute of Information & Communications Technology Planning & Evaluation (IITP) grant funded by the Korea government (MSIT) (No. 2016-0-00145, Smart Summary Report Generation from Big Data Related to a Topic).

© Springer Nature Switzerland AG 2019
A. C. Nayak and A. Sharma (Eds.): PRICAI 2019, LNAI 11670, pp. 338–350, 2019.
https://doi.org/10.1007/978-3-030-29908-8_27

and political ideology analysis, among other purposes [3,15]. Recently, deep learning approaches that employ recurrent neural networks (RNNs), convolutional neural networks (CNNs), and attention mechanisms have been shown to be effective for sentence classification. Among them, BLSTM2DCNN [18] employs a two-dimensional (2D) convolutional network on top of a bidirectional LSTM for text classification. BLSTM2DCNN has been shown to perform well on several text-classification tasks because the network can capture the dependency among feature vector dimensions via 2D convolutions and the bidirectional long-term contextual information via BLSTM. By contrast, DARLM [19] is another recently developed neural network for sentence classification. This model includes two attention subnets that attending different parts of a sentence with each other. Then, an example discriminator assigns a label to the given sentence by utilizing attention information from both subnets. The neural models proposed in these studies have network architectures and decision mechanisms optimized for sentence classification.

Despite these advances in deep learning for sentence classification, many sentence classifiers do not consider the out-of-vocabulary (OOV) problem, which appears in almost all sentence classification tasks. During training, neural classifiers have access to complete information about a sentence to be classified. By contrast, in practice, these classifiers may be applied to sentences containing multiple unseen words. This OOV problem interferes with the prediction of neural classifiers, and the problem becomes severe when the unseen words in a sentence deliver the key information that determines the class of the sentence. One possible solution for the OOV problem is using characters or subwords [13] as basic units for sentence classification instead of words [17]. Because characters and subwords are smaller units than words, sentence classifiers based on such small units can avoid the OOV problem naturally. However, the performance of character-level sentence classifiers is inferior to that of word-level models, even though character-level sentence classifiers have considerably deeper and more complex network structures [7]. This is because a character by itself has no meaning, whereas a word has a definite meaning.

In this paper, we propose a neural sentence classification model that is robust to the OOV problem, even if the proposed model utilizes words as the basic unit for classification. To this end, we introduce the *unknown word prediction* (UWP) task as an auxiliary task to train the proposed model. The UWP task predicts the would-be word when an unknown word is included in a sentence. To train a network for this task, some proportion of the words in training sentences are randomly selected and replaced with the $\langle unk \rangle$ token. Then, a network is trained to predict the words to be substituted instead of the $\langle unk \rangle$ tokens by considering all other words in a sentence. The objective of UWP is similar to that of the *masked language model* (MLM) [2], which has been proved to be useful for obtaining robust and contextual word representations of a given sentence.

The proposed neural network consists of three sub-networks, namely, a shared sentence encoder, sentence classification network, and an UWP network. The sentence encoder takes a sentence, that is, a sequence of words, as input and outputs a sequence of contextual word representations. Because the auxiliary UWP task is executed to generate robust word representations that can be used for sentence classification, the sentence encoder should be shared across both the main and the auxiliary task-specific networks. The UWP network is placed on top of the sentence encoder, and then the network predicts the word that is the original word of a given ⟨unk⟩ token based on all other known words in the sentence. Concurrently, the sentence classification network takes a sequence of word representations from the sentence encoder and performs classification based on the given sequence.

In multi-task learning, an auxiliary task can give the model useful hints, which are difficult to learn in the main task [1]. By applying the UWP task to a network as an auxiliary task for sentence classification, the proposed model obtains such hints for solving the OOV problem from two perspectives. The first perspective is that a neural classifier can be configured to predict the approximate meanings of unseen words. Thus, the UWP task provides a direct solution for the OOV problem encountered in sentence classification. Another perspective is that word representations of *known* words become more contextual. As a result, the meaning representation of the entire sentence becomes robust, even if a sentence includes a number of unseen words.

To demonstrate the effectiveness of the proposed model, a number of experiments are conducted on several sentence classification benchmarks including SST-1, SST-2, TREC-6, and TREC-50. In a comparison with the baselines without the UWP auxiliary task, the proposed model consistently outperforms the baselines over all four benchmark datasets; especially, on the SST-1 benchmark, the performance gain in terms of accuracy is up to 1.2% compared to that of the baselines.

The rest of this paper is organized as follows. In Sect. 2, we briefly introduce previous works on recent neural models for sentence classification. In Sects. 3 and 4, we describe the learning algorithm and the architecture of the proposed model. The experimental setting and the results are given in Sects. 5 and 6. Finally, we conclude the study in Sect. 7.

2   Related Work

In recent years, deep learning methods, including modern neural modules such as recurrent units, convolutions, and attention mechanisms, have yielded notable performance when applied to sentence classification. Even more recently, the developers of most deep learning models have blended more than two of the aforementioned modules to enhance performance. In BLSTM2DCNN [18], 2D convolutions are introduced, and their filters are defined across the feature vector dimension as well as the word sequence. These 2D convolutions summarize the contextual information generated by a bidirectional LSTM to classify a sentence. The DARLM [19] combines all three of the aforementioned modules; it is thus composed of a convolutional layer for text encoding, two different attention mechanisms for feature selection, and two LSTM layers on top of each attention mechanism for aggregating the contextual information. These studies show the importance of generating appropriate contextual representations and summarizing them.

Breaking words down into smaller units is one of solutions to the OOV problem. A number of character-level [8,17] and subword-level [5,13] classification models have been proposed. Essentially, these models seem to eliminate the OOV problem from sentence classification. However, the smaller units rarely convey meanings and increase the length of the input sequence. By contrast, the proposed model uses a word as the basic unit and infers contextual meanings by considering the surrounding words.

Multi-task learning is widely used to improve sentence classification performance [12]. To facilitate multi-task learning, the tasks to be performed should be related each other. When this condition is satisfied, an auxiliary task can help improve the performance of the associated main task through the provision of additional hints that can only be obtained from the auxiliary task. For instance, the execution of a word-level sentiment classification task can improve the performance of the associated sentence level sentiment classification task [16]. This makes sense because the polarity of each word in a sentence is crucial for determining the sentiment of the entire sentence. It is also intuitive that understanding the contextual meaning of each word in a sentence is very important for predicting the class of the sentence. In addition, human beings understand a sentence that contains words unknown to them by contextually approximating the meaning of those words. Based on this intuition, we set UWP as the auxiliary task for sentence classification.

The proposed UWP task was motivated by the masked language model (MLM), which is used for training BERT [2]. In the training procedure of MLM, a neural network is forced to predict the original words of masked words in a sentence. With this training, the network produces more robust word representations, even when a word is masked. Unlike the MLM in BERT, which applied to a very large-scale corpus to obtain robust word representations, we show that with relatively small data, the UWP task is adequately effective as an auxiliary task for sentence classification.

3 Sentence Classification with Auxiliary Word Predictor

Figure 1 describes the overall architecture of the proposed model. As shown in this figure, the proposed model follows a general network structure for multi-task learning, which comprises one shared sub-network and multiple task-specific sub-networks [12]. In the proposed model, the shared network is a sentence encoder, and two task-specific networks constitute a sentence classification network for the main task and an UWP network for the auxiliary task. Let $D = \{(x, \mathbf{y})\}$ be the training dataset for sentence classification, where $x = (w_1, \ldots, w_L)$ is a input sentence of length L, and $\mathbf{y} \in \{0, 1\}^C$ is a one-hot vector for the class label of a sentence x. Because D does not provide the training data for the auxiliary task, we first generate a training dataset $D' = \{(x', \mathbf{y})\}$ for both tasks. x' is a

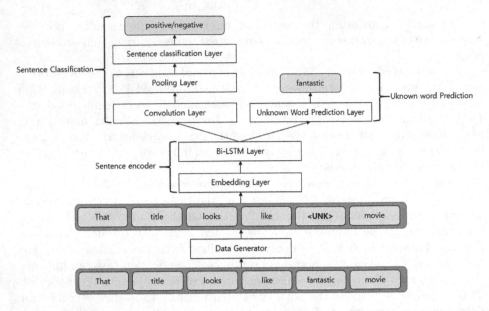

Fig. 1. Architecture of proposed sentence classification network with an auxiliary word predictor.

corrupted x obtained by replacing a few words with the unknown token $\langle unk \rangle$. In this study, 15% of the words in x were randomly replaced with the $\langle unk \rangle$ token.

After D' is prepared, x' with U unknown words is input to the RNN-based sentence encoder f_{enc} to produce a sequence of word vectors $H = (\mathbf{h}_1, \ldots, \mathbf{h}_L)$. Then, the CNN-based sentence classification network f_{cls} takes H and predicts the probability distribution of the class label $\mathbf{y}$ of x. Thus, two parameter sets θ_{cls} and θ_{enc} with respect to f_{enc} and f_{cls} are trained to minimize

$$\sum_{(x', \mathbf{y}) \in D'} \mathcal{L}_{cls}(\mathbf{y}, f_{cls}(f_{enc}(x'; \theta_{enc}); \theta_{cls})), \qquad (1)$$

where $\mathcal{L}_{cls}$ is the cross-entropy loss.

The UWP network f_{aux} for the auxiliary task is a feed-forward neural network, and it takes H as the input. Because more than one $\langle unk \rangle$ token can be included in x', f_{aux} predicts the original word for each $\langle unk \rangle$ token. Then, θ_{aux}, a set of all parameters of the auxiliary word predictor, is jointly trained with θ_{enc} to minimize

$$\sum_{(x', i: w_i = \langle unk \rangle) \in D'} \mathcal{L}_{aux}(\mathbf{w}_i, f_{aux}(f_{enc}(x'; \theta_{enc}); \theta_{aux})), \qquad (2)$$

where $\mathbf{w}_i$ is a one-hot vector for the i-th word in x, which is replaced by the $\langle unk \rangle$ token, and $\mathcal{L}_{aux}$ is the cross-entropy loss.

Algorithm 1. Training procedure of entire proposed model

---

    input   : Training set $D = (x, \mathbf{y})$, hyperparameters α and λ
    Parameters : $\Theta = (\theta_{enc}, \theta_{cls}, \theta_{aux})$

    initialize   : All parameters Θ are randomly initialized.
1 **repeat**
2     $D_{batch} \leftarrow$ sample(D, b) // **sample a minibatch of size** b
3     $D'_{batch} \leftarrow \emptyset$ // **initialize the corrupted training set**
4   **for** $(x, \mathbf{y}) \in D_{batch}$ **do**
5         $(x', \mathbf{y}) \leftarrow$ generate(D_{batch}) // **sample a corrupted tuple**
6         $D'_{batch} \leftarrow D'_{batch} \cup \{(x', \mathbf{y})\}$
7   **end**
    // **joint training of entire networks**
8   Train f_{enc} and f_{cls} by Eq. 1
9   Train f_{enc} and f_{aux} by Eq. 2
10 **until** *convergence*;

The goal of the auxiliary task is to help the sentence encoder produce a robust representation of x. Thus, f_{enc} should be optimized to jointly minimize both loss $\mathcal{L}_{cls}$ and $\mathcal{L}_{aux}$. As a result, the final loss of the proposed model is as follows.

$$\mathcal{L} = \mathcal{L}_{cls} + \alpha \mathcal{L}_{aux} + \lambda \|\Theta\|_2, \tag{3}$$

where $\Theta = (\theta_{enc}, \theta_{cls}, \theta_{aux})$ denote all parameters of the proposed model, and the hyperparameter α balances the main classification task and the auxiliary word prediction task. λ is an l_2 regularization hyperparameter.

Algorithm 1 describes the detailed procedure for training the proposed model. The proposed model contains three parameter sets, namely, θ_{enc}, θ_{cls}, and θ_{aux}, which come from f_{enc}, f_{cls}, and f_{aux}, respectively. All these parameters are initialized randomly before training. In each epoch of the algorithm, a small set of tuples is sampled from the training set and corrupted with $\langle unk \rangle$ tokens. Once a corrupted training set is prepared, the entire network is trained jointly with the loss given in Eq. 3 by lines 8–9 until the training converges.

4 Network Implementation

The proposed model begins with a shared sentence encoder f_{enc}. The shared sentence encoder consists of an embedding layer and a Bi-LSTM layer. The embedding layer converts an input sentence x of L words into a sequence of word vectors in the form of the matrix $X = [\mathbf{v}_1^T, \ldots, \mathbf{v}_L^T]$. The Bi-LSTM layer encodes X into a contextual representation H by reflecting the left and right contexts against each d-dimensional embedding vector $\mathbf{h}_i$. That is,

$$\mathbf{h}_i = \overrightarrow{\mathbf{h}_i} \oplus \overleftarrow{\mathbf{h}_i}, \tag{4}$$

where $\oplus$ is the element-wise sum. Thus, the output of the sentence encoder is $H = [\mathbf{h}_1^T, \ldots, \mathbf{h}_L^T]$, where $H \in \mathbb{R}^{L \times d}$. This H is fed to both the sentence classification network and the UWP network.

We employ a CNN as the sentence classification network f_{cls} because the performance of CNNs in sentence classification has been demonstrated [18]. Most CNNs used for sentence classification generally apply one-dimensional (1d) convolution and 1d pooling operations [4]. However, Zhou et al. [18] introduced 2d convolution and 2d pooling operations to sentence classification and showed the effectiveness of the 2d operations in practice. Following the work of Zhou et al., we use a 2d convolutional layer and a 2d max pooling layer for f_{cls}. The convolution operation of the convolutional layer involves a 2D filter $m \in \mathbb{R}^{k \times d_m}$, which is applied to a window of k words and d_m feature dimensions. After the convolution operation is applied to H, the convolutional layer outputs a feature matrix $O_{conv} \in \mathbb{R}^{(l-k+1) \times (d-d_m+1)}$. The 2d max pooling operation is then applied to obtain a summarized feature map. With the pooling size $p \in \mathbb{R}^{p_1 \times p_2}$, the operation is applied to O_{conv} for extracting the maximum value features. By flattening the max-pooled feature map, a fixed-sized feature vector $\mathbf{o} \in \mathbb{R}^{\lfloor (l-k+1)/p_1 \rfloor \cdot \lfloor (d-d_m+1)/p_2 \rfloor}$ is obtained. Finally, $\mathbf{o}$ is fed to the classification layer, and the target class label is determined by

$$\mathbf{y} = softmax(W_y \cdot \mathbf{o} + \mathbf{b}_y), \tag{5}$$

where W_y and $\mathbf{b}_y$ denote a weight matrix and a bias vector of the classification layer, respectively.

The UWP network f_{aux} consists of a fully-connected layer that serves as a word prediction layer. f_{aux} takes H, the output of f_{enc}, as its input. Then, the network computes the probability distribution of the output words at each position i by

$$[\mathbf{w}_1, \ldots, \mathbf{w}_L] = softmax(W_{aux} \cdot H^T + \mathbf{b}_{aux}), \tag{6}$$

where W_{aux} and $\mathbf{b}_{aux}$ are the weight matrix and the bias vector, respectively. The output of the word prediction layer includes all L predicted words, but only U words at the same positions as the $\langle unk \rangle$ tokens are words of interest for the auxiliary task. To solve this problem, a one-hot masking vector $\mathbf{m}_i \in \mathbb{R}^L$ that indicates the position of a $\langle unk \rangle$ token at the i-th position is used to generate the final output of f_{aux} as follows.

$$\mathbf{w}_i = [\mathbf{w}_1, \ldots, \mathbf{w}_L] \cdot \mathbf{m}_i^T. \tag{7}$$

Note that this operation is executed for all U $\langle unk \rangle$ tokens.

5   Experiments

To demonstrate the effectiveness of the proposed model, we conducted a number of experiments on four widely used benchmark datasets for sentence classification.

Table 1. Summary statistics of datasets. c: number of classes, l: average sentence length, m: max sentence length, train/dev/test: train/development/test set size, vocab: vocabulary size in training data, unk_num: number of sentences that include at least 1 unknown word, and unk_max: max number of unknown word in a sentence.

Data	c	l	m	train	dev	test	vocab	unk_num	unk_max
SST-1	5	19	56	8544	1101	2210	16581	1240	9
SST-2	2	19	56	6920	872	1821	14830	1080	9
TREC-6	6	7	17	5452	-	500	8679	266	4
TREC-50	50	7	17	5452	-	500	8679	266	4

- **SST-1:** Stanford Sentiment Treebank was introduced by Socher et al. [14]. This dataset includes reviews with fine-grained labels (very negative, negative, neutral, positive, very positive).
- **SST-2:** This dataset is a coarse-grained version of SST-1. Thus, this dataset contains only the sentences with positive and negative labels from SST-1.
- **TREC-6:** A question classification dataset [9]. This dataset contains questions of six types, namely, abbreviation, description, entity, human, location, and numeric value.
- **TREC-50:** Another question classification dataset [9]. This dataset was created to classify a question into one of the fine-grained 50 question types.

Table 1 summarizes the statistics of the four benchmark datasets. As shown in this table, over 50% of the test sentences contain unseen words during training time. Thus, we can infer that the unseen words may significantly influence the classification performance of the proposed model.

The classification performance of the proposed model is compared with that of two baseline models. The first baseline model is a neural network with the same architecture as that of the proposed model, except for the auxiliary word predictor. Thus, this baseline did not encounter the $\langle unk \rangle$ token during training. Note that this baseline is a re-implemented version of BLSTM2DCNN [18], which exhibits the state-of-the-art performances on several sentence classification benchmarks. The second baseline model has the same architecture as the first baseline model. However, this baseline model is trained with the corrupted dataset D'. The injection of some noises into the training dataset has the effect of network regularization, which often improves performance.

5.1   Training Details and Hyperparameters

In the experiments, the Word2Vec embeddings trained by [11] were utilized as the pretrained word vectors. We initialized the vectors of the words that appeared only in the benchmark training datasets through random sampling from a uniform distribution in the range of $[-0.1, 0.1]$. The dimensions of the word embedding vector $\mathbf{v}_i$ and the contextual word vector $\mathbf{h}_i$ from the sentence encoder f_{enc} were set to 300. We used 100 convolutional filters with the window

Table 2. Classification results obtained with four sentence classification benchmarks. **BLSTM2DCNN:** the performance reported in [18]. **BLSTM2DCNN baseline:** a re-implemented version of BLSTM2DCNN. **BLSTM2DCNN baseline w/ noise injection:** a re-implemented version of BLSTM2DCNN with ⟨unk⟩ tokens injected into the training dataset.

Model	SST-1	SST-2	TREC-6	TREC-50
BLSTM2DCNN [18]	52.4	89.5	96.1	-
BLSTM2DCNN baseline	47.2	86.6	95.0	86.6
BLSTM2DCNN baseline w/ noise injection	47.5	87.1	94.1	86.0
Proposed model	**48.4**	**87.1**	**95.6**	**87.0**

size of (3,3). The 2D pooling size was set to (2,2). We performed mini-batch training with a batch size of 10. AdaDelta was used as an optimizer with the default learning rate of 0.1. For regularization, we employed the dropout operation with a rate of 0.5 for word embeddings, 0.2 for the Bi-LSTM layer, and 0.4 for the output of the pooling layer. Moreover, we imposed the l_2 penalty with the coefficient 10^{-5} over all parameters.

6   Results and Analysis

Table 2 shows the classification results obtained with four benchmark datasets. Unfortunately, we could not reproduce the exact performance of BLSTM2DCNN because the accuracy of the BLSTM2DCNN baseline is 1%–5% lower on three datasets than the corresponding performance reported in the original paper. As a result, the proposed model failed to exceed the result reported in the work of [18]. However, the proposed model consistently outperformed two baseline models on all four benchmark datasets. The proposed method achieved the best accuracies of 48.4% on SST-1, 95.6% on TREC-6, and 84.0% on TREC-50 relative to the baseline models.

It is known that training a neural network with noise-injected data regularizes the network, which may improve network performance. In our experiments, this was true for the tasks of SST-1 and SST-2 but not for the tasks of TREC-6 and TREC-50. More specifically, noise injection into the training data increased the accuracy of the BLSTM2DCNN baseline by 0.3% on SST-1 and by 0.5% on SST-2, while it decreased the accuracy of the BLSTM2DCNN baseline by 0.9% on TREC-6 and 0.6% on TREC-50. However, the proposed model yielded additional performance gains by introducing UWP as an auxiliary training task. This can be ascribed to the fact that the proposed auxiliary word predictor ensures that the sentence encoder produces not only more robust contextualized word representations but also well-approximated meaning representations for ⟨unk⟩ tokens.

We can understand the reason for performance improvement by observing the sentence representations produced by different models. Figure 2 shows two visu-

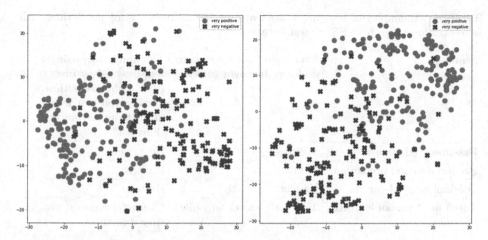

Fig. 2. Two visualizations of sentence representations by BLSTM2DCNN baseline (left) and proposed model (right) on SST-1 test dataset. All sentences in this figure contain ⟨*unk*⟩ tokens. (Color figure online)

Table 3. Performance comparison among the proposed model and BLSTM2DCNN baselines with different basic units on the benchmark datasets. Note that none of the models use any pretrained embeddings.

Model	SST-1	SST-2	TREC-6	TREC-50
BLSTM2DCNN using character	31.1 ± 0.5	63.7 ± 0.8	86.4 ± 0.5	76.4 ± 0.2
BLSTM2DCNN using subword2000	36.8 ± 0.4	76.2 ± 0.6	91.2 ± 0.2	80.2 ± 0.4
BLSTM2DCNN using subword4000	36.9 ± 1.1	76.0 ± 0.7	92.2 ± 0.6	81.9 ± 0.1
BLSTM2DCNN using word	39.7 ± 1.3	79.5 ± 0.9	93.0 ± 0.1	82.6 ± 0.5
Proposed model	$\mathbf{41.1 \pm 0.7}$	$\mathbf{81.3 \pm 1.0}$	$\mathbf{93.2 \pm 0.2}$	$\mathbf{83.7 \pm 0.3}$

alizations of sentence representations projected using T-SNE [10]. The left visualization in Fig. 2 shows sentence representations generated by the BLSTM2DCNN baseline, while the right one shows those generated by the proposed model. Because SST-1 is a difficult task, and the sentences in this figure contain more than one ⟨*unk*⟩ tokens, the red circles (very positive sentences) and blue crosses (very negative sentences) are jumbled in both figures. Nonetheless, in the right figure, the two areas of positive (top-right) and negative sentences (bottom-left) are more distinguishable than those in the left figure. These sentence representations were generated by summarizing $\mathbf{h}_i$'s in Eq. 4. Thus, the difference between the left and the right figures can be ascribed to the contextual representation power of the sentence encoder, which is jointly optimized for the UWP task.

Table 4. Examples of sentence classifications and unknown word predictions. All sentences are taken from SST-1 test dataset.

Sentence	Every good actor needs to do his or her own ⟨unk⟩	The film is surprisingly well-directed by brett ⟨unk⟩, who keeps things moving well – at least until the problematic third act
Baseline prediction	Positive	Negative
Proposed model prediction	Neutral	Positive
Original word of ⟨unk⟩	Hamlet	Ratner
Top-5 most probable words of ⟨unk⟩	Time . character way one	Character comedy way time director

Consequently, this different representation power inevitably contributes to the superior sentence classification performance of the proposed model.

Table 3 summarizes the classification results of the BLSTM2DCNN baselines with various basic units and those of the proposed model. The BLSTM2DCNNs using characters and subword units [6] eliminated the OOV problem by breaking down words into smaller units so that the vocabulary opened up. For the subword-level models, we limited the vocabulary size to 2,000 and 4,000. As can be seen, the results obtained with the BLSTM2DCNNs with the smaller units (character-level and two subword-level) are inferior to those achieved with the word-level BLSTM2DCNN over all benchmarks. These results indicate that the use of smaller units requires more complex and sophisticated architecture design. Finally, the proposed model achieved the best performance on all benchmarks because it replaced ⟨unk⟩ tokens with appropriate contextualized meaning representations.

Lastly, in Table 4, we introduce two example sentences that were correctly classified by the proposed model but misclassified by the baseline model. Each sentence in this table includes a ⟨unk⟩ token and the table shows the originals word of them as well as top-5 most likely words predicted by the proposed model. For the first sentence, the unknown word is 'hamlet' which means a representative character or acting methods. Thus words like 'character' and 'way' predicted by the proposed model are quite appropriate for the ⟨unk⟩ token. Similarly, the actual word of ⟨unk⟩ token in the second sentence is 'ratner' – the last name of the film director. Again, the proposed model correctly predicted the unknown word as the word 'director.' Although, unknown words in both sentences are not very critical for classification, the proposed model was able to make right decisions through appropriately contextualized word representations as well as properly estimated ⟨unk⟩ tokens.

7   Conclusion

In this paper, we propose a neural sentence classifier with an auxiliary UWP. To improve the classification performance of the model during testing, the proposed model was trained to predict not only the class label of the given sentences but also unknown words by considering all other words as contextual information. As a result, the proposed model generated robust representations of unknown words. In addition, the proposed auxiliary task enhanced the robustness of the entire sentence representation, which improved the classification performance of the proposed model. In the experiments, the proposed model consistently outperformed two baselines in terms of the sentence classification performance on four benchmark datasets.

References

1. Cheng, H., Fang, H., Ostendorf, M.: Open-domain name error detection using a multi-task RNN. In: Proceedings of the 2015 Conference on Empirical Methods in Natural Language Processing, pp. 737–746 (2015)
2. Devlin, J., Chang, M.W., Lee, K., Toutanova, K.: BERT: pre-training of deep bidirectional transformers for language understanding (2018). arXiv preprint arXiv:1810.04805
3. Iyyer, M., Enns, P., Boyd-Graber, J., Resnik, P.: Political ideology detection using recursive neural networks. In: Proceedings of the 52nd Annual Meeting of the Association for Computational Linguistics (Volume 1: Long Papers), pp. 1113–1122 (2014)
4. Kim, Y.: Convolutional neural networks for sentence classification. In: Proceedings of the 2014 Conference on Empirical Methods in Natural Language Processing (EMNLP), pp. 1746–1751 (2014)
5. Kudo, T.: Subword regularization: improving neural network translation models with multiple subword candidates. In: Proceedings of the 56th Annual Meeting of the Association for Computational Linguistics (Volume 1: Long Papers), pp. 66–75 (2018)
6. Kudo, T., Richardson, J.: Sentencepiece: a simple and language independent subword tokenizer and detokenizer for neural text processing. In: Proceedings of the 2018 Conference on Empirical Methods in Natural Language Processing: System Demonstrations, pp. 66–71 (2018)
7. Le, H.T., Cerisara, C., Denis, A.: Do convolutional networks need to be deep for text classification? In: Workshops at the Thirty-Second AAAI Conference on Artificial Intelligence (2018)
8. Lee, J., Cho, K., Hofmann, T.: Fully character-level neural machine translation without explicit segmentation. Trans. Assoc. Comput. Linguist. 5, 365–378 (2017)
9. Li, X., Roth, D.: Learning question classifiers. In: Proceedings of the 19th International Conference on Computational Linguistics, vol. 1, pp. 1–7. Association for Computational Linguistics (2002)
10. van der Maaten, L., Hinton, G.: Visualizing data using t-SNE. J. Mach. Learn. Res. 9(Nov), 2579–2605 (2008)
11. Mikolov, T., Sutskever, I., Chen, K., Corrado, G.S., Dean, J.: Distributed representations of words and phrases and their compositionality. In: Advances in Neural Information Processing Systems, pp. 3111–3119 (2013)

12. Ruder, S.: An overview of multi-task learning in deep neural networks (2017). arXiv preprint arXiv:1706.05098
13. Sennrich, R., Haddow, B., Birch, A.: Neural machine translation of rare words with subword units. In: Proceedings of the 54th Annual Meeting of the Association for Computational Linguistics (Volume 1: Long Papers), pp. 1715–1725 (2016)
14. Socher, R., et al.: Recursive deep models for semantic compositionality over a sentiment treebank. In: Proceedings of the 2013 Conference on Empirical Methods in Natural Language Processing, pp. 1631–1642 (2013)
15. Wang, S., Manning, C.D.: Baselines and bigrams: simple, good sentiment and topic classification. In: Proceedings of the 50th Annual Meeting of the Association for Computational Linguistics: Short Papers, vol. 2, pp. 90–94. Association for Computational Linguistics (2012)
16. Yu, J., Jiang, J.: Learning sentence embeddings with auxiliary tasks for cross-domain sentiment classification. In: Proceedings of the 2016 Conference on Empirical Methods in Natural Language Processing, pp. 236–246 (2016)
17. Zhang, X., Zhao, J., LeCun, Y.: Character-level convolutional networks for text classification. In: Advances in Neural Information Processing Systems, pp. 649–657 (2015)
18. Zhou, P., Qi, Z., Zheng, S., Xu, J., Bao, H., Xu, B.: Text classification improved by integrating bidirectional LSTM with two-dimensional max pooling. In: Proceedings of COLING 2016, the 26th International Conference on Computational Linguistics: Technical Papers, pp. 3485–3495 (2016)
19. Zhou, Q., Wang, X., Dong, X.: Differentiated attentive representation learning for sentence classification. In: Proceedings of the 27th International Joint Conference on Artificial Intelligence, pp. 4630–4636. AAAI Press (2018)

Effective Representation for Easy-First Dependency Parsing

Zuchao Li[1,2], Jiaxun Cai[1,2], and Hai Zhao[1,2(✉)]

[1] Department of Computer Science and Engineering,
Shanghai Jiao Tong University, Shanghai, China
`zhaohai@cs.sjtu.edu.cn`
[2] Key Laboratory of Shanghai Education Commission for Intelligent Interaction
and Cognitive Engineering, Shanghai Jiao Tong University, Shanghai, China
{`charlee,caijiaxun`}`@sjtu.edu.cn`

Abstract. Easy-first parsing relies on subtree re-ranking to build the complete parse tree. Whereas the intermediate state of parsing processing is represented by various subtrees, whose internal structural information is the key lead for later parsing action decisions, we explore a better representation for such subtrees. In detail, this work introduces a bottom-up subtree encoding method based on the child-sum tree-LSTM. Starting from an easy-first dependency parser without other handcraft features, we show that the effective subtree encoder does promote the parsing process, and can make a greedy search easy-first parser achieve promising results on benchmark treebanks compared to state-of-the-art baselines. Furthermore, with the help of the current pre-training language model, we further improve the state-of-the-art results of the easy-first approach.

Keywords: Easy-first algorithm · Dependency parsing ·
Effective representation

1 Introduction

Transition-based and graph-based parsers are two typical models used in dependency parsing. The former [27] can adopt rich features in the parsing process but are subject to limited searching space, while the latter [10,23,25] searches the entire tree space but limits to local features with higher computational costs. Besides, some other variants are proposed to overcome the shortcomings of both graph and transition based approaches. Easy-first parsing approach [11] is introduced by adopting ideas from the both models and is expected to benefit from the nature of the both. Ensemble method [19] was also proposed, which employs the parsing result of a parser to guide another in the parsing process.

This paper was partially supported by National Key Research and Development Program of China (No. 2017YFB0304100) and Key Projects of National Natural Science Foundation of China (U1836222 and 61733011).

© Springer Nature Switzerland AG 2019
A. C. Nayak and A. Sharma (Eds.): PRICAI 2019, LNAI 11670, pp. 351–363, 2019.
https://doi.org/10.1007/978-3-030-29908-8_28

Most recent works promote the parsing process by feature refinement. Instead, this work will explore the intermediate feature representation in the incremental easy-first parsing process. Easy-first dependency parser formalizes the parsing process as a sequence of attachments that build the dependency tree bottom-up. Inspired by the fact that humans always parse a natural language sentence starting from the easy and local attachment decisions and proceeding to the harder part instead of working in fixed left-to-right order, the easy-first parser learns its own notion of easy and hard, and defers the attachment decisions it considers to be harder until sufficient information is available. In the primitive easy-first parsing process, each attachment would simply delete the child node and leave the parent node unmodified. However, as the partially built dependency structures always carry rich information to guide the parsing process, effectively encoding those structures at each attachment would hopefully improve the performance of the parser (Fig. 1).

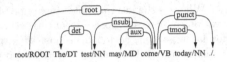

Fig. 1. A fully built dependency tree with part-of-speech (POS) tags and *root* token.

There exists a series of studies on encoding the tree structure created in different natural language processing (NLP) tasks using either recurrent neural network or recursive neural network [12,30]. However, most works require the encoded tree to have fixed maximum factors, and thus are unsuitable for encoding dependency tree where each node could have an arbitrary number of children. Other attempts allow arbitrary branching factors and have succeeded in particular NLP tasks.

[31] introduces a child-sum tree-structured Long Short-Term Memory (LSTM) to encode a completed dependency tree without limitation on branching factors, and shows that the proposed tree-LSTM is effective on semantic relatedness task and sentiment classification task. [50] proposes a recursive convolutional neural network (RCNN) architecture to capture syntactic and compositional-semantic representations of phrases and words in a dependency tree and then uses it to re-rank the k-best list of candidate dependency trees. [16] employs two vanilla LSTMs to encode a partially built dependency tree during parsing: one encodes the sequence of left-modifiers from the head outwards, and the other encodes the sequence of right-modifiers in the same manner.

In this paper, we inspect into the bottom-up building process of the easy-first parser and introduce pre-trained language model features and a subtree encoder for more effective representation to promote the parsing process[1]. Unlike the work in [16] that uses two standard LSTMs to encode the dependency subtree

---

[1] Our code is available at https://github.com/bcmi220/erefdp.

in a sequential manner (which we will refer to as HT-LSTM later in the paper), we employ a structural model that provides the flexibility to incorporate and drop an individual child node of the subtree. Further, we introduce a multilayer perceptron between depths of the subtree to encode other underlying structural information like relation and distance between nodes.

From the evaluation results on the benchmark treebanks, the proposed model gives results greatly better than the baseline parser and outperforms the neural easy-first parser presented by [16]. Besides, our greedy bottom-up parser achieves performance comparable to those parsers that use beam search or re-ranking method [1,50].

2   Easy-First Parsing Algorithm

Easy-first parsing could be considered as a variation of transition-based parsing method, which builds the dependency tree from easy to hard instead of working in a fixed left-to-right order. The parsing process starts by making easy attachment decisions to build several dependency structures, and then proceeds to the harder and harder ones until a well-formed dependency tree is built. During training, the parser learns its own notion of easy and hard, and learns to defer specific kinds of decisions until more information is available [11].

The main data structure in the easy-first parser is a list of unattached nodes called *pending*. The parser picks a series of actions from the allowed action set, and applies them upon the elements in the *pending* list. The parsing process stops until the *pending* solely contains the root node of the dependency tree.

At each step, the parser chooses a specific action $\hat{a}$ on position i using a scoring function $score(\cdot)$, which assigns scores to each possible action on each location based on the current state of the parser. Given an intermediate state of parsing process with *pending* $P = \{p_0, p_1, \cdots, p_N\}$, the attachment action is determined by

$$\hat{a} = \underset{act \in \mathcal{A},\ 1 \leq i \leq N}{\operatorname{argmax}}\ score(act(i)),$$

where $\mathcal{A}$ denotes the set of the allowed actions, i is the index of the node in the *pending*. Besides distinguishing the correct attachments from the incorrect ones, the scoring function is supposed to assign the "easiest" attachment with the highest score, which in fact determines the parsing order of an input sentence. [11] employs a linear model for the scorer:

$$score(act(i)) = \boldsymbol{w} \cdot \boldsymbol{\phi}_{act(i)},$$

where $\boldsymbol{\phi}_{act(i)}$ is the feature vector of attachment $act(i)$, and $\boldsymbol{w}$ is a parameter that can be learned jointly with other components in the model.

There are exactly two types of actions in the allowed action set: ATTACHLEFT(i) and ATTACHRIGHT(i) as shown in Fig. 2. Let p_i refer to i-th element in the *pending*, then the allowed actions can be formally defined as follows:

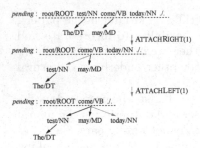

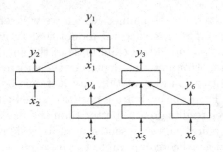

Fig. 2. Illustration of the *pending* states before and after the two type of attachment actions

Fig. 3. Tree-LSTM neural network with arbitrary number of child nodes

- ATTACHLEFT(i): attaching p_{i+1} to p_i which results in an arc (p_i, p_{i+1}) headed by p_i, and removing p_{i+1} from the *pending*.
- ATTACHRIGHT(i): attaching p_i to p_{i+1} which results in an arc (p_{i+1}, p_i) headed by p_{i+1}, and removing p_i from the *pending*.

3 Parsing with Subtree Encoding

3.1 Dependency Subtree

Easy-first parser builds up a dependency tree incrementally, so in intermediate state, the *pending* of the parser may contain two kinds of nodes:

- *subtree root:* the root of a partially built dependency tree;
- *unprocessed node:* the node that has not yet be attached to a parent or assigned a child.

Note that each processed node should become a subtree root (attached as a parent) or be removed from the *pending* (attached as a child). A subtree root in the *pending* actually stands for a dependency structure whose internal nodes are all processed, excluding the root itself. Therefore it is supposed to be more informative than the unprocessed nodes to guide the latter attachment decisions.

In the easy-first parsing process, each *pending* node is attached to its parent only after all its children have been collected. Thus, any structure produced in the parsing process is guaranteed to be a dependency subtree that is consistent with the above definition.

3.2 Recursive Subtree Encoding

In the primitive easy-first parsing process, the node that has been removed does not affect the parsing process anymore. Thus the subtree structure in the *pending* is simply represented by the root node. However, motivated by the success of encoding the tree structure properly for other NLP tasks [16,18,31], we employ the child-sum tree-LSTM to encode the dependency subtree in the hope of further parsing performance improvement.

Child-Sum Tree-LSTM. Child-sum tree-LSTM is an extension of standard LSTM proposed by [31] (hereafter referred to tree-LSTM). Like the standard LSTM unit [14], each tree-LSTM unit contains an input gate i_j, an output gate o_j, a memory cell c_j and a hidden state h_j. The major difference between tree-LSTM unit and the standard one is that the memory cell updating and the calculation of gating vectors are depended on multiple child units. As shown in Fig. 3, a tree-LSTM unit can be connected to numbers of child units and contains one forget gate for each child. This provides tree-LSTM the flexibility to incorporate or drop the information from each child unit.

Given a dependency tree, let $C(j)$ denote the children set of node j, x_j denote the input of node j. Tree-LSTM can be formulated as follow [31]:

$$\tilde{h}_j = \sum_{k \in C(j)} h_k, \tag{1}$$

$$
\begin{aligned}
i_j &= \sigma(W^{(i)} x_j + U^{(i)} \tilde{h}_j + b^{(i)}), \\
f_{jk} &= \sigma(W^{(f)} x_j + U^{(f)} h_k + b^{(f)}), \\
o_j &= \sigma(W^{(o)} x_j + U^{(o)} \tilde{h}_j + b^{(o)}), \\
u_j &= \tanh(W^{(u)} x_j + U^{(u)} \tilde{h}_j + b^{(u)}), \\
c_j &= i_j \odot u_j + \sum_{k \in C(j)} f_{jk} \odot c_k, \\
h_j &= o_j \odot \tanh(c_j).
\end{aligned}
\tag{2}
$$

where $k \in C(j)$, and h_k is the hidden state of the k-th child node, c_j is the memory cell of the head node j, and h_j is the hidden state of node j. Note that in Eq. (2), a single forget gate f_{jk} is computed for each hidden state h_k.

Our subtree encoder uses tree-LSTM as the basic building block incorporated with the distance and relation label.

Incorporating Distance and Relation Features. Distance embedding is a usual way to encode the distance information. In our model, we use vector $v^{(d)}_{h,m_k}$ to represent the relative distance of head word h and its k-th modifier m_k:

$$
\begin{aligned}
d_{h,m_k} &= index(h) - index(m_k), \\
v^{(d)}_{h,m_k} &= Embed^{(d)}(d_{h,m_k}),
\end{aligned}
$$

where $index(\cdot)$ is the index of the word in the original input sentence, and $Embed^{(d)}$ represents the distance embeddings lookup table.

Similarly, the relation label $v^{(rel)}_{h,m_k}$ between head-modifier pair (h, m_k) is encoded as a vector according to the relation embeddings lookup table $Embed^{(r)}$. Both of the two embeddings lookup tables are randomly initialized and learned jointly with other parameters in the neural network.

To incorporate the two features, our subtree encoder introduces an additional feature encoding layer between every connected tree-LSTM unit. Specifically, the two feature embeddings are first concatenated to the hidden state of the corresponding child node. Then we apply an affine transformation on the resulted vector g_k, and further pass the result through a tanh activation

$$g_k = \varphi(h_k \oplus v_{h,m_k}^{(d)} \oplus v_{h,m_k}^{(rel)}),$$

$$\varphi(x) = \tanh(W^{(\varphi)}x + b^{(\varphi)})$$

where $W^{(\varphi)}$ and $b^{(\varphi)}$ are learnable parameters. After getting g_k, it is fed into the next tree-LSTM unit. Therefore, the hidden state of child node h_k in Eqs. (1) and (2) is then replaced by g_k.

3.3   The Bottom-Up Constructing Process

In our model, a dependency subtree is encoded by performing the tree-LSTM transformation on its root node and computing the vector representation of its children recursively until reaching the leaf nodes. More formally, given a partially built dependency tree rooted at node h with children (modifiers): $h.m_1, h.m_2, h.m_3, \cdots$, which may be roots of some smaller subtree. Then the tree can be encoded like:

$$\tau_h = f(\omega_{h.m_1}, \omega_{h.m_2}, \omega_{h.m_3}, x_j), \tag{3}$$

$$\omega_{h.m_k} = \varphi(\tau_{h.m_k}, v_{h,m_k}^{(d)}, v_{h,m_k}^{(rel)}), k \in \{1, 2, 3, \cdots\}$$

where f is the tree-LSTM transformation, φ is the above-mentioned feature encoder, $\tau_{h.m_k}$ refers to the vector representation of subtree rooted at node m_k, and x_j denotes the embedding of the root node word h. In practice, x_j is always a combination of the word embedding and POS-tag embedding or the output of a bidirectional LSTM. We can see clearly that the representation of a fully parse tree can be computed via a recursive process.

When encountering the leaf nodes, the parser regards them as a subtree without any children and thus sets the initial hidden state and memory cell to a zeros vector respectively:

$$\tau^{(leaf)} = f(\mathbf{0}, x^{(leaf)}) \tag{4}$$

In the easy-first parsing process, each dependency structure in the *pending* is built incrementally. Namely, the parser builds several dependency subtrees separately and then combines them into some larger subtrees. So, when the parser builds a subtree rooted at h, all its children have been processed by some previous steps. The subtree encoding process can be naturally incorporated into the easy-first parsing process in a bottom-up manner using the dynamic programming technique.

Specifically, in the initial step, each node w_i in the input sentence is treated like a subtree without any children. The parser initializes the *pending* with the tree representation $\tau_{w_i}^{(leaf)}$ of those input nodes using Eq. (4). For each node in *pending*, the parser maintains an additional *children* set to hold their processed

Table 1. Comparison with baseline easy-first parser.

	Dev (%)		Test (%)	
	LAS	UAS	LAS	UAS
BiLSTM parser	90.73	92.87	90.67	92.83
RCNN	91.05	93.25	91.01	93.21
HT-LSTM	91.23	93.23	91.36	93.27
tree-LSTM	**92.32**	**94.27**	**92.33**	**94.31**
tree-LSTM + ELMo	**92.97**	**94.95**	**93.09**	**95.33**
tree-LSTM + BERT	**93.14**	**95.68**	**93.27**	**95.71**
tree-LSTM + ELMo + BERT	**93.44**	**95.87**	**93.49**	**95.87**

Table 2. Results under the same settings reported in [16].

	Dev (%)		Test (%)	
	LAS	UAS	LAS	UAS
Baseline parser	78.83	82.97	78.43	82.55
+tree-LSTM	91.10	92.98	91.08	92.94
+Bi-LSTM	91.62	93.49	91.52	93.46
+pre-train	92.01	93.97	91.95	93.95
Baseline parser*	79.0	83.3	78.6	82.7
+HT-LSTM*	90.1	92.4	89.8	92.0
+Bi-LSTM*	90.5	93.0	90.2	92.6
+pre-train *	90.8	93.3	90.9	93.0

children. Each time the parser performs an attachment on the nodes in the *pending*, the selected modifier is removed from *pending* and then added to the children set of the selected head. The vector representation of the subtree rooted at the selected head is recomputed using Eq. (3). The number of times that the easy-first parser performs updates on the subtree representations is equal to the number of actions required to build a dependency tree, namely, $N - 1$, where N is the input sentence length.

3.4 Incorporating HT-LSTM and RCNN

Both HT-LSTM and RCNN can be incorporated into our framework. However, since the RCNN model employs POS tag dependent parameters, its primitive form is incompatible with the incremental easy-first parser, for which we leave a detail discussion in Sect. 4.4. To address this problem, we simplify and reformulate the RCNN model by replacing the POS tag dependent parameters with a global one. Specifically, for each head-modifier pair (h, m_k), we first use a convolutional hidden layer to compute the combination representation:

$$z_k = \tanh(W^{(global)} p_k), 0 < k \leq K,$$
$$p_k = x_h \oplus g_k,$$
$$g_k = \varphi(\tau_k \oplus v_{h,m_k}^{(d)} \oplus v_{h,m_k}^{(rel)}),$$

where K is the size of the children set $C(h)$ of node h, $W^{(global)}$ is the global composition matrix, τ_k is the subtree representation of the child node m_k, which can be recursively computed using the RCNN transformation. After convolution, we stack all z_k into a matrix $Z^{(h)}$. Then to get the subtree representation for h, we apply a max pooling over $Z^{(h)}$ on rows:

$$\tau_h = \max_k Z_{j,k}^{(h)}, 0 < j \leq d, 0 < k \leq K,$$

where d is the dimensionality of z_k.

4 Experiments and Results

We evaluate our parsing model on English Penn Treebank (PTB) and Chinese Penn Treebank (CTB), using unlabeled attachment scores (UAS) and labeled attachment scores (LAS) as the metrics. Punctuations are ignored as in previous work [8,16]. Pre-trained word embeddings and language model have been shown useful in a lot of tasks. Therefore, we also add the latest ELMo [28] and BERT [7] pre-trained language model layer features to enhance our representation.

4.1 Treebanks

For English, we use the Stanford Dependency (SD 3.3.0) [6] conversion of the Penn Treebank [24], and follow the standard splitting convention for PTB, using sections 2–21 for training, Sect. 22 as development set and Sect. 23 as test set. Stanford POS tagger [32] is to give predicted POS tags.

For Chinese, we adopt the splitting convention for CTB described in [9,39, 45,48]. The dependencies are converted with the Penn2Malt converter. Gold segmentation and POS tags are used as in previous work [9].

4.2 Results

Improvement over Baseline Model. To explore the effectiveness of the proposed subtree encoding model, we implement a baseline easy-first parser without additional subtree encoders and conduct experiments on PTB. The baseline model contains a BiLSTM encoder and uses pre-trained word embedding, which we refer to BiLSTM parser. We also re-implement both HT-LSTM and RCNN and incorporate them into our framework for subtree encoding. All the four models share the same hyper-parameters settings and the same neural components except the subtree encoder.

The results in Table 1 show that our proposed tree-LSTM encoder model outperforms the BiLSTM parser with a margin of 1.48% in UAS and 1.66% in LAS on the test set. Though the RCNN model keeps simple by just using a single global matrix $W^{(global)}$, it draws with the HT-LSTM model in UAS on both the development set and the test set, and slightly underperforms the latter one in LAS. Note that the HT-LSTM is more complicated, which contains two LSTMs. Such results demonstrate that simply sequentializing the subtree fails to effectively incorporate the structural information. A further error analysis of the three models is given in the following section.

Besides, to make a fair comparison, we also run our model under the same setting as those reported in [16], and report the results in Table 2. Experiment results show that the performance of the tree-LSTM parser declines slightly but still outperforms the HT-LSTM parser. The "+" symbol denotes a specific extension over the previous line. The results with * is reported in [16]. It is worth noting that their weak baseline parser does not use Bi-LSTM and pre-trained embeddings.

Table 3. Comparison of results on the test sets. Acronyms used: (g) greedy, (b) beam search, (re) re-ranking, (3rd) 3rd-order, (1st) 1st-order. Because ELMo does not have a Chinese version, the "+ELMo" rows have no results.

System	Method	PTB-SD		CTB	
		LAS (%)	UAS (%)	LAS (%)	UAS (%)
Dyer et al. [9]	Transition (g)	90.9	93.1	85.5	87.1
Kiperwasser and Goldberg [17]	Transition (g)	91.9	93.9	**86.1**	**87.6**
Andor et al. [1]	Transition (b)	**92.79**	**94.61**	–	–
Zhu et al. [50]	Transition (re)	–	94.16	–	87.43
Zhang and McDonald [36]	Graph (3rd)	90.64	93.01	86.34	87.96
Wang and Chang [33]	Graph (1st)	91.82	94.08	86.23	87.55
Kiperwasser and Goldberg [17]	Graph (1st)	90.9	93.0	84.9	86.5
Dozat and Manning [8]	Graph (1st)	**94.08**	**95.74**	**88.23**	**89.30**
Wang et al. [34]	Graph (1st)	**94.54**	95.66	–	–
Wang et al. [34] + ELMo	Graph (1st)	**95.25**	**96.35**	–	–
Zhang et al. [37]	Seq2seq (b)	91.60	93.71	85.40	87.41
Li et al. [20]	Seq2seq (b)	92.08	94.11	86.23	88.78
Kiperwasser and Goldberg [16]	EasyFirst (g)	90.9	93.0	85.5	87.1
This work	EasyFirst (g)	**92.33**	**94.31**	**86.37**	**88.65**
This work + ELMo	EasyFirst (g)	**93.09**	**95.33**	–	–
This work + BERT	EasyFirst (g)	**93.27**	**95.71**	**87.44**	**89.52**
This work + ELMo + BERT	EasyFirst (g)	**93.49**	**95.87**	–	–

Comparison with Previous Parsers. We now compare our model with some other recently proposed parsers. The results are compared in Table 3. The work in [16] (HT-LSTM) is similar to ours and achieves the best result among the recently proposed easy-first parsers[2]. Our subtree encoding parser outperforms their model on both PTB and CTB. Besides, the proposed model also outperforms the RCNN based re-ranking model in [50], which introduces an RCNN to encode the dependency tree and re-ranks the k-best trees produced by the base model. Note that although our model is based on the greedy easy-first parsing algorithm, it is also competitive to the search-based parser in [1]. The model in [8] outperforms ours, however, their parser is graph-based and thus can enjoy the benefits of global optimization.

4.3   Error Analysis

To characterize the errors made by parsers and the performance enhancement by importing the subtree encoder, we present some analysis on the error rate with respect to the sentence length and POS tags. All analysis is conducted on the unlabeled attachment results from the PTB development set.

Error Distribution over Dependency Distance. Figure 4 shows the error rate of different subtree encoding methods with respect to sentence length. The

---

[2] Here we directly refer to the original results reported in [16].

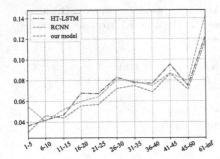
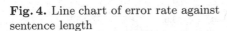

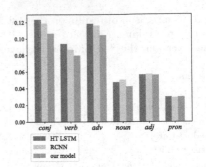

Fig. 4. Line chart of error rate against sentence length

Fig. 5. Error rate with respect to POS tags

error rate curves of the three models share the same tendency: as the sentence length grows, the error rate increases. In most of the cases, the curve of our model lies below the other two curves, except the case that the sentence length lies in 6-10 where the proposed model underperforms the other two with a margin smaller than 1%. The curve of HT-LSTM and that of RCNN cross with each other at several points. It is not surprising since the overall results of the two models are very close. The curves further show that tree-LSTM is more suitable for incorporating the structural information carried by the subtrees produced in the easy-first parsing process.

Error Distribution over POS Tags. [26] distinguishes *noun, verb, pronoun, adjective, adverb, conjunction* for POS tags to perform a linguistic factors analysis. To follow their works, we conduct a mapping on the PTB POS tags and skip those which cannot be mapped into one of the six above-mentioned POS tags. Then we evaluate the error rate with respected to the mapped POS tags and compare the performance of the three parsers in Fig. 5.

The results seem to be contradicted with the previous ones at first sight since the HT-LSTM model underperforms the RCNN one in most cases. This interesting result is caused by the overwhelming number of *noun*. According to statistics, the number of *noun* is roughly equal to the total number of *verb, adverb* and *conjunction*.

Typically, the *verb, conjunction* and *adverb* tend to be closer to the root in a parse tree, which leads to a longer-distance dependency and makes it more difficult to parse. The figure shows that our model copes better with those kinds of words than the other two models.

The other three categories of words are always attached lower in a parse tree and theoretically should be easier to parse. In the result, the three models perform similarly on *adjective* and *pronoun*. However, the RCNN model performs worse than the other two models on *noun*, which can be attributed to too simple RCNN model that is unable to cover different lengths of dependency.

4.4 Related Work

Recently, neural networks have been adopted for a wide range of traditional NLP tasks [41–44,46,47,49]. A recent line of studies including Chinese word segmentation [2,3], syntactic parsing [20,22,39], semantic role labeling [4,13,21] and other NLP applications [5,15,35,38,40] have drawn a lot of attention. Easy-first parsing has a special position in dependency parsing system. As mentioned above, to some extent, it is a kind of hybrid model that shares features with both transition and graph based models, though quite a lot of researchers still agree that it belongs to the transition-based type as it still builds parse tree step by step. Since easy-first parser was first proposed in [11], the most progress on this type of parsers is [16] who incorporated neural network for the first time.

Most of the RNNs are limited to a fixed maximum number of factors [29]. To release the constraint of the limitation of factors, [50] augments the RNN with a convolutional layer, resulting in a recursive convolutional neural network (RCNN). The RCNN is able to encode a tree structure with an arbitrary number of factors, and is used in a re-ranking model for dependency parsing.

Child-sum tree-LSTM [31] is a variant of the standard LSTM which is capable of getting rid of the arity restriction, and has been shown effective on semantic relatedness and the sentiment classification tasks. We adopt the child-sum tree-LSTM in our incremental easy-first parser to promote the parsing.

5 Conclusion

To enhance the easy-first dependency parsing, this paper proposes a tree encoder and integrates pre-trained language model features for a better representation of partially built dependency subtrees. Experiments on PTB and CTB verify the effectiveness of the proposed model.

References

1. Andor, D., et al.: Globally normalized transition-based neural networks. In: Proceedings of ACL (2016)
2. Cai, D., Zhao, H.: Neural word segmentation learning for Chinese. In: Proceedings of ACL (2016)
3. Cai, D., Zhao, H., Zhang, Z., Xin, Y., Wu, Y., Huang, F.: Fast and accurate neural word segmentation for Chinese. In: Proceedings of ACL (2017)
4. Cai, J., He, S., Li, Z., Zhao, H.: A full end-to-end semantic role labeler, syntactic-agnostic over syntactic-aware? In: Proceedings of COLING (2018)
5. Chen, K., Wang, R., Utiyama, M., Sumita, E., Zhao, T.: Syntax-directed attention for neural machine translation. In: Proceedings of AAAI (2018)
6. De Marneffe, M.C., Manning, C.D.: Stanford typed dependencies manual. Technical report (2008)
7. Devlin, J., Chang, M.W., Lee, K., Toutanova, K.: BERT: pre-training of deep bidirectional transformers for language understanding (2018). arXiv preprint arXiv:1810.04805

8. Dozat, T., Manning, C.D.: Deep biaffine attention for neural dependency parsing. In: Proceedings of ICLR (2017)
9. Dyer, C., Ballesteros, M., Ling, W., Matthews, A., Smith, N.A.: Transition-based dependency parsing with stack long short-term memory. In: Proceedings of ACL-IJCNLP (2015)
10. Eisner, J.: Efficient normal-form parsing for combinatory categorial grammar. In: Proceedings of ACL (1996)
11. Goldberg, Y., Elhadad, M.: An efficient algorithm for easy-first non-directional dependency parsing. In: Proceedings of HLT: NAACL (2010)
12. Goller, C., Kuchler, A.: Learning task-dependent distributed representations by backpropagation through structure. In: IEEE International Conference on Neural Networks (1996)
13. He, S., Li, Z., Zhao, H., Bai, H.: Syntax for semantic role labeling, to be, or not to be. In: Proceedings of ACL (2018)
14. Hochreiter, S., Schmidhuber, J.: Long short-term memory. Neural Comput. **9**, 1735–1780 (1997)
15. Huang, Y., Li, Z., Zhang, Z., Zhao, H.: Moon IME: neural-based Chinese pinyin aided input method with customizable association. In: Proceedings of ACL, Demo (2018)
16. Kiperwasser, E., Goldberg, Y.: Easy-first dependency parsing with hierarchical tree LSTMs. TACL **4**, 445–461 (2016)
17. Kiperwasser, E., Goldberg, Y.: Simple and accurate dependency parsing using bidirectional LSTM feature representations. TACL **4**, 313–327 (2016)
18. Kuncoro, A., Ballesteros, M., Kong, L., Dyer, C., Neubig, G., Smith, N.A.: What do recurrent neural network grammars learn about syntax? In: Proceedings of EACL (2017)
19. Kuncoro, A., Ballesteros, M., Kong, L., Dyer, C., Smith, N.A.: Distilling an ensemble of greedy dependency parsers into one MST parser. In: Proceedings of EMNLP (2016)
20. Li, Z., Cai, J., He, S., Zhao, H.: Seq2seq dependency parsing. In: Proceedings of COLING (2018)
21. Li, Z., et al.: A unified syntax-aware framework for semantic role labeling. In: Proceedings of EMNLP (2018)
22. Li, Z., He, S., Zhang, Z., Zhao, H.: Joint learning of POS and dependencies for multilingual universal dependency parsing. In: Proceedings of CoNLL (2018)
23. Ma, X., Zhao, H.: Fourth-order dependency parsing. In: Proceedings of COLING (2012)
24. Marcus, M.P., Marcinkiewicz, M.A., Santorini, B.: Building a large annotated corpus of English: the Penn treebank. Comput. Linguist. **19**, 313–330 (1993)
25. McDonald, R., Crammer, K., Pereira, F.: Online large-margin training of dependency parsers. In: Proceedings of ACL (2005)
26. McDonald, R., Nivre, J.: Characterizing the errors of data-driven dependency parsing models. In: Proceedings of EMNLP-CoNLL (2007)
27. Nivre, J.: An efficient algorithm for projective dependency parsing. In: Proceedings of IWPT (2003)
28. Peters, M., et al.: Deep contextualized word representations. In: Proceedings of NAACL-HLT (2018)
29. Socher, R.: Recursive deep learning for natural language processing and computer vision. Ph.D. thesis (2014)

30. Socher, R., Manning, C.D., Ng, A.Y.: Learning continuous phrase representations and syntactic parsing with recursive neural networks. In: Proceedings of the NIPS (2010)
31. Tai, K.S., Socher, R., Manning, C.D.: Improved semantic representations from tree-structured long short-term memory networks. In: Proceedings of ACL-IJCNLP (2015)
32. Toutanova, K., Klein, D., Manning, C.D., Singer, Y.: Feature-rich part-of-speech tagging with a cyclic dependency network. In: Proceedings of NAACL (2003)
33. Wang, W., Chang, B.: Graph-based dependency parsing with bidirectional LSTM. In: Proceedings of ACL (2016)
34. Wang, W., Chang, B., Mansur, M.: Improved dependency parsing using implicit word connections learned from unlabeled data. In: Proceedings of EMNLP (2018)
35. Xiao, Y., Cai, J., Yang, Y., Zhao, H., Shen, H.: Prediction of microrna subcellular localization by using a sequence-to-sequence model. In: Proceedings of ICDM (2018)
36. Zhang, H., McDonald, R.: Enforcing structural diversity in cube-pruned dependency parsing. In: Proceedings of ACL (2014)
37. Zhang, Z., Liu, S., Li, M., Zhou, M., Chen, E.: Stack-based multi-layer attention for transition-based dependency parsing. In: Proceedings of EMNLP (2017)
38. Zhang, Z., Wang, R., Utiyama, M., Sumita, E., Zhao, H.: Exploring recombination for efficient decoding of neural machine translation. In: Proceedings of EMNLP (2018)
39. Zhang, Z., Zhao, H., Qin, L.: Probabilistic graph-based dependency parsing with convolutional neural network. In: Proceedings of ACL (2016)
40. Zhang, Z., Li, J., Zhu, P., Zhao, H.: Modeling multi-turn conversation with deep utterance aggregation. In: Proceedings of COLING (2018)
41. Zhao, H., Chen, W., Kazama, J., Uchimoto, K., Torisawa, K.: Multilingual dependency learning: exploiting rich features for tagging syntactic and semantic dependencies. In: Proceedings of CoNLL (2009)
42. Zhao, H., Huang, C.N., Li, M.: An improved Chinese word segmentation system with conditional random field. In: Proceedings of SIGHAN, p. 5 (2006)
43. Zhao, H., Huang, C.N., Li, M., Lu, B.L.: A unified character-based tagging framework for Chinese word segmentation. ACM Trans. Asian Lang. Inf. Process. 9, 5 (2010)
44. Zhao, H., Huang, C.N., Li, M., Lu, B.L.: Effective tag set selection in Chinese word segmentation via conditional random field modeling. In: Proceedings of PACLIC, p. 20 (2006)
45. Zhao, H., Kit, C.: Parsing syntactic and semantic dependencies with two single-stage maximum entropy models. In: Proceedings of CoNLL (2008)
46. Zhao, H., Kit, C.: Unsupervised segmentation helps supervised learning of character tagging for word segmentation and named entity recognition. In: Proceedings of SIGHAN, p. 6 (2008)
47. Zhao, H., Kit, C.: Integrating unsupervised and supervised word segmentation: the role of goodness measures. Inf. Sci. 181, 163–183 (2011)
48. Zhao, H., Song, Y., Kit, C., Zhou, G.: Cross language dependency parsing using a bilingual lexicon. In: Proceedings of ACL-IJCNLP (2009)
49. Zhao, H., Zhang, X., Kit, C.: Integrative semantic dependency parsing via efficient large-scale feature selection. J. Artif. Intell. Res. 46, 203–233 (2013)
50. Zhu, C., Qiu, X., Chen, X., Huang, X.: A re-ranking model for dependency parser with recursive convolutional neural network. In: Proceedings of ACL-IJCNLP (2015)

HRCR: Hidden Markov-Based Reinforcement to Reduce Churn in Question Answering Forums

Reza Hadi Mogavi[1(✉)], Sujit Gujar[2], Xiaojuan Ma[1], and Pan Hui[1,3]

[1] Hong Kong University of Science and Technology, Clear Water Bay, Hong Kong
{rhadimogavi,mxj,panhui}@cse.ust.hk
[2] International Institute of Information Technology, Hyderabad, Hyderabad, India
sujit.gujar@iiit.ac.in
[3] University of Helsinki, Helsinki, Finland

Abstract. The high rate of *churning users* who abandon the Community Question Answering forums (CQAs) may be one of the crucial issues that hinder their development. More personalized question recommendation to users might help to manage this problem better. In this paper, we propose a new algorithm (we name HRCR) that recommends questions to users such *to reduce their churning probability*. We present our algorithm in a two-fold structure: First, we use Hidden Markov Models (HMMs) to uncover the users' engagement states inside a CQA. Second, we apply a Reinforcement Learning Model (RL) to recommend users the questions that match better with their engagement mood and thus help them get into a better engagement state (the one with the least churning probability). Experiments on a large-scale offline dataset from Stack Overflow show a meaningful reduction in the churning probability of the users who comply with HRCR's question recommendations.

Keywords: Question answering forum · Churn · Flow theory · Computational user engagement · Reinforcement learning

1 Introduction

Community Question Answering forums (CQAs) like Stack Overflow[1] facilitate knowledge sharing online [23]. CQAs are dependent on continuous user participation to preserve their sustainability [10,16]. Especially, retention of *contributing users* who provide answers to questions has priority and is challenging [13,15]. Related literature suggests *personalizing tasks* before crowdsourcing can grow users' willingness to make more and better quality contributions [21,25].

In this paper, we introduce a new question recommendation algorithm (we name HRCR) that aims to reduce churning (see [11,16]) of the contributing users through recommending more personalized questions. While literature is

---

[1] https://stackoverflow.com/.

© Springer Nature Switzerland AG 2019
A. C. Nayak and A. Sharma (Eds.): PRICAI 2019, LNAI 11670, pp. 364–376, 2019.
https://doi.org/10.1007/978-3-030-29908-8_29

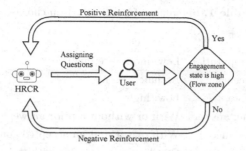

Fig. 1. High-level illustration of the HRCR. HRCR assigns questions to users. If the assigned questions lead users to a higher engagement state (i.e., Flow Zone), HRCR receives a positive reinforcement (feedback); otherwise, the agent receives a negative reinforcement. Through iterations, HRCR finds the best policy for assigning questions to users.

rich in question recommendation algorithms (see [4,9,18]), HRCR is distinctive in following aspects: First, HRCR considers churn tendency of contributing users before recommending questions. Second, HRCR is inspired by the psychological theory of *Flow* (see [8]) to increase user engagement.

HRCR, in essence, is a Markov Decision Process (MDP) based recommendation system. Although the idea of using MDP to build recommender systems is not new (see [6,20]), we are first to use them for building a more personalized question distributing algorithm. Figure 1 shows a high-level idea of how HRCR is working. After HRCR assigns a question to a user, it receives a positive reinforcement (feedback) if the users' engagement state is high (i.e., Flow Zone); otherwise, it receives a negative reinforcement. Our goal is to fill the gap of the existing question recommendation algorithms in term of considering churn.

More precisely, we first use Hidden Markov Models to uncover the *stochastic behavioral pattern* behind user participation. We are particularly inspired by the psychological theory of Flow to interpret the hidden states of our HMM. We then use the resolved hidden states of our HMM to reinforce a standard Markov Decision Process model. We make several simplifying assumptions to avoid the otherwise prohibitive number of parameters in our HRCR formulation. Through iterations on the MDP, HRCR finds the best recommendations for users. Experiments on a real-world dataset from a well-known CQA forum shows that HRCR can meaningfully help in churn reduction of the contributing users while preserving its simplicity.

2   Users' Action Choices

Users often have their own set of principles for deciding what questions they answer. Although knowledgeability in the field comes as an important factor, there are often more parameters involved. Table 1 provides a reference to such parameters. We use Stack Overflow, to characterize the CQA users [16,18]. How-

Table 1. Reference of possible action choices

Factors	Action choices
Questioner's reputation	Low, moderate, high
Question's recency	Archaic, recent
Question's score	Low, high
State of having answers	With or without a prior answer
State of acceptance	With or without an accepted contribution
Familiarity with topic	Familiar or unfamiliar with the asked topic
Bounty prize	With or without a bounty prize

ever, the features should be similar on the CQAs like Yahoo! Answers and Quora. In the rest, we elaborate on our list of factors.

Questioner's Reputation (QR). A contributing user can be selective on whose question she answers. For example, a contributing user who has got a lower reputation score might not have the sufficient skill or knowledge to answer to a question asked by a highly reputed user. We use, first and third quartiles of the user reputation values to define low, moderate and highly reputed users. Any user with a reputation between 26 and 1,580 is assumed to be a moderately reputed user. Reputation scores below 26 and above 1,580 are labeled as low and high reputation values respectively.

Question's Recency (QRC). CQAs like Stack Overflow are designed in a way to give a higher priority to the questions submitted most recently. Our dataset shows that the average response time for a typical question in the CQA of Stack Overflow is 6.23 days. In this paper, we name a question as "recent" if its time gap from being asked to the time that a user wants to contribute to it does not exceed seven days; otherwise, it is labeled as an "archaic" question.

Question Score (QS). It is routine for CQAs to ask users to express their opinions about the quality of a post. The CQA of Stack Overflow provides a voting system (up-votes or down-votes) to its users, to collect their opinions. A question's score is a measure which is found by the deduction of a question's number of down-votes from its number of up-votes. This measure helps a contributor to skip poor quality questions. We label a question "low-scoring" if its score is ≤ 1; otherwise, it is labeled a "high-scoring" question.

State of Having Answers (SA). Some of the CQA users waste a great deal of time to contribute to the questions which are already answered. Prior answers to a question are more likely to be appreciated (through receiving up-votes or being selected as the accepted answers) by the questioner [2]. A CQA user should select from contributing to a question with or without another answer.

State of Acceptance (SAC). It is an uneasy decision for many CQA users to decide whether to contribute to an already answered question or not. Some of the CQAs provide their questioners with a flag option to inform the others if they do not need a further contribution to their questions. In the CQA of Stack Overflow, if the questioner is satisfied by one of the answers received, she can flag that answer as an "accepted answer" to inform the other contributors. By considering this flag option, a user can decide to make a further contribution to that question or not.

Familiarity with Topic (FT). In many of the CQAs, users are demanded to choose related tags (or keywords) to summarize their post briefly. All of the tags that a user has created or is related to are stored in the user's profile. For a contributing user, a question is labeled as "familiar" if there is at least one common tag between the user's profile and that question; otherwise, we label that question as an "unfamiliar" question.

Bounty Prize (BP). A bounty is a special prize which is made up of the reputation scores. In a CQA like Stack Overflow, a questioner can offer a share of its reputation to the best answer contributor of a question. The questions which have bounties are typically more challenging than ordinary questions.

3   HRCR Architecture

HRCR is a question recommendation algorithm with the focus on the churn reduction of the contributing users. HRCR uses a two-fold structure to decide which questions to recommend users first. HRCR first trains a Hidden Markov Model (HMM) to elicit the engagement state of the users. The resolved engagement states of the HMM comply generally with the notion of the Flow theory in psychology. According to the Flow theory, users are highly engaged with their experiences if the challenges they face are in-line with their level of skills [8].

Next, we use a Reinforcement Learning Model (RL) to recommend users the action choices that increase the probability of getting into the Flow Zone with respect to the trained probabilities of the HMM. This way, our HMM and RL models complement each other: HMM by providing the engagement states of the users, and RL by using the HMM predictions for each contribution to update the value of the agent's reward function. We show that the action choices that put the users in the Flow Zone can meaningfully decline the churning probability of the users.

4   Implementation Details

Data Statistics. We use the public access data dump of Stack Overflow in Archive[2] to build our target dataset. Our target data is enclosed within a range from January 1st, 2014 to January 1st, 2017. The target data includes 12,390 contributing users. We have randomly excluded 3,810 users for testing purposes and the remaining data (8,580 users) are applied for training purposes. Table 2 summarizes the important features of our dataset.

---

[2] https://archive.org/details/stackexchange.

Table 2. Data statistics of the CQA of Stack Overflow

Measured statistics	Value
Average contribution of a user	6.93
Variance of user contributions	2.13
Average reputation of a user	481.81
Variance of user reputations	17.52
Total number of users	12,390
Number of churned users	4,821

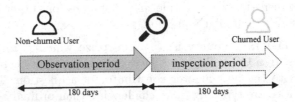

Fig. 2. A user is labeled as churned if her average participation rate within the inspection period is less than 20% of her prior participation rate in the observation period.

Data Preprocessing. In order to study churning behavior of the contributing users, we need a ground truth dataset of churned and non-churned users. Towards this end, we define churned users similar to [11]. We assume that a user has churned if her average rate of participation (of all measurable activities) in an inspection period (which is the immediate 180 days following the observation period) drops to less than 20% of her average participation rate in a prior time that we call the observation period (which is the immediate 180 days after a user's first contribution). Figure 2 is an illustration of the observation and inspection time periods.

Furthermore, we have carried out the following data preprocessing procedures to eliminate the unwanted consequences of noise: (1) The majority of the users of Stack Overflow are ephemeral users and thus churn the CQA before making enough contributions [16]. We thus remove the users who make no contributions and less than 5 posts in sum, from our study. (2) Users with a lower reputation measure (below 10) are sifted out of this study to resolve the problem of imbalanced class labels.

Methodologies. The major application of *Hidden Markov Models* (HMMs) is to render the sequence of observations that have an underlying stochastic process [17]. The HMM is formally defined with a tuple λ as follows:

$$\lambda = (S_{HMM}, V_{HMM}, E, F, \pi) \tag{1}$$

$S_{HMM} = \{s_1, s_2, ..., s_N\}$ is a set of N individual states which are hidden. $V_{HMM} = \{v_1, v_2, ..., v_M\}$ is a set comprised of M possible observations which are measured directly for a state. E is an $N \times N$ matrix to show the transition probabilities of the states. Entries of matrix E are derived from the following equation.

$$E = \{e_{ij}\}; \qquad e_{ij} = P(q_{t+1} = s_j | q_t = s_i) \tag{2}$$

Equation 2 shows the transition probability between states i and j. The variables of q_t and q_{t+1} show the hidden states before and after a new contribution is made respectively.

F is the emission probability of an observation k being generated from a hidden state j. Observation likelihood, F, is formally defined as follows:

$$F = \{f_j(k)\}; \qquad f_j(k) = P(v_k | s_j) \tag{3}$$

Finally, π is to show the probability of the initial states. In this research, we stick to the assumption that all of the hidden states have the same probability at the beginning. The HMM is learnt if only all of the parameters of λ are resolved.

Reinforcement Learning Models (RLs) are applied where an agent attempts to learn the best action policy by interacting with and receiving feedback from its stochastic environment [19]. First, the agent observes its state and then executes an action (or series of actions) which lead to another subsequent state. After getting to the next state, the agent assesses the value of its action according to a *reward function*. The agent considers the punishments and gains it has incurred over time to enhance its action policies iteratively [19]. The RL is formally defined with a tuple v as follows:

$$v = (S_{RL}, V_{RL}, R) \tag{4}$$

S_{RL} refers to the state space of the model and V_{RL} shows the set of all possible actions an RL agent can perform [19]. The feedback function, R, is used to guide an agent toward its goal. The goal of RL agents is to maximize their reward gains within a certain time period. In this research, we use the engagement states inspired by the Flow theory to define R and to recommend users the actions that put them into the Flow Zone.

5   Extraction of User Engagement State

Setup. Inspired by the flow theory measures, we use the user's skill and challenge levels to find out the user's engagement state. We modify a measure by [24], called "Z-score" to determine user skills in the CQA of Stack Overflow.

$$Z_{mod} = \frac{\alpha - \beta}{\sqrt{\alpha + \beta}} \tag{5}$$

Equation 5 shows the modified Z-score. The variables of α and β represent the total number of accepted answers, and the questions asked, respectively. The feature of "competitiveness" by [16], provided us with a basis to develop the measure of the perceived challenge of users through Eq. 6.

$$Challenge = \frac{\gamma}{\prod_{n=1}^{\gamma} Rank(c_n)} \tag{6}$$

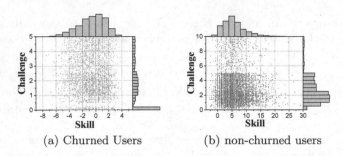

(a) Churned Users   (b) non-churned users

Fig. 3. Marginal skill and challenge distribution of the churned and non-churned

The variables of γ and c_n refer to the total number of answers and the nth contribution of a user, respectively. $Rank(.)$ is a function which returns the total number of contributions received for *a question* divided by the user's position within the contributions (sorted by the answer score). Although more parameters should be included to get a precise measurement of user skill and challenge, we stick to the introduced parameters to avoid complexity. Besides, our skill and challenge plots of churned and non-churned users comply with the psychological implications of the Flow theory. Figures 3a and b show the skill and challenge distributions of the churned and non-churned users respectively. As is shown, users with higher levels of skill and challenge are less likely to churn. There also exists a Pearson correlation of 0.163 with the significance of $P = 0.01$, between the user skill and challenge measures. This implies that the users' perceived level of challenge increases as users' level of skill increases.

We further conduct a Brown-Forsythe test [3] to find if there is a significant difference between skill and challenge levels of the users with different churning attitudes. The results show that the variations of user skill ($P = 0.05$) and challenge ($P = 0.05$) are statistically significant for the churned and non-churned users as we have hypothesized.

Inference of States. We train an HMM λ to infer the engagement state of the users. We feed the input of HMM with the skill and challenge measures of users. As true number of hidden states is typically determined by the Bayesian Information Criterion (BIC) and Akaike Information Criterion (AIC) (see [1,5,12]), we exhaustively search for the best AIC and BIC measures among two to four hidden states. As Fig. 4a shows, the best number of hidden states returned is four which complies with the simple quadrant framework of the Flow theory. Thereby, we name the hidden states of HMM to be $S_{HMM} = \{$Apathy, Anxiety, Boredom, Flow Zone$\}$. We use the Forward-Backward Algorithm by [17] to resolve the entries of the matrices E and F. After running 100 iterations of the algorithm with three different random starts, the Expectation Maximization value converges in the 23rd iteration as Fig. 4b implies. We show the transition probabilities between the hidden states in Fig. 4c. As is shown, the probabilities of the self-loops in all of the states are below half; thus, the users are more likely to change their states over time. However, among all of the hidden states, the

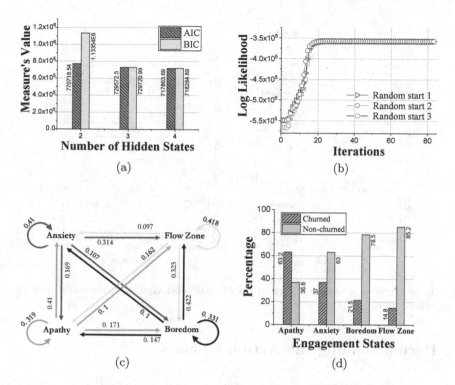

Fig. 4. (a) BIC and AIC measures (b) Convergence of HMM (c) Transition probabilities (d) Churning probabilities

state of Flow Zone has the largest self-loop value. This means that users are more likely to stay in the Flow Zone once they are in it. We notice that the users who are in the Apathy state are not as likely as the users in the Anxiety and Boredom states to enter the Flow Zone. Furthermore, as Fig. 4d depicts, the users who are in the Flow Zone are also less likely to churn the CQA. This makes the Flow Zone a perfect destination to steer users' behaviors to. We exploit this knowledge to devise our Reward function of RL.

Contrary to the Flow Zone, the Apathy state has the largest probability of churning. The Apathy state users are mostly new-comers to the CQA. The users who are in this state are mainly susceptible to entering to the Anxiety state.

We show the emission distributions F, for the skill and challenge measures, in Fig. 5. As is shown, the majority of the users have skills higher than zero, and the perceived challenges are relatively higher in the Flow Zone and the Anxiety state in comparison with the Apathy and Boredom states.

In the rest of this research, we use our built HMM to predict the engagement state of the users after each contribution.

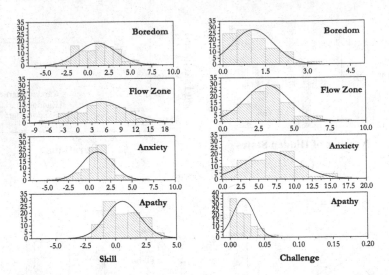

Fig. 5. Continuous emission distribution of skill and challenge measures per each engagement state

6   Recommending the Action Choices

Setup. From all possible combinations between the action choices of Table 1, we can get 144 types of authentic actions for a user to choose from that form our V_{RL} set. We use *1-of-k coding* to model the states of S_{RL} [7]. Each state of S_{RL} is a 144-dimensional column vector that has only one of its entries equal to 1 and the remaining entries are zeros. If we think of a user as an agent, the long-term goal would be to chose among the action choices in such a way to maximize its expected reward value over time. Equation 7 shows the discounted reward function that we use to train the RL.

$$Reward = \sum_{l=1}^{\infty} \delta P_{HMM}(FlowZone|.) \qquad (7)$$

P_{HMM} is the probability of getting into the Flow Zone (after making lth contribution) given the user's current satisfaction status. δ shows the discount factor which is a number between zero and 1 that shows the difference of importance between future and present rewards [19]. In this study, δ is set to 0.4. The inspiration behind our reward function is that as users choose action choices which increase the probability of getting into the Flow Zone, their probability of churning should drop.

Best Action Policy. We assume that an agent κ chooses from the authentic actions V_{RL} in contribution t so as to get a higher payoff. We recommend this choice of action with an RL as follows. We start with the Bellman's optimality equation [19], and initiate the Q-value of the agent κ ($Q_{t=0}^{\kappa}(v_{RL}, s_{RL})$) at $t = 0$

to a positive constant Q_0 for $\forall v_{RL} \in V_{RL}$ and $\forall s_{RL} \in S_{RL}$. We update the value of $Q_{t+1}^{\kappa}(v_{RL}, s_{RL})$ at the end of each contribution through Eq. 8 if s_{RL} is a new action choice; or else, we stick with $Q_{t+1}^{\kappa}(v_{RL}, s_{RL}) = Q_t^{\kappa}(v_{RL}, s_{RL})$.

$$Q_{t+1}^{\kappa}(v_{RL}, s_{RL}) = (1 - \theta)Q_t^{\kappa}(v_{RL}, s_{RL}) + \theta R_t \tag{8}$$

From Eq. 8, R_t is the Reward function which is attained through Eq. 7 at the end of each contribution. θ indicates the learning rate which is a number between zero and 1. In this research, we set $\theta = 0.2$. In the beginning of each contribution, the agent κ selects an action through $argmax_{s \in S_{RL}} Q_t^{\kappa}(v_{RL}, s_{RL})$ and running a majority vote over the possible actions. And if the results are tied, one action is randomly selected. Throughout the learning procedure, the agent uses the $\varepsilon - greedy$ approach within each time step to explore the new types of actions. In this study, the entries of the RL model (v) are learnt with $\varepsilon = 0.01$. Since our model applies a tabular version of Q-learning Algorithm (rather than the version which applies an approximation function), the solution should deterministically converge to an optimal policy [22]. We test ten random seeds, each with 5000 rounds of iterations to resolve the model parameters. It takes an average of 2230 iterations for each seed to converge.

7   Verification of Churn Reduction

The empirical evaluation of the personalized recommending systems appears to be challenging for many researchers who study the popular CQAs that do not provide direct control options. Hence, we use an off-line churn prediction model as a baseline to see if following the HRCR's recommended questions will help to reduce the churning probability of the contributing users.

Baseline Churn Predictor. We build a probabilistic classification model (i.e., a Logistic Regression (LR)) to predict the churning probability of each user over time as she contributes to answering the questions. We use the churn prediction features of [16] as the evidential features for predicting churn. As Fig. 6a summarizes, our baseline LR retrieves the Accuracy of 84% and the F1-score of 78% when it is applied to the test data. We get the Area Under Curve (AUC) value of 96%.

Comparison of the Churning Probabilities. Next we compare the average churn probability of the users based on the average similarity of their actions to the actions that are recommended by our HRCR algorithm over time. From Table 1, we know that there are 7 varying factors that the users can decide on before contributing to a question. At each contribution time, if the user totally complies with the HRCR's recommended action factors, the user gains the similarity value of 7. If there are no common factors between the user's action factors and the ones that the HRCR recommends, the user gets the similarity value of 0; thus, it is apparent that the similarity value should range between 0 and 7 based on the number of common activity factors between the HRCR's action recommendations and the user performance. Figure 6b shows the average

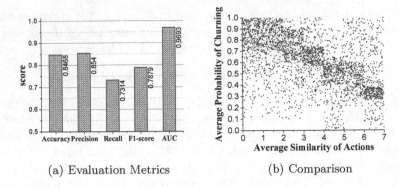

      (a) Evaluation Metrics   (b) Comparison

Fig. 6. (a) Performance of the base churn predictor (b) Users' churning probability for different ranges of similarity with the HRCR's recommended actions

similarity of actions along with the average churning probability of each user over time. We observe a decreasing trend of the average churning probability of the users over time as they comply with the HRCR's action recommendations.

Hypothesis Testing. We also run a statistical Brown-Forsythe test to see the effect of action similarities on the churning behavior of the users. The results show that the users whose actions share more than a 50% similarity with the HRCR's recommended action choices over time, have a significantly different churning probability ($P = 0.05$) from those who share less similarities. Thus the hypothesis that the users who share more similarity with our proposed model are less likely to churn is achieved.

8   Conclusion and Future Work

Takeaway Message. Flow theory implies that the users who get into the Flow Zone might be less likely to churn [8]. Our results comply with the Flow theory. HRCR also fills the gap of question recommendations to the contributing users by considering their tendency for churning. We observe that the churning probability of the users in the engagement state of the Flow Zone is respectively 22.2%, 48.4%, and 6.7% less than the churning probability in the engagement states of Anxiety, Apathy, and Boredom. We also observe that there exists approximately 42% difference between the average churning probability of those who do not follow the HRCR's recommendations at all and those who happen to contribute to the CQA fully in accordance to what the HRCR recommends.

Limitations and Future Work. In this research we investigated the users who have made at least more than or equal to 5 contributions and a reputation score above 10. Since these users are contributing more often to the CQA forums, reducing their churning probability was of higher interest for us in comparison with the ephemeral users. Besides, the ephemeral users make the dataset intensively imbalanced. Hence, our work is limited by the users we have studied.

Although there exists various versions of the flow theory [14], we stick with the quadrant framework of the flow theory for two main reasons: First, it matches with the best number of hidden states we retrieve for our HMM. In addition, if we increase the number of engagement states, the emission distributions of skills and challenges would become very similar to one another, and the efficiency of the predictions will drop. Our second reason is simplicity of the model and ease of understanding. Nevertheless, exploring the other variations of the flow theory remains as future work.

In this paper, we used the concept of Flow theory to recommend users better questions. We introduced an innovative two-fold algorithm (we called HRCR) using a *Hidden Markov Model* and a *Reinforcement Learning Model* to personalize the question recommendations on the CQAs. Experiment results on the popular CQA of Stack Overflow proves HRCR helpful in reducing churn.

Acknowledgement. This research has been supported in part by project 16214817 from the Research Grants Council of Hong Kong and the 5GEAR project from the Academy of Finland. We would like to also thank our reviewers and Mr. Young D. Kwon for their valuable comments.

References

1. Akaike, H.: A new look at the statistical model identification. IEEE Trans. Autom. Control **19**(6), 716–723 (1974)
2. Bosu, A., Corley, C.S., Heaton, D., Chatterji, D., Carver, J.C., Kraft, N.A.: Building reputation in stackoverflow: an empirical investigation. In: 2013 10th Working Conference on Mining Software Repositories (MSR), pp. 89–92. IEEE (2013)
3. Brown, M.B., Forsythe, A.B.: Robust tests for the equality of variances. J. Am. Stat. Assoc. **69**(346), 364–367 (1974)
4. Chang, S., Pal, A.: Routing questions for collaborative answering in community question answering. In: 2013 IEEE/ACM International Conference on Advances in Social Networks Analysis and Mining (ASONAM 2013), pp. 494–501. IEEE (2013)
5. Chen, S.S., Gopinath, R.A.: Model selection in acoustic modeling. In: Sixth European Conference on Speech Communication and Technology (1999)
6. Choi, S., Ha, H., Hwang, U., Kim, C., Ha, J.W., Yoon, S.: Reinforcement learning based recommender system using biclustering technique. arXiv preprint arXiv:1801.05532 (2018)
7. Christopher, M.B.: Pattern Recognition and Machine Learning, vol. 1. Springer, New York (2016)
8. Csikszentmihalyi, M.: Beyond Boredom and Anxiety. Josseybass, San Francisco. Well-Being: The Foundations of Hedonic Psychology, pp. 134–154 (1975)
9. Dror, G., Koren, Y., Maarek, Y., Szpektor, I.: I want to answer; who has a question?: Yahoo! answers recommender system. In: Proceedings of the 17th ACM SIGKDD International Conference on Knowledge Discovery and Data Mining, pp. 1109–1117. ACM (2011)
10. Dror, G., Pelleg, D., Rokhlenko, O., Szpektor, I.: Churn prediction in new users of Yahoo! answers. In: Proceedings of the 21st International Conference on World Wide Web, pp. 829–834. ACM (2012)

11. Karnstedt, M., Hennessy, T., Chan, J., Hayes, C.: Churn in social networks: a discussion boards case study. In: 2010 IEEE Second International Conference on Social Computing, pp. 233–240. IEEE (2010)
12. Li, C., Biswas, G.: A Bayesian approach to temporal data clustering using hidden Markov models. In: ICML, pp. 543–550 (2000)
13. Movshovitz-Attias, D., Movshovitz-Attias, Y., Steenkiste, P., Faloutsos, C.: Analysis of the reputation system and user contributions on a question answering website: stackoverflow. In: Proceedings of the 2013 IEEE/ACM International Conference on Advances in Social Networks Analysis and Mining, pp. 886–893. ACM (2013)
14. Nakamura, J., Csikszentmihalyi, M.: Flow theory and research. In: Handbook of Positive Psychology, pp. 195–206 (2009)
15. Oliveira, N., Muller, M., Andrade, N., Reinecke, K.: The exchange in stackexchange: divergences between stack overflow and its culturally diverse participants. Proc. ACM Hum.-Comput. Interact. 2(CSCW), 130 (2018)
16. Pudipeddi, J.S., Akoglu, L., Tong, H.: User churn in focused question answering sites: characterizations and prediction. In: Proceedings of the 23rd International Conference on World Wide Web, pp. 469–474. ACM (2014)
17. Rabiner, L.R., Juang, B.H.: An introduction to hidden Markov models. IEEE ASSP Mag. 3(1), 4–16 (1986)
18. San Pedro, J., Karatzoglou, A.: Question recommendation for collaborative question answering systems with RankSLDA. In: Proceedings of the 8th ACM Conference on Recommender Systems, pp. 193–200. ACM (2014)
19. Sutton, R.S., Barto, A.G.: Reinforcement Learning: An Introduction, vol. 1. MIT Press, Cambridge (1998)
20. Tang, X., Chen, Y., Li, X., Liu, J., Ying, Z.: A reinforcement learning approach to personalized learning recommendation systems. Br. J. Math. Stat. Psychol. 72(1), 108–135 (2019)
21. Trimponias, G., Ma, X., Yang, Q.: Rating worker skills and task strains in collaborative crowd computing: a competitive perspective. In: The World Wide Web Conference, pp. 1853–1863. ACM (2019)
22. Watkins, C.J., Dayan, P.: Q-learning. Mach. Learn. 8(3–4), 279–292 (1992)
23. Yuan, S., Zhang, Y., Tang, J., Hall, W., Cabotà, J.B.: Expert finding in community question answering: a review. Artif. Intell. Rev. 52, 1–32 (2019)
24. Zhang, J., Ackerman, M.S., Adamic, L.: Expertise networks in online communities: structure and algorithms. In: Proceedings of the 16th International Conference on World Wide Web, pp. 221–230. ACM (2007)
25. Zheng, L., Chen, L.: Mutual benefit aware task assignment in a bipartite labor market. In: 2016 IEEE 32nd International Conference on Data Engineering (ICDE). pp. 73–84. IEEE (2016)

Boosting Variational Generative Model via Condition Enhancing and Lexical-Editing

Zhengwei Tao(ID), Waiman Si, Juntao Li, Dongyan Zhao, and Rui Yan(✉)

Institute of Computer Science and Technology, Peking University, Beijing, China
{tttzw,lijuntao,zhaody,ruiyan}@pku.edu.cn
rs4wsi@gmail.com

Abstract. Conditional Variational Autoencoder (CVAE) has shown promising performance in text generation. However, CVAE is inadequate to generate sentences that are highly coherent to its condition due to error accumulation in decoding and KL-vanishing problem. In this paper, we propose an Edit-CVAE (ECVAE) in which we attempt to exploit information-related data to address the problem by (1) explicitly editing the generated sentence. (2) enriching the latent representation. While maintaining the diversity and information consistency. Experiment results on dialogue and Chinese poetry generation show that our method substantially increases generative coherence while maintaining the diversity and information consistency.

Keywords: Natural language processing ·
Open-domain conversation system · Variational inference

1 Introduction

Traditional neural network models [29, 34] have been applied in text generation for a long while. Recently, Variational Autoencoder (VAE) manifests great performance by introducing multivariate Gaussian or multi-way categorical latent variable [10, 12], which facilitates better generation in wording diversity and one-to-many diversity [5, 10]. CVAE, an extension of VAE, can generate sentences under certain attributes while maintaining the merits of VAE [4, 7–9]. However, despite its advantages, CVAE also presents some shortcomings. One salient problem is generative incoherence which is the incoherence between a generated sentence and its corresponding condition. Experimental results show that the generative coherence of CVAE is weaker than basic text generation models. We evaluate the generative coherence on both CVAE and Seq2Seq model with attention (AS2S) [29]. Our results shows that CVAE performs worse than AS2S on a large scale, see Table 2. The results show that CVAE incompetently generates coherence sentences.

Causes of this issue are two-fold: (1) errors in the generation can be accumulated, and the latent information would diminish as decoding forward [16, 31].

© Springer Nature Switzerland AG 2019
A. C. Nayak and A. Sharma (Eds.): PRICAI 2019, LNAI 11670, pp. 377–391, 2019.
https://doi.org/10.1007/978-3-030-29908-8_30

(2) CVAE tends to ignore the latent variable in order to reduce the optimization loss (aka. KL-vanishing) [11,13]. In order to solve the problem of incoherence generation, existing works attempt to enrich the capacity of the latent representation [11,13], augment the ability to extract the latent information of the decoder [14,15] and use new decoding process [16,33]. Although these methods aren't directly designed to solve the incoherence problem, they could alleviate it to some degree. However, these methods rely heavily on single data point itself which can be noisy and incapable. CVAE benefits from variational uncertainty, but this uncertainty also hurts CVAE in turn by interfering data information by the introduced bias. Under such circumstances, leveraging data itself may seem insufficient to solve this dilemma. Also, implicitly modifying latent representation could be inefficient to mitigate the problem due to the uncontrollability and unobservability.

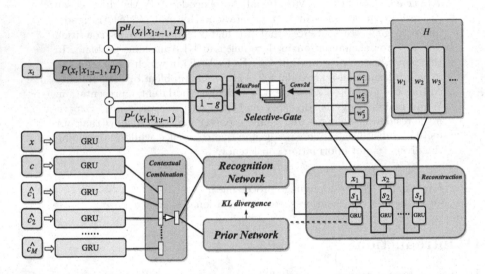

Fig. 1. Overview of ECVAE. Recognition network computes posterior and Prior network computes prior. Both networks utilize retrived data. Dashed lines represent inference.

To this end, we attempt to address the problem from different perspectives. Intuitively, utilizing additional similar data provides more information than relying on current data itself [22,26]. Similar data provides more comprehensive information of data distribution and data structure. To learn an abundant latent representation, we expect that fusing extra information-related data with original data should boost the generative coherence owing to enhancing similar information and reducing the bias. Besides, considering implicitly using extra data to enhance the latent representation may not enough in some cases, we propose to explicitly edit the generated sentence based on the extra information-related data to improve generative coherence, which effects together with implicitly modification. It allows to alleviate error accumulation and rewrite the words when the generation diverges from the correct trajectory. To the best of our knowledge, we are the first to combine editing strategy with variational inference model on text generation to boost generation coherence. We conduct experiments on two challenging text generation tasks with different languages and requisitions (genres), i.e., neural conversation generation and Chinese poetry generation, to evaluate the performance of our proposed model. Specifically, two large open datasets, STC and OpenSubtitle are used for training conversation systems, and a large open Chinese poetry corpus is utilized for comparison. Experimental results show that our proposed model mitigates the generation incoherence problem of CVAE and improves the performance of conversation generation over several baselines on both human evaluations and automatic metrics, such as RUBER, and distinct scores. Also, our model achieves substantial improvements on Chinese poetry generation on both quantitative and qualitative studies. Such results hint that our proposed model can effectively improve the performance of variational inference models regarding generation coherence.

2   Preliminaries

2.1   VAE and CVAE

In general, VAE consists of an encoder and a decoder, which relate to the encoding process where input x is mapped to a latent variable z, and the decoding process where the latent variable z is reconstructed into the original x. Formally, the encoding process computes the posterior distribution $q_\theta(z|x)$ while the decoding process can be formulated as $p_\theta(x|z)$ regarding as the condition distribution of input x conditioned on z. In VAE, we set z to multivariate Gaussian distribution i.e. $p_\theta(z) \sim N(\mu, \sigma)$. Here θ denotes the parameters of encoder and decoder. VAE supposes to maximize the distribution of the data x, i.e. $p_\theta(x)$. However, as represented in [1], consider large datasets and intractable integral of marginal likelihood $p(x)$, the true posterior $q_\theta(z|x)$ is simulated by a variational approximation $q_\phi(z|x)$ in modeling the encoding process, where ϕ represents the parameters for q.

The learning objective of VAE is transferred to maximize the log-likelihood $logp_\theta(x)$ over input x. Since intractability of the marginal distribution, to facilitate learning, one can target on pushing up the evidence lower bound of $logp_\theta(x)$ (ELBO):

$$ELBO(x; \theta, \phi) = E_{q_\phi(z|x)}[logp_\theta(x|z)] - KL(q_\phi(z|x)||p_\theta(z)) \qquad (1)$$

Alternatively, we maximize $ELBO(x; \theta, \phi)$ instead of $logp_\theta(x)$. In (1), the KL-divergence term $KL(\cdot)$ regards as the regularization for stimulating the approximated posterior $q_\phi(z|x)$ to be close to the prior $p_\theta(z)$. The $E[\cdot]$ is the expectation conditioned on the approximation prosterior $q_\phi(z|x)$.

CVAE extends VAE with an extra condition c, which is for supervision of the exact condition. In text generation, the condition c usually represents for sentiment, persona, politeness, dialogue act, and sentence function, etc. The objective of CVAE is thus to maximize the reconstruction log-likelihood of the input x under the condition of c. Thus, for CVAE, we have the ELBO formulated as

$$ELBO(x, c; \theta, \phi) = E_{q_\phi(z|x,c)}[logp_\theta(x|z,c)] - KL(q_\phi(z|x,c)||p_\theta(z|c)) \quad (2)$$

All items in ELBO of CVAE are conditioned on c. In testing mode, given a condition c, we first sample z according to $p_\theta(z|c)$, then use z to generate a sentence by utilizing $p_\theta(x|z)$.

2.2 Incoherent Generation in CVAE

CVAE tends to generate sentences which are incoherent to its condition, see results in Table 2, where the generation consistence of CVAE is inferior to some basic model (Seq2Seq with attention). It could be caused by two reasons:

KL Divergence Vanishing. During the training phase, the KL-divergence loss collapses quickly to zero [11,12], which degenerates the CVAE to autoencoder. This phenomenon incurs that the decoder would discard the latent variable and generate by only utilizing the language model causing to generating incoherent sentences.

Decoding Error Accumulation. When we use Recurrent Neural Network (RNN) to decode, the generative error will accumulate as encoding previously predicted words in the vector representations. This issue is caused by over-confidence problem in text generation as indicated in [16]. Information from z will become weaker as the decoding step increases. Besides, the training objective leads to this issue as well [31] and the generation would deviate from the correct trajectory if error accumulation occurs.

3   The Model

Suppose that we have a dataset D. Each $d, d \in D$, contains a variable-length input sequence $x = \{x_1, ..., x_{T_x}\}$ and its condition sequence $c = \{c_1, ..., c_{T_c}\}$, i.e. $d = (x, c)$. The task of text generation is to predict x given its condition c. Our proposed framework is illustrated in Fig. 1. It consists of two parts, contextual combination and editing generation, where their details are elaborated in the following subsections.

3.1   Contextual Combination

Given $d = (x, c)$, we attempt to obtain M information-related data $\hat{d}$ from the same corpus according to word-overlap and rerank the retrieved data by BM25 score via Whoosh[1]. For a data (x, c), we retrieve data $\hat{d} = (\hat{x}, \hat{c})$ whose $\hat{x}$ has the highest word-overlap with x. Then we compute the BM25 scores between x and

---

[1] https://whoosh.readthedocs.io.

each $\hat{x}$ and choose the highest M ones. These $\hat{d}$ contain similar context, semantic and syntactic information with d. To extract the information contains in a (x, c), we use bidirectional Gated Recurrent Units (GRU) as our encoder. Specifically, to encode x, for each word in x, the bidirectional GRU obtains forward hidden state $\overrightarrow{h_t^x}$ and backward hidden state $\overleftarrow{h_t^x}$, where $\overrightarrow{h_t^x}=$GRU $(x_t, \overrightarrow{h_{t-1}^x}); \overleftarrow{h_t^x}=$GRU $(x_t, \overleftarrow{h_{t-1}^x})$. The semantic of the i-word is represented by $h_i^x = [\overrightarrow{h_i^x}; \overleftarrow{h_i^x}]$, where ; is vector concatenation. We consider $h_{T_x}^x$ as the representation of the input x. We can utilize the same encoder to acquire sentence representation $h_{T_c}^c$ for condition sentence c. It is the same with $\hat{c}$ where we acquire $h_{T_i}^{\hat{c}_i}$ for each $\hat{c}_i, i = 1, ..., M$, T_i is the length of $\hat{c}_i$.

Considering that the noise introduced by variational inference would interfere the encoding information of c, more similar data would enhance the context information and denoise if we ensemble these data channel of $\hat{d}$. We achieve this by concatenating:

$$h^{\mathbf{c}} = [h_T^c; h_{T_1}^{\hat{c}_1}; ...; h_{T_M}^{\hat{c}_M}] \tag{3}$$

We consider $\mathbf{c}$ as ensembled condition and regard $h^{\mathbf{c}}$ as ensembled condition representation. We denote this operation as *contextual combination*.

In order to compute the posterior distribution $q_\phi(z|x, \hat{x}, \mathbf{c})$, we assume $q_\phi(z|x, \mathbf{c}) \sim N(\mu_q, \sigma_q I)$, μ and σ are the key parameters to be learned, and they are computed by

$$\begin{bmatrix} \mu_q \\ log(\sigma_q^2) \end{bmatrix} = \text{MLP}^q([h_{T_x}^x; h^{\mathbf{c}}]) \tag{4}$$

where MLP^q is the fully-connected neural network for posterior. Similarly, the prior $p_\theta(z|\mathbf{c})$ is formulated as another multivariate Gaussian distribution $N(\mu_p, \sigma_p I)$. μ_p, σ_p can be calculated by another multi-layer perception MLP^p.

$$\begin{bmatrix} \mu_p \\ log(\sigma_p^2) \end{bmatrix} = \text{MLP}^p(h^{\mathbf{c}}) \tag{5}$$

After acquiring the prior and posterior distribution of latent variable z, we can compute the KL-divergence in (6) as the following.

$$\widehat{ELBO}(x, \mathbf{c}; \theta, \phi) = E_{q_\phi(z|x,\mathbf{c})}[logp_\theta(x|z, \mathbf{c})] - KL(q_\phi(z|x, \mathbf{c})||p_\theta(z|\mathbf{c})) \tag{6}$$

3.2 Editing Generation

contextual combination process already have fused information-related data into the encoding. However, this combination seems implicit, which hinders the similar data from effecting better and lacks interpretability. Based on that, we design a novel reconstruction process to leverage on the same retrieved data more explicitly to enhance the generation consistence of CVAE, which effects together with *contextual combination*. We now elaborate the reconstruction process with our proposed editing mechanism.

The decoder of ECVAE is also a GRU. During training, we sample z from posterior $q_\phi(z|x, \hat{x}, \mathbf{c})$ and take it as the initial hidden state of the decoder. At each step, the decoder computes the hidden state based on the last output and last hidden: $s_t = \text{GRU}(x_{t-1}, s_{t-1})$. CVAE generates a word x_t according to decoder's language model(GRU) $P^L(x_t|x_{1:t-1}) = \text{Softmax}(s_t^T W^L + b^L)$, where $W^L \in R^{d_s \times V}$ is the projection matrix, $P^L \in R^V$ is the generation probability, and d_s is the dimension of the decoder hidden state and V is the total vocab size. $\text{Softmax}(x)_i = \frac{\exp x_i}{\Sigma \exp x_j}$ for an arbitrary vector x.

Considering that (1) decoding error may accumulate in the common generation process (2) variational noise may still interfere with the encoding process in some case even we use *contextual combination*. We wish to generate/copy a word from the retrieved data when at each step of reconstruction step when it needs. We design our model that can generate/copy a word from $(\hat{x}_1, ..., \hat{x}_M, \hat{c}_1, ..., \hat{c}_M)$ at each decoding step as another way to leverage the retrieved data.

We first look up to vocabulary and pick the word embeddings of these words. We concatenate all word embeddings of $(\hat{x}_1, ..., \hat{x}_M, \hat{c}_1, ..., \hat{c}_M)$ and, w.l.o.g we denote $H = (w_1, w_2, ...w_n)$, $H \in R^{d_w \times n}$, where n is the total number of items in H, d_w is the dimension of word embedding. H is the retrieval bank in which stores the similar information with (x, c). Given s_t, we compute the response of each item in H via a bilinear function and obtain the distribution according to the response by softmax activation. Then we sum up all items in H according to the distribution and get the contextual vector of H conditioned on s_t.

$$v = H \cdot \text{Softmax}(H^T W_1^H s_t + b^H)$$
$$\hat{P}^H = v^T W_2^H \tag{7}$$

where $W_1 \in R^{d_w \times d_s}$ and $W_2 \in R^{d_w \times v}$ is the projection matrix, $\hat{P}^H \in R_V$. In order to maintain the generation consistence with (x, c), we only copy words in $d = (x, c)$. We mask the probabilities of words that do not show up in d and use the contextual vector to compute the copy probability:

$$mask_i = \begin{cases} 0, & word_i \in d \\ -\infty, & otherwise \end{cases}$$
$$P^H(x_t|H) = \text{Softmax}(\hat{P}^H + mask) \tag{8}$$

After acquiring language model generating distribution $P^L(x_t|x_{1:t-1})$ and copying distribution $P^H(x_t|H)$, we compute the genuine generating $P(x_t|x_{1:t-1}, H)$. Similar to copy mechanism [24,25], we utilize a selective gate to determine the ratio between two distributions, which depends on the history generation $x_{1:t-1}$. On the one hand, if history $x_{1:t-1}$ matches well with condition c, it is reasonable to use the language model to generate; otherwise, it should copy a word from H to maintain the consistency with c.

We first look up the embedding table to get word embedding $C = [w_1^c; ...; w_{T_c}^c] \in R^{d_w \times T_c}$ and $X_{1:t-1} = [w_1^x; ...; w_{t-1}^x] \in R^{d_w \times (t-1)}$, w_i^c is the word embedding of c_i, T_c is the length of c. Then we compute semantic alignment between $x_{1:t-1}$ and c.

Table 1. Statistics of STC, OST and Poetry dataset.

	Train	Valid	Test	Vocab
STC	500,000	5,000	5,000	30,000
OST	500,000	10,000	10,000	30,000
Poetry	97,367	1,000	1,000	8461

$$A = X_{1:t-1}^T C \tag{9}$$

Then we utilize a convolutional neural network (CNN) to learn the matching score between $x_{1:t-1}$ and c, where the score is interpreted as the degree of semantic consistency.

$$F = \text{Conv2d}(A)$$
$$K = \text{MaxPool}(F)$$
$$g = \text{MLP}(\text{flat}(K)) \tag{10}$$

Conv2d is 2-dimensional convolution operation to the input, MaxPool takes the maximum value of the specific size of regions from the output of Conv2d, and flat is flatten operation which converts a matrix to a vector by concatenate all the columns. After acquiring flattened vector K, we compute the selective-gate g via an MLP.

Finally we acquire the genuine generating distribution $P(x_t|x_{1:t-1}, H)$ by:

$$P(x_t|x_{1:t-1}, H) = g \times P^L(x_t|x_{1:t-1}) + (1 - g) \times P^H(x_t|H) \tag{11}$$

4   Experiments

We conduct experiments on neural conversation and neural poetry generation. In neural conversation, c and x represent query and response respectively, but only condition c would be provided in the testing phase. For poetry generation, in the first iteration, the title and the first line of the poetry represent c and x respectively, and then the first and second line would be the c and x in the second iteration and such on. In the testing phase, only the title c would be supplied. We generate the first line given the title. And then generate the second sentence conditioned on the previously generated one.

4.1   Datesets

For dialogue generation we use two datasets. STC dataset (short text conversation) [28] and OpenSubTitle denoted as OST[2]. The original STC dataset consists of 4M dialogue pairs (query and response). We sample 0.5M, 5k, 5k of it for training validation and testing respectively. For OST dataset, we sample 0.5M, 10k, 10k for training, validation and testing.

---

[2] OST is clollected from www.opensubtitles.org.

Table 2. Automatic evaluations on STC dialogue generation dataset. Bold number represents the best result. We estimate the lower bound as the standard to measure the absolute value on the specific dataset by randomly choose a response for each query and calculate **Embedding Similarity**. The lower bound of each sub-metric: avg: 0.518, ext: 0.285 and grd: 0.322.

	RUBER					Diversity		Embedding similarity		
	Q-R	min	max	avg.G	avg.A	dist.1	dist.2	avg	ext	grd
AS2S	0.623	0.573	0.879	0.726	0.676	0.078	0.253	0.531	0.334	**0.385**
ESM	0.631	0.594	0.698	0.746	0.698	0.076	0.256	0.544	**0.345**	0.384
CVAE	0.552	0.536	0.914	0.725	0.655	**0.128**	0.523	0.560	0.333	0.359
ECVAE[†]	0.580	0.556	0.893	0.724	0.670	0.123	0.516	**0.567**	0.343	0.367
ECVAE	**0.643**	**0.615**	**0.918**	**0.766**	**0.717**	0.127	**0.553**	0.564	0.343	0.366

Table 3. Automatic evaluations on OPS dialogue generation dataset. Bold number represents best result. The lower bound of each metric is avg: 0.905, ext: 0.512 and grd: 0.710

	RUBER					Diversity		Embedding similarity		
	Q-R	min	max	avg.G	avg.A	dist.1	dist.2	avg	ext	grd
AS2S	0.650	0.635	**0.932**	0.741	0.784	0.041	0.166	0.912	**0.561**	**0.744**
ESM	0.688	0.662	0.914	0.755	0.788	0.038	0.148	0.914	0.560	0.743
CVAE	0.570	0.545	0.872	0.658	0.709	0.054	**0.291**	0.915	0.539	0.731
ECVAE[†]	0.590	0.575	0.905	0.690	0.740	**0.054**	0.289	0.915	0.544	0.734
ECVAE	**0.768**	**0.732**	0.927	**0.811**	**0.830**	0.052	0.286	**0.918**	0.549	0.739

For poetry generation, we collect a corpus contains a collection of Tang dynasty poetry and Song dynasty denoted as Poetry[3]. We filter out poetry that does not obey the genres which are quatrain and eight-line with 5 or 7 characters in each line. We randomly sample 1k for validation, 1k for testing and use rest fro train which contains 97367 items. Table 1 contains detail statistic information about all datasets including vocabulary size.

4.2 Baselines

We compare the proposed ECVAE with the following baseline models:

AS2S: Sequence-to-sequence model with attention [29] which is a prevailing framework in text generation.

CVAE: Conditional Variational Autoencoder for generating responses [6] which is our basic model. We use the same KL annealing strategy [10] and bag-of-word loss [9] for both baseline *CVAE* and our model.

ESM: Ensemble model which utilizes both retrieval-based dialogue model and generative dialogue model [26]. Similar to our model, ESM first retrieves

---

[3] Poetry is from https://github.com/chinese-poetry/chinese-poetry.

semantic-related query from the same training corpus and then generate the response based on two queries. For a fair comparison, we use the same retrieve engine for both our model and ESM, and we don't conduct re-ranking strategy in [26] since this approach takes effect for both models.

ECVAE[†]: Conditional Variational Autoencoder with our proposed editing mechanism which is our overall model without the *contextual combination*.

4.3 Experimental Setting

We implement all the models in PyTorch 0.4.0 with following hyperparameters:

We use single layer bidirectional GRU as the encoder and its hidden state sizes are set to 300. The decoder is GRU as well with the size of the hidden state set to 400. The word embedding has size 200. For VAEs model, we set the latent variable vector size to be 200. In all experiments, all the initial weights are sampled from a uniform distribution $[-0.08, 0.08]$. We use Adam optimizer with a learning rate of 0.001 and gradient clipping at 5. For selective-gate, we use 2×2 window size, 1 strip and 15 channels in the convolution network and 2×2 window size, 2 strip in max pooling operation.

Table 4. Automatic evaluations on Poetry dataset. Bold number represents the best result.

	RUBER	Diversity	
	Q-R	dist.1	dist.2
AS2S	0.795	0.054	0.251
ESM	0.804	0.057	0.321
CVAE	0.748	0.085	**0.651**
ECVAE[†]	0.762	**0.093**	0.608
ECVAE	**0.811**	0.088	0.605

4.4 Evaluation Metrics

To comprehensively evaluate the quality of generation from the different perspective, we employ metrics (we don't use well-known metric **BLEU** here since **Embedding Similarity** already considers the word-overlap in vector level.):

RUBER: RUBER (Referenced metric and Unreferenced metric Blended Evaluation) is a metric for evaluating the generative coherence and embedding similarity [30][4]. To generally evaluate the response quality based on referenced and unreferenced metrics mentioned above, RUBER provides four blended metrics: min and max calculates the minimum and maximum value; avg.A and

---

[4] We use **RUBER** from https://github.com/liming-vie/RUBER.

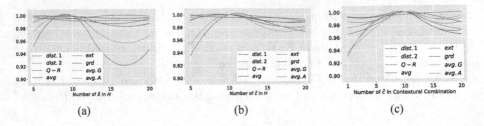

Fig. 2. Quantitative Analysis. Both (a) (b) are on 10 retrieved $\hat{c}$ in *contextual combination*. (a) contains 10 retrieved $\hat{c}$ in H and varies in number of $\hat{x}$ while (b) contains 10 retrieved $\hat{x}$ in H and varies in number of $\hat{c}$. (c) contains 10 $\hat{x}$ and 10 $\hat{c}$ in H while varies in number of $\hat{c}$ in *contextual combination*. y axis in all figures are scores normalized in $[0, 1]$, x axis is the number of sentences.

avg.G computes arithmetic and geometric mean. Also, we denote Q-R as the matching score to evaluate the relatedness between a generated response and its query/condition, which comes from a trained neural model. All metrics range from 0 to 1 demonstrate uncorrelated to correlated. This metric shows high correlation with human annotation.

Diversity: Distinct is a widely used metric to evaluate the diversity of generated sentence [31]. We use dist.1 and dist.2 to evaluate ability to generate distinct response. Dist.1 and Dist.2 represent the proportion of unique uni-grams and bi-grams in the generated result.

Embedding Similarity: Embedding similarity computes the cosine similarity between the ground-truth and generated sentence embedding. avg calculates the average of word embeddings in a sentence. ext takes the most extreme value of each dimension in word embeddings. grd finds the most similar word in two sentences using cosine similarity. We use Glove[5] to train the word embedding on corresponding datasets.

Human Evaluation: We invite three well-educated research students to rate the generated response by each model. We sample 100 test data of STC and poetry and use each model to generate the best responses. All generations are presented to evaluators simultaneously. The evaluators are asked to rate according to (1) coherence and relevance to query, (2) fluency and (3) informative. For each case, we ask each evaluator to pick out models outperform in these three aspects respectively. The picked out model would get 1 score in the that aspect. So we would obtain scores of all model in each aspect from each evaluator. We consider these three aspects are equally important, therefore, we average scores of three aspects. We present the average score of all evaluators for each model. The higher score means better performance.

---

[5] https://github.com/stanfordnlp/GloVe.

Table 5. Human evaluation result on STC and Poetry.

	STC	Poetry
AS2S	0.224	0.206
ESM	0.213	0.224
CVAE	0.139	0.189
ECVAE†	0.176	0.120
ECVAE	**0.245**	**0.258**

5 Results

We report our neural conversation generation results STC in Table 2 and OPS in Table 3, and neural poetry generation results in Table 4. In general, our proposed model outperforms all baselines in all three tasks, but the result also illustrates distinct distribution and characteristic of datasets. We conduct quantitative analysis to show the effectiveness of each component in our model and qualitative analysis to analyze generated case by our model.

5.1 Quantitative Analysis

Coherence. From the results, our ECVAE† has better RUBER scores than conventional CVAE especially in the Q-R score which indicates the effectiveness of the editing process. Building upon it, the proposed ECVAE achieves highest RUBER scores in all tasks as well, and it confirms our expectation that generating sentences with contextual combination and editing mechanism can improve the generative coherence significantly.

Diversity. Our ECVAE† achieves comparable results in diversity against CVAE. Interestingly, we note that ECVAE† excels the dist.1 of all models in OPS and Poetry. One reason for that is the editing mechanism forcing the model to use various words during the inference. Besides, ECVAE can generate sentences with more unique bi-gram ($Dist.2$) words compared to baselines.

Consistency. We can see that our ECVAE performs better in the avg of Embedding Similarity than baselines in both dialogue datasets, which is what we emphasize the most since it reflects the similarity between the ground truth and predicted sentence. It proofs that our model maintains generative consistency comparing to all baselines.

Contextual Combination. We fix 10 $\hat{x}$ and 10 $\hat{c}$ in H while altering the number of $\hat{c}$ for *contextual combination*. Results in Fig. 2(c) show that around 9 $\hat{c}$ for *contextual combination* are the best. More than 9 $\hat{c}$ are detrimental owing to the similarity of the retrieved data because excessive data that is unrelated to c would bring more noise in the encoder.

Selective Gate. We fix 10 retrieved $\hat{c}$ both in *contextual combination* and H while varying the number of $\hat{x}$ in H. From Fig. 2(a) we can see that 8 $\hat{x}$ in H

Table 6. Case study on STC and Poetry of each model. Underline denotes ECVAE copies words from its condition to enhance coherence.

condition	STC.1:<u>热火</u>vs步行者季后赛对决 2分36秒<u>热火</u> 85 比84 决胜时刻 It's in final 2 minutes 36seconds play-offs between Heat and Pacers.	STC.3:狮子座女生一直这么霸气 Leo girl is very strong.	Poetry.1:花朝溪上, 有愚普游其二二 On the river near flowers, a man recalls traveling time in the past.
	STC.2:柚子 吃得晚了吃得特别香 It's better to eat grapefruit later.	STC.4:有时候眼见 未必是现实 What the eye sees is not necessarily true.	
AS2S	STC.1:这是在挑战极限啊 It's a challenge of the of the limit.	STC.3:这是狮子座的狮子座 It's a Leo Leo.	Poetry.1:白鹭东风作色来,到头唯有水澄鲜 天末凉生伊陇鹤,老师犹放小参禅 Bird come with the east wind and the water is clean. Birds born in cold, a teacher meditates.
	STC.2:< unk > < unk > unknown words.	STC.4:我也想知道这是什么玩意 I want to know what it is as well.	
ESM	STC.1:我的弹跳力太强了 I'm good at springing force.	STC.3:这是什么电影啊 What movie is it?	Poetry.1:山川胜概是山川,老农不识农家乐 一雨春风满地花,今日春风无处着 Near beautiful the mountain, farmer loses the way. Flowers blossom after rain but can feel spring.
	STC.2:吃货的时候真的很难吃 It feels nasty when you're eating.	STC.4:看到这图就知道是真的 Knowing its true after seeing the picture.	
CVAE	STC.1:被评论的编辑是个傻 The commentator is an idiot.	STC.3:大拇趾想所以要抱抱 Wanna a hug.	Poetry.1:尽有廉轩不拼时,何如宝得看青莲 君心尽舞纷纷云, 无端不数人龙遊 Every thing is cheap, he has to look at Qinglian. The heart is full of clouds and can't find people.
	STC.2:我也想养一只猫猫 I wanna raise a cat.	STC.4:我不知道什么不同的图片是否可以看懂 I don't know if I understand different picture.	
ECVAE	STC.1:<u>热火</u>外线詹姆斯直接连续外线3分 James scored three points consecutively.	STC.3:这就是霸气 外露的行为 This is a strong behaviour.	Poetry.1:少年亲见草鞋游,人间随意随回首 只怪当年不足身,只恐当年书不足 A man travels with grass shoes, looking back casually. only to regret to learn when he was young.
	STC.2:这货是吃<u>柚子</u> 皮 He's eating grapefruit skin.	STC.4:眼见 未必是谣言也有过的 It can be true as well that what eye sees is rumor.	

reaches the peak. Insufficient $\hat{x}$ leads to a performance drop, which is probably due to the inadequate $\hat{x}$. It cannot provide enough supervision to train the gate and H. On the other hand, excessive $\hat{x}$ introduce noise to the model. Then, we conduct the same experiments as above but changing the number of $\hat{c}$ in H with fixed 10 retrieved $\hat{x}$, in which there are 10 retrieved $\hat{c}$ in *contextual combination* as well. It can be seen from the data in Fig. 2(b) that retrieved 10 $\hat{c}$ in H giving the best result. More or less $\hat{c}$ would incur performance drop which is similar to *contextual combination*.

Human Evaluate. Automatic evaluation metrics is not able to reflect the quality of the generation completely, so we need human expert to assess the performance that cannot be measured by machine such as fluency or coherence. Table 5 reflects that our model can generate meaningful and fluent sentences while keeping the generative coherent.

5.2 Qualitative Analysis

As shown in Table 6, we present some generated examples coming from each model. We pick some examples from STC and Poetry dataset. As we can see, in STC.1, ECVAE can copy the word "Heat" (partial name of an NBA team) and then use it to keep the generating on track. Then the topic of the whole sentence is about "Heat" team, and our model even generates "Lebron James", a player of "Heat". In another example, STC.2, although ESM generates with the meaning 'eat,' it can not detail the exact object. ECVAE copy the word 'grapefruit' and then generate a well-stated sentence according to it. In poetry generation, ECVAE can also utilize keyword in its condition and then generate the next line that is highly related to the condition. All the results reveal the effectiveness of contextual combination and editing mechanism in improving the quality of generation.

6   Related Work

CVAE in Text Generation. The VAE is introduced in [1,3], and variational inference family has shown promising performance in text generation: [10] propose to employ VAE to generate sentences from latent space; [5] propose generating utterances from continuous latent space by using hierarchical structure; [9] fuse external knowledge information to guide dialogue generation; [6] combined CVAE and keywords to generate poetry. Existing works suffer from lacking coherence between generation and its condition. This problem mainly comes from two phenomena: KL-vanishing and error accumulation in decoding. When the KL-vanishing happens, the decoder would discard the latent variable in training.

KL Vanishing and Decoding Incoherence. To alleviate the KL-vanishing, most methods fall into two main categories. One is to improve posterior distribution. [10] proposes the KL-annealing to increase KL divergence loss weight slowly. [11] utilizes adversarial training to learn the latent variable distribution. [17,18] combine information of future decoding step provided by a backward RNN. [12,13] change the latent viable distribution to increase density modes. Another group attempts to change decoder architecture to improve its ability to obtain information from latent representation. [10] proposes word dropout to weak decoder during training. [19] utilizes a diluted CNN to generate sentences. [15] adds residual connections to avoid KL vanishing. To enhance decoding, [32] argues that original attention only focuses on particular parts o input sentence and propose multi-head attention for seq2seq. The main difference of our model is that we attempt to mitigate in favor of extra synonymous data explicitly.

Retrieval Boosting. Our work is also related to retrieval boosting generation. [26,27] proposes to combine retrieval-based and generative-based dialogue generation. [21,22] introduce a paradigm which generates sentences based on prototypes.

7   Conclusions

We propose Edit-CVAE (ECVAE) in which we first combine information from extra semantic-related data to enrich the latent representation. We utilize additional data to edit the generated sentence. We conduct experiments on neural conversation generation and neural poetry generation, and then evaluate the effectiveness of our model comprehensively. All experimental results show that our model obtains improvement in generative consistence while maintaining generative diversity and semantic-consistency.

Acknowledgments. This work was supported by the National Key Research and Development Program of China (No. 2017YFC0804001), the National Science Foundation of China (NSFC No. 61672058; NSFC No. 61876196).

References

1. Kingma, D.P., Welling, M.: Auto-encoding variational bayes. arXiv:1312.6114 (2013)
2. Doersch, C.: Tutorial on variational autoencoders. arXiv:1606.05908 (2016)
3. Sohn, K., Lee, H., Yan, X.: Learning structured output representation using deep conditional generative models. In: NeurIPS (2015)
4. Li, J., et al.: Generating classical chinese poems via conditional variational autoencoder and adversarial training. In: EMNLP (2018)
5. Serban, I.V., et al.: A hierarchical latent variable encoder-decoder model for generating dialogues. In: AAAI (2017)
6. Yang, X., et al.: Generating thematic Chinese poetry with conditional variational autoencoder. CoRR (2017)
7. Hu, Z., et al.: Toward controlled generation of text. In: Proceedings of the 34th International Conference on Machine Learning-Volume 70. JMLR.org (2017)
8. Shen, X., et al.: A conditional variational framework for dialog generation. arXiv preprint arXiv:1705.00316 (2017)
9. Zhao, T., Zhao, R., Eskenazi, M.: Learning discourse-level diversity for neural dialog models using conditional variational autoencoders. arXiv preprint arXiv:1703.10960 (2017)
10. Bowman, S.R., et al.: Generating Sentences from a Continuous Space. In: CoNLL (2016)
11. Shen, X., et al.: Improving variational encoder-decoders in dialogue generation. In: AAAI (2018)
12. Xiao, Y., Zhao, T., Wang, W.Y.: Dirichlet variational autoencoder for text modeling. arXiv preprint arXiv:1811.00135 (2018)
13. Serban, I.V., et al.: Piecewise latent variables for neural variational text processing. In: Proceedings of the 2nd Workshop on Structured Prediction for Natural Language Processing (2017)
14. Xu, X., et al.: Better conversations by modeling, filtering, and optimizing for coherence and diversity. In: EMNLP (2018)
15. Dieng, A.B., et al.: Avoiding latent variable collapse with generative skip models. arXiv preprint arXiv:1807.04863 (2018)
16. Jiang, S., de Rijke, M.: Why are sequence-to-sequence models so dull? Understanding the low-diversity problem of chatbots. In: Proceedings of the 2018 EMNLP Workshop SCAI (2018)
17. Goyal, A., et al.: Z-forcing: training stochastic recurrent networks. In: NeurIPS (2017)
18. Du, J., et al.: Variational autoregressive decoder for neural response generation. In: EMNLP (2018)
19. Yang, Z., et al.: Improved variational autoencoders for text modeling using dilated convolutions. arXiv preprint arXiv:1702.08139 (2017)
20. Kim, Y., et al.: Semi-amortized variational autoencoders. arXiv preprint arXiv:1802.02550 (2018)
21. Guu, K., et al.: Generating sentences by editing prototypes. Trans. Assoc. Comput. Linguist. 6, 437–450 (2018)
22. Wu, Y., et al.: Response generation by context-aware prototype editing. arXiv preprint arXiv:1806.07042 (2018)
23. Cao, Z., et al.: Retrieve, rerank and rewrite: soft template based neural summarization. In: ACL (2018)

24. Gu, J., et al.: Incorporating copying mechanism in sequence-to-sequence learning. arXiv preprint arXiv:1603.06393 (2016)
25. See, A., Liu, P.J., Manning, C.D.: Get to the point: summarization with pointer-generator networks. In: Proceedings of the 55th Annual Meeting of the Association for Computational Linguistics (2017)
26. Song, Y., et al.: Two are better than one: an ensemble of retrieval-and generation-based dialog systems. arXiv preprint arXiv:1610.07149 (2016)
27. Song, Y., et al.: An Ensemble of Retrieval-Based and Generation-Based Human-Computer Conversation Systems (2018)
28. Wang, H., et al.: A dataset for research on short-text conversations. In: Proceedings of the 2013 Conference on Empirical Methods in Natural Language Processing (2013)
29. Bahdanau, D., Cho, K., Bengio, Y.: Neural machine translation by jointly learning to align and translate. arXiv preprint arXiv:1409.0473 (2014)
30. Tao, C., et al.: Ruber: an unsupervised method for automatic evaluation of open-domain dialog systems. In: AAAI (2018)
31. Li, J., et al.: A diversity-promoting objective function for neural conversation models. arXiv preprint arXiv:1510.03055 (2015)
32. Tao, C., et al.: Get the point of my utterance! learning towards effective responses with multi-head attention mechanism. In: IJCAI (2018)
33. Pereyra, G., et al.: Regularizing neural networks by penalizing confident output distributions. arXiv preprint arXiv:1701.06548 (2017)
34. Sutskever, I., Vinyals, O., Le, Q.V.: Sequence to sequence learning with neural networks. In: Neurips (2014)
35. Vaswani, A., et al.: Attention is all you need. In: Neurips (2017)

Noise-Based Adversarial Training for Enhancing Agglutinative Neural Machine Translation

Yatu Ji, Hongxu Hou$^{(\boxtimes)}$, Junjie Chen, and Nier Wu

Computer Science Department, Inner Mongolia University, Hohhot, China
jiyatu0@126.com, cshhx@imu.edu.cn, chenjj@imau.edu.cn, wunier04@126.com

Abstract. This study solves the problem of unknown(UNK) word in machine translation of agglutinative language in two ways. (1) a multi-granularity preprocessing based on morphological segmentation is used for the input of generative adversarial net. (2) a filtering mechanism is further used to identify the most suitable granularity for the current input sequence. The experimental results show that our approach has achieved significant improvement in the two representative agglutinative language machine translation tasks, including Mongolian→Chinese and Japanese→English.

Keywords: Agglutinative language machine translation · UNK · Generative adversarial network

1 Introduction

Neural machine translation (NMT) [2,3,8,9] systems have a major drawback in handing rare words, which is more prominent in agglutinative language tasks, due to the sparsity of the vocabulary. The varied morphology largely deceives the translation model directly resulting in a large amount of an $<unk>$ symbol. This can be illustrated by the following sentences.

1. 病院はどこですか？
2. 病院はどちら側ですか？

The word '∠' in the first sentence modifies the bject '病院' need add the suffix '∟'. However, it modifies the same object with a different suffix 'ちら'. This largely deceives the translation model directly resulting in a large amount of out-of-vocabulary (OOV) in the restricted vocabulary, and then it is crudely considered the same as an $<unk>$ symbol. In addition, assume '∠' is a rare word, if it is replaced by the $<unk>$ symbol, the two sentences will be same.

For NLP tasks, Generative adversarial network (GAN) [14] is immature. Some studies, such as [1,12], used GAN for semantic analysis and domain adaptation. [11,13] successfully applied GAN to sequence generation tasks, such as poem generation, speech language generation and machine translation.

© Springer Nature Switzerland AG 2019
A. C. Nayak and A. Sharma (Eds.): PRICAI 2019, LNAI 11670, pp. 392–396, 2019.
https://doi.org/10.1007/978-3-030-29908-8_31

In this study, we propose a novel training approach based on value filtering strategy. During training, corpus is divided into different granularity according to morphology as input of the model, which is transformed into a corresponding value by an additional value filter. Then the model selects the most suitable granularity for decoder according to this value. Experiments on Mongolian→Chinese and Japanese→English translation tasks show that more than 4 and 2 points in BLEU score can be gained with our approach over several strong baselines.

2   Filtering Mechanism for GAN

We consider a standard training process of our model composed of four basic components (Fig. 1).

(1) *training RL-based generator(G) by the XENT*: The selection of G is individualized and targeted. We focus on long short term memory with attention mechanism in this work, because the temporal structure enables it to capture dependency semantics in agglutinating language. We case our problem in the reinforcement framework [4,7,10]. Training G by XENT can ensure that an optimal initial state and provides a good search space for beam search. Note that the input to each module is a sample from the distribution over words produced at the precious timestep.

(2) *value generating and filtering.* The reward R of each sequence is generated by observing RL, then it feeds into a CNN implemented value iteration net. The convolutional layer in CNN corresponds to a particular action Q. The next-iteration value is then stacked with the reward and fed back into the convolutional layer N times, where N depends on the length of the sequence. Subsequently, a long-term value V_{update} is generated by decoding a sequence.

(3) *value discriminating.* The discriminator(D) is dedicated to distinguishing the filtered results with the target ground truth, which provides the probability p_D. The optimization target of D is to minimize the cross entropy loss for binary classification between generation and ground truth.

(4) *model optimization.* Through the analysis of the above components, we can formulate the optimization scheme of the model as follows:

$$J_\theta = E_{(x,y)}[logp_D(x,y)] + E_{(x,y')}[log(1 - p_D(x,y'))], \qquad (1)$$

where (x,y) is the ground truth sentence pair, (x,y') is the generated translation pair. $p_D(.,.)$ represents a probability which proved by D. J_θ can be regard as an opposite *game* process between maximum and minimum expectations, which is the maximum expectation for the generation for G, and the minimum expectation between generation and ground truth for D. We use $log(1 - p_D(x,y')$ obtained from D as a estimation of the reward, the corresponding gradient of :

$$\frac{\partial_{Loss}}{\partial_{\theta \sim G}} = E_{y'}[log(1 - p_D(x,y'))\frac{\partial}{\partial_{\theta \sim G}}logG(y'|x)], \qquad (2)$$

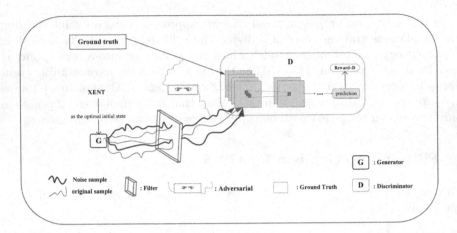

Fig. 1. A data flow presentation of model, in which different colors represent the training process of each component.

where $\frac{\partial}{\partial \theta_{\sim G}} logG(y'|x)$ represents the gradients specified with parameters of the RL-based model. Therefore, we can describe the gradient update of model as:

$$\theta_{\sim G} \leftarrow \theta_{\sim G} + l\frac{\partial}{\partial_{\theta_{\sim G}}}, \tag{3}$$

where l is the learning rate.

3   Experiment and Analysis

3.1   Dataset and Multi-granularity Preprocessing

We report the experimental results on two typical agglutinative language: Mongolian→Chinese (Mo-Ch) and Japanese→English (Ja-En). We use the data from CLDC & CWMT2017 (0.2M) and OPUS in LREC2016 (2.2M)[1]. In order to obtain input samples with multi granularity in two languages, we adopt independent-developed Mongolian affix segmenter and Japanese semantic segmenter JUMAN++[2]. Finally, we divide the training data into four categories:{Original, Original&Affixes, Original&Case, Original&Affixes&Case}. The test set (800 sentences) is composed of a subset of CWMT2017 test set.

We use the MIXER [6], Transformer [9] and BR-CSGAN [13] as the baseline system, and stop the pre-training of initial model until the accuracy of dev achieves at δ which is set to 0.7 in the LSTM. We set the G to generate 500 negative examples per iteration. Selecting a D's kernel width of not less than 3 is reasonable and valid for our vocabulary. All models are trained on up to single Titan-X GPU.

---

[1] https://object.pouta.csc.fi/OPUS-MultiUN/v1/moses/ar-en.txt.zip.
[2] http://nlp.ist.i.kyoto-u.ac.jp.

Table 1. BLEU score of systems under different noise modes

System		Original	Original&Affixes	Original&Case	Original&Affixes & Case
Mn-Ch	Transformer [9]	28.5	30.2	29.8	30.5
	MIXER [6]	29.7	30.4	28.6	31.3
	BR-CAGAN [13]	29.9 (15+17)	31.7 (22+25)	31.1 (15+19)	32.3 (27+32)
	Our	30.6 (15+*11*)	32.5 (22+*19*)	30.8 (15+*15*)	⋆**35.4** (27+*21*)
Ja-En	Transformer [9]	29.7	27.4	23.7	30.1
	MIXER [6]	24.5	26.2	22.3	24.7
	BR-CAGAN [13]	27.8 (38+30)	29.2 (47+41)	23.1 (43+39)	29.6 (66+52)
	Our	28.8 (38+*19*)	29.2 (47+*25*)	24.4 (43+*22*)	⋆**31.3** (66+*28*)

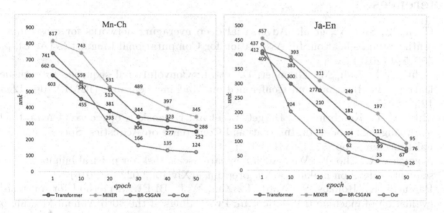

Fig. 2. Number of $<unk>$ symbols in the translations of different models in each epoch.

3.2   Analysis of Evaluation Metrics

We mainly analyze the experimental results in three aspects: BLEU score, training efficiency, unknown word tokens in translations.

BLEU and Training Efficiency. We use BLEU [5][3] score as an evaluation metric to measure the similarity degree between the generation and the human translation. Also, we compared the two GAN-based models by counting the time of pre-training and adversarial training, (e.g., $15 + 17$ indicates 15 h of pre-training and 17 h of adversarial training). From the Table 1, we can clearly observe that Adversarial-NMT obtains satisfactory BLEU score against baseline systems. In particular, the results show that the GAN-based model is obviously superior to baseline systems in any kind of noisy corpus.

A Discussion of Proposed Method for Redundant Tokens. Because of the restrict liberation on granularity, the mixed noises will cause under-fitting

---

[3] https://github.com/moses-smt/mosesdecoder/blob/master/scripts/generic/multi-bleu.perl.

or over-fitting, which motivates us to alleviate the confusion of decoder on granularity selection through VIN. A clear and credible evidence is the number of unknown words ($<unk>$) in the translation. Figure 2 shows a comparison of the number of word occurrences for each corpus and model.

4   Conclusion

The main contribution of this study is to show that NMT systems of agglutinative language are capable of generative adversarial translation by using morphological noises of corpus. This is both simpler and more effective than using a complex preprocessing method.

References

1. Chen, X., Sun, Y., et al.: Adversarial deep averaging networks for cross-lingual sentiment classification. In: Association for Computational Linguistics (ACL), pp. 557–570 (2016)
2. Gehring, J., Auli, M., Grangier, D., et al.: Convolutional sequence to sequence learning. In: International Conference on Machine Learning (ICML), pp. 1243–1252 (2017)
3. Mikolov, T., Kombrink, S., Burget, L., et al.: Extensions of recurrent neural network language model. In: International Conference on Acoustics, Speech, and Signal Processing, pp. 5528–5531 (2011)
4. Nogueira, R., Cho, K.: Webnav: a new large-scale task for natural language based sequential decision making. arXiv preprint arXiv:1602.02261 (2016)
5. Papineni, K., Roukos, S., Ward, T., Zhu, W.J.: BLEU: a method for automatic evaluation of machine translation. In: Proceedings of the 40th Annual Meeting on Association for Computational Linguistics (ACL), pp. 311–318 (2002)
6. Ranzato, M., Chopra, S., et al.: Sequence level training with recurrent neural networks. arXiv:1511.06732 (2015)
7. Sunmola, F.T., Wyatt, J.L.: Model transfer for Markov decision tasks via parameter matching. In: PlanSIG 2006, pp. 246–252 (2006)
8. Sutskever, I., Vinyals, O., Le, Q.V.: Sequence to sequence learning with neural networks. In: Conference and Workshop on Neural Information Processing Systems (NIPS), pp. 3104–3112 (2014)
9. Vaswani, A., Shazeer, N., et al.: Attention is all you need. In: Conference and Workshop on Neural Information Processing Systems (NIPS), pp. 5998–6008 (2017)
10. Volodymyr, M., Koray, K., et al.: Human-level control through deep reinforcement learning. Nature 518(7540), 529 (2015)
11. Yu, L., Zhang, W., Wang, J., Yu, Y.: SeqGAN: sequence generative adversarial nets with policy gradient. In: The Association for the Advancement of Artificial Intelligence (AAAI), pp. 2852–2858 (2016)
12. Zhang, Y., Barzilay, R., Jaakkola, T.: Aspect-augmented adversarial networks for domain adaptation. Trans. Assoc. Comput. Linguist. 5(1), 515–528 (2017)
13. Zhen, Y., Wei, C., Feng, W., Bo, X.: Improving neural machine translation with conditional sequence generative adversarial nets. In: The North American Chapter of the Association for Computational Linguistics (NAACL), pp. 1346–1355 (2018)
14. Zhu, J.Y., Park, T., et al.: Unpaired image-to-image translation using cycle-consistent adversarial networks. In: IEEE International Conference on Computer Vision (ICCV), pp. 2223–2232 (2017)

Concept Mining in Online Forums Using Self-corpus-Based Augmented Text Clustering

Wathsala Anupama Mohotti$^{(\boxtimes)}$, Darren Christopher Lukas, and Richi Nayak

Queensland University of Technology, 2 George Street, Brisbane, Australia
{mohotti,darren.lukas,r.nayak}@qut.edu.au

Abstract. This paper proposes a self-corpus-based text augmentation technique with clustering for concept mining in a discussion forum. Sparseness in text data, which challenges the distance and density measures in determining the concepts in a corpus, is handled through self-corpus-based document expansion via matrix factorization. Experiments with a real-world dataset show that the proposed method is able to infer useful concepts.

Keywords: Concept mining · Corpus-based augmentation · Clustering

1 Introduction

An online forum is a formal mechanism that community uses to exchange information through posted messages that are organized into "threads" [5]. The forums can reflect concepts, themes, and concerns of online societies in diverse fields such as education, marketing and politics [5,6]. A handful of studies have applied data and text mining methods to explore the predictive power of the forum data [4,6]. In the education domain, discussion forums have been analyzed to assess interactivity over a period of time to predict early warnings for students at-risk [6]. In marketing, online forum data is used to identify product defects [5] with predictive models. However, these works neglect the natural text content used in the online discussion. A few studies have applied text mining in online forums for sentiment analysis [4] with supervised approaches to classify forum threads. However, the unavailability of ground-truths in online forum data creates the demand for conducting the analysis in unsupervised setting [4].

In this paper, we propose a concept mining method that can extract concepts based on text discussions in the unsupervised setting. Concept mining of online forums data faces the same challenges as traditional text mining methods [1]. Sparse nature of text vectors and a higher number of dimensions make distance and density-based methods to perform poorly due to distance concentration [1]. Specifically, distance differences between far and near points become negligible in higher dimensions [1]. In addition, density based methods are unable to identify

© Springer Nature Switzerland AG 2019
A. C. Nayak and A. Sharma (Eds.): PRICAI 2019, LNAI 11670, pp. 397–402, 2019.
https://doi.org/10.1007/978-3-030-29908-8_32

dense patches in sparse text data. Moreover, forum data is usually homogeneous where a minor variation in the distance/density measures will determine groupings. Probabilistic and matrix factorization based approaches have been introduced to handle higher dimensions in text [1]. However, information loss in these dimensional reduction methods is evident.

Distinct from these works, we introduce a novel approach for content mining in online forums using clustering and document expansion, named as ConMine to understand the main concepts and themes present in user discussions. The self-corpus based document expansion [8] in ConMine, via Non-negative Matrix Factorization (NMF), learns virtual terms from the same corpus that semantically match the applied domain. A centroid-based clustering is then applied to the expanded text to differentiate the concepts. ConMine automatically learns the number of clusters to be produced within the augmentation process. Finally, we synthesize meaningful concepts with the help of experts via word-cloud visualization. ConMine approach is evaluated on real-world data taken from the Queensland University of Technology (QUT), Australia. The empirical analysis shows that ConMine is able to handle sparse and homogeneous nature of text in discussion forums and identify concepts more accurately than the benchmarks.

2   Concept Mining with Self-corpus-Based Augmentation

The proposed three-step ConMine Algorithm is outlined in Fig. 1. Consider an online forum corpora $\mathcal{D} = \{D_1, D_2, ..D_i, ...D_s\}$ over a time period s where D_i represents the corpus at time i. Let D_i be a collection of N distinct posts, $\{P_1, P_2, ...P_N\}$, that contain a total of M distinct terms $\{t_1, t_2, ...t_M\}$.

Self-corpus-Based Augmentation with Matrix Factorization: In contrast to using external knowledge bases [3], we conjecture that the self-corpus based augmentation is well suited for augmenting text as it follow forums' text patterns. Let A be the $M \times N$ matrix representation of D_i. We decompose A using NMF to have the lower rank matrices W and H which are non-negative and in the size of $M \times k$ and $k \times N$ respectively with the low-order rank k set as the number of topics. The k is learned using the intrinsic topic coherence measure. The matrix factorization process iteratively approximates W and H such that they can represent high-dimensional A with the least error as in Eq. 1.

$$\min_{W,H \geq 0} \frac{1}{2}\|A - WH\| = \sum_{i=1}^{M}\sum_{j=1}^{N}\left(A_{i,j} - (WH)_{i,j}\right)^2 \tag{1}$$

Topic membership of each post in D_i is obtained considering the maximum coefficient value in H for a post. This associated topic is used to identify the virtual terms for each post using W. The coefficients in W are sorted in decreasing order. The coefficients that yield higher value than *mean+standard deviation* of the distribution become the terms to represent a topic as in [8]. Each text post of D_i is expanded using the most probable terms as virtual terms that correspond to its topic vector and form D_i'.

```
Algorithm : Concept Mining with Augmented Text and Clustering (ConMine)
Input : Set of forum posts D_i = {P_1, P_2, ..., P_N}
Output : Meaningful set of concepts C = {c_1, c_2, ..., c_m}
Augmentation Process :
        D'_i = {}
        k, VirtualwordsMatrix ← NMFbasedTopicModeling(D_i)
        while each P_a ∈ D_i augmented
                P'_a ← NMFbasedAugmentation(P_a, VirtualwordsMatrix)
                D'_i ← D'_i ∪ P'_a
        return D'_i = {P'_1, P'_2, ..., P'_N}, k    //augmented media posts & number of clusters
Clustering Process :
        μ = {μ_1, μ_2, ..., μ_k}
        {s_1, s_2, ..., s_k} ← selectRandomSeeds(D'_i, k)  // Initial random value for centres
        for g ← 1 to k
            c_g ← {}
            μ_g ← s_g
        while each P'_a ∈ D'_i assign to a cluster
                for g ← 1 to k
                    j ← argmin|μ_g − P'_a|
                c_g ← j
                c_g ← c_g ∪ {P'_a}
                μ_g ← (1/|c_g|) Σ_{P'_b ∈ c_g} P'_b    //update centroid
        return {c_1, c_2, ..., c_k}    //k clusters
Knowledge Synthesis :
        C = {}
        for each c ∈ {c_1, c_2, ..., c_k}
            postprocessing(c)
            wordcloudVisualization(c)
            C ← C ∪ {expertValidation(c)}
        return C = {c_1, c_2, ..., c_m}    // m concepts
```

Fig. 1. Clustering algorithm for concept mining: ConMine

Augmented Text Clustering: The data matrix D'_i with augmented posts is represented with a weighted term × post matrix to partition into k clusters. We use the centroid-based clustering as it is reported to produce an accurate outcome for the homogeneous data [2]. As the online forum data shows the homogeneous nature, we partition the N posts into k clusters (obtained through the previous step) using k-means. Initial k cluster centers are randomly chosen. Then each post $P'_a \in D'_i$ is compared with each k center to decide on the closest to be assigned. This process updates the respective cluster center in each iteration.

Knowledge Synthesis for meaningful Communities: Within this step, we generate the m concepts that are meaningful to the domain in k clusters after doing further post-processing and consultation with domain experts. We analyze terms in each cluster through visualization and the highly occurring common words are removed. This is an iterative quality checking process that includes manual intervention. This process results in the m ($\leq k$) meaningful concepts discussed in a forum.

3   Empirical Analysis

Datasets: The dataset is obtained from the online forum, Essential Supervisory Practices (ESP), a 5-week training program for higher-degree research supervisors at QUT between 2015 to 2017. The posts from all years have been combined on a weekly basis, resulting in five datasets as in Table 1. We consider each post, regardless of its type (i.e, original or reply), as a single document after applying standard text pre-processing steps. After comparing the experimental results with multiple weighting schemes, posts are organized in vector space model (VSM) with the *tf*idf* weighting schema to derive the topics, while the augmented posts are represented using *tf* for clustering.

Table 1. Summary of the datasets used in the experiments

Dataset	Number of posts	Number of unique terms	Average post length (in terms)	
			Before augmentation	After augmentation
W1	1664	7090	154	165
W2	1495	7385	177	194
W3	1416	7145	155	165
W4	1402	7057	161	174
W5	1568	6893	145	153

Benchmarks and Evaluation Measures: The proposed NMF based approach for document expansion using topics in ConMine is evaluated against probabilistic LDA (pLDA) [1] and Latent Semantic Indexing (LSI) [1]. The state-of-the-art clustering methods of DBSCAN [8], LDA [1], LSI [1] and NMF [1] are used for benchmarking the concepts of clustering in ConMine. Accuracy of topic vector formation and clustering process were evaluated with the intrinsic measures topic coherence [7] and Silhouette score [7] respectively.

Accuracy: ConMine with NMF is found best in terms of topic coherence (Fig. 2(a)). LSI, which approximates factors with both positive and negative entries, is not able to provide stronger topic distribution in VSM which is represented with strictly positive entries. pLDA, which approximates topics using the probability of terms considers only the term count and neglects the context of the words and frequencies, has provided inferior results. We empirically learn the number of topics as shown in Fig. 2(b) which produces highly cohesive topics. This number is used in deriving topics for the post augmentation as well as it is set as k in the clustering process. Figure 2(c) compared clustering in ConMine with and without post augmentation. Increased tightness of the clusters, indicated by a higher silhouette score after augmentation in each method, confirms the benefit of augmentation by handling the sparseness in high-dimensional text via added terms. ConMine shows the highest increase in silhouette value compared to all the baselines. In the homogeneous data, the density concept (DBSCAN) creates contiguity-based clusters where very different

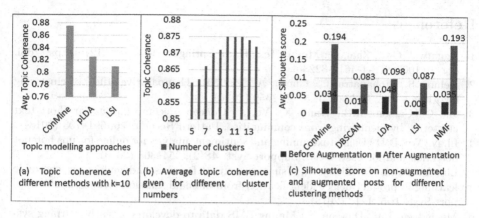

(a) Topic coherence of different methods with k=10

(b) Average topic coherence given for different cluster numbers

(c) Silhouette score on non-augmented and augmented posts for different clustering methods

Fig. 2. Results of the experiments

data items may end up in the same cluster giving the worst results. LDA which uses term counts-based probability is unable to predict the correct cluster due to the negligence of context of the terms. However, NMF as a clustering method performs similar to ConMine with a marginal difference showing the importance of mapping higher to lower-dimension space. The identified Concepts for each week are given in Table 2.

Table 2. Concepts identified for datasets

Dataset	Identified concepts by ConMine
W1	Research skill, Milestones, Supervisors, Meetings, Publications
W2	Experience in supervising, Relationship between student and supervisor
W3	Writing thesis, Writing literature review, Plagiarism and research issues
W4	Emotional issues, Completion, Strategy for unsatisfactory progress
W5	Examiner comments, Final submission and Seminar practice

4 Conclusion

We proposed and evaluated a concept mining method, ConMine, on a real-world forum data for understanding the discussions that are held on online forums. To handle the sparsity and high dimensionality in text, we use NMF (which approximates topic vectors in a linear manner considering the context of terms) to obtain virtual words for post-expansion. Leveraging the intrinsic measurements, we learn the optimal number of k topics that are further used in centroid-based clustering to obtain the clusters/concepts within the augmented text. Results show that ConMine can deal with the sparse and homogeneous nature of online forum data to obtain some useful concepts.

References

1. Aggarwal, C.C., Zhai, C.: Mining Text Data. Springer, Boston (2012). https://doi.org/10.1007/978-1-4614-3223-4
2. Dehuri, S., Mohapatra, C., Ghosh, A., Mall, R.: Comparative study of clustering algorithms. Inf. Technol. J. **5**, 551–559 (2006)
3. Jia, C., Carson, M.B., Wang, X., Yu, J.: Concept decompositions for short text clustering by identifying word communities. Pattern Recogn. **76**, 691–703 (2018)
4. Li, N., Wu, D.D.: Using text mining and sentiment analysis for online forums hotspot detection and forecast. Decis. Support Syst. **48**(2), 354–368 (2010)
5. Liu, Y., Jiang, C., Zhao, H.: Using contextual features and multi-view ensemble learning in product defect identification from online discussion forums. Decis. Support Syst. **105**, 1–12 (2018)
6. Macfadyen, L.P., Dawson, S.: Mining LMS data to develop an "early warning system" for educators: a proof of concept. Comput. Educ. **54**(2), 588–599 (2010)
7. Mehta, V., Caceres, R.S., Carter, K.M.: Evaluating topic quality using model clustering. In: CIDM, pp. 178–185. IEEE (2014)
8. Mohotti, W.A., Nayak, R.: Corpus-based augmented media posts with density-based clustering for community detection. In: ICTAI, pp. 379–386. IEEE (2018)

Knowledge Representation and Reasoning

Rational Inference Patterns

Lars-Phillip Spiegel[1], Gabriele Kern-Isberner[1]($\boxtimes$) (iD), and Marco Ragni[2] (iD)

[1] TU Dortmund, Dortmund, Germany
gabriele.kern-isberner@cs.tu-dortmund.de
[2] University of Freiburg, Freiburg, Germany

Abstract. Understanding, formalizing and modelling human reasoning is a core topic of artificial intelligence. In psychology, numerous fallacies and paradoxes have shown that classical logic is not a suitable logical framework for this. In a recent paper, Eichhorn, Kern-Isberner, and Ragni have succeeded in resolving paradoxes and modelling human reasoning consistently in a non-monotonic resp. conditional logic environment with so-called inference patterns. For further studies using inference patterns, however, it is mandatory to understand better how inference patterns are triggered by the characteristics of specific examples used in the empirical tests. The goal of this paper is to categorize empirical tasks by formal inference patterns and then find crucial features of the corresponding reasoning tasks in such a way that they can be used to predict the reasoning of human subjects according to the task. To this end a large amount of psychological studies dealing with human reasoning from the literature were investigated and classified according to the observed inference patterns. From this classification, we learnt a decision tree revealing which features of empirical tasks lead to which inference pattern in most cases. These results provide insights into the reasoning modes of humans which is important for choosing the right formal model, and help setting up proper tasks for testing inference patterns.

Keywords: Nonmonotonic reasoning · Conditional reasoning · Rational human inference · Psychological experiments

1 Introduction

Human rational reasoning has been a leading paradigm for formal approaches to knowledge representation and reasoning (see, e.g., [1,9]), and recently, interest has increased to evaluate formal theories of reasoning by empirical evidence, or the other way round, to find formal models of reasoning that are able to explain human reasoning (see, e.g., [12]). However, as many psychological experiments show, humans may systematically violate for a conditional (*if A then C*) the inference rules *modus ponens (MP)* (*Given the conditional and A, infer C*) and *modus tollens (MT)* (*Given the conditional and $\neg C$, infer $\neg A$*) that are valid according to classical logic, while also using the logically invalid inference rules *affirmation of consequent (AC)* (*Given the conditional and C, infer A*) and

© Springer Nature Switzerland AG 2019
A. C. Nayak and A. Sharma (Eds.): PRICAI 2019, LNAI 11670, pp. 405–417, 2019.
https://doi.org/10.1007/978-3-030-29908-8_33

denial of antecedent (DA) (Given the conditional and ¬A, infer ¬C). This has often lead to the conclusion that human reasoning is from a classical logic perspective (hopelessly) irrational, but also more recently to the insight that classical logic may not be the right norm to judge human reasoning [4,10,11,14]. In the paper [5], the authors showed that 3-valued conditional logic resp. basic semantics of nonmonotonic reasoning provide a much more coherent formal model of human reasoning that is not only apt to resolve prima facie irrationality but can also be used to elaborate on hypotheses for hidden assumptions and plausible knowledge that humans relied on when solving the reasoning tasks. The key idea in that paper was to consider all inference rules (MP, MT, AC, DA) together as so-called *inference patterns* and interpret them via formal conditionals.

In this paper which is based on the bachelor thesis [13], we continue this work by elaborating on links between inference patterns and the setting of experimental tasks with respect to wording, familiarity with the topic, suggested alternatives or disablers, and other features. The following small example illustrates how significantly reasoning depends on such settings: *If butter is heated, then it melts.* Does this also mean that butter is always heated when it melts? Not necessarily because this statement describes prima facie the logical implication: *Butter is heated ⇒ Butter melts.* In classical logic it is then a fallacy to say that butter is always heated when it melts because the implication is only defined to work one way. But how do you melt butter without heating it? In fact, usually the definition of melting is "Become fluid under the influence of heat". In the real world the conclusion that butter is heated when it melts is perfectly correct according to this definition. Nevertheless, according to the definition of implication it is logically incorrect. This statement about heated butter is taken from a real psychological study [15]. In this study, the reasoning behavior of subjects was examined and it was found that 95% of all subjects concluded from the statement *butter melts* the statement *it was heated.* In another study, subjects were given the statement *if Rex is a terrier, then he likes apples,* logically corresponding to the implication *Rex is a terrier ⇒ Rex likes apples* [8]. In this study, only 8% of the test subjects concluded from the statement *Rex likes apples* the statement *Rex is a terrier.* So what's the difference between these two studies? Obviously, of course, the assumption is that in the real world not anything that is called Rex and likes apples must also be a terrier. Although this difference is very obvious, it is also difficult to be worked out from studies because they often group several statements having the same logical structure into one experiment. In order to use inference patterns as a formal model of reasoning not only on an abstract, aggregated level, as it was done in [5], but also to explain and predict reasoning behavior of a specific individual in a given empirical task, characteristic features of such tasks must be taken into account. This paper elaborates on such features by investigating tasks from well-known empirical studies, and uses the most characteristic features for classifying inference behavior in the form of inference patterns. We describe inference patterns by their most salient features, and present a decision tree that can help classify empirical tasks by (expected) inference patterns. These results will pave the way to empirically validate formal inference patterns for individual reasoning, but are also interesting for psychologists in general when designing empirical tasks.

The rest of this paper is organized as follows: In the next section we will provide some formal preliminaries about conditionals and preferential models. In Sect. 3, we explain inference patterns and identify main features of psychological experiments. Moreover, we briefly discuss some features that we use in the following analysis of the experiments and for classification. In Sect. 4 we will analyze which features show up in most frequently drawn inference patterns across the experiments and present a decision tree based on three core features that visualizes concisely why some inference patterns are drawn. A general discussion in Sect. 5 concludes the paper.

2   Formal Preliminaries

We build upon a propositional logical language $\mathcal{L}$ defined over a set of atoms Σ and making use of the classical connectors conjunction ($\wedge$), disjunction ($\vee$), and negation ($\neg$); furthermore, $A \Rightarrow B$ means $\neg A \vee B$. For ease of reading, $\neg A$ is abbreviated by $\overline{A}$, and the conjunction symbol is often omitted, i.e., AB means $A \wedge B$. The set of interpretations or *worlds*, represented by conjunctions over literals, is Ω. A formula A is true in a world ω if and only if $\omega \models A$, ω is then called a model of A. The set of all models of A is denoted by $\text{Mod}(A)$. The total set of logical consequences of a formula A is $\text{Cn}(A) = \{B | A \models B\}$. In the following, we recall basics on conditionals and preferential semantics which are both necessary for setting up the non-classical logical framework used in this paper.

Conditionals $(B|A)$ are suitable to express commonsense statements, or rules with exceptions "If A then (usually) B".

Definition 1 (Conditionals [3,5]). *A conditional $(B|A)$ consists of two formulas $A, B \in \mathcal{L}$, which are combined by the conditional operator $|$. A is called the* premise *and B is the* consequence *of the conditional. There are three possible evaluations of a conditional $(B|A) \in (\mathcal{L}|\mathcal{L})$ in a world $\omega \in \Omega$:*

$$[\![(B|A)]\!]_\omega = \begin{cases} true & if \ \omega \models AB \\ false & if \ \omega \models A\overline{B} \\ undefined & if \ \omega \models \overline{A} \end{cases}$$

The conditional $(B|A)$ is verified *by a world ω if the formula AB is satisfied in the world ω. It is* falsified *by ω if the formula $A\overline{B}$ is satisfied in the world ω. If $\overline{A}$ is true in the world ω then the conditional is called* neutral *regarding ω.*

The set of all conditionals over $\mathcal{L}$ is denoted by $(\mathcal{L}|\mathcal{L})$. A *conditional knowledge base* Δ is a (non-empty, finite) set of conditionals $\Delta = \{(B_1|A_1), \ldots, (B_n|A_n)\} \subseteq (\mathcal{L}|\mathcal{L})$. Semantics of conditionals and consistency of conditional knowledge bases can be defined in terms of preferential models which provide a basic semantics for nonmonotonic logics.

Definition 2 (Preferential models [9]**).** *A* preferential model *is a triple* $(M, \preceq, \models)$ *consisting of a set of* states M, *a* preference relation *between these* states $\preceq \subseteq M \times M$, *and a* satisfiability relation *between models and formulas* $\models \subset M \times \mathfrak{L}$.

For a set of formulas $\mathcal{A}$ and a model m, $m \models \mathcal{A}$ iff $m \models a$ for all $a \in \mathcal{A}$. For two states m_1, m_2, $m_1 \approx m_2$ iff $m_1 \preceq m_2$ and $m_2 \preceq m_1$, and $m_1 \prec m_2$ if $m_1 \preceq m_2$ and not $m_2 \preceq m_1$. Furthermore, for a model m_1 and a formula A, A is *preferentially satisfied* by m_1, $(m_1 \models_\prec A)$, if $m_1 \models A$, and there is no m_2 with $m_2 \models A$ and $m_2 \prec m_1$. The *preferential inference relation* $\succ\!\!\!\sim_\prec$ is defined via $A \succ\!\!\!\sim_\prec B$ iff $m \models_\prec A$ implies $m \models B$, i.e., iff all minimal models of A satisfy B.

For the context used in this paper, $M = \Omega$ is the set of worlds, and $\preceq$ is a plausibility relation, i.e., a total preorder on worlds. That is, for two worlds ω_1, ω_2, we say $\omega_1 \preceq \omega_2$ if ω_1 is at most as plausible as ω_2. The satisfaction relation $\models$ is then simply defined as the normal satisfaction relation on worlds. The plausibility relation can also be lifted to formulas: For two formulas A and B, $A \preceq B$ means that for every model of B, there must be at least one model of A that is at least as plausible [5]. Since $\preceq$ is assumed to be a total preorder, this is equivalent to saying that all minimal models of A are as least as plausible as all minimal models of B. Then, it is easy to show that $A \succ\!\!\!\sim_\prec B$ iff $AB \prec A\overline{B}$ [5]. A state of knowledge, or *epistemic state* Ψ can now be represented by such a preferential model, more specifically, by such a plausibility relation $\preceq$ on worlds (the rest is classical logic), and can provide semantics for conditionals by interpreting the conditional operator via the nonmonotonic inference relation $\succ\!\!\!\sim_\prec$: Ψ *accepts* a conditional $(B|A)$, $\Psi \models (B|A)$, iff $A \succ\!\!\!\sim_\prec B$, iff $AB \prec A\overline{B}$. We also make use of so-called *weak conditionals* $(\!|B|A|\!)$ that are accepted by Ψ iff Ψ does not accept $(\neg B|A)$, i.e., $\Psi \models (\!|B|A|\!)$ iff $A \not\succ\!\!\!\sim_\prec \overline{B}$, iff $AB \preceq A\overline{B}$. The most plausible beliefs Bel(Ψ) of Ψ being represented by a total preorder $\preceq$ are all propositions which are implied by all most plausible (i.e., minimal wrt $\preceq$) worlds. A conditional knowledge base consisting of (weak) conditionals is *consistent* iff there is a total preorder that accepts all conditionals.

3 Human Reasoning and Inference Patterns

This section describes methods we developed and applied to investigate human reasoning patterns and covers some examples from a variety of psychological studies in the literature. Moreover, we discuss and illustrate features of experimental tasks, and show their relevance for inference patterns observed in empirical studies.

3.1 Inference Rules and Logical Fallacies

In psychology, human reasoning is often evaluated by inference rules: the logically valid rules of *modus ponens* (MP) and *modus tollens* (MT), and the logically invalid rules *denial of antecedent (DA)* and *affirmation of consequent (AC)*. Modus ponens assumes the premise to be true and deduces the consequence as true, modus tollens assumes the consequence as false and concludes that the premise must also be false. For denial of antecedent, the premise is assumed to be false, and it is inferred that the consequence must also be false. Affirmation of consequence perceives the consequence as true, and from that deduces the truth of the premise.

Example 1. To illustrate these rules, we use a classical example from [2], where the implication *if Lisa has an essay to write, then she (always) studies late in the library* is used.

- According to MP, the statement *Lisa has an essay to write* properly concludes the statement *Lisa studies late in the library.*
- According to MT, the statement *Lisa does not study late in the library* correctly concludes *Lisa has no essay to write* because otherwise she would study in the library until late in the evening.
- According to DA, the statement *Lisa has no essay to write* concludes *Lisa does not study late in the library*, but there may also be a lot of other reasons why Lisa studies late in the library, e.g., when an examination is pending.
- According to AC, the statement *Lisa studies late in the library* concludes the statement *Lisa has an essay to write*, but Lisa might not have to write an essay, but prepare herself for an examination.

3.2   Inference Patterns

An inference pattern indicates for each inference rule, modus ponens MP, modus tollens MT, affirmation of consequent AC, and denial of antecedent DA, whether the agent uses this rule to infer something in an application context or not.

Definition 3 (Inference pattern [5]). *An inference pattern ϱ is a 4-tuple which specifies for each inference rule MP, MT, AC, DA whether this inference rule is used (e. g., MP), or not (e. g., $\neg MP$). The set of all 16 resulting inference patterns is $\mathcal{R}$.*

In [5], we used the aggregated responses of all test persons for each task to decide whether an inference rule was applied by a majority of test persons, or not. This means, if the inference rule was applied by at least 50% of all test persons, then the inference pattern shows a positive occurring of the respective rule, otherwise, it shows a negative occurrence.

If we now use a total preorder $\preceq$ as a plausibility relation over all possible worlds, as is the case, for example, with preferential models, using or not using an inference rule induces an inference and thus an inequality, as shown in Table 1.

Thus, every inference pattern describes four inequalities. An inference pattern is called *rational* if the system of inequalities can be represented by a total preorder $\preceq$. Since every inference rule corresponds to a conditional (e.g., MP corresponds to $(B|A)$), this is equivalent to saying that the conditional knowledge

Table 1. Inference rules and induced inequalities on the models.

Rule	Inference	Inequality		Rule	Inference	Inequality
MP	$A \mathrel{|\!\sim} B$	$AB \prec A\bar{B}$		$\neg$ MP	$A \mathrel{|\!\not\sim} B$	$A\bar{B} \preceq AB$
MT	$\bar{A} \mathrel{|\!\sim} \bar{B}$	$\bar{A}\bar{B} \prec A\bar{B}$		$\neg$ MT	$\bar{A} \mathrel{|\!\not\sim} \bar{B}$	$A\bar{B} \preceq \bar{A}\bar{B}$
AC	$B \mathrel{|\!\sim} A$	$AB \prec \bar{A}B$		$\neg$ AC	$B \mathrel{|\!\not\sim} A$	$\bar{A}B \preceq AB$
DA	$\bar{A} \mathrel{|\!\sim} \bar{B}$	$\bar{A}B \prec \bar{A}\bar{B}$		$\neg$ DA	$\bar{A} \mathrel{|\!\not\sim} \bar{B}$	$\bar{A}B \preceq \bar{A}\bar{B}$

base consisting of the (weak) conditionals corresponding to the (not-satisfied) inference rules is consistent. This way, human rationality is formally characterized by consistence in a conditional logic.

Investigating the inequalities for all 16 inference patterns, it turns out that only two inference patterns describe non-solvable inequalities and are thus irrational. These are the patterns $(MP, \neg MT, \neg AC, DA)$ and $(\neg MP, MT, AC, \neg DA)$ [5]. In practice, it turns out that these irrational inference patterns are drawn very rarely. In the studies analyzed in this work they add up to just about 1% which reflects a very good overall rational reasoning behavior and thwarts the frequent findings of irrationality based on classical logic.

Moreover, the paper [5] also presented techniques to extract a suitable conditional knowledge base Δ generating the respective preorder, and the most plausible beliefs from the total preorder.

3.3 Features of Tasks in Empirical Studies

In this section, examples of some of the studies considered are presented, and their features and results are discussed in some detail. These features build the base for the later classification.

Alternatives and Disablers. In [2] the suppression of the logically valid and invalid inference rules is seperately examined. For this purpose, in Experiment 1 the subjects were presented with alternatives or additional conditions (disablers) together with the implication. For each inference rule, the subjects then received an assumption, possibly including an alternative or additional condition, as well as a choice of three conclusions. The subjects were instructed to accept the assumption as true and to choose which of the three conclusions follows from the given statements. Results of the experiment are shown in Table 2.

The Role of Negation. In [6] the effect of negation on reasoning tasks was examined. It was based on a known experiment in which a statement was made about the relationship between characters on the front and back of a card. These statements like *If there is (not) a P on one side of the card, then there is (not) a 1*

Table 2. Evaluation and inference pattern for Experiment 1, Table 1 in [2]

Argument type	MP	MT	AC	DA	Inference pattern
Simple	96	92	71	46	$(MP, MT, AC, \neg DA)$
Alternative	96	96	13	4	$(MP, MT, \neg AC, \neg DA)$
Additional condition	38	33	54	63	$(\neg MP, \neg MT, AC, DA)$

on the other side of the card are then varied in, among other things, the negated part of the implication, and whether the negation was given either explicitly or implicitly. Results of this experiment are shown in Table 3.

Table 3. Evaluation and inference pattern in [6] for explicit negation

Argument type	MP	MT	AC	DA	Inference pattern
If p, then q	95	60	60	35	$(MP, MT, AC, \neg DA)$
If p, then not q	100	75	40	20	$(MP, MT, \neg AC, \neg DA)$
If not p, then q	100	50	85	50	(MP, MT, AC, DA)
If not p, then not q	100	35	60	30	$(MP, \neg MT, AC, \neg DA)$

Counterfactual Implications. In [7], causal counterfactual statements were investigated. These are statements that speculate about a possibility of which it is uncertain whether it has taken place or not. To evoke such an interpretation in subjects, these statements are usually phrased in the subjunctive mood. In Experiment 3a of [7], the subjects were tasked with a normal implication, a counterfactual implication, and a fictional story. The stories dealt with different events, but all involved a counterfactual conditional, and the subjects were asked to speculate about a different outcome of the story by completing the sentence *"If only ..."* at the end of the story. Results of the experiment are shown in Table 4.

Table 4. Evaluation and inference pattern from Experiment 3a, Table 4 in [7]

Argument type	MP	MT	AC	DA	Inference pattern
Normal	80	58	40	20	$(MP, MT, \neg AC, \neg DA)$
Counterfactual	86	81	46	46	$(MP, MT, \neg AC, \neg DA)$
Fict. story	49	45	53	59	$(\neg MP, \neg MT, AC, DA)$

These examples show clearly which significant effect slight variations of experimental tasks may have on the reasoning behavior of the test persons. Therefore, categorizing experimental tasks with the help of suitable features is crucial to set up reasonable and justified hypotheses on the reasoning behavior of people.

3.4 Features for Classifying Experimental Tasks

For the classification, many possible features come to mind. The examples in the previous subsection show how negation, the presence of alternatives, as well as counterfactual phrasing or a fictionary story might influence the reasoning behavior of test persons.

In Table 5, an overview of all used features and their values that we used for classification can be found. *Negation*, *Alternatives*, and *(Counter)Factual* have been illustrated in the previous subsection. *Age Group* and *Task Type* are self-explanatory, the other features are explained below:

Meaning. Notwithstanding the type of tasks, each implication can be assigned a degree of good reason. For instance, the degree of good reason of the implication "terriers like apples" is lower than the degree of "terriers like meat".

Wording. Some studies also deal with the wording of an implication. They distinguish between *If ... then* and *Only ... if* conditionals which makes a big difference. *Only ... if* conditionals can be understood to express the implication "backwards".

Abstraction. While some tasks describe everyday situations, others abstract to the purely logical level with a $p \Rightarrow q$ implication.

Strictness. Often, the simple question is *what, if anything, follows?*. However, there are also studies that instruct the subjects, e.g., to draw only logical or only absolutely necessary conclusions explicitly. There are also studies that emphasize that the implication is true, while others do not mention it.

4 Describing Inference Patterns by Features

In the bachelor thesis [13], 22 studies with 35 experiments were investigated with respect to the inference pattern they induce. The number of participants is 29.65 on average in the studies considered, with a minimum of 8 and a maximum of 116. Only six inference patterns were ever drawn at a frequency of more than 5%. The proportion of irrational patterns is only 1.1%. For this paper, we reduced data and the number of investigated inference patterns in order to be able to focus on the most interesting aspects of our findings. We selected among the most frequent ones the following five inference patterns: (MP, MT, AC, DA), (MP, MT, AC, ¬DA), (MP, ¬MT, AC, DA), (MP, ¬MT, AC, ¬DA), and (MP, MT, ¬AC, ¬DA). For each considered inference pattern, the data for some features is analyzed and discussed in some detail here. Moreover, for each inference pattern and some features, Table 6 shows whether a feature has

Table 5. List of used features and their values

Feature	values	Feature	values
Age Group	Adults Children	Abstraction	Concrete Abstract
Task Type	Definition Causal Spatial Prevention Arbitrary Other	Familiarity	High Medium Low
		Meaning	High Low
Negation	None Premise	(Counter)Factual	Factual Counterfactual
	Consequence Both	Strictness	High Normal
Alternatives	None		Low
	Implicit Given	Wording	If Then Only If

been unusually frequent for this pattern. Regarding the single inferences, modus ponens was drawn in 100% of all considered cases followed by modus tollens with 86.1% and affirmation of consequent with 70.4%. Denial of antecedent was the least frequently drawn inference with only 52.8%.

Table 6. Inferential patterns and their number of occurrences in the considered dataset, as well as some unusually frequent features. Patterns are abbreviated in such a way that a T means a conclusion was drawn, and an F that is was not. The order is MP, MT, AC, DA.

Pattern	# occurr.	Unusually frequent features				
		Type	Negation	Abstraction	Familiarity	Strict.
TTTT	52	Prevention, Spatial	None	Concrete	Med., High	Normal
TTTF	9	Arbitrary	Cons., Both	Abstract	Low	High
TTFF	32	Def., Causal	Consequent	-	High	Low
TFTF	10	Arbitrary	Präm., Both	Abstract	Low	High
TFTT	5	Arbitrary	Prämise	Abstract	Low	High

Moreover, as explained in Sect. 3.2, each of the inference patterns induce a total preorder. Table 7 shows for each inference pattern a generating conditional knowledge base and the most plausible beliefs of the appertaining total preorder.

For describing the inference patterns in the following, we put a focus on their most salient features negation, alternatives, and abstraction.

Table 7. Inference patterns with a generating conditional knowledge base and most plausible beliefs of their appertaining total preorder

Inference pattern	Δ example	$\mathrm{Bel}(\Delta)$
(MP, MT, AC, DA)	$\{(B|A), (A|B)\}$	$\mathrm{Cn}(A \Leftrightarrow B)$
$(MP, \neg MT, AC, DA)$	$\{(B|A), (|A|\overline{B}|), (\overline{B}|\overline{A})\}$	$\mathrm{Cn}(AB)$
$(MP, MT, AC, \neg DA)$	$\{(\overline{A}|\overline{B}), (A|B), (|B|\overline{A}|)\}$	$\mathrm{Cn}(AB)$
$(MP, \neg MT, AC, \neg DA)$	$\{(B|A), (|A|\overline{B}|), (A|B)\}$	$\mathrm{Cn}(AB)$
$(MP, MT, \neg AC, \neg DA)$	$\{(B|A)\}$	$\mathrm{Cn}(A \Rightarrow B)$

Most of the examined tasks yield the (MP, MT, AC, DA) pattern, namely 52. In this case, test persons drew all inferences which is not logically correct. It is important to note that no experiments with Negation or Alternatives yielded this pattern, and it was only rarely found in abstract cases. Therefore one may assume that this pattern describes the "normal" commonsense reasoning behavior for everyday reasoning tasks best, where people understand an implication as a biconditional (which validates all four inference rules). Indeed, regarding Table 7, the most plausible beliefs show a logical equivalence, and the generating knowledge base contains the conditional in both ways.

There are only a few examples for the (MP, MT, AC, ¬DA) inference pattern, with 9 (8.3%) in total. This pattern only occurs in abstract experiments with no alternatives. It is also noteworthy that a negated consequence, possibly accompanied by a negated premise, appears to underline this pattern. Additionally the *Only if* wording is a very good indicator for this pattern. As the conditional knowledge base for this case reveals (see Table 7), people seem to take the consequence as a requirement for the premise, which is consistent with the idea that *Only if* wordings are effective "backwards".

(MP, MT, ¬AC, ¬DA) is the logically correct inference pattern. It is the second most frequently found one in the data set occurring in 32 reasoning tasks in total. This means that although the subjects do not reason logically correct in general, they still do so with a good share of almost 30%. All considered experiments that contained either a given or an implied alternative yielded this pattern. This is consistent with the findings in [2] that alternatives suppress logical fallacies. Surprisingly, a negated consequence also supports this pattern, while negated premises do not appear at all here. The level of abstraction does not seem to have any effect. These settings seem to prevent the biconditional understanding, people are inclined to assign a clear direction to the conditional. This is also reflected by the findings in Table 7, where the most plausible beliefs in this case are given by a logical implication, and the conditional knowledge base contains just a single conditional.

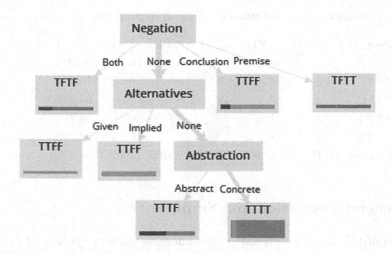

Fig. 1. Decision tree based on three core features: negation, alternatives, abstraction

The (MP, ¬MT, AC, ¬DA) inference pattern is the third most frequently used one. Almost all experiments that yielded this pattern were abstract and none of them included any alternatives. Negated premises, usually accompanied by negated consequences, support this pattern, while experiments with only negated consequences do not. Both conditional knowledge base and most plausible beliefs in Table 7 show a very strong connection between premise and conclusion which are both deemed to be plausible.

(MP, ¬MT, AC, DA) is the inference pattern which is least frequently found, in 5 tasks only. It is noteworthy, however, that all five of these experiments have exactly the same characteristics. All experiments yielding this pattern are abstract and do not contain any alternatives. Moreover, all of them have a negated premise and most experiments that have a negated premise yield this pattern. The conditional knowledge base for this pattern in Table 7 reveals that the premise is seen as a necessary prerequisite for the conclusion, matching the finding that no alternatives are indicated in the task.

A decision tree (see Fig. 1) using only the most salient features negation, alternatives, and abstraction was created using the Rapidminer program[1] by using cross-validation and decision-tree operators, and tested by cross-validation. Results can be seen in Table 8. This small decision tree achieved 81% accuracy overall, but only with lower precision and recall for the rarer inference patterns compared to larger trees with more features. Bigger trees with all ten features increase the accuracy to 90.84%, mostly increasing the values of rarer patterns [13].

---

[1] https://rapidminer.com/.

Table 8. Results of cross-validation on minimal decision tree with three features.

Pred\True	TTTF	TFTF	TTFF	TTTT	TFTT	Class precision
TTTF	1	0	0	4	0	20.00%
TFTF	2	6	0	0	0	75.00%
TTFF	2	0	28	0	0	93.33%
TTTT	4	1	4	48	0	84.21%
TFTT	0	3	0	0	5	62.50%
Class recall	11.11%	60.00%	87.50%	92.31%	100%	

5   General Discussion and Summary

The aim of this work was, based on the inference patterns presented in [5], to investigate the relationships between observed inference behavior of human subjects in psychological studies and characteristics of the corresponding empirical tasks, and to understand and explain the significant differences in observed inference patterns caused by slight variations of the task by considering conditional assumptions and most plausible beliefs that the inference patterns reveal. We described the most frequently used inference patterns and classified them with a decision tree by means of most salient features found in the tasks presented to the subjects.

Prominent features for classification are *negation, alternatives* and *abstraction*. The results presented in [2] confirm that alternatives suppress the logically invalid inferences. It was also noted that only one example in concrete tasks showed neither the inference pattern (MP, MT, AC, DA) nor $(MP, MT, \neg AC, \neg DA)$. It can therefore be assumed that people in everyday life, where they are often dealing with non-abstract and non-negated problems, mainly reason in these two patterns. Likewise, these results suggest that inference behavior of people may be influenced by deliberate application of negation and wording.

The results of this paper are valuable (at least) in two respects: From a formal perspective, the shown relationships between inference patterns and characteristics of the corresponding empirical tasks support a better commonsense interpretation of the abstract inference patterns and provide insights into the cognitive relevance of these patterns. From a psychological perspective, whether inference patterns are used or not, the influence of seemingly slight variations of test scenarios on inference behavior was revealed very clearly in this paper, and hypotheses for explaining these crucial differences were proposed that may trigger further empirical studies in the future.

Acknowledgements. This work was possible due to the DFG-projects KI1413/5-1 to G. Kern-Isberner and RA1934/2-1 as part of the priority program "New Frameworks of Rationality" (SPP 1516), and a Heisenberg DFG fellowship RA1934/3-1, /4-1 to M. Ragni.

References

1. Brachman, R., Levesque, H.: Knowledge Representation and Reasoning. Morgan Kaufmann Publishers, San Francisco (2004)
2. Byrne, R.M.: Suppressing valid inferences with conditionals. Cognition **31**(1), 61–83 (1989)
3. DeFinetti, B.: Theory of Probability: A Critical Introductory Treatment (Translated by Antonio Machí and Adrian Smith), vol. 1–2. Wiley, Hoboken (1975)
4. Dietz, E.A., Hölldobler, S., Ragni, M.: A computational approach to the suppression task. In: Miyake, N., Peebles, D., Cooper, R. (eds.) Proceedings of the 34th Annual Conference of the Cognitive Science Society, pp. 1500–1505. Cognitive Science Society, Austin (2012)
5. Eichhorn, C., Kern-Isberner, G., Ragni, M.: Rational inference patterns based on conditional logic. In: Proceedings of the Thirty-Second AAAI Conference on Artificial Intelligence (2018)
6. Evans, J.S.B.T., Handley, S.J.: The role of negation in conditional inference. Q. J. Exp. Psychol. Sect. A Hum. Exp. Psychol. **52**(3), 739–769 (1999)
7. Frosch, C.A., Byrne, R.M.: Causal conditionals and counterfactuals. Acta Psychol. **141**(1), 54–66 (2012)
8. Kern, L.H., Mirels, H.L., Hinshaw, V.G.: Scientists' understanding of propositional logic: an experimental investigation. Soc. Stud. Sci. **13**(1), 131–146 (1983)
9. Makinson, D.: General theory of cumulative inference. In: Reinfrank, M., de Kleer, J., Ginsberg, M.L., Sandewall, E. (eds.) NMR 1988. LNCS, vol. 346, pp. 1–18. Springer, Heidelberg (1989). https://doi.org/10.1007/3-540-50701-9_16
10. Ragni, M., Eichhorn, C., Bock, T., Kern-Isberner, G., Tse, A.P.P.: Formal non-monotonic theories and properties of human defeasible reasoning. Mind. Mach. **27**(1), 79–117 (2017)
11. Ragni, M., Kola, I., Johnson-Laird, P.N.: On selecting evidence to test hypotheses: a theory of selection tasks. Psychol. Bull. **144**(8), 779–796 (2018)
12. Ragni, M., Eichhorn, C., Kern-Isberner, G.: Simulating human inferences in the light of new information: a formal analysis. In: Proceedings of the Twenty-Fifth International Joint Conference on Artificial Intelligence, IJCAI 2016, New York, NY, USA, 9–15 July 2016, pp. 2604–2610 (2016). http://www.ijcai.org/Abstract/16/370
13. Spiegel, L.P.: Klassifikation von Beispielen aus kognitiven Studien mit Hilfe von Inferenzmustern. Bachelor thesis, Technische Universität Dortmund (2018)
14. Stenning, K., Lambalgen, M.: Human Reasoning and Cognitive Science. Bradford Books, MIT Press, Cambridge (2008)
15. Thompson, V.A.: Conditional reasoning: the necessary and sufficient conditions. Can. J. Exp. Psychol./Revue canadienne de psychologie expérimentale **49**(1), 1–60 (1995)

Harnessing Higher-Order (Meta-)Logic to Represent and Reason with Complex Ethical Theories

David Fuenmayor[1(✉)] and Christoph Benzmüller[1,2]

[1] Freie Universität Berlin, Berlin, Germany
david.fuenmayor@fu-berlin.de
[2] University of Luxembourg, Esch-sur-Alzette, Luxembourg

Abstract. The computer-mechanization of an ambitious explicit ethical theory, Gewirth's Principle of Generic Consistency, is used to showcase an approach for representing and reasoning with ethical theories exhibiting complex logical features like alethic and deontic modalities, indexicals, higher-order quantification, among others. Harnessing the high expressive power of Church's type theory as a meta-logic to semantically embed a combination of quantified non-classical logics, our work pushes existing boundaries in knowledge representation and reasoning. We demonstrate that intuitive encodings of complex ethical theories and their automation on the computer are no longer antipodes.

1 Introduction

Hybrid architectures for ethical autonomous agents that integrate both bottom-up learning and top-down deliberation from upper principles are receiving increased attention; cf. Dignum (2017, 2018); Scheutz (2017); Malle (2016); Dennis et al. (2016); Anderson and Anderson (2014); Wallach et al. (2008) and the references therein. Irrespective of the preferred direction, it is becoming increasingly evident that adequate explicit representations of ethical knowledge are beneficial, if not mandatory, to obtain satisfactory solutions.Bottom-up approaches may benefit from expressive languages to *explicitly* represent the learned ethical knowledge in an scrutable, communicable and transferable manner. Top-down approaches usually rely on expressive logic languages to enable an *intuitive and accurate representation and reasoning* with ethical theories. Unfortunately, however, very few approaches are currently available that enable adequate and realistic, explicit formal encodings of non-trivialized ethical theories, and that at the same time support intuitive interactive-automated reasoning with them.

In this paper *we demonstrate a methodology and implementation of such an ambitious ethical reasoning machinery.* Our approach is based on classical higher-order logic (HOL), aka Church's type theory (Benzmüller and Andrews 2019),

Supported by VolkswagenStiftung, grant *Consistent, Rational Arguments in Politics (CRAP).*

© Springer Nature Switzerland AG 2019
A. C. Nayak and A. Sharma (Eds.): PRICAI 2019, LNAI 11670, pp. 418–432, 2019.
https://doi.org/10.1007/978-3-030-29908-8_34

which we exploit as a meta-logic to encode combinations of non-classical logics for normative reasoning as suited for a given application context. The methodology and techniques we present, cf. also Benzmüller et al. (2019), can bring many benefits to the design of ethically-critical systems aiming at scrutability, verifiability, and the ability to provide justification for its decision-making. They are particularly relevant to the design of explicit ethical agents (Moor 2009). In particular, this area faces tough philosophical and practical challenges. No consensus is currently in sight, if possible at all, concerning the choice of upper moral values and principles that constitute a generally agreed normative ethics for intelligent autonomous agents. For example, utilitarianism and deontology have both been critically discussed in this context.

We exemplarily study another relevant and ambitious theory in normative ethics: Alan Gewirth's "Principle of Generic Consistency (PGC)" (Gewirth 1981; Beyleveld 1991), which has been proposed as an emendation of the *Golden Rule.* Our aim is not to defend or assess Gewirth's work in comparison to other approaches. We instead present a methodology and technique enabling the intuitive and accurate representation of ambitious ethical theories, and for this we take the PGC as a showcase and exemplarily assess its logical validity. Such an ambitious ethical theory has never before been assessed on the computer at such a level of detail (i.e. without trivializing it by abstraction).

Our method enables the reuse of modern interactive and automated higher-order theorem proving technology, and in this sense it establishes a *relevant bridge between different research communities.* On a practical level our work also addresses what we consider one of the biggest challenges in the area: to *represent complex ethical theories in both a machine and human interpretable manner and to carry out complex reasoning in real-time with incomplete and inconsistent information.* And finally, as a side-effect, we have *revealed and fixed some (minor) issues in Gewirth's PGC.*

Our choice of HOL at the meta-level is motivated by the goal of flexibly combining expressive non-classical logics as required for the formal encoding of complex ethical theories. Current theories in normative and machine ethics are, quite understandably, formulated predominantly in natural language. While this supports human deliberation and agreement about what kind of moral beings we want future intelligent agents to be, it also hampers their implementation in machines. Hence expressive formal languages are required, which enable flexible combinations of different types of non-classical logics. This is because ethical theories are usually challenged by complex linguistic expressions, including modalities (alethic, epistemic, temporal, etc.), counterfactual conditionals, generalized quantifiers, (un-)conditional obligations, among many others.

The meta-logical approach we exploit and demonstrate grounds on a technique known as *(shallow) semantical embedding.* The approach will be addressed in Sect. 2, where we present an extended embedding of a dyadic deontic logic (DDL) by Carmo and Jones (2002) in HOL and *combine, among others, conditional obligations with further modalities and quantifiers.* The combined logic is immune to known paradoxes in deontic logic, in particular, the so-called contrary-to-duty scenarios, in which a 'secondary' obligation must come into effect when a 'primary' obligation is violated (contradicted). Moreover, conditional (dyadic) obligations in DDL are of a defeasible and paraconsistent nature and thus lend themselves to normative reasoning with incomplete and inconsistent information. In Sects. 3 and 4 we will represent and formally assess Gewirth's PGC using this expressive logic combination. We also demonstrate how our technique has been utilized to reveal and fix some (minor) issues in Gewirth's work. Related work and short summary are presented in Sect. 5, and a formally-verified, unabridged version of our formal encoding of Gewirth's theory and argument is provided in Fuenmayor and Benzmüller (2018).

2   Combining Expressive Logics in HOL

We utilize the *shallow semantical embeddings* (SSE) approach to combining logics. SSE exploits HOL as a meta-logic in order to embed the syntax and semantics of some target logics, thereby turning theorem proving systems for HOL into universal reasoning engines (Benzmüller 2019). Moreover, an approach drawing upon SSE has beenproposed as the foundation for a flexible deontic logic reasoning infrastructure (Benzmüller et al. 2019). We thus assess, in some sense, the promises of this framework at hand of a non-trivial, concrete example.

In the following, we present *an extract* of the embedding of (extended) DDL in HOL. Our work thereby extends previous work by Benzmüller et al. (2018): Besides adding higher-order quantification, we also extend this embedding to a two-dimensional semantics (Schroeter 2017) by additionally adding contextual information; for this we use Kaplanian *contexts of use*, cf. Kaplan (1989a,b). The system platform used to implement this ambitious logic combination is the Isabelle proof assistant (Nipkow et al. 2002). In what follows, we are using Isabelle/HOL syntax to render axioms, theorems and definitions (providing the appropriate indications when needed).[1]

2.1   Definition of Types

The type w corresponds to the original type for possible worlds/situations in DDL, cf. Benzmüller et al. (2018). We draw in this work upon David Kaplan's *logic of indexicals/demonstratives* as originally presented in Kaplan (1989a). In Kaplan's logical theory, entities of the aforementioned type w would correspond to his so-called "circumstances of evaluation". Moreover, Kaplan introduces an additional dimension c, so-called "contexts of use", which allow for the modelling of particular context-dependent linguistic expressions, i.e. *indexicals* (see Sect. 2.4). We additionally introduce some type aliases: *wo* for intensions (also called "contents" or "propositions" in Kaplan's work), which are identified with their truth-sets i.e. the set of worlds at which the proposition is true, and *cwo*

---

[1] The formal content of this paper has been generated directly by Isabelle from our source files. A benefit is the prevention of typos. As a side contribution we showcase the usability of modern proof assistants for the non-initiated in order to foster their application.

(aliased m) for sentence meanings (also called "characters" in Kaplan's theory), which are modelled as functions from contexts to intensions. Moreover, a type e for individuals is introduced to e.g. enable quantification over individuals.

> **typedecl** w — Type for possible worlds (Kaplan's "circumstances of evaluation")
> **typedecl** c — Type for Kaplan's "contexts of use"
> **typedecl** e — Type for individuals
> **type-synonym** wo = w⇒bool — Type for contents/propositions
> **type-synonym** cwo = c⇒wo — Type for sentence meanings (Kaplan's "characters")
> **type-synonym** m = cwo   — Type alias 'm' for characters

2.2 Embedding of DDL Modal and Deontic Operators

The semantics of DDL draws on Kripke semantics for its (normal) alethic modal operators and on a neighbourhood semantics[2] for its (non-normal) deontic operators. In order to embed those, we need to introduce the operators av and pv (which can be seen as accessibility relations between worlds), and ob (denoting a neighborhood function operating on sets of worlds) at the meta-logical level. Several axioms, not shown here, adequately constraint the interpretations of av, pv and ob (e.g. $av(w)$ is always a subset of $pv(w)$). See Carmo and Jones (2002) and Benzmüller et al. (2018) for further details.

The following Isabelle/HOL commands illustrate the way logical operators in the target logic (enhanced DDL) can be defined as metalogical predicates using lambda expressions of the appropriate arity/type. The two definitions below, introduced using Isabelle's keyword "abbreviation", realize the embedding of the different modal box and diamond operators (shown here only for $\Box_a$ and $\Diamond_a$). Each of them is embedded as a function from sentence meanings to sentence meanings (type "$m \Rightarrow m$"), and they employ (restricted) quantification over possible worlds, following a Kripke semantics.[3]

> **abbreviation** cjboxa :: m⇒m ($\Box_a$-) **where** $\Box_a\varphi \equiv \lambda$c w. $\forall$v. (av w) v ⟶ (φ c v)
> **abbreviation** cjdiaa :: m⇒m ($\Diamond_a$-) **where** $\Diamond_a\varphi \equiv \lambda$c w. $\exists$v. (av w) v $\land$ (φ c v)

The following definitions correspond to the semantical embedding of DDL deontic operators in Isabelle/HOL. The first one represents conditional obligations of the form "φ must be the case given σ" and is embedded as a dyadic relation (type "$m \Rightarrow m \Rightarrow m$"). The second and third represent the so-called "actual" and "ideal" obligations.

---

[2] Neighbourhood semantics is a generalisation of Kripke semantics, developed independently by Dana Scott and Richard Montague. Whereas a Kripke frame features an accessibility relation $R : W \rightarrow 2^W$ indicating which worlds are alternatives to (or, accessible from) others, a neighborhood frame $N : W \rightarrow 2^{2^W}$ (or, as in our case, $N : 2^W \rightarrow 2^{2^W}$) features a neighbourhood function assigning to each world (or set of worlds) a set of sets of worlds.

[3] Note that in addition to the ASCII name "cjboxa", Isabelle/HOL supports graphical notation "($\Box_a$-)". This is essential for obtaining intuitive mathematical representations.

abbreviation cjod :: m$\Rightarrow$m$\Rightarrow$m $(\mathbf{O}\langle\text{-}|\text{-}\rangle)$ **where** $\mathbf{O}\langle\varphi|\sigma\rangle \equiv \lambda$c w. ob ($\sigma$ c) (φ c)
abbreviation cjoa :: m$\Rightarrow$m $(\mathbf{O}_a\text{-})$ **where**
$\quad \mathbf{O}_a\varphi \equiv \lambda$c w. (ob (av w)) ($\varphi$ c) $\wedge$ ($\exists$x. (av w) x $\wedge$ $\neg$(φ c x))
abbreviation cjop :: m$\Rightarrow$m $(\mathbf{O}_i\text{-})$ **where**
$\quad \mathbf{O}_i\varphi \equiv \lambda$c w. (ob (pv w)) ($\varphi$ c) $\wedge$ ($\exists$x. (pv w) x $\wedge$ $\neg$(φ c x))

2.3   Logical Validity (Classical)

The SSE technique also allows us to embed different notions of logical validity: context-dependent modal validity and general validity (modal validity in each context).

$\quad$**abbreviation** modvalidctx :: m$\Rightarrow$c$\Rightarrow$bool $(\lfloor\text{-}\rfloor^M)$ **where** $\lfloor\varphi\rfloor^M \equiv \lambda$c. $\forall$w. φ c w
$\quad$**abbreviation** modvalid :: m$\Rightarrow$bool $(\lfloor\text{-}\rfloor)$ **where** $\lfloor\varphi\rfloor \equiv \forall$c. $\lfloor\varphi\rfloor^M$ c

2.4   Kaplan's Context Features

Kaplan's theory, originally named "Logic of Demonstratives (LD)" (Kaplan 1989a,b), aims at modeling the behavior of certain context-sensitive linguistic expressions like the pronouns 'I', 'my', 'it', the demonstrative pronouns 'that', 'this', the adverbs 'here', 'now', 'tomorrow', the adjectives 'actual', 'present', and others. Such expressions are known as *indexicals* and so Kaplan's logical system, among others, is usually referred to as a "logic of indexicals".

It is characteristic of an indexical that its content varies with context, i.e. they have a context-sensitive character. Non-indexicals have a fixed character. LD models context-sensitivity by representing contexts as quadruples of features: $\langle Agent(c), Position(c), World(c), Time(c)\rangle$. The agent and the position of context c can be seen as the actual speaker and place of the utterance respectively, while c's world and time stand for the circumstances of evaluation of the expression's content and allow for the interaction of indexicals with alethic and tense modalities respectively. To keep things simple, we restrict ourselves to representing a context c as the pair: $\langle Agent(c), World(c)\rangle$ and model the functional concepts "Agent" and "World" as uninterpreted logical constants. An extension of our work to operate on Kaplan's context quadruples is straightforward.

$\quad$**consts** Agent::c$\Rightarrow$e — function retrieving the agent corresponding to context c
$\quad$**consts** World::c$\Rightarrow$w — function retrieving the world corresponding to context c

2.5   Indexical Validity

Kaplan's notion of (context-dependent) logical truth for a sentence corresponds to its context-sensitive formula (of type "m", i.e. "$c\Rightarrow w\Rightarrow bool$") being true in the given context and at its corresponding world. Kaplan's notion of logical validity for a sentence requires its truth in all contexts. This notion is also known as indexical validity.

$\quad$**abbreviation** ldtruectx::m$\Rightarrow$c$\Rightarrow$bool $(\lfloor\text{-}\rfloor\text{-})$ **where** $\lfloor\varphi\rfloor_c \equiv \varphi$ c (World c)
$\quad$**abbreviation** ldvalid::m$\Rightarrow$bool $(\lfloor\text{-}\rfloor^D)$ **where** $\lfloor\varphi\rfloor^D \equiv \forall$c. $\lfloor\varphi\rfloor_c$

The following lemmas show that indexical validity is indeed weaker than its classical modal counterpart (truth at all worlds for all contexts).

lemma $\lfloor A \rfloor \implies \lfloor A \rfloor^D$ **by** simp — proven using Isabelle's term-rewriting engine (simp)
lemma $\lfloor A \rfloor^D \implies \lfloor A \rfloor$ **nitpick oops** — countermodel

The *countermodel* computed by the model finder *Nitpick* (Blanchette and Nipkow 2010) for the latter lemma consists of one context c_1 and two worlds w_1 and w_2; where $\text{World}(c_1) = w_1$ and where A holds for c_1 and w_1, but not for c_1 and w_2 (*Nitpick* returns further insightful details which we omit here). Below we use *Nitpick* to show that the interplay between indexical validity and the DDL modal and deontic operators does not result in *modal collapse*. Moreover, we show that the necessitation rule does not work for indexical validity (in contrast to classical modal validity as defined for DDL).

lemma $\lfloor P \rightarrow O_a P \rfloor^D$ **nitpick oops** — countermodel for deontic modal collapse found
lemma $\lfloor P \rightarrow \Box_a P \rfloor^D$ **nitpick oops** — countermodel for alethic modal collapse found
lemma $\lfloor A \rfloor^D \implies \lfloor \Box_a A \rfloor^D$ **nitpick oops** — countermodel for necessitation rule found

Below we introduce a kind of "a priori necessity" operator (to be contrasted to the more traditional alethic necessity). This operator satisfies the necessitation rule for indexical validity.[4] In Kaplan's framework, a sentence being logically (i.e. indexically) valid means its being true *a priori*: It is guaranteed to be true in every possible context in which it is uttered, even though it may express distinct propositions (i.e. contents or intensions) in different contexts. This correlation between indexical validity and *a prioricity* has also been claimed in other two-dimensional semantic frameworks (Schroeter 2017).

abbreviation ldvalidbox :: m$\Rightarrow$m ($\Box^D$-) **where** $\Box^D \varphi \equiv \lambda c$ w. $\lfloor \varphi \rfloor^D$
lemma NecLD: $\lfloor A \rfloor^D \implies \lfloor \Box^D A \rfloor^D$ **by** simp — necessitation rule proven (term-rewriting)

2.6   Quantification

By utilizing Isabelle/HOL's parameterized types (rank-1 polymorphism), we can easily enrich our logic with (first-order and higher-order) quantifiers.

abbreviation mforall::('t$\Rightarrow$m)$\Rightarrow$m ($\forall$) **where** $\forall \Phi \equiv \lambda c$ w.$\forall$x. (Φ x c w)
abbreviation mexists::('t$\Rightarrow$m)$\Rightarrow$m ($\exists$) **where** $\exists \Phi \equiv \lambda c$ w.$\exists$x. (Φ x c w)

This definition of embedded parametric quantifiers (which reuses λ-abstraction to avoid the explicit introduction of a new binding mechanism) follows earlier work (Benzmüller and Paulson 2013). However, it is defined here for Kaplan's sentence meanings and in this sense constitutes another relevant extension of previous work.

3   Representing Gewirth's Ethical Theory

In this section we encode and mechanize Gewirth's (1981) ethical theory—respectively, ethical argument—which aims at justifying an upper moral principle called the "Principle of Generic Consistency" (PGC). In a nutshell, according

---

[4] Note that $\Box^D$ is not part of Kaplan's original system. It has been added by us in order to better highlight some semantic features of our formalization of Gewirth's theory in the next section and for enabling the use of the necessitation rule for drawing inferences.

to this principle, any intelligent agent (by virtue of its self-understanding as an agent) is rationally committed to asserting that (i) it has rights to freedom and well-being, and (ii) all other agents have those same rights. The argument used by Gewirth to derive the PGC (presented in detail in Gewirth (1981); Beyleveld (1991)) is by no means trivial and has stirred much controversy in legal and moral philosophy during the last decades. It has also been discussed in political philosophy as an argument for the *a priori* necessity of human rights (Beyleveld 2012). Perhaps more relevant for us, the PGC has lately been proposed as a means to bound the impact of artificial general intelligence (AGI) by Kornai (2014).

Kornai draws on Gewirth's PGC as the paradigmatic principle which, assuming it can reliably be represented in a machine, will enable the design of a safety mechanism of a mathematical nature that ensures that an AGI will always respect basic human's rights over all other things. This is based on the assumption that such an intelligent agent is able to recognize itself, as well as humans, as agents acting voluntarily on self-chosen purposes, i.e. as what Gewirth calls: prospective purposive agents (PPA). Every agent designed to follow the PGC will thus be deductively committed, on pain of self-contradiction, to acting in accord with the *generic* rights (i.e. to freedom and well-being) of all agents.[5]

3.1   Gewirth's Ethical Theory

Gewirth's meta-ethical position is known as moral (or ethical) rationalism. According to it, moral principles are knowable *a priori*, by reason alone. Immanuel Kant is the most famous figure who has defended such a position. He argued for the existence of upper moral principles (e.g. his "categorical imperative") from which we can reason in a top-down fashion to deduce and evaluate other more concrete maxims and actions. In contrast to Kant, Gewirth derives such upper moral principles by starting from purely logical (i.e. non-moral) considerations alone. The argument for the PGC employs what Gewirth calls "the dialectically necessary method" within the "internal viewpoint" of an agent. Although the logical inferences leading to the PGC are drawn relative to the reasoning agent, Gewirth (1981) further argues that *"the dialectically necessary method propounds the contents of this relativity as necessary ones, since the statements it presents reflect judgements all agents necessarily make on the basis of what is necessarily involved in their actions ... The statements the method attributes to the agent are set forth as necessary ones in that they reflect what is conceptually necessary*

---

5 Our work constitutes a most relevant first step for further assessment of Kornai's claim. E.g. we plan to embody our encoding of Gewirth's theory in virtual agents and devise and conduct respective empirical studies. The merits of the work presented here are however not tied to the validity of Kornai's claim. We illustrate that representation and reasoning with complex ethical theories is meanwhile feasible to an extent unmatched before; and this is highly relevant for implementing explicit ethical intelligent systems. In the following, we will present some commented extracts of our formal encoding of Gewirth's theory and of the computer-supported verification of the argument leading to the PGC.

to being an agent who voluntarily or freely acts for purposes he wants to attain." In other words, the "dialectical necessity" of the assertions and inferences made in the argument comes from the definitional features (i.e. conceptual analysis) of the involved notions of agency, purposeful action, obligation, rights, etc. In order to adequately represent this informal notion of *a priori* dialectical/analytic necessity, we resorted to the formal notion of *indexical validity* as developed in David Kaplan's logical framework LD (Kaplan 1989a,b).

The cogency of Gewirth's theory will be put to the test in Sect. 4 by using it to reconstruct his argument (with minor fixes) for the PGC as logically valid. However, we first need to introduce the basic theory itself. To get some inspiration we study the main steps of Gewirth's argument (with original numbering from Beyleveld (1991)):

(1) [**Premise**] I act voluntarily for some (freely chosen) purpose E—equivalent by definition to: I am a prospective purposive agent (PPA).
(2) E is (subjectively) good—i.e. I value E proactively.
(3) My freedom and well-being (FWB) are generically necessary conditions of my agency—i.e. I need them to achieve any purpose whatsoever.
(4) My FWB are necessary goods (at least for me).
(5) I have (maybe nobody else does) a claim right to my FWB.
(13) [**Conclusion**] Every PPA has a claim right to their FWB.

In his informal proof, Gewirth claims that the latter generalization step (from "I" to all agents) is done on purely logical grounds and does not presuppose any kind of universal moral principle, and his result is meant to hold with some kind of necessity.[6] In this respect, Deryck Beyleveld, author of an authoritative book on Gewirth's theory (1991), comments on its first page: *"[Gewirth's] argument purports to establish the PGC as a rationally necessary proposition with an apodictic status for any PPA equivalent to that enjoyed by the logical principle of noncontradiction itself."*

In what follows, we provide some *meaning postulates*[7] for the core ethical concepts used to articulate both the PGC and the argument leading to it (as outlined above). We illustrate how to exploit the expressivity of our embedded object logic (DDL enhanced with quantifiers and contexts) to *intuitively* represent and mechanize such a complex ethical theory for the first time in a computer. We also illustrate the utilization of interactive proof assistants (Isabelle/HOL) to assess the argument and to reason with Gewirth's theory.

3.2 Agency

Since Isabelle/HOL is a based on a Church's functional type theory, we need to assign all terms a type. We give "purposes" the same type as sentence meanings

---

[6] We were indeed able to formally verify Gewirth's claim, on condition of committing to an alternative notion of (logical) necessity: Kaplan's "indexical validity".

[7] Definitions and axiomatized conceptual interrelations framing the inferential role of terms. We also refer to them as "explications". Meaning postulates were introduced in Carnap (1952).

(type '$c\Rightarrow w\Rightarrow bool$' aliased 'm'), so that "acting on a purpose" is represented analogously to having a certain propositional attitude (like "desiring that so and so ... "). The terms "ActsOnPurpose" and "NeedsForPurpose" obtain functional types, and thus expressions like "(ActsOnPurpose A E)" and "(NeedsForPurpose A P E)" are read as "agent A acts on purpose E" and "agent A needs to have property P in order to reach purpose E". We also define a type alias p for properties (functions mapping individuals to characters).

> **type-synonym** p = e⇒m — function from individuals to sentence meanings (characters)
> **consts** ActsOnPurpose:: e⇒m⇒m
> **consts** NeedsForPurpose:: e⇒p⇒m⇒m

In Gewirth's argument, an individual with agency (i.e. capable of purposive action) is said to be a PPA (prospective purposive agent). This definition is supplemented with a meaning postulate stating that being a PPA is an essential (i.e. identity-constitutive) property of an individual. Quite interestingly, this postulate entails a kind of ability for a PPA to recognize other PPAs.[8] For instance, if some individual holds itself as a PPA (seen from its own perspective/context'd') then this individual 'Agent(d)' is considered a PPA from any other agent's perspective/context 'c'.

> **definition** PPA:: p **where** — Definition of PPA
> **axiomatization where** essentialPPA: $\lfloor \forall$a. PPA a $\rightarrow \Box^D(\text{PPA a})\rfloor^D$
> **lemma** recognizeOtherPPA: $\forall$c d. $\lfloor$PPA (Agent d)$\rfloor_d \longrightarrow \lfloor$PPA (Agent d)$\rfloor_c$
>   **using** essentialPPA **by** blast — proven using Isabelle blast tactic (tableaux)

3.3   Goodness

Gewirth's concept of (subjective) goodness applies to purposes and is relative to some agent. It is thus modeled as a binary relation relating an individual (of type 'e') with a purpose (of type 'm'). The axioms below are meaning postulates interrelating the concept of goodness with agency and are given as indexically valid sentences (in Kaplan's sense).[9] In particular, we have noticed the need to postulate a further axiom (*explGoodness3*), which represents the intuitive notion of "seeking the good" by asserting that, from an agent's perspective, necessarily good purposes are not only action motivating, but also entail an instrumental obligation to their realization (but only where possible).

> **consts** Good::e⇒m⇒m
> **axiomatization where**
>   explGoodness1: $\lfloor \forall$a P. ActsOnPurpose a P $\rightarrow$ Good a P$\rfloor^D$
>   explGoodness2: $\lfloor \forall$P M a. Good a P $\wedge$ NeedsForPurpose a M P $\rightarrow$ Good a (M a)$\rfloor^D$
>   explGoodness3: $\lfloor \forall \varphi$ a. $\Diamond_p\varphi \rightarrow \mathbf{O}\langle\varphi \mid \Box^D$Good a $\varphi\rangle\rfloor^D$

---

[8] Lemma "recognizeOtherPPA" below is indeed inferred from axiom "essentialPPA" using Isabelle's *blast* tactic (a tableaux prover).

[9] Their higher-order and modal nature well illustrates the need for expressive knowledge representation and reasoning techniques.

3.4   Freedom and Well-Being

According to Gewirth, enjoying freedom and well-being (which we take together as the predicate "FWB") is the *contingent* property which represents the "necessary conditions" or "generic features" of agency (i.e. FWB is *always* required in order to be able to act on *any* purpose whatsoever). As before, we take this as an *a priori* characteristic of FWB and therefore axiomatize it as an indexically valid sentence. The last two axioms postulate that FWB is a contingent property.

> **consts** FWB::p — FWB is a property (has type $e{\Rightarrow}m$)
> **axiomatization where**
>   explicationFWB1: $\lfloor \forall P\ a.\ \text{NeedsForPurpose a FWB P} \rfloor^D$
>   explicationFWB2: $\lfloor \forall a.\ \Diamond_p\ \text{FWB a} \rfloor^D$
>   explicationFWB3: $\lfloor \forall a.\ \Diamond_p\ \neg\text{FWB a} \rfloor^D$

3.5   Obligation and Interference

Kant's Law ("ought implies can") plays an important role in Gewirth's argument.[10] We have noticed the need to slightly amend it in order to render the argument as logically valid. The new variant reads as: "ought implies *ought to* can". Our variation is indeed closer to Gewirth's (1981, pp. 91–95) textual description, that having an obligation to do X implies that *"I ought (in the same sense and the same criterion) to be free to do X, that I ought not to be prevented from doing X, that my capacity to do X ought not to be interfered with."*[11]

> **lemma** $\lfloor O_i\varphi \rightarrow \Diamond_p\varphi \rfloor$ **using** sem-5ab **by** simp
> **axiomatization where** OIOAC: $\lfloor O_i\varphi \rightarrow O_i(\Diamond_a\varphi) \rfloor^D$

Concerning the concept of interference, we have noticed the need to presume that the existence of an individual b (successfully) interfering with some state of affairs φ implies that φ cannot possibly be obtained in any of the actually possible situations (and the other way round). This axiom implies that if someone (successfully) interferes with agent a having FWB, then a can no longer possibly enjoy its FWB (and the converse).

> **consts** InterferesWith::$e{\Rightarrow}m{\Rightarrow}m$
> **axiomatization where** explicationInterference: $\lfloor (\exists b.\ \text{InterferesWith b } \varphi) \leftrightarrow \neg\Diamond_a\varphi \rfloor$
> **lemma** InterferenceWithFWB: $\lfloor \forall a.\ (\exists b.\ \text{InterferesWith b (FWB a)}) \leftrightarrow \neg\Diamond_a(\text{FWB a}) \rfloor$
>   **using** explicationInterference **by** blast

3.6   Rights and Other-Directed Obligations

Gewirth (1981, p. 66) points out the existence of a correlation between an agent's own claim rights and other-referring obligations. A claim right is a right which entails duties or obligations for other agents regarding the right-holder (so-called

---

[10] This theorem is indeed derivable directly in DDL from the definition of obligations: If φ oughts to obtain then φ is possible.

[11] Below we use Isabelle's *simp* tool to prove that Kant's lemma follows from one of the DDL semantic conditions (not shown here).

Hohfeldian claim rights in legal theory). We model this concept of claim rights in such a way that an individual a has a (claim) right to having some property φ if and only if it is obligatory that every (other) individual b does not interfere with the state of affairs (φ a). Since there is no particular individual to whom this directive is addressed, this obligation has been referred to by Gewirth as being "other-directed" (aka. "other-referring") in contrast to "other-directing" obligations which entail a moral obligation for some particular subject (Beyleveld 1991, p. 41, 51). This latter distinction is essential to Gewirth's argument.

definition RightTo::e⇒(e⇒m)⇒m **where** RightTo a φ ≡ $\mathbf{O}_i$(∀b. ¬InterferesWith b (φ a))

Now that all axioms of the theory are in place, we need to show that they are indeed logically consistent. For this we use Isabelle's model finder *Nitpick* to compute a corresponding model (not shown here) having one context, one individual and two worlds.

lemma True **nitpick**[satisfy, card c = 1, card e = 1, card w = 2] **oops** — model found

4   Reasoning with Gewirth's Ethical Theory

The PGC can be seen as a particular variant (or emendation) of the *golden rule*: treating others as one's self would wish to be treated. A self-acknowledged agent (i.e. a PPA) would read the PGC as a moral commandment: "I ought to act in accord with the generic rights of my recipients as well as of myself" (Gewirth 1981, p. 153). Urging a fellow human being to obey such a principle without having explained its deeper rationale will presumably at best elicit an absent-minded, cursory acknowledgment. The difficulty here lies not only in the lack of understanding or agreement of what the given words mean (what is a "generic right"?), but also in the addressee's lack of 'immersion' in the underlying conceptual framework and the inferential practices behind such a principle (an unaware addressee would not be able to infer a third-party obligation from a right claim). In short, any moral principle *qua sentence* makes best sense in the context of the background theory from which it is obtained as a well-founded part; this has been argued e.g. by the philosopher Quine in his holistic view of meaning (cf. 1960).

This situation is not much different for machines. In order to correctly interpret and apply an ethical principle, we need to (i) determine the meaning of its constituent concepts (action/agency, right, freedom and well-being, etc.); and (ii) determine the meaning of other relevant concepts (goodness, necessity, interference, obligation, etc.) playing a role in its articulation (and justification) within the underlying theory. Talk of meanings can be obscure, so let us put it in model-theoretical terms: The set of models of the logical theory has to be constrained to properly fit the target conceptualization (i.e. to only entail intended models). These constraints are set by meaning postulates, i.e. axioms and definitions. Their adequacy can be assessed by studying the extent to which they enable the validation (or invalidation) of candidate theorems (or non-theorems). As is already known, the main theorem we aim at validating here is the PGC, suitably

```
 55    ● ● ●              ☰ Isabelle2018/HOL – GewirthArgument.thy
 56  (**The following is a formalized proof for the main conclusion of Gewirth's argument, which
 57  asserts that the following sentence is valid from every PPA's standpoint: "Every PPA has a
 58  claim right to its freedom and well-being (FWB)" *)
 59  theorem PGC: shows "⌊∀x. PPA x → (RightTo x FWB)⌋ᵘ"
 60  proof - {
 61    fix C::c (**'C' is some arbitrarily chosen context (agent's perspective)*)
 62    {
 63      fix I::"e" (**'I' is some arbitrarily chosen individual (agent's perspective)*)
 64      {
 65        fix E::m (**'E' is some arbitrarily chosen purpose*)
 66        {
 67          (**(1) I act voluntarily on purpose E:*)
 68          assume P1: "⌊ActsOnPurpose I E⌋c"
 69          (**(1a) I am a PPA:*)
 70          from P1 have P1a: "⌊PPA I⌋c" using PPA_def by auto
 71          (**(2) purpose E is good for me:*)
 72          from P1 have C2: "⌊Good I E⌋c" using explGoodness1 essentialPPA by meson
 73          (**(3) I need FWB for any purpose whatsoever:*)
 74          from explicationFWB1 have C3: "⌊∀P. NeedsForPurpose I FWB P⌋ᵘ" by simp
 75          hence "∃P.⌊Good I P ∧ NeedsForPurpose I FWB P⌋ᵘ"
 76            using explicationFWB2 explGoodness3 sem_5ab by blast
 77          (**FWB is (a priori) good for me (in a kind of definitional sense):*)
 78          hence "⌊Good I (FWB I)⌋ᵘ" using explGoodness2 by blast
 79          (**(4) FWB is an (a priori) necessary good for me:*)
 80          hence C4: "⌊□ᵘ(Good I (FWB I))⌋c" by simp
 81          (**I ought to pursue my FWB on the condition that I consider it to be a necessary good:*)
 82          have "⌊O(FWB I | □ᵘ(Good I) (FWB I))⌋c" using explGoodness3 explicationFWB2 by blast
 83          (**There is an (other-directed) obligation to my FWB:*)
 84          hence "⌊Oᵢ(FWB I)⌋c" using explicationFWB2 explicationFWB3 C4 CJ_14p by fastforce
 85          (**It must therefore be the case that my FWB is possible:*)
 86          hence "⌊Oᵢ(◇ₐ(FWB I))⌋c" using OIOAC by simp
 87          (**There is an obligation for others not to interfere with my FWB:*)
 88          hence "⌊Oᵢ(∀a. ¬InterferesWith a (FWB I))⌋c" using InterferenceWithFWB by simp
 89          (**(5) I have a claim right to my FWB:*)
 90          hence C5: "⌊RightTo I FWB⌋c" using RightTo_def by simp
 91        }
 92        (**I have a claim right to my FWB (since I act on some purpose E):*)
 93        hence "⌊ActsOnPurpose I E → RightTo I FWB⌋c" by (rule impI)
 94      }
 95      (**In the following "allI" is the logical generalization rule: all-quantifier introduction*)
 96      hence "⌊∀P. ActsOnPurpose I P → RightTo I FWB⌋c" by (rule allI)
 97      (**I have a claim right to my FWB since I am a PPA:*)
 98      hence "⌊PPA I → RightTo I FWB⌋c" using PPA_def by simp
 99    }
100    (**Every agent has a claim right to its FWB since it is a PPA:*)
101    hence "∀x. ⌊PPA x → RightTo x FWB⌋c" by simp
102  }
103  (**(13) For every perspective C: every agent has a claim right to its FWB:*)
104  thus C13: "∀C. ⌊∀x. PPA x → (RightTo x FWB)⌋c" by (rule allI)
105  qed
```

Fig. 1. Gewirth's proof encoded in the Isabelle/HOL proof assistant.

paraphrased as: *Every PPA has a claim right to its freedom and well-being.* The reconstructed proof in Isabelle/HOL of the theorem below is shown in Fig. 1.

theorem PGC: **shows** $\forall C.\ \lfloor PPA\ (Agent\ C) \to (RightTo\ (Agent\ C)\ FWB)\rfloor_C$

In Sects. 2 and 3, besides from formally articulating Gewirth's theory, we have used some of Isabelle's proof methods (simp, blast, etc.) and the *Nitpick* model finder to verify some relevant inferences and to guarantee consistency, thus the theory's adequacy has already partly been assessed. In addition, we have used a combination of interactive and automated theorem proving to reconstruct Gewirth's argument for the PGC as logically valid by formally proving it within the complex logical framework built so far. We thus contribute an exemplary case study illustrating how to reason with highly-expressive formal representations of complex, natural-language ethical theories by harnessing the power of higher-order theorem provers (drawing on the SSE approach). In the argument's reconstruction as displayed in Fig. 1, some of the intermediate inference steps leading to the main conclusion (PGC) have indeed been hinted at by automated tools; cf. Fuenmayor and Benzmüller (2019, 2018) for further details. In particular, some missing implicit premises (not considered in Gewirth's original argument) have been uncovered, namely the explications of the concepts of *goodness* and *interference* and the amendment to Kant's Law: "ought implies *ought to* can". Note that the mechanized argument matches the granularity-level as can also be found in human constructed informal arguments, and all the sub-arguments (sub-proofs) can automatically be found by automated theorem proving technology. Moreover, the whole proof as presented can be automatically verified using a standard laptop in under a second.

5   Related Work and Summary

We achieve several improvements over related work such as Bringsjord et al. (2006) and Furbach and Schon (2015): (i) Due the use of enriched DDL (enabled by our higher-order meta-logic) we are not suffering from contrary-to-duty issues; (ii) we make use of truly higher-order encodings as required for the adequate modeling of the PGC; (iii) we overcome unintuitive, machine-oriented formula representations; and (iv) we do not stop with supporting proof automation, but combine it with intuitive user interaction. Combinations of (i)–(iv) also apply to more recent related work by Govindarajulu and Bringsjord (2017), Hooker and Kim (2018) and Pereira and Saptawijaya (2016), which are not applicable to complex theories like Gewirth's PGC without considering significant simplifications (accepting e.g. contrary-to-duty issues is potentially dangerous).

Utilizing a semantical embedding of a suitable combination of expressive non-classical logics in meta-logic HOL, an ambitious ethical theory, Gewirth's PGC, has exemplarily been encoded and mechanized on the computer. Our methodology supports both highly intuitive representation of and interactive-automated reasoning with the encoded theory. Automated theorem provers have even helped to reveal some hidden issues in Gewirth's argument. The presented methodology is motivating research in different, albeit related, directions: (i) for conducting analogous formal assessments of further ambitious ethical theories, and (ii) for progressing with the implantation of explicit ethical reasoning competencies in future intelligent autonomous systems *by adapting state-of-the-art theorem proving technology and by combining the expertise of different research communities.*

References

Anderson, M., Anderson, S.L.: GenEth: a general ethical dilemma analyzer. In: Twenty-Eighth AAAI Conference on Artificial Intelligence (2014)

Benzmüller, C.: Universal (meta-)logical reasoning: recent successes. Sci. Comput. Program. **172**, 48–62 (2019). https://doi.org/10.1016/j.scico.2018.10.008. Url (preprint): http://doi.org/10.13140/RG.2.2.11039.61609/2

Benzmüller, C., Andrews, P.: Church's type theory. In: Zalta, E.N. (ed.) The Stanford Encyclopedia of Philosophy. Metaphysics Research Lab, Stanford University (2019). https://plato.stanford.edu/entries/type-theory-church/

Benzmüller, C., Paulson, L.: Quantified multimodal logics in simple type theory. Logica Univers. **7**(1), 7–20 (2013). https://doi.org/10.1007/s11787-012-0052-y. (Special Issue on Multimodal Logics)

Benzmüller, C., Farjami, A., Parent, X.: A dyadic deontic logic in HOL. In: Broersen, J., Condoravdi, C., Nair, S., Pigozzi, G. (eds.) Deontic Logic and Normative Systems – 14th International Conference, DEON 2018, Utrecht, The Netherlands, 3–6 July 2018, pp. 33–50. College Publications (2018). ISBN 978-1-84890-278-7. John-Jules Meyer Best Paper Award

Benzmüller, C., Parent, X., van der Torre, L.W.N.: Designing normative theories of ethical reasoning: formal framework, methodology, and tool support. CoRR, abs/1903.10187 (2019). http://arxiv.org/abs/1903.10187

Beyleveld, D.: The dialectical necessity of morality: an analysis and defense of Alan Gewirth's argument to the principle of generic consistency. University of Chicago Press (1991)

Beyleveld, D.: The principle of generic consistency as the supreme principle of human rights. Hum. Rights Rev. **13**(1), 1–18 (2012). ISSN 1874-6306

Blanchette, J.C., Nipkow, T.: Nitpick: a counterexample generator for higher-order logic based on a relational model finder. In: Kaufmann, M., Paulson, L.C. (eds.) ITP 2010. LNCS, vol. 6172, pp. 131–146. Springer, Heidelberg (2010). https://doi.org/10.1007/978-3-642-14052-5_11. ISBN 978-3-642-14051-8

Bringsjord, S., Arkoudas, K., Bello, P.: Toward a general logicist methodology for engineering ethically correct robots. IEEE Intell. Syst. **21**(4), 38–44 (2006)

Carmo, J., Jones, A.J.I.: Deontic logic and contrary-to-duties. In: Gabbay, D.M., Guenthner, F. (eds.) Handbook of Philosophical Logic, pp. 265–343. Springer, Dordrecht (2002). https://doi.org/10.1007/978-94-010-0387-2_4

Carnap, R.: Meaning postulates. Philos. Stud. **3**(5), 65–73 (1952)

Dennis, L.A., Fisher, M., Slavkovik, M., Webster, M.: Formal verification of ethical choices in autonomous systems. Robot. Auton. Syst. **77**, 1–14 (2016). https://doi.org/10.1016/j.robot.2015.11.012

Dignum, V.: Responsible autonomy. In: IJCAI 2017, pp. 4698–4704 (2017)

Dignum, V: Special issue: ethics and artificial intelligence. Ethics Inf. Technol. **20**(1) (2018)

Fuenmayor, D., Benzmüller, C.: Formalisation and evaluation of Alan Gewirth's proof for the principle of generic consistency in Isabelle/HOL. Archive of Formal Proofs (2018). https://www.isa-afp.org/entries/GewirthPGCProof.html

Fuenmayor, D., Benzmüller, C.: Isabelle/HOL sources associated with this PRICAI-2019 paper (2019). http://bit.ly/Appendix-PRICAI-19

Furbach, U., Schon, C.: Deontic logic for human reasoning. In: Eiter, T., Strass, H., Truszczyński, M., Woltran, S. (eds.) Advances in Knowledge Representation, Logic Programming, and Abstract Argumentation. LNCS (LNAI), vol. 9060, pp. 63–80. Springer, Cham (2015). https://doi.org/10.1007/978-3-319-14726-0_5

Gewirth, A.: Reason and Morality. University of Chicago Press, Chicago (1981)

Govindarajulu, N.S., Bringsjord, S.: On automating the doctrine of double effect. In: Proceedings of the Twenty-Sixth International Joint Conference on Artificial Intelligence, IJCAI 2017, pp. 4722–4730 (2017). https://doi.org/10.24963/ijcai.2017/658

Hooker, J.N., Kim, T.W.N.: Toward non-intuition-based machine and artificial intelligence ethics: a deontological approach based on modal logic. In: Proceedings of the 2018 AAAI/ACM Conference on AI, Ethics, and Society, pp. 130–136. ACM (2018)

Kaplan, D.: Demonstratives. In: Almog, J., Perry, J., Wettstein, H. (eds.) Themes from Kaplan, pp. 481–563. Oxford University Press, Oxford (1989a)

Kaplan, D.: Afterthoughts. In: Almog, J., Perry, J., Wettstein, H. (eds.) Themes from Kaplan, pp. 565–612. Oxford University Press, Oxford (1989b)

Kornai, A.: Bounding the impact of AGI. J. Exp. Theor. Artif. Intell. **26**(3), 417–438 (2014)

Malle, B.F.: Integrating robot ethics and machine morality: the study and design of moral competence in robots. Ethics Inf. Technol. **18**(4), 243–256 (2016)

Moor, J.: Four kinds of ethical robots. Philos. Now **72**, 12–14 (2009)

Nipkow, T., Wenzel, M., Paulson, L.C. (eds.): Isabelle/HOL: A Proof Assistant for Higher-Order Logic. LNCS, vol. 2283. Springer, Heidelberg (2002). https://doi.org/10.1007/3-540-45949-9

Pereira, L.M., Saptawijaya, A.: Programming Machine Ethics. SAPERE, vol. 26. Springer, Cham (2016). https://doi.org/10.1007/978-3-319-29354-7

Van Orman Quine, W.: Word and Object. MIT Press, Cambridge (1960)

Scheutz, M.: The case for explicit ethical agents. AI Mag. **38**(4), 57–64 (2017)

Schroeter, L.: Two-dimensional semantics. In: Zalta, E.N. (eds.) The Stanford Encyclopedia of Philosophy. Metaphysics Research Lab, Stanford University (2017)

Wallach, W., Allen, C., Smit, I.: Machine morality: bottom-up and top-down approaches for modelling human moral faculties. AI Soc. **22**(4), 565–582 (2008)

Aleatoric Dynamic Epistemic Logic for Learning Agents

Tim French[(✉)], Andrew Gozzard, and Mark Reynolds

The University of Western Australia, Crawley, Western Australia
{tim.french,mark.reynolds,andrew.gozzard}@uwa.edu.au

Abstract. We propose a generalisation of dynamic epistemic logic, where propositions are aleatoric: that is, rather than having true/false values, propositions have odds of being true. Agents in such a system suppose a probability distribution of possible worlds, and based on observations are able to refine this probability distribution to match their observations. We demonstrate this logic with respect to some games of chance.

Keywords: Probabilsitic logic · Game playing ·
Dynamic epistemic logic

1 Introduction

Dynamic epistemic logic (DEL) has been widely applied for reasoning about games and security [7,8]. In practical applications in these domains, agents' belief models have an element of probability, and there has been considerable work investigating probabilistic extensions to DEL [1,3,15]. Here, rather than extending a propositional modal logic with the capability to represent and reason about probabilities, we apply the recent development of the *modal aleatoric calculus* [11] to revise all logical operators so that they are interpreted probabilistically. This subtle difference takes us from *reasoning about probabilities* to *reasoning probabilistically*, and is a core principal of Bayesian epistemiology [4].

The games we are interested in are games of chance and bluffing. Typically these games have a hidden epistemic state, so that the knowledge of all agents is not equal. There is also an element of chance, either coming through an initial deal of cards, or some random element such as a dice or coin. Finally there should be a strategic advantage to having knowledge, so players have an incentive to discover what their opponent knows, and to hide their knowledge from an opponent. Such games include traditional games such as Poker, and Bridge, and more recent games such as Clue, Werewolf or Love Letter. *Aleatoric* comes from the Latin word for dice and literally means "depending on the throw of dice". This describes both explicit elements of such games (card deals, dice rolls for example) as well as the policies and strategies of players in the game (so a player may bluff 10% of the time).

© Springer Nature Switzerland AG 2019
A. C. Nayak and A. Sharma (Eds.): PRICAI 2019, LNAI 11670, pp. 433–445, 2019.
https://doi.org/10.1007/978-3-030-29908-8_35

Our aim is to provide a lightweight logic, *aleatoric dynamic epistemic logic*, for formalising reasoning processes in games of chance. This has broader applications in reasoning in multi-agent systems with a degree of uncertainty. The aleatoric dynamic epistemic logic allows agents to express strategies or theories of how other agents will act. By observing the actions of other agents, the logic uses Bayesian conditioning to update the agents' belief models.

To demonstrate the logic, we will use the game, *The Resistance*, which is a card game where players are required to sabotage one another without revealing their true purpose, and the *Dining Cryptographers Problem* [5] which is a well known puzzle in epistemic reasoning.

1.1 The Resistance

The Resistance[1] by Don Eskridge, is a bluffing game for five to ten players and is similar to the games Werewolf and Mafia. Approximately one third of the players are allocated as being government spies, while the rest are true members of the resistance. The spies know each others' identity, but the true members of the resistance do not know who is a spy. The game consists of a number of rounds. Each round proceeds as follows:

1. A leader is allocated (randomly, or the person to the left of the previous leader)
2. The leader proposes a group of players to go on a "mission". The size of the group is given (depending on the number of players and round), and the leader may include themselves.
3. All players vote publicly on whether they support the choice. If a majority support it the mission proceeds. Otherwise, the leadership moves to the left, and the process starts again. If five missions are voted against in a row, the spies are declared the winner.
4. The mission succeeds only if no one betrays the mission. Each player on the mission plays a token (face down) to indicate whether they betray the mission. These are shuffled and then revealed to everyone. If a betrayal token was played, the spies win, otherwise the resistance wins.

The first group to win three rounds wins the game. The true members of the resistance would like a majority of missions to succeed, whilst the spies would like a majority to fail. As the spies are in a minority, they must do everything they can to hide their true identity, and the identity of the other spies. However, they also need to influence the debate and vote so that the spies are sent on enough missions to achieve their goal.

This game actually has relatively little uncertainty in it. The only randomness is in the initial assignment of spies, and the spies themselves have generally got perfect information about the state of the game (the only exception is that the decision to betray is taken simultaneously by all spies on a mission). However, the limited uncertainty for the non-spies is enough to make a compelling game,

---

[1] http://www.indieboardsandcards.com/resistance.php.

and the relatively simple sets of actions available to players makes the game ideal to analyse. In fact all actions can be modelled as public announcements [16].

2 Related Work

Modal and epistemic logics have been applied for reasoning about uncertainty in multi-agent systems [14], and more recent work on dynamic epistemic logic [8,16] has looked at how agents incorporate new information into thier belief structures. There are explicit probabilistic extensions of these logics, that maintain the Boolean interpretation of formulas, but include probabilistic terms [9,10,15].

These logics use the many possible worlds interpretation of uncertainty, with a probability distribution over the worlds. Then the modality becomes the expected likelihood of a world satisfying the proposition, coupled with a comparative operator ($>$ or $\geq$), so we can express that the likelihood of a proposition holding is greater than another. Such logics are able to reason *about* probabilities, so an agent may reason "It is more likely to rain than it is likely to snow, but both are less than 50%". The consequence of this is that these logics can not apply marginalisation on an observation, as all formulas are either true or false). In these papers the dynamic component only removes impossible state and normalizes the probabilities. Not being able to apply Bayesian conditioning on observations makes them very weak for reasoning about games such as The Resistance, where another players actions can reveal a lot about the likelihood of the hidden state.

This is quite different to the many valued approach, where probabilities are not explicit parts of a formula, but intrinsic in the semantics. Halpern's book [13] gives an excellent overview of these approaches, and the representation of uncertainty in multi-agent systems based on Dempster-Shafer models of belief [18]. Of particular note is the work of Kooi [15] and van Benthem [3] extending dynamic epistemic logic with explicit probabilities. In these cases, the informative updates such as public announcements are realised as Bayesian conditioning. Baltag and Smets [2] have provided similar extensions in the context of belief revision.

Recently the paper [11] has presented a variation on modal logic, where variables and formulas are aleatoric, rather than Boolean. That is, they may be modelled as independent random events, like to roll of dice. The paper also presented the *modal aleatoric calculus* for computing probability preserving transformations.

3 Syntax and Semantics

Probabilistic uncertainty is difficult to model and hard to reason about, because sets of events or variables have dependencies that are complex to represent. So for an agent to reason about probabilities, we have the challenges of determining what dependencies exists between variables, what the agent knows about these dependencies, and how observations effect what the agent knows about these dependencies.

Following [11], we present *Aleatoric Dynamic Epistemic Logic* (ADEL), which is a generalisation of dynamic epistemic logic to apply to aleatoric variables. The difference with probabilistic dynamic epistemic logic (PDEL) [15] is subtle: In PDEL it is possible to express that the statement "Alice thinks X has probability 0.5" is true; whereas the language here simply has a term "Alice's expectation of X" which may have a value that is greater than 0.5. We present a syntax for constructing complex terms in this logic, and a semantics for assignment values to terms, given a particular interpretation or model.

3.1   Syntax

The syntax is given for a set of random variables X, and a set of agents N. We also have constants $\top$ and $\bot$. The syntax of aleatoric dynamic epistemic logic, ADEL, is as follows:

$$\alpha ::= x \mid \top \mid \bot \mid (\alpha?\alpha:\alpha) \mid (\alpha|\alpha)_i \mid [\alpha]\alpha$$

where $x \in X$ is a random variable and $i \in N$ is an agent. As usual, we let $v(\alpha)$ refer to the set of variables that appear in α. We refer to $\top$ as *always* and $\bot$ as *never*. The *if-then-else* operator $(\alpha?\beta:\gamma)$ is read *if α then β else γ* and uses the ternary conditional syntax of programming languages such as C. The *marginal expectation* operator $(\alpha|\beta)_i$ is *agents i's expectation of α given β* (the marginal probability i assigns to α given β). The *global observation* operator $[\alpha]\beta$ is *the expectation of β once α is observed by all agents*. This corresponds to Bayesian conditioning on a public announcement of α.

Some abbreviations we can define in ADEL are as follows:

$$\begin{array}{ll}
\alpha \wedge \beta = (\alpha?\beta:\bot) & \alpha \vee \beta = (\alpha?\top:\beta) \\
\neg\alpha = (\alpha?\bot:\top) & \alpha \to \beta = (\alpha?\beta:\top) \\
\alpha^{\frac{0}{b}} = \top & \alpha \leftrightarrow \beta = (\alpha?\beta:\neg\beta) \\
\alpha^{\frac{a}{b}} = \bot \text{ if } b < a \neq 0 & E_i\alpha = (\alpha|\top)_i \\
\alpha^{\frac{a}{b}} = (\alpha?\alpha^{\frac{a-1}{b-1}}:\alpha^{\frac{a}{b-1}}) \text{ if } b \geq a \neq 0 & B_i\alpha = (\bot|\neg\alpha)_i
\end{array}$$

where a and b are natural numbers. The boolean abbreviations are correspond to fuzzy modal logic with the product norm [19] but $\alpha^{\frac{a}{b}}$, $E_i\alpha$ and $B_i\alpha$ are new. The modality $E_i\alpha$ is agent i's expectation of α being true, which is just α conditioned on the uniformly true $\top$. The operator $B_i\alpha$ uses a property of the conditional operator: it evaluates $(\alpha|\beta)_i$ as vacuously true if and only if there is no expectation that β can ever be true. Therefore, $(\bot|\neg\alpha)_i$ can only be true if agent i always expects $\neg\alpha$ to be false, and thus agent i *believes* α. The formula $\alpha^{\frac{a}{b}}$ (α *a out of b*) allows us to explicitly represent degrees of belief in the language. It is interpreted as α *is true at least a times out of b*. Note that this is not a statement saying what the frequency of α is. Rather it describes the event of α being true a times out of b. Therefore, if α was unlikely (say true 5%

of the time) then $\alpha^{\frac{9}{9}}$ describes a very unlikely event. This allows us to encode degrees of belief, which is seen most clearly in the context of the if-then-else operator: $(\alpha^{\frac{4}{5}}?\beta : \gamma)$ represents "if α is very likely, then return the expectation for β, otherwise give the expectation for γ", where the α being "very likely" is the expectation that α will be true in 4 out of 5 times.

3.2   Semantics

Aleatoric dynamic epistemic logic is interpreted over *probability models* similar to the probability structures defined in [13], although they have aleatoric variables in place of propositional assignments.

Definition 1. *Given a set S, we use the notation $PD(S)$ to notate the set of probability distributions over S, where $\mu \in PD(S)$ implies: $\mu : S \longrightarrow [0,1]$; and $\Sigma_{s \in S} \mu(s) = 1$.*

Definition 2. *Given a set of variables X and a set of agents N, a probability model is specified by the tuple $P = (W, \pi, f)$, where:*

- *W is a set of possible worlds.*
- *$\pi : N \longrightarrow W \longrightarrow PD(W)$ assigns for each agent, for each world $w \in W$, a probability distribution $\pi_i(w)$ over W such that for all i, for all $u, v \in W$ $\pi_i(u, v) > 0$ implies $\pi_i(u) = \pi_i(v)$. We will write $\pi_i(w, v)$ in place of $\pi(i)(w)(v)$.*
- *$f : W \longrightarrow X \longrightarrow [0,1]$ is a probability assignment so for each world w, for each variable x, $f_w(x)$ is the probability of x being true.*

A pointed probability model, $P_w = (W, \pi, f, w)$, specifies a world in the model as the point of evaluation.

We note the condition on π enforces the property: $\forall i \in N$, $\forall w, u, v \in W$, $\pi_i(w, u) > 0$ implies $\forall v$, $\pi_i(u, v) = \pi_i(w, v)$, so if $\pi_i(w, u) > 0$, $\pi_i(u, u) = \pi_i(w, u)$. However, it is still possible that $\pi_i(w, w) = 0$. This aligns with the modal logic KD45, which is transitive, Euclidean and serial, and is often applied for reasoning about belief [6].

Given a pointed model P_w, the semantic interpretation of a ADEL formula α is $P_w(\alpha) \in [0,1]$ which is the expectation of the formula being supported by a sampling of the model, where the sampling is done with respect to the distributions specified by π and f.

Definition 3. *The semantics of aleatoric dynamic epistemic logic take a pointed probability model, f_w, and a proposition defined in ADEL, α, and calculate the expectation of α holding at P_w. Given an agent i, a world w and a ADEL formula α, we define i's expectation of α at w as*

$$E_w^i(\alpha) = \sum_{u \in W} \pi_i(w, u).P_u(\alpha).$$

Then the semantics of ADEL *are as follows:*

$$P_w(\top) = 1 \quad P_w(\bot) = 0 \quad \overset{\bullet}{P_w}(x) = f_w(x)$$
$$P_w((\alpha?\beta:\gamma)) = P_w(\alpha).P_w(\beta) + (1 - P_w(\alpha)).P_w(\gamma)$$
$$P_w((\alpha | \beta)_i) = \frac{E_w^i(\alpha \wedge \beta)}{E_w^i(\beta)} \text{ if } E_w^i(\beta) > 0 \text{ and } 1 \text{ otherwise}$$
$$P_w([\alpha]\beta) = P_w^\alpha(\beta)$$

where P^α *is the model* (W, π', f) *such that for all* $u, v \in W$, $\pi_i'(u, v) = \frac{P_v(\alpha).\pi_i(u,v)}{E_u^i(\alpha)}$ *if* $E_u^i(\alpha) > 0$, *and* $\pi_i(u, v)$ *otherwise.*

These semantics deserve some discussion, but first we should show that they are well formed. The interpretation of $P_w([\alpha]\beta)$ is given with respect to the model P_w^α, so it is required that P_w^α is a probability model.

Lemma 1. *Given a probability model* P *and some formula of* ADEL, α, *the structure* P^α *is a probability model.*

Proof. As P^α only varies from P in the definition of π', it is sufficient to show that for all $w \in W$ and all $i \in A$, $\pi_i^\alpha(w)$ is a probability distribution of W. In the case that $E_w^i(\alpha) = 0$, we have $\pi_i^\alpha(w) = \pi_i(w)$, so the result follows immediately. When $E_w^i(\alpha) > 0$, we must show:

1. For all v, $\pi_i^\alpha(w, v) \in [0, 1]$. This follows since we have $\pi_i^\alpha(w, v) = P_v(\alpha).\pi_i(w, v)/E_w^i(\alpha)$ and also $E_w^i(\alpha) \geq P_v(\alpha).\pi_i(w, v) \geq 0$.
2. $\sum_{u \in W} \pi_i^\alpha(w, u) = 1$. This follows since:

$$\sum_{u \in W} \pi_i^\alpha(w, u) = \sum_{u \in W} \frac{P_u(\alpha).\pi_i(w, u)}{\sum_{v \in W} P_v(\alpha)\pi_i(w, v)} = 1.$$

3. For all $u, v \in W$, $\pi_i^\alpha(w, v) > 0$ implies $\pi_i^\alpha(w, u) = \pi_i^\alpha(v, u)$. Expanding these definitions, we have

$$\pi_i^\alpha(w, u) = \frac{P_u(\alpha).\pi_i(w, u)}{E_w^i(\alpha)} = \frac{P_u(\alpha).\pi_i(v, u)}{E_v^i(\alpha)} = \pi_i^\alpha(v, u).$$

Therefore, P^α is a probability model.

The concept of *sampling* is intrinsic in the rationale of these semantics. The word *aleatory* has its origins in the Latin for dice-player (*aleator*), and we suppose that our agents are committed aleators, in that they use dice (or sample probability distributions) for everything. We imagine these semantics being interpreted by agents armed with a set of labelled coins. If we ask "is x true" the agent will take the coin marked x, flip it and if it lands heads, reply "yes". Every formula is evaluated as a sampling process this way. To interpret $(\alpha?\beta : \gamma)$, the agent will execute the sampling procedure for α and if it returns true, the agent will proceed with the sampling procedure for β, otherwise the agent will continue to sample γ.

The marginal operator $(\alpha \mid \beta)_i$ expresses agent i's expectation of α *marginalised* by β. The intuition for these semantics corresponds to a sampling protocol. The agent i samples a world from their probability distribution and sample β. If β is true, then i samples α at that world and returns the result. Otherwise agent i resamples a world from their probability distribution, and repeats the process. In the case that β is never true, we assign $(\alpha \mid \beta)_i$ probability 1, as being vacuously true.

The observation operation $[\alpha]\beta$ is the expectation of β after α is *observed by all agents* (or *publicly announced* in the terminology of dynamic epistemic logic). The interpretation of α is also stochastic, so we imagine that as before, the mental model of the universe is sampled, and α is true in that sampling. Further, we suppose that all agents are told that α was true in that sampling. Now every agent updates their mental model of the universe, taking this new information into account. The pointed model P_w is their prior expectation of the universe, and we apply Bayesian conditioning to determine the new (posterior) model of the universe. The Bayesian conditioning is applied only to π (the probability distribution of different worlds) and not to f, (the probability distribution of random variables in a single world). The reason for this is that in a single world, all propositions are independent, so conditioning would have no effect. Finally, note that if an agent assigns zero probability to α at a world, they do not modify their probability distribution for that world. That is, the agent refuses to accept new information that contradicts its current beliefs. However, the agent still recognises that α was publicly announced, and other agents who do not consider α impossible would have accepted the information and updated their beliefs accordingly.

Finally, we note the distinction between the marginal operator and the observation operator. The marginal operator $(\alpha \mid \beta)_i$ is modal and allows the agent to speculate about the likelihood of α being sampled given that β was sampled. This captures the concept of dependence: each sampling of an aleatoric variable is an independent event, but it does depend on the possible world in which the sampling is done.

The observation operator, on the other hand, models belief change. Rather than an agent speculating about the relationship between two formula, it involves the agent observing a sampling event, and using this observation to update their distribution of worlds. Whilst both operators reflect the concept of dependence between formulas, the marginal operator is passive, whilst the observation operation is not.

4   Example

Here we present a simple analysis of a small version of the game, The Resistance. We suppose that there are four players, $\{1, 2, 3, 4\}$, and two of them are spies. This gives six possible configurations at the start of the game The spies know the identity of all the other spies, but the other players do not. The non-spies only know that they are not spies, and therefore assign equal probability to the

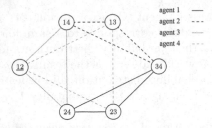

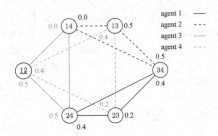

(a) The resistance common prior. The worlds are labelled with the agents who are spies, and each agent considers all linked worlds equally likely.

(b) The probability model after one of agents 2 and 3 betrays the mission (some rounding has been applied).

Fig. 1. A model of agents' knowledge before an after an action in the game, The Resistance

three worlds in which they are not spies, and zero probability to the worlds in which they are spies. Finally, we will suppose that all players have common assumptions about how other players will behave. Particularly for every agent, $i \in \{1, 2, 3, 4\}$, there is a variable x_i, which is the probability of agent i betraying a mission, if i is a spy on that mission. We will suppose that for all i, for all worlds where i is a spy, x_i has initial value $\frac{3}{4}$. There are also variables s_i to dictate who is a spy, so s_1 has probability 1 in worlds w_{12}, w_{13} and w_{14} and probability 0 in all other worlds. This gives the model depicted in Fig. 1a, which is a common prior for all players if the game. The left-most world is underlined, as that is the actual world, where agents 1 and 2 are spies.

Now suppose that 2 and 3 are sent on a mission, and 2 betrays the mission. All agents are informed that exactly one agent betrayed the mission, which is equivalent to the announcement $(x_2 \wedge \neg x_3) \vee (\neg x_2 \wedge x_3)$. We can calculate this event has 0 probability in the world (14), since neither 2 nor 3 are spies in that world. The event has $\frac{3}{4}$ probability in worlds (12), (13), (24) and (34), and probability $\frac{87}{256}$ in world (23).

Every agent can infer different information from this announcement. Agent 3 will know 2 is a spy, and assigns equal probability to 1 and 4 being spies. Agent 4 does not know who is a spy, but the fact that only one agent betrayed the mission makes it less likely that both agents 2 and 3 are spies, so agent 4's expectation that agent 1 is a spy actually increases. That is,

$$E_4 s_1 = \frac{2}{3} \text{ and } [(x_2 \wedge \neg x_3) \vee (\neg x_2 \wedge x_3)]E_4 s_1 = \frac{384}{481}.$$

The ADEL allows us to express more complex policies for agents. In the instance described above an agent simply flips a biased coin (with probability $\frac{3}{4}$ of coming up betray). However, we could also specify a policy whereby agent 2 will betray, if both non-spies think 2 is a spy, or if both non-spies think 2 is not a spy. That is, in world w_{12} of Fig. 1a agent 2's likelihood of betraying a mission could be

$$P_{w_{12}}(E_2(E_3s_2 \wedge E_4s_2) \vee E_2(E_3\neg s_2 \wedge E_4\neg s_2)) = \frac{46}{81}. \qquad (1)$$

We note that this policy doesn't use any aleatoric variables (there is no flipping of coins) and the action depends entirely on the uncertainty of other agents. More complex policies could combine aleatoric variable and agent uncertainty. If another agent assumes that player 2 is using the policy represented by (1), then after observing player 2's actions the player could apply Bayesian conditioning to the uncertainty functions. Thus the players' policies and actions naturally evolve and respond to the information inherent in the actions of other players.

5 Actions

In this section we generalise the concept of global observation, with the concept of an *action*, which corresponds to a move in a game. In dynamic epistemic logic, such change is achieved through action models [8], and the probabilistic dynamic epistemic logic of Kooi [15] has been extended by Sack with action models [17]. Here we give an account of a similar extension in the context of Aleatoric Dynamic Epistemic Logic.

An action differs from a global observation in two ways: it can change the likelihood of a random variable, and it can model effects that are not symmetric for all agents. Actions are composed of *possible events* and for each agent i, we include a *null event*, $\aleph_i$, to model consequence for agents who believe an event is impossible.

Definition 4. *An action is described by a tuple $\mathcal{A} = (E, \sim_i, pre, post)$, where*

- *E is a finite set of possible* events, *including the null events $\aleph_i$.*
- *$\sim_i \subseteq E \times E$ is the uncertainty relation for each agent i, such that*
  - *$\sim_i$ is Euclidean, serial and transitive,*
  - *$\forall e \in E \setminus \{\aleph_j \mid j \neq i\}(\aleph_i, e) \in \sim_i$, and*
  - *$\forall e \in E, (e, \aleph_i) \notin \sim_i$.*
- *$pre : E \longrightarrow$ ADEL assigns a pre-condition to each event, where $pre(\aleph_i) = \top$*
- *$post : E \longrightarrow X \hookrightarrow$ ADEL is a post condition that reassigns the likelihood for some variables, where $dom(post(\aleph_i)) = \emptyset$.*

A pointed action is $\mathcal{A}_e$ is a action with a specific event specified. Given an event $e \in E$, we let $[e]_i$ be the set $\{e' \mid e \sim_i e'\}$.

For comparison we see that are very similar to the action models of [8], but allowing for the unsuccessful announcements of [12], and ontic change. The semantics for action execution are as follows:

Definition 5. *Given a probability model $P = (W, \pi, f)$ and an action $\mathcal{A} = (E, \sim_i, pre, post)$, the execution of $\mathcal{A}$ on P is the probability model $\mathcal{A} \otimes P = (W', \pi', f')$, where:*

- $W' = \{(e,w) \in E \times W \mid P_w(pre(e)) > 0\}$,

$$\pi'_i((e_1,u),(e_2,v)) = \begin{cases} 0, & if\ e_2 \notin [e_1]_i,\ and\ e_2 \neq \aleph_i \\ \frac{P_v(pre(e_2)).\pi_i(u,v)}{\Sigma_{e \in [e_1]_i} P_v(E_i pre(e))}, & if\ e_2 \in [e_1]_i\ and\ \Sigma_{e \in [e_1]_i} P_u(E_i pre(e)) > 0 \\ \pi_i(u,v)\ if\ e_2 = \aleph_i\ and\ \Sigma_{e \in [e_1]_i} P_u(E_i pre(e)) = 0 \end{cases}$$

- if $x \in dom(post(e))$, then $f'_{(e,w)}(x) = P_w(post(e)(x))$ and $f'_{(e,w)}(x) = f_w(x)$ otherwise.

If P_w is a pointed probability model and $\mathcal{A}_e$ is a pointed action, then $(\mathcal{A} \otimes P)_{(e,w)}$ is the execution of $\mathcal{A}_e$ on P_w (also written $\mathcal{A}_e \otimes P_w$).

As with Lemma 1 we can show $\mathcal{A}_e \otimes P_w$ is always a pointed probability model. The null events $\aleph_i$ model the effect of an agent witnessing an event that is inconsistent with their beliefs. For example, if an agent witnessed a coin land tails, when they believed the coined to be double headed, belief revision is typically required. Here we duck the issue, assuming that agents are presidentially resolute in their beliefs, and if they witness a contradictory event they will refuse to believe it. This is modelled by the agent believing the null event occurred, which did not have any ontic effect, although it may have impacted the beliefs of other agents.

An action is a generalisation of a global observation. Ignoring the post function or multiple agents, an action $\mathcal{A}_e$ can be thought of as stochastic global observations of pre(e') for each $e' \in [e]_i$, normalised by agents i's expectation of e'. In fact, every global observation is an action with a single event, and a trivial post function, along with the corresponding null events for each agent.

It is also a little strange that we do not require any stochastic element for an action, even though many games have stochastic actions (e.g. rolling a die, or drawing a card from a deck). Instead of encoding this stochastic choice we model it as an element of the probability model, so a fair coin would be modelled by a random variable h, with value 0.5 in every world, and an event predicated on the coin landing heads would have precondition h, and an event predicated on tails would have precondition $\neg h$. This allows for stochastic elements of actions to vary with worlds (e.g. if it is possible the coin is biased, or if we may have a "hot deck" containing mostly high value cards).

We can include actions as a syntactic element in the same way that global observations are include in ADEL.

Definition 6. *The* Aleatoric Action Model Logic *consists of formulas*

$$\alpha ::= x \mid \top \mid \bot \mid (\alpha?\alpha:\alpha) \mid (\alpha|\beta)_i \mid [\mathcal{A}_e]\alpha$$

where $x \in X$, $i \in N$ *and* $\mathcal{A}_e$ *is a pointed action model. Given a pointed probability model* P_w, *the semantics are as in Definition 3.2, with the additional clause:*

$$P_w([\mathcal{A}_e]\alpha) = (\mathcal{A}_e \otimes P_w)(\alpha)\ if\ P_w(pre^{\mathcal{A}}(e)) > 0$$
$$and\ P_w([\mathcal{A}_e]\alpha) = 0\ otherwise.$$

6 Example

As a simple example we consider the *dining cryptographers problem* [5], a well known example of epistemic reasoning. In this problem three cryptographers are dining and their meal is anonymously paid for. They wish to know if one of them paid for it without infringing the benefactor's right to anonymity. The protocol is that each pair flip a coin hidden from the third, and then they publicly announce whether the coins they saw were the same, but they invert their response if they were the one who paid for the meal. If an even number (possibly 0) claim that the coins were the same, then no one inverted their response, so the meal must have been paid for by an external party.

This problem has been extensively studied in the context of epistemic logics, assuming the knowledge state created by the coin flips. With aleatoric action model logic we are also able to model the coin flips as actions, as in Fig. 2. The action to reveal whether the coins are the same is a standard public announcement, as in Fig. 3. The probability model P, starts as four states **a**, **b**, **c** and **none** (respectively, a paid, b paid, c paid and no one paid), with variables for the coin c, whether some agent at the table paid p, and the observed coin flips (all initialised to 0.0). The only uncertainty between the states is in whether the meal was paid for by one of the agents. In this instance we will assume that there are three agents a, b, and c.

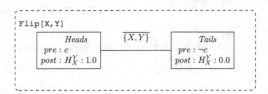

Fig. 2. An action model capturing the coin flip of the dining cryptographers problem. The action is specified for an arbitrary pair of agents X and Y, and has states *Heads* and *Tails* with the pre-condition depending on an random variable c (the coin). The random variable H_X^Y is set to 1.0 iff the coin flipped was heads. Any agent not in $\{X,Y\}$ will not know H_X^Y, but X and Y will. The null actions $\aleph_i$ have been omitted.

The application of $\texttt{Flip}[\texttt{X},\texttt{Y}]$ for each pair of agents builds a probability model, where a fragment of the first step is shown in Fig. 4b. The full protocol requires a further 2 coin flips leading to a 32 state model which we will not show here. The subsequent actions of $\texttt{Reveal}$ eliminates either all but 6 states (in the case one agent paid), or all but 2 states (in the case no agent paid).

This is a simple 3 agent instance of the dining cryptographers problem, but we can see that the use of aleatoric reasoning allows us to investigate richer aspects of the problem. In the original problem, the coin is simply used to initialise uncertainty, and then non-probabilistic epistemic reasoning is applied. However, if the coin was biased, and one of the agents (say a) knew the bias, aleatoric

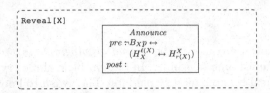

Fig. 3. The action of a player, X, revealing if the coins were the same or different, where $\ell(X)$ is the agent to X's left and $r(X)$ is the agent to X's right. The precondition is predicated on whether X believes that the meal was paid for by someone at the table.

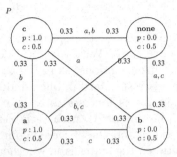

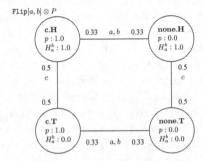

(a) The initial state of the model. It is assumed that there is a fair coin (c) that maybe sampled at any time. As none of the coins have been flipped at this stage, their value is not reported.

(b) A fragment of the model after a and b have observe a common coin flip, recorded as H_a^b. We only include the states where either c paid or no one paid.

Fig. 4. A sketch of a probabilsitic action in the dining cryptographers problem

action model logic would give the probability a assigns to each other agent paying, diminishing the anonymity in the protocol.

7 Conclusion

This project aims to resolve the discrepancies between the reasoning capabilities of dynamic epistemic logic, and the pragmatics of building game playing agents. Aleatoric Dynamic Epistemic Logic is a true generalisation of public announcement logic, but gives a much richer language that allows agents to weigh the observations they make against past experience a converge on a set of beliefs that accurately models the agents' experience.

References

1. Baltag, A., Smets, S.: Group belief dynamics under iterated revision: fixed points and cycles of joint upgrades. In: Proceedings of the 12th TARK, pp. 41–50 (2009)
2. Baltag, A., Smets, S.: Probabilistic dynamic belief revision. Synthese **165**(2), 179 (2008)

3. van Benthem, J., Gerbrandy, J., Kooi, B.: Dynamic update with probabilities. Stud. Logica **93**(1), 67–96 (2009)
4. Bovens, L., Hartmann, S.: Bayesian Epistemology. Oxford University Press on Demand, Oxford (2003)
5. Chaum, D.: The dining cryptographers problem: unconditional sender and recipient untraceability. J. Cryptol. **1**(1), 65–75 (1988)
6. Cocchiarella, N.B., Freund, M.A.: Modal Logic: An Introduction to Its Syntax and Semantics. Oxford University Press, Oxford (2008)
7. Dechesne, F., Wang, Y.: To know or not to know: epistemic approaches to security protocol verification. Synthese **177**(1), 51–76 (2010)
8. van Ditmarsch, H., van der Hoek, W., Kooi, B.: Dynamic Epistemic Logic. Synthese Library, vol. 337. Springer, Heidelberg (2007). https://doi.org/10.1007/978-1-4020-5839-4
9. van Eijck, J., Schwarzentruber, F.: Epistemic probability logic simplified. In: Proceedings of AiML, pp. 158–177 (2014)
10. Fagin, R., Halpern, J.Y., Megiddo, N.: A logic for reasoning about probabilities. Inf. Comput. **87**(1–2), 78–128 (1990)
11. French, T., Gozzard, A., Reynolds, M.: A modal aleatoric calculus for probabilistic reasoning. In: Khan, M.A., Manuel, A. (eds.) ICLA 2019. LNCS, vol. 11600, pp. 52–63. Springer, Heidelberg (2019). https://doi.org/10.1007/978-3-662-58771-3_6
12. Gerbrandy, J.: The surprise examination. Synthese **155**(1), 21–33 (2007)
13. Halpern, J.Y.: Reasoning About Uncertainty. MIT Press, Cambridge (2017)
14. Hintikka, J.: Knowledge and Belief. Cornell University Press, Ithaca (1962)
15. Kooi, B.P.: Probabilistic dynamic epistemic logic. J. Logic Lang. Inform. **12**(4), 381–408 (2003)
16. Plaza, J.: Logics of public communications. In: Proceedings of the 4th ISMIS, pp. 201–216. Oak Ridge National Laboratory (1989)
17. Sack, J.: Extending probabilistic dynamic epistemic logic. Synthese **169**(2), 241–257 (2009)
18. Shafer, G.: Dempster-shafer theory. In: Encyclopedia of Artificial Intelligence, pp. 330–331 (1992)
19. Zadeh, L.A.: Fuzzy sets. In: Zadeh, L.A. (ed.) Fuzzy Sets, Fuzzy Logic, and Fuzzy Systems: Selected Papers, pp. 394–432. World Scientific (1996)

Mastering Uncertainty: Towards Robust Multistage Optimization with Decision Dependent Uncertainty

Michael Hartisch[✉] and Ulf Lorenz

Chair of Technology Management, University of Siegen, Siegen, Germany
{michael.hartisch,ulf.lorenz}@uni-siegen.de

Abstract. We investigate, as a special case of robust optimization, integer linear programs with variables being either existentially or universally quantified. They can be interpreted as two-person zero-sum games between an existential and a universal player. In this setting the existential player must ensure the fulfillment of a system of linear constraints, while the universal variables can range within given intervals, trying to make the fulfillment impossible. We extend this approach by adding a linear constraint system the universal player must obey. Consequently, existential and universal variable assignments in early decision stages now can restrain possible universal variable assignments later on and vice versa resulting in a multistage optimization problem with *decision dependent uncertainty*. We present novel insights in structure and complexity.

1 Introduction

Mixed-integer linear programming (MIP) [24] is the state-of-the art technique for computer aided optimization of real world problems. Nowadays, commercial top solvers are able to solve large MIPs of practical size, but companies observe an increasing danger of disruptions, which prevent them from acting as planned. Thus, there is a need for planning and deciding under uncertainty. Uncertainty, however, often pushes the complexity of problems that are in the complexity class P or NP, to PSPACE [7,21]. Prominent solution paradigms for optimization under uncertainty are Stochastic Programming [5], Robust Optimization [3], Dynamic Programming [2], Sampling [13] and others, e.g. approximation techniques [17] and on-line optimization [8]. Relatively unexplored are the abilities of linear programming extensions for PSPACE-complete problems. In the early 2000s the idea of universally quantified variables, as they are used in quantified constraint satisfaction problems [11], was picked up again [25] - coining the term quantified integer program (QIP) - and further examined [10,19]. Quantified integer programming gives the opportunity to combine traditional linear

Partially supported by the German Research Foundation (DFG) project "Advanced algorithms and heuristics for solving quantified mixed - integer linear programs".

© Springer Nature Switzerland AG 2019
A. C. Nayak and A. Sharma (Eds.): PRICAI 2019, LNAI 11670, pp. 446–458, 2019.
https://doi.org/10.1007/978-3-030-29908-8_36

programming formulations with some uncertainty bits. Hence, a solution of a
QIP is a strategy for assigning existentially quantified variables such that some
linear constraint system is fulfilled. By adding a minmax objective function one
further must find the best strategy [10]. As not unusual in the context of optimization under uncertainty [3,4] a polyhedral uncertainty set can be used [15].
Recently, we made our solver for quantified integer programs – which combines
techniques known from game tree search, linear programming and (quantified)
boolean formula – available as open source [9].

Results from optimization under uncertainty sometimes tend to be too pessimistic and the uncertainty must be implemented carefully. In robust optimization uncertain parameters are bound within a given uncertainty set, often having
polyhedral, ellipsoidal or some other convex shape [4]. However, after specifying
the uncertainty set, this domain is fixed and no endogenous effects can be considered. Within the scope of the multistage optimization problem QIP, we will
introduce the possibility to influence the uncertainty set, making it dependent
on decisions made in previous stages. To the best of our knowledge only a few
results in the area of stochastic programming [1,12,14,16] and robust optimization [18,20,23,26] can be found regarding such an influence on uncertainty. This
neglection is probably due to worries regarding even higher complexity.

We will introduce decision dependent uncertainty sets in quantified integer
programs, making the domains for both variable types interdependent. These
Quantified Integer Programs with Interdependent Domains (QIP^{ID}) have the
following properties:

1. even a local information whether a variable of a QIP^{ID} is allowed to be set
   to a specific value demands the solution of an NP-complete problem. This is
   very different to classic robust optimization and also different to games like
   chess or go, where it is simple to check whether a certain move is legal.
2. despite the heavy intuitive differences between QIP and QIP^{ID}
   (a) QIP^{ID} is still in PSPACE.
   (b) QIP and QIP^{ID} are surprisingly close to each other: a polynomial time
       reduction from QIP^{ID} to QIP (the other way is trivial) is presented.

The paper is organized as follows: In Sect. 2 basic definitions and notations
regarding QIPs are introduces. In Sect. 3 an extension of QIPs is presented allowing the manipulation of the domain of uncertain variables and an extended minimax value is outlined. In Sect. 4 a parameterized polynomial reduction function
is presented mapping instances of the newly defined problem to QIP instances.

2 Basics of Quantified Integer Programming

Let $n \in \mathbb{N}$ be the number of variables and $x = (x_1,\ldots,x_n)^\top \in \mathbb{Z}^n$ a vector[1]
of variables.[2] For each variable x_k its domain $\mathcal{L}_k$ with $l_k, u_k \in \mathbb{Z}$, $l_k \leq u_k$,

---

[1] Henceforth, transposes are suppresses when they are clear from the context.
[2] $\mathbb{Z}$, $\mathbb{N}$ and $\mathbb{Q}$ are the set of integers, natural, and rational numbers, respectively.

$1 \leq k \leq n$, is given by $\mathcal{L}_k = \{y \in \mathbb{Z} \mid l_k \leq y \leq u_k\}$. The domain of the entire variable vector is described by $\mathcal{L} = \{y \in \mathbb{Z}^n \mid \forall k \in \{1, \ldots, n\} : y_k \in \mathcal{L}_k\}$. Let $Q \in \{\exists, \forall\}^n$ denote the vector of quantifiers. We call each maximal consecutive subsequence in Q consisting of identical quantifiers a *quantifier block* and denote the i-th block as $B_i \subseteq \{1, \ldots, n\}$ and the corresponding quantifier as $Q^{(i)} \in \{\exists, \forall\}$. Let $\beta \in \mathbb{N}$, $\beta \leq n$, denote the number of blocks and thus $\beta - 1$ is the number of quantifier changes. Let $\mu(i, k) = \sum_{j=1}^{i-1} |B_j| + k$ which maps the k-th variable of block i to its original index. The variable vector of variable block B_i will be referred to as $x^{(i)}$ and its range is given by $\mathcal{L}^{(i)} = \{y \in \mathbb{Z}^{|B_i|} \mid y_k \in \mathcal{L}_{\mu(i,k)}\}$. We call $\mathcal{E} = \{i \in \{1, \ldots, \beta\} \mid Q^{(i)} = \exists\}$ the set of existential variable blocks and $\mathcal{A} = \{i \in \{1, \ldots, \beta\} \mid Q^{(i)} = \forall\}$ the set of universal variable blocks.

Definition 1 (Quantified Integer Linear Program (QIP)). *Let* $A^{\exists} \in \mathbb{Q}^{m_{\exists} \times n}$ *and* $b^{\exists} \in \mathbb{Q}^{m_{\exists}}$ *for* $m_{\exists} \in \mathbb{N}$ *and let* $\mathcal{L}$ *and* Q *be given as described above. Let* $c \in \mathbb{Q}^n$ *be the vector of objective coefficients and let* $c^{(i)}$ *denote the vector of coefficients belonging to block* B_i*. Let the term* $Q \circ x \in \mathcal{L}$ *with the component wise binding operator* $\circ$ *denote the* quantification vector $(Q^{(1)}x^{(1)} \in \mathcal{L}^{(1)}, \ldots, Q^{(\beta)}x^{(\beta)} \in \mathcal{L}^{(\beta)})$ *such that every quantifier* $Q^{(i)}$ *binds the variables* $x^{(i)}$ *of block* i *to its domain* $\mathcal{L}^{(i)}$*. We call*

$$z = \min_{x^{(1)} \in \mathcal{L}^{(1)}} \left(c^{(1)} x^{(1)} + \max_{x^{(2)} \in \mathcal{L}^{(2)}} \left(c^{(2)} x^{(2)} + \ldots \min_{x^{(\beta)} \in \mathcal{L}^{(\beta)}} c^{(\beta)} x^{(\beta)} \right) \right)$$
$$s.t. \quad Q \circ x \in \mathcal{L}: \ A^{\exists} x \leq b^{\exists} \qquad\qquad (\star)$$

a QIP with objective function (for a minimizing existential player), given by $(A^{\exists}, b^{\exists}, c, \mathcal{L}, Q)$*.*

A QIP instance can be interpreted as a two-person zero-sum game between an *existential player* setting the existentially quantified variables and a *universal player* setting the universally quantified variables with payoff z. The variables are set in consecutive order according to the variable sequence. Consequently, we say that a player makes the move $x_k = y$ if she fixes the variable x_k to $y \in \mathcal{L}_k$. At each such move, the corresponding player knows the settings of $x_1, \ldots, x_{k-1}$ before taking her decision x_k. If the completely assigned vector $x \in \mathcal{L}$ satisfies the linear constraint system $A^{\exists} x \leq b^{\exists}$, the existential player pays $z = c^{\top} x$ to the universal player. If x does not satisfy $A^{\exists} x \leq b^{\exists}$, we say *the existential player loses* and the payoff will be $+\infty$. This is a small deviation from conventional zero-sum games but using[3] $\infty + (-\infty) = 0$ also fits for zero-sum games. The chronological order of the variable blocks given by Q can be represented using a game tree consisting of existential, universal and leaf nodes.

Definition 2 (Game Tree). *Let* $G = (V, E, c)$ *be the edge-labeled finite directed tree with a set of nodes* $V = V_{\exists} \cup V_{\forall} \cup V_L$*, a set of edges* E *and a vector of edge labels* $c \in \mathbb{Z}^{|E|}$*. Each inner level either consists of only nodes from* $V_{\exists}$ *or only of nodes from* $V_{\forall}$*, with the root node at level 0 being from* $V_{\exists}$ *and* V_L *being the set*

---

[3] This is only a matter of interpretation and consequences are not discussed further.

of leaves. The j-th variable is represented by the inner nodes at depth $j-1$ and outgoing edges from such a node represent moves from $\mathcal{L}_j$. The corresponding edge labels encode the variable assignments of the move.

Thus, a path from the root to a leaf represents a play of the QIP and the sequence of edge labels encodes its moves, i.e. the corresponding variable assignments. Solutions of a QIP are strategies [9]. A strategy is called a *winning strategy* if all paths from the root node to a leaf represent a vector x such that $A^\exists x \leq b^\exists$. A QIP is called *feasible* if $(\star)$ is true (see Definition 1), i.e. if a winning strategy exists. If there is more than one winning strategy, the objective function aims for a certain (the "best") one. The value of a strategy is given by its minimax value which is the maximum value at its leaves [22]. Note that a leaf not fulfilling $A^\exists x \leq b^\exists$ can be represented by the value $+\infty$. The objective value of a feasible QIP is the minimax value at the root, i.e. the minimax value of the optimal winning strategy, defined by the *principal variation* (PV) [6]: the sequence of variable assignments being chosen during optimal play. For any $v \in V$ we call $f(v)$ the outcome of optimal play by both players starting at v.

Example 1. *Let us consider a QIP with $n = 4$ binary variables, $Q = (\exists, \forall, \exists, \forall)$, $c = (2, -2, -3, -2)$ and let the constraint system $A^\exists x \leq b^\exists$ given by*

$$\begin{aligned}
x_1 + x_2 + x_3 &\leq 2 \\
-x_1 \qquad + x_3 - x_4 &\leq 0 \\
- x_2 + x_3 - x_4 &\leq 0 \\
-x_1 + x_2 - x_3 + x_4 &\leq 1 \ .
\end{aligned}$$

The minimax value of the root node of the game tree is 2 and the principal variation is given by $x_1 = 1$, $x_2 = 0$, $x_3 = 0$ and $x_4 = 0$. The inner node at level 1 resulting from setting $x_1 = 0$ has the minimax value $+\infty$, i.e. after setting $x_1 = 0$ there exists no winning strategy.

3   The Extension: QIP with Interdependent Domain

In a QIP the universally quantified variables are only restricted to the domain $\mathcal{L}$, whereas the existential variables also must aim at fulfilling the constraint system $A^\exists x \leq b^\exists$. This results in an asymmetry, as – even though the min-max semantic is symmetric – only the existential player has to cope with a polytope influenced by the opponent: the interdependence between existential and universal decisions is only represented in one direction through the restriction $A^\exists x \leq b^\exists$. Thus, in this setting it is difficult to model most two-person games, since moves by any player almost always depend on previous own and opponent decisions. But also from the viewpoint of operations research this non-symmetric behavior can be inadequate for certain problems, e.g. the maintenance of a machine could prevent its failure – an active restriction of the anticipated uncertainty– at the expense of the required maintenance time.

We introduce a second constraint system $A^\forall x \leq b^\forall$, $A^\forall \in \mathbb{Q}^{m_\forall \times n}$ and $b^\forall \in \mathbb{Q}^{m_\forall}$, $m_\forall \in \mathbb{N}$, the universal player must satisfy. Both players have the superordinate goal to fulfill their system. In particular, if $A^\forall x \not\leq b^\forall$ and $A^\exists x \leq b^\exists$ for a fixed variable vector $x \in \mathcal{L}$ the universal player loses and the payoff will be $-\infty$. The odd situation may now occur, that for a completely assigned variable vector $x \in \mathcal{L}$ both systems are not met, i.e. $A^\exists x \not\leq b^\exists$ and $A^\forall x \not\leq b^\forall$. In such a situation, however, one of the players must have been the first to make an illegal move: this player loses immediately, just like in games like Chess or Go. Since the superordinate goal for each player is to fulfill her own constraint system, the set of legal variable allocations is defined as follows:

Definition 3 (Legal Variable Allocation). *For variable block $i \in \{1, \ldots, \beta\}$ the set of legal variable allocations $\mathcal{F}^{(i)}(\tilde{x}^{(1)}, \ldots, \tilde{x}^{(i-1)})$ depends on the assignment of previous variable blocks $\tilde{x}^{(1)}, \ldots, \tilde{x}^{(i-1)}$ and is given by*

$$\mathcal{F}^{(i)} = \left\{ \hat{x}^{(i)} \in \mathcal{L}^{(i)} \mid \exists x = (\tilde{x}^{(1)}, \ldots, \tilde{x}^{(i-1)}, \hat{x}^{(i)}, x^{(i+1)}, \ldots, x^{(\beta)}) \in \mathcal{L} : A^{Q^{(i)}} x \leq b^{Q^{(i)}} \right\}$$

i.e. after assigning the variables of block i there still must exist an assignment of x such that the system of $Q^{(i)} \in \{\exists, \forall\}$ is fulfilled. The dependence on the previous variables $\tilde{x}^{(1)}, \ldots, \tilde{x}^{(i-1)}$ will be omitted when clear.

Hence, moves that will wipe out any chance of fulfilling the own constraint system are forbidden explicitly. In particular, if $\mathcal{F}^{(i)} = \emptyset$ there is no move at all such that the constraint system of the player responsible for block i can be satisfied and since there are no legal moves left, the player in turn loses.

The subordinate goal for both players is still trying to optimize the objective function: The existential player is trying to minimize and the universal player is trying to maximize the objective value.

Definition 4 (QIP with Interdependent Domains (QIP^{ID})). *For given $A^\forall$, $A^\exists$, $b^\forall$, $b^\exists$, c, $\mathcal{L}$ and Q we call*

$$\min_{x^{(1)} \in \mathcal{F}^{(1)}} \left(c^{(1)} x^{(1)} + \max_{x^{(2)} \in \mathcal{F}^{(2)}} \left(c^{(2)} x^{(2)} + \ldots \max_{x^{(\beta)} \in \mathcal{F}^{(\beta)}} c^{(\beta)} x^{(\beta)} \right) \right)$$

$$s.t. \quad \exists x^{(1)} \in \mathcal{F}^{(1)} \; \forall x^{(2)} \in \mathcal{F}^{(2)} \ldots \forall x^{(\beta)} \in \mathcal{F}^{(\beta)} : \; A^\exists x \leq b^\exists \qquad (\star\star)$$

a Quantified Integer Program with Interdependent Domains (QIP^{ID}) given by $P = (A^\forall, A^\exists, b^\forall, b^\exists, c, \mathcal{L}, Q)$. For the following considerations we require the first variable block to consist of existential and the final variable block to consists of universal variables, i.e. $Q^{(1)} = \exists$ and $Q^{(\beta)} = \forall$. Further, we demand $\{x \in \mathcal{L} \mid A^\forall x \leq b^\forall\} \neq \emptyset$.

Using this definition the case of both systems being violated is bypassed: if both systems are not satisfied there must have been an illegal variable allocation at some early stage i with $F^{(i)} = \emptyset$. If $i \in \mathcal{E}$ this results in the noncompliance of $(\star\star)$ for this path in the game tree and hence a loss for the existential player. If $i \in \mathcal{A}$ Constraint $(\star\star)$ is trivially satisfied for this path, since $\forall x \in \emptyset$ any expression is true, and the objective value will be $-\infty$, as $\max_{x \in \emptyset} f(x) = -\infty$ for any function f.

In order to play optimally in each stage the player in turn not only has to figure out the set of legal moves $\mathcal{F}^{(i)}$, but also must ensure that a strategy exists such that she can satisfy her constraint system. Determining, whether a move is legal is equivalent to the question, whether a leaf in that part of the game tree below this move exists, representing a fulfilling variable assignment: finding such a path implies that the system of the player in turn is not "broken" yet. This, however itself is an NP-complete problem. As you can see in Example 2 and in Fig. 1 the allocation $x_1 = 2$ is legal, since a path $(x_2 = 1, x_3 = 0)$ exists fulfilling the existential system, whereas $x_1 = 3$ is illegal, since all leaves in the subtree below do not fulfill $A^\exists x \le b^\exists$. However, even though $x_1 = 2$ is legal, allocating x_1 this way would be a poor decision, as no strategy exists to fulfill the existential system.

Again, a game tree as in Definition 2 can be built for any QIP^{ID} instance. However, both the term strategy and minimax value must be adjusted in order to describe and find solutions of this problem type: A QIP^{ID} game might end before all variables are allocated, since one player might have no legal move left. Further, strategies for QIP^{ID} must not consider all possible moves from $\mathcal{L}^{(i)}$ but only all legal moves $\mathcal{F}^{(i)}$.

Definition 5 (Truncated Existential Strategy). *A* truncated strategy *(for the allocation of existential variables) $T = (V', E', c')$ is a subtree of a game tree $G = (V, E, c)$ of a QIP^{ID}. Each node $v_\exists \in V' \cap V_\exists$ has at most one child for which at least one leaf in the underlying sub-tree of G exists with $A^\exists x \le x^\exists$, i.e. the variable allocation represented by this child must be legal according to Definition 3. Each node $v_\forall \in V' \cap V_\forall$ has all the children as in G for which at least one leaf in their corresponding sub-tree in G exists with $A^\forall x \le b^\forall$, i.e. as many as there are in the corresponding domain $\mathcal{F}^{(i)}$.*

Definition 6 (Winning Truncated Existential Strategy). *Let $x_v \in \mathcal{L}$ denote the variable assignment corresponding to a leaf node $v \in V_L$ defined by the path from the root to v. A truncated existential strategy $T = (V', E', c')$ is called a winning truncated existential strategy, if for all nodes $\hat{v} \in V'$ without children (the leaves of the strategy; not necessarily $\hat{v} \in V_L$) it holds*

$$(\hat{v} \in V_L \wedge A^\exists x_{\hat{v}} \le b^\exists) \vee \hat{v} \in V_\forall ,$$

i.e. leaves either represents a fixed vector $x \in \mathcal{L}$ with $A^\exists x \le b^\exists$ or a partially filled vector $x^{(1)}, \ldots, x^{(i-1)}$ with $i \in \mathcal{A}$ and hence $\mathcal{F}^{(i)}(x^{(1)}, \ldots, x^{(i-1)}) = \emptyset$.

Since the outcome "universal player loses" can occur for QIP^{ID} the value $-\infty$ is added as possible outcome. Further, if both constraint systems are violated for a fixed variable vector $x \in \mathcal{L}$, the corresponding leaf node does not hold the information who made the first illegal move, i.e. who lost the game. The symbolic value $\pm\infty$ is introduced for such nodes.

Definition 7 (Extended Minimax Value). *Let $G = (V, E, c)$ be a game tree for a QIP^{ID} and let $w(v) : V_L \to \mathbb{Q} \cup \{+\infty, -\infty, \pm\infty\}$ be the weighting function of the leaf nodes $v \in V_L$ with*

$$w(v) = \begin{cases} c^\top x_v & , \ A^\exists x_v \leq b^\exists \ and \ A^\forall x_v \leq b^\forall \\ +\infty & , \ A^\exists x_v \not\leq b^\exists \ and \ A^\forall x_v \leq b^\forall \\ -\infty & , \ A^\exists x_v \leq b^\exists \ and \ A^\forall x_v \not\leq b^\forall \\ \pm\infty & , \ A^\exists x_v \not\leq b^\exists \ and \ A^\forall x_v \not\leq b^\forall \ . \end{cases}$$

For any node $v \in V$ the extended minimax value $f_e(v)$ is defined recursively by

$$f_e(v) = \begin{cases} w(v) & , \ v \in V_L \\ \min\{f_e(v') \mid (v,v') \in E, \ f_e(v') \neq \pm\infty\} & , \ v \in V_\exists \setminus V_{\pm\infty} \\ \max\{f_e(v') \mid (v,v') \in E, \ f_e(v') \neq \pm\infty\} & , \ v \in V_\forall \setminus V_{\pm\infty} \\ \pm\infty & , \ v \in V_{\pm\infty} \ . \end{cases}$$

with $V_{\pm\infty} = \{v \in V \setminus V_L \mid \forall v' \in V : (v,v') \in E \Rightarrow f_e(v') = \pm\infty\}$ being the set of nodes for which any leaf in the part of the game tree below has the value $\pm\infty$.

Since we demanded $\{x \in \mathcal{L} \mid A^\forall x \leq b^\forall\} \neq \emptyset$, after using the extended minimax value the root node lies within $\mathbb{Q} \cup \{+\infty, -\infty\}$. This can be interpreted as follows:

- Root node has extended minimax value $z = c^\top x \in \mathbb{Q}$:
  Both the universal as well as the existential player have a winning truncated strategy to satisfy their system. The outcome if both players play optimal is z.
- Root node has extended minimax value $+\infty$:
  The existential player has no winning truncated strategy to satisfy her system: The universal player can enforce $A^\exists x \not\leq b^\exists$ and hence the QIP^{ID} is infeasible.
- Root node has extended minimax value $-\infty$:
  The universal player has no winning truncated strategy to satisfy her system: The existential player can enforce $A^\forall x \not\leq b^\forall$.

Example 2. Let $c = (-1, -1, -2)^\top$, $Q = (\exists, \forall, \forall)$, $\mathcal{L}_1 = \{1, 2, 3\}$ and $\mathcal{L}_2 = \mathcal{L}_3 = \{0, 1\}$ and the two constraint systems given as follows:

$$A^\exists x \leq b^\exists : \quad \begin{array}{r} x_1 + x_2 + x_3 \leq 3 \\ 2x_1 - 3x_2 \quad\ \leq 3 \end{array} \qquad \Big| \qquad A^\forall x \leq b^\forall : \quad -x_1 + 2x_2 + x_3 \geq 0$$

The game tree of this QIP^{ID} instance is give in Fig. 1. The values at the leaves are assigned using the weighting function $w(v)$ as described above. The values at inner nodes are received by applying the extended minimax value. Hence, node values show the value according to optimal play starting from this node. In the first existential stage $\mathcal{F}^{(1)} = \{1, 2\}$ and in particular $3 \notin \mathcal{F}^{(1)}$ since the existential system cannot be fulfilled for $x_3 = 3$ (all leaves in the subtree beneath the decision $x_3 = 3$ have the property $+\infty$ or $\pm\infty$). Note, that with $x_1 = 2$ the universal variable x_2 cannot be set to 0, since a violation of the universal system would become inevitable. The optimal course of play (the PV) is $x_1 = 1$, $x_2 = 1$, $x_3 = 0$ with objective value -2.

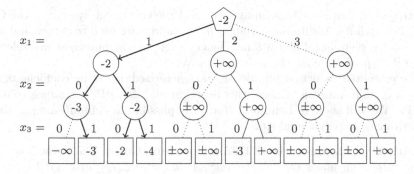

Fig. 1. Game tree for the given QIP^{ID} with rectangular leaves, circular universal nodes, and pentagonial existential root node. Leaves have either the property that some constraint system is not fulfilled ($+\infty$, $-\infty$, $\pm\infty$) or, if both systems are met, the resulting objective value. Illegal moves (not in $\mathcal{F}^{(i)}$) are indicated as dotted lines. The optimal truncated strategy is indicated by thicker arrows. Values at inner nodes indicate the outcome of optimal play.

Clearly $\text{QIP} \leq_p \text{QIP}^{\text{ID}}$ via polynomial-time reduction. Also, quite easily, QIP^{ID} is in PSPACE: a depth first-search can compute the presented extended minimax value, dealing with the crucial $\pm\infty$ nodes. However, PSPACE is a large class, and we are even able to present the non-trivial direct reduction $\text{QIP}^{\text{ID}} \leq_p \text{QIP}$: we reveal their close relationship by presenting a beautiful polynomial reduction. This gives us further structural insights into the nature of QIP^{ID}.

4 Parameterized Reduction Function

QIP^{ID} has one major difficulty compared to the QIP: In order to know whether a variable allocation is legal one has to solve an NP-complete subproblem, i.e. one has to check the feasibility of several integer programs in order to create the set $\mathcal{F}^{(i)}\left(x^{(1)}, \ldots, x^{(i-1)}\right)$, instead of simply ensuring compliance with the variable bounds. The resulting QIP from the reduction will ensure that setting one of the original variables illegally (i.e. outside of the specific $\mathcal{F}$-domain) would allow the other player to receive a much better payoff. Hence it is in each player's best interest to make moves that stay within the feasible region $\mathcal{F}$. The reduction function roughly works as follows: The original variables $x^{(i)}$ are also used in the arising QIP and in between some further variables are added in order to ensure that illegal moves (not element of $\mathcal{F}^{(i)}$) are disadvantageous compared to legal moves. For an existential variable block $x^{(i)}$ a verification vector $v^{(i)}$ is added to check, whether the existential system still can be satisfied. If no allocation of $v^{(i)}$ exists the constraint system of the QIP cannot be satisfied, which is to the detriment of the existential player. Hence, a legal variable assignment $x^{(i)} \in \mathcal{F}^{(i)}$ will be preferred for which a fulfilling allocation of $v^{(i)}$ will exist. A similar approach is used for universal variable blocks: If there is no allocation of the verification vector $v^{(i)}$ that fulfills the universal system, i.e. if $x^{(i)} \notin \mathcal{F}^{(i)}$, the objective value can be reduced massively and the constraint system of the QIP is fulfilled trivially. Additional variables $y^{(i)}$, t_i and p are used to detect, indicate and penalize such illegal allocations. Hence, selecting the universal variable $x^{(i)}$ from $\mathcal{F}^{(i)}$ is preferable for the universal player.

The presented reduction function is parameterized, i.e. some coefficients can range in certain intervals. Hence, this function maps a QIP^{ID} instance to a set of QIPs. We will show in Lemma 1 that it is possible to compute one of those QIP instances in polynomial time.

Definition 8 (Parameterized Reduction Function). *The reduction function r maps any given QIP^{ID} $P = (A^\forall, A^\exists, b^\forall, b^\exists, c, \mathcal{L}, Q)$ to a QIP with the following form:*

$$\min_{x^{(1)} \in \mathcal{L}^{(1)}} \left(c^{(1)} x^{(1)} + \ldots \max_{x^{(\beta)} \in \mathcal{L}^{(\beta)}} \left(c^{(\beta)} x^{(\beta)} - \min_{p \in \{0,1\}} \tilde{M} p \right) \right)$$

$$
\begin{aligned}
s.t. \quad & \exists x^{(1)} \in \mathcal{L}^{(1)} \quad \exists v^{(1)} \in \mathcal{L}^{(2)} \times \ldots \times \mathcal{L}^{(\beta)} \\
& \forall x^{(2)} \in \mathcal{L}^{(2)} \quad \forall v^{(2)} \in \mathcal{L}^{(3)} \times \ldots \times \mathcal{L}^{(\beta)} \\
& \exists y^{(2)} \in \{0,1\}^{m_\forall} \quad \exists t_2 \in \{0,1\} \\
& \exists x^{(3)} \in \mathcal{L}^{(3)} \quad \exists v^{(3)} \in \mathcal{L}^{(4)} \times \ldots \times \mathcal{L}^{(\beta)} \\
& \forall x^{(4)} \in \mathcal{L}^{(4)} \quad \forall v^{(4)} \in \mathcal{L}^{(5)} \times \ldots \times \mathcal{L}^{(\beta)} \\
& \exists y^{(4)} \in \{0,1\}^{m_\forall} \quad \exists t_4 \in \{0,1\} \\
& \quad \ldots \\
& \forall x^{(\beta)} \in \mathcal{L}^{(\beta)} \quad \exists y^{(\beta)} \in \{0,1\}^{m_\forall} \\
& \exists t_\beta \in \{0,1\} \quad \exists p \in \{0,1\} :
\end{aligned}
$$

$$A^\exists s^{(i)} - M t_{i-1} \leq b^\exists \qquad \forall i \in \mathcal{E} \qquad (1)$$

$$A^\exists x - M p \leq b^\exists \qquad (2)$$

$$-A^\forall s^{(i)} - (L - b^\forall - R^{LCD}) y^{(i)} \leq -L \qquad \forall i \in \mathcal{A} \qquad (3)$$

$$p - \sum_{i \in \mathcal{A}} t_i \leq 0 \qquad (4)$$

$$t_i - t_{i-2} - \sum_{k=1}^{m_\forall} y_k^{(i)} \leq 0 \qquad \forall i \in \mathcal{A} \qquad (5)$$

with $s^{(i)}$ being the abbreviation for the vector $(x^{(1)}, \ldots, x^{(i)}, v^{(i)})$. M, $\tilde{M}$, L, R^{LCD} and t_0 are parameters that must fulfill the following criteria: $t_0 = 0$, $L \in \mathbb{Q}^{m_\forall}$ with[4]

$$L_k \leq \min_{x \in \mathcal{L}} A_{k,*}^\forall x \qquad \forall k \in \{1, \ldots, m_\forall\}, \qquad (6)$$

---

[4] $A_{k,*}$ denotes the k-th row of matrix A.

being smaller than the smallest possible value of the left-hand side of line k of the universal system resulting in $A^\forall x \geq L$ being valid for any $x \in \mathcal{L}$. $M \in \mathbb{Q}^{m_\exists}$ must fulfill

$$M_k \geq \max_{x \in \mathcal{L}} A^\exists_{k,*} x - b^\exists_k \qquad \forall k \in \{1, \ldots, m_\exists\}, \tag{7}$$

being larger than the largest possible violation of line k of the existential system resulting in $A^\exists x \leq b^\exists + M$ being valid for all $x \in \mathcal{L}$. $\tilde{M} \in \mathbb{Q}$ is set such that

$$\max_{x \in \mathcal{L}} c^\top x - \tilde{M} < \min_{x \in \mathcal{L}} c^\top x \tag{8}$$

for example

$$\tilde{M} \geq \sum_{\substack{1 \leq i \leq n \\ c_i < 0}} c_i \cdot (l_i - u_i) + \sum_{\substack{1 \leq i \leq n \\ c_i \geq 0}} c_i \cdot (u_i - l_i) + 1. \tag{9}$$

$R^{LCD} \in \mathbb{Q}^{m_\forall}$ *is a vector with positive entries smaller than or equal to the reciprocals of the lowest common denominators of the rows of $A^\forall$ and $b^\forall$, ensuring for any line $k \in \{1, \ldots, m_\forall\}$ $A^\forall_{k,*} x \nleq b_k \Leftrightarrow A^\forall_{k,*} x \geq b_k + R^{LCD}_k$. In particular $0 < R^{LCD}_k \leq 1$, if all entries in row k are integer. Note, that for the entries of L and R^{LCD} only upper bounds are given, whereas for M and $\tilde{M}$ lower bounds are specified. Therefore, the result of the reduction is not unique, since the parameters L, M, $\tilde{M}$ and R^{LCD} only must satisfy the given bounds. Thus, r is a parameterized reduction function and we call $r(P)$ the set of QIPs created by the above mechanism.*

The close relationship between a QIP^{ID} and the corresponding QIPs resulting from the reduction function is given in the following theorem:

Theorem 1. *The following equivalences hold for any QIP^{ID} instance P:*

1. *P is feasible with optimal objective value $z \in \mathbb{Q}$*
   $\Leftrightarrow \forall R \in r(P)$: The QIP instance R is feasible with optimal objective value z
2. *P is feasible with optimal objective value $z = -\infty$*
   $\Leftrightarrow \forall R \in r(P)$: R is feasible with optimal objective value $\bar{z} < \min\limits_{x \in \mathcal{L}} c^\top x$
3. *P is infeasible*
   $\Leftrightarrow \forall R \in r(P)$: R is infeasible

In the following we present the proof idea of Theorem 1.

Proof Idea 1. *Let us assume $R \in r(P)$ with fixed variables $\tilde{p}$, $\tilde{x}^{(i)}$, $\tilde{v}^{(i)}$, $\tilde{y}^{(i)}$ and $\tilde{t}_i$, $i \in \{1, \ldots, \beta\}$. Assume $\tilde{x}^{(1)} \notin \mathcal{F}^{(1)}$, and hence this move should constitute a loss for the (starting) existential player. For $s^{(1)} = (\tilde{x}^{(1)}, \tilde{v}^{(1)})$ Constraint (1) for $i = 1$ is violated, since - by definition of $\mathcal{F}^{(1)}$ - no fulfilling $\tilde{v}^{(1)}$ can exists. Therefore, any illegal variable assignment of $\tilde{x}^{(1)}$ - in terms of P - results in an immediate loss for the existential player in R, regardless of the assignment of the remaining variables. In particular, if $\mathcal{F}^{(1)} = \emptyset$ the existential player will lose, i.e. R is infeasible.*

Let us now assume $\tilde{x}^{(1)} \in \mathcal{F}^{(1)}$ and $\tilde{v}^{(1)}$ is set such that $A^\exists s^{(1)} \leq b^\exists$, i.e. the fulfillment of Constraint (1) for $i = 1$ is ensured. Assume the universal variables $\tilde{x}^{(2)}$ and $\tilde{v}^{(2)}$ are selected such that for some line $k \in \{1, \ldots, m_\forall\}$ $A^\forall_{k,*} s^{(2)} \not\leq b^\forall_k$ and hence $A^\forall_{k,*} s^{(2)} \geq b^\forall_k + R^{LCD}_k$. Constraint (3) is constructed such that $y^{(2)}_k$ can be set to 1 only if line k is violated and hence $y^{(2)}_k$ is an indicator of such a violation. With $y^{(2)}_k = 1$ we further see that $t_2 = 1$ is valid for $i = 2$ in Constraint (5). This allows $t_i = 1$ for $2 < i \in \mathcal{A}$ according to Constraint (5) and after selection of M as in (7) Constraint (1) is trivially fulfilled for $2 < i \in \mathcal{E}$ and $p = 1$ is valid according to (4). Consequently, if the allocation of the universal variables $x^{(2)}$ and $v^{(2)}$ do not satisfy $A^\forall s^{(2)} \leq b^\forall$ the existential variable t_2 can be set to 1 and the entire remaining constraint system (1)–(5) can be satisfied easily. Additionally, since $p = 1$ and after the selection of $\tilde{M}$ according to (9) the objective value will be smaller than $\min_{x \in \mathcal{L}} c^\top x$. Hence, for $\tilde{x}^{(2)} \notin \mathcal{F}^{(2)}(\tilde{x}^{(1)})$ the objective value of optimal play will be smaller than $\min_{x \in \mathcal{L}} c^\top x$, which is to the detriment of the maximizing universal player and will be avoided if possible. For $\tilde{x}^{(2)} \in \mathcal{F}^{(2)}(\tilde{x}^{(1)})$ there exists some $\tilde{v}^{(2)}$ such that $A^\forall s^{(2)} \leq b^\forall$. Selecting $\tilde{v}^{(2)}$ otherwise once again is to the detriment of the maximizing universal player and will be avoided. For $\tilde{v}^{(2)}$ such that $A^\forall s^{(2)} \leq b^\forall$ the subsequent existential variables must hold $y^{(2)} = 0$ and $t_2 = 0$. This argumentation can be continued for the subsequent stages. Assuming that the variables up to stage i were selected such that for each $j < i$ it holds $x^{(j)} \in \mathcal{F}^{(j)}$, $v^{(j)}$ such that $A^{Q^{(j)}} s^{(j)} \leq b^{Q^{(j)}}$, $y^{(j)} = 0$ and $t_j = 0$ the following holds:

- If $i \in \mathcal{E}$ selecting $x^{(i)} \notin \mathcal{F}^{(i)}$ immediately results in an existential player's loss as Constraint (1) is violated.
- If $i \in \mathcal{A}$ selecting $x^{(i)} \notin \mathcal{F}^{(i)}$ allows $t_i = 1$ and hence the entire constraint system of R can be fulfilled trivially with objective value smaller than $\min_{x \in \mathcal{L}} c^\top x$.
- If $i \in \mathcal{A}$ and $x^{(i)} \in \mathcal{F}^{(i)}$ and $v^{(i)}$ such that $A^\forall s^{(i)} \leq b^\forall$ then the existential variables $y^{(i)}$ and t_i must be zero in order to fulfill Constraints (3) and (5).

Hence, illegal moves (not in $\mathcal{F}^{(i)}$) according to the QIP^{ID} P are punished in $R \in r(P)$ and therefore are disadvantageous. Therefore, legal allocations (taken from $\mathcal{F}^{(i)}$) for the $x^{(i)}$ variables are preferred, if possible.

In order to constitute a polynomial reduction function we further must show the following:

Lemma 1. Let $P = (A^\forall, A^\exists, b^\forall, b^\exists, c, \mathcal{L}, Q)$ be a given QIP^{ID}. An element of the set $r(P)$ can be computed in polynomial time with respect to the input size.

Proof. The size of the input P only depends on the number of variables n and the number of constraints $m_\exists$ and $m_\forall$. Obviously, values for the vectors L and M and the value for $\tilde{M}$ can be computed in polynomial time with respect to n, $m_\exists$ and $m_\forall$ by computing the bounds given in (6), (7) and (9), respectively. For the entries of R^{LCD} it is not necessary to find the *lowest* common denominator: it suffices to multiply the denominators of the non-zero entries of row k and take its reciprocal for R^{LDC}_k, which can be computed in polynomial time. The number of variables in the resulting QIP is in $\mathcal{O}(n + n^2 + n \cdot m_\forall)$: $\mathcal{O}(n)$ x and t variables, $\mathcal{O}(n^2)$ v variables, $\mathcal{O}(n \cdot m_\forall)$ y variables and one p variable. The number of constraints is $\mathcal{O}(n \cdot m_\exists + n \cdot m_\forall)$: $\mathcal{O}(n \cdot m_\exists)$ constraints in (1), $m_\exists$ constraints in (2), $\mathcal{O}(n \cdot m_\forall)$ in (3) and 1 and $\mathcal{O}(n)$ constraints in (4) and (5), respectively.

5   Conclusion

We examined the complex entanglements arising from allowing an exertion of influence by the optimizer on the uncertainty set in a robust multistage optimization problem. We focused on quantified integer programs, which itself cover a nearly unlimited field of applications. We introduced the formal problem QIP^{ID} where the modeling constraints of the optimizer and the uncertainty set interact: the domain of universal variables is restricted through a second system of linear constraints that can be affected by previous decisions. Solutions are truncated strategies, which can be depicted as subgraphs of the corresponding game tree, and an enhanced minimax algorithm can be used to find the best possible value if the variables are assigned optimally. Intuitively QIP^{ID} is a much more difficult problem than QIP since the complex subproblem of determining the action sets at decision nodes is itself an NP-complete problem. Nevertheless, the QIP^{ID} and the QIP are closely related, since the QIP^{ID} does not only stay in the same complexity class PSPACE as the QIP, but how close they are coupled can be seen with help of the presented parameterized reduction function, mapping QIP^{ID} instances to QIPs (in polynomial time). A proof of concept or practicability is still open, but in the past, technological progress has pushed the threshold of applicability in other domains as well, e.g. the famous rise of mixed integer linear programming.

References

1. Apap, R., Grossmann, I.: Models and computational strategies for multistage stochastic programming under endogenous and exogenous uncertainties. Comput. Chem. Eng. **103**, 233–274 (2017)
2. Bellman, R.: Dynamic Programming. Dover Publications, New York (2003)
3. Ben-Tal, A., Ghaoui, L.E., Nemirovski, A.: Robust Optimization. Princeton University Press, Princeton (2009)
4. Bertsimas, D., Brown, D., Caramanis, C.: Theory and applications of robust optimization. SIAM Rev. **53**(3), 464–501 (2011)
5. Birge, J., Louveaux, F.: Introduction to Stochastic Programming, 2nd edn. Springer, New York (2011). https://doi.org/10.1007/978-1-4614-0237-4
6. Campbell, M., Marsland, T.: A comparison of minimax tree search algorithms. Artif. Intell. **20**(4), 347–367 (1983)
7. Condon, A.: Space-bounded probabilistic game automata. J. ACM **38**(2), 472–494 (1991)
8. De Filippo, A., Lombardi, M., Milano, M.: Methods for off-line/on-line optimization under uncertainty. In: IJCAI, pp. 1270–1276 (2018)

9. Ederer, T., Hartisch, M., Lorenz, U., Opfer, T., Wolf, J.: Yasol: an open source solver for quantified mixed integer programs. In: Winands, M.H.M., van den Herik, H.J., Kosters, W.A. (eds.) ACG 2017. LNCS, vol. 10664, pp. 224–233. Springer, Cham (2017). https://doi.org/10.1007/978-3-319-71649-7_19
10. Ederer, T., Lorenz, U., Martin, A., Wolf, J.: Quantified linear programs: a computational study. In: Demetrescu, C., Halldórsson, M.M. (eds.) ESA 2011. LNCS, vol. 6942, pp. 203–214. Springer, Heidelberg (2011). https://doi.org/10.1007/978-3-642-23719-5_18
11. Gerber, R., Pugh, W., Saksena, M.: Parametric dispatching of hard real-time tasks. IEEE Trans. Comput. **44**(3), 471–479 (1995)
12. Goel, V., Grossmann, I.: A class of stochastic programs with decision dependent uncertainty. Math. Program. **108**(2), 355–394 (2006)
13. Gupta, A., Pál, M., Ravi, R., Sinha, A.: Boosted sampling: approximation algorithms for stochastic optimization. In: Proceedings of the Thirty-Sixth Annual ACM Symposium on Theory of Computing, STOC 2004, pp. 417–426. ACM, New York (2004)
14. Gupta, V., Grossmann, I.: A new decomposition algorithm for multistage stochastic programs with endogenous uncertainties. Comput. Chem. Eng. **62**, 62–79 (2014)
15. Hartisch, M., Ederer, T., Lorenz, U., Wolf, J.: Quantified integer programs with polyhedral uncertainty set. In: Plaat, A., Kosters, W., van den Herik, J. (eds.) CG 2016. LNCS, vol. 10068, pp. 156–166. Springer, Cham (2016). https://doi.org/10.1007/978-3-319-50935-8_15
16. Jonsbråten, T., Wets, R.J.B., Woodruff, D.: A class of stochastic programs with decision dependent random elements. Ann. Oper. Res. **82**, 83–106 (1998)
17. König, F.G., Lübbecke, M., Möhring, R., Schäfer, G., Spenke, I.: Solutions to real-world instances of PSPACE-complete stacking. In: Arge, L., Hoffmann, M., Welzl, E. (eds.) ESA 2007. LNCS, vol. 4698, pp. 729–740. Springer, Heidelberg (2007). https://doi.org/10.1007/978-3-540-75520-3_64
18. Lappas, N., Gounaris, C.: Multi-stage adjustable robust optimization for process scheduling under uncertainty. AIChE J. **62**(5), 1646–1667 (2016)
19. Lorenz, U., Martin, A., Wolf, J.: Polyhedral and algorithmic properties of quantified linear programs. In: de Berg, M., Meyer, U. (eds.) ESA 2010. LNCS, vol. 6346, pp. 512–523. Springer, Heidelberg (2010). https://doi.org/10.1007/978-3-642-15775-2_44
20. Nohadani, O., Roy, A.: Robust optimization with time-dependent uncertainty in radiation therapy. IISE Trans. Healthc. Syst. Eng. **7**(2), 81–92 (2017)
21. Papadimitriou, C.: Games against nature. J. Comput. Syst. Sci. **31**(2), 288–301 (1985)
22. Pijls, W., de Bruin, A.: Game tree algorithms and solution trees. Theoret. Comput. Sci. **252**(1), 197–215 (2001)
23. Poss, M.: Robust combinatorial optimization with variable cost uncertainty. Eur. J. Oper. Res. **237**(3), 836–845 (2014)
24. Schrijver, A.: Theory of Linear and Integer Programming. Wiley, New York (1986)
25. Subramani, K.: Analyzing selected quantified integer programs. In: Basin, D., Rusinowitch, M. (eds.) IJCAR 2004. LNCS (LNAI), vol. 3097, pp. 342–356. Springer, Heidelberg (2004). https://doi.org/10.1007/978-3-540-25984-8_26
26. Vujanic, R., Goulart, P., Morari, M.: Robust optimization of schedules affected by uncertain events. J. Optim. Theory Appl. **171**(3), 1033–1054 (2016)

Belief Change Properties of Forgetting Operations over Ranking Functions

Gabriele Kern-Isberner[1]([✉]), Tanja Bock[1]([✉]), Kai Sauerwald[2]([✉]) [iD],
and Christoph Beierle[2]([✉])

[1] Technische Universität Dortmund, 44227 Dortmund, Germany
{gabriele.kern-isberner,tanja.bock}@cs.tu-dortmund.de
[2] FernUniversität in Hagen, 58084 Hagen, Germany
{kai.sauerwald,christoph.beierle}@fernuni-hagen.de

Abstract. Intentional forgetting means to deliberately give up information and is a crucial part of change or consolidation processes, or to make knowledge more compact. Two well-known forgetting operations are contraction in the AGM theory of belief change, and various types of variable elimination in logic programming. While previous work dealt with postulates being inspired from logic programming, in this paper we focus on evaluating forgetting in epistemic states according to postulates coming from AGM belief change theory. We consider different forms of contraction, marginalization, and conditionalization as major representatives of forgetting operators to be evaluated. We use Spohn's ranking functions as a common semantic base to show that all operations can be realized in one logical framework, thereby exploring the richness of forgetting operations in a comparable way.

1 Introduction

Reasoning tasks in knowledge representation usually focus on how inferences are drawn (as in nonmonotonic logics), or changed (as in belief revision). Crucial constraints of reasoning processes such as focussing on relevant details only, leaving irrelevant aspects aside, are often left implicit. Hardly any example on reasoning that can be found in the literature explicitly considers what happens to the reasoning results if one, two, or a hundred irrelevant variables were added. Likewise, assumptions related to a specific context are often left implicit which means that reasoning results only hold in that specific context. Most people would take statements like "water boils at 100 centigrades" for granted, but this only holds in the context of the water being at sea level. Therefore, abstraction and omitting details which are found to be irrelevant or self-evident are important companions of reasoning the role of which, however, is often neglected; for a recent publication that focusses on abstraction, cf. [15].

We address abstraction and similar operations by the term *(intentional) forgetting* in this paper with the aim of making explicit and exploring formally those parts of reasoning tasks that are due to aspects of forgetting. There are

© Springer Nature Switzerland AG 2019
A. C. Nayak and A. Sharma (Eds.): PRICAI 2019, LNAI 11670, pp. 459–472, 2019.
https://doi.org/10.1007/978-3-030-29908-8_37

two major research fields for which formal properties of forgetting operations have been presented: in logic programming, especially in the field of answer set programming, forgetting atoms, literals, and sets of atoms or literals has been well investigated (for a recent survey, see [7]), and postulates have been formalized (see, in particular, [10]). The other field that explicitly deals with formal properties of deliberately giving up information is AGM belief change theory [1] that presents a formal framework of postulates for contraction operations that can deal with the omission of propositional formulas; this has been generalized to multiple contraction that allows for giving up sets of formulas [8]. Both fields are principal reference points when formal frameworks of forgetting are to be elaborated. While different types and operators of forgetting according to formal standards inspired by logic programming are evaluated in [13], in this paper we focus on investigating epistemic forgetting operations according to postulates of AGM-like iterated contraction [4,13].

In order to evaluate forgetting operations from the point of view of iterated AGM contraction, we re-interpret the postulates from [4,14] as postulates for forgetting on epistemic states, and apply them to forgetting operations defined for Spohn's ranking functions [16,17] (aka ordinal conditional functions, OCF) as representatives of epistemic states. We show how types of forgetting like variable elimination (aka marginalization), AGM-like contraction, and others can be distinguished along the lines of postulates, and that the framework of forgetting is richer than is suggested by the dichotomy of forgetting in logic programming, and contraction in AGM theory. We also emphasize the duality between reasoning and forgetting by re-interpreting conditionalization, an operation typically used for reasoning and belief change, as a forgetting operation. Since the framework of OCF provides established operations for marginalization and conditionalization, it seems to be a perfect, unifying environment for illustrating and evaluating different forgetting operations (cf. [2]).

In summary, the main contributions of this paper are a study of key forgetting operations; the identification of a new subclass of c-changes, the minimal c-contractions; and formal investigations of marginalization and conditionalization, c-contractions and their subclasses of c-ignoration and minimal c-contractions, with respect to AGM-like contraction postulates.

2   Basics on Conditionals and Iterated Contraction

Let $\mathcal{L}_\Sigma$ denote a finitely generated propositional language, with atoms $\Sigma = \{a, b, c, \ldots\}$, and with formulas $A, B, C, \ldots$. For conciseness of notation, we will omit the logical *and*-connector, writing AB instead of $A \wedge B$, and overlining formulas will indicate negation, i.e., $\overline{A}$ means $\neg A$. Let Ω_Σ denote the set of all possible worlds (propositional interpretations) over Σ. As usual, $\omega \models A$ means that the propositional formula $A \in \mathcal{L}$ holds in the possible world $\omega \in \Omega$, and $Mod(A) = \{\omega \mid \omega \models A\}$ denotes the set of all such possible worlds. By slight abuse of notation, we will use ω both for the model and the corresponding complete conjunction containing all atoms either in positive or negative form.

For a total preorder $\preceq$ on Ω we denote the set of minimal models of A with respect to $\preceq$ by $\min(\preceq, Mod(A))$. If $A \in \mathcal{L}$ is a formula, then the minimal set of signature elements from Σ to represent a formula which is equivalent to A is denoted by Σ_A. For a subset of signature elements $\Sigma' \subseteq \Sigma$ and a world $\omega \in \Omega_\Sigma$ we denote the Σ'-part of ω with $\omega^{\Sigma'} \in \Omega_{\Sigma'}$, mentioning exactly the atoms from Σ' such that $\omega \models \omega^{\Sigma'}$. For illustration, consider $\Sigma = \{a, b, c\}$ with $\omega_1 = \overline{a}b\overline{c} \in \Omega_\Sigma$ and $\Sigma' = \{a, b\}$. The Σ'-part of ω_1 is $\omega_1^{\Sigma'} = \overline{a}b$. For $\omega_2 = \overline{a}bc$ the Σ'-part would be the same. The marginalization of a propositional formula A to a signature $\Sigma' \subseteq \Sigma$ is defined iteratively by variable elimination: For a variable $V \in \Sigma$, the formula $A|_{\Sigma - \{V\}} = A^+ \vee A^-$ arises from A by replacing all occurrences of V by $\top$, yielding A^+, or by $\bot$, yielding A^-. Then, $A|_{\Sigma'}$ is the formula obtained from A by successively eliminating all variables in $\Sigma - \Sigma'$.

By introducing a new binary operator $|$, we obtain the set $(\mathcal{L}|\mathcal{L}) = \{(B|A) \mid A, B \in \mathcal{L}\}$ of conditionals over $\mathcal{L}$. $(B|A)$ formalizes "*if A then usually B*" and establishes a plausible connection between the *antecedent A* and the *consequent B*. Conditionals with tautological antecedents are taken as plausible statements about the world. Following De Finetti [5], a conditional $(B|A)$ can be *verified* (*falsified*) by a possible world ω iff $\omega \models AB$ ($\omega \models A\overline{B}$). If $\omega \not\models A$, then we say the conditional is *not applicable* to ω. Because conditionals go well beyond classical logic, they require a richer setting for their semantics than classical logic.

Ordinal conditional functions (OCFs), (also called *ranking functions*) $\kappa : \Omega \to \mathbb{N} \cup \{\infty\}$ with $\kappa^{-1}(0) \neq \varnothing$, were introduced (in a more general form) first by [16]. They express degrees of plausibility of propositional formulas A by specifying degrees of disbelief of their negations $\overline{A}$. More formally, we have $\kappa(A) := \min\{\kappa(\omega) \mid \omega \models A\}$, so that $\kappa(A \vee B) = \min\{\kappa(A), \kappa(B)\}$. Hence, due to $\kappa^{-1}(0) \neq \varnothing$, at least one of $\kappa(A), \kappa(\overline{A})$ must be 0. Every OCF κ induces a canonical total preorder $\preceq_\kappa$ on Ω, where $\omega_1 \preceq_\kappa \omega_2$ iff $\kappa(\omega_1) \leqslant \kappa(\omega_2)$. With $[\![\kappa]\!] = \{\omega \mid \kappa(\omega) = 0\}$, we denote the minimal models of κ, and $Bel(\kappa)$ denotes the theory of propositional formulas that hold in all $\omega \in [\![\kappa]\!]$. OCFs can serve as representations of epistemic states providing semantics for conditionals: A conditional $(B|A)$ is accepted in the epistemic state represented by κ, written as $\kappa \models (B|A)$, iff $\kappa(AB) < \kappa(A\overline{B})$, i.e., iff the verification AB of the conditional is more plausible than its falsification $A\overline{B}$. For a propositional formula A, we have $\kappa \models A$ iff $\kappa \models (A|\top)$ iff $\kappa(A) < \kappa(\overline{A})$ iff $\kappa(\overline{A}) > 0$, since at least one of $\kappa(A), \kappa(\overline{A})$ must be 0 due to $\kappa^{-1}(0) \neq \varnothing$.

AGM theory [1] deals with belief revision in the context of belief sets, i.e., deductively closed sets of propositions. Chopra, Ghose, Meyer and Wong [4], and later Caridroit, Konieczny, Marquis and Pino Pérez [3,14], argued that iterated contraction should fulfill postulates beyond the classical AGM ones and therefore adapted a set of AGM contraction postulates for epistemic states Ψ which are equipped with a total preorder $\preceq_\Psi$. For such epistemic states, the most plausible beliefs are denoted by $Bel(\Psi)$, having the minimal models of $\preceq_\Psi$ as their models. Note that this generalizes the definition of $Bel(\kappa)$ for OCFs from above. Using the underlying propositional language, $Bel(\Psi)$ can be represented by a single formula $\psi \in \mathcal{L}$ which is unique up to equivalence. Any such formula is denoted by

$\mathbf{B}(\Psi)$, and we have $Bel(\Psi) = Cn(\mathbf{B}(\Psi))$. These notations also apply to epistemic states which are represented by OCFs.

Let $-$ be a contraction operator that assigns a posterior epistemic state $\Psi - A$ to a prior state Ψ and a proposition A. The epistemic postulates extending the AGM contraction postulates for epistemic states are given as follows in [14]:

$$\mathbf{B}(\Psi) \models \mathbf{B}(\Psi - A) \qquad\qquad \text{(AGMes-1)}$$

$$\text{If } \mathbf{B}(\Psi) \not\models A, \text{ then } \mathbf{B}(\Psi - A) \models \mathbf{B}(\Psi) \qquad\qquad \text{(AGMes-2)}$$

$$\text{If } \mathbf{B}(\Psi - A) \models A, \text{ then } A \equiv \top \qquad\qquad \text{(AGMes-3)}$$

$$\mathbf{B}(\Psi - A) \wedge A \models \mathbf{B}(\Psi) \qquad\qquad \text{(AGMes-4)}$$

$$\text{If } A \equiv C, \text{ then } \mathbf{B}(\Psi - A) \equiv \mathbf{B}(\Psi - C) \qquad\qquad \text{(AGMes-5)}$$

$$\mathbf{B}(\Psi - (A \wedge C)) \models \mathbf{B}(\Psi - A) \vee \mathbf{B}(\Psi - C) \qquad\qquad \text{(AGMes-6)}$$

$$\text{If } \mathbf{B}(\Psi - (A \wedge C)) \not\models A, \text{ then } \mathbf{B}(\Psi - A) \models \mathbf{B}(\Psi - (A \wedge C)) \qquad \text{(AGMes-7)}$$

The translation of the AGM contraction postulates on belief sets for the propositional case is explained in detail in [3].

3   Forgetting in Epistemic States

In this section, we describe the general framework in which we consider forgetting in epistemic states, specify the forgetting operators to be considered in this paper, and recall technical details for realizing forgetting operators in epistemic states represented by OCFs.

Epistemic States and Forgetting Operators. The paper [2] discusses various kinds of forgetting and specifies characteristic properties for each of the forgetting operations on epistemic states. In that paper, an abstract model of epistemic states (also called belief states) is used in which each epistemic state Ψ is presupposed to be make use of a logical language $\mathcal{L}$ over a signature Σ and equipped with an inference relation $\mathrel{|\!\approx}$. The relation $\Psi \mathrel{|\!\approx} \varphi$ holds if an agent with belief state Ψ infers/believes/accepts φ, where φ can be a statement from $\mathcal{L}$, or a conditional from $(\mathcal{L}|\mathcal{L})$. The connection between both types of inferences is given by $\Psi \mathrel{|\!\approx} A$ iff $\Psi \mathrel{|\!\approx} (A|\top)$ iff $\mathbf{B}(\Psi) \models A$. Indeed, validation of conditionals (or rules) is a crucial characteristic feature of epistemic states that distinguishes them from flat belief sets, i.e., logical theories. One may even think of an epistemic state as being (basically) specified by the conditionals that are believed on its base; this view can be found in works on nonmonotonic reasoning and belief revision [9], and logic programming [6]. Moreover, conditionals can be easily related to total preorders which play an important role in belief revision [11].

The set of all (conditional) inferences will be denoted by $C(\Psi)$ with $C(\Psi) = \{(B|A) \in (\mathcal{L} \mid \mathcal{L}) \mid \Psi \mathrel{|\!\approx} (B|A)\}$. Using the conditional inference relation, we define the entailment relation between belief states Ψ_1, Ψ_2 by $\Psi_1 \mathrel{|\!\approx} \Psi_2$ iff $C(\Psi_2) \subseteq$

$C(\Psi_1)$. Equivalence among epistemic states is defined by the entailment relation $\approx\!\!\!| $ on epistemic states with $\Psi_1 \cong \Psi_2$ iff $\Psi_1 \approx\!\!\!| \Psi_2$ and $\Psi_2 \approx\!\!\!| \Psi_1$.

We further assume in this paper that epistemic states can be marginalized by restricting the signature Σ, and that they can be conditionalized by considering only models of a given proposition A. To be more precise, let $\Psi|_{\Sigma'}$ be the (unique) marginalized belief state for a subset Σ' of signature elements with $\Psi|_{\Sigma'} \approx\!\!\!| (B|A)$ iff $\Psi \approx\!\!\!| (B|A)$ and $(B|A) \in (\mathcal{L}_{\Sigma'}|\mathcal{L}_{\Sigma'})$. For conditionalization via the operator $|$, $\Psi|A$ has the intended meaning that Ψ should be interpreted under the assumption that $A \in \mathcal{L}$ holds; hence $\Psi|A \approx\!\!\!| (C|B)$ iff $\Psi \approx\!\!\!| (C|B)$ for $B \models A$ and $C \models A$. We further suppose that $\Psi|A \approx\!\!\!| A$ holds for every A.

If Ψ is a prior state, then we denote with Ψ_A° the posterior belief state of the agent after forgetting A, or after applying the change operation $\circ$ to Ψ with input A. In this context the object A that we want to forget can be a formula from $\mathcal{L}$, or a variable from Σ. We focus on the base case of forgetting a single logical item, but the approaches and results presented in this paper can be extended to deal with forgetting of several items.

The key idea of this paper is that different notions of forgetting can be specified by the inferences an agent can or can no longer draw after forgetting [2]. In the following, we focus on three major types of forgetting - contraction, marginalization and conditionalization - and additionally on ignoration as this is a special kind of contraction but covers a different aspect that is of interest for various applications as well. These four forgetting operations are described on a high level as follows:

$$
\begin{aligned}
\text{Contraction} :\;& \Psi_A^\circ \not\approx\!\!\!| A \\
\text{Ignoration} :\;& \Psi_A^\circ \not\approx\!\!\!| A \text{ and } \Psi^\circ \not\approx\!\!\!| \overline{A} \\
\text{Marginalization} :\;& \Psi_A^\circ = \Psi|_{\Sigma \setminus \Sigma_A} \\
\text{Conditionalization} :\;& \Psi_A^\circ = \Psi|\overline{A} \text{ with a conditionalization operator } | \text{ on } \Psi
\end{aligned}
$$

Contraction refers to the intention to directly give up information A, as it is known from the AGM framework [1]. Ignoration is a special contraction that enforces undecidedness between A and $\overline{A}$, thus giving up the judgement on A. Marginalization and conditionalization are well-established operators; they are reinterpreted here for defining forgetting operators that take the information A to be forgotten as their argument. For further discussion on these operators please see [2]. In the rest of this section, we will instantiate this general framework with ordinal conditional functions and specific forgetting operators based on them.

Instantiations of Forgetting with OCF. OCFs, or ranking functions κ over Σ provide all technical features that we expect from epistemic states. The *conditional inference relation* is defined as $\mathbf{C}(\kappa) = \{(B|A) \in (\mathcal{L}|\mathcal{L}) \mid \kappa \approx\!\!\!| (B|A)\}$ and a conditional $(B|A)$ is an inference of a ranking function κ, denoted by $\kappa \approx\!\!\!| (B|A)$, iff $\kappa(AB) < \kappa(A\overline{B})$.

In the following, we recall the instantiations of the four forgetting operators described above within the framework of OCFs from [2]. The object A which is to be forgotten can be a formula from $\mathcal{L}$, or a variable from Σ. Throughout this paper, we assume that in case that A is a formula, A is neither $\bot$ nor $\top$ because we are considering general strategies of forgetting here, leaving out limit cases for the moment.

Definition 1 (c-contraction by a single proposition [2]**).** *A change from κ to κ° is called a c-contraction with A, if there exist integers γ^+, γ^- such that the following equation holds*

$$\kappa^\circ_A(\omega) = -\gamma^- - \kappa(\overline{A}) + \kappa(\omega) + \begin{cases} \gamma^+ & \text{if } \omega \models A \\ \gamma^- & \text{if } \omega \models \overline{A} \end{cases} \tag{1}$$

and the following condition is satisfied:

$$\gamma^- - \gamma^+ \leqslant \kappa(A) - \kappa(\overline{A}) \tag{2}$$

The integers γ^+, γ^- are shifting factors the role of which is to make models of A resp. $\overline{A}$ more or less plausible in a uniform way. Equation (2) ensures that after the c-contraction $\kappa^\circ_A \not\models A$ holds and it is $\kappa^\circ_A(A) \geqslant \kappa^\circ_A(\overline{A})$, so especially that $\kappa^\circ_A(\overline{A}) = 0$ holds.

c-Contractions form a large family of diverse contraction operations which deliberately go beyond AGM [13]. Therefore, we choose special instances of c-contractions to elaborate on the potential of c-contractions for forgetting operations in more depth. In general, c-contractions are parametrized by γ^+, γ^- satisfying (2). There are two obvious ways to choose the difference $\gamma^- - \gamma^+$ that are directly related to information provided by the prior κ and ensure (2). First, setting $\gamma^- - \gamma^+ = -\kappa(\overline{A})$ guarantees the validity of (2) because $\kappa(A) \geqslant 0$ in any case. This first choice provides *minimal c-contraction* with a minimal amount of change among c-contractions because A-worlds are not shifted, and $\overline{A}$-worlds are shifted minimally (see Definition 2).

Definition 2 (minimal c-contraction by a single proposition). *A change from κ to κ° is called a minimal c-contraction with A, if there exist integers γ^+, γ^- such that the following equation holds*

$$\kappa^\circ_A(\omega) = -\gamma^- - \kappa(\overline{A}) + \kappa(\omega) + \begin{cases} \gamma^+ & \text{if } \omega \models A \\ \gamma^- & \text{if } \omega \models \overline{A} \end{cases} \tag{3}$$

and the following condition is satisfied:

$$\gamma^- - \gamma^+ = -\kappa(\overline{A}) \tag{4}$$

Note that minimal c-contractions can be described compactly by

$$\kappa^\circ_A(\omega) = \kappa(\omega) + \begin{cases} 0 & \text{if } \omega \models A \\ -\kappa(\overline{A}) & \text{if } \omega \models \overline{A} \end{cases}, \tag{5}$$

where (5) results from (3) and (4) by an easy calculation.

As a second obvious choice, setting $\gamma^- - \gamma^+ = \kappa(A) - \kappa(\overline{A})$ guarantees the validity of (2) for trivial reasons, yielding the class of *c-ignorations* (see Definition 3).

Definition 3 (c-ignoration by a single proposition [2]). *A change from κ to κ° is called a c-ignoration with A, if there exist integers γ^+, γ^- such that*

$$\kappa_A^\circ(\omega) = -\gamma^- - \kappa(\overline{A}) + \kappa(\omega) + \begin{cases} \gamma^+ & \text{if } \omega \models A \\ \gamma^- & \text{if } \omega \models \overline{A} \end{cases} \tag{6}$$

holds and the following condition is satisfied:

$$\gamma^- - \gamma^+ = \kappa(A) - \kappa(\overline{A}) \tag{7}$$

Equation (7) in Definition 3 ensures that $\kappa_A^\circ \not\models A$ and $\kappa_A^\circ \not\models \overline{A}$ holds. It is $\kappa_A^\circ(A) = \kappa_A^\circ(\overline{A}) = 0$ after the change operation.

Proposition 1. *For every c-ignoration, $Mod(\mathbf{B}(\kappa_A^\circ)) = \min(\preceq_\kappa, Mod(A)) \cup \min(\preceq_\kappa, Mod(\overline{A}))$ holds.*

Proof. The success condition of c-ignoration ensures that $\kappa_A^\circ(A) = \kappa_A^\circ(\overline{A}) = 0$ holds. For models of A, $\omega \models A$, the rank after the c-ignoration is computed by $-\gamma^- - \kappa(\overline{A}) + \kappa(\omega) + \gamma^+ = 0$. With Eq. 7 we get the condition $\kappa(\omega) - \kappa(\overline{A}) + \kappa(\overline{A}) - \kappa(A) = 0$, leading to $\kappa(\omega) = \kappa(A)$. For $\omega' \models \overline{A}$ we have $-\gamma^- - \kappa(\overline{A}) + \kappa(\omega') + \gamma^- = 0$ which leads to $\kappa(\omega') - \kappa(\overline{A}) = 0$ and $\kappa(\omega') = \kappa(\overline{A})$. So we get $Mod(\mathbf{B}(\kappa_A^\circ)) = \min(\preceq_\kappa, Mod(A)) \cup \min(\preceq_\kappa, Mod(\overline{A}))$. $\square$

OCFs can be seen as qualitative abstractions of probabilities [9]. Therefore, very similar to probabilities, they also allow for marginalization and conditionalization while observing their specific arithmetic characteristics (cf. [16]):

Definition 4 (marginalization of κ to Σ' [2]). *Let κ be an OCF over Σ and $\Sigma' \subseteq \Sigma$. The marginalization of κ to Σ', denoted by $\kappa|_{\Sigma'} : \Omega_{\Sigma'} \to \mathbb{N}$, is given by*

$$\kappa|_{\Sigma'}(\omega') = \min\{\kappa(\omega) \mid \omega \in \Omega_\Sigma \text{ and } \omega \models \omega'\}. \tag{8}$$

Applying this to implement marginalization as a forgetting operation, we obtain the operation of *forgetting by OCF-marginalization* $\kappa_A^\circ = \kappa|_{\Sigma \setminus \Sigma_A}$.

Definition 5 (conditionalization of κ by A [16]). *Let κ be a ranking function and A a proposition, then the conditionalization of κ by A is the ranking function $\kappa|A : Mod(A) \to \mathbb{N}$, defined on the models of A as follows:*

$$\kappa|A(\omega) = \kappa(\omega) - \kappa(A) \tag{9}$$

This yields the operation of *forgetting by OCF-conditionalization* $\kappa_A^\circ = \kappa|\overline{A}$.

4   Evaluation of Different Kinds of Forgetting by AGMes

In this section, we evaluate the forgetting operators that we defined in Sect. 3 according to the AGM postulates for contraction in epistemic states shown in Sect. 2. Table 1 gives an overview which postulates are fulfilled for each forgetting

Table 1. Evaluation of AGMes postulates for the different forgetting operations

AGMes	(-1)	(-2)	(-3)	(-4)	(-5)	(-6)	(-7)
C-Contraction	✗	✗	✓	✗	✓	✗	✗
Minimal C-Contraction	✓	✓	✓	✓	✓	✓	✓
C-Ignoration	✓	✗	✓	✗	✓	✗	✗
OCF-Marginalization	✓	✗	✓	✗	✓	✗	✓
OCF-Conditionalization	✗	✓	✓	✓	✓	✓	✓

operator. It is clearly seen that the epistemic AGM-postulates provide a useful standard for the five forgetting operations because each operation satisfies at least two postulates, but no two operations show the same pattern. Only minimal c-contraction satisfies all seven postulates. Ignoration fulfills (AGMes-1) while contraction does not. Another noticeable thing is that each operation fulfills the same two postulates, namely (AGMes-3) and (AGMes-5). (AGMes-3) is basically the success condition that each forgetting operator has to satisfy. (AGMes-5) is fulfilled by the forgetting operators as well, since every operator based on models ensures this condition. Therefore, we will omit proofs for these two postulates in the following subsections. Moreover, the full AGMes-compliance of minimal c-contractions is covered by the results of [13]. The remaining results for contraction, ignoration, marginalization and conditionalization will be discussed and proved in the following subsections.

4.1 Contraction

We start with considering c-contraction which seems to be the closest to AGM iterated contraction.

Proposition 2. *c-Contraction in general as defined in Definition 1 does not fulfill (AGMes-1), (AGMes-2), (AGMes-4), (AGMes-6) and (AGMes-7), but fulfills (AGMes-3) and (AGMes-5).*

Proof. c-Contraction does not fulfill (AGMes-1), as the counterexample in Fig. 1a shows. Let κ be a ranking function over the signature $\Sigma = \{a, b\}$ with $\kappa(ab) = 0$, $\kappa(a\bar{b}) = \kappa(\bar{a}\,\bar{b}) = 1$, and $\kappa(\bar{a}b) = 2$. By a c-contraction by a with $\gamma^+ = 1$ and $\gamma^- = -1$ we get κ_a° with $\kappa_a^\circ(\bar{a}\,\bar{b}) = 0$, $\kappa_a^\circ(ab) = \kappa_a^\circ(\bar{a}b) = 1$, and $\kappa_a^\circ(a\bar{b}) = 2$. We had $ab \equiv \mathbf{B}(\kappa)$ before the contraction and $\bar{a}\bar{b} \equiv \mathbf{B}(\kappa_a^\circ)$ after the contraction. Thus, $ab \not\equiv \bar{a}\bar{b}$ shows that (AGMes-1) is not fulfilled.

For (AGMes-2) let κ be a ranking function with $\kappa(\bar{a}\,\bar{b}) = 0$, $\kappa(ab) = \kappa(\bar{a}b) = 1$, and $\kappa(a\bar{b}) = 2$ (see Fig. 1c). After a c-contraction of κ by a with $\gamma^+ = 0$ and $\gamma^- = 1$ we get κ_a° with $\kappa_a^\circ(ab) = \kappa_a^\circ(\bar{a}\,\bar{b}) = 0$ and $\kappa_a^\circ(a\bar{b}) = \kappa_a^\circ(\bar{a}b) = 1$. This leads to $\mathbf{B}(\kappa_a^\circ) \equiv ab \vee \bar{a}\bar{b} \not\equiv \bar{a}b \equiv \mathbf{B}(\kappa)$, showing that (AGMes-2) is not fulfilled.

Figure 1c is also a counterexample for (AGMes-4). We have $\mathbf{B}(\kappa) \equiv \bar{a}\,\bar{b}$, but with a c-Contraction it is just as possible as with a c-ignoration that afterwards we believe neither a nor $\neg a$. In Fig. 1c we get $\mathbf{B}(\kappa_a^\circ) \equiv ab \vee \bar{a}\,\bar{b}$ which means that

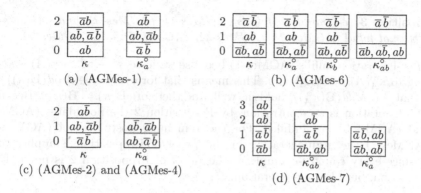

(a) (AGMes-1)

(b) (AGMes-6)

(c) (AGMes-2) and (AGMes-4)

(d) (AGMes-7)

Fig. 1. Counterexamples for c-contraction

$\mathbf{B}(\kappa_a^\circ) \not\models \overline{a}$ and by a conjunction with a we can infer a and cannot model all the beliefs of the previous ranking function. We have $\mathbf{B}(\kappa) \models \overline{a}$ but $\mathbf{B}(\kappa_a^\circ) \wedge a \models a$ which leads to $\mathbf{B}(\kappa_a^\circ) \wedge a \not\models \overline{a}$.

c-Contraction does not fulfill (AGMes-6). Let κ be the ranking function over the signature $\Sigma = \{a, b\}$ with $\kappa(\overline{a}b) = \kappa(a\overline{b}) = 0$, $\kappa(ab) = 1$, and $\kappa(\overline{a}\overline{b}) = 2$ as shown in Fig. 1b. Because $\kappa(a) = \kappa(\overline{a}) = \kappa(b) = \kappa(\overline{b}) = 0$, a c-contraction of κ with a or b, respectively, and by choosing $\gamma^+ = 0$ and $\gamma^- = 0$, does not change the ranking function, leading to $\kappa_a^\circ = \kappa$ and $\kappa_b^\circ = \kappa$. For κ, we have $\mathbf{B}(\kappa) \equiv \overline{a}b \vee a\overline{b}$ and the same for κ_a° and also for κ_b°, i.e., $\mathbf{B}(\kappa) \equiv \mathbf{B}(\kappa_a^\circ) \equiv \mathbf{B}(\kappa_b^\circ)$. A c-contraction of κ by ab and with $\gamma^+ = 0$, $\gamma^- = 1$ results in κ_{ab}° with $\kappa_{ab}^\circ(\overline{a}b) = \kappa_{ab}^\circ(a\overline{b}) = \kappa_{ab}^\circ(ab) = 0$ and $\kappa_{ab}^\circ(\overline{a}\overline{b}) = 2$. Hence, $\mathbf{B}(\kappa_{ab}^\circ) \equiv \overline{a}b \vee a\overline{b} \vee ab$ and therefore $ab \in Mod(\mathbf{B}(\kappa_{ab}^\circ))$ but $ab \notin Mod(\mathbf{B}(\kappa_a^\circ) \vee \mathbf{B}(\kappa_b^\circ))$.

For (AGMes-7), let κ be the ranking function with $\kappa(\overline{a}b) = 0$, $\kappa(\overline{a}\overline{b}) = 1$, $\kappa(a\overline{b}) = 2$, and $\kappa(a\,b) = 3$, cf. Fig. 1d. The belief of κ is $\mathbf{B}(\kappa) \equiv \overline{a}b$. By forgetting $a \wedge b$ with a c-contraction we have to choose γ^-/γ^+ according to Equation (2). The second table in Fig. 1d shows the result of κ_{ab}° with $\gamma^- = 3$ and $\gamma^+ = 0$. The result κ_a° of a c-contraction of κ by a and with $\gamma^- = 2$, $\gamma^+ = 0$ is shown in Fig. 1d. We get $\mathbf{B}(\kappa_{ab}^\circ) \equiv \overline{a}b \vee ab$ and $\mathbf{B}(\kappa_a^\circ) \equiv \overline{a}b \vee a\overline{b}$ which means that $\mathbf{B}(\kappa_{ab}^\circ) \not\models a$ but $\mathbf{B}(\kappa_a^\circ) \not\models \mathbf{B}(\kappa_{ab}^\circ)$.   □

The general concept of c-contractions allows for a broad range of contractions that, in this generality, only fulfill (AGMes-3) and (AGMes-5). While every c-contraction by A effectively forgets A, it is still possible to violate the AGMes postulates which concern only the most plausible beliefs.

4.2   Ignoration and Minimal c-Contraction

Ignoration is similar to contraction except for postulating explicitly undecided-ness between A and $\neg A$. In the framework of ranking functions, a c-ignoration is a special kind of c-contraction [2] which makes at least one model of A and one of $\neg A$ maximally plausible so that afterwards we believe neither A nor $\neg A$. This characteristic makes it possible to fulfill (AGMes-1).

Proposition 3. *c-Ignoration fulfills (AGMes-1), (AGMes-3), and (AGMes-5) but does not fulfill (AGMes-2), (AGMes-4), (AGMes-6), and (AGMes-7).*

Proof. c-Ignoration fulfills (AGMes-1) because we have $\gamma^- - \gamma^+ = \kappa(A) - \kappa(\overline{A})$ leading to $\kappa_A^\circ(A) = \kappa_A^\circ(\overline{A}) = 0$. This means that for every $\omega \in Mod(\mathbf{B}(\kappa))$ we have that $\omega \in Mod(\mathbf{B}(\kappa_A^\circ))$ holds as well, and therefore $\mathbf{B}(\kappa) \models \mathbf{B}(\kappa_A^\circ)$. Because every c-ignoration is a c-contration, by Proposition 2 the postulates (AGMes-3) and (AGMes-5) are fulfilled. For the remaining postulates of (AGMes-2) to (AGMes-7) the argumentation in the proof of Proposition 2 applies also here, since every counter-example in the proof of Proposition 2 is not only a c-contraction, but also a c-ignoration.   □

Another specific form of c-contractions are the new subclass of minimal c-contractions (cf. Sect. 3), which satisfy all AGMes postulates.

Proposition 4. *Minimal c-contractions fulfill (AGMes-1) to (AGMes-7).*

The proof of Proposition 4 can be derived from the results given in [13, Theorem 8]. Minimal c-contractions shift worlds in a minimal way, thus fully complying with the minimal change paradigm of AGM.

4.3 Marginalization

Let us now turn to a substantially different forgetting operation. Forgetting by OCF-marginalization has the goal to remove certain aspects of the language and only take the remaining signature elements into account.

Proposition 5. *Forgetting by OCF-marginalization fulfills (AGMes-1), (AGMes-3), (AGMes-5), and (AGMes-7). Forgetting by OCF-marginalization does not fulfill (AGMes-2), (AGMes-4), (AGMes-6).*

Proof. We start by showing that (AGMes-1) is fulfilled for forgetting by OCF-marginalization. The rank of the worlds with the reduced signature is computed by $\kappa|_{\Sigma'}(\omega') = \min\{\kappa(\omega) \mid \omega \in \Omega_\Sigma \text{ and } \omega \models \omega'\}$. Thus, for all $\omega \in Mod(\mathbf{B}(\kappa))$ there exists a $\omega' \in Mod(\mathbf{B}(\kappa_A^\circ))$ with $\omega \models \omega'$, implying $\mathbf{B}(\kappa) \models \mathbf{B}(\kappa_A^\circ)$.

Let κ be a ranking function over the signature $\Sigma = \{a, b, c\}$ as shown on the left side in Fig. 2a. For (AGMes-2), we only consider the second table of Fig. 2a, representing the forgetting of a in κ, namely $\kappa_a^\circ = \kappa|_{\{b,c\}}$. From $\mathbf{B}(\kappa) \equiv \overline{a}bc$ and $\mathbf{B}(\kappa_a^\circ) \equiv bc$ but $bc \not\models \overline{a}bc$.

The example in Fig. 2a also shows that forgetting by OCF-marginalization does not fulfill (AGMes-4). We have $\mathbf{B}(\kappa) \equiv \overline{a}bc$ and $\mathbf{B}(\kappa_a^\circ) \equiv bc$, leading to $\mathbf{B}(\kappa_a^\circ) \wedge a \models a$. Therefore, $\mathbf{B}(\kappa_a^\circ) \wedge a \models \mathbf{B}(\kappa)$ cannot be fulfilled because $abc \not\models \overline{a}$.

For (AGMes-6), consider all ranking functions as shown in Fig. 2a. The two tables in the middle show the forgetting of a in κ (second table), leading to $\kappa_a^\circ = \kappa|_{\{b,c\}}$, and $\kappa_b^\circ = \kappa|_{\{a,c\}}$ (third table) as the result of the forgetting of b. We have $\mathbf{B}(\kappa_a^\circ) \equiv bc$ and $\mathbf{B}(\kappa_b^\circ) \equiv \overline{a}c$. The forgetting of ab in κ leads to the ranking function $\kappa_{ab}^\circ = \kappa|_{\{c\}}$ over the reduced signature $\Sigma' = \{c\}$ as shown in

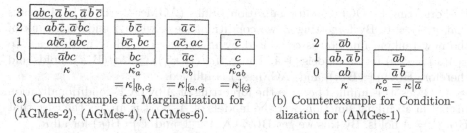

(a) Counterexample for Marginalization for (AGMes-2), (AGMes-4), (AGMes-6).

(b) Counterexample for Conditionalization for (AMGes-1)

Fig. 2. Counterexamples for marginalization and conditionalization.

Fig. 2a on the right side. There we get $\mathbf{B}(\kappa_{ab}^\circ) \equiv c$. Now $c \not\models bc \vee \bar{a}c$ implies that $\mathbf{B}(\kappa_{ab}^\circ) \not\models \mathbf{B}(\kappa_a^\circ) \vee \mathbf{B}(\kappa_b^\circ)$ and therefore (AGMes-6) is not fulfilled.

Forgetting by OCF-marginalization fulfills (AGMes-7). The forgetting of a ranking function κ by A leads to $\kappa_A^\circ = \kappa|_{\Sigma_1}$ with $\Sigma_1 = \Sigma \setminus \Sigma_A$ and we get $\kappa_{AB}^\circ = \kappa|_{\Sigma'}$ with $\Sigma' = \Sigma \setminus (\Sigma_A \cup \Sigma_B)$ for the forgetting by AB. Because $\Sigma' \subseteq \Sigma_1$, we can consider $\mathbf{B}(\kappa_A^\circ)$ and $\mathbf{B}(\kappa_{AB}^\circ)$ over $\mathcal{L}(\Sigma_1)$. It is clear that $\mathbf{B}(\kappa_{AB}^\circ) \models A$ can never hold. Because $\omega_1 \in Mod_{\Sigma_1}(\mathbf{B}(\kappa_A^\circ))$ iff $\kappa_A^\circ(\omega_1) = 0$, there exists a $\omega \in \Omega$ such that $\omega \models \omega_1$ and $\kappa(\omega) = 0$. This leads to $\kappa(\omega_1) = 0$. Accordingly, it is the case that $\omega' \in Mod_{\Sigma'}(\mathbf{B}(\kappa_{AB}^\circ))$ iff $\kappa_{AB}^\circ(\omega') = 0$ iff $\kappa(\omega') = 0$. Further it is the case that $\omega_1 \models \mathbf{B}(\kappa_{AB}^\circ)$ iff $\omega_1^{\Sigma'} \models \mathbf{B}(\kappa_{AB}^\circ)$ iff $\kappa(\omega_1^{\Sigma'}) = 0$. For (AGMes-7) we now assume that $\omega_1 \models \mathbf{B}(\kappa_A^\circ)$ holds. This is the case iff $\kappa_A^\circ(\omega_1) = 0$ iff $\kappa(\omega_1) = 0$. From this we can conclude that $\kappa(\omega_1^{\Sigma'}) = 0$ holds iff $\kappa_{AB}^\circ(\omega_1^{\Sigma'}) = 0$ iff $\omega_1^{\Sigma'} \models \mathbf{B}(\kappa_{AB}^\circ)$ iff $\omega_1 \models \mathbf{B}(\kappa_{AB}^\circ)$. This means that $\mathbf{B}(\kappa_A^\circ) \models \mathbf{B}(\kappa_{AB}^\circ)$ holds. □

Forgetting by OCF-marginalization ensures that no new beliefs are added but fails to recover the original beliefs when the negation of A was believed before. This information is completely lost under forgetting by OCF-marginalization. However, a bit surprisingly, forgetting by OCF-marginalization respects a kind of coherence, as expressed by (AGMes-7), even without the explicit prerequisite $\mathbf{B}(\kappa_{AB}^\circ) \not\models A$.

4.4 Conditionalization

Another form of forgetting is forgetting by OCF-conditionalization which restricts the models of a ranking function to a specific context, forgetting any model outside of the context.

Proposition 6. *Forgetting by OCF-conditionalization does not fulfill (AGMes-1) but fulfills (AGMes-2), (AGMes-3), (AGMes-4), (AGMes-5), (AGMes-6), and (AGMes-7).*

Proof. For (AGMes-1) let κ be the OCF over $\Sigma = \{a, b\}$ with $\kappa(ab) = 0$, $\kappa(a\bar{b}) = \kappa(\bar{a}\bar{b}) = 1$, and $\kappa(\bar{a}b) = 2$ as shown in Fig. 2b. The forgetting of κ by a results in $\kappa_a^\circ = \kappa|\bar{a}$ with $\kappa_a^\circ(\bar{a}\bar{b}) = 0$ and $\kappa_a^\circ(\bar{a}b) = 1$ (cf. Fig. 2b). We get $\mathbf{B}(\kappa) \equiv ab$ but $\mathbf{B}(\kappa_a^\circ) \equiv \bar{a}\bar{b}$; hence, $ab \not\models \bar{a}\bar{b}$ shows that (AGMes-1) is not fulfilled.

Forgetting by OCF-conditionalization fulfills (AGMes-2). $\mathbf{B}(\kappa) \not\models A$ means that $\kappa(\overline{A}) = 0$. By forgetting A we conditionalize κ with $\overline{A}$ and each rank of the new ranking function is computed by $\kappa(\omega) - \kappa(\overline{A})$ leading in this case to $\kappa|\overline{A}(\omega) = \kappa(\omega)$ for all models of $\overline{A}$. Thus, $Mod(\mathbf{B}(\kappa_A^\circ)) \subseteq Mod(\mathbf{B}(\kappa))$ holds and therefore $\mathbf{B}(\kappa_A^\circ) \models \mathbf{B}(\kappa)$, and (AGMes-2) is fulfilled.

(AGMes-4) is fulfilled because the forgetting of A by OCF-conditionalization results in a ranking function over the models of $\overline{A}$. So it is always the case that $\mathbf{B}(\kappa_A^\circ) \models \overline{A}$ holds. By this we get $\mathbf{B}(\kappa_A^\circ) \wedge A \equiv \bot$ and $\bot \models \mathbf{B}(\kappa)$ for all κ.

Forgetting by OCF-conditionalization fulfills (AGMes-6). It is the case that $\kappa_{A \wedge B}^\circ = \kappa|(\overline{A} \vee \overline{B})$ which means that the rank of all models of $\overline{A} \vee \overline{B}$ are reduced by $\kappa(\overline{A} \vee \overline{B}) = \min\{\kappa(\overline{A}), \kappa(\overline{B})\}$. Because every $\omega \in Mod(\mathbf{B}(\kappa_{A \wedge B}^\circ))$ is a belief of κ_A° or of κ_B°, we observe that $\omega \in Mod(\mathbf{B}(\kappa_A^\circ)) \cup Mod(\mathbf{B}(\kappa_B^\circ))$ holds and therefore $\mathbf{B}(\kappa_{A \wedge B}^\circ) \models \mathbf{B}(\kappa_A^\circ) \vee \mathbf{B}(\kappa_B^\circ)$ holds.

For (AGMes-7) we conclude $\kappa_{A \wedge B}^\circ(\overline{A}) = 0$ from $\mathbf{B}(\kappa_{A \wedge B}^\circ) \not\models A$. With this and the considerations from above we get $\kappa(\overline{A}) = \min\{\kappa(\overline{A}), \kappa(\overline{B})\}$. Every minimal model of κ_A° is a minimal model of $\kappa_{A \wedge B}^\circ$. Therefore, $Mod(\mathbf{B}(\kappa_A^\circ)) \subseteq Mod(\mathbf{B}(\kappa_{A \wedge B}^\circ))$ and $\mathbf{B}(\kappa_A^\circ) \models \mathbf{B}(\kappa_{A \wedge B}^\circ)$. Thus, (AGMes-7) is fulfilled. □

Forgetting by OCF-conditionalization fulfills almost all AGMes postulates. Only (AGMes-1) cannot be fulfilled because it is possible that some of the most plausible beliefs do not refer to the context after forgetting and hence are removed during forgetting by OCF-conditionalization.

5   Discussion, Conclusion, and Further Work

In this paper, we evaluated four key forgetting operators defined for ranking functions according to the standards of AGM contraction. We analysed the large class of c-contractions and their special subclass of c-ignorations [2] and the new minimal c-contractions, investigated the classical marginalization operator, and used conditionalization as a forgetting operator. Conditionalization fulfils nearly all postulates except for the very first one, which is due to the fact that conditionalization establishes belief in $\neg A$ if A is to be forgotten. This seems to be unintentionally strong, in particular when compared to c-ignoration that demands for being undecided between A and $\neg A$ after forgetting, but ensures the best compatibility with AGM contraction theory otherwise.

The paper [12] evaluates the forgetting operators mentioned above, except for c-ignoration and minimal c-contractions, from the point of view of forgetting in answer set programming (ASP). In that paper, ASP postulates for forgetting are generalized to epistemic states and their satisfaction for each forgetting operator is investigated. Together with the observations of [12], our results obtained here show clearly the type difference between forgetting in logic programming (which is basically marginalization) and forgetting through contraction in AGM theory, but also reveals that this distinction is not exclusive: Both forgetting by marginalization and by conditionalization satisfy postulates from both areas, encouraging further studies of formal properties of forgetting as an operator that is interesting in itself. Next steps will be to generalize our studies to deal with forgetting in epistemic states that are equipped just with total preorders. However, while any total preorder can be considered as a ranking function by numbering consecutively the layers of the total preorder, the full arithmetics allowing us to realize forgetting in ranking functions by addition and subtraction conveniently will not be available and also not justifiable any more in a total preorder setting.

Acknowledgements. The research reported here was supported by the German Research Society (DFG) within the Priority Research Program *Intentional Forgetting in Organisations* (DFG-SPP 1921; grants BE 1700/9-1, KE 1413/10-1).

References

1. Alchourrón, C.E., Gärdenfors, P., Makinson, D.: On the logic of theory change: partial meet contraction and revision functions. J. Symb. Logic **50**(2), 510–530 (1985)
2. Beierle, C., Kern-Isberner, G., Sauerwald, K., Bock, T., Ragni, M.: Towards a general framework for kinds of forgetting in common-sense belief management. KI - Künstliche Intelligenz **33**(1), 57–68 (2019)
3. Caridroit, T., Konieczny, S., Marquis, P.: Contraction in propositional logic. Int. J. Approx. Reason. **80**, 428–442 (2017)
4. Chopra, S., Ghose, A., Meyer, T.A., Wong, K.-S.: Iterated belief change and the recovery axiom. J. Philos. Logic **37**(5), 501–520 (2008)
5. de Finetti, B.: La prévision, ses lois logiques et ses sources subjectives. Ann. Inst. H. Poincaré **7**(1), 1–68 (1937). Engl. transl. Theory Probab. J. (1974). Wiley
6. Eiter, T., Fink, M., Sabbatini, G., Tompits, H.: On updates of logic programs: semantics and properties. Technical report, TU Vienna (2001)
7. Eiter, T., Kern-Isberner, G.: A brief survey on forgetting from a knowledge representation and reasoning perspective. KI **33**(1), 9–33 (2019)
8. Fermé, E.L., Saez, K., Sanz, P.: Multiple kernel contraction. Stud. Logica **73**(2), 183–195 (2003)
9. Goldszmidt, M., Pearl, J.: Qualitative probabilities for default reasoning, belief revision, and causal modeling. Artif. Intell. **84**, 57–112 (1996)
10. Gonçalves, R., Knorr, M., Leite, J.: The ultimate guide to forgetting in answer set programming. In: Principles of Knowledge Representation and Reasoning: Proceedings of the Fifteenth International Conference, KR 2016, pp. 135–144. AAAI Press (2016)
11. Katsuno, H., Mendelzon, A.: Propositional knowledge base revision and minimal change. Artif. Intell. **52**, 263–294 (1991)
12. Kern-Isberner, G., Bock, T., Beierle, C., Sauerwald, K.: Axiomatic evaluation of epistemic forgetting operators. In: Proceedings of the 32nd International FLAIRS Conference, FLAIRS-32, pp. 470–475. AAAI Press (2019)
13. Kern-Isberner, G., Bock, T., Sauerwald, K., Beierle, C.: Iterated contraction of propositions and conditionals under the principle of conditional preservation. In: 3rd Global Conference on Artificial Intelligence, GCAI 2017. EPiC Series in Computing, vol. 50, pp. 78–92. EasyChair (2017)
14. Konieczny, S., Pino Pérez, R.: On iterated contraction: syntactic characterization, representation theorem and limitations of the levi identity. In: Moral, S., Pivert, O., Sánchez, D., Marín, N. (eds.) SUM 2017. LNCS (LNAI), vol. 10564, pp. 348–362. Springer, Cham (2017). https://doi.org/10.1007/978-3-319-67582-4_25

15. Saribatur, Z.G., Eiter, T.: Omission-based abstraction for answer set programs. In: Principles of Knowledge Representation and Reasoning: Proceedings of the Sixteenth International Conference, KR 2018, pp. 42–51. AAAI Press (2018)
16. Spohn, W.: Ordinal conditional functions: a dynamic theory of epistemic states. In: Harper, W.L., Skyrms, B. (eds.) Causation in Decision, Belief Change, and Statistics, II, pp. 105–134. Kluwer Academic Publishers (1988)
17. Spohn, W.: The Laws of Belief: Ranking Theory and Its Philosophical Applications. Oxford University Press, Oxford (2012)

Identity Resolution in Ontology Based Data Access to Structured Data Sources

David Toman$^{(\boxtimes)}$ and Grant Weddell

Cheriton School of Computer Science, University of Waterloo, Waterloo, Canada
{david,gweddell}@uwaterloo.ca

Abstract. Earlier work has proposed a notion of referring expressions and types in first order knowledge bases as a way of more effectively answering conjunctive queries in ontology based data access (OBDA). We consider how PTIME description logics can be combined with referring expressions to provide a more effective virtual front-end to nested relational data sources via OBDA. In particular, we consider replacing the standard notion of an assertion box, or ABox, with a more general notion of a concept box, or CBox, and show how this can serve as a front-end to such data sources.

1 Introduction

In a query answer $(a_1, \ldots, a_n)$ over structured data sources viewed as a first order knowledge base $\mathcal{K}$, the common assumption is that each a_i will correspond to some constant symbol occurring in $\mathcal{K}$. A more general option has been proposed in [1] in which each a_i can now be a *referring expression*, in particular, a well-formed formulae ψ that is free in one variable and that satisfies a number of additional conditions for any interpretation $\mathcal{I}$ of $\mathcal{K}$. First, ψ should not be *vacuous*: it should hold of at least one individual in $\triangle^{\mathcal{I}}$. Second, ψ should be *singular*: it should hold of *at most one* individual in $\triangle^{\mathcal{I}}$. And third, the singularity property of ψ should be ensured by the ontological component of $\mathcal{K}$.

In this paper, we consider query answering in which the ontological component of $\mathcal{K}$ consists of a TBox $\mathcal{T}$ expressed in terms of a *description logic* (DL), and in which the remaining part of $\mathcal{K}$ consists of a CBox $\mathcal{C}$ instead of an ABox, where $\mathcal{C}$ consists of a finite set of referring expressions in the form of concept descriptions in the DL, and for which each must be singular and non-empty in all models of $\mathcal{K}$.

The DL we consider is $partial-\mathcal{CFDI}_{nc}^{\forall-}$ [4,9], a dialect of the PTIME feature-based $\mathcal{CFD}$ family designed for capturing structured data sources, and our main focus is on query answering over a knowledge base $\mathcal{K}$ consisting of a TBox and CBox pair $(\mathcal{T}, \mathcal{C})$ expressed in terms of $partial-\mathcal{CFDI}_{nc}^{\forall-}$. The main technical difficulty is on mapping $\mathcal{C}$ to a combination of an ABox $\mathcal{A}$ and a way of distinguishing the constant symbols occurring in $\mathcal{A}$ that "stand in place" of referring expressions in $\mathcal{C}$. This must be done in a way that ensures off-the-shelf query answering over $(\mathcal{T}, \mathcal{A})$ can be used to compute the certain answers to

© Springer Nature Switzerland AG 2019
A. C. Nayak and A. Sharma (Eds.): PRICAI 2019, LNAI 11670, pp. 473–485, 2019.
https://doi.org/10.1007/978-3-030-29908-8_38

queries over the original $\mathcal{K} = (\mathcal{T}, \mathcal{C})$ by a simple substitution of the distinguished constants by their referring expressions.

A core problem in deriving the ABox relates to identity issues when introducing new constants. Of particular significance is that fix-point computations are necessary when such constants are introduced. Indeed, this can be necessary when a TBox derives from relational data sources with tables that have uniqueness constraints as well as primary keys, or for which primary keys themselves are not minimal. In database parlance, one would say in this case that primary keys are *superkeys* but not *candidate keys*. Such "key conversion" tables can serve to map between alternative primary keys and thereby lead to additional query answers.

While several approaches to integrating information in settings in which the same individual can be identified in several (even syntactically incomparable) ways have been considered in the past [2], we show how the integration can be achieved naturally within $partial-\mathcal{CFDI}_{nc}^{\forall-}$ by reducing the problem to existing ABox completion procedures for query answering over $partial-\mathcal{CFDI}_{nc}^{\forall-}$. In particular, using concepts and procedures developed in [1] and [4], we define a natural way of capturing (perhaps multiple) external identities of objects. Subsequently, we show how query answering can be achieved in such a setting via an embedding into standard $partial-\mathcal{CFDI}_{nc}^{\forall-}$ reasoning and query answering problems. We also present examples that show how this technique can apply in an OBDA setting to structured knowledge bases, both in the relational setting and in the setting of nested relational or document databases such as MongoDB.

The remainder of the paper is organized as follows. We begin with the necessary background material in Sect. 2 in which we introduce $partial-\mathcal{CFDI}_{nc}^{\forall-}$ concepts, and "standard" knowledge bases consisting of a TBox of inclusion dependencies over such concepts and an ABox of assertions. Our main results then follow in Sect. 3 in which an ABox is replaced with a CBox of $partial-\mathcal{CFDI}_{nc}^{\forall-}$ concepts called *referring expressions*. We then define a mapping of CBoxes to ABoxes and show how each of the following can be resolved with the use of this mapping: (1) diagnosing an *admissibility* condition for a CBox, (2) satisfiability of knowledge bases with a CBox, and (3) query answering over knowledge bases with a CBox. The admissibility condition requires that the TBox ensures each referring expression occurring in the CBox is singular in the sense outlined above. Throughout, we introduce examples to illustrate why CBoxes are useful and how identification issues become far more complicated as a consequence. We conclude with summary comments in Sect. 4.

2   Background

The description logic $partial-\mathcal{CFDI}_{nc}^{\forall-}$ is a member of the $\mathcal{CFD}$ family of DLs which are fragments of FOL with underlying signatures based on disjoint sets of unary predicate symbols called *primitive concepts*, constant symbols called *individuals* and unary function symbols called *features*. Note that features deviate from the normal practice of admitting *roles* denoting binary predicate symbols.

SYNTAX	SEMANTICS: DEFN OF "$(\cdot)^{\mathcal{I}}$"	
$C ::= A$	$A^{\mathcal{I}} \subseteq \triangle$	(primitive concept; $A \in PC$)
$\mid \forall Pf.C$	$\{x \mid Pf^{\mathcal{I}}(x) \in C^{\mathcal{I}}\}$	(value restriction)
$\mid \exists Pf$	$\{x \mid Pf^{\mathcal{I}}(x) \text{ exists}\}$	(existential restriction)
$\mid \neg C$	$\triangle \setminus C^{\mathcal{I}}$	(negation)
$\mid \exists f^{-1}$	$\{f^{\mathcal{I}}(x) \mid x \in \triangle\}$	(inverse feature)
$\mid C : Pf_1, ..., Pf_k \to Pf_0$	(see text)	(PFD)
$\mid \{a\}$	$\{a^{\mathcal{I}}\}$	(nominal)
$\mid C_1 \sqcap C_2$	$C_1^{\mathcal{I}} \cap C_2^{\mathcal{I}}$	(conjunction)
$\mid \exists f^{-1}.C$	$\{f^{\mathcal{I}}(x) \mid x \in C^{\mathcal{I}}\}$	(qualified inverse feature)

Fig. 1. Syntax and semantics of concept descriptions.

However, features make it easier to incorporate concept constructors that are better suited to the capture of (possibly nested) relational data sources, particularly so when they include dependencies such as primary keys, uniqueness constraints, functional dependencies and foreign keys. This is achieved by a straightforward reification of n-ary predicates and by using a concept constructor peculiar to the $\mathcal{CFD}$ family called a *path functional dependency* (PFD). Consider the case of a role R. It can be reified as a primitive concept R_C, two features $R\text{-}dom$ and $R\text{-}ran$ and an inclusion dependency of the form

$$R_C \sqsubseteq R_C : R\text{-}dom, R\text{-}ran \to self$$

in $partial\text{-}\mathcal{CFDI}_{nc}^{\forall-}$. By introducing a PFD on the right-hand-side, the dependency ensures any combination of $R\text{-}dom$ and $R\text{-}ran$ values uniquely determine an R 2-tuple. Note that an $\mathcal{ALC}$ inclusion dependency mentioning R of the form "$A \sqsubseteq \forall R.B$", can also be captured in $partial\text{-}\mathcal{CFDI}_{nc}^{\forall-}$ as the inclusion dependency

$$\forall R\text{-}dom.A \sqsubseteq \forall R\text{-}ran.B.$$

Concepts and TBoxes in $partial\text{-}\mathcal{CFDI}_{nc}^{\forall-}$ are defined as follows:

Definition 1 ($partial\text{-}\mathcal{CFDI}_{nc}^{\forall-}$ **Concepts and TBoxes**). Let F and PC be sets of feature names and primitive concept names, respectively. A *partial path expression* is defined by the grammar "$Pf :: = f.Pf \mid self$" for $f \in F$. We define derived *concept descriptions* by the grammar on the left-hand-side of Fig. 1. A *path functional dependency* concept, or PFD, is obtained by using the sixth production of this grammar.

An *inclusion dependency* $\mathcal{C}$ is an expression of the form $C_1 \sqsubseteq C_2$. A *terminology* (TBox) $\mathcal{T}$ consists of a finite set of inclusion dependencies. A *posed question* $\mathcal{Q}$ is a single inclusion dependency.

The *semantics* of expressions is defined with respect to a structure $\mathcal{I} = (\triangle, \cdot^{\mathcal{I}})$, where $\triangle$ is a domain of "objects" and $\cdot^{\mathcal{I}}$ an interpretation function that fixes the interpretations of primitive concepts A to be subsets of $\triangle$ and primitive features f to be partial functions $f^{\mathcal{I}} : \triangle \rightarrow \triangle$. The interpretation is extended to partial path expressions, $self^{\mathcal{I}} = \lambda x.x$, $(f.\mathsf{Pf})^{\mathcal{I}} = \mathsf{Pf}^{\mathcal{I}} \circ f^{\mathcal{I}}$, in the natural way, and derived concept descriptions C not including PFDs as defined in the centre column of Fig. 1.

Note that $partial\text{--}\mathcal{CFDI}_{nc}^{\forall-}$ adopts the *strict* interpretation of undefined values, which means that argument terms *must* be defined whenever equality and set membership do hold. This implies, for any concept C, that $(\forall\,\mathsf{Pf}.C)^{\mathcal{I}}$ must be a subset of $(\exists\mathsf{Pf})^{\mathcal{I}}$. This also suggests the following interpretation of concepts that are PFDs:[1]

$$(C : \mathsf{Pf}_1, \ldots, \mathsf{Pf}_k \rightarrow \mathsf{Pf}_0)^{\mathcal{I}} = \{x \mid \forall y.(y \in C^{\mathcal{I}} \wedge x \in (\exists\mathsf{Pf}_0)^{\mathcal{I}} \wedge y \in (\exists\mathsf{Pf}_0)^{\mathcal{I}} \wedge$$
$$\bigwedge_{i=1}^{k}(x \in (\exists\mathsf{Pf}_i)^{\mathcal{I}} \wedge y \in (\exists\mathsf{Pf}_i)^{\mathcal{I}} \wedge \mathsf{Pf}_i^{\mathcal{I}}(x) = \mathsf{Pf}_i^{\mathcal{I}}(y))) \rightarrow \mathsf{Pf}_0^{\mathcal{I}}(x) = \mathsf{Pf}_0^{\mathcal{I}}(y) \}.$$

A $partial\text{--}\mathcal{CFDI}_{nc}^{\forall-}$ TBox $\mathcal{T}$ consists of a set of *inclusion dependencies* of the form $C_1 \sqsubseteq C_2$. An interpretation $\mathcal{I}$ satisfies an inclusion dependency if $C_1^{\mathcal{I}} \subseteq C_2^{\mathcal{I}}$, and is a *model of* $\mathcal{T}$ ($\mathcal{I} \models \mathcal{T}$) if it satisfies all inclusion dependencies in $\mathcal{T}$. The *logical implication problem* asks if $\mathcal{T} \models \mathcal{Q}$ holds, that is, if $\mathcal{Q}$ is satisfied in all models of $\mathcal{T}$.   □

Since features are still *functional* there is no need for a qualified existential restriction of the form "$\exists f.C$". Such restrictions can be equivalently written as "$\forall f.C \sqcap \exists f$". Hence, the use of qualified existential restrictions in the rest of the paper should be considered to be syntactic sugar.

To ensure PTIME reasoning in $partial\text{--}\mathcal{CFDI}_{nc}^{\forall-}$, we require that a TBox $\mathcal{T}$ has a conservative extension $\mathcal{T}'$ in which the structure of concepts C and D in each inclusion dependency $C \sqsubseteq D$ are given by the following grammars (for more general TBoxes and normalization see [9]):

$$C ::= A \mid \forall f.A \mid \exists f$$
$$D ::= A \mid \neg A \mid \forall f.A \mid \exists f^{-1} \mid \exists f \mid A : \mathsf{Pf}_1, \ldots, \mathsf{Pf}_k \rightarrow \mathsf{Pf}$$

Observe how this effectively requires the left-hand-side of inclusion dependencies to employ only the first three concept constructors in Fig. 1, and disallows entirely the use of the last three concept constructors in Fig. 1 to occur at all in a TBox. Indeed, allowing the use of qualified inverse features or nominals, or the occurrence of a PFD on the left-hand-side leads to undecidability [5,6]. (We have included these constructors in preparation for the introduction of *referring expressions* introduced in the next section.)

A TBox must also satisfy two additional syntactic conditions. First, as a consequence of inverse feature and value restriction interaction, whenever both $A \sqsubseteq \exists f^{-1}$ and $\forall f.A' \sqsubseteq B$ occur in $\mathcal{T}$, then at least one of $A \sqsubseteq A'$, $A' \sqsubseteq A$,

---

[1] This constitutes a minimal condition for capturing when one violates an inclusion dependency of the form "$C_1 \sqsubseteq C_2 : \mathsf{Pf}_1, \ldots, \mathsf{Pf}_k \rightarrow \mathsf{Pf}_0$".

or $A \sqsubseteq \neg A'$ also occurs in $\mathcal{T}$. Relaxing this condition leads to intractability [8]. And finally, as a consequence of inverse feature and PFD interaction, any PFD occurring in $\mathcal{T}$ must adhere to one of the following two forms to yet again avoid undecidability [7]:

$$C : \mathsf{Pf}_1, \ldots, \mathsf{Pf} \,.\, \mathsf{Pf}_i, \ldots, \mathsf{Pf}_k \to \mathsf{Pf} \ \text{ or } \ C : \mathsf{Pf}_1, \ldots, \mathsf{Pf} \,.\, g, \ldots, \mathsf{Pf}_k \to \mathsf{Pf} \,.\, f.$$

Definition 2 (*partial*$-\mathcal{CFDI}_{nc}^{\forall-}$ **Knowledge Bases**). Let IN be a set of constant symbols. A *partial*$-\mathcal{CFDI}_{nc}^{\forall-}$ ABox $\mathcal{A}$ consists of a set of *assertions* of the form "$A(a)$", "$a = b$", "$a \neq b$", and "$f(a) = b$", with the usual interpretation mapping constant symbols to domain elements, and interpreting the assertions as set membership and an equality/inequality between a constant and another constant or a function application to a constant, respectively. A *partial*$-\mathcal{CFDI}_{nc}^{\forall-}$ knowledge base $\mathcal{K}$ consists of a TBox $\mathcal{T}$ and ABox $\mathcal{A}$. $\square$

Proposition 3 (*partial*$-\mathcal{CFDI}_{nc}^{\forall-}$ **KB Satisfiability** [9]). Satisfiability of *partial*$-\mathcal{CFDI}_{nc}^{\forall-}$ knowledge bases is complete for PTIME. $\square$

Conjunctive queries are formed as usual from atomic queries (or *atoms*), corresponding to concept descriptions, and equalities between variables and applications of features to variables, using conjunction and existential quantification. To simplify notation, we conflate conjunctive queries with the set of its constituent atoms and a set of *answer variables*:

Definition 4 (Conjunctive Query). Let φ be a set of atoms (representing a conjunction) $A(x_i)$ and $f(x_{i_1}) = x_{i_2}$, where A is a primitive concept description, f a feature (including *self*), and $\bar{x}$ a tuple of variables. We call the expression $\{\bar{x} \mid \varphi\}$ a *conjunctive query* (CQ). $\square$

A conjunctive query $\{\bar{x} \mid \varphi\}$ is therefore a notational variant of the formula $\exists \bar{y}. \bigwedge_{\psi \in \varphi} \psi$ in which $\bar{y}$ contains all variables appearing in φ but not in $\bar{x}$. The usual definition of certain answers is given by the following:

Definition 5 (Certain Answer). Let $\mathcal{K}$ be a *partial*$-\mathcal{CFDI}_{nc}^{\forall-}$ KB and $Q = \{\bar{x} \mid \varphi\}$ a CQ. A *certain answer* to Q over $\mathcal{K}$ is a substitution of constant symbols $\bar{a}$, $[\bar{x} \mapsto \bar{a}]$, such that $\mathcal{K} \models \varphi[\bar{x} \mapsto \bar{a}]$. $\square$

Proposition 6 (*partial*$-\mathcal{CFDI}_{nc}^{\forall-}$ **Query Answering** [4]). Query answering over *partial*$-\mathcal{CFDI}_{nc}^{\forall-}$ knowledge bases is complete for PTIME (data complexity). $\square$

The query answering algorithm presented in [4] requires both ABox completion and query reformulation. The former is needed to propagate concept memberships along feature chains present in the data when implied by a TBox, and the latter is needed to avoid the need for potentially exponentially many witnesses of anonymous objects. Note that the ABox completion also deals with *equalities* stipulated in the ABox and/or generated by PFD-based inclusion dependencies in the knowledge base TBox.

3   Referring Expressions and CBoxes

In this section, we introduce *referring expressions*: concept descriptions that will serve as external identifiers of objects in $partial-\mathcal{CFDI}_{nc}^{\forall-}$ knowledge bases.

We will require that these concept descriptions *behave* the same way *constant symbols* behave in the traditional setting: we expect their interpretations to be *singular* in every model of a given knowledge base.

Definition 7 (Referring Expressions and Singularity). Let $\mathcal{T}$ be a TBox and let C be a $partial-\mathcal{CFDI}_{nc}^{\forall-}$ concept description conforming to the grammar

$$C ::= A \mid C_1 \sqcap C_2 \mid \exists f.C \mid \exists f^{-1}.C \mid \{a\},$$

where a is a constant symbol. We say that C is a *referring expression*, and say in addition that C is *singular with respect to* $\mathcal{T}$ if $|C^{\mathcal{I}}| \leq 1$ for all interpretations $\mathcal{I}$ that are models of $\mathcal{T}$.   □

We use referring expressions to define the counterpart of *assertions* in traditional knowledge bases.

Example 8. Consider where the interpretation of primitive concept PERSON is intended to be all people, and where each person is identified by *ssn*, a social security number. The following referring expressions might be used to identity two individuals:

$$\text{PERSON} \sqcap \exists ssn.\{123\}, \text{ and PERSON} \sqcap \exists ssn.\{456\}.$$

The referring expressions would qualify as singular if the underlying TBox ensured that *ssn*-values can indeed serve as a way of identifying people, e.g., by having the inclusion dependency

$$\text{PERSON} \sqsubseteq \text{PERSON} : ssn \to self.$$

The two concepts can replace the usual ABox assertions of the form $\text{PERSON}(c_1)$ and $\text{PERSON}(c_2)$ (which require the introduction of additional constant symbols for the two individuals). Moreover, ABox assertions of the form $f(c_1) = c_2$ can also be captured by concepts, in our example:

$$\text{PERSON} \sqcap \exists ssn.\{123\} \sqcap \exists mother.(\text{PERSON} \sqcap \exists ssn.\{456\})$$

□

Note that such descriptions naturally arise when the assertion part of a knowledge base is captured in various database back-ends, e.g., in relational databases, via keys and foreign keys, or in nested relational or document databases, such as MongoDB[2], in which the structure of the referring expressions correspond to JSON[3].

---

[2] https://www.mongodb.com/.
[3] https://www.json.org/.

Example 9. To illustrate the flexibility of referring expressions in the nested relational or document setting, consider where PERSON is the name of a JSON collection that contains the following two documents:

```
{ "fname" : "John", "lname" : "Smith", "age" : 25,
  "wife" : { "fname" : "Mary" },
  "phone" : [
    {"loc" : "home", "dnum" : "212 555-1234"}
  ] }

{ "fname" : "Mary", "lname" : "Smith", "age" : 27,
  "husband" : { "fname" : "John" },
  "phone" : [
    {"loc" : "home", "dnum" : "212 555-1234"},
    {"loc" : "work", "dnum" : "212 666-4567"}
  ] }
```

In our setting, the documents can be naturally and directly represented by the following pair of referring expressions:[4]

$\text{PERSON} \sqcap (\exists \mathit{fname}.\{\text{"John"}\}) \sqcap (\exists \mathit{lname}.\{\text{"Smith"}\}) \sqcap (\exists \mathit{age}.\{25\})$
$\sqcap \exists \mathit{wife}.(\exists \mathit{fname}.\{\text{"Mary"}\})$
$\sqcap \exists \mathit{phone\text{-}dom}^{-1}.\exists \mathit{phone\text{-}ran}.((\exists \mathit{loc}.\{\text{"home"}\}) \sqcap (\exists \mathit{dnum}.\{\text{"212 555-1234"}\}))$

$\text{PERSON} \sqcap (\exists \mathit{fname}.\{\text{"Mary"}\}) \sqcap (\exists \mathit{lname}.\{\text{"Smith"}\}) \sqcap (\exists \mathit{age}.\{27\})$
$\sqcap \exists \mathit{husband}.(\exists \mathit{fname}.\{\text{"John"}\})$
$\sqcap \exists \mathit{phone\text{-}dom}^{-1}.\exists \mathit{phone\text{-}ran}.((\exists \mathit{loc}.\{\text{"home"}\}) \sqcap (\exists \mathit{dnum}.\{\text{"212 555-1234"}\}))$
$\sqcap \exists \mathit{phone\text{-}dom}^{-1}.\exists \mathit{phone\text{-}ran}.((\exists \mathit{loc}.\{\text{"work"}\}) \sqcap (\exists \mathit{dnum}.\{\text{"212 666-4567"}\}))$

Each would quality as a referring expression if, e.g., the underlying TBox were to contain the inclusion dependency

$$\text{PERSON} \sqsubseteq \text{PERSON} : \mathit{fname}, \mathit{lname} \to \mathit{self},$$

that is, if the combination of an fname and an lname identifies a PERSON. □

Thus, identities of entities referenced in documents (and sub-documents) are now captured using referring expressions. This potentially allows for join operations on document databases that are not typically supported by such systems. Queries navigating JSON documents can now be expressed as conjunctive queries over the $\mathit{partial\text{-}CFDI}_{nc}^{\forall-}$ representation.

This development leads to a revision of the definition of $\mathit{partial\text{-}CFDI}_{nc}^{\forall-}$ knowledge bases in which the traditional ABox is replaced by what we call a CBox, that is, by a set of concept descriptions that are referring expressions for individuals the knowledge base *knows* about.

---

[4] Here, we translate **phone** fields of documents as roles that are reified, in the sense outline above, in order to enable different people to share phones.

Definition 10 (CBoxes, Knowledge Bases, and Query Answers). Let $\mathcal{T}$ be a *partial*$-\mathcal{CFDI}_{nc}^{\forall-}$ TBox and $Q = \{(x_1,\ldots,x_k) \mid \varphi\}$ a conjunctive query. We define a *CBox* $\mathcal{C}$ to be a set of *partial*$-\mathcal{CFDI}_{nc}^{\forall-}$ concept descriptions. A *partial*$-\mathcal{CFDI}_{nc}^{\forall-}$ *knowledge base* $\mathcal{K}$ is a pair $(\mathcal{T},\mathcal{C})$. We say that the CBox $\mathcal{C}$ is *admissible* for $\mathcal{T}$ if each $C \in \mathcal{C}$ is a referring expression that is singular with respect to $\mathcal{T}$, and that $\mathcal{I}$ is a model of $\mathcal{K}$ if $\mathcal{I} \models \mathcal{T}$ and, for every $C \in \mathcal{C}$, $|C^{\mathcal{I}}| = 1$. Thus, $\mathcal{K}$ is *consistent* if such a model exists. Finally, we say that $(C_1,\ldots,C_k)$ is a certain answer to Q in $\mathcal{K}$ if

$$\mathcal{K} \models \exists x_1,\ldots,x_k.(\varphi \wedge C_1(x_1) \wedge \ldots \wedge C_k(x_k))$$

for $\{C_1,\ldots,C_k\} \subseteq \mathcal{C}$. □

3.1 Identity Resolution

CBoxes inherently represent information about how objects in a knowledge base are identified in which there are no restrictions on how such identification must be captured. In particular, there are no *uniformity conditions* on identification of objects that must hold, such as requiring each object to have a single global identifier in all assertions in the knowledge base.

However, CBoxes allow one to *capture* various resolutions of the heterogeneity of identification, e.g., through *translation tables* or *cross-links* [2]. As the following illustrates, these can be captured using TBox/CBox assertions:

Example 11. Consider the following *partial*$-\mathcal{CFDI}_{nc}^{\forall-}$ knowledge base:

$$\mathcal{T} = \{ \text{FRIEND} \sqsubseteq \text{PERSON},$$
$$\text{FRIEND} \sqsubseteq \text{PERSON} : \textit{fname} \rightarrow \textit{self},$$
$$\text{MATRIARCH} \sqsubseteq \text{PERSON},$$
$$\text{MATRIARCH} \sqsubseteq \text{PERSON} : \textit{lname} \rightarrow \textit{self},$$
$$\text{PERSON} \sqsubseteq \text{PERSON} : \textit{fname}, \textit{lname} \rightarrow \textit{self}, \ldots\}$$

$$\mathcal{C} = \{ \text{FRIEND} \sqcap \exists \textit{fname}.\{\text{``Mary''}\},$$
$$\text{PERSON} \sqcap (\exists \textit{fname}.\{\text{``Mary''}\}) \sqcap (\exists \textit{lname}.\{\text{``Smith''}\}),$$
$$\text{MATRIARCH} \sqcap \exists \textit{lname}.\{\text{``Smith''}\}, \ldots\}$$

Observe that the three referring expressions in $\mathcal{C}$ are each singular with respect to $\mathcal{T}$. On co-reference, note that the first two inclusion dependencies in $\mathcal{T}$ imply that the first referring expression and the second

$$\text{PERSON} \sqcap (\exists \textit{fname}.\{\text{``Mary''}\}) \sqcap (\exists \textit{lname}.\{\text{``Smith''}\})$$

must refer to the same object, and that the next two inclusion dependencies in $\mathcal{T}$ imply the same for the second and the third referring expression. Thus, the object referred to by

$$\text{FRIEND} \sqcap \exists \textit{fname}.\{\text{``Mary''}\}$$

should be a certain answer to the conjunctive query

$$\{x \mid \text{MATRIARCH}(x)\}.$$

The same happens for all pairs of referring expressions in $\mathcal{C}$ subsumed by FRIEND and MATRIARCH, respectively, for which there is a PERSON *cross-link*. □

3.2 On Minimal Referring Expressions

In relational databases the notion of *candidate key*, a key that has a *minimal* set of attributes of a relation, is typically used as an external identifier of objects stored in the database.

Our development of a *referring expression* strictly generalizes the notion of a *superkey* in the relational setting: sets of attributes, not necessarily minimal, that identifies an object or entity. We now present a procedure that (syntactically) minimizes a referring expression to obtain minimal co-referring referring expressions that are counterparts to relational candidate keys.

Theorem 12 (Minimal Referring Expressions). Let $\mathcal{T}$ be a *partial*$-$ $\mathcal{CFDI}_{nc}^{\forall-}$ TBox and C a referring expression w.r.t. $\mathcal{T}$. We say that subconcepts of C of the form A, $\{a\}$, $\exists f.\top$, $\exists f^{-1}.\top$, and $\top \sqcap \top$ are *leaves* of C and write $C[L \mapsto \top]$ for a description C in which a leaf L was replaced by $\top$. Assuming "first-leaf" and "next-leaf" denote functions that successively enumerate all leaves of C, the procedure

1. $L := \text{first-leaf}(C);$
2. **while** $C[L \mapsto \top]$ is singular w.r.t. $\mathcal{T}$ **do**
3. $C := C[L \mapsto \top]; L := \text{next-leaf}(C);$
4. **done**
5. **return** $C;$

computes a syntactically-minimal co-referring expression for C. (Note that replacing a leaf by $\top$ may create additional leaves.)

Proof (sketch): Since the $C[L \mapsto \top]$ operation weakens the concept description C, it preserves satisfiability. The algorithm tests for singularity at every step. Hence the result is a minimal referring expression equivalent to C since no additional leaves can be removed. □

The algorithm finds a minimal referring expression in time linear in $|C|$. Analogous to the relational setting, backtracking this algorithm facilitates the discovery of alternative minimal referring expressions, and, also analogous to the relational setting, there can be exponentially many of these.

3.3 Reasoning with CBoxes

Our technique crucially depends on mapping CBoxes to (standard) ABoxes as follows. We begin by defining how individual concepts corresponding to referring expressions are transformed:

$$\text{ToABox}(a : C_1 \sqcap C_2) \;\; \mapsto \;\; \text{ToABox}(a : C_1) \cup \text{ToABox}(a : C_2)$$
$$\text{ToABox}(a : \exists f.C) \;\; \mapsto \;\; \{f(a) = b\} \cup \text{ToABox}(b : C), b \text{ fresh}$$
$$\text{ToABox}(a : \exists f^{-1}.C) \;\; \mapsto \;\; \{f(b) = a\} \cup \text{ToABox}(b : C), b \text{ fresh}$$
$$\text{ToABox}(a : \{b\}) \;\; \mapsto \;\; \{a = b\}$$
$$\text{ToABox}(a : A) \;\; \mapsto \;\; \{A(a)\}, A \text{ primitive}$$

The ToABox function converts a CBox assertion C to a set of ABox assertions by introducing constant names for all necessary individuals, in particular a constant a_C for the (witness of satisfiability of) C itself. The mapping is then lifted to CBoxes by applying it on all referring expressions in the CBox as follows:

$$\text{ToABox}(\mathcal{C}) = \bigcup_{C \in \mathcal{C}} \text{ToABox}(a_C : C) \cup \{a_i \neq a_j \mid a_i, a_j \text{ individuals in } \mathcal{C}, i \neq j\}$$

Note that we make nominals distinct since they correspond to values from a structured data source, such as a relational database, for which the *unique name assumption* (UNA) will usually apply. Also, one could reuse the textual representation of the concepts to serve as the invented constant names.

Theorem 13 (CBox Admissibility). Let $\mathcal{T}$ be a *partial*$-\mathcal{CFDI}_{nc}^{\forall-}$ TBox and C a concept description. Then C is a singular referring expression w.r.t. $\mathcal{T}$ if and only if the knowledge base

$$(\mathcal{T} \cup \{A \sqsubseteq \neg B\}, \text{ToABox}(a : C) \cup \text{ToABox}(b : C) \cup \{A(a), B(b)\})$$

is inconsistent, where A and B are primitive concepts not occurring in $\mathcal{T}$ and C and a and b are distinct constant symbols.

Proof (sketch): The ToABox mapping expands complex concepts in a CBox to sets of assertions in a corresponding ABox. By case analysis we can show that a model of $(\mathcal{T} \cup \{A \sqsubseteq \neg B\}, \text{ToABox}(a : C) \cup \text{ToABox}(b : C) \cup \{A(a), B(b)\})$ provides a counterexample to C's singularity (w.r.t. $\mathcal{T}$). Moreover, whenever C is not singular, such a model can be constructed by appropriately naming additional individuals in a counterexample to C's singularity. □

It is easy to verify that the CBox in Example 11 is admissible w.r.t. the given TBox since *fname*, *lname*, and the combination of *fname* and *lname* are respective keys of FRIEND, MATRIARCH and *PERSON*.

Theorem 14 (Satisfiability of KBs with CBoxes). Let $\mathcal{K} = (\mathcal{T}, \mathcal{C})$ be a knowledge base with an admissible CBox $\mathcal{C}$. Then $\mathcal{K}$ is consistent if $(\mathcal{T}, \text{ToABox}(\mathcal{C}))$ is consistent.

Proof (sketch): Similar to the argument in the proof sketch in Theorem 13. □

3.4   Query Answering over CBoxes

We now show how query answering over CBox-based knowledge bases can be reduced to the standard case of ABoxes. We also show an example of the utility of CBoxes in capturing distinct co-references to a particular object, and how $partial\text{-}\mathcal{CFDI}_{nc}^{\forall-}$ based TBoxes can account for such co-references.

Theorem 15 (Query Answering). Let $\mathcal{K} = (\mathcal{T}, \mathcal{C})$ be a consistent knowledge base and $Q = \{(x_1, \ldots, x_k) \mid \varphi\}$ a conjunctive query over $\mathcal{K}$. Then $(C_1, \ldots, C_k)$ is a certain answer to Q in $\mathcal{K}$ if and only if $\{C_1, \ldots, C_k\} \subseteq \mathcal{C}$ and $(a_{C_1}, \ldots, a_{C_k})$ is a certain answer to Q over $(\mathcal{T}, \text{ToABox}(\mathcal{C}))$.

<u>Proof (sketch):</u> Since $C_1, \ldots, C_k$ are singular referring expressions, there must be individuals $o_1, \ldots, o_k$ witnessing non-emptiness of $C_1, \ldots, C_k$, respectively, that make the query true in every model of $\mathcal{K}$. Case analysis shows that, in the corresponding models of $(\mathcal{T}, \text{ToABox}(\mathcal{C}))$, these individuals will be the interpretations of the constant symbols $(a_{C_1}, \ldots, a_{C_k})$ (and vice versa).   □

Note that ABox completion [4,9] will make constant symbols belonging to *co-referring* referring expressions equal automatically. This, in turn, realizes all reasoning needed to capture the effects of *translation tables* in a TBox/CBox:

Example 16. Consider again the $partial\text{-}\mathcal{CFDI}_{nc}^{\forall-}$ knowledge base $(\mathcal{T}, \mathcal{C})$ in Example 11. An ABox $\mathcal{A}$ generated by $\text{ToABox}(\mathcal{C})$ would introduce three constant symbols in generated assertions, as in the following:

$$\mathcal{A} = \{\ \text{FRIEND}(a_1), fname(a_1) = \text{``Mary''},$$
$$\text{PERSON}(a_2), fname(a_2) = \text{``Mary''}, lname(a_2) = \text{``Smith''},$$
$$\text{MATRIARCH}(a_3), lname(a_3) = \text{``Smith''}, \ldots\}$$

The standard ABox completion [4,9] will then ultimately generate the equality "$a_1 = a_3$" by virtue of the inclusion dependencies in $\mathcal{T}$, including the facts that both FRIEND and MATRIARCH are subsumed by PERSON. (Here, PERSON might serve the role of a translation concept corresponding to a translation table available in some structured data source.)   □

On Query Answers. The definition of *certain answers* asks for all tuples of constants—in our setting proxied by referring expressions—for which the query is entailed by the knowledge base. Thus the selection of referring expressions in the CBox determines components of query answers presented to the user. There are two considerations:

1. Additional answers may be needed; these can be obtained by considering additional referring expressions describing, e.g., sub-documents, to the CBox (as long as the CBox remains admissible);
2. Simpler answers may be desired, i.e., simpler referring expressions denoting the answers; these can be obtained by appropriate selection of minimal referring expressions (and removing all the more complex referring expressions from answers).

Both of these goals can be achieved by a simple housekeeping that determines which referring expressions are eligible to appear in query answers. This step can be easily combined with the CBox-to-ABox mapping by appropriately marking the generated constant symbols. Indeed, similar marking is commonly used, e.g., when ABoxes are *normalized* in most OBDA settings.

4 Summary

We have considered how referring expressions corresponding to concepts in a description logic can serve the role of constant symbols in both assertion boxes and in query answering, and how doing so leads to a more effective and direct way of achieving an integration of structured data sources via OBDA, as well as more descriptive and meaningful answers to queries.

Admitting referring expressions leads naturally to a notion of a concept box or CBox in place of an ABox in a knowledge base. This in turn raises a number of technical issues: how to ensure referring expressions in a CBox refer to a single individual, how to check for knowledge base consistency, and how to evaluate conjunctive queries over the knowledge base. We have shown how all these issues can be resolved by a mapping of CBoxes to ABoxes. The mapping enables off-the-shelf procedures for TBox completion, for consistency checking, and for ABox completion and query rewriting over standard knowledge bases consisting of a TBox and ABox.

In [1], the notion of a *referring expression type* was also introduced. For future work, we plan to explore how such a typing discipline can be used to push parts of the mapping of CBoxes to ABoxes to backend database sources along the lines outline in [3]. Our new ability of detecting co-reference to objects by referring expressions can also lead to an ability to detect duplicate answers in query results. Future work along this line can enable additional capabilities in query formulation and answering, such as an ability for "limit k" operators in queries.

References

1. Borgida, A., Toman, D., Weddell, G.: On referring expressions in query answering over first order knowledge bases. In: Proceedings of KR 2016, pp. 319–328 (2016)
2. Calvanese, D., Giese, M., Hovland, D., Rezk, M.: Ontology-based integration of cross-linked datasets. In: Arenas, M., et al. (eds.) ISWC 2015, Part I. LNCS, vol. 9366, pp. 199–216. Springer, Cham (2015). https://doi.org/10.1007/978-3-319-25007-6_12
3. Jacques, J.S., Toman, D., Weddell, G.E.: Object-relational queries over $\mathcal{CFDI}_{nc}$ knowledge bases: OBDA for the SQL-Literate. In: Proceedings of International Joint Conference on Artificial Intelligence, IJCAI, pp. 1258–1264 (2016)
4. McIntyre, S., Borgida, A., Toman, D., Weddell, G.: On limited conjunctions and partial features in parameter-tractable feature logics. In: Proceedings of AAAI Conference on Artificial Intelligence 2019, Honolulu, HI, USA (2019, in press)

5. Toman, D., Weddell, G.: On keys and functional dependencies as first-class citizens in description logics. In: Furbach, U., Shankar, N. (eds.) IJCAR 2006. LNCS (LNAI), vol. 4130, pp. 647–661. Springer, Heidelberg (2006). https://doi.org/10.1007/11814771_52
6. Toman, D., Weddell, G.E.: On the interaction between inverse features and path-functional dependencies in description logics. In: Proceedings of the International Joint Conference on Artificial Intelligence (IJCAI), pp. 603–608 (2005)
7. Toman, D., Weddell, G.E.: On keys and functional dependencies as first-class citizens in description logics. J. Autom. Reason. **40**(2–3), 117–132 (2008)
8. Toman, D., Weddell, G.: On adding inverse features to the description logic $\mathcal{CFD}_{nc}^{\forall}$. In: Pham, D.-N., Park, S.-B. (eds.) PRICAI 2014. LNCS (LNAI), vol. 8862, pp. 587–599. Springer, Cham (2014). https://doi.org/10.1007/978-3-319-13560-1_47
9. Toman, D., Weddell, G.: On partial features in the $\mathcal{DLF}$ family of description logics. In: Booth, R., Zhang, M.-L. (eds.) PRICAI 2016. LNCS (LNAI), vol. 9810, pp. 529–542. Springer, Cham (2016). https://doi.org/10.1007/978-3-319-42911-3_44

SPARQL Queries over Ontologies Under the Fixed-Domain Semantics

Sebastian Rudolph$^{(\boxtimes)}$ (iD), Lukas Schweizer$^{(\boxtimes)}$ (iD), and Zhihao Yao$^{(\boxtimes)}$

Computational Logic Group, TU Dresden, Dresden, Germany
{sebastian.rudolph,lukas.schweizer,zhihao.yao}@tu-dresden.de

Abstract. Fixed-domain reasoning over OWL ontologies is adequate in certain closed-world scenarios and has been shown to be both useful and feasible in practice. However, the reasoning modes hitherto supported by available tools do not include querying. We provide the formal foundations of querying under the fixed domain semantics, based on the principle of certain answers, and show how fixed-domain querying can be incorporated in existing reasoning methods using answer set programming (ASP).

1 Introduction

Semantic web technologies [13] are widely adopted for knowledge representation on the Web or in other scenarios requiring intelligent data management. For expressing sophisticated background knowledge, the ontology language OWL 2 and its profiles are the standard [17,30]. OWL 2 is based on expressive description logics [4,21] and supported by optimized engines for reasoning and querying [12,28,29].

The success of OWL 2 has led to its usage also in scenarios that actually go against its standard semantics, which operates under the open-world assumption. In many such scenarios, the involved elements (the "domain") are actually known upfront. In order to better account for such scenarios, an alternative, "fixed-domain" semantics has been proposed and tools providing reasoning support have been implemented on top of answer-set solvers [9,24,25].

While the existing reasoning support is helpful for standard reasoning tasks such as satisfiability testing and also for non-standard ones such as model enumeration, sometimes more elaborate information needs must be addressed. For sophisticated querying tasks in the Semantic Web setting, SPARQL has been established as the query language of choice [31], originally designed as querying formalism for RDF graphs [27]. The recent SPARQL 1.1 standard, however, supports queries over OWL ontologies by means of the so called *entailment regimes* [5]. Given that querying OWL ontologies even under very basic queries is not known to be decidable [23], the proposed approach constitutes a compromise, implementing what is practically feasible under the open world semantics.

Under the fixed-domain semantics, however, a tighter integration of OWL background knowledge and querying can be realized without risking decidability.

© Springer Nature Switzerland AG 2019
A. C. Nayak and A. Sharma (Eds.): PRICAI 2019, LNAI 11670, pp. 486–499, 2019.
https://doi.org/10.1007/978-3-030-29908-8_39

Table 1. Syntax and semantics of role and concept constructors in $\mathcal{SROIQ}$, where $a_1, \ldots a_n$ denote individual names, s a role name, r a role expression and C and D concept expressions.

Name	Syntax	Semantics
Inverse role	s^-	$\{(x, y) \in \Delta^{\mathcal{I}} \times \Delta^{\mathcal{I}} \mid (y, x) \in s^{\mathcal{I}}\}$
Universal role	u	$\Delta^{\mathcal{I}} \times \Delta^{\mathcal{I}}$
Top	$\top$	$\Delta^{\mathcal{I}}$
Bottom	$\bot$	$\emptyset$
Negation	$\neg C$	$\Delta^{\mathcal{I}} \setminus C^{\mathcal{I}}$
Conjunction	$C \sqcap D$	$C^{\mathcal{I}} \cap D^{\mathcal{I}}$
Disjunction	$C \sqcup D$	$C^{\mathcal{I}} \cup D^{\mathcal{I}}$
Nominals	$\{a_1, \ldots, a_n\}$	$\{a_1^{\mathcal{I}}, \ldots, a_n^{\mathcal{I}}\}$
Univ. restriction	$\forall r.C$	$\{x \mid \forall y.(x, y) \in r^{\mathcal{I}} \rightarrow y \in C^{\mathcal{I}}\}$
Exist. restriction	$\exists r.C$	$\{x \mid \exists y.(x, y) \in r^{\mathcal{I}} \wedge y \in C^{\mathcal{I}}\}$
Self concept	$\exists r.Self$	$\{x \mid (x, x) \in r^{\mathcal{I}}\}$
Qualified number	$\leqslant n\, r.C$	$\{x \mid \#\{y \in C^{\mathcal{I}} \mid (x, y) \in r^{\mathcal{I}}\} \leq n\}$
Restriction	$\geqslant n\, r.C$	$\{x \mid \#\{y \in C^{\mathcal{I}} \mid (x, y) \in r^{\mathcal{I}}\} \geq n\}$

Under these circumstances we can realize querying following the principle of *certain answers*: each fixed-domain model of a given ontology can be conceived as an RDF graph which can be SPARQL-queried in separation. Only if a query answer is returned when querying each and every model, it qualifies as query answer for the corresponding ontology.

Since model enumeration is a task readily provided by existing fixed-domain reasoners, the above definition immediately gives rise to a brute-force algorithm for fixed-domain ontological querying. However, the combinatorial explosion typically occurring in model-enumeration makes the feasibility of such an approach appear highly doubtful. We therefore propose an alternative method based on a tighter integration with existing reasoning technology, where SPARQL query evaluation is encoded in the same answer set program that produces the models. By means of this tight integration, we can leverage the structural similarity of certain answers and skeptical consequences.

2   Description Logics

OWL 2 DL, the version of the Web Ontology Language we focus on, is based on description logics (DLs, [4,21]). We briefly recap the description logic $\mathcal{SROIQ}$ (for details see [14]). Let N_I, N_C, and N_R be finite, disjoint sets called *individual names*, *concept names*, and *role names*, respectively.[1] These atomic entities can be used to form complex ones as displayed in Table 1.

---

[1] To ensure compatibility with their later usage in RDF and SPARQL, we silently presume that all these vocabulary elements are Internationalized Resource Identifiers (IRIs).

Table 2. Syntax and semantics of $\mathcal{SROIQ}$ axioms.

Axiom α	$\mathcal{I} \models \alpha$, if	
$r_1 \circ \cdots \circ r_n \sqsubseteq r$	$r_1^{\mathcal{I}} \circ \cdots \circ r_n^{\mathcal{I}} \subseteq r^{\mathcal{I}}$	RBox $\mathcal{R}$
$\mathsf{Dis}(s, r)$	$s^{\mathcal{I}} \cap r^{\mathcal{I}} = \emptyset$	
$C \sqsubseteq D$	$C^{\mathcal{I}} \subseteq D^{\mathcal{I}}$	TBox $\mathcal{T}$
$C(a)$	$a^{\mathcal{I}} \in C^{\mathcal{I}}$	ABox $\mathcal{A}$
$r(a, b)$	$(a^{\mathcal{I}}, b^{\mathcal{I}}) \in r^{\mathcal{I}}$	
$a \doteq b$	$a^{\mathcal{I}} = b^{\mathcal{I}}$	
$a \not\doteq b$	$a^{\mathcal{I}} \neq b^{\mathcal{I}}$	

A $\mathcal{SROIQ}$ *knowledge base* $\mathcal{K}$ is a tuple $(\mathcal{A}, \mathcal{T}, \mathcal{R})$ where $\mathcal{A}$ is a $\mathcal{SROIQ}$ ABox, $\mathcal{T}$ is a $\mathcal{SROIQ}$ TBox and $\mathcal{R}$ is a $\mathcal{SROIQ}$ RBox. Table 2 presents the respective axiom types available in the three parts.[2] We use $N_I(\mathcal{K})$, $N_C(\mathcal{K})$, and $N_R(\mathcal{K})$ to denote the sets of individual names, concept names, and role names occurring in $\mathcal{K}$, respectively.

The semantics of $\mathcal{SROIQ}$ is defined via interpretations $\mathcal{I} = (\Delta^{\mathcal{I}}, \cdot^{\mathcal{I}})$ composed of a non-empty set $\Delta^{\mathcal{I}}$ called the *domain of* $\mathcal{I}$ and a function $\cdot^{\mathcal{I}}$ mapping individual names to elements of $\Delta^{\mathcal{I}}$, concept names to subsets of $\Delta^{\mathcal{I}}$, and role names to subsets of $\Delta^{\mathcal{I}} \times \Delta^{\mathcal{I}}$. This mapping is extended to complex role and concept expressions (cf. Table 1) and finally used to define satisfaction of axioms (see Table 2). We say that $\mathcal{I}$ *satisfies* a knowledge base $\mathcal{K} = (\mathcal{A}, \mathcal{T}, \mathcal{R})$ (or $\mathcal{I}$ is a *model* of $\mathcal{K}$, written: $\mathcal{I} \models \mathcal{K}$) if it satisfies all axioms of $\mathcal{A}$, $\mathcal{T}$, and $\mathcal{R}$. We say that a knowledge base $\mathcal{K}$ *entails* an axiom α (written $\mathcal{K} \models \alpha$) if all models of $\mathcal{K}$ are models of α.

Example 1. Consider a knowledge base $\mathcal{K} = (\mathcal{A}, \mathcal{T}, \mathcal{R})$. Let $\mathcal{A}$ contain the assertions `Aca(alice)`, `Aca(bob)`, `Aca(claire)`, `Aca(david)`, `Aca(eve)`, stating that the mentioned individuals are all academics and the assertions `supervises(alice, bob)`, `supervises(bob, claire)`, and `supervises(david, eve)` indicating supervision relationships and `inProject(bob, projectX)`, `inProject(david, projectY)`, as well as `inProject(eve, projectY)` to indicate research project affiliations.

Let $\mathcal{T}$ contain the axioms `Aca` $\sqsubseteq$ `Masterstudent` $\sqcup$ `PhDstudent` $\sqcup$ `Professor` as well as `Masterstudent` $\sqsubseteq$ `¬PhDstudent`, `Masterstudent` $\sqsubseteq$ `¬Professor`, and `PhDstudent` $\sqsubseteq$ `¬Professor` to indicate that every academic must be in exactly one of the three categories. Moreover, we

---

[2] The original definition of $\mathcal{SROIQ}$ contained more RBox axioms (expressing transitivity, (a)symmetry, (ir)reflexivity of roles), but these can be shown to be syntactic sugar. Moreover, the definition of $\mathcal{SROIQ}$ contains so-called *global restrictions* which prevents certain axioms from occurring together. These complicated restrictions, while crucial for the decidability of classical reasoning in $\mathcal{SROIQ}$ are not necessary for fixed-domain reasoning considered here, hence we omit them for the sake of brevity.

impose some constraints on supervision relationships: $\exists$supervises.$\top$ $\sqsubseteq$ (Professor $\sqcup$ PhDstudent) $\sqcap$ $\forall$supervises.(Masterstudent $\sqcup$ PhDstudent) as well as $\exists$supervises.PhDstudent $\sqsubseteq$ Professor and PhDstudent $\sqsubseteq$ $\forall$supervises.Masterstudent.

It can be readily checked that $\mathcal{K}$ is satisfiable. It would, however, become unsatisfiable upon adding the assertion supervises(finn, alice). Note also that, e.g., $\mathcal{K} \models \neg$Masterstudent(david).

3   Fixed-Domain Semantics

In DLs, models can be of arbitrary cardinality – for a satisfiability check, for example, all what matters is the mere existence of a model. Yet, in many applications, the domain of interest is known to be finite. Restricting reasoning to models of finite domain size (called *finite model reasoning*, a natural assumption in database theory), has been intensively studied in DLs [7,16,20,22]. As opposed to assuming the domain to be merely finite (but of arbitrary, unknown size), one can consider the case where the domain has an *a priori known cardinality* and use the term *fixed domain* [9].

Definition 1 (Fixed-Domain Semantics). *Given a non-empty finite set $\Delta \subseteq N_I$, called* fixed domain, *an interpretation $\mathcal{I} = (\Delta^{\mathcal{I}}, \cdot^{\mathcal{I}})$ is said to be Δ-fixed (or just fixed, if Δ is clear from the context), if $\Delta^{\mathcal{I}} = \Delta$ and $a^{\mathcal{I}} = a$ for all $a \in \Delta$. Accordingly, for a DL knowledge base $\mathcal{K}$, we call an interpretation $\mathcal{I}$ a Δ-model of $\mathcal{K}$, if $\mathcal{I}$ is a Δ-fixed interpretation and $\mathcal{I} \models \mathcal{K}$. A knowledge base $\mathcal{K}$ is called Δ-satisfiable if it has a Δ-model. We say $\mathcal{K}$ Δ-entails an axiom α ($\mathcal{K} \models_\Delta \alpha$) if every Δ-model of $\mathcal{K}$ is also a model of α.*

Example 2. Consider the knowledge base $\mathcal{K}$ from Example 1. Assume, we let $\Delta = \{$alice, bob, claire, david, eve, projectX, projectY$\}$. It is not hard to see that $\mathcal{K}$ is Δ-satisfiable. Moreover, $\mathcal{K}$ Δ-entails the axiom $\neg$Aca $\sqsubseteq$ {projectX, projectY}, whereas this axiom is not generally entailed.

4   RDF

We will now very briefly introduce RDF [8], and show how to represent a Δ-fixed interpretation as *RDF graph* which in our setting will serve as essential data structure over which SPARQL queries are evaluated. We will omit named graphs from our presentation as they are not meaningful in our context.

Let I, B, L be countably infinite, pairwise disjoint sets, called *IRIs, blank nodes*, and *RDF literals*, respectively. A tuple $(v_1, v_2, v_3) \in (I \cup B) \times I \times (I \cup B \cup L)$ is called an *RDF triple*, where v_1 is called the *subject*, v_2 the *predicate*, and v_3 the *object*. An *RDF graph G* (or just *graph*) is a set of RDF triples, and we use term(G) as the set of all elements from $I \cup B \cup L$ occurring in G, and blank(G) $\subseteq B$ to denote the set blank nodes occurring in G. We will make later use of Definition 2 that defines the construction of an RDF graph given a Δ-fixed interpretation, promoting the interpretation as queryable artifact.

Definition 2. *Let $\mathcal{I}$ be a Δ-fixed interpretation. Then the RDF graph $G(\mathcal{I})$ induced by $\mathcal{I}$ consists of the triples $(a, \text{rdf:type}, C)$ for all $a \in C^{\mathcal{I}}$, and (a, r, b) for all $(a, b) \in r^{\mathcal{I}}$.*

5   SPARQL

We will give a very compact introduction on the core elements of SPARQL [31], similar to [3,19]. For reasons of space and relevance, we will focus on SELECT queries and omit aggregates and solution modifiers.

Let V be a countably infinite set of available variables, where $V \cap (I \cup B \cup L) = \emptyset$. A tuple from $(I \cup L \cup V) \times (I \cup L \cup V) \times (I \cup V)$ is called *triple pattern*, and we call a finite set of triple patterns a *basic graph pattern*. Complex *graph patterns* are now inductively defined: (i) every basic graph pattern is a graph pattern, (ii) for graph patterns P_1 and P_2, the expressions P_1 AND P_2, P_1 UNION P_2, P_1 MINUS P_2, and P_1 OPT P_2 are graph patterns and (iii) for P a graph pattern and C a filter constraint (defined below), P FILTER C is a graph pattern. The set of variables occurring in a graph pattern P is denoted with $\text{var}(P)$. A *filter constraint* is defined recursively as follows: (i) if $?X, ?Y \in V$ and $u \in I \cup L$ then $?X = u$, $?X = ?Y$, bound($?X$), isIRI($?X$), isLiteral($?X$), and isBlank($?X$) are *atomic filter constraints*; (ii) if C_1 and C_2 are filter constraints then $(\neg C_1)$, $(C_1 \wedge C_2)$, and $(C_1 \vee C_2)$ are *complex filter constraints*.

Finally, a SPARQL *query* q is a structure SELECT $?X_1 \ldots ?X_n$ WHERE P with $?X_1, \ldots, ?X_n$ variables and P a graph pattern. We use $\text{avar}(q) = \{?X_1, \ldots, ?X_n\}$ to denote the set of *answer variables*.

Example 3. In the following, a simple SPARQL query q_1 asks for all projects in which some PhD student is involved.

```
SELECT ?Y
WHERE { ?X rdf:type PhDStudent. ?X inProject ?Y }
```

The next SPARQL query q_2 retrieves employees who are PhD students or professors together with their projects.

```
SELECT ?X ?Y
WHERE { { ?X rdf:type PhDStudent. UNION ?X rdf:type Professor. }
        AND ?X inProject ?Y. }
```

A *mapping* μ is a partial function $\mu : V \to (I \cup B \cup L)$. The domain of μ, $\text{dom}(\mu) \subseteq V$, are the variables for which μ is defined. Two mappings μ_1, μ_2 are *compatible*, written $\mu_1 \sim \mu_2$, if for all $?X \in \text{dom}(\mu_1) \cap \text{dom}(\mu_2)$, it holds that $\mu_1(?X) = \mu_2(?X)$. Given a triple pattern t, we let $t\mu$ denote the triple obtained by replacing every variable $?X \in \text{dom}(\mu)$ in t by $\mu(?X)$.

Definition 3. *Let t be a triple pattern, P, P_1, P_2 graph patterns, and G an RDF graph, then the evaluation $\langle\!\langle \cdot \rangle\!\rangle_G$ is defined as:*

$$\langle\!\langle \{t_1, ..., t_k\} \rangle\!\rangle_G = \{\mu \mid \mathsf{dom}(\mu) = \bigcup_{1 \le i \le k} \mathsf{var}(t_i) \text{ and } \{t_1\mu, ..., t_k\mu\} \subseteq G\}$$

$$\langle\!\langle P_1 \text{ AND } P_2 \rangle\!\rangle_G = \{\mu_1 \cup \mu_2 \mid \mu_1 \in \langle\!\langle P_1 \rangle\!\rangle_G, \mu_2 \in \langle\!\langle P_2 \rangle\!\rangle_G, \mu_1 \sim \mu_2\}$$

$$\langle\!\langle P_1 \text{ UNION } P_2 \rangle\!\rangle_G = \langle\!\langle P_1 \rangle\!\rangle_G \cup \langle\!\langle P_2 \rangle\!\rangle_G$$

$$\langle\!\langle P_1 \text{ MINUS } P_2 \rangle\!\rangle_G = \langle\!\langle P_1 \rangle\!\rangle_G \setminus \langle\!\langle P_2 \rangle\!\rangle_G$$

$$\langle\!\langle P_1 \text{ OPT } P_2 \rangle\!\rangle_G = \{\mu_1 \cup \mu_2 \mid \mu_1 \in \langle\!\langle P_1 \rangle\!\rangle_G, \mu_2 \in \langle\!\langle P_2 \rangle\!\rangle_G, \mu_1 \sim \mu_2\}$$
$$\cup \{\mu_1 \mid \mu_1 \in \langle\!\langle P_1 \rangle\!\rangle_G, \forall \mu_2 \in \langle\!\langle P_2 \rangle\!\rangle_G. \mu_1 \not\sim \mu_2\}$$

$$\langle\!\langle P \text{ FILTER } C \rangle\!\rangle_G = \{\mu \in \langle\!\langle P \rangle\!\rangle_G \mid C\mu = \top\}$$

$$\langle\!\langle \text{SELECT } ?X_1...?X_n \text{ WHERE } P \rangle\!\rangle_G = \{\mu|_{\{?X_1,...,?X_n\}} \mid \mu \in \langle\!\langle P \rangle\!\rangle_G\}$$

Let C, C_1, C_2 be filter constraints, $?X, ?Y \in V$, $a \in I \cup B \cup L$. The valuation of C on a mapping μ, written $C\mu$ takes one of the three values $\{\top, \bot, \epsilon\}$ and is defined as follows. $C\mu = \epsilon$, if:

$$C = \mathsf{isBlank}(?X), \ C = \mathsf{isIRI}(?X), \ C = \mathsf{isLiteral}(?X), \text{ or} \tag{1}$$
$$C = (?X = a) \text{ with } ?X \notin \mathsf{dom}(\mu);$$
$$C = (?X = ?Y) \text{ with } ?X \notin \mathsf{dom}(\mu) \text{ or } ?Y \notin \mathsf{dom}(\mu); \tag{2}$$
$$C = (\neg C_1) \text{ where } C_1\mu = \epsilon; \tag{3}$$
$$C = (C_1 \vee C_2) \text{ with } \top \notin \{C_1\mu, C_2\mu\} \text{ and } \epsilon \in \{C_1\mu, C_2\mu\}; \tag{4}$$
$$C = (C_1 \wedge C_2) \text{ with } \bot \notin \{C_1\mu, C_2\mu\} \text{ and } \epsilon \in \{C_1\mu, C_2\mu\}. \tag{5}$$

$C\mu = \top$, if:

$$C = \mathsf{bound}(?X) \text{ with } ?X \in \mathsf{dom}(\mu); \tag{1}$$
$$C = \mathsf{isBlank}(?X) \text{ with } ?X \in \mathsf{dom}(\mu) \text{ and } \mu(?X) \in B; \tag{2}$$
$$C = \mathsf{isIRI}(?X) \text{ with } ?X \in \mathsf{dom}(\mu) \text{ and } \mu(?X) \in I; \tag{3}$$
$$C = \mathsf{isLiteral}(?X) \text{ with } ?X \in \mathsf{dom}(\mu) \text{ and } \mu(?X) \in L; \tag{4}$$
$$C = (?X = a) \text{ with } ?X \in \mathsf{dom}(\mu) \text{ and } \mu(?X) = a; \tag{5}$$
$$C = (?X = ?Y) \text{ with } ?X, ?Y \in \mathsf{dom}(\mu) \text{ and } \mu(?X) = \mu(?Y); \tag{6}$$
$$C = (\neg C_1) \text{ with } C_1\mu = \bot; \tag{7}$$
$$C = (C_1 \vee C_2) \text{ with } C_1\mu = \top \text{ or } C_2\mu = \top; \tag{8}$$
$$C = (C_1 \wedge C_2) \text{ with } C_1\mu = \top \text{ and } C_2\mu = \top. \tag{9}$$

$C\mu = \bot$, otherwise.

6   SPARQL over Knowledge Bases Under Fixed Domain Semantics

In database theory, as it is the case for SPARQL, a database instance is typically conceived to be complete in terms of knowledge, and thus queries are answered under the closed-world assumption (e.g. a person not listed in an employee database is not an employee) [2]. In contrast, a DL knowledge base represents incomplete knowledge, thus the mere absence of a fact does not allow to assume its truth value to be *false*. Alike the notion of axiom entailment, this has coined the notion of *certain query answers* [1], where (intuitively) a tuple is considered to be an answer if it is the result of evaluating the query over every model of the knowledge base. Thus, each interpretation $\mathcal{I}$ is seen as database instance, over which the query is evaluated. For the evaluation of a SPARQL query over some model $\mathcal{I}$, we will therefore use the RDF graph $G(\mathcal{I})$ induced by $\mathcal{I}$, as introduced in Sect. 4. To obtain the certain answers to a SPARQL query, we collect only those answers that are returned upon executing the query over the RDF graph $G(\mathcal{I})$ of each and every model $\mathcal{I}$ of the queried knowledge base $\mathcal{K}$.

Definition 4. *The set of* certain answers *to a SPARQL query q over a DL knowledge base $\mathcal{K}$ and a fixed domain Δ, is defined by* $\mathsf{cert}_\Delta(\mathcal{K}, q) = \{\mu \mid \mu \in \langle\!\langle q \rangle\!\rangle_{G(\mathcal{I})}$ *for all* $\mathcal{I} \models_\Delta \mathcal{K}\}$.

Example 4. Consider the knowledge base $\mathcal{K}$ from Example 1. Like in Example 2 we let $\Delta = \{\mathtt{alice}, \mathtt{bob}, \mathtt{claire}, \mathtt{david}, \mathtt{eve}, \mathtt{projectX}, \mathtt{projectY}\}$. For q_1 from Example 3 we obtain $\mathsf{cert}_\Delta(q_1, \mathcal{K}) = \{?Y \mapsto \mathtt{projectX}\}$. For q_2 we get $\mathsf{cert}_\Delta(q_2, \mathcal{K}) = \{(?X \mapsto \mathtt{bob}, ?Y \mapsto \mathtt{projectX}), (?X \mapsto \mathtt{david}, ?Y \mapsto \mathtt{projectY})\}$.

7   Practical SPARQL Answering

Practical fixed-domain reasoning for DL knowledge bases has been realized via a translation-based approach [9]. The given finite domain allows to translate DL axioms into ASP rules, and thereby make use of modern solvers to evaluate the resulting program in order to check satisfiability, as well as enumerating models – which in turn correspond to answer sets.

In consequence, it is a straightforward idea to build on top of this translation to answer SPARQL queries, in particular since translating SPARQL to datalog rules has already been proposed [3,19]; in fact, it was shown that SPARQL is equally expressive as non-recursive safe datalog with default negation.

We essentially combine both approaches (ASP-based model enumeration and ASP-based query evaluation) and adapt them to make them compatible. After providing a short introduction of answer set programming, we will sketch the translation of DL knowledge bases into answer set programs [9]. In more detail, the translation of SPARQL queries into a stratified answer set program is given thereafter.

7.1   Answer Set Programming

We review the basic notions of answer set programming [18] under the stable-model semantics [11], for further details we refer to [6,10].

We fix a countable set $\mathcal{U}$ of *(domain) elements*, also called *constants*; and presume a total order $<$ over the domain elements. An *atom* is an expression $p(t_1, \ldots, t_n)$, where p is a *predicate* of arity $n \geq 0$ and each t_i is either a variable or an element from $\mathcal{U}$. An atom is *ground* if it is free of variables. $B_{\mathcal{U}}$ denotes the set of all ground atoms over $\mathcal{U}$. A *(normal) rule* ρ is of the form

$$a \leftarrow b_1, \ldots, b_k, \ \text{not}\, b_{k+1}, \ldots, \ \text{not}\, b_m.$$

with $m \geq k \geq 0$, where a is an atom or empty (in the latter case the rule is called *integrity constraint*), $b_1, \ldots, b_m$ are atoms, and "not" denotes *default negation*. The *head* of ρ is the singleton set $H(\rho) = \{a\}$ if a is an atom and $H(\rho) = \emptyset$ otherwise, and the *body* of ρ is $B(\rho) = \{b_1, \ldots, b_k, \text{not}\, b_{k+1}, \ldots, \text{not}\, b_m\}$. Furthermore, $B^+(\rho) = \{b_1, \ldots, b_k\}$ and $B^-(\rho) = \{b_{k+1}, \ldots, b_m\}$. A rule ρ is *safe* if each variable in ρ occurs in $B^+(r)$. A rule ρ is *ground* if no variable occurs in ρ. A *fact* is a ground rule with empty body. An *(input) database* is a set of facts. A (normal) *program* is a finite set of normal rules. For a program Π and an input database D, we often write $\Pi(D)$ instead of $D \cup \Pi$. For any program Π, let $\mathcal{U}_\Pi$ be the set of all constants appearing in Π. $Gr(\Pi)$ is the set of rules $\rho\sigma$ obtained by applying, to each rule $\rho \in \Pi$, all possible substitutions σ from the variables in ρ to elements of $\mathcal{U}_\Pi$.

An *interpretation* $I \subseteq B_{\mathcal{U}}$ *satisfies* a ground rule ρ iff $H(\rho) \cap I \neq \emptyset$ whenever $B^+(\rho) \subseteq I$, $B^-(\rho) \cap I = \emptyset$. I satisfies a ground program Π, if each $\rho \in \Pi$ is satisfied by I. A non-ground rule ρ (resp., a program Π) is satisfied by an interpretation I iff I satisfies all groundings of ρ (resp., $Gr(\Pi)$). $I \subseteq B_{\mathcal{U}}$ is an *answer set* (also called *stable model*) of Π iff it is the subset-minimal set satisfying the *Gelfond-Lifschitz reduct* $\Pi^I = \{H(\rho) \leftarrow B^+(\rho) \mid I \cap B^-(\rho) = \emptyset, \rho \in Gr(\Pi)\}$. For a program Π, we denote the set of its answer sets by $\mathcal{S}(\Pi)$.

Consequences. We rely on two notions of consequence: Given a program Π and a ground atom α, we say that Π *cautiously entails* α, written $\Pi \models_\forall \alpha$, if $\alpha \in S$ for every answer set $S \in \mathcal{S}(\Pi)$. Likewise, we say that Π *bravely entails* α, written $\Pi \models_\exists \alpha$, if there exists an answer set $S \in \mathcal{S}(\Pi)$ with $\alpha \in S$. The set of all cautious consequences of Π is denoted $\mathsf{Cn}^\forall(\Pi)$ and the set of its brave consequences $\mathsf{Cn}^\exists(\Pi)$.

7.2 Translating DL Knowledge Bases

An ASP translation of $\mathcal{SROIQ}$ knowledge bases has been proposed in [9,26]. Intuitively, given a fixed domain, one can guess an interpretation and verify modelhood with appropriate constraints (resulting from the axioms). Thus, the key idea of the translation is that every axiom is turned into an integrity constraint, and the only rules with nonempty head are so-called "guessing rules" for the extensions of every concept and role. Following this *guess and check* approach, the translation is rather direct, for example, a simple concept subsumption $A \sqsubseteq B$ becomes a constraint of the form $\leftarrow A(X), \text{not}\, B(X)$; i.e. ruling out interpretations where X is an instance of A but not of B, and hence not satisfying the subsumption.

For a DL knowledge base $\mathcal{K}$ and fixed domain Δ, let $\Pi(\mathcal{K}, \Delta)$ denote the answer set program resulting from translating $\mathcal{K}$ with respect to Δ. It is shown that every answer set $S \in \mathcal{S}(\Pi(\mathcal{K}, \Delta))$, corresponds to a Δ-model of $\mathcal{K}$, and vice versa. Hence, it is possible to obtain the corresponding RDF graph $G(\mathcal{I})$ of every model via the answer sets and evaluate a SPARQL query on it. Since the translation has been implemented and is available in the tool WOLPERTINGER [25], which is able to enumerate Δ-models, SPARQL query evaluation could be realized with only little implementation effort; i.e. retrieve all models and evaluate the query on each of the induced graphs, and compute the intersection of all answers – that would be taking Definition 4 literally. However, as the sets of enumerated models tend to be very large due to combinatorial explosion, we are certain that this approach would not be feasible. Therefore, we will propose another translation-based approach.

In $\Pi(\mathcal{K}, \Delta)$, predicate names directly correspond to concept and role names in $\mathcal{K}$. This translation can syntactically be lifted to a triple notation, such that, e.g. , translating $A \sqsubseteq B$ results in the constraint $\leftarrow triple(X, \mathsf{rdf:type}, A), not\ triple(X, \mathsf{rdf:type}, B)$. We let $\Pi_{\mathrm{RDF}}(\mathcal{K}, \Delta)$ denote this lifted program. Now by letting $\mathrm{RDF}(S) = \{(v1, v2, v3) \mid triple(v1, v2, v3) \in S\}$, we obtain the following correspondence.

Lemma 1. *Let $\mathcal{K}$ be a DL knowledge base, Δ a fixed domain, and $\mathcal{I}$ a Δ-fixed interpretation. Then $\mathcal{I} \models_\Delta \mathcal{K}$ if and only if there exists some answer set $S \in \mathcal{S}(\Pi_{\mathrm{RDF}}(\mathcal{K}, \Delta))$ such that $G(\mathcal{I}) = \mathrm{RDF}(S)$.*

This correspondence now provides us with the right starting point for applying the SPARQL querying – again via a translation into ASP.

7.3   Translating SPARQL Queries

We let $\Pi(q)$ denote the answer set program resulting from the translation of a SPARQL query q, into rules, closely following [19]. Intuitively, the translation follows the recursive definition of $\langle\!\langle q \rangle\!\rangle_G$ (cf. Definition 3), evaluating the graph pattern P_q of q inside out. For a set of variables $V = \{X_1, \ldots, X_n\}$, we denote with $\overline{V} = (X_1, \ldots, X_n)$ the sequence of variables obtained relying on some lexicographic ordering. $\Pi(q)$ is then obtained with the initial call $\tau(\mathsf{avar}(q), P_q, 1)$ of the translation τ defined in the following. Thereby, the dedicated atom $answer_i$ represents the result of evaluating the sub-graph pattern at position i in the query graph pattern seen as binary tree; thus, alike the definition of $\langle\!\langle q \rangle\!\rangle_G$ (cf. Definition 3), the translation τ traverses the binary tree. For the translation of filter expressions via the function Φ we refer the reader to [19].

$$\tau(V, \{T_1, \ldots, T_n\}, i) = \{answer_i(\overline{V}) \leftarrow triple(T_1), \ldots, triple(T_n)\}$$
$$\text{where } T_i = (v_i, v_i', v_i'') \text{ is a triple pattern.} \tag{1}$$

$$\tau(V, P_1 \text{ AND } P_2, i) = \{answer_i(\overline{V}) \leftarrow answer_{2i}(\overline{V_{P_1}'}), answer_{2i+1}(\overline{V_{P_2}'}),$$

$$join_{|S_{P_{1,2}}|}(\overline{S'_{P_{1,2}}}, \overline{S''_{P_{1,2}}}, \overline{S_{P_{1,2}}})\}$$

$$\cup \tau(\mathsf{var}(P_1), P_1, 2i) \cup \tau(\mathsf{var}(P_2), P_2, 2i+1) \cup \mathsf{Join}(|S_{P_{1,2}}|)$$
$$\text{with } V'_{P_1} = \mathsf{var}(P_1)[S_{P_{1,2}} \to S'_{P_{1,2}}]$$
$$\text{and } V'_{P_2} = \mathsf{var}(P_2)[S_{P_{1,2}} \to S''_{P_{1,2}}] \tag{2}$$

$$\tau(V, P_1 \text{ UNION } P_2, i) = \{answer_i(\overline{V[(V\setminus\mathsf{var}(P_1)) \to \mathsf{null}]}) \leftarrow answer_{2i}(\overline{\mathsf{var}(P_1)}),$$
$$answer_i(\overline{V[(V\setminus\mathsf{var}(P_2)) \to \mathsf{null}]}) \leftarrow answer_{2i+1}(\overline{\mathsf{var}(P_2)})\}$$
$$\cup \tau(\mathsf{var}(P_1), P_1, 2i) \cup \tau(\mathsf{var}(P_2), P_2, 2i+1) \tag{3}$$

$$\tau(V, P_1 \text{ MINUS } P_2, i) = \{answer_i(\overline{V[(V\setminus\mathsf{var}(P_1)) \to \mathsf{null}]}) \leftarrow answer_{2i}(\overline{\mathsf{var}(P_1)}),$$
$$\text{not } answer_{2i+1}(\overline{\mathsf{var}(P_1) \cap \mathsf{var}(P_2)})\}$$
$$\cup \tau(\mathsf{var}(P_1), P_1, 2i) \cup \tau(\mathsf{var}(P_2), P_2, 2i+1) \tag{4}$$

$$\tau(V, P_1 \text{ OPT } P_2, i) = \tau(V, P_1 \text{ AND } P_2, i) \cup \tau(V, P_1 \text{ MINUS } P_2, i) \tag{5}$$

$$\tau(V, P \text{ FILTER } C, i) = \tau(\mathsf{var}(P), P, 2i) \cup \Phi(answer_i(\overline{V}) \leftarrow answer_{2i}(\overline{\mathsf{var}(P)}), C) \tag{6}$$

The translation of AND (joins), realized in Rule (2) requires some more explanation. First, the variables to join on are determined via $S_{P_{1,2}} = \mathsf{var}(P_1) \cap \mathsf{var}(P_2)$ (*shared variables*), and we denote with $S'_{P_{1,2}}$ and $S''_{P_{1,2}}$ the renamed copies of the shared variables $S_{P_{1,2}}$. For example, $S'_{P_{1,2}} = \{X'_1, \ldots, X'_n\}$ for $S_{P_{1,2}} = \{X_1, \ldots, X_n\}$. Thus, in Rule (2), the shared variables in $answer_{2i}$ are replaced by their singly primed version, and the shared variables in $answer_{2i}$ to their doubly primed version, respectively. The non-primed version is bound by $join_n(\ldots)$, which basically ensures that any value joins with null, for n shared variables. To implement this, we define the rule set $\mathsf{Join}(n)$ as follows:

$join(\mathsf{null}, \mathsf{null}, \mathsf{null})$
$join(X, X, X) \leftarrow term(X).$
$join(X, \mathsf{null}, X) \leftarrow term(X). \quad join(\mathsf{null}, X, X) \leftarrow term(X).$
$join_1(X'_1, X''_1, X_1) \leftarrow join(X'_1, X''_1, X_1)$
$join_2(X'_1, X'_2, X''_1, X''_2, X_1, X_2) \leftarrow join_1(X'_1, X''_1, X_1), join(X'_2, X''_2, X_2)$
$join_3(X'_1, X'_2, X'_3, X''_1, X''_2, X''_3, X_1, X_2, X_3) \leftarrow join_2(X'_1, X'_2, X''_1, X''_2, X_1, X_2), join(X'_3, X''_3, X_3)$
$\vdots$
$join_n(X'_1, \ldots, X'_m, \ldots, X_1, \ldots, X_n) \leftarrow join_{n-1}(X'_1, \ldots, X'_{n-1}, \ldots, X_1, \ldots, X_{n-1}),$
$join(X'_n, X''_n, X_n)$

Example 5. The following rule is the result of applying $\tau(\{?Y\}, P_{q_1}, 1)$ for the query q_1 in Example 3.

$$answer_1(Y) \leftarrow triple(X, \mathsf{rdf:type}, \mathsf{PhDStudent}), triple(X, \mathsf{inProject}, Y).$$

For the query q_2 we obtain the following result for the computation of $\tau(\{?X, ?Y\}, P_{q_1}, 1)$ (omitting the rules defining the *join* predicate).

$$answer_1(X, Y) \leftarrow answer_2(X'), answer_3(X'', Y), join_1(X', X'', X).$$
$$answer_2(X) \leftarrow answer_4(X).$$
$$answer_2(X) \leftarrow answer_5(X).$$
$$answer_3(X, Y) \leftarrow triple(X, \mathsf{inProject}, Y).$$
$$answer_4(X) \leftarrow triple(X, \mathsf{rdf{:}type}, \mathsf{PhDStudent}).$$
$$answer_5(X) \leftarrow triple(X, \mathsf{rdf{:}type}, \mathsf{Professor}).$$

Note that $\Pi(q)$ is stratified and hence can have only one answer set. Moreover, observe that the answer set of $\Pi(q)$ might contain instances of $answer_i$ with the null constant, with the intuitive meaning that the corresponding answer variables do not have a value assigned. In contrast the mapping μ is not defined to map variables onto null. Therefore, for some $V \supseteq \mathsf{dom}(\mu)$ let μ_V be the total function with domain V such that $\mu_V(?X) = \mu(?X)$ if $?X \in \mathsf{dom}(\mu)$ and $\mu_V(?X) = \mathsf{null}$ otherwise. Now, for an RDF graph G, let $\mathrm{ASP}(G)$ denote the translation of G into a database of *triple* atoms. Then we obtain the following lemma.

Lemma 2. *Let G be an RDF graph and let q be a SPARQL query with $V = \mathsf{avar}(q)$. Then $\mu \in \langle\!\langle q \rangle\!\rangle_G$ if and only if $\mu : V \to \mathsf{terms}(G)$ and $answer_1(\overline{V}\mu_V)$ is an element of the one and only answer set of $\Pi(q) \cup \mathrm{ASP}(G)$.*

7.4   Combining Model Generation and Querying

It is straightforward to reformulate Lemma 2 for models of our knowledge base.

Lemma 3. *Let $\mathcal{I}$ be a Δ-model for the DL knowledge base $\mathcal{K}$ and let q be a SPARQL query with $V = \mathsf{avar}(q)$. Then $\mu \in \langle\!\langle q \rangle\!\rangle_{G(\mathcal{I})}$ if and only if $\mu : V \to \mathsf{terms}(G)$ and $answer_1(\overline{V}\mu_V)$ is an element of the one and only answer set of $\Pi(q) \cup \mathrm{ASP}(G(\mathcal{I}))$.*

Now we are ready to "plug together" the results from Lemmas 1 and 3 to obtain the correctness result for the described translation.

Theorem 1. *For a DL knowledge base $\mathcal{K}$ over a fixed domain Δ, and a SPARQL query q with $V = \mathsf{avar}(q)$, it holds that $\mu \in \mathsf{cert}_\Delta(\mathcal{K}, q)$ if and only if $\mu : V \to \mathsf{terms}(G)$ and $answer_1(\overline{V}\mu_V) \in \mathsf{Cn}^\forall(\Pi_{\mathrm{RDF}}(\mathcal{K}, \Delta) \cup \Pi(q))$.*

Proof (Sketch). First, we observe that no predicate from $\Pi_{\mathrm{RDF}}(\mathcal{K}, \Delta)$ occurs in the head of any rule of $\Pi(q)$. Hence, by an application of the well-known splitting theorem [15], we can establish the following correspondence:

S is an answer set of $\Pi_{\mathrm{RDF}}(\mathcal{K}, \Delta) \cup \Pi(q)$ if and only if $S = S' \cup S''$ where S' is an answer set of $\Pi_{\mathrm{RDF}}(\mathcal{K}, \Delta)$ and S'' is an answer set of $\Pi(q) \cup S'$.   (†)

Now consider a mapping μ such that $answer_1(\overline{V}\mu_V) \in \mathsf{Cn}^\forall(\Pi_{\mathrm{RDF}}(\mathcal{K}, \Delta) \cup \Pi(q))$. By the definition of cautious consequences, this means that $answer_1(\overline{V}\mu_V) \in S$ for every answer set S of $\Pi_{\mathrm{RDF}}(\mathcal{K}, \Delta) \cup \Pi(q)$. By (†), this means that $answer_1(\overline{V}\mu_V) \in S''$ for the one and only answer set S'' of $\Pi(q) \cup S'$ for every answer set S' of $\Pi_{\mathrm{RDF}}(\mathcal{K}, \Delta)$. Now, using Lemmas 1 and 3 we find that this is the case exactly if for every Δ-model $\mathcal{I}$ of $\mathcal{K}$ (represented by $S' = \mathrm{ASP}(G(\mathcal{I}))$), we find that $\mu \in \langle\!\langle q \rangle\!\rangle_{G(\mathcal{I})}$. Now, by the definition of certain answers, the latter is the case exactly if $\mu \in \mathsf{cert}_\Delta(\mathcal{K}, q)$.   □

8   Conclusion

In this paper, we introduced the formal underpinnings for answering SPARQL queries over OWL ontologies under the fixed domain semantics. As usual for query answering over expressive logics, we employ the principle of certain answers. We also proposed a way to realize this task by means of cautious inferencing over answer set programs, allowing to employ existing, highly optimized off-the-shelf machinery for that purpose.

As next steps in our research, we will evaluate the approach over synthetic and real-world data sets in order to verify the (albeit very plausible) assumption that our proposed approach is superior to the brute-force approach of enumerating and querying all models.

Beyond that, our initial work raises many interesting conceptual questions.

Adequacy of the Certain Answer Principle. On the one hand it is natural to ask for "guaranteed" results applying to all scenarios complying with the knowledge base. On the other hand, fixed-domain reasoning is often employed in the search for solutions to some sort of constraint satisfaction problem, and the enumerated models represent the solutions to that problem. In such a setting, one might also ask for *possible answers*, i.e., answers obtained from some model (rather than all of them). We foresee that such a setting can be captured by our approach in a straightforward way by considering brave consequences rather than cautious ones.

On yet another note, an alternative approach would be to conceive the set of models of a knowledge base as a collection of RDF graphs, stored together in an RDF dataset using named graphs. SPARQL queries could then be executed over this "super-model".

Aggregates. For space reasons, we refrained from addressing aggregates. The technically most straightforward (and readily implementable) way to define answers for queries featuring aggregates would again be to fully execute the query over each model separately and then intersect the result sets over all models. Such strategy might, however lead to unintuitive results. If we queried the knowledge base from Example 1 asking for academics and the number of projects each of them is in, we would get an empty result, since there exist models where every person is working in each of the project, where in the "standard model" one or no project would be assigned to every person. This observation suggests that under certain circumstances we might perform other operations than just intersection when accumulating certain answers.

Acknowledgments. We are grateful for the valuable feedback from the anonymous reviewers, which helped greatly to improve this work. This work has been funded by the European Research Council via the ERC Consolidator Grant No. 771779 (DeciGUT).

References

1. Abiteboul, S., Duschka, O.M.: Complexity of answering queries using materialized views. In: Proceedings of the 7th Symposium on Principles of Database Systems (PODS), pp. 254–263. ACM Press (1998)
2. Abiteboul, S., Hull, R., Vianu, V.: Foundations of Databases. Addison-Wesley, Boston (1995)
3. Angles, R., Gutierrez, C.: The expressive power of SPARQL. In: Sheth, A., et al. (eds.) ISWC 2008. LNCS, vol. 5318, pp. 114–129. Springer, Heidelberg (2008). https://doi.org/10.1007/978-3-540-88564-1_8
4. Baader, F., Calvanese, D., McGuinness, D., Nardi, D., Patel-Schneider, P.: The Description Logic Handbook: Theory, Implementation, and Applications, 2nd edn. Cambridge University Press, Cambridge (2007)
5. Birte Glimm, C.O. (ed.): SPARQL 1.1 Entailment Regimes. W3C Working Draft, 21 March 2013. http://www.w3.org/TR/sparql11-entailment/
6. Brewka, G., Eiter, T., Truszczyński, M.: Answer set programming at a glance. Commun. ACM **54**(12), 92–103 (2011)
7. Calvanese, D.: Finite model reasoning in description logics. In: Proceedings of Description Logic Workshop, 1996. AAAI Technical Report, vol. WS-96-05, pp. 25–36. AAAI Press (1996)
8. Cyganiak, R., Wood, D., Lanthaler, M. (eds.): RDF 1.1 Concepts and Abstract Syntax. W3C Recommendation, 25 February 2014. http://www.w3.org/TR/rdf11-concepts/
9. Gaggl, S.A., Rudolph, S., Schweizer, L.: Fixed-domain reasoning for description logics. In: Proceedings of European Conference on AI (ECAI), 2016. Frontiers in Artificial Intelligence and Applications, vol. 285, pp. 819–827. IOS Press (2016)
10. Gebser, M., Kaminski, R., Kaufmann, B., Schaub, T.: Answer Set Solving in Practice. Synthesis Lectures on Artificial Intelligence and Machine Learning, Morgan & Claypool Publishers, San Rafael (2012)
11. Gelfond, M., Lifschitz, V.: Classical negation in logic programs and disjunctive databases. New Gener. Comput. **9**(3/4), 365–386 (1991)
12. Glimm, B., Horrocks, I., Motik, B., Stoilos, G., Wang, Z.: HermiT: an OWL 2 reasoner. J. Autom. Reason. **53**(3), 245–269 (2014)
13. Hitzler, P., Krötzsch, M., Rudolph, S.: Foundations of Semantic Web Technologies. Chapman & Hall/CRC, Boca Raton (2009)
14. Horrocks, I., Kutz, O., Sattler, U.: The even more irresistible $\mathcal{SROIQ}$. In: Proceedings of the 10th International Conference on Principles of Knowledge Representation and Reasoning (KR), pp. 57–67. AAAI Press (2006)
15. Lifschitz, V., Turner, H.: Splitting a logic program. In: Proceedings of the 11th International Conference on Logic Programming (ICLP), pp. 23–37. MIT Press (1994)
16. Lutz, C., Sattler, U., Tendera, L.: The complexity of finite model reasoning in description logics. Inf. Comput. **199**(1–2), 132–171 (2005)
17. Motik, B., Cuenca Grau, B., Horrocks, I., Wu, Z., Fokoue, A., Lutz, C. (eds.): OWL 2 Web Ontology Language: Profiles. W3C Recommendation, 27 October 2009. http://www.w3.org/TR/owl2-profiles/

18. Niemelä, I.: Logic programs with stable model semantics as a constraint programming paradigm. Ann. Math. Artif. Intell. **25**(3–4), 241–273 (1999)
19. Polleres, A., Wallner, J.P.: On the relation between SPARQL1.1 and answer set programming. J. Appl. Non-Class. Logics **23**(1–2), 159–212 (2013)
20. Rosati, R.: Finite model reasoning in *DL-Lite*. In: Bechhofer, S., Hauswirth, M., Hoffmann, J., Koubarakis, M. (eds.) ESWC 2008. LNCS, vol. 5021, pp. 215–229. Springer, Heidelberg (2008). https://doi.org/10.1007/978-3-540-68234-9_18
21. Rudolph, S.: Foundations of description logics. In: Polleres, A., et al. (eds.) Reasoning Web 2011. LNCS, vol. 6848, pp. 76–136. Springer, Heidelberg (2011). https://doi.org/10.1007/978-3-642-23032-5_2
22. Rudolph, S.: Undecidability results for database-inspired reasoning problems in very expressive description logics. In: Proceedings of the 15th International Conference on the Principles of Knowledge Representation and Reasoning (KR), pp. 247–257. AAAI Press (2016)
23. Rudolph, S., Glimm, B.: Nominals, inverses, counting, and conjunctive queries or: why infinity is your friend!. J. Artif. Intell. Res. **39**, 429–481 (2010)
24. Rudolph, S., Schweizer, L.: Not too big, not too small... complexities of fixed-domain reasoning in first-order and description logics. In: Oliveira, E., Gama, J., Vale, Z., Lopes Cardoso, H. (eds.) EPIA 2017. LNCS (LNAI), vol. 10423, pp. 695–708. Springer, Cham (2017). https://doi.org/10.1007/978-3-319-65340-2_57
25. Rudolph, S., Schweizer, L., Tirtarasa, S.: Wolpertinger: a fixed-domain reasoner. In: Proceedings of the 16th International Semantic Web Conference (ISWC), Posters & Demonstrations. CEUR, vol. 1963. CEUR-WS.org (2017)
26. Rudolph, S., Schweizer, L., Tirtarasa, S.: Justifications for description logic knowledge bases under the fixed-domain semantics. In: Benzmüller, C., Ricca, F., Parent, X., Roman, D. (eds.) RuleML+RR 2018. LNCS, vol. 11092, pp. 185–200. Springer, Cham (2018). https://doi.org/10.1007/978-3-319-99906-7_12
27. Schreiber, G., Raimond, Y. (eds.): RDF 1.1 Primer. W3C Recommendation, 24 February 2014. http://www.w3.org/TR/rdf11-primer/
28. Sirin, E., Parsia, B., Grau, B.C., Kalyanpur, A., Katz, Y.: Pellet: a practical OWL-DL reasoner. J. Web Semant. **5**(2), 51–53 (2007)
29. Steigmiller, A., Liebig, T., Glimm, B.: Konclude: system description. J. Web Semant. **27**, 78–85 (2014)
30. W3C OWL Working Group: OWL 2 Web Ontology Language: Document Overview. W3C Recommendation (2009). https://www.w3.org/TR/owl2-overview/
31. W3C SPARQL Working Group: SPARQL 1.1 Overview. W3C Recommendation, 21 March 2013. http://www.w3.org/TR/sparql11-overview/

Augmenting an Answer Set Based Controlled Natural Language with Temporal Expressions

Rolf Schwitter[✉]

Department of Computing, Macquarie University, Sydney, NSW 2109, Australia
Rolf.Schwitter@mq.edu.au

Abstract. In this paper we discuss how we can augment an existing controlled natural language in a systematic way with temporal expressions in order to write high-level temporal specifications which require reasoning about action and change. We show that domain-dependent axioms which are necessary to specify time-varying properties, deal with the commonsense law of inertia, and with continuous change can be expressed directly and in a transparent way on the level of the controlled natural language. The resulting temporal specification including the corresponding axioms and the required terminological knowledge can be translated automatically into an executable answer set program and then be used by a linguistically motivated version of the Event Calculus, implemented as an answer set program, for temporal reasoning and question answering.

Keywords: Controlled natural language · Answer set programming · Event calculus

1 Introduction

Despite the impressive progress made in the domain of data-driven applications, there is still a strong need for mechanisms that support the manual acquisition and encoding of fine-grained commonsense and domain knowledge that can be used for automated reasoning. Controlled natural languages offer such a mechanism that allows subject matter experts to specify knowledge in a human-readable and machine-processable way [10,16]. Controlled natural languages are simplified forms of natural languages; they are constructed by restricting the size of the grammar and the vocabulary of a natural language in order to reduce or eliminate ambiguity and complexity so that these languages can be automatically translated into a formal target language. These characteristics are interesting and make controlled natural languages useful as high-level interface languages to knowledge systems; in particular, if the writing process of these languages is supported by a suitable authoring tool that facilitates and guides the construction of a textual specification [6,9].

© Springer Nature Switzerland AG 2019
A. C. Nayak and A. Sharma (Eds.): PRICAI 2019, LNAI 11670, pp. 500–513, 2019.
https://doi.org/10.1007/978-3-030-29908-8_40

Most existing controlled natural languages deal with static knowledge. We believe that there exist only two controlled languages: Computer Processable English (CPL) [1] and PENG Light [17] that support reasoning with dynamic commonsense knowledge. In the case of CPL, sentences are parsed and then translated via intermediate logical forms into statements of the KM knowledge representation language that relies on a situation calculus mechanism to process the effects of actions and to update existing situations. In contrast to CPL, a textual specification in PENG Light is translated during the parsing process into discourse representation structures, the basic meaning-carrying units of discourse representation theory [7], and then further into a Prolog-based version of the simplified Event Calculus [8,20] in order to reason about events, fluents and time points. From a linguistic point of view, discourse representation theory is interesting, since it has been designed to handle anaphoric expressions in a uniform way, but discourse representation theory is equivalent to first-order logic and has therefore the same computational problems and limitations as knowledge representation language as a first-order logic.

In this paper, we explore a different avenue for writing temporal specifications that is based on Answer Set Programming (ASP) [3,4,13] as knowledge representation formalism and combine ASP with Event Calculus reasoning [8,14,15,20]. Previously, the author [18] showed that a bi-directional grammar can be used to specify and verbalise ASP programs in controlled natural language and to resolve anaphoric references in a similar way as in discourse representation theory by enforcing structural constraints in the data structure of the grammar. Taking this work as a starting point, we show how a temporal specification can be written in controlled natural language and translated into an ASP program that uses a linguistically motivated version of the Event Calculus for temporal reasoning and question answering. We will illustrate that a temporal specification can be written entirely in controlled natural language together with all domain-dependent effect axioms and relevant terminological knowledge that are necessary for temporal reasoning.

2   Answer Set Programming

Answer Set Programming (ASP) is a form of declarative programming and has its roots in logic programming, disjunctive databases and non-monotonic reasoning [4,13]. ASP is an expressive formal language for knowledge representation and automated reasoning and is based on the answer set semantics for logic programs [4]. In ASP, problems are represented in terms of finite logic theories and are then solved by reducing these problems to finding answer sets which declaratively describe the solutions to these problems. An ASP program consists of a set of rules of the form:

$$L_1 \; ; \; \ldots \; ; \; L_k \; :- \; L_{k+1}, \; \ldots, \; L_m, \; \text{not } L_{m+1}, \; \ldots, \; \text{not } L_n.$$

where all L_i are classical literals. A classical literal L is either an atom a or a negated atom $\neg a$. A literal of the form $\text{not } L$ stands for a negation as failure literal. The disjunction (;) is an epistemic disjunction [5]. The arrow (:-) that points to the left stands for *if*; the part on the left of that arrow is the head of the rule and the part on the right is its body. If the body is empty (n=0), then we omit the arrow and end up with a *fact*. If the head is empty (k=0), then we keep the arrow and end up with an *integrity constraint*. ASP is interesting as a knowledge representation language, since it allows us to combine strong negation and negation as failure to specify non-monotonic theories. Furthermore, ASP is supported by powerful programming tools; for example, the *clingo* system [3] combines a grounder and a solver, and integrates scripting languages.

3   Towards an ASP-Based Event Calculus

The basic notation of the Event Calculus distinguishes between events, fluents and time points [8,15,20]. An event represents an action that may occur in a particular domain and can be described on the linguistic level by an event verb; a fluent represents a time-varying property in a domain and can be described by an adjective or a stative verb, and a time point represents an instant of time and can be described by a date/time. An event may happen at a particular time point and initiate or terminate a fluent; a fluent has a truth value and may hold or not hold at that time point. After an event occurs, the truth value of a fluent may change and require an update. Using the Event Calculus, we can specify commonsense knowledge about events that initiate fluents, terminate fluents or release them from the commonsense law of inertia [15]. Additionally, we have trajectory and anti-trajectory axioms that describe the behaviour of continuous change. For example, the following are two core axioms of the simplified Event Calculus [20] in logic programming notation:

```
holds_at(F, T2) :-
   happens(E, T1), initiates(E, F, T1), T1 < T2, not clipped(T1, F, T2).
clipped(T1, F, T2) :-
   happens(E, T), T1 < T, T < T2, terminates(E, F, T).
```

The first axiom states that a fluent F holds at time point T2, if an event E happens at time point T1 and initiates the fluent at T1, and T1 is before T2, and it is not provable that the fluent has been clipped between these two time points. The second axiom states that a fluent F is clipped between the time points T1 and T2, if an event E happens at T and terminates the fluent at T and T is after T1 and before T2. To illustrate our approach, we modify this version of the Event Calculus in order to make it compatible with ASP and prepare it for the formal output that is generated by the controlled natural language processor. As the following two rules illustrate, we applied three main modifications:

```
holds_at(F, T2) :-
   initiated_at(F, T1), time_point(T2), T1 < T2, not clipped(T1, F, T2).
clipped(T1, F, T2) :-
   terminated_at(F, T), time_point(T1), time_point(T2), T1 <= T, T < T2.
```

(1) Many ASP solvers require a safety condition on rules: a rule is safe, if every variable occurs in a positive body literal of the rule: therefore, we have to add the literal `time_point(T2)` to the body of the first rule and the literals `time_point(T1)` and `time_point(T2)` to the body of the second rule to guarantee rule safety. (2) The literal `happens/2` is not necessary in the body of these rules, since we can check if an event happens at a specific time point in the body of the effect axioms and end up with the two literals: `initiated_at/2` and `terminated_at/2` instead of `initiates/3` and `terminates/3`. (3) Rather than specifying that a fluent is clipped, if an event terminates that fluent after T1 and before T2, we specify that the fluent was terminated at or after T1 and before T2 (more about this in Sect. 5).

4   The Controlled Natural Language PENG^{ASP}

PENG^{ASP} is a controlled natural language designed to support subject matter experts who do not necessarily have a formal background in logic to write high-level specifications [18]. The language processor of the PENG^{ASP} system uses a chart parser that relies on a unification-based grammar and lexicon. The language processor generates an internal representation of an ASP program, resolves anaphoric expressions and generates lookahead information during the parsing process. The lookahead information informs the author about how a sentence can be completed and enforces the structure of the controlled natural language on the user interface level [6]. The grammar of the PENG^{ASP} system is bi-directional and parametrised so that it can be used for processing a specification and translating the specification into an ASP program as well as generating a semantically equivalent specification from an ASP program [18].

4.1   A Temporal Specification in PENG^{ASP}

The static version of the controlled natural language PENG^{ASP} did not support the writing of temporal specifications that require reasoning about events, fluents and time points. In this section, we present an extension of the controlled natural language where the grammar is partitioned away from the existing static grammar and where the extension can be used as required. The extension consists of temporal modifiers that allow us to specify when an particular event occurs or does not occur and when a fluent holds or does not hold. These modifiers include date and time expressions and references to time points. A date is introduced by the preposition *on* and has the format *YYYY-MM-DD*, a time is introduced by the preposition *at* and has the format *HH:MM:SS* or shorter *HH:MM*. Once a date has been introduced all subsequent time points refer to that date, until a new date is introduced, for example:

1. The train AV8504 is located at Roma Termini on 2019-02-20 at 06:30.
2. The train departs from Roma Termini at 06:45.
3. The train arrives at Firenze Campo di Marte at 08:03 and departs from Firenze Campo di Marte at 08:10.
4. The train arrives at Bozen/Bolzano at 11:17.

The first sentence (1) describes a fluent that holds at a given time point (*on 2019-02-20 at 06:30*). The next three sentences (2–4) describe four events that occur at subsequent time points (*06:45, 08:03, 08:10* and *11:17*) on the same day. However, note that this specification does not tell us, for example, that the train is no longer at Roma Termini after 06:45 or that the train is located at Bozen/Bolzano after 11:17. Our own commonsense knowledge fills in these gaps, but this knowledge is not available by default for automated reasoning and needs to be added to the specification in the form of positive and negative effect axioms. Below are four effect axioms (5–8) that describe which fluents are initiated and terminated as consequence of one or more events that occur (or do not occur) at a particular time point.

5. If a vehicle departs from a location at a time point then the vehicle will no longer be located at that location after that time point.
6. If a vehicle arrives at a location at a time point then the vehicle will be located at that location after that time point.
7. If a vehicle departs from a location at a time point then the vehicle will be in transit after that time point.
8. If a vehicle arrives at a location at a time point and the vehicle does not provably make a stopover at that time point then the vehicle will no longer be in transit after that time point.

These effect axioms add new linguistic constructions to the controlled language PENG^{ASP}; for example, the construction *will be in transit* serves a trigger to construct an initiating effect axiom and the construction *will no longer be located at* servers as a trigger to construct a terminating effect axiom. Note that the expression *does not provably ...* stands for weak negation in contrast to *does not ...* which stands for strong negation in PENG^{ASP}. Note also that the above effect axioms are quite general and speak about *vehicle* and *location* and could therefore also be used for different scenarios. In our case, we have to connect the terminology used in these effect axioms with the terminology in our temporal specification. We can achieve this directly on the level of the controlled natural language and make therefore the following ontological commitments:

9. Roma Termini is a railway station.
10. Firenze Campo di Marte is a railway station.
11. Bozen/Bolzano is a railway station.
12. Every railway station is a location.
13. Every train is a vehicle.
14. If a vehicle arrives at a location at a time point T1 and departs from that location at a time point T2 and T1 is before T2 then the vehicle makes a stopover between T1 and T2.

The sentences (9–11) assign an instance to a class and the sentences (12) and (13) specify general class inclusion axioms. The sentence (14) defines a composite event (*makes a stopover at*) that consists of two subevents (*arrives at* and *departs from*) and that occurs between two time points (*T1* and *T2*). Given this additional background knowledge, the ASP system *clingo* [3] can figure out which fluents hold or do not hold at a given point in time; for example, that the train 8504 is in transit at 07:30 since it is a vehicle, located at Firenze Campo di Marte at 08:05 and in transit at the same time since Firenze Campo di Marte is a location, but not located at the same railway station after 08:10.

4.2 Translating the Temporal Specification

The translation of the temporal specification results in the set of ASP atoms displayed below which is based on a flat notation that uses a small number of typed literals (e.g. `class/2`, `named/2`, `fluent/3`, `event/3`, `data_prop/3`) which are associated with linguistic categories (e.g. noun, verb, adjective, proper name, date and time) in the lexicon, together with those literals (`holds_at/2` and `happens/2`) which belong to the language of the Event Calculus.

```
class(1,train).
named(1, av8504).
holds_at(fluent(1, 2, located_at), 1550644200).
named(2, roma_termini).
data_prop(3, 1550644200, date_time).
happens(event(1, 2, depart_from), 1550645100).
data_prop(4, 1550645100, date_time).
happens(event(1, 5, arrive_at), 1550649780).
named(5, firenze_campo_di_marte).
data_prop(6, 1550649780, date_time).
happens(event(1, 5, depart_from), 1550650200).
data_prop(7, 1550650200, date_time).
happens(event(1, 8, arrive_at), 1550661420).
named(8, bozen_bolzano).
data_prop(9, 1550661420, date_time).
```

The numbers 1-9 are Skolem constants and the time points are represented as integers and stand for Unix timestamps. Note that whenever the anaphora resolution algorithm detects a time, for example 06:45 without an immediately preceding date, then it looks for the closest date (2019-02-20) so that a suitable Unix timestamp can be constructed (in our case: 1550645100). If no date is available then the system date is taken as reference date.

4.3 Translating the Effect Axioms

The sentences (5–8) that specify effect axioms are translated into ASP rules that contain the Event Calculus literals `initiated_at/2` and `terminated_at/2` as head. The first argument of these literals is a fluent type and the second argument is a time point.

```
terminated_at(fluent(A, B, located_at), T) :-
  class(A, vehicle),  happens(event(A, B, depart_from), T),
  class(B, location), class(C, time_point), data_prop(C, T, date_time).
initiated_at(fluent(A, B, located_at), T) :-
  class(A, vehicle), happens(event(A, B, arrive_at), T),
  class(B, location), class(C, time_point), data_prop(C, T, date_time).
initiated_at(fluent(A, in_transit), T) :-
  class(A, vehicle),  happens(event(A, B, depart_from), T),
  class(B, location), class(C, time_point), data_prop(C, T, date_time).
terminated_at(fluent(A, in_transit), T) :-
  class(A, vehicle), happens(event(A, B, arrive_at), T),
  class(B, location), class(C, time_point), data_prop(C, T, date_time),
  not happens(event(A, make_stopover), T).
```

Note that the Event Calculus literal happens/2 that triggers the initiation or termination of a fluent at a given time point occurs in the body of these rules. This is a consequence of the linguistic structure of the effect axioms that we used as a starting point and has an impact on the linguistically motivated reconstruction of the Event Calculus (see Sect. 5).

4.4   Translating the Terminological Knowledge

The terminological knowledge that connects the temporal specification with the effect axioms is translated into the following ASP facts and rules:

```
class(2, railway_station).
class(5, railway_station).
class(8, railway_station).
class(X, vehicle) :- class(X, train).
class(X, location) :- class(X, railway_station).
```

The definition that introduces a compound event (make_stopover) via two subevents (arrive_at and depart_from) results in the following ASP rule:

```
happens(event(A, make_stopover), T1, T2) :-
  class(A, vehicle), happens(event(A, B, arrive_at), T1),
  class(B, location), time_point(T1),
  happens(event(A, B, depart_from), T2), time_point(T2),   T1 < T2.
```

Note that the entire temporal specification can be verbalised again in controlled natural language since our grammar is bi-directional, but verbalisation requires sentence planning as discussed in [18].

5   A Linguistically Motivated ASP-Based Event Calculus

Recently, Lee and Palla [11] showed that the circumscriptive Event Calculus can be reformulated in first-order stable model semantics [2], and represented as an ASP program under the assumption that the domain is given and finite. The axioms of our linguistically motivated ASP-based version of the Event Calculus trace back to those of the circumscriptive Event Calculus of Miller and Shanahan [14] and to the work of Mueller [15]. Below we discuss our version of the ASP representation of the 17 core axioms of the Event Calculus.

EC1. A fluent F is clipped between time points T1 and T2, if the fluent was terminated at time point T, and T was at or after T1 and before T2:

```
clipped(T1, F, T2) :-
    terminated_at(F, T), time_point(T1), time_point(T2), T1 <= T, T < T2.
```

EC2. A fluent F is declipped between time points T1 and T2, if the fluent was initiated at time point T, and T was at or after T1 and before T2:

```
declipped(T1, F, T2) :-
    initiated_at(F, T), time_point(T1), time_point(T2), T1 <= T, T < T2.
```

EC3. A fluent F is stopped between time points T1 and T2, if the fluent was terminated at time point T, and T was after T1 and before T2:

```
stopped_in(T1, F, T2) :-
    terminated_at(F, T), time_point(T1), time_point(T2), T1 < T, T < T2.
```

EC4. A fluent F is started between time points T1 and T2, if the fluent was initiated at time point T, and T was after T1 and before T2:

```
started_in(T1, F, T2) :-
    initiated_at(F, T), time_point(T1), time_point(T2), T1 < T, T < T2.
```

EC5. A fluent F2 holds at time point T2, if there is a trajectory defined where a fluent F1 was initiated at time point T1 and the value of a fluent F2 becomes true at time point T2 while the fluent F1 was not provably stopped between time points T1 and T2:

```
holds_at(F2, T2) :-
    trajectory(F1, T1, F2, T2), not stopped_in(T1, F1, T2).
```

EC6. A fluent F2 holds at time point T2, if there is an anti-trajectory defined where a fluent F1 was terminated at time point T1 and the value of a fluent F2 becomes true at time point T2 while the fluent F1 was not provably started between time points T1 and T2:

```
holds_at(F2, T2) :-
    anti_trajectory(F1, T1, F2, T2), not started_in(T1, F1, T2).
```

EC7. A fluent F is released from the commonsense law of inertia between time points T1 and T2, if the fluent was released at time point T, and T was after T1 and before or at T2:

```
released_at_between(T1, F, T2) :-
    released_at(F, T), time_point(T1), time_point(T2), T1 < T, T <= T2.
```

EC8. A fluent F is released from the commonsense law of inertia between time points T1 and T2, if the fluent was initially released at time point T, and T was at or after T1 and before T2:

```
released_between(T1, F, T2) :-
    init_released_at(F, T), time_point(T1), time_point(T2),
    T1 <= T, T < T2.
```

EC9. A fluent F holds at time point T2, if the fluent did hold at time point T1 and was not provably released and not provably clipped between time point T1 and some later time point T2:

```
holds_at(F, T2) :-
    holds_at(F, T1), time_point(T1), time_point(T2), T1 < T2,
    not released_at_between(T1, F, T2), not clipped(T1, F, T2).
```

EC10. A fluent F does not hold at time point T2, if the fluent did not hold at time point T1 and was not provably released and not provably declipped between time point T1 and some later time point T2:

```
-holds_at(F, T2) :-
    -holds_at(F, T1), time_point(T1), time_point(T2), T1 < T2,
    not released_at_between(T1, F, T2), not declipped(T1, F, T2).
```

EC11. A fluent F is released from the commonsense law of inertia at time point T2, if the fluent was released at time point T1 and was not provably clipped and not provably declipped between time point T1 and some later time point T2:

```
released_at(F, T2) :-
    released_at(F, T1), time_point(T2), T1 < T2,
    not clipped(T1, F, T2), not declipped(T1, F, T2).
```

EC12. A fluent F is not released from the commonsense law of inertia at time point T2, if the fluent was not released at time point T1 and was not provably released between time point T1 and some later time point T2:

```
-released_at(F, T2) :-
    -released_at(F, T1), time_point(T2), T1 < T2,
    not released_between(T1, F, T2).
```

EC13. A fluent F is released between time points T1 and T2, if the fluent was initially released at time point T and T is after T1 and before T2:

```
released_in(T1, F, T2) :-
    init_released_at(F, T), time_point(T1), time_point(T2),
    T1 < T, T < T2.
```

EC14. A fluent F holds at time point T2, if the fluent was initiated at time point T1 and was not provably stopped and not provably released between time point T1 and some later time point T2:

```
holds_at(F, T2) :-
    initiated_at(F, T1), time_point(T2), T1 < T2,
    not stopped_in(T1, F, T2), not released_in(T1, F, T2).
```

EC15. A fluent F does not hold at time point T2, if the fluent was terminated at time point T1 and was not provably started and not provably released between time point T1 and some later time point T2:

```
-holds_at(F, T2) :-
    terminated_at(F, T1), time_point(T2), T1 < T2,
    not started_in(T1, F, T2), not released_in(T1, F, T2).
```

EC16. A fluent F is released from the commonsense law of inertia at time point T2, if the fluent was initially released from that law at time point T1 and was not provably stopped and not provably started between time point T1 and some later time point T2:

```
released_at(F, T2) :-
    init_released_at(F, T1), time_point(T2), T1 < T2,
    not stopped_in(T1, F, T2), not started_in(T1, F, T2).
```

EC17. A fluent F is not released from the commonsense law of inertia at time point T2, if the fluent was either initiated at time point T1 or terminated at time point T1, and was not provably released between time point T1 and some later time point T2:

```
-released_at(F, T2) :-
    initiated_at(F, T1), time_point(T2), T1 < T2,
    not released_in(T1, F, T2).
-released_at(F, T2) :-
    terminated_at(F, T1), time_point(T2), T1 < T2,
    not released_in(T1, F, T2).
```

In contrast to Mueller [15], our implementation of the core axioms is more suitable for controlled natural language processing, since we do not need to check whether an event happens in the following axioms: EC1-6, 8, 13-18. This is done in a natural way in the effect axioms (see Sect. 4.3). Similarly, we do not need to check in EC5 and EC6 whether an event initiated or terminated a fluent, since the trajectory/anti-trajectory axiom takes care of this. In addition to these core axioms of the Event Calculus, the PENG^{ASP} system uses a number of bridging axioms to interface the domain-dependent axioms derived from the textual specification with the core axioms.

6   Releasing Fluents from Inertia

The commonsense law of inertia is a default assumption that objects tend to stay at the same location unless they are affected by events [12]. A large number of the core axioms of the Event Calculus deal with the commonsense law of inertia. The axioms EC9 and EC10, for example, specify that at any given time point a fluent is or is not subject to the commonsense law of inertia. The axioms EC14 and EC15 enforce the commonsense law of inertia after a fluent was initiated or terminated. The axiom EC16 specifies when a fluent is released from inertia and the axiom EC17 when inertia for a fluent is restored. Finally, the axioms EC11 and EC12 specify when the truth value of the fluent does not change. We can trigger the initial release of a fluent from the commonsense law of inertia and specify the conditions on the level of the controlled natural language, for example:

15. If a signal light is activated at a time point and a lightning strikes the train network at that time point then the signal light may or may not be activated after that time point.

When a fluent is released from inertia its truth value can fluctuate and may or may not be the same as before. The sentence (15) is translated into the following choice rule:

```
{ init_released_at(fluent(A, activated), T) } :-
  class(A, signal_light), holds_at(fluent(A, activated), T),
  class(B, time_point), data_prop(B, T, date_time),
  class(C, lightning), happens(event(C, D, strike), T),
  class(D, train_network).
```

Given additional effect axioms – here in controlled natural language – such as:

16. If a signal light is activated at a time point and the train passes the signal light at that time point then the train will be safe after that time point.
17. If a signal light is not provably activated at a time point and the train passes the signal light at that time point then the train will no longer be safe after that time point.

and suitable events that trigger these axioms, we end up with two answer sets after execution: one in which the train is safe and one in which the train is not safe.

7   Working with Continuous Change

The controlled natural language PENG[ASP] can also describe forms of continuous change; continuous change is ubiquitous in commonsense domains. Let us assume that a train is moving with an average speed of 120 km/h, then we can calculate for each point in time how far the train is away from its starting point using the Event Calculus and *clingo*'s built-in support for arithmetic functions. We can specify domain-specific axioms for continuous change directly in controlled language, for example:

18. If a vehicle is X km away from a location at a time point T1 and the vehicle will be moving after T1 and T1 is before a time point T2 and Y = X + (120 * (T2−T1)/3600) then the vehicle will be Y km away from that location at T2.

Sentence (18) contains a formula that calculates the distance covered by a vehicle and is translated into a trajectory axiom [19]. This trajectory axiom has the form `trajectory(F1, T1, F2, T2)` and is used by the core axiom EC5 and makes sure that if a fluent F1 was initiated at time point T1, then the fluent F2 becomes true at time point T2 (depending on a particular formula).

8 Evaluation

In order to evaluate our linguistically motivated version of the Event Calculus, we add a number of additional time points (06:30, 07:30, 08:05, 11:15, and 11:18) to the temporal specification introduced in Sect. 4.1, translate it into an ASP program, execute the Event Calculus/ASP program and observe the following results:

```
holds_at(fluent(1, 2, located_at), 1550644200)    % 06:30
holds_at(fluent(1, 2, located_at), 1550645100)    % 06:45
holds_at(fluent(1, in_transit), 1550647800)       % 07:30
holds_at(fluent(1, in_transit), 1550649780)       % 08:03
holds_at(fluent(1, 5, located_at), 1550649900)    % 08:05
holds_at(fluent(1, in_transit), 1550649900)       % 08:05
holds_at(fluent(1, 5, located_at), 1550650200)    % 08:10
holds_at(fluent(1, in_transit), 1550650200)       % 08:10
holds_at(fluent(1, in_transit), 1550661300)       % 11:15
holds_at(fluent(1, in_transit), 1550661420)       % 11:17
holds_at(fluent(1, 8, located_at), 1550661480)    % 11:18

-holds_at(fluent(1, 2, located_at), 1550647800)   % 07:30
...
-holds_at(fluent(1, in_transit), 1550661480)      % 11:18
```

These results tell us, for example, that the train is located at Roma Termini at 06:45, in transit at 07:30 but not located anymore at Roma Termini at the same time. Furthermore, the train is still in transit at 08:03 and located at Fierenze Campo di Marte after that time point until 08:10. The train is not in transit anymore at 11:18 after it arrived at Bozen/Bolzano and is located there at the same time. Alternatively, we can interrogate the resulting answer set using questions in controlled natural language, for example:

19. When is the train located at a railway station?

20. Is the train in transit at 08:05?
21. How far away is the train from Roma Termini at 10:30?

These questions are translated in a similar way into ASP rules as discussed in [18] and contain a specific answer literal in the rule head.

9   Conclusion

In this paper, we started from the controlled natural language PENG^{ASP} that was designed as a high-level interface language to write specifications and terminological knowledge that can be translated into and executed as ASP programs. We showed how this controlled natural language can be augmented with temporal expressions and be used to write temporal specifications. The controlled language PENG^{ASP} uses a small number of specific linguistic expressions that trigger the construction of positive and negative effect axioms, axioms that release fluents from the commonsense law of inertia, and axioms that deal with continuous change. We showed that the form of these domain-dependent axioms that are automatically derived from a controlled natural language specification has an impact on the form of the axioms of the Event Calculus. This observation prompted the introduction of a linguistically motivated ASP-based version of the Event Calculus. The resulting specifications are automatically translated into an ASP program, and if temporal literals are detected, our linguistically motivated version of the Event Calculus is used for temporal reasoning and question answering.

References

1. Clark, P., Harrison, P., Jenkins, T., Thompson, J., Wojcik, R.: Acquiring and using world knowledge using a restricted subset of English. The 18th International FLAIRS Conference (FLAIRS 2005), pp. 506–511 (2005)
2. Ferraris, P., Joohyung Lee, J., Lifschitz, V.: Stable models and circumscription. Artif. Intell. **175**, 236–263 (2011)
3. Gebser, M., et al.: Potassco User Guide, Version 2.2.0 (2019). https://github.com/potassco/guide/releases/
4. Gelfond, M., Lifschitz, V.: The stable model semantics for logic programming. In: Proceedings of the Fifth International Conference on Logic Programming (ICLP), pp. 1070–1080 (1988)
5. Gelfond, M., Kahl, Y.: Knowledge Representation, Reasoning, and the Design of Intelligent Agents, The Answer-Set Programming Approach. Cambridge University Press, Cambridge (2014)
6. Guy, S., Schwitter, R.: The PENG^{ASP} system: architecture, language and authoring tool. J. Lang. Resour. Eval. Spec. Issue: Control. Nat. Lang. **51**(1), 67–92 (2017)
7. Kamp, H., van Genabith, J., Reyle, U.: Discourse representation theory. In: Gabbay, D., Guenthner, F. (eds.) Handbook of Philosophical Logic, vol. 15, pp. 125–394. Springer, Dordrecht (2011). https://doi.org/10.1007/978-94-007-0485-5_3

8. Kowalski, R., Sergot, M.: A logic-based calculus of events. New Gener. Comput. **4**, 67–94 (1986)
9. Kuhn, T., Schwitter, R.: Writing support for controlled natural languages. In: Proceedings of ALTA, pp. 46–54 (2008)
10. Kuhn, T.: A survey and classification of controlled natural languages. Comput. Linguist. **40**(1), 121–170 (2014)
11. Lee, J., Palla, R.: Reformulating temporal action logics in answer set programming. In: Proceedings of the AAAI Conference on Artificial Intelligence (AAAI), pp. 786–792 (2012)
12. Lifschitz, V.: Formal theories of action. the frame problem in artificial intelligence. In: Proceedings of the 1987 Workshop, Los Altos, CA, pp. 35–57 (1987)
13. Lifschitz, V.: What is answer set programming? In: Proceedings of AAAI, pp. 1594–1597 (2008)
14. Miller, R., Shanahan, M.: Some alternative formulations of the event calculus. In: Kakas, A.C., Sadri, F. (eds.) Computational Logic: Logic Programming and Beyond. LNCS (LNAI), vol. 2408, pp. 452–490. Springer, Heidelberg (2002). https://doi.org/10.1007/3-540-45632-5_17
15. Mueller, E.T.: Commonsense Reasoning: An Event Calculus Based Approach, 2nd edn. Morgan Kaufmann, Burlington (2015)
16. Schwitter, R.: Controlled natural language for knowledge representation. In: Proceedings of COLING 2010, Stroudsburg, PA, USA, pp. 1113–1121. Association for Computational Linguistics (2010)
17. Schwitter, R.: Specifying events and their effects in controlled natural language. Comput. Linguist. Relat. Fields Procedia - Soc. Behav. Sci. **27**, 12–21 (2011)
18. Schwitter, R.: Specifying and verbalising answer set programs in controlled natural language. J. Theory Pract. Log. Program. **18**(3–4), 691–705 (2018)
19. Shanahan, M.: Representing continuous change in the event calculus. In: Proceedings of ECAI 1990, pp. 598–603 (1990)
20. Shanahan, M.: Solving the Frame Problem. MIT Press, Cambridge (1997)

Predictive Systems: The Game Rock-Paper-Scissors as an Example

Mathias Zink[ID], Paulina Friemann[ID], and Marco Ragni[✉][ID]

Cognitive Computation Lab, University of Freiburg, 79110 Freiburg, Germany
zinkmathias@web.de, {friemanp,ragni}@cs.uni-freiburg.de

Abstract. In simple-decision-making scenarios such as in repeated two-person games human behavior is to some extend predictable. To investigate this research question, we focused on developing a system for the Rock-Paper-Scissor (RPS) game. Our approach included three steps: (i) To generate a large data-base of experimental data, (ii) to analyze the data to detect systematic patterns and deviations from rational behavior within the test persons, and (iii) to employ methods from machine learning to identify patterns and predict the next throw of the opponent. We identified as the best current approach a *Gated-Reccurent-Unit using User Statistics*, which is able to predict the next throw and hence win in about 50% of the cases, beating state-of-the-art approaches. Potentials and limitations of our approach are discussed.

Keywords: Predictive systems · Two-person games · Cognitive AI

1 Introduction

The rise of AI systems and companion systems in our everyday life make a successful interaction with humans a core focus. One important part is that such systems can adapt themselves to the cognition of its human partner. This, however, requires to determine how humans makes decision. In this article, we combine a simple cognitive experiment with techniques of machine learning. Stepping into the footsteps of Shannon's 'mind-reading machine' [23], we investigated the applicability of standard Machine Learning techniques, namely Artificial Neural Networks, Support-Vector Machines and Recommender systems, to the task of predicting a human's decision. Particularly, we use the well known and studied game of Rock-Paper-Scissors (RPS). Due to its simplicity we are able to transfer the players decisions almost exclusively onto underlying cognitive processes and their perception of the opponent, making it an interesting game to study. Sequence production has been studied and predicted for the canonical version of RPS, Matching Pennies [5,16]. RPS, however, provides a more challenging subject for prediction, because it encourages the use of strategies and theory of mind, and due to its familiarity to a broad population, it simplified the acquisition of participants.

© Springer Nature Switzerland AG 2019
A. C. Nayak and A. Sharma (Eds.): PRICAI 2019, LNAI 11670, pp. 514–526, 2019.
https://doi.org/10.1007/978-3-030-29908-8_41

We applied the different methods to find and predict patterns in human decision making behavior in RPS and to conclude which features are the most important to look out for. We first describe the theory of the game. Afterwards, we introduce the conducted experiment and present the most important findings. We then use the behavioral data to evaluate the different models.

Rock-Paper-Scissors. A formal description of the Rock-Paper-Scissors game family was introduced by [18]. They established the format $RPS(n,b,s,r)$, which means that n players simultaneously show a move among b possible moves with possible s winning regulations at each round out of r round matches in total.In the basic variant $RPS(2,3,1,r)$, the $n = 2$ players can choose from $b = 3$ different actions: R (Rock), P (Paper) and S (Scissors), all depicted by a specific motion of their hand. Depending on cultural background the players will count to three ('Ro-sham-bo', 'Rock-Paper-Scissors', etc.) while swinging their fist simultaneously and revealing their chosen throw on three. Conditioned on the following rules player one or two receives a point (or none if their chosen throws match): Action 'Rock' beats action 'Scissors', which in turn beats 'Paper', which in turn is better than 'Rock'. It has been shown that from a game refinement point of view our classical model should be played for $r = 9$ rounds, resulting in a very balanced game refinement score [24]. This is the reason why in the following we will use RPS(2,3,1,9).

Rock-Paper-Scissors in Psychology. It has been shown that human players exhibit a significant cycling behavior in their throws of RPS, depending on the player's success or failure in the previous round of the game [27]. A winning round increases the probability of subjects to use the same throw again, while losing makes them more likely to swap. This corresponds to the well known win-stay-lose-shift strategy. Users were matched against a random user after every round, therefore they only have knowledge about their own last throw. [10] claims that negative reinforced cycling has a much bigger impact than the positive one and show that cycle behavior depends on the used throw as well as the outcome. Additionally, they propose that the cycle's direction tends to continue onto the next turns. It is also suspected that the distribution at which the three possible actions are chosen is not uniform: 'Rock' is chosen most frequently, obtaining a pick rate around 35.5%, followed by 'Paper' and 'Scissors', with the latter one being the least played (cf. [27, 29]).

Computational RPS. Reported metrics in the literature are very diverse (e.g., [2, 19]). Most works only focus on *beating* the human player at RPS. Win rates are reported there as either

$$\frac{wins}{wins + losses} \quad \text{or as} \quad \frac{wins}{wins + losses + draws}$$

We only aimed at *predicting* the player. However, these different approaches are still partially comparable, as in every round where the system can successfully predict the player, it is also able to beat them. This means that the win rate would be at least as high as the prediction accuracy:

$$\frac{correct\ predictions}{incorrect\ predictions + correct\ predictions} \tag{1}$$

One of the first appearances of a strategy-guided RPS playing machine was in the year 2000 by means of a genetic algorithm [2]. It achieved a win rate around 35.5%, i.e., in 35.5% of rounds did the algorithm win versus human players. A more recent approach was presented by [20], using a machine learning algorithm based on Gaussian mixture models. In contrast to the last approach, both, the opponent's and the robot's turns were captured to train the algorithm. The authors achieved a win rate of up to 36.6%. The most recent approaches used a large data set to train their algorithms [4,19]. Both treated the game as Markov Processes. The latest approach used up to 10-th order Markov chains with a win rate of almost 40% [19].

Rationality. The study of reasoning and rationality in cognitive science is concerned with the investigation of the relation between normative rationality principles (e.g., formal logics) and inference processes in humans. It has been observed for a long time that human reasoning deviates systematically from normative approaches (e.g., [3]). In RPS, this can be observed by inspecting the relative frequencies at which the actions are picked. A rational agent would pick all actions at the same rate, since the payoffs are equal. Humans however tend to prefer actions over one another, and are unable to create random sequences [15,26]. Humans act rather irrationally, especially following negative or disappointing events [11]. But they mostly still converge to the rational equilibrium in the long run when facing the problem repeatedly, as they adjust their strategies based on past experience [14]. Even though single decisions might be very different to what game theory suggests, humans may ultimately learn from errors and return to a more rational behavior again. If we learn more about the factors that drive us to commit to specific patterns, we can gain insights into the workings of the human mind. Learning algorithms should be able to pick up on those behavioral peculiarities and predict a participant's decisions above chance level. Speaking in terms of game theory, we can gain information about a human's *strategy*, which allows us to exploit it to win at RPS.

2   Experiment

2.1   Method

Participants. For our first experiment we collected data of 185 different participants (145 male, 40 female) with a total of 2273 played games with 28841 turns. Recruitment was done via a University mailing list. Users played on average $\mu = 12.07$ games ($\sigma = 43.07$).

Materials and Procedure. We set up a web page where participants could play the game of RPS online. The layout of the website is shown in Fig. 1. Participants played exclusively against bots. Following every game of RPS we gave the users a short questionnaire in which they could specify which strategy they used and which strategy they thought their opponent to use. Options included for example that there was no strategy involved, that users used gambits, i.e., having a predefined patterns of three throws, or using an explicit strategy, such as reading the opponent or using the throw that beats the user's own last throw. During data acquisition, we tried to give the user the impression of playing against a human opponent, as this has been shown to influence playing experience [28]. For this, we used features that would occur while playing against a human opponent, i.e., a waiting period to find an opponent, and for them to get ready for the next turn. We used a total of seven different bots to oppose our human players. The bots get selected randomly:

- RandomBot: Chooses one of the three actions at random
- CounterOwnBot: Chooses the action that would have won against the Bot's last throw in 50% and the others in 25% respectively
- CounterLastBot: Chooses the action that counters the user's last throw in 50% and the others in 25% respectively
- Sequence3/4/5Bot: Chooses the action of a generated random sequence of 3/4/5 throws and repeats it
- GameFrequencyBot: Chooses the action that counters the last throw of the player

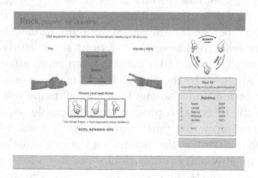

Fig. 1. Layout of the experiment-website

All bots also react to players who repeat a single action multiple times, countering it after three repeated occurrences. Based on previous studies in the literature we can formulate two hypotheses: Hypotheses 1 (H1): The probablilities of the three actions are not uniform. Hypotheses 2 (H2): Users tend to continue their behavioral pattern after a win and change strategy after a loss.

2.2   Results and Discussions

Throw distribution. As in most other studies action 'Rock' was chosen most, while 'Scissors' was chosen the least. All statistical evaluations were done using a one-sample z-Test. The null hypothesis, that the distribution is uniformly distributed can be rejected (Rock: $\mu = 34.91$, $z = 3.49$, $p < .001$; Paper: $\mu = 33.02$, $z = .70$, $p = .48$; Scissors: $\mu = 32.07$, $z = 2.82$, $p = .0048$).

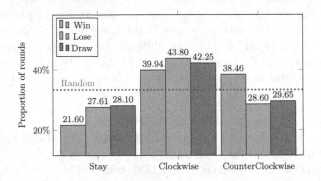

Fig. 2. Cycling behavior of the participants after winning, losing or drawing. Uniform distribution marked in red. (Color figure online)

Cycling behavior. Figure 2 depicts the cycling behavior during the experiment. Clockwise rotation in this case means R → S → P → R, i.e., choosing the action that would have lost against their own last action. Participants chose the clockwise cycling behavior well above random chance level. Contrary to the literature, users tend to change their actions significantly more often than expected (Stay: $z = 16.9$, $p < .001$), notably even more after winning the last throw. This might be caused by the players fear of making themselves predictable by repeating their action and also the fact that humans tend to change their selection too often when trying to produce unpredictable sequences.

Both these factors imply that participants tried to counter the win-stay-lose-shift strategy. Probabilities for the remaining options (clockwise and counter-clockwise after a win, stay and counter-clockwise after a loss) do not differ decisively, which further supports the claim. The increased probability of the users to cycle clockwise after a draw shows that the players expect their opponent to shift action when there was no winner in the last round. Some of the tendencies could be explained by the selection of bots. The bots cycling behavior shows a bias in the bots' behavior towards counter-clockwise cycling. Furthermore, the high percentage of the 'Stay' behavior in the case of a win could be learned by the user and thus be countered with a clockwise cycle. The comparison showed that the cycling behavior of the human participants was influenced by our choice of bots. Therefore it is difficult to draw conclusions on human cycle preferences in the general case.

Continued cycling. When considering the cycling behavior for another turn, as [10] proposed, we obtain the proportions visualized in Fig. 3. This revealed a significant trend towards maintaining a strategy once started (38.40%, $z = 16.96$, $p < .001$). This coincides well with the findings reported in [10].

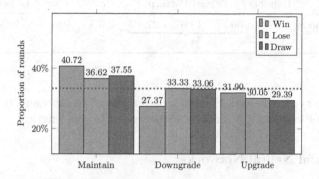

Fig. 3. Continued cycling two rounds after the initial win, lose or draw, expected uniform distribution marked in red. 'Maintain' denotes a continuation of the same strategy, 'Downgrade' a clockwise switch, i.e., staying in the first round and cycle clockwise in the second, cycling clockwise and then counter-clockwise, or cycling counter-clockwise and then staying. 'Update' denotes the contrary. (Color figure online)

Strategy and Strategy Detection. Evaluation of the questionnaires revealed that participants claimed in 52.4% of all games that they tried to read their opponent's strategy. In around 37.3% of the cases on the other hand they claimed to have ignored their opponents actions. In the remaining cases the users opted for a strategy in between. This relates to the just mentioned dependencies in the cycle selection. The detection accuracy of the opposing bot's strategy is noted down in Table 1. We found that only the Sequence Bots (Seq3, Seq4 & Seq5), corresponding to the 'Fixed Pattern' answer, and the GameFrequency Bots were detected correctly above chance level (14.3%). The increased proportion for the Sequence Bots is not surprising, as a static repeating pattern is easier to detect than random movement or specific cycling behavior. Moreover, the Sequence3/4/5 bots have a fairly high probability of 77.8%, 55.5% and 38.3%, respectively, to not even use all three available throws, which makes them even more predictable. The good detection for the 'GameFrequency Bot', however, is due to the corresponding answer 'ReadOpponent' being the most used answer overall (27.8%). Overall we can say that the specific strategies were not reliably identified by our participants.

3   Prediction of Throwing Behavior

The analysis of the experimental data shows that there exist clear tendencies in throwing behavior. This means that a prediction above chance level should be

Table 1. Proportions of correctly identified strategies for the various Bots.

Bot	Random	Seq3	Seq4	Seq5	Counter Last	Counter Own	Game Freq	Overall
Corresponding Answer	Random	Fixed Pattern			Counter Last	Counter Own	Read Opponent	
Correct Answers	7.4%	30.3%	19.6%	21.4%	2.3%	10.2%	28.0%	15.1%
			21.0%					

possible. The cycling behavior implies that a model to represent the data must have a form of *memory*, i.e., encode sequences of longer length.

3.1 Artificial Neural Networks

Firstly we examined Artificial Neural Networks [30] (ANNs) as means of predicting the throw of the human player in the next round based on their past behavior. Inputs for the ANNs were the last throws by the two players. In all our networks we used a fully connected layer with three output neurons, representing the possible actions, and a *softmax* activation function as last layer before the output to map the output to the range [0,1]. All networks were trained using *Keras* [8]. It builds upon Google's open-source software *Tensorflow* [1]. For faster learning, we use Nvidia's *CUDA* [17] architecture. We compared the different network types described a standard recurrent network (RNN) [22], an implementation of a Long-Short-Term-Memory network (LSTM) [13] and an RNN with gated recurrent units (GRU) [7].[1] The training process was implemented using the *holdout method*, i.e., we randomly split our data set in a training set with 80% of the data, and a validation set containing the remaining 20%. We encoded the actions of 'Rock', 'Paper' and 'Scissors' using a *one-hot encoding* to prevent unwanted dependencies. The same method is used for the user ids added during the input variation later in this section. To find the best set of initial parameters for the model, we used a genetic algorithm [12]. We optimized the parameters on the LSTM and adopted them for all network types. Tested parameter combinations can be found in Table 2, whereas chosen settings are printed in bold. We optimized for prediction accuracy as well as simplicity of the model.

Results and Discussion. Results can be found in Table 4. It is evident that the LSTM and the GRU outperform the simple recurrent network. This also demonstrates the additional dependencies that can be picked up during a game of Rock-Paper-Scissors. The better learning times led to the usage of GRU as our continuation network type, as they converged faster than standard RNN and the LSTM, and yielded slightly better results.

---

[1] Originally, we tried prediction using a standard Feed Forward network, however, this did not converge.

Table 2. Tested configurations, best combination marked in bold.

Parameter	Selections
Neurons	2, 4, 8, 16, 32, 64, **96**, 128, 256
Layers	1, 2, 3, 4, 5
Activation function	Linear, **ReLU**, Tanh, Sigmoid
Optimizer	RMSprop, **Adam**, AdaMax, Nadam
Loss	**Categorical cross-entropy**, Mean squared error

Sequence Learning. In this approach we increased the size of a single input interval by using a variable amount of history steps n. Figure 4 shows the accuracy means for a fixed GRU network with 96 neurons in 1 layer for $n = 1,...,8$.

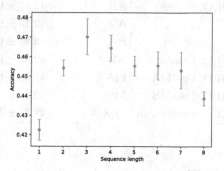

Fig. 4. Accuracy means and standard deviations for varying sequence length, averaged over five random seeds.

Results and Discussion. Accuracy improved for $n > 1$, with the best results for $n = 3$. If training on sequence lengths longer than $n = 3$, then the accuracy starts to drop. This may be due to a larger heterogeneity in the sequences. Another point that could have played into this result, is the fact that most humans are not capable to include information that lies more than two steps back (see e.g., the N-Back Task [9]).

After we found a suitable network structure for our problem, we started to evaluate input information configurations. Due to the randomization during learning, we lost many of the discovered peculiarities in the data. By varying the input features of our networks, we were able to partially prevent this and improve the results considerably. Firstly we changed the length of the input sequence we fed into the network. Adding these sequential dependencies, allowed the network to learn on more, and less contradictory, information. The network performance peaked for $n = 3$ at around 47%. This implied, that history more than 3 turns away did not add more behavioral dependencies, which means that human behavior in a game of RPS, that lies back more than three turns, does not influence the current decision considerably anymore.

Input Variation. Finally, we evaluated different networks by varying their input features. For this purpose we again fixed the network parameters, using a GRU with 96 neurons in 1 layer, as this structure showed to be the most effective, especially considering learning times. We then changed the input from the standard $A(t), B(t)$ to those listed in Table 3 and compare the maximum accuracy reached with this feature combination.

Table 3. Variation of features used as input for the neural networks. A(t) denotes the throw of the human player at time t. B(t) is the throw of the Bot at time t.

Input	Abbreviation	Accuracy (%)
A(t), B(t)	AB	42.17 ± 0.66
A(t)	A	42.15 ± 0.80
B(t)	B	35.97 ± 0.27
A(t), B(t), Gender	ABG	42.90 ± 0.30
A(t), B(t), UserIdA	IdA	47.75 ± 0.30
A(t), B(t), UserIdB	IdB	42.28 ± 0.46
A(t), B(t), UserStatsA	StatsA	49.57 ± 0.46

Results and Discussion. The input of the human player's action alone led to a similar accuracy as providing the input of the human player as well as the Bot's. If only providing the action of the Bot, accuracy drops down to almost chance level. This indicates that the opponent's actions have almost no influence on the choice of actions if you consider all participants at once. The large improvement of the results by adding the unique ID of the user, or other stats, i.e., the user's throw distribution as well as their cycling tendencies, implies that there is a non-negligible difference between users. It allows us to reach a prediction accuracy of almost 50%, marking the highest value for this data set for all used methods and inputs. Results can again be found in Table 4.

3.2   Recommender Systems

We used an adapted version of Recommender Systems (RS) [21], in which we mapped the last n throws for both players to the items to rate. Additionally we had to take into account that the same user can rate the same sequence multiple times with different outputs, because they occurred multiple times in his games and he reacted differently. As there exist exactly $3^2 \cdot n$ different sequences, and similar sequences represent totally different behavior in the game, item-based filtering approaches do not offer valuable results. This left us with user-based collaborative filtering concepts to represent the Recommender methods.

Sequence-Based Approach. In the first approach, we ignored the fact for which user a given sequence occurred, but counted the times a specific output followed this sequence. The one with the most votes was then chosen as predicted next throw. In case of two- or three-way ties, we added this sequence two to three times with the different outputs. Afterwards we verified for each sequence if the predicted result coincided with the actually observed ones, leaving us with a prediction accuracy similar to the ones issued by the neural networks. This technically is still a user-based approach, as we compare how other players react in the same situation. Since it is heavily centered around the given sequence however, we will refer to it as sequence-based in the continuation of this work.

User-Based Approach. In our second approach we addressed the single user more, giving every user exactly one vote. The main difference is that in the sequence-based approach, users who played a large amount of games strongly influenced the voting. This is not the case for this approach. It is important to note that for both recommender approaches we did not use the holdout method to validate the model, as this is not needed for this type.

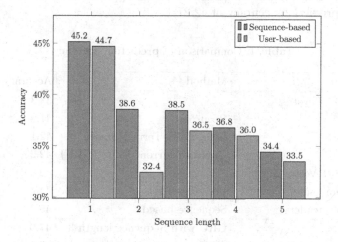

Fig. 5. Performance of recommender systems for varying input length

Results and Discussion. Both recommender approaches (see Fig. 5) achieved a surprisingly good prediction accuracy of around 45% using a single sequence input, indicating that many users react very similar in comparable situations. The sequence-based approach expectedly outperformed the user-based approach in all scenarios. This is not surprising as the amount of votes is significantly higher (thousands compared to exactly 185-1), reducing outliers and improving the inferred prediction accuracy all together. Since the users only voted on other peoples' behavior, this also means that reoccurring sequences in a single user are not taken into account. The drop in accuracy from sequence length 1 to 2 implies that secondary behavior, meaning reactions that do not occur as direct response, are less comparable for different users. This may coincide with the secondary cycling behavior we examined in 2.2, where the cycling probabilities did not show strong tendencies considering the whole group of participants. The even higher drop in the user-based case demonstrates that the same individual seems to repeat behavior while different users show only little affinity. This is somewhat balanced out again for $n \geq 3$, which implies that the secondary cycling behavior is particularly diverse for the different users. The overall descending accuracy for growing input length can again be attributed to the linearly decreasing amount of data, paired with an increase of input combinations by a factor of $2n$. This consequentially leads to a lack of sequence matches.

3.3 Support Vector Machines

We trained a Support Vector Machine (SVM) [25] using *LIBSVM* [6]. By using grid search and cross validation techniques we compared different parameter combinations. The best results were yielded by a radial basis function kernel. Results can again compared against the other approaches in Table 4. The SVM achieved a prediction accuracy of 42.27%.

Table 4. Comparison of prediction accuracy.

System	Method	Accuracy (%)
Chance Level		33.3
ANN	Recurrent	41.9
ANN	Long-Short-Term Memory	42.1
ANN	Gated Recurrent Units (GRU)	42.2
Support Vector Machine		42.3
Recommender	User-based	44.7
Recommender	Sequence-based	45.2
ANN	GRU with Sequence length 3	47.0
ANN	GRU with User ID	47.8
ANN	**GRU with User Statistics**	**49.6**

4   Discussion and Conclusion

This article focuses on analyzing, if human behavior in games can be predicted using computational tools. In order to examine this, we conducted experimental studies. On the basis of 28841 rounds of RPS versus probabilistic bots, we were able to predict the players' decision-making at an accuracy that surpasses state-of-the-art approaches. Evaluating the data, we were able to detect significant deviations from rational behavior within our test persons. We were able to replicate characteristics found in prior studies, such as the biased choice of throw and the primary and secondary cycling behavior of humans. During evaluation of the primary cycle behavior we noticed a bias in the cycling behavior of the bots. We managed to show that many of our users were able to detect this deviation from rationality and counter it accordingly, or even sometimes overcompensated this behavior. We then used different machine learning methods on the task of predicting the players – neural networks, support vector machines and collaborative filtering approaches. Formalizing RPS as a supervised learning problem held several difficulties. Mainly, the input feature count is relatively small due to the simple structure of the game. This can lead to contradictory mappings which limit the prediction accuracy which the different methods can achieve.

We were able to support the claim that the average player did not identify the opposing bot's strategy in their games by adding said information to the network inputs. This also implied that the opponent's choice of action did not have as big of an impact on a player's decision making as expected. The overall best accuracy was achieved by adding the unique ID of the user (47.7%), or when adding additional stats about the players, such as their throw preferences or cycling behavior (49.5%). Compared to the win rates of the state of the art, which do not exceed 40%, a RPS playing system equipped with these models could well beat a human player in over 50% of games. The experiment showed that single users exhibit strong reoccurring patterns. It is therefore necessary to tailor machine learning approaches to handle small data on single users.

Acknowledgements. This paper was supported by DFG grants RA 1934/3-1, RA 1934/2-1, RA 1934/4-1, and RA 1934/9-1 to MR.

References

1. Abadi, M., et al.: TensorFlow: large-scale machine learning on heterogeneous systems (2015). https://www.tensorflow.org/
2. Ali, F.F., Nakao, Z., Chen, Y.W.: Playing the rock-paper-scissors game with a genetic algorithm. In: Proceedings of the Congress on Evolutionary Computation, vol. 1, pp. 741–745 (2000)
3. Allais, M.: L'extension des théories de l'équilibre économique général et du rendement social au cas du risque. Econometrica, J. Econ. Soc. 269–290 (1953)
4. Bayern, S.: Rock, paper, scissors: humans against AI (2001). http://www.essentially.net/rsp/index.jsp
5. Bowling, M., Veloso, M.: Rational and convergent learning in stochastic games. In: Lang, J. (ed.) International Joint Conference on Artificial Intelligence, vol. 17, pp. 1021–1026. Lawrence Erlbaum Associates Ltd. (2001)
6. Chang, C.C., Lin, C.J.: LIBSVM: a library for support vector machines. ACM Trans. Intell. Syst. Technol. **2**, 27:1–27:27 (2011). http://www.csie.ntu.edu.tw/~cjlin/libsvm
7. Cho, K., Van Merriënboer, B., Bahdanau, D., Bengio, Y.: On the properties of neural machine translation: encoder-decoder approaches. arXiv preprint arXiv:1409.1259 (2014)
8. Chollet, F., et al.: Keras (2015). https://github.com/fchollet/keras

9. Cohen, J.D., et al.: Temporal dynamics of brain activation during a working memory task. Nature **386**(6625), 604 (1997)

10. Dyson, B.J., Wilbiks, J.M.P., Sandhu, R., Papanicolaou, G., Lintag, J.: Negative outcomes evoke cyclic irrational decisions in rock, paper, scissors. Sci. Rep. **6**, 20479 (2016)

11. Epstein, S., Lipson, A., Holstein, C., Huh, E.: Irrational reactions to negative outcomes: evidence for two conceptual systems. J. Pers. Soc. Psychol. **62**(2), 328 (1992)

12. Harvey, M.: Let's evolve a neural network with a genetic algorithm. Coastline automation, posted on April (2017)

13. Hochreiter, S., Schmidhuber, J.: Long short-term memory. Neural Comput. **9**(8), 1735–1780 (1997)

14. Lee, D., McGreevy, B.P., Barraclough, D.J.: Learning and decision making in monkeys during a rock-paper-scissors game. Cognit. Brain Res. **25**(2), 416–430 (2005)

15. Lopes, L.L., Oden, G.C.: Distinguishing between random and nonrandom events. J. Exp. Psychol.: Learn. Mem. Cognit. **13**(3), 392 (1987)

16. Mookherjee, D., Sopher, B.: Learning behavior in an experimental matching pennies game. Games Econ. Behav. **7**(1), 62–91 (1994)

17. Nickolls, J., Buck, I., Garland, M., Skadron, K.: Scalable parallel programming with cuda. Queue **6**(2), 40–53 (2008)

18. Panumate, C., Iida, H., Terrillon, J.C.: A game informatical analysis of RoShamBo (2016)

19. Pomerleau, N.: Rock paper scissors. https://www.neilpomerleau.com/posts/wp-content/uploads/rps-report.pdf

20. Pozzato, G., Michieletto, S., Menegatti, E.: Towards smart robots: rock-paper-scissors gaming versus human players. In: PAI@ AI* IA, pp. 89–95. Citeseer (2013)

21. Resnick, P., Varian, H.R.: Recommender systems. Commun. ACM **40**(3), 56–59 (1997)

22. Rumelhart, D.E., Hinton, G.E., Williams, R.J.: Learning internal representations by error propagation. Technical report. California Univ San Diego La Jolla Inst for Cognitive Science (1985)

23. Shannon, C.E.: A mind-reading machine. Bell Laboratories memorandum (1953)

24. Sutiono, A.P., Ramadan, R., Jarukasetporn, P., Takeuchi, J., Purwarianti, A., Iida, H.: A mathematical model of game refinement and its applications to sports games. EAI Endorsed Trans. Creat. Technol. **15**, 1–7 (2015)

25. Vapnik, V.: The support vector method of function estimation. In: Suykens, J.A.K., Vandewalle, J. (eds.) Nonlinear Modeling, pp. 55–85. Springer, Boston (1998). https://doi.org/10.1007/978-1-4615-5703-6_3

26. Wagenaar, W.A.: Generation of random sequences by human subjects: a critical survey of literature. Psychol. Bull. **77**(1), 65 (1972)

27. Wang, Z., Xu, B., Zhou, H.J.: Social cycling and conditional responses in the rock-paper-scissors game. arXiv preprint arXiv:1404.5199 (2014)

28. Weibel, D., Wissmath, B., Habegger, S., Steiner, Y., Groner, R.: Playing online games against computer-vs. human-controlled opponents: effects on presence, flow, and enjoyment. Comput. Human Behav. **24**(5), 2274–2291 (2008)

29. Xu, B., Zhou, H.J., Wang, Z.: Cycle frequency in standard rock-paper-scissors games: evidence from experimental economics. Physica A: Stat. Mech. Appl. **392**(20), 4997–5005 (2013)

30. Zurada, J.M.: Introduction to Artificial Neural Systems, vol. 8. West Publishing Company, St. Paul (1992)

Strategy-Proofness, Envy-Freeness and Pareto Efficiency in Online Fair Division with Additive Utilities

Martin Aleksandrov[✉] and Toby Walsh

Technical University Berlin, Berlin, Germany
{martin.aleksandrov,toby.walsh}@tu-berlin.de

Abstract. We consider fair division problems where indivisible items arrive one by one in an online fashion and are allocated immediately to agents who have additive utilities over these items. Many existing offline mechanisms do not work in this online setting. In addition, many existing axiomatic results often do not transfer from the offline to the online setting. For this reason, we propose here three *new* online mechanisms, as well as consider the axiomatic properties of three previously proposed online mechanisms. In this paper, we use these mechanisms and characterize classes of online mechanisms that are strategy-proof, and return envy-free and Pareto efficient allocations, as well as combinations of these properties. Finally, we identify an important impossibility result.

Keywords: Online fair division · Strategy-proofness · Envy-freeness · Pareto efficiency · Additive utilities

1 Introduction

Fair division is an important problem facing our society today as increasing economical, environmental, and other pressures require us to try to do more with limited resources. An especially challenging form of fair division is when we are allocating available resources in an *online* fashion with only partial knowledge of the future resources and agent's preferences for these resources. There are many applications of online fair division for *social good*. For example, when a kidney is donated, it must be allocated to a patient within a few hours. As a second example, food items arrive at a food bank and must be allocated and distributed to charities promptly. As a third example, when allocating charging slots to electric cars, we may not know when or where cars will arrive for charging. As a fourth example, when managing a river, we might start allocating irrigation water to farmers today, not knowing how much it will rain the next month. As a fifth example, when allocating memory to cloud services, we may not know what and how many services are requested in the next moment.

Funded by the European Research Council under the Horizon 2020 Programme via AMPLify 670077.

© Springer Nature Switzerland AG 2019
A. C. Nayak and A. Sharma (Eds.): PRICAI 2019, LNAI 11670, pp. 527–541, 2019.
https://doi.org/10.1007/978-3-030-29908-8_42

The online nature of such fair division problems changes the mechanisms available to allocate items. For example, with the well-known (offline) *sequential allocation* mechanism, agents pick their most preferred remaining items in turns. In an online setting, an agent's most preferred item may not be currently (or even ever) available. To tackle this, we propose three *new* - ONLINE SERIAL DICTATOR, ONLINE RANDOM PRIORITY and PARETO LIKE - as well as study three existing - LIKE, BALANCED LIKE and MAXIMUM LIKE- online mechanisms. The online nature also means we may need to consider *new* axiomatic properties. For example, in deciding if agents have any incentive to misreport preferences in an online setting, we may consider the past fixed but the future unknown. This leads to a *new* and weaker form of *online strategy-proofness* (OSP). Therefore, it might be easier to achieve strategy-proofness in an online than in an offline setting. Also, we give a *new* and stronger form of envy-freeness, called *shared envy-freeness* (SEF), in which agents might be envious of each other but only over the items that they like in common. For example, in the paper assignment problem, reviewers tend to bid for papers in their field of expertise and not for papers outside this field [13]. In this context, SEF guarantees envy-freeness across the different fields.

We provide characterization results for strategy-proofness (SP), envy-freeness (EF) and Pareto efficiency (PE). For example, we characterize completely the class of online mechanisms that are SP, and the class of online mechanisms that are PE ex post. We also characterize the class of SP and EF mechanisms. Thus, a mechanism for online fair division is SP and EF ex ante iff it returns the same random assignment as LIKE. The same holds for SEF ex ante mechanisms. Also, we prove that a mechanism is SP, PE ex post and EF ex ante iff it returns the same probability distribution of allocations as ONLINE RANDOM PRIORITY. We further give an important impossibility result. In offline fair division, stochastic Pareto efficiency and envy-freeness are always possible simultaneously (e.g. the probabilistic serial mechanism [5]). However, we prove that no online mechanism can be both Pareto efficient ex ante and envy-free ex ante.

2   Related Work

We consider the model of online fair division from [17] in which items are indivisible and arrive one-by-one over time. We primarily contrast our characterization results with similar results in (offline) fair division. For example, we prove that *no* online mechanism can be both PE and EF ex ante. By comparison, the (offline) probabilistic serial mechanism satisfies both stochastic PE and EF [5]. In fact, it follows from our results that there could be an unbounded number of mechanisms that are just PE ex ante or EF ex ante. We can show that other (offline) characterizations (e.g. [6,14]) break in the online setting as well. By comparison, as online mechanisms can be applied to offline problems by picking a sequence of the items, our results can be mapped into such settings. For example, our PARETO LIKE mechanism returns all possible PE ex post allocations in the offline problem. As a result, this mechanism characterizes the set of offline such mechanisms. As another example, we prove that ONLINE RANDOM PRIORITY is SP and PE ex post, but not PE ex ante. With this mechanism, agents with the same cardinal utilities receive the same expected utilities (i.e. it is symmetric). This is in-line with the impossibility result that *no* (offline or online) mechanism for offline matching is SP, PE ex ante and symmetric [19]. Yet more related results are shown in many other fair division (e.g. [7,10,12,16]), voting (e.g. [11,18]) and kidney exchange (e.g. [8,9]) settings. Our results can also be mapped to such settings.

3 Online and Additive Fair Division

An online fair division *instance* consists of a set of *agents* $N = \{1, \ldots, n\}$, and an ordered set of indivisible *items* $O = \{o_1, \ldots, o_m\}$. We suppose that item o_j arrives at round j when each agent $i \in N$ becomes aware of their sincere *utility* $u_{ij} \in \mathbb{R}_{\geq 0}$ and places a possibly strategic *bid* $v_{ij} \in \mathbb{R}_{\geq 0}$ for o_j. We suppose at least one agent has positive utility for every item as, otherwise, we can simply discard the item. We use *online* mechanisms that allocate o_j immediately, supposing the allocation of o_1 to o_{j-1} is fixed and there is *no information* of o_{j+1} to o_m. We consider only *non-wasteful* mechanisms that share the probability of 1 for o_j only among agents that bid positively for it if there is at least one such agent and, otherwise, discard o_j.

An *allocation* π_j of o_1 to o_j gives a bundle of items π_{ji} to each agent $i \in N$ such that $\bigcup_{i \in N} \pi_{ji} = \{o_1, \ldots, o_j\}$ and $\pi_{ji} \cap \pi_{jk} = \emptyset$ for each $i \neq k$. We write $u_{ik}(\pi_j)$ for the *utility* of agent $i \in N$ for π_{jk}. We write $u_i(\pi_j)$ for $u_{ii}(\pi_j)$. A mechanism induces a probability distribution over the set Π_j of all allocations of items o_1 to o_j. We write $\overline{u}_{ik}(\Pi_j)$ for the *expected utility* of agent $i \in N$ for the expected allocation of agent $k \in N$ and $p_{ik}(\Pi_j)$ for the *probability* of agent $i \in N$ for item o_k in this distribution. We write $\overline{u}_i(\Pi_j)$ for $\overline{u}_{ii}(\Pi_j)$ and $p_i(\Pi_j)$ for $p_{ij}(\Pi_j)$. We suppose *additive* utilities and expected utilities.

$$u_{ik}(\pi_j) = \sum_{o_h \in \pi_{jk}} u_{ih} \qquad \overline{u}_{ik}(\Pi_j) = \sum_{h=1}^{j} p_{kh}(\Pi_j) \cdot u_{ih}$$

We consider three common properties of mechanisms: strategy-proofness, envy-freeness and Pareto efficiency.

Definition 1. *(SP) A mechanism is* strategy-proof (SP) *if, for each instance with $m \in \mathbb{N}$ items, no agent $i \in N$ can strictly increase $\overline{u}_i(\Pi_m)$ by reporting any sequence $v_{i1}, \ldots, v_{im}$ other than $u_{i1}, \ldots, u_{im}$, supposing all other agents bid sincerely for items o_1 to o_m.*

Definition 2. *(EF) A mechanism is* envy-free ex post (EFP) *iff, for each instance with $m \in \mathbb{N}$ items and allocation $\pi_m \in \Pi_m$ returned by the mechanism with positive probability, $\forall i, k \in N : u_{ii}(\pi_m) \geq u_{ik}(\pi_m)$. A mechanism is* envy-free ex ante (EFA) *iff, for each instance with $m \in \mathbb{N}$ items, $\forall i, k \in N : \overline{u}_{ii}(\Pi_m) \geq \overline{u}_{ik}(\Pi_m)$.*

Definition 3. *(PE) A mechanism is* Pareto efficient ex post (PEP) *iff, for each instance with $m \in \mathbb{N}$ items and allocation $\pi_m \in \Pi_m$ returned by the mechanism with positive probability, no $\pi'_m \in \Pi_m$ is such that $\forall i \in N : u_i(\pi'_m) \geq u_i(\pi_m)$ and $\exists k \in N : u_k(\pi'_m) > u_k(\pi_m)$. Also, it is* Pareto efficient ex ante (PEA) *iff, no mechanism gives at least $\bar{u}_i(\Pi_m)$ to each $i \in N$ and more than $\bar{u}_k(\Pi_m)$ to some $k \in N$.*

To characterize SP, EF and PE mechanisms, we will use two equivalence relations between outcomes of mechanisms. We say that two mechanisms are *ex ante equivalent* iff, for each instance of $m \in \mathbb{N}$ items, agent $i \in N$ and item $o_j \in O$, the probabilities of i for o_j under both mechanisms are equal, whilst these mechanisms are *ex post equivalent* iff, for each instance of $m \in \mathbb{N}$ items and allocation $\pi_m \in \Pi_m$, the probabilities of π_m under both mechanisms are equal (i.e. each of the two mechanisms returns an identical distribution of allocations).

4   Six Cardinal Mechanisms

Many offline mechanisms cannot be used in the online setting because only one item is available at any time. For this reason, we propose three *new* as well as study three existing online mechanisms. For every arriving item o_j, each mechanism first computes a set of agents feasible for o_j given an allocation $\pi_{j-1} \in \Pi_{j-1}$. An agent that is feasible for o_j then receives it with *conditional probability* that is uniform with respect to the other agents that are feasible for o_j. Thus, for the first j items, each mechanism returns a probability distribution over Π_j and an actual allocation with some positive probability that is obtained as a product of j conditional randomizations.

- ONLINE SERIAL DICTATOR: it has a strict priority order σ of the agents prior to round one, and the unique feasible agent for o_j is the first agent in σ that bids positively for o_j.
- ONLINE RANDOM PRIORITY: it draws uniformly at random a strict priority order σ of the agents prior to round one, and runs ONLINE SERIAL DICTATOR with it.
- PARETO LIKE: agent $i \in N$ is feasible for o_j if extending π_{j-1} by allocating o_j to i is Pareto efficient ex post.
- LIKE: agent $i \in N$ is feasible for o_j if $v_{ij} > 0$ [2].
- BALANCED LIKE: agent $i \in N$ is feasible for o_j if $v_{ij} > 0$ and i has the fewest items in π_{j-1} among those with positive bids for o_j [2].
- MAXIMUM LIKE: agent $i \in N$ is feasible for o_j if $v_{ij} = \max_{k \in N} v_{kj}$ [3].

In Example 1, we demonstrate that these mechanisms may return distributions of allocations that are different from each other.

Example 1. Let us consider an instance with $N = \{1, 2\}$ and $O = \{o_1, o_2\}$. The utilities of agents for items are given in the below table.

	item o_1	item o_2
agent 1	1	2
agent 2	2	1

In this instance, supposing sincere bidding, there are 4 possible allocations: $\pi^1 = (\{o_1, o_2\}, \emptyset)$, $\pi^2 = (\emptyset, \{o_1, o_2\})$, $\pi^3 = (\{o_1\}, \{o_2\})$, and $\pi^4 = (\{o_2\}, \{o_1\})$. ONLINE SERIAL DICTATOR with fixed $\sigma = (1, 2)$ returns π^1 with probability 1, ONLINE RANDOM PRIORITY returns π^1 and π^2 with probabilities $1/2$, PARETO LIKE returns π^1 with probability $1/2$, π^2 and π^4 with probabilities $1/4$, LIKE returns π^1 to π^4 with probabilities $1/4$, BALANCED LIKE returns π^3 and π^4 with probabilities $1/2$, and MAXIMUM LIKE returns π^4 with probability 1. $\qquad\square$

We note that the ONLINE SERIAL DICTATOR mechanism is similar to the (offline) *serial dictatorship* mechanism [15]. However, agents have no quota on the number of items they receive with ONLINE SERIAL DICTATOR, and only take items for which they declare non-zero utility. The ONLINE RANDOM PRIORITY mechanism is also similar to the (offline) *random priority* mechanism [1]. Finally, the LIKE mechanism can be seen as the online analog of the (offline) *probabilistic serial* mechanism (see [5]) with agents "eating" each next item which they like.

5   Strategy-Proofness

We begin by considering strategic behavior of agents. We provide a simple characterization of mechanisms that are strategy-proof. For $i \in N$, we say that $p_i(\Pi_j)$ is a *step* function iff it is 0 if $v_{ij} = 0$ and it admits the same value for any bid $v_{ij} > 0$ supposing the bids of the other agents for o_1 to o_j, and the bids of agent i for o_1 to o_{j-1} are fixed. A mechanism is a *step* mechanism iff, for each instance with $m \in \mathbb{N}$ items, $i \in N$ and $o_j \in O$, $p_i(\Pi_j)$ is a step function. For $i \in N$, we say that $p_i(\Pi_j)$ is a *memoryless* function iff it takes the same value for all possible bids v_{i1} to $v_{i(j-1)}$ of agent i for items o_1 to o_{j-1} given fixed bid v_{ij} of agent i for item o_j and fixed bids of the other agents for items o_1 to o_j. A mechanism is a *memoryless* mechanism iff, for each instance with $m \in \mathbb{N}$ items, $i \in N$ and $o_j \in O$, $p_i(\Pi_j)$ is a memoryless function.

With a step mechanism, $p_i(\Pi_j)$ does not depend on the size of an agent's non-zero bid for item o_j but it may depend on the allocation history. By comparison, with a memoryless mechanism, $p_i(\Pi_j)$ may depend on the size of their non-zero bid for item o_j but not on the allocation history. As a consequence, with a memoryless step mechanism, $p_i(\Pi_j)$ depends only on the combination of the non-zero bids for item o_j.

Theorem 1. *A non-wasteful mechanism for online fair division is strategy-proof iff it is a memoryless step mechanism.*

Proof. Pick $i \in N$ in an instance. Let us view $\overline{u}_i(\Pi_j)$ and $p_i(\Pi_j)$ as functions of v_{i1} to v_{ij}. That is, we write $\overline{u}_i(\Pi_j) = \overline{u}_i(v_{i1}, \ldots, v_{ij})$ and $p_i(\Pi_j) = p_i(v_{i1}, \ldots, v_{ij})$. Consider a memoryless step mechanism. Suppose now that all agents bid sincerely. Then, $\overline{u}_i(u_{i1}, \ldots, u_{im}) = \sum_{j=1}^{m} p_i(u_{i1}, \ldots, u_{ij}) \cdot u_{ij}$. Suppose next that only i bids strategically v_{i1} to v_{im}. Then, $\overline{u}_i(v_{i1}, \ldots, v_{im}) = \sum_{j=1}^{m} p_i(v_{i1}, \ldots, v_{ij}) \cdot u_{ij}$. For each o_j with $v_{ij} = u_{ij}$, $p_i(v_{i1}, \ldots, v_{ij}) \cdot u_{ij} = p_i(u_{i1}, \ldots, u_{ij}) \cdot u_{ij}$ as the mechanism is a memoryless step. For each o_j with $v_{ij} > 0$ and $u_{ij} = 0$, $p_i(v_{i1}, \ldots, v_{ij}) \cdot u_{ij} = p_i(u_{i1}, \ldots, u_{ij}) \cdot u_{ij} = 0$. For each o_j with $v_{ij} = 0$ and $u_{ij} > 0$, $p_i(v_{i1}, \ldots, v_{ij}) \cdot u_{ij} = 0$ and $p_i(u_{i1}, \ldots, u_{ij}) \cdot u_{ij} \geq 0$ as the mechanism is non-wasteful. Consequently, the mechanism is strategy-proof.

Consider a strategy-proof mechanism. First, assume that it is not a step and $p_i(u_{i1}, \ldots, u_{i(j-1)}, v_{ij})$ admits different values for different positive values of v_{ij} supposing that the bids of other agents for items o_1 to o_j are fixed. WLOG, we can suppose that item o_j is the last item to arrive. We can also suppose $u_{ij} > 0$ as the case $u_{ij} = 0$ is trivial. Agent i has an incentive to report $v_{ij} > u_{ij}$ (or $v_{ij} < u_{ij}$) and, thus, strictly increase $p_i(u_{i1}, \ldots, u_{i(j-1)}, u_{ij})$ and $\overline{u}_i(u_{i1}, \ldots, u_{i(j-1)}, u_{ij})$. Second, assume that the mechanism is a step but not memoryless. Suppose that agent i gets different probabilities for item o_j for alternative bids v_{ik} compared to their sincere bids u_{ik} with $k < j$. WLOG, for each o_k with $k < j$, we suppose that $p_i(v_{i1}, \ldots, v_{ik}) = p_i(u_{i1}, \ldots, u_{ik})$. Otherwise, we truncate the problem to the first such round j. WLOG, we also suppose that $p_i(v_{i1}, \ldots, v_{i(j-1)}, u_{ij}) > p_i(u_{i1}, \ldots, u_{i(j-1)}, u_{ij})$. Otherwise, we swap v_{ik} for u_{ik} for $k < j$. We let agent i have utility 1 for all items except o_j and utility j for o_j. Thus, the bids v_{ik} increase the expected utility of agent i compared to the bids u_{ik}. We reached contradictions under both assumptions.   □

The LIKE mechanism is a memoryless step and so is strategy-proof. We observe that the ONLINE SERIAL DICTATOR and ONLINE RANDOM PRIORITY mechanisms are also memoryless steps and, hence, are also both strategy-proof. On the other hand, the BALANCED LIKE mechanism is just a step mechanism and is neither memoryless nor strategy-proof. Furthermore, the MAXIMUM LIKE mechanism is only memoryless and the PARETO LIKE mechanism is neither a step nor a memoryless mechanism. Consequently, these two mechanisms are not strategy-proof.

Thus far, we have made the strong assumption that an agent has complete knowledge of any future items. In practice, agents may have limited or even no knowledge about the future. We next capture this formally in terms of a definition of a weaker form of strategy-proofness.

Definition 4. *(OSP) A mechanism is* online strategy-proof (OSP) *if, for each instance with $m \in \mathbb{N}$ items and $j \in \{1, \ldots, m\}$, no agent $i \in N$ can strictly increase $\overline{u}_i(\Pi_j)$ by reporting any bid v_{ij} other than u_{ij}, supposing agent i bids sincerely for o_1 to o_{j-1} and all other agents bid sincerely for items o_1 to o_j.*

Indeed, it is harder for an agent to benefit from a strategic bidding with only partial information of the future. For this reason, many mechanisms that are not strategy-proof are online strategy-proof. For example, the BALANCED LIKE mechanism is online strategy-proof with no knowledge of future items, but stops being strategy-proof with complete knowledge of these future items even if all utilities are just 0 or 1 [2]. In the other direction, it is easy to show that a mechanism that is strategy-proof is also online strategy-proof. The reason for this is simple. If an agent cannot increase their expected utility by misreporting their utilities for any subset of items, then they cannot do it by misreporting their utility for any individual item, including the last one. We give a simple characterization of mechanisms that are online strategy-proof.

Theorem 2. *A non-wasteful mechanism for online fair division is online strategy-proof iff it is a step mechanism.*

Proof. We show the "if" direction. Suppose the mechanism is a step. Consider an instance, an agent $i \in N$ and an item o_j. The allocation of this item does not have an impact on the allocation of earlier items as this is now fixed. If $u_{ij} > 0$, then agent i has no incentive to report 0 for it as their expected utility can only decrease, and also has no incentive to report any positive value $v_{ij} \neq u_{ij}$ as their probability for item o_j is a step function. If $u_{ij} = 0$, then agent i has no incentive to report $v_{ij} > 0$ as their expected utility cannot increase. Hence, i cannot increase $\overline{u}_i(\Pi_j)$. The mechanism is online strategy-proof. We next sketch the "only if" direction. Suppose the mechanism is not a step. The result follows by the second part of the proof of Theorem 1. $\square$

It follows immediately that the ONLINE SERIAL DICTATOR, ONLINE RANDOM PRIORITY, LIKE and BALANCED LIKE mechanisms are all online strategy-proof. In contrast, the MAXIMUM LIKE and PARETO LIKE mechanisms are not as they are not steps and agents have an incentive to report a larger bid for an item.

To sum up, we might use the ONLINE SERIAL DICTATOR, ONLINE RANDOM PRIORITY, or LIKE mechanism for strategy-proofness with complete information. However, for online strategy-proofness with no information about future items, we can also use the BALANCED LIKE mechanism.

6   Envy-Freeness

We continue with envy-freeness. We suppose agents bid sincerely. This might be because we use a mechanism that is strategy-proof or online strategy-proof. There is *no* envy-free ex post mechanism [2]. We, therefore, mainly focus on fairness in expectation. Uncertainty about the future means that envy-freeness ex ante is now harder to achieve than in the offline setting. Nevertheless, it is always *possible* as the LIKE mechanism is envy-free ex ante.

By Example 1, the ONLINE RANDOM PRIORITY and LIKE mechanisms can return different ex post allocations. Nevertheless, they are ex ante equivalent and, therefore, envy-free ex ante. Unfortunately, ex ante equivalence to the LIKE mechanism only provides a partial characterization as there is an unbounded number of envy-free ex ante mechanisms that are *not* ex ante equivalent to it. We show this in Example 2.

Example 2. Let us consider the fair division of items o_1 and o_2 to agents 1 and 2 with utilities as follows: $u_{11} = 1$, $u_{12} = 1$, $u_{21} = 0$ and $u_{22} = 1$. Further, consider the mechanism that works as LIKE on each instance except on this one in which it gives item o_2 to agent 2 with some probability in $(1/2, 1]$. This mechanism is envy-free ex ante but it is not ex ante equivalent to LIKE.   □

In Example 2, the mechanism is neither memoryless, nor a step. Therefore, by Theorem 1, it is not strategy-proof. However, we can give a complete characterization of *all* strategy-proof and envy-free ex ante mechanisms.

Theorem 3. *A non-wasteful mechanism for online fair division is strategy-proof and envy-free ex ante iff it is ex ante equivalent to the LIKE mechanism.*

Proof. If a mechanism is ex ante equivalent to LIKE, then it is envy-free ex ante and a memoryless step by the definition of LIKE. By Theorem 2, the mechanism is strategy-proof. If a mechanism is envy-free ex ante and strategy-proof, then it is a memoryless step. We show that it is ex ante equivalent to LIKE by induction on the round number j. In the base case, the mechanism is clearly ex ante equivalent to LIKE. In the step case, suppose that the mechanism is ex ante equivalent to LIKE for items o_1 to o_{j-1} (i.e. hypothesis) but not for item o_j. That is, there are two agents $i, k \in N$ that like item o_j with $p_i(\Pi_j) < p_k(\Pi_j)$. As the mechanism is envy-free ex ante up to round $(j-1)$, we have that $\overline{u}_{ii}(\Pi_{j-1}) \geq \overline{u}_{ik}(\Pi_{j-1})$. As the mechanism is memoryless step, we can suppose that $u_{ij} = 1 - (\overline{u}_{ik}(\Pi_{j-1}) - \overline{u}_{ii}(\Pi_{j-1}))/(p_k(\Pi_j) - p_i(\Pi_j)) > 0$. We, hence, obtain that $\overline{u}_{ik}(\Pi_{j-1}) - \overline{u}_{ii}(\Pi_{j-1}) + (p_k(\Pi_j) - p_i(\Pi_j)) \cdot u_{ij} > 0$, or i envies ex ante k for o_1 to o_j. This contradicts the fact that the mechanism is envy-free ex ante up to round j. Consequently, $p_i(\Pi_j) = p_k(\Pi_j)$. The result follows.   □

We can give similar results if we weaken strategy-proof mechanisms to memoryless or step mechanisms. We omit these proofs for reasons of space.

Proposition 1. *A step mechanism for online fair division is envy-free ex ante iff it is ex ante equivalent to the LIKE mechanism.*

Proposition 2. *A memoryless mechanism for online fair division is envy-free ex ante iff it is ex ante equivalent to the LIKE mechanism.*

On a restricted preference domain, the LIKE mechanism characterizes all envy-free ex ante mechanisms, even without the assumption of strategy-proofness. The following result applies to common domains of positive cardinal, identical cardinal, identical ordinal, Borda (e.g. $1, 2, \ldots, m$) or lexicographic (e.g. $2^0, 2^1, \ldots, 2^m$) utilities. This result holds for *wasteful* (i.e. not non-wasteful) mechanisms as well.

Theorem 4. *With non-zero cardinal utilities, a mechanism for online fair division is envy-free ex ante iff it is ex ante equivalent to the LIKE mechanism.*

Proof. We first show the "if" direction. If a mechanism is ex ante equivalent to LIKE, then it is envy-free ex ante as LIKE. We next show the "only if" direction. The proof is by induction as in Theorem 3. In the step case, we consider $i, k \in N$

that like o_j. We have that $\overline{u}_{ii}(\Pi_{j-1}) = \overline{u}_{ik}(\Pi_{j-1})$ and $\overline{u}_{kk}(\Pi_{j-1}) = \overline{u}_{ki}(\Pi_{j-1})$ as the cardinal utilities are non-zero and the mechanism is ex ante equivalent to LIKE for o_1 to o_{j-1} by the hypothesis. Hence, $p_i(\Pi_j) = p_k(\Pi_j)$ as the mechanism is envy-free ex ante up to round j. □

We can also completely characterize a stronger notion of envy-freeness even with general utilities. Shared envy-freeness requires that each pair of agents are envy-free of each other only over the items that both agents in the pair like in common. We write $u_{ik}^{\mathrm{SEFP}}(\pi_j)$ for the utility of agent $i \in N$ over the items in π_{ji} that both agents i and $k \in N$ like. We write $\overline{u}_{ik}^{\mathrm{SEFA}}(\Pi_j)$ for the expected utility of agent $i \in N$ over the items o_1 to o_j that both agents i and $k \in N$ like.

$$u_{ik}^{\mathrm{SEFP}}(\pi_j) = \sum_{\substack{o_h \in \pi_{ji} \\ u_{kh} > 0}} u_{ih} \qquad \overline{u}_{ik}^{\mathrm{SEFA}}(\Pi_j) = \sum_{\substack{h=1 \\ u_{kh} > 0}}^{j} p_{ih}(\Pi_j) \cdot u_{ih}$$

We note $u_{ik}^{\mathrm{SEFP}}(\pi_j) \leq u_{ii}(\pi_j)$ and $\overline{u}_{ik}^{\mathrm{SEFA}}(\Pi_j) \leq \overline{u}_{ii}(\Pi_j)$. A mechanism is *shared envy-free ex post (SEFP)* iff, for each instance with $m \in \mathbb{N}$ items and allocation $\pi_m \in \Pi_m$ returned by the mechanism with positive probability, $\forall i, k \in N : u_{ik}^{\mathrm{SEFP}}(\pi_m) \geq u_{ik}(\pi_m)$. A mechanism is *shared envy-free ex ante (SEFA)* iff, for each instance of $m \in \mathbb{N}$ items, $\forall i, k \in N : \overline{u}_{ik}^{\mathrm{SEFA}}(\Pi_m) \geq \overline{u}_{ik}(\Pi_m)$. Shared envy-freeness coincides with envy-freeness with non-zero cardinal utilities. For this reason, shared envy-freeness is only possible in expectation.

Theorem 5. *A non-wasteful mechanism for online fair division is shared envy-free ex ante iff it is ex ante equivalent to the LIKE mechanism.*

Proof. If a mechanism is ex ante equivalent to LIKE, then it is envy-free ex ante. Every pair of agents receive each of their commonly liked item with the same probability. The mechanism is, therefore, shared envy-free ex ante. If a mechanism is shared envy-free ex ante, then the proof resembles the one of Theorem 3. In the step case, we consider round j and agents i, k that like item o_j. WLOG, assume that the mechanism is not ex ante equivalent to LIKE for item o_j and $p_i(\Pi_j) < p_k(\Pi_j)$. By the hypothesis, the mechanism is ex ante equivalent to LIKE up to round $(j-1)$. Hence, $\overline{u}_{ik}(\Pi_{j-1}) = \overline{u}_{ik}^{\mathrm{SEFA}}(\Pi_{j-1})$ and $\overline{u}_{ki}(\Pi_{j-1}) = \overline{u}_{ki}^{\mathrm{SEFA}}(\Pi_{j-1})$. As the mechanism is shared envy-free ex ante up to round j, $p_i(\Pi_j) = p_k(\Pi_j)$. This contradicts our assumption. □

If we limit ourselves to 0/1 utilities, we say that a mechanism is *bounded envy-free ex post with 1 (BEFP)* iff, for each instance of $m \in \mathbb{N}$ items and $\pi_m \in \Pi_m$ returned by the mechanism with positive probability, $\forall i, k \in N : u_{ii}(\pi_m) + 1 \geq u_{ik}(\pi_m)$. For example, the BALANCED LIKE mechanism is bounded envy-free ex post with 1 [2]. In fact, we can immediately conclude the following partial characterization.

Corollary 1. *With 0/1 cardinal utilities, a non-wasteful mechanism for online fair division is bounded envy-free ex post with 1 if it returns a subset of the allocations returned by the BALANCED LIKE mechanism.*

Benade et al. [4] showed that the random assignment of each next item (i.e. LIKE) is asymptotically optimal in the ex post sense, with a bound of the (maximum) envy that increases as the number of rounds increases. Unfortunately, this means that we cannot put any trivial bound on the envy ex post in general.

To sum up, we can use the LIKE or ONLINE RANDOM PRIORITY mechanism if we want envy-freeness ex ante. With 0/1 utilities, we can bound the ex post envy between agents to at most one unit of utility with the BALANCED LIKE mechanism which also happens to be envy-free ex ante in this domain [2].

7   Pareto Efficiency

We consider lastly Pareto efficiency supposing agents act sincerely. With 0/1 utilities, each mechanism is Pareto efficient as the sum of agents' utilities in each returned allocation is m. This is not true in general. We start with Pareto efficiency ex post. The ONLINE SERIAL DICTATOR, ONLINE RANDOM PRIORITY and MAXIMUM LIKE mechanisms are all Pareto efficient ex post. We might hope that a given Pareto efficient ex post mechanism returns some of the allocations returned by these three mechanisms. However, this does not hold as they may return only some of the Pareto efficient allocations. We illustrate this in Example 3.

Example 3. Let us consider the fair division of items o_1 and o_2 to agents 1 and 2 with utilities as in the below table.

	item o_1	item o_2
agent 1	1	4
agent 2	2	3

The allocation that gives o_1 to 1 and o_2 to 2 is Pareto efficient ex post. None of ONLINE SERIAL DICTATOR, ONLINE RANDOM PRIORITY or MAXIMUM LIKE returns this allocation. Note that PARETO LIKE does return it.   □

By Example 3, we conclude that we cannot characterize all Pareto efficient ex post mechanisms in terms of allocations returned by the ONLINE SERIAL DICTATOR, ONLINE RANDOM PRIORITY and MAXIMUM LIKE mechanisms. However, we can use the PARETO LIKE mechanism for this purpose.

Theorem 6. *The* PARETO LIKE *mechanism returns only and all Pareto efficient ex post allocations.*

Proof. By definition, the mechanism returns only PE ex post allocations. For this reason, we next only show that it returns all such allocations. Consider such an allocation π_m. Assume π_m is not returned by it. Run the mechanism and follow π_m until the first round $j \in (1, m]$ when some agent $i \in N$ gets o_j in π_m but i is not feasible for o_j given the sub-allocation π_{j-1} of π_m of o_1 to o_{j-1}.

Such a round exists as π_m is not returned by the mechanism. Further, π_{j-1} is Pareto efficient ex post for o_1 to o_{j-1}. Otherwise, the mechanism would not get to round j by following π_m. Also, the allocation extending π_{j-1} by allocating o_j to i is Pareto efficient ex post. Otherwise, this allocation can be Pareto improved for o_1 to o_j and together with the allocations of o_{j+1} to o_m in π_m can Pareto improve π_m. This contradicts the Pareto efficiency of π_m. Hence, the allocation extending π_{j-1} is Pareto efficient ex post. By the definition of the mechanism, it then follows that i is feasible for o_j which contradicts our assumption. Hence, π_m is returned by the mechanism with positive probability. $\square$

By Theorem 6, we conclude that a non-wasteful mechanism for online fair division is Pareto efficient ex post iff it returns a subset of the allocations of the PARETO LIKE mechanism. Such a mechanism may not be strategy-proof. However, we can characterize *all* mechanisms that are strategy-proof and Pareto efficient ex post.

Theorem 7. *A non-wasteful mechanism for online fair division is strategy-proof and Pareto efficient ex post iff it is ex post equivalent to a probability distribution of the* ONLINE SERIAL DICTATOR *mechanisms.*

Proof. We start with the "if" direction. If a mechanism is ex post equivalent to a probability distribution of ONLINE SERIAL DICTATORS, then it is strategy-proof and Pareto efficient ex post as each ONLINE SERIAL DICTATOR. We next prove the "only if" direction. Consider a strategy-proof and Pareto efficient ex post mechanism and assume that it is not ex post equivalent to any probability distribution of ONLINE SERIAL DICTATORS. Hence, there is an instance, an allocation and $j \in [1, m]$ such that the mechanism and ONLINE SERIAL DICTATOR with some priority ordering σ agree on o_1 to o_{j-1} but the mechanism and any such ONLINE SERIAL DICTATOR disagree on o_j. WLOG, let the mechanism give o_j to 1 and ONLINE SERIAL DICTATOR with σ give o_j to 2 such that 2 is immediately before 1 in σ. Both agents like item o_j. We can show that there is o_k with $k < j$ such that 1 and 2 like o_k, and that o_k is allocated to agent 2 with both mechanisms. By Theorem 1, with the mechanism, the probabilities of 2 for o_k and 1 for o_j do not change for any positive bids of these agents for these items. WLOG, let then $u_{1j} = 1, u_{1k} = 2, u_{2j} = 2, u_{2k} = 1$. Hence, the allocation that extends π_{j-1} by allocating o_j to agent 1 is not Pareto efficient ex post. $\square$

Let us next add the ex ante properties. There is an unbounded number of Pareto efficient ex post and envy-free ex ante (or Pareto efficient ex ante) mechanisms that are not strategy-proof. To see this, consider the mechanism for the instance in Example 2, that runs the ONLINE RANDOM PRIORITY (or MAXIMUM LIKE) mechanism on each other instance. Nevertheless, by Theorems 3 and 7, the only strategy-proof such mechanism is the ONLINE RANDOM PRIORITY mechanism.

Corollary 2. *A non-wasteful mechanism for online fair division is strategy-proof, Pareto efficient ex post and envy-free ex ante iff it is ex post equivalent to the* ONLINE RANDOM PRIORITY *mechanism.*

A mechanism that is Pareto efficient ex post might not be Pareto efficient ex ante. For example, the ONLINE RANDOM PRIORITY mechanism is Pareto efficient ex post but not ex ante. To see this, consider the instance in Example 1. The reverse direction may also not hold. That is, a mechanism that is Pareto efficient ex ante may not necessarily be Pareto efficient ex post. We show this in Example 4.

Example 4. Consider the mechanism that runs MAXIMUM LIKE on each instance except on the instance from Example 1. In this instance, the mechanism works as follows: agent 1 gets o_1 and o_2 with probabilities 1 and $1 - \epsilon$, and agent 2 gets these items with probabilities 0 and ϵ where $\epsilon > 0$. With this mechanism, agent 1 gets expected utility $3 - 2\epsilon$, whilst agent 2 gets expected utility ϵ. This outcome is Pareto efficient ex ante for any $\epsilon < 1/2$. But, there is one returned allocation that gives o_1 to agent 1 and o_2 to agent 2. This outcome is not Pareto efficient ex post.   □

It is easy to see that the mechanism in Example 4 is not strategy-proof. Interestingly, we can give a complete characterization of mechanisms that are strategy-proof, Pareto efficient ex post and Pareto efficient ex ante.

Theorem 8. *A non-wasteful mechanism for online fair division is strategy-proof, Pareto efficient ex post and ex ante iff it is ex post equivalent to the* ONLINE SERIAL DICTATOR *mechanism.*

Proof. We show the "if" direction. The mechanism returns the same allocation as ONLINE SERIAL DICTATOR. Hence, it is strategy-proof, Pareto efficient ex post and Pareto efficient ex ante. We next show the "only if" direction. By Theorem 7, the mechanism is a probability distribution of ONLINE SERIAL DICTATORS. Suppose that there are at least two different allocations which are the result of different ONLINE SERIAL DICTATORS in this distribution. WLOG, assume that agent 1 have the highest priority with probability $p_1 \in (0, 1)$, agent 2 with $p_2 \in (0, 1 - p_1]$ and agent $k \in N \setminus \{1, 2\}$ with $p_k \in [0, 1 - p_1 - p_2]$. Suppose that agent $i \in \{1, 2\}$ likes all items with 1 except o_i which they like with u, and agent $k \in N \setminus \{1, 2\}$ likes items positively. The expected utility of agent $i \in \{1, 2\}$ is $p_i \cdot (n - 1 + u)$ and the one of agent $k \in N \setminus \{1, 2\}$ is p_k multiplied by the sum of their utilities. Consider now another distribution of allocations, in which agent $i \in \{1, 2\}$ gets p_i for each item they like with 1 except items o_1, o_2, $p_1 + p_2$ for item o_i and 0 for $o \in \{o_1, o_2\} \setminus \{o_i\}$ whereas agent $k \in N \setminus \{1, 2\}$ gets p_k for each item. This allocation Pareto improves the allocation of the mechanism for $u > \max\{(p_1/p_2), (p_2/p_1)\}$. Hence, the mechanism is not Pareto efficient ex ante. Therefore, p_1 and p_2 cannot be both positive and, for this reason, each mechanism in the distribution gives the highest priority to the same agent. We can inductively show this for each priority.   □

We next observe one last difference to the offline setting where stochastic Pareto efficiency and envy-freeness are always possible [5]. In online fair division, *no* mechanism (even wasteful) satisfies Pareto efficiency ex ante and envy-freeness ex ante unless we consider simple 0/1 utilities (e.g. the BALANCED LIKE mechanism).

Theorem 9. *With general cardinal utilities, no mechanism for online fair division is envy-free ex ante and Pareto efficient ex ante.*

Proof. Consider an envy-free ex ante mechanism and the instance with non-zero utilities in Example 1. By Theorem 4, to ensure envy-freeness ex ante for o_1, the mechanism should give it to each agent with $1/2$. By Theorem 4, to ensure envy-freeness for both o_1 and o_2, the mechanism then should give o_2 to each agent with $1/2$. The expected utility of each agent is $3/2$. This expected allocation is Pareto dominated by the allocation in which each agent gets the item they value with 2. Hence, the mechanism is not Pareto efficient ex ante.   □

To sum up, we might use the ONLINE RANDOM PRIORITY or PARETO LIKE mechanism for Pareto efficiency ex post, or the MAXIMUM LIKE or ONLINE SERIAL DICTATOR mechanism for Pareto efficiency ex ante. With 0/1 utilities, we may also use the LIKE or BALANCED LIKE mechanism.

8   Conclusions

We summarize all results in Table 1 and Fig. 1. For completeness, we add some simple results for the case of identical utilities when the PARETO LIKE and MAXIMUM LIKE mechanisms become ex post equivalent to the LIKE mechanism, the BALANCED LIKE mechanism becomes ex ante equivalent to the LIKE mechanism, and each of these becomes Pareto efficient as the sum of agents' utilities is a constant in each allocation.

Table 1. Axiomatic results. Key: $\star$ - the result follows from [Aleksandrov *et al.* 2015].

Mechanism	SP	OSP	EFA	SEFA	EFP	SEFP	BEFP	PEA	PEP
General cardinal utilities									
ONLINE RP	✓	✓	✓	✓	×	×	×	×	✓
ONLINE SD	✓	✓	×	×	×	×	×	✓	✓
MAXIMUM LIKE	×	×	×	×	×	×	×	✓	✓
PARETO LIKE	×	×	×	×	×	×	×	×	✓
LIKE	✓*	✓	✓*	✓	×*	×	×*	×	×
BALANCED LIKE	×*	✓	×*	×	×*	×	×*	×	×
Identical cardinal utilities									
LIKE	✓*	✓	✓*	✓	×*	×	×	✓	✓
BALANCED LIKE	×	✓	✓	✓	×*	×	×	✓	✓
Binary cardinal utilities									
LIKE	✓*	✓	✓*	✓	×*	×	×*	✓	✓
BALANCED LIKE	×*	✓	✓*	×	×*	×	✓*	✓	✓

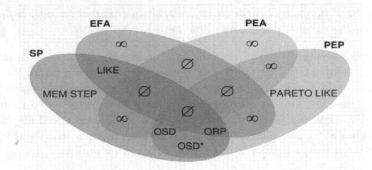

Fig. 1. General characterization results. Key: $\emptyset$ - no mechanisms, ∞ - inf. many mechanisms.

In future work, we will add quotas to our setting. And, we will extend our results to approximations of envy-freeness and general monotone utilities.

References

1. Abdulkadiroglu, A., Sönmez, T.: Random serial dictatorship and the core from random endowments in house allocation problems. Econometrica **66**(3), 689–702 (1998)
2. Aleksandrov, M., Aziz, H., Gaspers, S., Walsh, T.: Online fair division: analysing a food bank problem. In: Proceedings of the Twenty-Fourth International Joint Conference on Artificial Intelligence (IJCAI 2015), Buenos Aires, Argentina, 25–31 July 2015, pp. 2540–2546 (2015)
3. Aleksandrov, M., Walsh, T.: Most competitive mechanisms in online fair division. In: Kern-Isberner, G., Fürnkranz, J., Thimm, M. (eds.) KI 2017. LNCS (LNAI), vol. 10505, pp. 44–57. Springer, Cham (2017). https://doi.org/10.1007/978-3-319-67190-1_4
4. Benade, G., Kazachkov, A.M., Procaccia, A.D., Psomas, C.A.: How to make envy vanish over time. In: Proceedings of the 2018 ACM Conference on Economics and Computation, EC 2018, pp. 593–610. ACM, New York (2018)
5. Bogomolnaia, A., Moulin, H.: A new solution to the random assignment problem. J. Econ. Theory **100**(2), 295–328 (2001)
6. Brams, S.J., King, D.L.: Efficient fair division: help the worst off or avoid envy? Rat. Soc. **17**(4), 387–421 (2005)
7. Chevaleyre, Y., Endriss, U., Estivie, S., Maudet, N.: Multiagent resource allocation in k-additive domains: preference representation and complexity. Ann. Oper. Res. **163**(1), 49–62 (2008)
8. Dickerson, J.P., Procaccia, A.D., Sandholm, T.: Dynamic matching via weighted myopia with application to kidney exchange. In: Proceedings of the Twenty-Sixth AAAI Conference on Artificial Intelligence (2012)
9. Dickerson, J.P., Sandholm, T.: Futurematch: combining human value judgments and machine learning to match in dynamic environments. In: Proceedings of the Twenty-Ninth AAAI Conference, pp. 622–628 (2015)

10. Freeman, R., Zahedi, S.M., Conitzer, V., Lee, B.C.: Dynamic proportional sharing: a game-theoretic approach. In: Proceedings of the ACM on Measurement and Analysis of Computing Systems - SIGMETRICS, vol. 2, no. 1, pp. 3:1–3:36, April 2018
11. Gibbard, A.: Manipulation of voting schemes: a general result. Econometrica **41**(4), 587–601 (1973)
12. Kash, I.A., Procaccia, A.D., Shah, N.: No agent left behind: dynamic fair division of multiple resources. JAIR **51**, 579–603 (2014). https://doi.org/10.1613/jair.4405
13. Lian, J.W., Mattei, N., Noble, R., Walsh, T.: The conference paper assignment problem: using order weighted averages to assign indivisible goods. In: Proceedings of the Thirty-Second AAAI Conference on Artificial Intelligence, (AAAI-18), New Orleans, Louisiana, USA, 2–7 February 2018, pp. 1138–1145 (2018)
14. Manea, M.: Serial dictatorship and pareto optimality. Games Econ. Behav. **61**(2), 316–330 (2007). https://doi.org/10.1016/j.geb.2007.01.003
15. Svensson, L.G.: Strategy-proof allocation of indivisible goods. Soc. Choice Welf. **16**(4), 557–567 (1999)
16. Walsh, T.: Online cake cutting. In: Brafman, R.I., Roberts, F.S., Tsoukiàs, A. (eds.) ADT 2011. LNCS (LNAI), vol. 6992, pp. 292–305. Springer, Heidelberg (2011). https://doi.org/10.1007/978-3-642-24873-3_22
17. Walsh, T.: Allocation in practice. In: Lutz, C., Thielscher, M. (eds.) KI 2014. LNCS (LNAI), vol. 8736, pp. 13–24. Springer, Cham (2014). https://doi.org/10.1007/978-3-319-11206-0_2
18. Xia, L., Conitzer, V.: Strategy-proof voting rules over multi-issue domains with restricted preferences. In: Saberi, A. (ed.) WINE 2010. LNCS, vol. 6484, pp. 402–414. Springer, Heidelberg (2010). https://doi.org/10.1007/978-3-642-17572-5_33
19. Zhou, L.: On a conjecture by Gale about one-sided matching problems. J. Econ. Theory **52**(1), 123–135 (1990)

Knowledge Enhanced Neural Networks

Alessandro Daniele[1,2(✉)] and Luciano Serafini[1]

[1] Fondazione Bruno Kessler, Trento, Italy
{daniele,serafini}@fbk.eu
[2] University of Florence, Florence, Italy

Abstract. We propose Knowledge Enhanced Neural Networks (KENN), an architecture for injecting prior knowledge, codified by a set of logical clauses, into a neural network.

In KENN clauses are directly incorporated in the structure of the neural network as a new layer that includes a set of additional learnable parameters, called *clause weights*. As a consequence, KENN can learn the level of satisfiability to impose in the final classification. When training data contradicts a constraint, KENN learns to ignore it, making the system robust to the presence of wrong knowledge. Moreover, the method returns learned clause weights, which gives us informations about the influence of each constraint in the final predictions, increasing the interpretability of the model. We evaluated KENN on two standard datasets for multi-label classification, showing that the injection of clauses automatically extracted from the training data sensibly improves the performances. Furthermore, we apply KENN to solve the problem of finding relationship between detected objects in images by adopting manually curated clauses. The evaluation shows that KENN outperforms the state of the art methods on this task.

Keywords: Neural-symbolic integration · Neural networks · Fuzzy logic · Visual Relationship Detection

1 Introduction

In the last decades, there have been an increased interest on Neural-Symbolic systems, i.e., systems that integrates neural networks and symbolic reasoning (Besold et al. 2017). In this paper, we propose *Knowledge Enhanced Neural Network (KENN)*, a neural network model that exploits prior knowledge about the domain of interest. Such knowledge is expressed by a set of logical rules (or clauses). For instance, in an image classification task the rule $\forall(Dog(x) \rightarrow Animal(x))$, stating that dogs are animals, is used by KENN to predict the two labels *Dog* and *Animal*.

We would like to thank Alessandro Sperduti and Marco Gori for useful suggestions and feedbacks. We also thank Ivan Donadello for providing us with the knowledge base for VRD Dataset produced during his Ph.D.

© Springer Nature Switzerland AG 2019
A. C. Nayak and A. Sharma (Eds.): PRICAI 2019, LNAI 11670, pp. 542–554, 2019.
https://doi.org/10.1007/978-3-030-29908-8_43

Suppose we have a neural network (NN) for some relational classification task, that we call *basic NN*. KENN *enhances* the prediction done by the basic NN by injecting the background knowledge. This is obtained by adding a final layer in the basic NN. This additional layer is designed to increase the satisfaction of the clauses by modifying the predictions. It uses *clause weights* parameters that represent the influence of each clause on the final predictions. Differently from other neuro-symbolic integration approaches, clauses weights are not given, but they are learned. By changing the weight of clause, KENN can ignore clauses in the prior knowledge that are not fully satisfied inside training data. Although KENN can be used in principle for relational data, in this paper we focus our attention on multi-label classification problems. Notice however that motivations, general framework and theoretical results hold for relational data as well.

We evaluate KENN on two aspects. First we evaluate the ability of KENN to learn *clause weights*, which make KENN robust to less reliable knowledge bases. For this reasons, we made some experiments using automatically generated knowledge bases. We tested KENN on Yeast dataset (Elisseeff and Weston 2001) and Emotions dataset (Trohidis et al. 2008) using clauses generated directly from their training data. Results showed that KENN can make efficient usage of this kind of rules. In a second experiment we evaluate KENN on Predicate Detection task of Visual Relationship Detection Dataset (VRD Dataset) (Lu et al. 2016) using a manually curated prior knowledge proposed by Donadello (2018). KENN outperformed state of the art methods, with best results on the *Zero Shot Learning* variant of the task. This last experiment confirms the fact that background knowledge plays a key role in machine learning when there is a scarcity of training data.

2 Related Work

Many previous works attempt to combine learning models with logical knowledge. Among them, there are (Hybrid) Markov Logic Networks (Wang and Domingos 2008; Richardson and Domingos 2006) and *Probabilistic Soft Logic* (PSL) (Bach et al. 2017). However, in each of these systems the functions that bind logic rules with predictions are very simple while in KENN could be a general Neural Network. Moreover, they can not deal with real valued features.

A different line of research, called Neural-Symbolic systems, has focused on combining neural networks models with logical knowledge (Besold et al. 2017). Early proposals, like KBANN (Towell and Shavlik 1994, were restricted on propositional logic. More recently, systems working with *First Order Logic* were developed. In this category there are *Logic Tensor Network* (LTN) (Serafini and d'Avila Garcez 2016) and *Semantic Based Regularization* (SBR) (Diligenti et al. 2017). The two methods have a similar approach: they deals with logical constraints by maximize their satisfaction during training. In (Demeester et al. 2016), rules are used to constraint the learnt embeddings. However, they restrict the type of rules to implications with a singular literal in the body. Another approach consists on using a distillation mechanism for injecting logic (Hu et al. 2016).

The main difference between KENN and its major competitors lies on the way logic formulas are used: in KENN they become part of the classifier instead of being enforced during training. More precisely, methods like LTN and SBR force the constraints satisfaction during training making the assumption that the knowledge is in general correct. Instead, we assume there is a relationship between clauses and correct results, but this relationship is not known. The logical constraints are seen as a prior belief rather than prior knowledge. More in details, KENN has internal learnable parameters associated to the logic formulas. To the best of our knowledge, there are no previous methods that can inject logical constraints into a neural network while being able to learn clause weights. This make KENN suitable for scenarios where the given knowledge contains errors or when rules are softly satisfied in the real world but it is not known the extent on which they are correct.

Current implementation of KENN can deal with multi-label classification problems. Although the theory is more general, so far we performed evaluations only on those kind of problems. Multi-label classification is a supervised learning task relevant in many disciplines, e.g., bioinformatics (Elisseeff and Weston 2001), scene classification (Boutell et al. 2004) and text categorization (Loza Mencía and Fürnkranz 2010). In multi-label classification we are interested in mapping specific observations to subsets of all the possible labels (Tsoumakas and Katakis 2007; Park and Fürnkranz 2008). It differs from binary classification and, more in general, from multi-class classification, because the classes are not mutually exclusive, i.e. multiple labels can be associated to a single instance. Formally, it is given a set of labels $\mathcal{L} = \{\lambda_i | i = 1...m\}$ and a training set $\mathcal{T} = \{(x_i, y_i) | i = 1...n\}$, where x_i denotes features of the ith observation and the classification $y_i \in 2^{\mathcal{L}}$ is a subset of labels associated to such observation. We want to find a classifier $\phi : \mathcal{X} \to 2^{\mathcal{L}}$ which, given features x of an observation, returns the associated set of labels. KENN is extremely suited to formulate multi-label classification with constraints. Indeed KENN uses vectorial representation for both features and corresponding classification, with $y_{i,j}$ equal to one if λ_j is associated to observation i, zero otherwise. While it is possible to train many different binary classifiers for each label, such method does not take into account relationships among them. Indeed, in real-world applications, labels are often not independent. For this reason, it could be useful to exploit knowledge about labels relationships provided by some human expert. More formally, it is given a prior knowledge $\mathcal{K}_{C,R}$ composed of a set of clauses. Each clause is a disjunction of (positive or negated) literals, each of which is a class (i.e., a label) of the multi-label classification problem. For instance, a possible clause could be $\lambda_1 \vee \neg \lambda_3 \vee \lambda_4$. Such clause tells the system that at least one of the specified literal should be true.

3   KENN: Overview of the Model

Let NN be a neural network (called base NN) that takes in input the feature vectors $\mathbf{x}_1, \ldots, \mathbf{x}_N$ of N objects and returns an initial output $\mathbf{y} = (\mathbf{y}_C, \mathbf{y}_R)$

that contains the predictions about the classes of these objects, i.e., $\mathbf{y}_C$, and the predictions about their relations $\mathbf{y}_R$ (in this work we restrict to binary relations). The predictions $\mathbf{y}$ of the base NN are revised by a function, called *Knowledge Enhancer* (KE), to force the satisfaction of the logical constraints on the classes C and relations R that are contained in the knowledge base $\mathcal{K}_{C,R}$.

KE must be differentiable and it can be seen as a new final layer for the base NN that encodes the background knowledge. The entire network is still differentiable end-to-end, making it possible to train the model with back-propagation algorithm. KE contains additional parameters that can be learned as well. In particular *clause weights* w_c determine the strength of each clause $c \in \mathcal{K}_{C,R}$. Figure 1 shows a high level overview of the model.

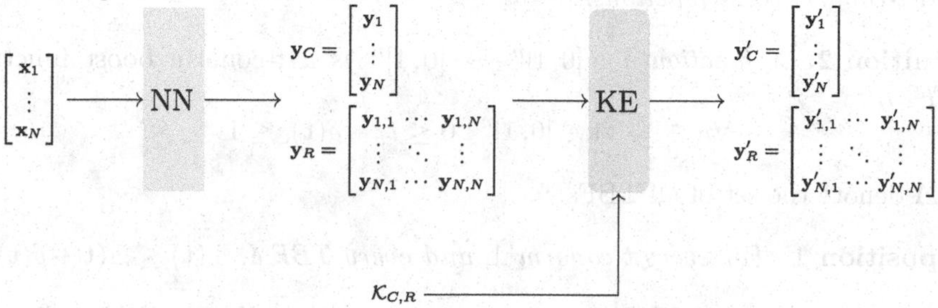

Fig. 1. KENN model: features are given as input to a neural network (NN) and predictions on predicates values are returned. Knowledge Enhancer modifies the predictions based on logical constraints ($\mathcal{K}_{C,R}$)

For each clause $c \in \mathcal{K}_{C,R}$, KE internally produces the change to be applied to the predictions $\mathbf{y}$ of the base NN in order to increase c satisfaction. It then calculates the final predictions $\mathbf{y}'$ by adding to the NN outputs the weighted sum of such changes using as weights parameters w_c. In order to increase satisfaction of a clause, we first need to define what satisfaction mean. We rely on fuzzy logic, where the satisfaction of a disjunction of literals is represented with a t-conorm function.

Definition 1. *A t-conorm* $\perp : [0,1] \times [0,1] \to [0,1]$ *is a binary function which satisfies the following properties:*

1. $\perp(a,b) = \perp(b,a)$
2. $\perp(a,b) \leq \perp(c,d)$ *if* $a \leq c$ *and* $b \leq d$
3. $\perp(a, \perp(b,c)) = \perp(\perp(a,b), c)$
4. $\perp(a,0) = a$

We represent a t-conorm as a unary function over vectors ($\mathbf{t} = \langle t_1, t_2...t_n \rangle$):

$$\perp(\mathbf{t}) = \perp(t_1, \perp(t_2, \perp(t_3...\perp(t_{n-1}, t_n))))$$

4   Boost Functions

The KE should implement a function that changes the values of the predictions **y** of the basic NN into **y**′ in order to increase the truth value of each clause c contained in the knowledge base $\mathcal{K}_{C,R}$.

Given a clause c composed of n literals, if the ith component of **t** is the truth value of the ith literal of c, then $\perp(\mathbf{t})$ is the truth value of c. Intuitively, the KE should implement a function $\delta : [0,1]^n \to [0,1]^n$ that increases the value of $\perp(\mathbf{t})$.

4.1   T-Conorm Boost Functions

Let us first define the class of functions that increases the values of a t-conorm, called *t-conorm boost functions*.

Definition 2. *A function* $\delta : [0,1]^n \to [0,1]^n$ *is a* t-conorm boost function *(TBF) iff:*

$$\forall n \in \mathbb{N} \ \ \forall \mathbf{t} \in [0,1]^n \ \ 0 \le t_i + \delta(\mathbf{t})_i \le 1$$

Let Δ denote the set of all TBFs.

Proposition 1. *For every t-conorm* $\perp$ *and every TBF* δ, $\perp(\mathbf{t}) \le \perp(\mathbf{t} + \delta(\mathbf{t}))$

Proof. By definition of TBF, $\forall i \in [1,n]$, $t_i \le t_i + \delta(\mathbf{t})_i$; the conclusion directly follows from the Property 2. of t-conorm.

TBFs are used in the KE to update the initial predictions **y** done by the base NN. We want to keep this change as minimal as possible. Therefore we look at TBF's that improve the t-conorm value *in a minimal way*, so that it is not possible to obtain a higher improvement with smaller modifications on literals values. We define the concept of minimality for a TBFs.

Definition 3. *A function* $\delta \in \Delta$ *is minimal with respect to a norm* $\| \cdot \|$ *and a t-conorm* $\perp$ *iff:*

$$\forall \delta' \in \Delta \ \ \forall n \in \mathbb{N} \ \ \forall \mathbf{t} \in [0,1]^n \tag{1}$$
$$\|\delta'(\mathbf{t})\| < \|\delta(\mathbf{t})\| \to \perp(\mathbf{t} + \delta'(\mathbf{t})) < \perp(\mathbf{t} + \delta(\mathbf{t}))$$

For any function $f : \mathbb{R}^n \to \mathbb{R}$ we define $\delta^f : \mathbb{R}^n \to \mathbb{R}^n$ as

$$\delta^f(\mathbf{t})_i = \begin{cases} f(\mathbf{t}) & \text{if } i = \operatorname{argmax}_{j=1}^n t_j \\ 0 & \text{otherwise} \end{cases} \tag{2}$$

Theorem 1. *If we choose function* f *such that* $0 \le f(\mathbf{t}) \le 1 - \max_{j=1}^n t_j$, *then* δ^f *functions are minimal TBFs for the Gödel t-conorm and* l_p-norm.

Proof. Gödel t-conorm $\bot(\cdot)$ is defined as $\bot(\mathbf{t}) = \max_{i=1}^{n}(t_i)$; and l_p-norm is defined as $\|\mathbf{t}\|_p = \left(\sum_{k=1}^{n}|t_k|^p\right)^{1/p}$. Suppose that $\delta \in \Delta$ is such that

$$\|\delta(\mathbf{t})\|_p < \|\delta^f(\mathbf{t})\|_p$$

If $j = \operatorname{argmax}_{k=1}^{n}(t_k + \delta(\mathbf{t})_k)$, we can derive:

$$\bot(\mathbf{t} + \delta(\mathbf{t})) = t_j + \delta(\mathbf{t})_j$$

and, if $i = \operatorname{argmax}_{k=1}^{n} t_k$, we have that

$$\bot(\mathbf{t} + \delta^f(\mathbf{t})) = t_i + f(\mathbf{t})$$

Since $t_i \geq t_j$, we just need to demonstrate that $\delta(\mathbf{t})_j < f(\mathbf{t})$. Notice that:

$$\delta(\mathbf{t})_j = (|\delta(\mathbf{t})_j|^p)^{1/p} \leq \left(\sum_{k=1}^{n}|\delta(\mathbf{t})_k|^p\right)^{1/p} = \|\delta(\mathbf{t})\|_p < \|\delta^f(\mathbf{t})\|_p$$

Since $\delta^f(\mathbf{t})$ changes only the value of the ith component of $\mathbf{t}$ we have that $\|\delta^f(\mathbf{t})\|_p = f(\mathbf{t})$.

4.2 Boosting Preactivations

Applying directly a TBF to the final prediction $\mathbf{y}$ of the base NN could be problematic since we have to respect the constraint that the improved value should remain in $[0,1]$. This implies that f cannot be a linear function. Notice that the outputs $\mathbf{y}$ of NN are calculated by applying the sigmoid activation function over the preactivations $\mathbf{z}$ generated in the last layer. I.e.:

$$y_i = \sigma(z_i) = \frac{1}{1 + e^{-z_i}}$$

where y_i is the activation of the ith predicate and z_i the corresponding preactivation. In Sect. 4.1 we showed that if we increase only the value of the highest literal (highest activation), such a change is minimal for Gödel t-conorm. We can apply the same strategy to preactivations and still have a minimal change that increase the t-conorm. In other words, we can increase z_i instead of y_i, when $i = \operatorname{argmax}_{j=1}^{n} z_j$ and the previously proven properties still hold. Applying changes on preactivations has the advantage of guaranteeing to be in $[0,1]$.

Proposition 2. *For any function $f : \mathbb{R}^n \to \mathbb{R}^+$, the function*

$$\delta^g(\mathbf{y}) = \sigma(\mathbf{z} + \delta^f(\mathbf{z})) - \sigma(\mathbf{z}) \tag{3}$$

with $\sigma(\mathbf{z}) = \mathbf{y}$, is a minimal TBF.

Proof. Notice that $\mathbf{y} + \delta^g(\mathbf{y}) = \sigma(\mathbf{z} + \delta^f(\mathbf{z}))$ which is in $[0,1]^n$. Furthermore, for every $i \neq \operatorname{argmax}_{j=1}^{n} y_j$, $\delta^g(\mathbf{y})_i = 0$. Theorem 1 guarantees that δ^g is a TBF.

The function δ^g is not directly used by KENN; it is implicitly induced by the application of δ^f on $\mathbf{z}$. Therefore, by showing that it is a minimal TBF, we prove that applying δ^f on $\mathbf{z}$ is indeed equivalent to apply a minimal TBF on the NN predictions. We allow a distinct f_c function for every clause c. This is motivated by the fact that we want a δ^{f_c} that is proportional to *clause weight* w_c (learnable parameter) that expresses the strengths of the clause. The simplest function that conforms to this property is the constant function w_c (with $w_c \in [0, \infty]$). The function applied to $\mathbf{z}$ to increase c satisfaction is therefore δ^{w_c} defined as in 2.

4.3   Soft Approximation of δ^{w_c}

Although δ^{w_c} respects our minimality property, there are two problems when using it inside a neural network: first, it is not differentiable; second, it is too strict when multiple literals have close values. In those cases, it increases just one of the values even if the difference is minimal. To obviate these problems, in our implementation we substitute δ^{w_c} with the *softmax* function $(sm(\cdot))$ multiplied by w_c, that can be seen as a soft differentiable approximation of δ^{w_c}:

$$\delta_s^{w_c}(\mathbf{z})_i = w_c \cdot sm(\mathbf{z})_i = w_c \cdot \frac{e^{z_i}}{\sum_{j=1}^n e^{z_j}}$$

Intuitively, the idea is that, in order to satisfy a clause, at least one of its literal must be true. The softmax function act as a selector for the most promising true literal, that is the one with highest supporting evidences (biggest preactivation).

5   KENN Architecture

Figure 2 shows in details the architecture of KENN's last layer. KE is a function that takes as input the preactivations $\mathbf{z}$ of the base NN and produces the final activations $\mathbf{y}'$. Internally, for each clause $c \in \mathcal{K}_{C,R}$, it has a submodule called *Clause Enhancer* (CE), which returns the adjustment to apply on preactivations to enforce c satisfaction. The CEs outputs are then combined linearly using *clauses weights* and summed to the initial preactivations. Lastly, the final predictions are calculated by applying the logistic function:

$$y_i' = \sigma\left(z_i + \sum_{\substack{c \in \mathcal{K}_{C,R} \\ i \in c}} p_{c,i} \cdot \delta_s^{w_c}(\mathbf{p}_c \odot \mathbf{z}_c)_i\right) \tag{4}$$

where $i \in c$ means that the clause c contains a positive or negative literal corresponding to the preactivations z_i, $\mathbf{z}_c$ are all the preactivations corresponding to the literals in c, $\odot$ is the element-wise multiplication, and $\mathbf{p}_c$ is a polarity vector of 1 and -1, with $p_i = 1$ if the ith literal of the clause is positive and $p_i = -1$ if it is negative. The term w_c is a positive weight associated to clause c; it is a parameter of the model, which is learned during training. Notice that setting w_c to zero make clause c irrelevant for the final predictions, making it possible for KENN to learn to ignore clauses.

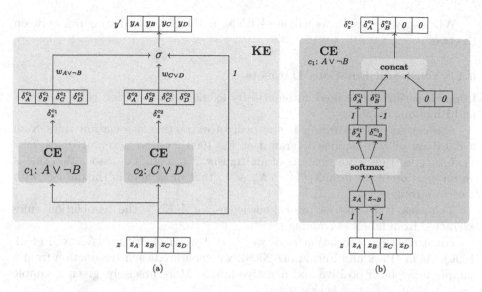

Fig. 2. KENN: (a) Knowledge enhancer, (b) Clause enhancer

6 Experimental Evaluation

We tested KENN on three datasets: Yeast (Elisseeff and Weston 2001), Emotions (Trohidis et al. 2008) and VRD Dataset (Lu et al. 2016). Table 1 reports some figures of the three datasets. Please note that for VRD the features are two bounding boxes of an image together with the class of the contained object (there are 100 possible classes and we used as features the one hot encoding of them plus the 4 coordinates of each bounding box).

Table 1. Datasets statistics

Dataset	Features	Labels	Train	Test
Yeast	103	14	1500	917
Emotions	72	6	392	202
VRD	208	70	4000	1000

We implemented KENN using TensorFlow and used RMSProp (Tijmen and Hinton 2012) as learning algorithm. For all the experiments we used as NN a neural network with zero hidden layers, i.e., a logistic regression (LR). We tried two learning strategies: in the **end-to-end** strategy, we trained KENN (base NN + KE) end-to-end using the entire training set; with **greedy** strategy we split the training set into two subset. With the first one we trained the base NN, with the second the KE (freezing the NN parameters).

When showing results we will use KENN$_e$ and KENN$_g$ to distinguish between the two.

6.1   Yeast and Emotions Datasets

In our experiments we used automatically extracted association rules for Yeast and Emotions datasets.

Association rules learning is the task of extracting association rules from a database of transactions (Agrawal et al. 1993). Given a set of items $\mathcal{I} = \mathcal{I}_1, \mathcal{I}_2...\mathcal{I}_n$, a transaction consists of an itemset $\mathcal{T} \subseteq \mathcal{I}$ and an association rule is an implication of the form $\lambda_1 \wedge ... \wedge \lambda_n \rightarrow \lambda_j$ that must hold in the dataset with a certain confidence.

The idea is to use as prior knowledge for KENN the association rules extracted from labels of training set.

For extracting association rules we used Apriori algorithm (Agrawal et al. 1996). As in (Park and Fürnkranz 2008), we generated each transaction from a sample using both positive and negative labels. More precisely, given a sample (x_i, y_i), the transaction is calculated as:

$$\left\{\lambda_j | y_{i,j} = 1\right\} \bigcup \left\{\neg\lambda_j | y_{i,j} = 0\right\}$$

where λ_j is the name of jth predicate.

To select the best values of support and confidence for the Apriori algorithm we tried different combinations of values. We generated the rules and trained KENN using 2/3 of the Training and evaluate the model on the remain samples. We finally kept the values with highest accuracy. Then we used the optimal value for extracting rules from the entire Training Set and trained KENN on it. It is worth noticing that, when using greedy approach of learning, for every configuration tried for Apriori parameters the accuracy of KENN was greater or equal than the one of the corresponding neural network, confirming the robustness of our method to poorly written knowledge bases. The found values for support and confidence are respectively 0.2 and 0.99 for Yeast Dataset and 0.2 and 0.7 for Emotions.

Table 2. Results with and without prior knowledge on yeast and emotions datasets. The table shows hamming loss and accuracy.

	Yeast		Emotions	
	HL	Acc	HL	Acc
LR	22.38	37.52	27.89	33.70
KENN$_e$	22.24	46.43	34.57	25.74
KENN$_g$	**20.86**	**48.56**	**24.59**	**38.78**

In Table 2 the final comparison between the logistic regression and KENN: the usage of prior knowledge improved both HL and accuracy when using the greedy approach of learning, while the end-to-end one results in smaller improvements in Yeast dataset and a degradation in Emotions.

In principle, we could expect better results training jointly the NN and KE, because there could be some combinations of parameters for the entire model such that the NN performs poorly while the entire system gives good results. This type of configurations cannot be learned by the greedy approach. In other words, when training with greedy approach, some solutions in the hypothesis space can not be reached. For this reason, greedy approach has a smaller ability to fit the data. However, this could also imply an increased risk of overfitting. Therefore, we can expect better results of the end-to-end version when the amount of data is big enough to overcome overfitting. Indeed, the problem with the end-to-end version is visible in particular on Emotions dataset, which has less than 400 samples for its training.

6.2 Predicate Detection

Visual Relationship Detection (VRD) is the task of finding objects in an image and capture their interactions (Donadello 2018; Lu et al. 2016; Zhang et al. 2017). It is composed of three subtask: Relationship Detection, Phrase Detection and Predicate Detection (Lu et al. 2016). The VRD Dataset contains a total amount of 6672 triplets types. Among them 1877 can be find only in the Test Set and predicting them is the goal of the Zero Shot Learning variant of the task.

We trained both LR and KENN using RMSProp (Tijmen and Hinton 2012) and cross entropy as loss function. For evaluating the results we used the $Recall@n$ ($n \in \{50, 100\}$) metric proposed by Lu et al. (2016) that is the percentage of times a correct relationship is found on the n predictions with highest score. We evaluated KENN on the Predicate Detection task using the manually curated knowledge base described in (Donadello 2018). This knowledge base contains 206 clauses divided in three groups, namely: domain (resp. range) clauses, that restrict the domain (range) of a binary relation (e.g., $\neg ParkOn(x,y) \vee Street(y) \vee Road(y) \vee Grass(y)$); mutual exclusivity clauses (e.g. $\neg Behind(x,y) \vee \neg SitOn(x,y)$); and subrelation clauses, that state the containment between relations (e.g., $\neg Ride(x,y) \vee On(x,y)$).

Results on Predicate Detection task are shown in Table 3. KENN outperformed other methods on all the metrics except for Recall@100 of the standard variant of the task where it is surpassed by Yu et al. (2017). Moreover, the best results of KENN can be seen on Zero Shot Learning version where the difference between KENN and the second best system is more than 10%. In *Zero Shot Learning* the aim is to predict previously unseen triplets, therefore it is rather difficult to learn to predict them from the Training Set. This confirms the ability of KENN to use the Knowledge Base.

Another interesting result is the value obtained by KENN compared to LTN (Donadello 2018). In particular considering that the two works used the same Prior Knowledge. A possible explanation is given by the ability of KENN to learn *clause weights*. Indeed, many weights results to be zero after learning. An example of a zero weighted clause is: $\neg Ride(x,y) \vee On(x,y)$.

Table 3. Results on VRD Predicate Detection task

	Standard		Zero Shot	
	R@50	R@100	R@50	R@100
Lu et al. (2016)	47.87	47.87	8.45	8.45
Dai et al. (2017)	80.78	81.90	–	–
Yu et al. (2017)	85.64	**94.65**	54.20	74.65
Donadello (2018)	78.63	91.88	46.28	70.15
LR	54.58	60.55	33.88	44.91
$KENN_g$	59.87	71.42	43.88	63.99
$KENN_e$	**86.02**	91.91	**68.95**	**83.83**

Although the rule seems correct it is not in general satisfied on training and test set. This is because labels have been added manually, therefore there are plenty of missing relations. Our hypothesis is that people have a tendency to add the most informative labels making some of the clauses unsatisfied.

Additionally, as in previous experiments, we increased the results of LR, the basic NN used by KENN. This time, although greedy version brought great improvements over the logistic regression, the best results are achieved by the end-to-end approach. This provide an additional support for the overfitting hypothesis made in Sect. 6.1, since VRD has a much bigger training set than Emotions and Yeast. Moreover, in previous experiments, prior knowledge was automatically extracted from the training data. Clauses could be satisfied in the training set just by chance. In such cases, Apriori algorithm extracts misleading rules that are not satisfied at test time. This implies a further increase in the chances of overfitting.

7   Conclusions and Future Work

With its results on the three analyzed datasets, KENN showed to be able to efficiently exploit prior knowledge improving performances of the base neural network. Moreover, experiments on VRD dataset showed that KENN is competitive against other approaches, in particular tanks to its ability to effectively learn *clauses weights*.

In future work, investigating alternatives to Gödel t-conorm might be useful. In addition, we would like to perform experiments with more complex neural networks rather than adding rules to a simple logistic regression. Moreover, we are interested in extend KENN implementation for relational data. Finally, further investigations could prove helpful to fully understand the different scenarios in which the two learning approaches (greedy and end-to-end) are more suited.

References

Agrawal, R., Imieliński, T., Swami, A.: Mining association rules between sets of items in large databases. In: ACM SIGMOD Record, vol. 22, no. 2, pp. 207–216. ACM (1993)

Agrawal, R., Mannila, H., Srikant, R., Toivonen, H., Verkamo, A.I.: Fast discovery of association rules. In: Advances in Knowledge Discovery and Data Mining, pp. 307–328. American Association for Artificial Intelligence, Menlo Park (1996). http://dl.acm.org/citation.cfm?id=257938.257975

Bach, S.H., Broecheler, M., Huang, B., Getoor, L.: Hinge-loss Markov random fields and probabilistic soft logic. J. Mach. Learn. Res. **18**(109), 1–67 (2017)

Besold, T.R., et al.: Neural-symbolic learning and reasoning: a survey and interpretation. CoRR abs/1711.03902 (2017). http://arxiv.org/abs/1711.03902

Boutell, M.R., Luo, J., Shen, X., Brown, C.M.: Learning multi-label scene classification (2004)

Dai, B., Zhang, Y., Lin, D.: Detecting visual relationships with deep relational networks. In: Proceedings of the IEEE Conference on Computer Vision and Pattern Recognition (2017)

Demeester, T., Rocktäschel, T., Riedel, S.: Lifted rule injection for relation embeddings. In: Proceedings of the 2016 Conference on Empirical Methods in Natural Language Processing, EMNLP 2016, Austin, Texas, USA, 1–4 November 2016, pp. 1389–1399 (2016). http://aclweb.org/anthology/D/D16/D16-1146.pdf

Diligenti, M., Gori, M., Saccà, C.: Semantic-based regularization for learning and inference. Artif. Intell. **244**, 143–165 (2017)

Donadello, I.: Semantic image interpretation - integration of numerical data and logical knowledge for cognitive vision. Ph.D. thesis, Trento Univ., Italy (2018)

Elisseeff, A., Weston, J.: A kernel method for multi-labelled classification. In: Proceedings of the 14th International Conference on Neural Information Processing Systems: Natural and Synthetic, NIPS 2001, pp 681–687. MIT Press, Cambridge (2001). http://dl.acm.org/citation.cfm?id=2980539.2980628

Hu, Z., Ma, X., Liu, Z., Hovy, E.H., Xing, E.P.: Harnessing deep neural networks with logic rules. In: Proceedings of the 54th Annual Meeting of the Association for Computational Linguistics, ACL 2016, 7–12 August 2016, Berlin, Germany, Volume 1: Long Papers. The Association for Computer Linguistics (2016). http://aclweb.org/anthology/P/P16/P16-1228.pdf

Loza Mencía, E., Fürnkranz, J.: Efficient multilabel classification algorithms for large-scale problems in the legal domain. In: Francesconi, E., Montemagni, S., Peters, W., Tiscornia, D. (eds.) Semantic Processing of Legal Texts. LNCS (LNAI), vol. 6036, pp. 192–215. Springer, Heidelberg (2010). https://doi.org/10.1007/978-3-642-12837-0_11

Lu, C., Krishna, R., Bernstein, M., Fei-Fei, L.: Visual relationship detection with language priors. In: Leibe, B., Matas, J., Sebe, N., Welling, M. (eds.) ECCV 2016. LNCS, vol. 9905, pp. 852–869. Springer, Cham (2016). https://doi.org/10.1007/978-3-319-46448-0_51

Park, S.H., Fürnkranz, J.: Multi-label classification with label constraints. In: Hüllermeier, E., Fürnkranz, J. (eds.) Proceedings of the ECML PKDD 2008 Workshop on Preference Learning (PL 2008, Antwerp, Belgium), pp 157–171 (2008). http://www.mathematik.uni-marburg.de/~kebi/ws-ecml-08/12.pdf

Richardson, M., Domingos, P.: Markov logic networks. Mach. Learn. **62**(1–2), 107–136 (2006)

Serafini, L., d'Avila Garcez, A.S.: Logic tensor networks: deep learning and logical reasoning from data and knowledge. CoRR abs/1606.04422 (2016)

Tijmen, T., Hinton, G.: Lecture 6.5-rmsprop: divide the gradient by a running average of its recent magnitude. In: COURSERA: Neural Networks for Machine Learning (2012)

Towell, G.G., Shavlik, J.W.: Knowledge-based artificial neural networks. Artif. Intell. **70**(1–2), 119–165 (1994). https://doi.org/10.1016/0004-3702(94)90105-8

Trohidis, K., Tsoumakas, G., Kalliris, G., Vlahavas, I.: Multi-label classification of music into emotions. In: Proceedings of the 9th International Conference on Music Information Retrieval, Philadelphia, USA, pp 325–330 (2008)

Tsoumakas, G., Katakis, I.: Multi-label classification: an overview. Int. J. Data Warehouse. Mining (IJDWM) **3**(3), 1–13 (2007)

Wang, J., Domingos, P.: Hybrid Markov logic networks. In: Proceedings of the 23rd National Conference on Artificial Intelligence, AAAI 2008, vol. 2, pp. 1106–1111. AAAI Press (2008)

Yu, R., Li, A., Morariu, V.I., Davis, L.S.: Visual relationship detection with internal and external linguistic knowledge distillation. CoRR abs/1707.09423 (2017)

Zhang, H., Kyaw, Z., Chang, S.F., Chua, T.S.: Visual translation embedding network for visual relation detection. In: 2017 IEEE Conference on Computer Vision and Pattern Recognition (CVPR), pp. 3107–3115 (2017)

Encoding Epistemic Strategies for General Game Playing

Shawn Manuel, David Rajaratnam$^{(\boxtimes)}$, and Michael Thielscher

School of Computer Science and Engineering, University of New South Wales,
Sydney, Australia
shawn_manuel_000@yahoo.com, {David.Rajaratnam,mit}@unsw.edu.au

Abstract. We propose a general approach for encoding epistemic strategies for playing incomplete information games. A game strategy involves selecting actions in order to maximise an outcome (e.g., winning the game). In an epistemic strategy the selection of actions is based on reasoning about the knowledge of other players. We show how epistemic strategies can be encoded by supplementing a GDL-II game description with a set of epistemic rules to produce a GDL-III game that an appropriate reasoner can use to play the original GDL-II game. We prove the formal correctness of this approach and provide a practical evaluation to show its efficacy for playing the co-operative multi-player game of Hanabi. It was found that the encoded epistemic rules were able to provide players with a strategy that allowed them to play Hanabi near optimally.

1 Introduction

General Game Playing (GGP) is a sub-field within AI aimed at creating systems that can learn to play a variety of strategy games when given only the game rules at runtime [11]. Unlike specialised systems, a general game player cannot rely on game specific algorithms that have been designed in advance. Instead it requires a form of general intelligence that enables the player to autonomously adapt to new games. This is exemplified by the annual international GGP competition [13].

A feature of GGP is the Game Description Language (GDL) used to specify complete information games [11], and subsequently extended to deal with *imperfect information* games (GDL-II) [18]. However, while GDL-II can be used to specify the rules of imperfect information games, it lacks the expressive power to describe the *strategy* a player should follow to actually play such a game. In particular, a multi-agent, imperfect information game typically requires players to reason about the knowledge, or *epistemic state*, of other players in order to play effectively [1]. Such a player is said to be following an *epistemic strategy*.

While epistemic strategies cannot be encoded in GDL-II, a more recent language extension, GDL-III, does allow for the specification of epistemic goals and rules (GDL-III; for GDL with *imperfect information* and *introspection*) [22]. While intended as a language for describing epistemic games, it has also been used to model epistemic puzzles, such as the 'Muddy Children' puzzle, where the goal of each child is to know whether or not she has mud on her forehead [21].

© Springer Nature Switzerland AG 2019
A. C. Nayak and A. Sharma (Eds.): PRICAI 2019, LNAI 11670, pp. 555–567, 2019.
https://doi.org/10.1007/978-3-030-29908-8_44

In this paper we introduce a further application of GDL-III; as a language for representing epistemic strategies. In particular, we provide a framework for encoding epistemic strategies for GDL-II games within the GDL-III language. Any GDL-III reasoner (i.e., a logical reasoner that can track the state of a GDL-III game) can then be co-opted into being an effective GDL-II game player.

To motivate the use of epistemic strategies, and to study the potential efficacy of our approach, we consider the game of Hanabi[1]. Hanabi has been the subject of recent interest [4,7,9,10,17,20,24], and has been proposed as a new frontier for AI research in a similar league to games such as poker and Go [5]. Hanabi is well-suited for our purposes, as the game rules require no epistemic properties (i.e. it is GDL-II representable), yet it is a multi-player, imperfect information game that requires players to reason about the knowledge of other players in order to play effectively. Existing AI players are specialised [9,10,17,20] with epistemic strategies that can be abstracted from the underlying search algorithms. We investigate the application of two of these strategies encoded in GDL-III.

The rest of the paper proceeds as follows: Sect. 2 provides background to GGP and Hanabi, Sect. 3 formalises the encoding of epistemic strategies in GDL-III, Sect. 4 outlines the modeling of strategies for Hanabi, and Sect. 5 provides an experimental evaluation of these strategies.

2   Background

The Game Description Language (GDL) [12], and its extension GDL-II for imperfect-information games [19], is a formal language for specifying the rules of strategy games to a general game-playing system. GDL uses a prefix-variant of the syntax of logic programs along with the following special keywords:

(**role** ?r)	?r is a player
(**init** ?f)	feature ?f holds in the initial position
(**true** ?f)	feature ?f holds in the current position
(**legal** ?r ?m)	?r has move M in the current position
(**does** ?r ?m)	player ?r does move M
(**next** ?f)	feature ?f holds in the next position
terminal	the current position is terminal
(**goal** ?r ?v)	player ?r gets payoff ?v
(**sees** ?r ?p)	player ?r observes ?p in the next position
random	the random player (aka. Nature)

GDL-II can be used to describe a variety of commonly played imperfect-information games (see http://ggpserver.general-game-playing.de).

Example. Hanabi is a fully cooperative, incomplete information game where a team of two to five players work together to play cards from a deck in order to complete up to five stacks of sequentially numbered cards. Crucially, each player

---

[1] https://en.wikipedia.org/wiki/Hanabi_(card_game).

only sees the cards of the other players' and must therefore rely on those other players to inform them about cards in their own hand in order to play correctly.

The game uses a special deck of 50 cards consisting of five colours where each colour has 10 ranked cards from 1 to 5. Each player begins with a randomly dealt hand of four or five cards where a player's hand is held with the cards facing away such that only the other players can see their colour and rank. Play proceeds with players taking turns to select one of three types of actions:

- **Play**: Select any one of the player's own cards to reveal; and add it to a stack of the same colour if its number is the next in sequence for that stack.
- **Discard**: Select any one of the player's own cards to discard from the game.
- **Hint**: Select another player and declare the positions of all cards in their hand that share the same colour or rank.

The game also features two types of tokens, *information* and *life* tokens. Information tokens restrict the number of hint actions that can be made and can only be regained when a discard action is taken. Life tokens limit the number of unsuccessful play actions where all players lose if the last life token is lost.

An example set of rules for a Hanabi version with just 2 players, colours and ranks and a hand size with 1 card position is shown below[2]:

```
(role random) (role player1) (role player2)

(cardCol ukn) (cardCol red) (cardCol grn)
(cardNum ukn) (cardNum 1) (cardNum 2)
(position 1) (succ 0 1) (succ 1 2)

(<= (card ?colour ?number)
    (cardCol ?colour) (cardNum ?number))

(<= (legal ?r (play ?pos))
    (true (control ?r))
    (true (hand ?r ?pos (card ?col ?num))))

(<= (legal ?r noop)
    (role ?r) (not (true (control ?r))))

(<= (sees ?r1 (does ?r2
                   (play ?pos (card ?col ?num))))
    (role ?r1) (does ?r2 (play ?pos))
    (true (hand ?r2 ?pos (card ?col ?num))))

(<= (next (stacksize ?col ?x))
    (does ?r (play ?pos (card ?col ?num)))
    (correct_play ?r ?pos))

(<= (correct_play ?r ?pos)
```

---

[2] For the full Hanabi GDL-II encoding see: https://git.io/fhbVz.

```
(true (hand ?r ?pos (card ?col ?num)))
(true (stacksize ?col ?prev))
(succ ?prev ?num))
```

Semantics. A game description Σ that obeys GDL syntactic restrictions [16] determines a state transition system as follows. A move m is *legal* for role r in state $s = \{f1...fn\}$ if (**legal** r m) follows from Σ and the facts $s^{\text{true}} = \{(\textbf{true}\ f1)...(\textbf{true}\ fn)\}$. Given a state s and a joint move M (i.e. a legal move m for every player r), the *updated state* $u(M, s)$ consists of all f for which (**next** f) follows from Σ and the facts $s^{\text{true}} = \{(\textbf{true}\ f1)...(\textbf{true}\ fn)\}$ and $M^{\text{does}} = \{(\textbf{does}\ r1\ m1)...(\textbf{does}\ rk\ mk)\}$. The *observations* for player r after joint move M in state s are given by the derivable instances of (**sees** r p) in the same way. For example, consider $s = \{(\text{hand player1 1 (card red 2))},$ (control player1), (stacksize red 1)}. Then the move (play 1) is legal for player1 and noop is legal for both player2 and **random**. The state resulting from this legal move is {(stacksize red 2)}, where both players will observe (**does** player1 (play 1 (card red 2))).

Definition 1 ([19]). *The* semantics *of a valid GDL-II game description G is given by*

- $R = \{r\colon G \models (\textbf{role}\ r)\}$
- $s_0 = \{f\colon G \models (\textbf{init}\ f)\}$
- $t = \{S\colon G \cup s^{\text{true}} \models \textbf{terminal}\}$
- $l = \{(r, m, S)\colon G \cup s^{\text{true}} \models (\textbf{legal}\ r\ m)\}$
- $u(M, S) = \{f\colon G \cup M^{\text{does}} \cup s^{\text{true}} \models (\textbf{next}\ f)\}$
- $\mathcal{I} = \{(r, M, S, p)\colon G \cup M^{\text{does}} \cup s^{\text{true}} \models (\textbf{sees}\ r\ p)\}$
- $g = \{(r, v, S)\colon G \cup s^{\text{true}} \models (\textbf{goal}\ r\ v)\}$

Legal play sequences are sequences of joint moves, beginning in the initial state, in which all players always select a legal move. Legal play sequences δ and δ' are *indistinguishable* by player r (i.e., are in the same information set), written $\delta \sim_r \delta'$, iff r's moves and observations are identical in δ and δ'.

GDL-III. Hanabi can be sufficiently described as a GDL-II game with incomplete information where the value of the colour and number of cards in each position in a player's hand is hidden until they play or discard it. However, playing optimally requires players to maintain a knowledge base for each player of the known facts about each card in their hand to determine which cards are correct plays and what information other players need to identify correct plays. This provides an opportunity to use an epistemic strategy encoded in a recent extension of GDL to reason about a player's knowledge when selecting an action. GDL-III [22] introduces the following keywords in order to support the axiomatisation of game rules that depend on the knowledge of players:

(**knows** ?r ?p)	player ?r knows ?p in the current state
(**knows** ?p)	?p is common knowledge in the current state

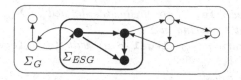

Fig. 1. Epistemic strategy game Σ_{ESG} is a sub-game of an existing GDL-II game Σ_G.

The semantics for GDL-III is more involved since the transition system is now also shaped by what players know: Let (s, K) be an arbitrary *knowledge state*, where s is a state and K a set of ground **knows**-instances. Move m for role r is *legal* in (s, K) if (**legal** r m) can be derived from $\Sigma \cup s^{\mathrm{true}} \cup K$. The *resulting state* when joint move M is executed in (s, K) consists of all f such that (**next** f) can be derived from $\Sigma \cup s^{\mathrm{true}} \cup K \cup M^{\mathrm{does}}$. Player r *observes* p when M is executed in (s, K) iff (**sees** r p) can be derived from $G \cup s^{\mathrm{true}} \cup K \cup M^{\mathrm{does}}$. By definition, the initial state is common knowledge among the agents. A *legal play sequence* is a sequence of joint moves, beginning in the initial knowledge state, in which all players always select a legal action. Legal play sequences δ and δ' are *indistinguishable* by r, written $\delta \sim_r \delta'$, iff r's moves and observations are the same in δ and δ'. Player r *knows* a property ϕ after a legal play sequence δ iff ϕ is true in all δ' that r cannot distinguish from δ. Finally, ϕ is *common knowledge* after δ if it holds after all δ' in the transitive closure $\sim^C$ of $\bigcup_{r \in R} \sim_r$.

3   A Framework for Epistemic Strategies

In this section we formally introduce the framework for representing epistemic strategies as GDL-III games. We provide a general method to transform an arbitrary GDL-II game into a GDL-III game through the addition of legal move definitions based on the knowledge of players. We then prove a number of properties; starting with the correctness of the framework, through to establishing that the framework defines a computationally interesting GDL-III fragment.

3.1   Defining the Transformation

The general approach of enriching a given "source" game with a set of epistemic strategy rules is illustrated in Fig. 1. A game can be viewed as a state transition system, with the joint moves defining the transition between states. Adding a set of epistemic strategy rules limits the available moves to only those that follow the strategy; effectively pruning the state space and defining a sub-game.

We define the GDL axiomatisation of this process in two stages. We first define a transformation function for arbitrary GDL-II games and then define how this transformation is applied to create a GDL-III game.

Definition 2. *Let Σ_G be a GDL axiomatisation of a GDL-II game. The set of rules $\tau(\Sigma_G)$ is obtained from Σ_G as follows:*

- *if* ep_legal *or* src_legal *are predicates defined in* Σ_G, *replace them with unique names not occurring in* Σ_G,
- *replace every occurrence of* (**legal** r a) *with* (src_legal r a), *for arbitrary* r *and* a; *and*
- *add the following rules:*

```
(<= (legal ?r ?a) (ep_legal ?r ?a))
(<= (ep_legal ?r) (ep_legal ?r ?a))
(<= (legal ?r ?a)
    (not (ep_legal ?r)) (src_legal ?r ?a))
```

Definition 2 provides a purely syntactic transformation of a GDL-II game. It should also be noted that the resulting transformation is itself a legal GDL-II game. Furthermore, because (ep_legal ?r ?a) is used but not defined in the game $\tau(\Sigma_G)$, the non-monotonicity of default negation in the third additional rule ensures logical equivalent to the original game.

However, the default negation of the third rule can also be used to provide a mechanism for incorporating an epistemic strategy, which we outline now. For the following definition we rely on the notion of the *dependency graph* of a logic program and the related concepts of a logic program being *stratified* [3] and *safe* (or *allowed*) [15]. It should be noted that the GDL specification requires that all game descriptions are stratified and safe [16].

Definition 3 (Epistemic Enrichment). *For a GDL-II game* Σ_G *and a transformation function* τ *satisfying Definition 2, let* Σ_{ESG} *be the GDL-III game:*

$$\Sigma_{ESG} = \tau(\Sigma_G) \cup \Sigma_{strategy} , where$$

$\Sigma_{strategy}$ *is a set of GDL-III rules satisfying the following requirements:*

- $\Sigma_{strategy}$ *must be both stratified and safe,*
- *rule heads cannot include the GDL keyword predicates (e.g.,* **next**, **true***),*
- *rule heads cannot include auxiliary predicates that have been defined in* Σ_G,
- *any rule containing* (ep_legal r a) *in the head, for arbitrary* r *and* a, *must also include* (src_legal r a) *as a positive dependency in the dependency graph for* $\Sigma_{strategy}$.

Definition 3 turns a GDL-II game into a GDL-III game by enriching the original game with a set of rules where the **knows** predicate can occur in the body of the rule. This new GDL-III game can be used by a GDL-III reasoner to play the original game; where it chooses actions if there is an epistemic strategy for the given state or simply plays a legal move otherwise. The effectiveness of the resulting player will be dependent on the extent to which the epistemic rules are able to encode a winning strategy.

3.2   Properties

In this section we establish some basic properties of our framework. In particularly, we show that, despite the difference in semantics between GDL-II and GDL-III, there is a clear link between the GDL-III game constructed through the method defined in Definition 3 and the original GDL-II game.

Proposition 1. *For any syntactically correct GDL-II game Σ_G and set of epistemic strategy rules $\Sigma_{strategy}$, the resulting epistemically enriched game Σ_{ESG} is a syntactically correct GDL-III game.*

Proof. Σ_G satisfies GDL keyword restrictions, is safe and stratified (see [16]). These properties are preserved by $\tau(\Sigma_G)$. The restriction on $\Sigma_{strategy}$ also satisfies these properties such that $\tau(\Sigma_G) \cup \Sigma_{strategy}$ is also stratified and safe.   □

In order to establish further properties we use the notion of *legal play sequences*, where a sequence of moves $M_1, \ldots, M_n$ corresponds to legal moves from a starting state s_0, such that $s_i = u(M_i, s_{i-1})$, for state update function u. In particular, we establish that the game Σ_{ESG} is a restriction over the game Σ_G.

Proposition 2. *Given any GDL-II game Σ_G and set of epistemic strategy rules $\Sigma_{strategy}$, then for the resulting GDL-III game Σ_{ESG}, and legal play sequence of Σ_{ESG}, $M_1, \ldots, M_n$ with corresponding states $s_0, s_1, \ldots, s_n$ it holds that:*

- *the sequence $M_1, \ldots, M_n$ is a legal play sequence of Σ_G with corresponding states $s'_0, s'_1, \ldots, s'_n$.*
- *if s_n is a terminal state of Σ_{ESG} then s'_n is a terminal state of Σ_G.*
- *for each player r the goal value $g(r, s_n)$ of Σ_{ESG} in state s_n will be the same as the goal value $g(r, s'_n)$ in game Σ_G and corresponding state s'_n.*

Proof. By induction on the states in the play sequence. The initial state of a game in GDL-II/III corresponds to the fluents defined by `init`. Since $\tau(\Sigma_G)$ and $\Sigma_{strategy}$ do not introduce any new fluents or modify `init` therefore the objective fluents of the initial state of Σ_G will be identical to Σ_{ESG}; furthermore $\tau(\Sigma_G)$ does not change `goal` values or how `terminal` is determined.

Consider the first move M_1 and an arbitrary role r of Σ_{ESG}; $M_1(r) = a$ is a legal action for r in s_0 for game Σ_{ESG}. Hence either (`ep_legal r a`) or (`src_legal r a`) is true in s_0 (Definition 2). But if (`ep_legal r a`) is true then (`src_legal r a`) must also be true (by the dependency restriction in Definition 3). But (`src_legal r a`) is simply the rewrite of (`legal r a`) from the original game, hence (`legal r a`) must also be a legal move for role r in s'_0 for game Σ_G. Hence M_1 is also a legal move in Σ_G and so the transition from s'_0 by M_1 will also be a state s'_1 of Σ_G. Furthermore if s'_1 is terminal in Σ_{ESG} it will also be a terminal state of Σ_G with identical goal values for each player.

The same argument holds for the induction step; where assuming $M_1, \ldots, M_i$ ($i < n$) is also a legal sequence of Σ_G with states $s'_0, \ldots, s'_i$, then s'_{i+1} is also a state of Σ_G with the correct termination and goal values.   □

Now, Proposition 2 establishes that Σ_{ESG} represents a restriction over Σ_G and allows this restriction to be determined by a player's knowledge or the common knowledge of all players. However, in general GDL-III is a strictly more expressive language than GDL-II, since, determining if a game terminates in GDL-III is in general undecidable, even when subject to the usual syntactic restrictions that ensure the finiteness of the state space in GDL-II [22]. Consequently, the syntactic restrictions of Definition 3 result in the identification of an interesting fragment of GDL-III.

Proposition 3. *Given any GDL-II game Σ_G and set of epistemic strategy rules $\Sigma_{strategy}$, then for the resulting epistemically enriched game Σ_{ESG} determining if Σ_{ESG} terminates is decidable.*

Proof. A direct consequence of Proposition 2. Any legal play sequence of Σ_{ESG}, $M_1, \ldots, M_n$ with corresponding states $s_0, s_1, \ldots, s_n$ where s_n is a terminal state, is also a legal play sequence of Σ_G and corresponding state s'_n is also a terminal state of Σ_G. But Σ_G is a GDL-II game so determining termination is decidable, hence determining termination of Σ_{ESG} is also decidable.   □

The key to the decidability of Σ_{ESG} is that the syntactic restrictions ensures that the truth of fluents in a state or the termination of a state is independent of the knowledge of players. This is not true of GDL-III in general.

Hence not only does the proposed framework allow for the encoding of epistemic strategies for playing specific games, but it does so in a manner that preserves the decidability of the original game. This means that the encoding of epistemic strategies in GDL-III is both of theoretical and potentially of practical interest. In the following section we apply this theory to encoding strategies for the game of Hanabi and show the efficacy of the approach.

4 Encoding Epistemic Strategies in Hanabi

This section outlines the GDL-III encoding of two epistemic rule-based strategies that can be used to extend the original GDL-II Hanabi source game to allow players to reason about game knowledge when selecting a move.

The two strategies considered are the *information strategy* and the *implicit strategy*. The information strategy takes a conservative approach to playing Hanabi. Hint moves are given to inform other players of card numbers or colours that are unknown to them, and cards are only played if its holder knows that it is playable based on knowing both its colour and number. This represents an individualistic strategy since it is agnostic to the strategies of other players.

In contrast, the implicit strategy represents a more optimistic approach to playing Hanabi. It selects moves based on players' knowledge, but with the additional implicit assumption that all players are following the same, or at least a pre-agreed, strategy. Hence, this strategy resembles that of an experienced group of players with an agreed convention on how to play the game.

4.1   The Information Strategy

This strategy models the use of hint moves to convey information of card properties that are not known to other players. It is adapted from the *Outer-State* strategy presented in Osawa's original paper on solving Hanabi [17], where players inform each other of properties that have not yet been stated. A player uses this strategy by selecting the first rule whose antecedent is satisfied from a list of epistemic rules. We provide a description of these rules, but also show a sample of the GDL encodings to illustrate the correspondence between the rule explanations and their GDL-III instantiations[3]:

1. If a known playable card is in our hand, play that card.

```
(<= (ep_legal ?r (play ?pos))
    (src_legal ?r (play ?pos))
    (knows ?r (correct_play ?r ?pos)))
```

2. If a known dead card is in our hand and there are clue tokens to be gained, discard that card.

```
(<= (ep_legal ?r (discard ?pos))
    (src_legal ?r (discard ?pos))
    (not (has_legal_play ?r))
    (knows ?r (has_dead_card ?r ?pos)))
```

3. If no known playable cards is in any hand, discard the card with the highest known number.

```
(<= (ep_legal ?r (discard ?pos))
    (src_legal ?r (discard ?pos))
    (not (has_legal_play ?r))
    (not (has_legal_discard ?r))
    (not (has_legal_hint_num ?r))
    (not (has_unknown_card ?r))
    (knows_highest_card ?r ?pos ?num))
```

4. If there are clue tokens, and a player has a playable card, hint its number if not known.

```
(<= (ep_legal ?r (hint ?r1 ?num))
    (src_legal ?r (hint ?r1 ?num))
    (not (has_legal_play ?r))
    (not (has_legal_discard ?r))
    (true (hand ?r1 ?pos (card ?col ?num)))
    (correct_play ?r1 ?pos)
    (not (knows ?r1 (hand ?r1 ?pos (cardNum ?num))))))
```

5. If there are clue tokens, and a player has a playable card, hint its colour if not known.
6. If there are clue tokens, hint a random card's number that is not known.
7. If there are clue tokens, hint a random card's colour.
8. If no clue tokens available, discard the highest known number card in hand.

---

[3] For the complete GDL-III information strategy encoding see: https://git.io/fhbVo.

4.2   The Implicit Strategy

This strategy aims to encode additional facts in certain moves which other players following the same strategy can infer when they observe those moves. This differs from the information strategy to use the hint moves to imply playability of a card based on the property hinted for that card. In the implicit strategy, the number property of a card is hinted if the card is playable otherwise the colour is hinted instead. The resulting strategy rules are as follows[4]:

1. If a known playable card is in our hand, play that card.
2. If a known dead card is in our hand and clue tokens to be gained, discard that card.
3. If no known playable cards are in any hand, then discard the card with the highest known number.
4. If there are clue tokens, and a player has a playable card, hint its number.
5. If there are clue tokens, and no known playable cards in any hands, hint a random card's colour.
6. If no clue tokens available, discard a random card.

5   Evaluation

Four experiments were conducted to evaluate the increased performance of the GDL-II game Hanabi extended with a GDL-III epistemic strategy. The first experiment investigated the practical lower bound of playing Hanabi without any strategy, where players select (legal) actions randomly. The second and third experiments evaluated the performance of the information and the implicit strategy respectively. Finally, the fourth experiment provided a crude upper bound by modelling the case of playing Hanabi where the cards in every players' hand is common knowledge, with the only unknowns being the cards in the deck.

Each experiment was run for 50 games each with six configurations of number of players and cards per player and a play clock of 10 s. Table 1 below shows the

Table 1. Outline of Hanabi game configurations

Id	nPl	Colours	Numbers	nHand	MaxScore
1	2	R,G	1,2	1	4
2	2	R,G	1,2,3	2	6
3	2	R,G	1,1,2,3	2	6
4	3	R,G,B	1,2	1	6
5	3	R,G,B	1,2,3	2	9
6	4	R,G,B,Y	1,2	1	8

---

[4] For the complete GDL-III implicit strategy encoding see: https://git.io/fhbVK.

details of each configuration of number of players, **nPl**, with hand size **nHand** and number card counts for each colour. The **Colours** and **Numbers** column indicate the colour and number values used to build the deck.

The experiments were run on a 2.5GHz MacOS laptop with 16GB memory. A GDL-III knowledge reasoner was implemented in ASP according to the formula in [21] to calculate knowledge at each timestep. This reasoner was adapted for time-restrained GGP matches to incorporate a timeout equal to the play clock that only returns an approximation of the knowledge state. As a result, games were played within a reasonable amount of time ranging from 5 s for games with a decksize of 4 to 4 min at the maximum decksize of 9.

Figure 2 below displays the comparison of the average scores for each experiment from the above table grouped by each of the six configurations. There is a clear increase in the performance of the players following an epistemic strategy to select moves based on their own knowledge and that of other players. For the random players without a strategy, multiple games were lost due to too many incorrect plays resulting in an average score almost always less than 2. On the other hand, the information and implicit strategies provided a similar increase in performance. The information strategy was seen to achieve a more consistent score, where legal play moves tend to follow from a sequence of two hint moves. While this strategy is less variable, it is also highly dependent on clue tokens for increasing its scalability to larger game configurations. The implicit strategy experiences more variance in achievable scores but is able to scale to games with more players due its ability to signal playability of a card with a single hint move. It is also worth noting that for configurations 5 and 6, the knowledge calculation for most time steps exceeded the timeout and resulted in players relying on an incomplete set of knowledge facts. Despite this, games using epistemic strategies were still able to outperform those with players playing randomly.

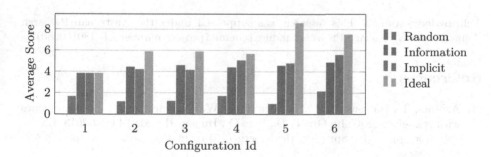

Fig. 2. Comparison of endgame results

6 Conclusion

We developed a framework for modelling epistemic games in GGP allowing players to reason about their knowledge of the game state using epistemic strategies.

We presented a formal approach to represent these strategies for GDL-II games as specialised GDL-III games. This approach was evaluated for the game of Hanabi where two epistemic strategies, the information and implicit strategies, were used to select player actions. From the evaluation, it was observed that the experimental upper bound for reasoning with complete knowledge was able to achieve a perfect score most of the time, although not guaranteed for all configurations. We then found that the information strategy achieved a more consistent score in contrast to the implicit strategy which achieved a higher maximum score. This was done by encoding an implicit recommendation to play a card if a player was hinted its number, which resulted in more effective use of limited hint actions. Yet this advantage was lost when multiple hints were given for a combination of both playable and unplayable cards.

In terms of related work, a number of logical frameworks exist for reasoning about the strategic abilities of players in games [8,14,23], mostly based on Alternating-time Temporal Logic [2]. However, these logics are based on modalities for the existence of strategies and do not provide means for specifying them [6]. An exception is a recent special-purpose modal logic for reasoning about strategies [25]. The main differences to our framework are that their strategies can only be conditioned on state properties and not on players' knowledge, and that using their logic would first require the development of a special-purpose automated theorem prover.

As future work, probabilistic reasoning could be incorporated within the implicit strategy to allow the same action to indicate either playable or non-playable cards with an associated probability, which could also be used to condition hint moves in some cases. Another avenue would consider optimal action selection using epistemic strategies combined with game independent strategies such as Monte-Carlo Tree Search to reduce the reliance on game specific strategy rules.

Acknowledgements. This research was supported under the Australian Research Council's (ARC) Linkage Projects funding scheme (project number LP 180100080).

References

1. Ågotnes, T., Harrenstein, P., van der Hoek, W., Wooldridge, M.: Boolean games with epistemic goals. In: Grossi, D., Roy, O., Huang, H. (eds.) LORI 2013. LNCS, vol. 8196, pp. 1–14. Springer, Heidelberg (2013). https://doi.org/10.1007/978-3-642-40948-6_1
2. Alur, R., Henzinger, T.A., Kupferman, O.: Alternating-time temporal logic. J. ACM **49**(5), 672–713 (2002)
3. Apt, K.R., Blair, H.A., Walker, A.: Towards a theory of declarative knowledge. In: Minker, J. (ed.) Foundations of Deductive Databases and Logic Programming, pp. 89–148. Morgan Kaufmann Publishers Inc., San Francisco (1987)
4. Baffier, J., et al.: Hanabi is NP-complete, even for cheaters who look at their cards. In: FUN. LIPIcs, Schloss Dagstuhl - Leibniz-Zentrum fuer Informatik, vol. 49, pp. 4:1–4:17 (2016)
5. Bard, N., et al.: The Hanabi challenge: a new frontier for AI research. CoRR abs/1902.00506 (2019). http://arxiv.org/abs/1902.00506

6. van Benthem, J.: In praise of strategies. In: Eijck, J.V., Verbrugge, R. (eds.) Games, Actions, and Social Software. ILLC Scientific Publications, Institute for Logic, Language and Computation (ILLC), University of Amsterdam (2008)
7. Bouzy, B.: Playing Hanabi near-optimally. In: Winands, M.H.M., van den Herik, H.J., Kosters, W.A. (eds.) ACG 2017. LNCS, vol. 10664, pp. 51–62. Springer, Cham (2017). https://doi.org/10.1007/978-3-319-71649-7_5
8. Chatterjee, K., Henzinger, T.A., Piterman, N.: Strategy logic. Inf. Comput. **208**(6), 677–693 (2010)
9. Cox, C., De Silva, J., Deorsey, P., Kenter, F.H., Retter, T., Tobin, J.: How to make the perfect fireworks display: two strategies for Hanabi. Math. Mag. **88**(5), 323–336 (2015)
10. Eger, M., Martens, C., Córdoba, M.A.: An intentional AI for Hanabi. In: IEEE Conference on Computational Intelligence and Games (CIG), pp. 68–75. IEEE (2017)
11. Genesereth, M., Love, N., Pell, B.: General game playing: overview of the AAAI competition. AI Mag. **26**(2), 62 (2005)
12. Genesereth, M., Thielscher, M.: General game playing. Synth. Lect. Artif. Intell. Mach. Learn. **8**(2), 20, 172–178 (2014)
13. Genesereth, M.R., Björnsson, Y.: The international general game playing competition. AI Mag. **34**(2), 107–111 (2013)
14. van der Hoek, W., Jamroga, W., Wooldridge, M.: A logic for strategic reasoning. In: Proceedings of the Fourth International Joint Conference on Autonomous Agents and Multi Agent Systems (AAMAS 2005), pp. 157–164. ACM (2005)
15. Lloyd, J., Topor, R.: A basis for deductive database systems II. J. Logic Program. **3**(1), 55–67 (1986)
16. Love, N., Henrichs, T., Haley, D., Schkufza, E., Genesereth, M.: General game playing: game description language specification. Technical report, Stanford Logic Group Computer Science Department Stanford University (2006)
17. Osawa, H.: Solving Hanabi: estimating hands by opponent's actions in cooperative game with incomplete information. In: AAAI Workshop: Computer Poker and Imperfect Information, pp. 37–43 (2015)
18. Schiffel, S., Thielscher, M.: Reasoning about general games described in GDL-II. In: AAAI, vol. 11, pp. 846–851 (2011)
19. Schiffel, S., Thielscher, M.: Representing and reasoning about the rules of general games with imperfect information. JAIR **49**, 171–206 (2014)
20. van den Bergh, M.J.H., Hommelberg, A., Kosters, W.A., Spieksma, F.M.: Aspects of the cooperative card game Hanabi. In: Bosse, T., Bredeweg, B. (eds.) BNAIC 2016. CCIS, vol. 765, pp. 93–105. Springer, Cham (2017). https://doi.org/10.1007/978-3-319-67468-1_7
21. Thielscher, M.: A formal description language for epistemic games. Technical report, UNSW-CSE-TR-201708, The University of New South Wales (2017)
22. Thielscher, M.: GDL-III: a description language for epistemic general game playing. In: Sierra, C. (ed.) IJCAI, pp. 1276–1282. AAAI Press, Melbourne, August 2017
23. Walther, D., van der Hoek, W., Wooldridge, M.: Alternating-time temporal logic with explicit strategies. In: Proceedings of the 11th Conference on Theoretical Aspects of Rationality and Knowledge, pp. 269–278. ACM (2007)
24. Walton-Rivers, J., Williams, P.R., Bartle, R., Perez-Liebana, D., Lucas, S.M.: Evaluating and modelling Hanabi-playing agents. In: 2017 IEEE Congress on Evolutionary Computation (CEC), pp. 1382–1389. IEEE (2017)
25. Zhang, D., Thielscher, M.: A logic for reasoning about game strategies. In: AAAI, vol. 15, pp. 1671–1677 (2015)

Non-zero-sum Stackelberg Budget Allocation Game for Computational Advertising

Daisuke Hatano[1]($\boxtimes$), Yuko Kuroki[2]($\boxtimes$), Yasushi Kawase[3]($\boxtimes$),
Hanna Sumita[4]($\boxtimes$), Naonori Kakimura[5]($\boxtimes$), and Ken-ichi Kawarabayashi[6]($\boxtimes$)

[1] RIKEN AIP, Tokyo, Japan
`daisuke.hatano@riken.jp`
[2] The University of Tokyo, Tokyo, Japan
`ykuroki@ms.k.u-tokyo.ac.jp`
[3] Tokyo Institute of Technology and RIKEN AIP, Tokyo, Japan
`kawase.y.ab@m.titech.ac.jp`
[4] Tokyo Metropolitan University, Hachioji, Japan
`sumita@tmu.ac.jp`
[5] Keio University, Tokyo, Japan
`kakimura@math.keio.ac.jp`
[6] National Institute of Informatics, Tokyo, Japan
`k_keniti@nii.ac.jp`

Abstract. Computational advertising has been studied to design efficient marketing strategies that maximize the number of acquired customers. In an increased competitive market, however, a market leader (a *leader*) requires the acquisition of new customers as well as the retention of her loyal customers because there often exists a competitor (a *follower*) who tries to attract customers away from the market leader. In this paper, we formalize a new model called the *Stackelberg budget allocation game with a bipartite influence model* by extending a budget allocation problem over a bipartite graph to a Stackelberg game. To find a *strong Stackelberg equilibrium*, a solution concept of the Stackelberg game, we propose two algorithms: an approximation algorithm with provable guarantees and an efficient heuristic algorithm. In addition, for a special case where customers are disjoint, we propose an exact algorithm based on linear programming. Our experiments using real-world datasets demonstrate that our algorithms outperform a baseline algorithm even when the follower is a powerful competitor.

Keywords: Stackelberg game · Budget allocation problem · Submodular

1 Introduction

An aim of *computational advertising* is to find the best advertisement that can help build customers loyalty. More specifically, the purpose of advertisers is to

© Springer Nature Switzerland AG 2019
A. C. Nayak and A. Sharma (Eds.): PRICAI 2019, LNAI 11670, pp. 568–582, 2019.
https://doi.org/10.1007/978-3-030-29908-8_45

devise an optimum allocation of budgets to *media*, such as newspapers, radio stations, TV, and websites, in order to maximize the number of activated customers. Recently, Alon *et al.* [1] proposed a model to deal with a simple case of the problem, called a *bipartite influence model*. In this study, we shall extend the model by integrating a game-theoretic framework, called the non-zero-sum *Stackelberg game* framework. Let us explain the model more precisely below.

In the bipartite influence model, we consider a bipartite graph where one side is a set of *media*, the other is a set of *customers*, and each edge is associated with a probability. Intuitively, each edge between a medium and a customer indicates that the customer is influenced by the medium with some given probability that depends on the budget allocated to the medium. We aim to allocate budgets on media so that the expected number of activated customers is maximized. The problem can be formulated as a combinatorial optimization problem. Constant-factor approximation algorithms for the problem have been developed in a framework of submodularity [1,11,12].

In this paper, we shall try to extend the above-mentioned model to deal with a situation of a duopoly where a market leader has occupied the market of a certain product for a long time and a competitor tries to break into the market. The competitor tries to grab the share of the market by aggressively marketing its product. On the other hand, the market leader wants to gain customers and retain her loyal customers simultaneously. This implies that the leader's gain does not necessarily result in the competitor's loss. In order to capture the dynamics of this market, we exploit a *Stackelberg game* [13] framework to model the interactions between the market leader and the competitor. The Stackelberg game is a two-player two-period game, in which one player (a *leader*) can commit to an action before the other player (a *follower*) plays an action. A standard solution concept of this game is the *strong Stackelberg equilibrium*, which is an optimal solution maximizing the leader's utility under the constraint that the follower plays a best response to the leader's action (i.e., intended to maximize the follower's utility).

The Stackelberg game matches to model our problem setting because the leader wants to increase the number of activated customers, and at the same time, prevent the outflow of her customers, which is achieved by finding a strong Stackelberg equilibrium. In a strong Stackelberg equilibrium, the leader plays a *mixed strategy* and the follower plays a *pure strategy*, where pure strategy and mixed strategy correspond to a budget allocation and a probability distribution over the pure strategies, respectively.

In this paper, we propose a new model called the *Stackelberg budget allocation game with a bipartite influence model*, which is an extension of the budget allocation problem presented in [1]. The difficulties of our game lie in the leader's utility function. Our game belongs to a non-zero-sum game, and the utility function is a submodular (nonlinear) function even when the follower's action is fixed. It is hard to construct an approximation algorithms by the following reasons: (i) the cumbersome constraint that the follower optimally responds and (ii) the leader's utility may be non-linearly changed by a follower's strategy. Thus, existing techniques for submodular functions cannot be directly applied to our problem. Furthermore, the leader's utility function is not necessarily monotone, that is, the utility does not always increase in the number of allocated budgets. This entails the increment of the number of pure strategies. To design an efficient algorithm is an arduous task.

In this paper, we propose three efficient algorithms:

- We design an approximation algorithm with theoretical guarantee. The key idea to construct an approximation algorithm is to create a zero-sum game close to the original non-zero-sum game, and to find an approximate minimax strategy of the zero-sum game with the aid of submodularity.
- We give an efficient heuristic algorithm that repeatedly finds a leader's pure strategy greedily and uniformly picks from the pure strategies. The running time is polynomial in the leader's budget. This heuristic can deal with a situation that the leader should not spend up her whole budget due to the non-monotonicity of the utility function. We also evaluate its performance by numerical experiments.
- If the customers are disjoint, we prove that a strong Stackelberg equilibrium can be found efficiently even when the leader has exponentially many pure strategies by using the multiple linear programming (LP for short) formulations. The point in the disjoint case is that we can aggregate a leader's mixed strategy to a fractional budget allocation. At the same time we can recover a mixed strategy in a compact representation without loss of the leader's utility. This enables us to save memories to keep a mixed strategy and reduce the size of LP instances.

The rest of the paper is organized as follows: We describe related work in Sect. 2 and define notations in Sect. 3. We formalize our model and analyze its (mathematical) properties in Sect. 4. We then provide an approximation and a heuristic algorithms in Sect. 5, and provide an exact algorithm for the disjoint customers in Sect. 6. In Sect. 7, we empirically show the performance of our algorithm, and finally we conclude the study in Sect. 8.

2 Related Work

Our problem setting can be viewed as a non-monotone non-zero-sum Stackelberg game with submodular functions. Vanek *et al.* [16] modeled a non-zero-sum Stackelberg game with submodular functions where the defender (the leader) cares about minimizing the loss of her utility. In our game, the leader maximizes her utility incorporating her loss against the follower's action. Thus, the goal of the leader is different. Moreover, direct application of their technique to find a Stackelberg equilibrium does not seem to work well in our setting. Recently, Wilder *et al.* [17] extended a bipartite influence model to a zero-sum Stackelberg game, which is closely related to our problem setting. They proved that the problem is APX-hard, while it has FPTAS for some special cases.

In combinatorial optimization and machine learning, approximation algorithms for maximizing submodular functions under certain constraints have been extensively studied [9]. Our problem can be viewed as a submodular maximization under a best-response constraint, which is more cumbersome than typical constraints in the submodular maximization literature (e.g., cardinality constraint and knapsack constraint).

The budget allocation problem with the bipartite influence model has been extended in [6,10,11,15]. In particular, some formulations have incorporated the view of the multi-agent system. Maehara et al. [10] extended a budget allocation in the bipartite influence model to a strategic form game, called *the budget allocation game with a bipartite influence model*. Hatano et al. [6] extended the budget allocation problem to the problem with two participants; advertiser and match maker. In the problem, there exist multiple advertisers who cooperatively maximize the influence on customers and single match maker who allocates slots of media to advertisers.

3 Preliminary

Let $\mathbb{Z}_+$ be the set of non-negative integers. For an integer $k \in \mathbb{Z}_+$, let $[k]$ be the set $\{1, 2, \ldots, k\}$. In this section, we describe the budget allocation problem with a bipartite influence model and the Stackelberg game.

3.1 Bipartite Influence Model

Let $G = (U, V; E)$ be a bipartite graph, where (U, V) is a bipartition and $E \subseteq U \times V$ is a set of edges. Each vertex $u \in U$ corresponds to a medium and $v \in V$ corresponds to a customer. Let n and m be the sizes of U and V, respectively. Each edge $uv \in E$ is associated with a probability $p_{uv} \in [0, 1]$, which means that allocating a budget to medium $u \in U$ activates customer $v \in V$ with probability p_{uv}. We assume that the activation events are independent. The advertiser has a total available budget of $k \in \mathbb{Z}_+$, and each medium $u \in U$ has a slot to which the advertiser can allocate her budget. The goal is to find the optimal budget allocation $z \in \{0, 1\}^U$ with $\sum_{u \in U} z_u \leq k$ that maximizes the number of activated customers. Throughout this paper, we identify a set S of media with its characteristic vector $z_S \in \{0, 1\}^U$. A probability that a customer $v \in V$ is activated by the advertiser's trial from media in U is given by

$$P_v(z) = 1 - \prod_{u \in N_v : z_u = 1}(1 - p_{uv}), \tag{1}$$

where $N_v = \{u \mid uv \in E\}$ is the set of the neighbors of v. The expected number of customers activated through the budget allocation z is given by $\sum_{v \in V} P_v(z)$. The objective of the budget allocation problem with a bipartite influence model is to find z that maximizes $\sum_{v \in V} P_v(z)$ subject to $\sum_{u \in U} z_u \leq k$.

The function $P_v(z)$ is shown to be a monotone submodular function [14]. Here, a function $f : \{0, 1\}^n \to \mathbb{R}$ is *submodular* if it satisfies $f(x) + f(y) \geq$

$f(x \vee y) + f(x \wedge y)$ for all $x, y \in \{0, 1\}^n$, where $x \vee y$ and $x \wedge y$ denote the vector of component-wise maxima and minima, respectively, i.e., $(x \vee y)_i = \max\{x_i, y_i\}$ and $(x \wedge y)_i = \min\{x_i, y_i\}$. A function f is *monotone* if it satisfies $f(x) \leq f(y)$ for all $x \leq y$, i.e., $x_i \leq y_i$ for all $i \in [n]$. Thus the budget allocation problem is a special case of the submodular maximization problem with a cardinality constraint, and it is well-known that the problem is NP-hard [3] and has a $(1 - 1/e)$-approximation algorithm [12].

3.2   Stackelberg Game

The *Stackelberg game* is played between two players: the leader and the follower. Both players can play a mixed strategy, but it is sufficient to consider that the follower plays a pure strategy. Let S_L and D_F be the sets of pure strategies of the leader and the follower, respectively. We denote the set of mixed strategies of the leader by $D_L = \{x \in [0, 1]^{S_L} \mid \sum_{s \in S_L} x_s = 1\}$, each of which is a probability distribution on pure strategies in S_L. We define $f : D_L \times D_F \to \mathbb{R}$ and $g : D_L \times D_F \to \mathbb{R}$ as utility functions of the leader and the follower, respectively. We define an instance of the game as $\mathcal{G} = (D_L, D_F, f, g)$. Let $\mathrm{BR}(x) = \arg\max_{y \in D_F} g(x, y)$ be the set of best responses of the follower against x. In this game, the leader will commit to play a mixed strategy before the follower plays his strategy. Thus the leader needs to find a mixed strategy x maximizing $f(x, y)$ under the constraint that the follower would choose a best-response pure strategy $y \in \mathrm{BR}(x)$. More precisely, the goal of this game is to find a leader's mixed strategy that forms a strong Stackelberg equilibrium, as indicated below.

Definition 1. *A strong Stackelberg equilibrium of $\mathcal{G}$ is a pair (x^*, y^*) that satisfies $f(x^*, y^*) \geq f(x, y)$ for all $x \in D_L$, $y \in \mathrm{BR}(x)$, and $y^* \in \mathrm{BR}(x^*)$.*

4   Stackelberg Budget Allocation Game

In this section, we extend the budget allocation problem with a bipartite influence model to a Stackelberg game. For any set S_L and $s \in S_L$, we denote by χ_s a characteristic vector in $\{0, 1\}^{S_L}$ such that $(\chi_s)_{s'} = 1$ for $s' = s$ and $(\chi_s)_{s'} = 0$ for $s' \neq s$ ($s' \in S_L$). For a mixed strategy $x \in D_L$, the support of x is the set of pure strategies that is played with non-zero probability under x, i.e., $\mathrm{supp}(x) = \{s \in S_L \mid x_s > 0\}$.

4.1   Definition

Let $G = (U, V; E)$ be a bipartite graph consisting of a set U of n media, a set V of m customers, and a set E of edges between them. For each $uv \in E$, we denote by p_{uv} a probability that a customer v is activated through a medium u by a leader's or a follower's trial, and by $p_{F,uv}$ a probability that a medium u activates a customer v who has been already activated by the leader. Two probabilities intuitively mean that p_{uv} is a basic activation probability in the market, and $p_{F,uv}$ is a probability that the follower recaptures customers who were activated by the leader. Let k_L and k_F be the budgets of the leader and the follower, respectively. An instance of the *Stackelberg budget allocation game with a bipartite influence model* is parameterized by $\phi = (G = (U, V; E), \{p_{uv}\}_{uv \in E}, \{p_{F,uv}\}_{uv \in E}, k_L, k_F)$.

We construct a Stackelberg game $\mathcal{G}$ from an instance ϕ as follows. A pure strategy for the leader (respectively the follower) is a set of at most k_L media (respectively k_F media). D_L and D_F of the game $\mathcal{G}$ are defined by setting $S_L = \{z \in \{0,1\}^U \mid \sum_{u \in U} z_u \leq k_L\}$ (or equivalently $S_L = \{S \subseteq U \mid |S| \leq k_L\}$) and $D_F = \{y \in \{0,1\}^U \mid \sum_{u \in U} y_u \leq k_F\}$.

Let $v \in V$ be any customer. Let z and y be a leader's and a follower's pure strategies, respectively. The probability that the leader activates v is given by the Eq. (1). If v is not activated by the leader, then the activation probability for the follower is given by the same basic probability, that is $P_v(y)$. If v is activated by the leader, then the probability that the follower attracts a customer $v \in V$ away from the leader is $P_{F,v}(y) = 1 - \prod_{u \in N_v : y_u = 1}(1 - p_{F,uv})$.

Example 1. We explain the difference between p_{uv} and $p_{F,uv}$. Consider a game instance illustrated in Fig. 1. There are three media u_1, u_2, u_3 and four customers v_1, v_2, v_3, v_4. For an arbitrary edge uv, $p_{uv} = 0.8$ and $p_{F,uv} = 0.5$. The budget for the leader and the follower is $k_L = 2$ and $k_F = 1$, respectively. At first, the leader plays a mixed strategy x that chooses $\{u_1, u_2\}$ w.p. 1. Suppose the situation in Fig. 1(a) where $\{u_1, u_2\}$ is chosen and v_1, v_2, and v_3 are activated w.p. 0.8, who are shown in gray. After that, the follower plays a pure strategy that chooses $\{u_3\}$. In Fig. 1(b), the customer v_2 switches to the follower w.p. 0.5 if v_2 is activated by the leader, and otherwise v_2 is activated w.p. 0.8. Thus, the probability to activate v_2 is $0.96 \cdot 0.5 + 0.04 \cdot 0.8 = 0.512$. In addition, v_4 is activated w.p. 0.8 because v_4 is non-activated.

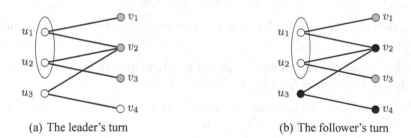

(a) The leader's turn   (b) The follower's turn

Fig. 1. The difference between p_{uv} and $p_{F,uv}$.

The utility functions f and g are given as follows. The expected number of customers that are activated by the leader but do not shift to the follower is given by

$$f(z, y) = \sum_{v \in V} P_v(z)(1 - P_{F,v}(y)).$$

The expected number of activated customers for the follower is given by

$$g(z, y) = \sum_{v \in V} \left(P_v(z) P_{F,v}(y) + (1 - P_v(z)) P_v(y) \right).$$

When the leader uses a mixed strategy x, we abuse the notation and write $P_v(x) = \mathbb{E}_{z \sim x}[P_v(z)]$. Here, for a probability distribution x over a domain D, $z \sim x$ means that we sample $z \in D$ from the distribution x. Similarly, we write $f(x, y) = \mathbb{E}_{z \sim x}[f(z, y)] = \sum_{v \in V} P_v(x)(1 - P_{F,v}(y))$, and $g(x, y) = \mathbb{E}_{z \sim x}[g(z, y)] = \sum_{v \in V} \left(P_v(x) P_{F,v}(y) + (1 - P_v(x)) P_v(y) \right)$.

The goal of the Stackelberg budget allocation game with a bipartite influence model is to find a leader's mixed strategy x in a strong Stackelberg equilibrium of the game $\mathcal{G}$. We define a function f_{BR} that receives a mixed strategy $x \in D_L$ and returns the leader's utility when the follower takes a best response, i.e.,

$$f_{\mathrm{BR}}(x) = \max \left\{ f(x, y) \mid y \in \mathrm{BR}(x) \right\}.$$

We aim to solve

$$\max \quad f_{\mathrm{BR}}(x) \qquad \text{s.t.} \quad x \in D_L. \tag{2}$$

Note that x is an optimal solution to (2) if and only if (x, y) is a strong Stackelberg equilibrium, where y is a best response against x. We can evaluate $f_{\mathrm{BR}}(x)$ for $x \in S_L$ in $O(|D_F| \cdot |E| \cdot |\mathrm{supp}(x)|)$ time by evaluating $f(x, y) |D_F|$ times. To obtain the value of $f(x, y)$, we evaluate $f(z, y)$ for $z \in \mathrm{supp}(x)$, which takes $O(|E|)$ time.

We now see that the leader's optimal strategy may not be a pure strategy.

Example 2. Consider an instance depicted in Fig. 2(a) with $k_L = k_F = 1$. In this case, an optimal strategy for the leader is $x^* = 0.5\chi_{\{u_1\}} + 0.5\chi_{\{u_2\}}$ and $f_{\mathrm{BR}}(x^*) = 1.1$ where the best response of the follower is $\{u_1\}$. However, $f_{\mathrm{BR}}(\chi_{\{u_1\}}) = f_{\mathrm{BR}}(\chi_{\{u_2\}}) = 0.6$ and $f_{\mathrm{BR}}(\chi_{\{u_3\}}) = 0.599$.

We next see that the leader may not use the whole budget in her optimal strategy.

Example 3. Consider an instance depicted in Fig. 2(b) with $k_L = 3$ and $k_F = 1$. Then $f_{\mathrm{BR}}(\chi_U) = 0$ while $f_{\mathrm{BR}}(\chi_{\{u_1\}}) = f_{\mathrm{BR}}(\chi_{\{u_3\}}) = 1$.

There also exists an instance without a pure Stackelberg equilibrium (see Example 4.4 in the upcoming full version).

4.2   Hardness

In this subsection, we show hardness results. We observe that finding a leader's optimal pure strategy when $k_F = 0$ is equivalent to the optimal budget allocation problem. Thus, it is NP-hard to find the leader's mixed strategy that forms a Stackelberg equilibrium even if $k_F = 0$, since our problem (2) when $k_F = 0$ always has the leader's optimal strategy that is pure. It is also known that the approximation ratio $1 - 1/e$ is best possible for the maximum coverage

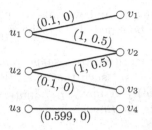

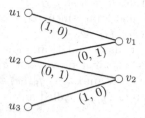

(a) An instance without leader's pure optimal strategies

(b) An instance where the leader should not spend whole her budget

Fig. 2. Examples of the Stackelberg budget allocation game, where the pair of numbers on each edge uv represents the activation probabilities p_{uv} and $p_{F,uv}$.

problem under the assumption that $P \neq NP$ [4]. Hence, our problem (2) is also inapproximable within ratio $1 - 1/e$ unless $P = NP$.

Moreover, when k_F is not a fixed constant, it is even NP-hard to evaluate $f_{BR}(x)$ for a given $x \in D_L$. The proof is reducing from the maximum coverage problem, which is shown to be NP-hard (see e.g., [7]). Given an integer k and a collection of sets $\mathcal{S} = \{S_1, S_2, \ldots, S_n\}$, the *maximum coverage problem* is to find a subset $\mathcal{S}' \subseteq \mathcal{S}$ of at most k sets such that the number of covered elements $\left| \bigcup_{S_i \in \mathcal{S}'} S_i \right|$ is maximized. See the upcoming full version for the proof.

Theorem 1. *It is NP-hard to compute $f_{BR}(x)$ for $x \in D_L$.*

5 Algorithms for Non-disjoint Customers

In this section, for the non-disjoint customers setting, which has no assumption about the graph structure, we propose two types of algorithms for (2). Let $\mathcal{G}$ be a game instance created from an instance $\phi = (G = (U, V; E), \{p_{uv}\}_{uv \in E}, \{p_{F,uv}\}_{uv \in E}, k_L, k_F)$ and let Λ be its data size. Due to the hardness result (Theorem 1), in this section we assume that k_F is a constant.

5.1 Approximation Algorithm via Zero-Sum Game

We shall approximately solve a game $\mathcal{G}$ by solving a zero-sum game close to $\mathcal{G}$. The core idea of constructing such a zero-sum game is to keep the same set of best-responses of the follower for any strategy of the leader as $\mathcal{G}$. Let us focus on the structure of f and g, which include the term $- \sum_{v \in V} P_v(x) P_{F,v}(y)$ and its negation, respectively. We define a utility function for the leader as

$$\Phi(x, y) = -g(x, y) + \sum_{v \in V} P_v(x)$$
$$= \sum_{v \in V} [P_v(x)(1 - P_{F,v}(y)) - P_v(y)(1 - P_v(x))].$$

Note that $C := \max_{y \in D_F} \sum_{v \in V} P_v(y) \geq -\Phi(x, y)$ and we can compute C in polynomial time since $|D_F|$ is polynomially bounded. Let $\mathcal{G}_\Phi$ be a zero-sum game $(D_L, D_F, \Phi, -\Phi)$.

For reals $\alpha \in [0, 1]$ and $\epsilon \geq 0$, we call an algorithm (α, ϵ)-approximation for $\mathcal{G}$ (resp. $\mathcal{G}_\Phi$) if it provides a strategy profile (x', y') such that $y' \in \mathrm{BR}(x')$ and $f(x', y') \geq \alpha \cdot \max_{x \in D_L} f_{\mathrm{BR}}(x) - \epsilon$ (resp. $\Phi(x', y') \geq \alpha \cdot \max_{x \in D_L, y \in \mathrm{BR}(x)} \Phi(x, y) - \epsilon$). Such (x', y') is called an (α, ϵ)-approximate solution.

Lemma 1. *Let (x', y') be an (α, ϵ)-approximate solution of a zero-sum game $\mathcal{G}_\Phi$, and let (x^*, y^*) be a strong Stackelberg equilibrium of the original game $\mathcal{G}$. Let $\epsilon_1 := \sum_{v \in V}(1 - P_v(x'))P_v(y')$ and $\epsilon_2 := \sum_{v \in V}(1 - P_v(x^*))P_v(y^*)$. Then (x', y') is an $(\alpha, \alpha\epsilon_2 - \epsilon_1 + \epsilon)$-approximate solution for the game $\mathcal{G}$.*

Proof. We remark that $f(x, y)$ can be rewritten by $\Phi(x, y)$ as $f(x, y) = \Phi(x, y) + \sum_{v \in V}(1 - P_v(x))P_v(y)$. Let $(\tilde{x}, \tilde{y})$ be the minimax strategy of $\mathcal{G}_\Phi$. We have

$$f(x', y') = \Phi(x', y') + \epsilon_1 \geq \alpha\Phi(\tilde{x}, \tilde{y}) - \epsilon + \epsilon_1$$
$$\geq \alpha\Phi(x^*, y^*) - \epsilon + \epsilon_1 = \alpha f(x^*, y^*) - (\alpha\epsilon_2 - \epsilon_1 + \epsilon),$$

where the second inequality holds by $\Phi(\tilde{x}, \tilde{y}) = \max_{x \in D_L, y \in \mathrm{BR}(x)} \Phi(x, y) \geq \Phi(x^*, y^*)$. ∎

To find an approximate strong Stackelberg equilibrium, it suffices to find an approximate minimax strategy for $\mathcal{G}_\Phi$. Note that since $|S_L|$ is an exponential size, finding a minimax strategy for $\mathcal{G}_\Phi$ is still intractable.

To this end, we use the *multiplicative weight update method* [2]. Based on this method, Kawase and Sumita [8] showed that, for any nonnegative monotone submodular functions $h_1, \ldots, h_\nu : \{0, 1\}^n \to \mathbb{R}_+$ and $\epsilon > 0$, there exists an algorithm that finds a $(1 - 1/e - \epsilon)$-approximate solution of $\max_{x \in D_L} \min_{i \in [\nu]} \mathbb{E}_{s \sim x}[h_i(s)]$ in polynomial time in n, ν and $1/\epsilon$. We set $h_y(z) = \Phi(z, y) + C$ for all pure strategies $z \in S_L$ and $y \in D_F$. By the definition, h_y is nonnegative monotone submodular for any $y \in D_F$. Thus, we see that we can compute a $(1 - 1/e - \epsilon)$-approximate solution for $\max_{x \in D_L} \min_{y \in D_F}(\Phi(x, y) + C)$ in polynomial time in Λ and $1/\epsilon$. This solution is $(1 - 1/e - \epsilon, (1/e + \epsilon)C)$-approximate for $\mathcal{G}_\Phi$. Therefore, by Lemma 1, we observe the following result.

Theorem 2. *For any $\epsilon > 0$, there exists a $(1 - 1/e - \epsilon, \beta)$-approximation algorithm where $\beta = (1 - 1/e)\epsilon_2 - \epsilon_1 + (1/e + \epsilon)C$ and the running time is polynomial with respect to Λ and $1/\epsilon$.*

5.2   Heuristic Algorithm

In this subsection, we propose a heuristic algorithm. Intuitively, in the algorithm, the players fictitiously play a game ℓ times. Here ℓ is a parameter. Let us assume that the leader would know that the follower estimates the leader's mixed strategy by observing the past budget allocations. In every phase, the leader needs to allocate her budgets so that the mixed strategy estimated by the

---

Algorithm 1. Proposed heuristic

---

   input: a parameter $\ell \in \mathbb{Z}_+$ **output:** a mixed strategy
1 $x, x^* \leftarrow \chi_\emptyset$;
2 **for** $i = 1$ *to* ℓ **do**
3      $S \leftarrow \emptyset$;
4   **for** $j = 1$ *to* k_L **do**
5          $x^u \leftarrow \frac{i-1}{i} x + \frac{1}{i} \chi_{S \cup \{u\}} \ (u \in U)$;
6          $r \in \arg\max_{u \in U \setminus S} f_{\text{BR}}(x^u)$;
7   **if** $f_{\text{BR}}(x^r) \geq f_{\text{BR}}(x)$ **then** $S \leftarrow S \cup \{r\}$;
8   **else** break;
9      $x \leftarrow \frac{i-1}{i} x + \frac{1}{i} \chi_S$;
10   **if** $f_{\text{BR}}(x^*) < f_{\text{BR}}(x)$ **then** $x^* \leftarrow x$;
11 **return** x^*;

---

follower maximizes the leader's utility. The algorithm outputs a mixed strategy by repeating this phase ℓ times.

We describe informally our algorithm, which is summarized in Algorithm 1. The algorithm repeatedly computes ℓ pure strategies $\chi_{S_1}, \ldots, \chi_{S_\ell} \in D_L$, and outputs the best mixed strategy among $\frac{1}{i}(\chi_{S_1} + \cdots + \chi_{S_i})$ $(i = 1, \ldots, \ell)$. At first round, χ_{S_1} is chosen to maximize $f_{\text{BR}}(x)$. Each χ_{S_i} is computed greedily (lines 4–9).

In each round i, we evaluate f_{BR} $O(n \cdot k_L)$ times, and each evaluation of f_{BR} takes $O(|D_F| \cdot |E| \cdot i)$ time. Thus the total running time is $O(|D_F| \cdot |E| \cdot n \cdot k_L \cdot \ell^2)$.

6   Algorithm for Disjoint Customers

In this section, we focus on the disjoint customers setting where each customer is interested in only one medium, i.e., $|N_v| = 1$ for all $v \in V$. This means that the utility functions f, g are bilinear. In this special case, we propose an LP-based algorithm, and modify it so that it runs fast when $|D_F|$ is small. We denote by Λ the data size of an input game instance $(G, \{p_{uv}\}_{uv \in E}, \{p_{F,uv}\}_{uv \in E}, k_L, k_F)$. The following proposition is the main result in this section.

Proposition 1. *When $|N_v| = 1$ for all $v \in V$, we can find a strong Stackelberg equilibrium (x, y) in polynomial time with respect to $|D_F|$ and Λ.*

As we will see in Sect. 6.1, it is easy to compute a strong Stackelberg equilibrium by a multiple LP formulation. The running time is polynomial with respect to λ, $|S_L|$, and $|D_F|$. However, this is not sufficient since $|S_L|$ could be exponentially large with respect to λ and $|D_F|$. To remove the dependency on $|S_L|$, we reduce the size of each LP in Sect. 6.2. The idea is a projection of a leader's mixed strategy $x \in [0, 1]^{S_L}$ onto a fractional budget allocation $r \in [0, 1]^U$.

6.1   Multiple LP Formulation

We first describe a simple exact algorithm to solve (2). The problem (2) is rewritten as

$$\max f(x, y)$$
$$\text{s.t. } x \in D_L, \tag{3}$$
$$g(x, y) \geq g(x, y') \quad \forall y' \in D_F.$$

When we fix $y = y^*$, LP (3) is equivalent to the following LP:

$$\max \sum_{z \in S_L} f(z, y^*) x_z$$
$$\text{s.t. } \sum_{z \in S_L} (g(z, y^*) - g(z, y')) x_z \geq 0 \quad \forall y' \in D_F, \tag{4}$$
$$\sum_{z \in S_L} x_z = 1,$$
$$x_z \geq 0 \quad \forall z \in S_L.$$

The simple algorithm solves (3) exactly by solving (4) for each $y^* \in D_F$. Each LP (4) is solvable in polynomial time with respect to Λ, $|S_L|$ and $|D_F|$, and the algorithm produces $|D_F|$ instances of LP (4). Thus this algorithm runs in polynomial time with respect to Λ, $|S_L|$, and $|D_F|$.

6.2   Reduced Formulation

Let A be a matrix in $\{0, 1\}^{U \times S_L}$ whose rows are all pure strategies. For notational convenience, we denote $p'_{uv} = p_{uv} - p_{F,uv}$ for each $uv \in E$. We denote by a fractional budget allocation $r \in [0, 1]^U$ with $\sum_{u \in U} r_u \leq k_L$. We remark that a fractional budget allocation is a different notion from a mixed strategy $x \in D_L$; the former is uniquely defined from the latter as $r_u = \sum_{S:u \in S} x_S$ ($u \in U$), but the converse may not hold.

We first observe that A projects a mixed strategy x to a fractional budget allocation $Ax \in [0, 1]^U$. Let $Q = \{r \in [0, 1]^U \mid r = Ax, x \in D_L\}$. See the upcoming full version for the proof.

Lemma 2. *For any vector z, it holds that $z \in Q$ if and only if*

$$0 \leq z \leq 1, \ \sum_{u \in U} z_u \leq k_L. \tag{5}$$

We can rewrite P_v and $P_{F,v}$ as $P_v(z) = p_{uv} z_u$ and $P_{F,v}(y) = p_{F,uv} y_u$, where u is the only neighbor of v. Then f and g are simplified as

$$f(x, y) = \sum_{u \in U} \sum_{v \in N_u} p_{uv} (Ax)_u (1 - p_{F,uv} y_u),$$
$$g(x, y) = \sum_{u \in U} \sum_{v \in N_u} p_{uv} y_u (1 - p'_{uv} (Ax)_u).$$

The utility functions $f(x, y)$ and $g(x, y)$ are bilinear. Moreover, they depend on a fractional budget allocation $Ax \in [0, 1]^U$ rather than x.

Lemma 3. *Assume that $|N_v| = 1$ for all $v \in V$. For each $x \in D_L$ and $y \in D_F$, it holds that $f(x, y) = f(x', y)$ and $g(x, y) = g(x'', y)$ for any $x \in D_L$ such that $Ax = Ax'$.*

This lemma gives us an intuition that we solve (4) for a fractional budget allocation r and recover a mixed strategy x. We claim that LP (4) is polynomially equivalent to

$$
\begin{aligned}
\max \quad & \sum_{u \in U} \sum_{v \in N_u} p_{uv}(1 - p_{F,uv} y_u^*) r_u \\
\text{s.t.} \quad & \sum_{u \in U} \sum_{v \in N_u} p_{uv} y_u' r_u' \geq 0, \\
& y_u' = y_u^* - y_u \quad \forall u \in U, y \in D_F, \\
& r_u' = 1 - p_{uv}' r_u \quad \forall u \in U, \\
& \sum_{u \in U} r_u \leq k_L, \\
& r_u \in [0, 1] \quad \forall u \in U.
\end{aligned}
\tag{6}
$$

Indeed, if (x, y) is an optimal solution for (4), then we obtain an optimal solution (r, y) for (6) by setting $r = Ax$. Conversely, let (r, y) be any optimal solution for (6). We observe that $r \in Q$ by Lemma 2. If we can construct a mixed strategy $x \in D_L$ such that $r = Ax$, then we see that (x, y) is an optimal solution for (4) by Lemma 3. In the following, we show that we can recover $x \in D_L$ such that $r = Ax$ in polynomial time with respect to Λ. See the upcoming full version for the proof.

Lemma 4. *For any $r^* \in Q$, there exists a polynomial-time algorithm that finds a mixed strategy $x \in D_L$ such that $|\mathrm{supp}(x)| \leq n + 1$ and $r^* = \sum_{z \in \mathrm{supp}(x)} x_z z$.*

Note that the mixed strategy x in the statement always exists by Carathéodory's theorem. Lemma 4 holds even if some $|N_v|$ is not necessarily equal to one. A leader's strategy in a Stackelberg equilibrium may have the support of a large size.

Therefore, we can solve (3) by solving (6) and recovering a mixed strategy $x \in D_L$ for each $y^* \in D_F$. This algorithm generates $|D_F|$ instances of LP (6) and each instance can be solved in polynomial time in Λ and $|D_F|$. Note that the data size of the LP (6) is bounded by polynomial in Λ and $|D_F|$. The recovered mixed strategy x has polynomial size in Λ. By summarizing the above arguments, Proposition 1 is proved.

7   Experiments

In this section, we evaluate the performance of the proposed approximation algorithm and the heuristic algorithm on real-world datasets. We execute the approximation algorithm Approx (the algorithm based on MWU described in Sect. 5.1 with 100 iterations and $\epsilon = 0.5$), and the heuristic algorithm Prop. (Algorithm 1 with $\ell = 10$). We compare the above algorithms with a baseline algorithm Greedy, which greedily chooses k_L media to maximize $\sum_{v \in V} P_v(z)$. We conduct a series of experiments on Movielens [5] and Yahoo! webscope [18] datasets to examine the leader's utility. The dataset MovieLens is constructed from MovieLens 100K Dataset[1] with 100,000 ratings (1 to 5) to 1,700 movies by 1,000 users. From

---

[1] http://grouplens.org/datasets/movielens/100k/.

the dataset, we select top n frequently rated movies and constructed a bipartite graph G with $n = 20$ media (movies) and $m = 844$ customers (users) with $|E| = 3506$ edges. The dataset Yahoo! Webscope is constructed from Yahoo! Search Marketing Advertiser Bidding Data[2], which contains a bipartite graph between 1,000 search keywords and 10,475 accounts, where each edge represents one bid to advertisement on the keyword with the bid price. From the dataset, we select top n frequently bidden keywords and constructed a bipartite graph G, which has $n = 50$ media (keywords) and $m = 447$ customers (accounts) with $|E| = 871$ edges. $\mathcal{U}(a, b)$ denotes an uniform distribution with maximum and minimum values a and b. For the above bipartite graphs, we set each basic activation probability as $p_{uv} \in \mathcal{U}(0, 0.2)$ for $uv \in E$ as in Wilder and Vorobeychik [17]. We generate two types of instances; for each edge $uv \in E$, the activation probability $p_{F,uv}$ is drawn from a distribution $\mathcal{D}_\mathcal{F} = \mathcal{U}(0, 0.2)$ in the first type of instances, whereas that is drawn from a distribution $\mathcal{D}_\mathcal{F} = \mathcal{U}(0.1, 0.9)$ in the second type of instances that models a scenario where the follower aims to take customers away from the leader. We set the leader's budget as $k_L = 1, 2, 4$, whereas the follower's budget is set to be $k_F = 2$. The results reported in Table 1 indicate that our algorithms clearly outperform Greedy especially when the follower is eager to strip the leader of her customers; that is, when $\mathcal{D}_\mathcal{F} = \mathcal{U}(0.1, 0.9)$.

Table 1. Results averaged over 30 instances for real-world datasets.

MovieLens $((n, m, |E|) = (20, 844, 3506))$					Yahoo! Webscope $((n, m, |E|) = (50, 447, 871))$				
$\mathcal{D}_\mathcal{F}$	(k_L, k_F)	Greedy	Approx	Prop.	$\mathcal{D}_\mathcal{F}$	(k_L, k_F)	Greedy	Approx	Prop.
$\mathcal{U}(0, 0.2)$	$(1, 2)$	37.05	37.05	37.05	$\mathcal{U}(0, 0.2)$	$(1, 2)$	5.42	5.42	5.46
$\mathcal{U}(0, 0.2)$	$(2, 2)$	65.10	65.10	65.24	$\mathcal{U}(0, 0.2)$	$(2, 2)$	10.68	10.68	10.72
$\mathcal{U}(0, 0.2)$	$(4, 2)$	114.22	114.22	114.22	$\mathcal{U}(0, 0.2)$	$(4, 2)$	19.56	19.56	19.55
$\mathcal{U}(0.1, 0.9)$	$(1, 2)$	14.34	14.55	17.22	$\mathcal{U}(0.1, 0.9)$	$(1, 2)$	2.20	3.00	3.67
$\mathcal{U}(0.1, 0.9)$	$(2, 2)$	24.22	29.97	31.87	$\mathcal{U}(0.1, 0.9)$	$(2, 2)$	5.03	6.36	7.05
$\mathcal{U}(0.1, 0.9)$	$(4, 2)$	54.46	54.46	56.12	$\mathcal{U}(0.1, 0.9)$	$(4, 2)$	11.72	12.68	13.31

8 Conclusion

We formalized a new model called the *Stackelberg budget allocation game with a bipartite influence model*. For the general case of our model, we proposed two algorithms: an approximation algorithm which has provable guarantee and a heuristic algorithm empirically outputs a better solution. We remark that, to the best of our knowledge, our approximation algorithm is the first algorithm with a provable guarantee for the non-zero sum submodular Stackelberg game. When the utility functions are bilinear, we proposed our LP-based algorithm and showed that it runs in polynomial time when the follower's budget is constant. We remark that in this case, we can generalize the budget constraint to a

---

[2] https://webscope.sandbox.yahoo.com/catalog.php?datatype=a.

matroid constraint and show a similar result. Finally, experimental results indicate that our approximation and heuristic algorithms empirically output good quality solutions especially in the setting that the follower is a powerful competitor.

Acknowledgments. This work was partially supported by JST ERATO Grant Number JPMJER1201, Japan, and JSPS KAKENHI Grant Numbers JP17K12744, JP18J23034, JP16K16005, JP17K12646, JP17K00028 and JP18H05291, Japan.

References

1. Alon, N., Gamzu, I., Tennenholtz, M.: Optimizing budget allocation among channels and influencers. In: Proceedings of the 21st World Wide Web Conference 2012, WWW 2012, pp. 381–388 (2012)
2. Arora, S., Hazan, E., Kale, S.: The multiplicative weights update method: a meta-algorithm and applications. Theor. Comput. **8**(1), 121–164 (2012)
3. Cornuejols, G., Fisher, M.L., Nemhauser, G.L.: Location of bank accounts to optimize float: an analytic study of exact and approximate algorithms. Manag. Sci. **23**(8), 789–810 (1977)
4. Feige, U.: A threshold of ln n for approximating set cover. J. ACM **45**(4), 634–652 (1998)
5. Harper, M., Konstan, J.: The movielens datasets: history and context. ACM Trans. Interact. Intell. Syst. **5**(4), 19 (2015). https://doi.org/10.1145/2827872
6. Hatano, D., Fukunaga, T., Maehara, T., Kawarabayashi, K.: Lagrangian decomposition algorithm for allocating marketing channels. In: Proceedings of the 29th AAAI Conference on Artificial Intelligence, AAAI 2015, pp. 1144–1150 (2015)
7. Hochbaum, D.S. (ed.): Approximation Algorithms for NP-hard Problems. PWS Publishing Co., Boston (1997)
8. Kawase, Y., Sumita, H.: Randomized strategies for robust combinatorial optimization. In: Proceedings of the 33rd AAAI Conference on Artificial Intelligence AAAI 2019 (2019)
9. Krause, A., Golovin, D.: Submodular function maximization. In: Tractability: Practical Approaches to Hard Problems, pp. 71–104 (2014)
10. Maehara, T., Yabe, A., Kawarabayashi, K.: Budget allocation problem with multiple advertisers: a game theoretic view. In: Proceedings of the 32nd International Conference on Machine Learning, ICML 2015, pp. 428–437 (2015)
11. Miyauchi, A., Iwamasa, Y., Fukunaga, T., Kakimura, N.: Threshold influence model for allocating advertising budgets. In: Proceedings of the 32nd International Conference on Machine Learning, ICML 2015, pp. 1395–1404 (2015)
12. Nemhauser, G.L., Wolsey, L.A., Fisher, M.L.: An analysis of approximations for maximizing submodular set functions - I. Math. Program. **14**(1), 265–294 (1978)
13. Simaan, M.A., Cruz, J.B.: On the stackelberg strategy in nonzero-sum games. J. Optim. Theor. Appl. **11**(5), 533–555 (1973)
14. Soma, T., Kakimura, N., Inaba, K., Kawarabayashi, K.: Optimal budget allocation: theoretical guarantee and efficient algorithm. In: Proceedings of the 31st International Conference on Machine Learning, ICML 2014, pp. 351–359 (2014)
15. Staib, M., Jegelka, S.: Robust budget allocation via continuous submodular functions. In: Proceedings of the 34th International Conference on Machine Learning, ICML 2017, pp. 3230–3240 (2017)

16. Vanek, O., Yin, Z., Jain, M., Bosanský, B., Tambe, M., Pechoucek, M.: Game-theoretic resource allocation for malicious packet detection in computer networks. In: Proceeding of the 11th International Conference on Autonomous Agents and Multiagent Systems, AAMAS 2012, pp. 905–912 (2012)
17. Wilder, B., Vorobeychik, Y.: Defending elections against malicious spread of misinformation. In: Proceedings of the 33rd AAAI Conference on Artificial Intelligence AAAI 2019 (2019)
18. Yahoo: Yahoo! search marketing advertiser bidding data (2007). https://webscope.sandbox.yahoo.com/catalog.php?datatype=a

Game Equivalence and Bisimulation
for Game Description Language

Guifei Jiang[1]([⊠]), Laurent Perrussel[2], Dongmo Zhang[3], Heng Zhang[4],
and Yuzhi Zhang[1]

[1] Nankai University, Tianjin, China
G.Jiang@nankai.edu.cn
[2] University of Toulouse, Toulouse, France
[3] Western Sydney University, Penrith, Australia
[4] Tianjin University, Tianjin, China

Abstract. This paper investigates the equivalence between games represented by state transition models and its applications. We first define a notion of bisimulation equivalence between state transition models and prove that it can be logically characterized by Game Description Language (GDL). Then we introduce a concept of quotient state transition model. As the minimum equivalent of the original model, it allows us to improve the efficiency of model checking for GDL. Finally, we demonstrate with real games that bisimulation equivalence can be generalized to characterize more general game equivalence.

Keywords: Game equivalence · Bisimulation equivalence ·
General Game Playing

1 Introduction

General Game Playing (GGP) is concerned with creating intelligent agents that understand the rules of previously unknown games and learn to play these games without human intervention [9]. To represent the rules of arbitrary games, a formal game description language (GDL) was introduced as an official language for GGP in 2005. GDL is originally a machine-processable, logic programming language [14]. Most recently, it has been adapted as a logical language for game specification and strategic reasoning [24]. Moreover, its epistemic and dynamic extensions have also been developed [13,23].

As a logical language for representing game rules and specifying game properties, the logical properties, especially the expressive power of GDL have not been fully investigated yet. For instance, which game properties are definable in GDL? When two GDL game descriptions are equivalent? How to distinguish two GDL-defined games? In this paper, we will address these questions through a *bisimulation* approach.

The notion of *bisimulation* plays a pivotal role to identify the expressive power of a logic. It was independently defined and developed in the areas of

© Springer Nature Switzerland AG 2019
A. C. Nayak and A. Sharma (Eds.): PRICAI 2019, LNAI 11670, pp. 583–596, 2019.
https://doi.org/10.1007/978-3-030-29908-8_46

theoretical computer science [12,16] and the model theory of modal logic [3,4]. Since bisimulation-equivalent structures can simulate each other in a stepwise manner, they cannot be distinguished by the concerned logic. An appropriate notion of bisimulation for a logic allows us to study the expressive power of that logic in terms of structural invariance and language indistinguishability [11].

Besides identifying the expressivity of a logic, bisimulation equivalence also allows us to obtain the minimum equivalent of the original model, called *the quotient model*, which can be used to improve the efficiency of model checking [2]. Moreover, in terms of GDL, bisimulation equivalence tells us when two game structures are essentially the *same*, and thus gives us a natural criterion on the equivalence between games. Exploiting game equivalence may provide a bridge for knowledge transfer between a new game and a well-studied game in GGP. In particular, Zhang *et al.* considered that two games are equivalent exactly if the state machines described them are identical (isomorphic) [25]. Such definition might be too strong, as it would rule out many non-identical but essentially equivalent games, such as bisimulation-equivalent games.

Based on the above considerations, we will use in this paper a concept of bisimulation as a tool to investigate the expressive power of GDL and capture the equivalence between games represented by state transition models. We first define a concept of bisimulation equivalence between state transition models and prove that it coincides with the invariance of GDL-formulas on state transition models. This justifies that the notion of bisimulation equivalence is appropriate for GDL. Then we introduce a concept of quotient state transition model and show that it is bisimulation-equivalent to its original model. Considering its smaller size, this provides a way to improve the efficiency of model checking. Finally we demonstrate with real games that bisimulation equivalence can be generalized to capture a wider range of game equivalence.

The rest of this paper is structured as follows: Sect. 2 introduces the framework for game description. Section 3 defines the concept of bisimulation equivalence and introduces the notion of quotient model. Section 4 generalizes bisimulation equivalence to characterize more general game equivalence. Finally, we conclude with related work and future work.

2   The Framework

Let us now introduce the GDL-based framework from [24]. All games are assumed to be played in multi-agent environments. Each game is associated with a *game signature*. A *game signature* S is a triple $(N, \mathcal{A}, \Phi)$, where

- $N = \{1, 2, \cdots, m\}$ is a non-empty finite set of agents,
- $\mathcal{A}$ is a non-empty finite set of *actions* such that it contains *noop*, an action without any effect , and
- $\Phi = \{p, q, \cdots\}$ is a finite set of propositional atoms for specifying individual features of a game state.

Through the rest of the paper, we will consider a fixed game signature S, and all concepts are based on the game signature unless otherwise specified.

2.1   State Transition Models

This paper focuses on synchronous games where all players move simultaneously. These games can be specified by *state transition models* defined as follows:

Definition 1. *A state transition (ST) model M is a tuple $(W, w_0, T, L, U, g, \pi)$, where*

- W *is a non-empty finite set of possible states.*
- $w_0 \in W$, *representing the unique initial state.*
- $T \subseteq W$, *representing a set of terminal states.*
- $L \subseteq W \times N \times \mathcal{A}$ *is a legality relation, specifying legal actions for each agent at game states. Let $L_r(w) = \{a \in \mathcal{A} : (w, r, a) \in L\}$ be the set of all legal actions for agent r at state w. To make a game playable, we assume that (i) each agent has at least one available action at each state, i.e., $L_r(w) \neq \emptyset$ for any $r \in N$ and $w \in W$, and (ii) each agent can only do action noop at terminal states, i.e., $L_r(w) = \{noop\}$ for any $r \in N$ and $w \in T$.*
- $U : W \times \mathcal{A}^{|N|} \rightarrow W \setminus \{w_0\}$ *is an update function, specifying the state transition for each state and legal joint action, such that $U(w, \langle noop^r \rangle_{r \in N}) = w$ for any $w \in W \setminus \{w_0\}$.*
- $g : N \rightarrow 2^W$ *is a goal function, specifying the winning states of each agent.*
- $\pi : W \rightarrow 2^\Phi$ *is a standard valuation function.*

Note that to make the framework as general as possible, we use the concurrent game structure and the turn-based game structure involved in [24] is a special case by allowing a player only to do "noop" when it is not her turn. For convenience, let D denote the set of all joint actions $\mathcal{A}^{|N|}$. Given $d \in D$, we use $d(r)$ to specify the action taken by agent r.

The following notion specifies all possible ways in which a game can develop.

Definition 2. *Let $M = (W, w_0, T, L, U, g, \pi)$ be an ST-model. A path δ is an infinite sequence of states and joint actions $w_0 \xrightarrow{d_1} w_1 \xrightarrow{d_2} \cdots \xrightarrow{d_j} \cdots$ such that for any $j \geq 1$ and $r \in N$,*

1. $w_j \neq w_0$ *(that is, only the first state is initial.);*
2. $d_j(r) \in L_r(w_{j-1})$ *(that is, any action that is taken by each agent must be legal.)*
3. $w_j = U(w_{j-1}, d_j)$ *(state update)*
4. *if $w_{j-1} \in T$, then $w_{j-1} = w_j$ (self-loop after reaching a terminal state.)*

Let $\mathcal{P}(M)$ denote the set of all paths in M. For $\delta \in \mathcal{P}(M)$ and a stage $j \geq 0$, we use $\delta[j]$ to denote the j-th state of δ and $\theta_r(\delta, j)$ to denote the action taken by agent r at stage j of δ.

2.2   The Language

Let us now introduce the GDL-based language from [24] for game specification.

Definition 3. *The language $\mathcal{L}$ for game description is generated by the following BNF:*

$$\varphi ::= p \mid initial \mid terminal \mid legal(r,a) \mid wins(r) \mid does(r,a) \mid \neg\varphi \mid \varphi \wedge \psi \mid \bigcirc\varphi$$

where $p \in \Phi$, $r \in N$ and $a \in \mathcal{A}$.

Other connectives $\vee$, $\rightarrow$, $\leftrightarrow$, $\top$, $\bot$ are defined by $\neg$ and $\wedge$ in the standard way. Intuitively, *initial* and *terminal* specify the initial state and the terminal states of a game, respectively; $does(r,a)$ asserts that agent r takes action a at the current state; $legal(r,a)$ asserts that agent r is allowed to take action a at the current state, and $wins(r)$ asserts that agent r wins at the current state. Finally, the formula $\bigcirc\varphi$ means that φ holds in the next state.

We use the following abbreviations in the rest of paper. For $d = \langle a_r \rangle_{r \in N}$, $does(d) =_{def} \bigwedge_{r \in N} does(r, a_r)$, and $\bigcirc^k \varphi =_{def} \underbrace{\bigcirc \cdots \bigcirc}_{k} \varphi$. Note that our language is slightly different from [24] by introducing the agent parameter in $legal(\cdot)$ and $does(\cdot)$. To help the reader capture the intuition of the language, let us consider the following example.

Example 1 (Number Scrabble). Two players take turns to select numbers from 1 to 9 without repeating any numbers previously used. The first player who selects three numbers that add up to 15 wins.

The game signature $\mathcal{S}_{NS}$ is given as follows: $N_{NS} = \{b, w\}$ denoting two game players; $\mathcal{A}_{NS} = \{\alpha(n) \mid 1 \leq n \leq 9\} \cup \{noop\}$, where $\alpha(n)$ denotes selecting number n, and $\Phi_{NS} = \{s(r,n), turn(r) \mid r \in \{b, w\}$ and $1 \leq n \leq 9\}$, where $s(r,n)$ represents the fact that number n is selected by player r, and $turn(r)$ says that player r has the turn now. The rules of Number Scrabble can be naturally formulated by GDL-formulas as shown in Fig. 1 (where $r \in \{b, w\}$ and $-r$ represents r's opponent).

The formulas are intuitive. Formula 1 says at the initial state, player b has the first turn and all numbers are not selected. The next two formulas specify winning states of each player and the terminal states, respectively. The preconditions of each action (legality) are specified by Formula 4 and Formula 5. Formula 6 is the combination of the frame axioms and the effect axioms [19]. The last formula specifies the turn-taking.

2.3   The Semantics

The semantics of this language is based on ST-models with respect to a path and a stage of the path.

Definition 4. *Let $M = (W, w_0, T, L, U, g, \pi)$ be an ST-model. Given a path δ of M, a stage $j \geq 0$ and a formula $\varphi \in \mathcal{L}$, we say φ is true (or satisfied) at j of δ under M, denoted $M, \delta, j \models \varphi$, according to the following definition:*

1. $initial \leftrightarrow turn(\mathbf{b}) \wedge \neg turn(\mathbf{w}) \wedge \bigwedge_{i=1}^{9} \neg(s(\mathbf{b}, i) \vee s(\mathbf{w}, i))$

2. $wins(r) \leftrightarrow \left(\bigvee_{i=2}^{3}(s(r, i) \wedge s(r, 4) \wedge s(r, 11 - i)) \vee \bigvee_{i=1}^{2}(s(r, i) \wedge s(r, 6) \wedge s(r, 9 - i)) \vee \right.$
   $\left. \bigvee_{l=1}^{4}(s(r, 5 - l) \wedge s(r, 5) \wedge s(r, 5 + l)) \right)$

3. $teminal \leftrightarrow wins(\mathbf{b}) \vee wins(\mathbf{w}) \vee \bigwedge_{i=1}^{9}(s(\mathbf{b}, i) \vee s(\mathbf{w}, i))$

4. $legal(r, \alpha(n)) \leftrightarrow \neg(s(\mathbf{b}, n) \vee s(\mathbf{w}, n)) \wedge turn(r) \wedge \neg terminal$

5. $legal(r, noop) \leftrightarrow turn(-r) \vee terminal$

6. $\bigcirc s(r, n) \leftrightarrow s(r, n) \vee (\neg(s(\mathbf{b}, n) \vee s(\mathbf{w}, n)) \wedge does(r, \alpha(n)))$

7. $turn(r) \wedge \neg terminal \rightarrow \bigcirc \neg turn(r) \wedge \bigcirc turn(-r)$

Fig. 1. A GDL description of Number Scrabble.

$M, \delta, j \models p$	iff	$p \in \pi(\delta[j])$
$M, \delta, j \models \neg\varphi$	iff	$M, \delta, j \not\models \varphi$
$M, \delta, j \models \varphi_1 \wedge \varphi_2$	iff	$M, \delta, j \models \varphi_1$ and $M, \delta, j \models \varphi_2$
$M, \delta, j \models initial$	iff	$\delta[j] = w_0$
$M, \delta, j \models terminal$	iff	$\delta[j] \in T$
$M, \delta, j \models wins(r)$	iff	$\delta[j] \in g(r)$
$M, \delta, j \models legal(r, a)$	iff	$a \in L_r(\delta[j])$
$M, \delta, j \models does(r, a)$	iff	$\theta_r(\delta, j) = a$
$M, \delta, j \models \bigcirc\varphi$	iff	$M, \delta, j + 1 \models \varphi$

A formula φ is *valid* in an ST-model M, written $M \models \varphi$, if $M, \delta, j \models \varphi$ for any $\delta \in \mathcal{P}(M)$ and $j \geq 0$. A formula φ is called *satisfied at a state w* in M, written $M, w \models \varphi$, if it is true for all paths going through w, i.e., $M, \delta, j \models \varphi$ for any $\delta \in \mathcal{P}(M)$ and any $j \geq 0$ with $\delta[j] = w$. It follows that $M, w_0 \models \varphi$ iff $M, \delta, 0 \models \varphi$ for all $\delta \in \mathcal{P}(M)$.

3   Bisimulation Equivalence

In this section, we define the concept of bisimulation equivalence over state transition models and show it coincides with the invariance of GDL-formulas. We also introduce the quotient state transition model in terms of such relation.

3.1   Bisimulation and Invariance

Inspired by the notion of bisimulation in [7], we define the concept of bisimulation equivalence between ST-models as follows:

Definition 5. *Let $M = (W, w_0, T, L, U, g, \pi)$ and $M' = (W', w_0', T', L', U', g', \pi')$ be two ST-models. We say M and M' are bisimulation-equivalent, (bisimilar, for short), written $M \approx M'$, if there is a binary relation $Z \subseteq W \times W'$ such that $w_0 Z w_0'$, and for all states $w \in W$ and $w' \in W'$ with wZw', the following conditions hold:*

1. $\pi(w) = \pi'(w')$;
2. $w = w_0$ iff $w' = w_0'$;
3. $w \in T$ iff $w' \in T'$;
4. $a \in L_r(w)$ iff $a \in L_r'(w')$ for any $r \in N$ and $a \in \mathcal{A}$;
5. $w \in g(r)$ iff $w' \in g'(r)$ for any $r \in N$;
6. If $U(w,d) = u$, then there is $u' \in W'$ s.t. $U'(w',d) = u'$ and uZu';
7. If $U'(w',d) = u'$, then there is $u \in W$ s.t. $U(w,d) = u$ and uZu'.

Note that $\approx$ is an equivalence relation over ST-models. When Z is a bisimulation linking two states w in M and w' in M', we say that w and w' are *bisimilar*, written $M, w \leftrightarrow M', w'$. In particular, if $M \approx M'$, then their initial states are bisimilar, i.e., $M, w_0 \leftrightarrow M', w_0'$.

Another way to understand bisimulation equivalence is to observe that M is bisimilar to M' iff each path that can be developed in one model can also be induced in the other. To formalize this idea, we need generalize the notion of bisimilar over states to paths as follows:

Definition 6. *Consider two ST-models M and M'. Given two paths $\delta := w_0 \xrightarrow{d_1} w_1 \xrightarrow{d_2} \cdots$ in M and $\delta' := w_0' \xrightarrow{d_1'} w_1' \xrightarrow{d_2'} \cdots$ in M', we say δ and δ' are bisimilar, written $M, \delta \leftrightarrow M', \delta'$, iff for every $j \geq 0$ and $r \in N$, $M, \delta[j] \leftrightarrow M', \delta'[j]$ and $\theta_r(\delta, j) = \theta_r(\delta', j)$.*

That is, two paths are bisimilar if (i) all the corresponding states are bisimilar, and (ii) each agent takes the same action at every stage. With this, the above idea is restated as follows:

Lemma 1. *Given two ST-models M and M', $M \approx M'$ iff for every $\delta \in \mathcal{P}(M)$, there is $\delta' \in \mathcal{P}(M')$ such that $M, \delta \leftrightarrow M', \delta'$, and vice versa.*

Proof. ($\Rightarrow$) This direction holds directly by Condition 6 & 7 of Definition 5.

($\Leftarrow$) Let $Z = \{(w, w') \mid$ there are $\delta \in \mathcal{P}(M), \delta' \in \mathcal{P}(M')$ and $j \geq 0$ such that $\delta[j] = w, \delta'[j] = w'$, the local properties Condition 1–5 in Definition 5 hold for $\delta[j]$ and $\delta'[j]$, and $\theta_r(\delta, j) = \theta_r(\delta', j)\}$. Such a relation Z exists due to the assumption. It is easy to show that Z is a bisimulation between M and M'. ($\square$)

Let us now turn to the logical characterization of bisimulation equivalence. We begin with the invariance of GDL-formulas under path-bisimulation.

Proposition 1. *Let M, M' be two ST-models. For every $\delta \in \mathcal{P}(M)$ and $\delta' \in \mathcal{P}(M')$, if $M, \delta \leftrightarrow M', \delta'$, then $(M, \delta, j \models \varphi$ iff $M', \delta', j \models \varphi)$ for any $j \geq 0$ and $\varphi \in \mathcal{L}$.*

It is routine to prove this by induction on φ. This result asserts that two bisimilar paths preserve GDL-formulas at each stage. Note that the other direction does not hold. Here is a simple counter-example. Let M and M' be two ST-model depicted in Fig. 2, where $N = \{r\}$ and $\Phi = \emptyset$. Now consider two paths $\delta = w_0 \xrightarrow{a} w_1 \xrightarrow{b} \cdots$ in M and $\delta' = w_0' \xrightarrow{a} w_1' \xrightarrow{b} \cdots$ in M'. As $w_3 \notin T$ and $w_3' \in T'$, then $M, w_3 \not\approx M', w_3'$, i.e., the successors of w_1 and w_1' are not bisimilar, so

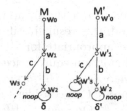

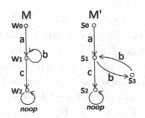

Fig. 2. δ and δ' are not bisimilar.

Fig. 3. M and M' are not bisimulation-equivalent.

$M, w_1 \not\approx M', w_1'$. Thus, δ and δ' are not bisimilar, i.e., $M, \delta \not\approx M', \delta'$. But it is easy to check that at each stage, δ and δ' satisfy the same GDL-formulas.

To prove the characterization result, we need one additional notion. For each path $\delta := w_0 \xrightarrow{d_1} w_1 \xrightarrow{d_2} \cdots \xrightarrow{d_j} \cdots$ in M, we induce a *trace* $V(\delta) = V(w_0) \cdot does(d_1) \cdot V(w_1) \cdots does(d_{j-1}) \cdot V(w_j) \cdots$, where $V(w) = \{p \in \Phi \mid p \in \pi(w)\} \cup \{initial \mid w = w_0\} \cup \{terminal \mid w \in T\} \cup \{wins(r) \mid w \in g(r) \text{ for } r \in N\} \cup \{legal(r, a) \mid a \in L_r(w) \text{ for } r \in N, a \in \mathcal{A}\}$. Let $trace(M)$ denote the set of all traces in M, i.e., $trace(M) = \{V(\delta) \mid \delta \in \mathcal{P}(M)\}$. Then it holds that

Lemma 2. *For two ST-models M and M', $M \approx M'$ iff $trace(M) = trace(M')$.*

We now provide the logical characterization of bisimulation equivalence.

Theorem 1. *Let M and M' be any two ST-models. Then $M \approx M'$ iff they satisfy the same GDL-formulas.*

Proof. Assume $M \approx M'$. For symmetry, it suffices to prove one case. For every $\varphi \in \mathcal{L}$, assume φ is satisfied in M. then there is $\delta \in \mathcal{P}(M)$ and stage $j \geq 0$ such that $M, \delta, j \models \varphi$. By the assumption and Lemma 1, there is $\delta' \in \mathcal{P}(M')$ such that $M, \delta \leftrightarrow M', \delta'$. And by Proposition 1, we have $M', \delta', j \models \varphi$. Thus, φ is satisfied in M'.

Now assume $M \not\approx M'$, then by Lemma 2 there is $\delta \in \mathcal{P}(M)$ for all $\delta' \in \mathcal{P}(M')$ $V(\delta) \neq V(\delta')$. It follows that for each $\delta' \in \mathcal{P}(M)$ there is $k \geq 0$ such that either $does(d_{k+1}) \neq does(d'_{k+1})$ or $V(\delta[k]) \neq V(\delta'[k])$. From the former, we obtain a formula $does(r, a_{k+1})$ (for $r \in N$ and $a_{k+1} \in \mathcal{A}$) such that $M, \delta, k \models does(r, a_{k+1})$ and $M', \delta', k \not\models does(r, a_{k+1})$. From the latter, we obtain a formula $\chi \in Atm$ such that either (i) $M, \delta, k \models \chi$ and $M', \delta', k \not\models \chi$, or (ii) $M, \delta, k \models \neg\chi$ and $M', \delta', k \not\models \neg\chi$. Let $\varphi_{\delta'}$ be the formula of the form $\bigcirc^k does(r, a_{k+1})$, $\bigcirc^k \chi$ or $\bigcirc^k \neg\chi$ to distinguish δ from δ'. It follows by the construction that $M, \delta, 0 \models \varphi_{\delta'}$ and $M', \delta', 0 \not\models \varphi_{\delta'}$. Let Δ be the conjunctions of all such obtained formulas for all paths in M', i.e., $\Delta := \bigwedge_{\delta' \in \mathcal{P}(M')} \varphi_{\delta'}$. Note that Δ is well-formed due to the fact that M' is finite-branching. Let us now consider formula $initial \wedge \Delta$. Then it is satisfied in M, i.e., $M, \delta, 0 \models initial \wedge \Delta$. But it is unsatisfied in M'. Otherwise, there are some $\delta' \in \mathcal{P}(M')$ and $j \geq 0$ such that $M', \delta', j \models initial \wedge \Delta$, then $M', \delta', 0 \models \Delta$, so $M', \delta', 0 \models \varphi_{\delta'}$, contradicting with $M', \delta', 0 \not\models \varphi_{\delta'}$. Thus, M and M' fail to satisfy the same set of GDL-formulas. $\hfill(\square)$

This theorem asserts that bisimulation equivalence and the invariance of GDL-formulas match on ST-models. On the one hand, this result justifies that the notion of bisimulation equivalence is natural and appropriate for GDL; On the other hand, it allows us to show the failure of bisimulation-equivalence easily. *Two ST-models are not bisimulation-equivalent if there is a GDL-formula that holds in one model and fails in the other.* For instance, let us consider two ST-models depicted in Fig. 3, where $N = \{r\}$ and $\Phi = \emptyset$. One can find formula $initial \wedge \bigcirc^2(does(r, c) \wedge \bigcirc terminal)$ that holds in M, but fails in M'. This leads to $M \not\approx M'$. Thus, two ST-models are bisimulation-equivalent if and only if they enjoy exactly the same properties. Alternatively, two ST-models are not bisimulation-equivalent if one has a property that the other does not have.

3.2 Bisimulation Quotient

In this subsection, we provide an alternative perspective to consider bisimulation as a relation between states within a single ST-model. Then we introduce the quotient ST-model under such relation.

Definition 7. *Let $M = (W, w_0, T, L, U, g, \pi)$ be an ST-models. A bisimulation is a binary relation $Z \subseteq W \times W$ s.t. for all states $w_1, w_2 \in W$ with $w_1 Z w_2$,*

1. *$\pi(w_1) = \pi(w_2)$;*
2. *$w_1 = w_0$ iff $w_2 = w_0$;*
3. *$w_1 \in T$ iff $w_2 \in T$;*
4. *$a \in L_r(w_1)$ iff $a \in L_r(w_2)$ for any $r \in N$ and $a \in \mathcal{A}$;*
5. *$w_1 \in g(r)$ iff $w_2 \in g(r)$ for any $r \in N$;*
6. *If $U(w_1, d) = u_1$, then there is $u_2 \in W$ s.t. $U(w_2, d) = u_2$ and $u_1 Z u_2$;*
7. *If $U(w_2, d) = u_2$, then there is $u_1 \in W$ s.t. $U(w_1, d) = u_1$ and $u_1 Z u_2$.*

States w_1 and w_2 are bisimulation-equivalent, denoted by $w_1 \sim_M w_2$, if there is a bisimulation Z for M with $w_1 Z w_2$.

It follows that a bisimulation over states for ST-model M is a bisimulation over ST-models for the pair (M, M). Clearly, $\sim_M$ is an equivalence relation on W. For $w \in W$, let $[w]_{\sim_M}$ be the equivalence class of state w under $\sim_M$, i.e., $[w]_{\sim_M} = \{w' \in W \mid w \sim_M w'\}$. We next define the quotient ST-model under such bisimulation equivalence.

Definition 8. *For an ST-model $M = (W, w_0, T, L, U, g, \pi)$ and a bisimulation equivalence $\sim_M$, the quotient ST-model $M/\sim_M = (W', w_0', T', L', U', g', \pi')$ is defined as follows:*

- *$W' = \{[w]_{\sim_M} \mid w \in W\}$ is the set of all $\sim_M$-equivalence classes;*
- *$w_0' = [w_0]_{\sim_M}$;*
- *$T' = \{[w]_{\sim_M} \mid w \in T\}$;*
- *$a \in L_r'([w]_{\sim_M})$ iff $a \in L_r(w)$ for any $r \in N$ and $a \in \mathcal{A}$;*
- *$U'([w]_{\sim_M}, d) = [u]_{\sim_M}$ iff $U(w, d) = u'$ for some $u' \in [u]_{\sim_M}$;*
- *$[w]_{\sim_M} \in g'(r)$ iff $w \in g(r)$ for any $r \in N$;*
- *$p \in \pi'([w]_{\sim_M})$ iff $p \in \pi(w)$ for any $p \in \Phi$.*

Note that the defined quotient ST-model is indeed a state transition model, and it is minimum as $\sim_M$ is the coarsest bisimulation for M. Moreover, an ST-model and its quotient ST-model are bisimulation-equivalent.

Proposition 2. *For any ST-model M, $M \approx M/\sim_M$.*

This follows from the fact that $Z = \{(w, [w]_{\sim_M}) \mid w \in W\}$ is a bisimulation between M and $M/\sim_M$. Combining this result and Theorem 1 allows us to perform model checking on the bisimulation-equivalent quotient ST-model. *A GDL-formula holds for the quotient if and only if it also holds for the original ST-model.* This provides a way to improve the efficiency of model checking for GDL in [13, 20]. Note that an adaption of bisimulation-quotienting algorithms for a finite transition system in [2] can be used to compute the quotient ST-model.

4 Bisimulation and Game Equivalence

State transition models may be viewed as representations of games, and bisimulation equivalence tells us when two state transition models are essentially the same. Thus, bisimulation equivalence provides a criterion on the equivalence between games, i.e., two games are equivalent if their state transition models are bisimulation-equivalent. In this section, we generalize this concept to capture more general game equivalence.

Let us first consider the following two games: Number Scrabble in Example 1 and Tic-Tac-Toe specified as follows:

Example 2 (Tic-Tac-Toe). Two players take turns in marking either a cross 'x'or a nought 'o' on a 3×3 board. The player who first gets three consecutive marks of her own symbol in a row wins this game.

The game signature for Tic-Tac-Toe, written $\mathcal{S}_{TT}$, is given as follows: $N_{TT} = \{x, o\}$ denoting the two game players; $\mathcal{A}_{TT} = \{a_{i,j} \mid 1 \leq i, j \leq 3\} \cup \{noop\}$, where $a_{i,j}$ denotes filling cell (i, j), and $\Phi_{TT} = \{p_{i,j}^r, turn(r) \mid r \in \{x, o\}$ and $1 \leq i, j \leq 3\}$, where $p_{i,j}^r$ represents the fact that cell (i, j) is filled by player r. The rules of this game is given in Fig. 4.

The initial state, each player's winning states, the terminal states and the turn-taking are given by formulas 1-3 and 7, respectively. The preconditions of each action (legality) are specified by Formula 4 and 5. Formula 6 specifies the state transitions.

Although the two games appear different in their game descriptions, they are actually equivalent (isomorphic) [15, 18]. Unfortunately, bisimulation equivalence is not able to capture such game equivalence as they are based on different game signatures. Then we generalize the notion of bisimulation equivalence as follows:

Definition 9. *Consider two ST-models $M_{\mathcal{S}} = (W, w_0, T, L, U, g, \pi)$ with $\mathcal{S} = (N, \mathcal{A}, \Phi)$ and $M'_{\mathcal{S}'} = (W', w'_0, T', L', U', g', \pi')$ with $\mathcal{S}' = (N', \mathcal{A}', \Phi')$. $M_{\mathcal{S}}$ and $M'_{\mathcal{S}'}$ are structure-equivalent, written $M_{\mathcal{S}} \sim M'_{\mathcal{S}'}$, if there are bijections $f_1 : N \mapsto N'$, $f_2 : \mathcal{A} \mapsto \mathcal{A}'$, $f_3 : \Phi \mapsto \Phi'$, and a relation $Z \subseteq W \times W'$ such that $w_0 Z w'_0$ and for all states $w \in W$ and $w' \in W'$ with wZw', the following conditions hold:*

1. $initial \leftrightarrow turn(\mathsf{x}) \wedge \neg turn(\mathsf{o}) \wedge \bigwedge\limits_{i,j=1}^{3} \neg(p_{i,j}^{\mathsf{x}} \vee p_{i,j}^{\mathsf{o}})$

2. $wins(r) \leftrightarrow \bigvee\limits_{i=1}^{3} \bigwedge\limits_{l=0}^{2} p_{i,1+l}^{r} \vee \bigvee\limits_{j=1}^{3} \bigwedge\limits_{l=0}^{2} p_{1+l,j}^{r} \vee \bigwedge\limits_{l=0}^{2} p_{1+l,1+l}^{r} \vee \bigwedge\limits_{l=0}^{2} p_{1+l,3-l}^{r}$

3. $teminal \leftrightarrow wins(\mathsf{x}) \vee wins(\mathsf{o}) \vee \bigwedge\limits_{i,j=1}^{3} (p_{i,j}^{\mathsf{x}} \vee p_{i,j}^{\mathsf{o}})$

4. $legal(r, a_{i,j}) \leftrightarrow \neg(p_{i,j}^{\mathsf{x}} \vee p_{i,j}^{\mathsf{o}}) \wedge turn(r) \wedge \neg terminal$

5. $legal(r, noop) \leftrightarrow turn(-r) \vee terminal$

6. $\bigcirc p_{i,j}^{r} \leftrightarrow p_{i,j}^{r} \vee (does(r, a_{i,j}) \wedge \neg(p_{i,j}^{\mathsf{x}} \vee p_{i,j}^{\mathsf{o}}))$

7. $turn(r) \wedge \neg terminal \rightarrow \bigcirc \neg turn(r) \wedge \bigcirc turn(-r)$

Fig. 4. A GDL description of Tic-Tac-Toe.

1. $p \in \pi(w)$ iff $f_3(p) \in \pi'(w')$;
2. $w = w_0$ iff $w' = w_0'$;
3. $w \in T$ iff $w' \in T'$;
4. $a \in L_r(w)$ iff $f_2(a) \in L'_{f_1(r)}(w')$ for any $r \in N$ and $a \in \mathcal{A}$;
5. $w \in g(r)$ iff $w' \in g'(f_1(r))$ for any $r \in N$;
6. If $U(w, d) = u$, then there is $u' \in W'$ s.t. $U'(w', \langle f_2(d(r)) \rangle_{r \in N}) = u'$ and uZu';
7. If $U'(w', d') = u'$, then there is $u \in W$ s.t. $U(w, \langle f_2^{-1}(d'(r')) \rangle_{r' \in N'}) = u$ and uZu'.

Note that $\sim$ is an equivalence relation over ST-models (with different game signatures). Clearly, $\approx$ is a special case of $\sim$ when $\mathcal{S} = \mathcal{S}'$. We say that two games are *equivalent* if their ST-models are structure-equivalent.

Let us back to the examples. As we expected, the equivalence between Number Scrabble and Tic-Tac-Toe can be captured by the structure equivalence. The mapping between their state transition models is demonstrated in Table 1. The basic idea is that *filling a cell corresponds to selecting the number in the cell, and the fact that a cell is filled amounts to the fact that the corresponding number is selected*. For instance, filling the left-bottom cell corresponds to selecting number 4, i.e., $f_2(a_{1,1}) = \alpha(4)$, and the fact that the center is filled by player x maps the fact that number 5 is selected by player b, i.e., $f_3(p_{2,2}^{\mathsf{x}}) = s(\mathsf{b}, 5)$. And the structure-bisimulation relation starts from their initial states and can be constructed step by step according to the mapping. For example, the states depicted in Table 2 and Fig. 5 are structure-bisimilar.

Table 1. The mapping.   **Table 2.** Filling the center.

2	7	6
9	5	1
4	3	8

	X	

$\{1,2,3,4,\underline{5},6,7,8,9\}$

b

Fig. 5. Selecting number 5.

Similarly, we say that w and w' are structure-bisimilar, written $M_S, w \leftrightarroweq_s M'_{S'}, w'$, if Z links two states w in M_S and w' in $M'_{S'}$. In particular, for two paths $\delta := w_0 \xrightarrow{d_1} w_1 \xrightarrow{d_2} \cdots \xrightarrow{d_j} \cdots$ in M_S and $\delta' := w'_0 \xrightarrow{d'_1} w'_1 \xrightarrow{d'_2} \cdots \xrightarrow{d'_j} \cdots$ in $M'_{S'}$, we say that δ and δ' are *structure-bisimilar*, written $M_S, \delta \leftrightarroweq_s M'_{S'}, \delta'$, iff for every $j \geq 0$ and $r \in N$, $M_S, \delta[j] \leftrightarroweq_s M'_{S'}, \delta'[j]$ and $f_2(\theta_r(\delta, j)) = \theta_{f_1(r)}(\delta', j)$. Similar to Bisimulation Equivalence, the following result displays that two ST-models are structure-equivalent iff each path that can be developed in one model can be also *simulated* in the other.

Lemma 3. *Given two ST-models M_S and $M'_{S'}$, $M_S \sim M'_{S'}$ iff for every $\delta \in \mathcal{P}(M_S)$, there is $\delta' \in \mathcal{P}(M'_{S'})$ such that $M_S, \delta \leftrightarroweq_s M'_{S'}, \delta'$, and vice versa.*

Let us turn to the logical characterization of structure equivalence. To this end, we begin with the transformation of GDL-formulas. The translation between languages is defined as follows:

Definition 10. *Consider two game signatures $S = (N, \mathcal{A}, \Phi)$ and $S' = (N', \mathcal{A}', \Phi')$ with the same bijections $f_1 : N \mapsto N'$, $f_2 : \mathcal{A} \mapsto \mathcal{A}'$ and $f_3 : \Phi \mapsto \Phi'$ of Definition 9. A translation tr is a bijective mapping from $\mathcal{L}_S$ onto $\mathcal{L}_{S'}$ such that for $p \in \Phi$, $r \in N$ and $a \in \mathcal{A}$,*

$$\text{tr}(p) = f_3(p) \qquad\qquad \text{tr}(initial) = initial$$
$$\text{tr}(terminal) = terminal \qquad \text{tr}(wins(r)) = wins(f_1(r))$$
$$\text{tr}(legal(r, a)) = legal(f_1(r), f_2(a)) \quad \text{tr}(does(r, a)) = does(f_1(r), f_2(a))$$
$$\text{tr}(\neg\varphi) = \neg\text{tr}(\varphi) \qquad\qquad \text{tr}(\varphi \wedge \psi) = \text{tr}(\varphi) \wedge \text{tr}(\psi)$$
$$\text{tr}(\bigcirc\varphi) = \bigcirc\text{tr}(\varphi)$$

Note that such a translation exists as there is a bijective mapping between the game signatures. The following result holds that if two paths are structure-bisimilar, then they preserve the corresponding GDL-formulas at each stage.

Lemma 4. *Let M_S, $M'_{S'}$ be two ST-models. For every $\delta \in \mathcal{P}(M_S)$ and $\delta' \in \mathcal{P}(M'_{S'})$, if $M_S, \delta \leftrightarroweq_s M'_{S'}, \delta'$, then $M_S, \delta, j \models \varphi$ iff $M'_{S'}, \delta', j \models \text{tr}(\varphi)$ for any $\varphi \in \mathcal{L}_S$ and $j \geq 0$.*

Note that the converse to this proposition does not hold. Please refer to Fig. 2 for a counter-example. We now provide the following logical characterization result that structure equivalence and the invariance of the corresponding GDL-formulas coincide on ST-models.

Proposition 3. *Let M_S, $M'_{S'}$ be two ST-models. The following are equivalent.*

1. $M_S \sim M'_{S'}$
2. *for every $\varphi \in \mathcal{L}_S$, φ is satisfied in M_S iff $\text{tr}(\varphi)$ is satisfied in $M'_{S'}$.*

Proof. The direction from Clause 1 to Clause 2 follows from Lemmas 3 and 4. To prove the other direction, we need the following notion.

Consider two ST-models $M_S = (W, w_0, T, L, U, g, \pi)$ with $S = (N, \mathcal{A}, \Phi)$ and $M'_{S'} = (W', w'_0, T', L', U', g', \pi')$ with $S' = (N', \mathcal{A}', \Phi')$. Given bijections

$f_1 : N \mapsto N'$, $f_2 : \mathcal{A} \mapsto \mathcal{A}'$ and $f_3 : \Phi \mapsto \Phi'$, let tr be a translation defined in Definition 10. A translation Tr is a bijection from $trace(M)$ to $trace(M')$. For every $V(\delta) \in trace(M)$, $\mathrm{Tr}(V(\delta)) = \mathrm{tr}(V(w_0)) \cdot \mathrm{tr}(does(d_1)) \cdot \mathrm{tr}(V(w_1)) \cdots \mathrm{tr}(does(d_{e-1})) \cdot \mathrm{tr}(V(w_e))$ where $\mathrm{tr}(V(w_j)) = \{\mathrm{tr}(\varphi) \in \mathcal{L}_{S'} : \varphi \in V(w_j)\}$ for any $0 \le j \le e$, and $\mathrm{tr}(does(d_j)) = \bigwedge_{r \in N} \mathrm{tr}(does(r, d_j(r)))$ for any $1 \le j \le e$. Let $\mathrm{Tr}(trace(M)) = \{\mathrm{Tr}(V(\delta)) \mid V(\delta) \in trace(M)\}$. Then we have that the fact holds that $M_S \sim M'_{S'}$ iff $\mathrm{Tr}(trace(M)) = trace(M')$. With this, the proof of the direction from Clause 2 to Clause 1 is similar to Theorem 1.($\square$)

We end this section with the interesting observation that the GDL-descriptions of Tic-Tac-Toe and Number Scrabble are logically equivalent in terms of the translation.

Observation 1. *Let Σ_{TT} and Σ_{NS} denote the GDL-descriptions of Tic-Tac-Toe (Fig. 4) and Number Scrabble (Fig. 1), respectively. Then $\models \bigwedge \mathrm{tr}(\Sigma_{TT}) \leftrightarrow \bigwedge \Sigma_{NS}$, where $\mathrm{tr}(\Sigma_{TT}) = \{\mathrm{tr}(\varphi) \in \mathcal{L}_{NS} \mid \varphi \in \Sigma_{TT}\}$.*

5   Conclusion

We have defined the notion of bisimulation for GDL and showed that it coincides with the invariance of GDL-formulas. We have also introduced the quotient model to improve the efficiency of model checking for GDL. Moreover, we have generalized the notion of bisimulation to capture more general game equivalence.

Although various game equivalence have been proposed in economics, mathematics and logic [1,5,8,22], few work has been done in the domain of GGP. To the best of our knowledge, Zhang *et al.* investigated game equivalence for knowledge transfer in GGP. They consider two games are equivalent if their state transition models are isomorphic [25]. While our notion of game equivalence is more general as it is based on bisimulation relation.

Directions of future research are manifold. We intend to explore the van Benthem Characterization Theorem for GDL [4]. More recently, GDL has been extended to GDL-II and epistemic GDL for representing and reasoning about imperfect information games [13,21]. We plan to study the expressiveness of these extended languages. Besides structure equivalence, it would be also interesting to investigate different types of game equivalence in GGP, such as strategic equivalence, subgame equivalence [6,10,17].

Acknowledgments. We are grateful to three anonymous referees for their insightful comments. Guifei Jiang acknowledges the support of the National Natural Science Foundation of China (NO.61806102), the Fundamental Research Funds for the Central Universities, and the Major Program of the National Social Science Foundation of China (NO.17ZDA026). Laurent Perrussel acknowledges the support of the ANR project AGAPE ANR-18-CE23-0013.

References

1. Alur, R., Henzinger, T.A., Kupferman, O., Vardi, M.Y.: Alternating refinement relations. In: Sangiorgi, D., de Simone, R. (eds.) CONCUR 1998. LNCS, vol. 1466, pp. 163–178. Springer, Heidelberg (1998). https://doi.org/10.1007/BFb0055622
2. Baier, C., Katoen, J.P., Larsen, K.G.: Principles of Model Checking. MIT Press, Cambridge (2008)
3. van Benthem, J.: Modal Correspondence Theory. Ph.D. thesis, University of Amsterdam (1977)
4. van Benthem, J.: Correspondence theory. In: Gabbay, D., Guenthner, F. (eds.) Handbook of Philosophical Logic. SYLI, pp. 167–247. Springer, Dordrecht (1984). https://doi.org/10.1007/978-94-009-6259-0_4
5. van Benthem, J.: Logic in Games. MIT Press, Cambridge (2014)
6. van Benthem, J., Bezhanishvili, N., Enqvist, S.: A new game equivalence and its modal logic. In: Proceedings Sixteenth Conference on Theoretical Aspects of Rationality and Knowledge (TARK 2017), pp. 57–74 (2017)
7. Blackburn, P., van Benthem, J., Wolter, F.: Handbook of Modal Logic, vol. 3. Elsevier, Amsterdam (2006)
8. Elmes, S., Reny, P.J.: On the strategic equivalence of extensive form games. J. Econ. Theor. **62**(1), 1–23 (1994)
9. Genesereth, M., Love, N., Pell, B.: General game playing: overview of the AAAI competition. AI Mag. **26**(2), 62–72 (2005)
10. Goranko, V.: The basic algebra of game equivalences. Studia Logica **75**(2), 221–238 (2003)
11. Grädel, E., Otto, M.: The freedoms of (guarded) bisimulation. In: Baltag, A., Smets, S. (eds.) Johan van Benthem on Logic and Information Dynamics. OCL, vol. 5, pp. 3–31. Springer, Cham (2014). https://doi.org/10.1007/978-3-319-06025-5_1
12. Hennessy, M., Milner, R.: Algebraic laws for nondeterminism and concurrency. J. ACM (JACM) **32**(1), 137–161 (1985)
13. Jiang, G., Zhang, D., Perrussel, L., Zhang, H.: Epistemic GDL: a logic for representing and reasoning about imperfect information games. In: Proceedings of the 25th International Joint Conference on Artificial Intelligence (IJCAI 2016), pp. 1138–1144 (2016)
14. Love, N., Hinrichs, T., Haley, D., Schkufza, E., Genesereth, M.: General game playing: game description language specification. Stanford Logic Group Computer Science Department Stanford University (2006). http://logic.stanford.edu/reports/LG-2006-01.pdf
15. Michon, J.A.: The game of jam: an isomorph of tic-tac-toe. Am. J. Psychol. **80**(1), 137–140 (1967). http://www.jstor.org/stable/1420555
16. Park, D.: Concurrency and automata on infinite sequences. In: Deussen, P. (ed.) GI-TCS 1981. LNCS, vol. 104, pp. 167–183. Springer, Heidelberg (1981). https://doi.org/10.1007/BFb0017309
17. Pauly, M.: Logic for Social Software. Ph.D. thesis. University of Amsterdam (2001). ILLC Dissertation Series 2001-10
18. Pell, B.: Strategy generation and evaluation for meta-game playing. Ph.D. thesis. University of Cambridge (1993)
19. Reiter, R.: The frame problem in the situation calculus: a simple solution (sometimes) and a completeness result for goal regression. Artif. Intell. Math. Theory Comput.: Pap. Honor. John McCarthy **27**, 359–380 (1991)

20. Ruan, J., van Der Hoek, W., Wooldridge, M.: Verification of games in the game description language. J. Log. Comput. **19**(6), 1127–1156 (2009)
21. Thielscher, M.: A general game description language for incomplete information games. In: Proceedings of the 24th AAAI Conference on Artificial Intelligence (AAAI 2010), pp. 994–999 (2010)
22. Thompson, F.: Equivalence of games in extensive form. Class. Game Theory, **36** (1997)
23. Zhang, D., Thielscher, M.: A logic for reasoning about game strategies. In: Proceedings of the 29th AAAI Conference on Artificial Intelligence (AAAI 2015), pp. 1671–1677 (2015)
24. Zhang, D., Thielscher, M.: Representing and reasoning about game strategies. J. Philos. Log. **44**(2), 203–236 (2015)
25. Zhang, H., Liu, D., Li, W.: Space-consistent game equivalence detection in general game playing. In: Cazenave, T., Winands, M.H.M., Edelkamp, S., Schiffel, S., Thielscher, M., Togelius, J. (eds.) CGW/GIGA -2015. CCIS, vol. 614, pp. 165–177. Springer, Cham (2016). https://doi.org/10.1007/978-3-319-39402-2_12

Characterizing the Expressivity of Game Description Languages

Guifei Jiang[1]([✉]), Laurent Perrussel[2], Dongmo Zhang[3], Heng Zhang[4], and Yuzhi Zhang[1]

[1] Nankai University, Tianjin, China
g.jiang@nankai.edu.cn
[2] University of Toulouse, Toulouse, France
[3] Western Sydney University, Penrith, Australia
[4] Tianjin University, Tianjin, China

Abstract. Bisimulations are a key notion to study the expressive power of a modal language. This paper studies the expressiveness of Game Description Language (GDL) and its epistemic extension EGDL through a bisimulations approach. We first define a notion of bisimulation for GDL and prove that it coincides with the indistinguishability of GDL-formulas. Based on it, we establish a characterization of the definability of GDL in terms of k-bisimulations. Then we define a novel notion of bisimulation for EGDL, and obtain a characterization of the expressive power of EGDL. In particular, we show that a special case of the bisimulation for EGDL can be used to characterize the expressivity of GDL. These characterizations not only justify the notions of bisimulation are appropriate for game description languages, but also provide a powerful tool to identify their expressive power.

Keywords: Bisimulation equivalence · Expressive power · Game description languages

1 Introduction

General Game Playing (GGP) is concerned with creating intelligent agents that understand the rules of previously unknown games and learn to play these games without human intervention [8]. To represent the rules of arbitrary games, a formal game description language (GDL) was introduced as an official language for GGP in 2005. GDL is originally a machine-processable, logic programming language [14]. Most recently, it has been adapted as a minimal logical language for game specification and strategic reasoning [21]. The epistemic extension EGDL has been also developed to incorporate imperfect information games [12].

Although GDL and EGDL are logical languages for representing game rules and specifying game properties, their logical properties, especially their expressive power have not been fully investigated yet. For instance, which game properties are definable or non-definable in GDL and EGDL? How to show a game

© Springer Nature Switzerland AG 2019
A. C. Nayak and A. Sharma (Eds.): PRICAI 2019, LNAI 11670, pp. 597–611, 2019.
https://doi.org/10.1007/978-3-030-29908-8_47

property is not definable in GDL and EGDL? When two game descriptions are equivalent? The existing work about the expressiveness of game description languages is rare, mostly investigating the relationships of these languages with other strategic logics. In particular, Ruan *et al.* study the relationship between GDL and Alternating-time Temporal Logic (ATL) by transferring a GDL game specification into an ATL specification [18]. Lorini and Schwarzentruber investigate the relation between GDL and Seeing-to-it-that Logics (STITs) by providing a polynomial embedding of GDL into STIT [13]. In this paper, we propose a different approach to address these questions via *bisimulation*.

The notion of *bisimulation* plays a pivotal role to identify the expressive power of a logic. It was independently defined and developed in the areas of theoretical computer science [11,15] and the model theory of modal logic [2,3]. Since bisimulation-equivalent structures can simulate each other in a stepwise manner, they cannot be distinguished by the concerned logic. An appropriate notion of bisimulation for a logic allows us to study the expressive power of that logic in terms of structural invariance and language indistinguishability [10].

On the basis of the above consideration, we use in this paper a bisimulation approach to investigate the expressive power of GDL and EGDL. To this end, we first define a notion of *bisimulation equivalence* for GDL and prove that *it preserves the invariance of GDL-formulas*. Based on this, we provide a characterization for the definability of GDL, and show that *a class of state transition models is definable in GDL iff they are closed under k-bisimulations*. This allows us to establish the non-definability of a property in GDL. For instance, we show that GDL does not allow to express the property that a player has a winning strategy. More importantly, to characterize the expressivity of EGDL, we define a novel notion of bisimulation, called *(m,n)-bisimulation*. We not only prove that (m,n)-bisimulation can be logically characterized by EGDL, but also establish a characterization of the definability of EGDL. These characterizations not only justify that the notions of bisimulation are appropriate for GDL and EGDL, but also provide a powerful tool to identify their expressive power. To the best of our knowledge, this work is the first to conduct a systematic study on the expressive power of Game Description Languages. This would help us to identify the roles of Game Description Languages compared to other existing strategic logics, such as ATL, STIT, and choose the right language for the intended application.

The rest of this paper is structured as follows: Sect. 2 introduces the framework of GDL. Section 3 defines the notion of bisimulation equivalence for GDL and characterizes its definability. Section 4 defines the notion of bisimulation for EGDL and characterizes its expressivity. Finally, we conclude with future work.

2   The GDL-Based Framework

Let us now introduce the GDL-based framework from [21]. Each game is associated with a *game signature*. A *game signature* $\mathcal{S}$ is a triple $(N, \mathcal{A}, \Phi)$, where

- $N = \{1, 2, \cdots, m\}$ is a non-empty finite set of agents,

- $\mathcal{A}$ is a non-empty finite set of *actions* such that it contains *noop*, an action without any effect, and
- $\Phi = \{p, q, \cdots\}$ is a finite set of propositional atoms for specifying individual features of a game state.

Through the rest of the paper, we will consider a fixed game signature $\mathcal{S}$, and all concepts are based on the game signature unless otherwise specified.

2.1   State Transition Models

The structure for modelling games is defined as follows:

Definition 1. *A state transition (ST) model M is a tuple $(W, w_0, T, L, U, g, \pi)$, where*

- W *is a non-empty finite set of* possible states.
- $w_0 \in W$, *representing the unique initial state.*
- $T \subseteq W$, *representing a set of* terminal states.
- $L \subseteq W \times N \times \mathcal{A}$ *is a legality relation, specifying legal actions for each agent at game states. Let $L_r(w) = \{a \in \mathcal{A} \ : \ (w, r, a) \in L\}$ be the set of all legal actions for agent r at state w. To make a game playable, it is assumed that (i) $L_r(w) \neq \emptyset$ for any $r \in N$ and $w \in W$, and (ii) $L_r(w) = \{noop\}$ for any $r \in N$ and $w \in T$.*
- $U : W \times \mathcal{A}^{|N|} \hookrightarrow W \backslash \{w_0\}$ *is a partial* update *function, specifying the state transition for each state and legal joint action, such that $U(w, \langle noop^r \rangle_{r \in N}) = w$ for any $w \in W \setminus \{w_0\}$.*
- $g : N \to 2^W$ *is a goal function, specifying the winning states of each agent.*
- $\pi : W \to 2^\Phi$ *is a standard valuation function.*

Note that to make the framework as general as possible, here we consider synchronous games and as demonstrated by Example 1, turn-based games involved in [21] are special cases by allowing a player only to do "noop" when it is not her turn. For convenience, let D denote the set of all joint actions $\mathcal{A}^{|N|}$. Given $d \in D$, we use $d(r)$ to specify the action taken by agent r.

The following notion specifies all possible ways in which a game can develop.

Definition 2. *Let $M = (W, w_0, T, L, U, g, \pi)$ be an ST-model. A path δ is an infinite sequence of states and joint actions $w_0 \xrightarrow{d_1} w_1 \xrightarrow{d_2} \cdots \xrightarrow{d_j} \cdots$ s.t. for any $j \geq 1$ and $r \in N$,*

1. *$w_j \neq w_0$ (that is, only the first state is initial.)*
2. *$d_j(r) \in L_r(w_{j-1})$ (that is, any action taken by each agent must be legal.)*
3. *$w_j = U(w_{j-1}, d_j)$ (state update)*
4. *if $w_{j-1} \in T$, then $w_{j-1} = w_j$ (self-loop after reaching a terminal state.)*

Let $\mathcal{P}(M)$ denote the set of all paths in M. For $\delta \in \mathcal{P}(M)$ and a stage $j \geq 0$, we use $\delta[j]$ to denote the j-th state on δ, and $\theta_r(\delta, j)$ the action taken by agent r at stage j of δ.

2.2   The Language

The language for game specification is given as follows:

Definition 3. *The language $\mathcal{L}_{GDL}$ for game description is generated by the following BNF:*

$$\varphi ::= \ p \mid initial \mid terminal \mid legal(r,a) \mid wins(r) \mid does(r,a) \mid \neg\varphi \mid \varphi \wedge \psi \mid \bigcirc\varphi$$

where $p \in \Phi$, $r \in N$ and $a \in \mathcal{A}$.

Other connectives $\vee$, $\rightarrow$, $\leftrightarrow$, $\top$, $\bot$ are defined by $\neg$ and $\wedge$ in the standard way. Intuitively, *initial* and *terminal* specify the initial state and the terminal states of a game, respectively; $does(r,a)$ asserts that agent r takes action a at the current state; $legal(r,a)$ asserts that agent r is allowed to take action a at the current state, and $wins(r)$ asserts that agent r wins at the current state. Finally, the formula $\bigcirc\varphi$ means that φ holds in the next state. We use the following abbreviations in the rest of paper. For $d = \langle a_r\rangle_{r\in N}$, $does(d) =_{def} \bigwedge_{r\in N} does(r,a_r)$, and $\bigcirc^k\varphi =_{def} \underbrace{\bigcirc\cdots\bigcirc}_{k}\varphi$.

Note that our language is slightly different from [21] by introducing the agent parameter in $legal(\cdot)$ and $does(\cdot)$. To help the reader capture the intuition of the language, let us consider the following example.

Example 1 (Tic-Tac-Toe). Two players take turns in marking either a cross 'x'or a nought 'o' on an 3×3 board. The player who first gets three consecutive marks of her own symbol in a row wins this game.

The game signature for Tic-Tac-Toe, written $\mathcal{S}_{TT}$, is given as follows: $N_{TT} = \{x, o\}$ denoting the two game players; $\mathcal{A}_{TT} = \{a_{i,j} \mid 1 \leq i,j \leq 3\} \cup \{noop\}$, where $a_{i,j}$ denotes filling cell (i,j), and $\Phi_{TT} = \{p_{i,j}^r, turn(r) \mid r \in \{x, o\}$ and $1 \leq i,j \leq 3\}$, where $p_{i,j}^r$ represents the fact that cell (i,j) is filled by player r. The rules of this game is given in Fig. 1.

1. $initial \leftrightarrow turn(x) \wedge \neg turn(o) \wedge \bigwedge\limits_{i,j=1}^{3} \neg(p_{i,j}^x \vee p_{i,j}^o)$

2. $wins(r) \leftrightarrow \bigvee\limits_{i=1}^{3}\bigwedge\limits_{j=0}^{2} p_{i,1+j}^r \vee \bigvee\limits_{j=1}^{3}\bigwedge\limits_{i=0}^{2} p_{1+i,j}^r \vee \bigwedge\limits_{i=0}^{2} p_{1+i,1+i}^r \vee \bigwedge\limits_{i=0}^{2} p_{1+i,3-i}^r$

3. $teminal \leftrightarrow wins(x) \vee wins(o) \vee \bigwedge\limits_{i,j=1}^{3} (p_{i,j}^x \vee p_{i,j}^o)$

4. $legal(r, a_{i,j}) \leftrightarrow \neg(p_{i,j}^x \vee p_{i,j}^o) \wedge turn(r) \wedge \neg terminal$

5. $legal(r, noop) \leftrightarrow turn(-r) \vee terminal$

6. $\bigcirc p_{i,j}^r \leftrightarrow p_{i,j}^r \vee (does(r, a_{i,j}) \wedge \neg(p_{i,j}^x \vee p_{i,j}^o))$

7. $turn(r) \wedge \neg terminal \rightarrow \bigcirc\neg turn(r) \wedge \bigcirc turn(-r)$

Fig. 1. A GDL description of Tic-Tac-Toe.

The initial state, each player's winning states, the terminal states and the turn-taking are given by formulas 1–3 and 7, respectively. The preconditions of each action (legality) are specified by Formula 4 and 5. Formula 6 is the combination of the frame axioms and the effect axioms [17].

2.3   The Semantics

The semantics of this language is specified as follows:

Definition 4. *Let $M = (W, w_0, T, L, U, g, \pi)$ be an ST-model. Given a path δ of M, a stage $j \in \mathbb{N}$ and a formula $\varphi \in \mathcal{L}_{GDL}$, we say φ is true (or satisfied) at j of δ under M, denoted $M, \delta, j \models \varphi$, according to the following definition:*

$$
\begin{array}{llll}
M, \delta, j \models p & \quad iff & \quad p \in \pi(\delta[j]) \\
M, \delta, j \models \neg\varphi & \quad iff & \quad M, \delta, j \not\models \varphi \\
M, \delta, j \models \varphi_1 \wedge \varphi_2 & \quad iff & \quad M, \delta, j \models \varphi_1 \text{ and } M, \delta, j \models \varphi_2 \\
M, \delta, j \models initial & \quad iff & \quad \delta[j] = w_0 \text{ and } j = 0 \\
M, \delta, j \models terminal & \quad iff & \quad \delta[j] \in T \\
M, \delta, j \models wins(r) & \quad iff & \quad \delta[j] \in g(r) \\
M, \delta, j \models legal(r, a) & \quad iff & \quad a \in L_r(\delta[j]) \\
M, \delta, j \models does(r, a) & \quad iff & \quad \theta_r(\delta, j) = a \\
M, \delta, j \models \bigcirc\varphi & \quad iff & \quad M, \delta, j + 1 \models \varphi
\end{array}
$$

An ST-model M *satisfies* a formula φ if there are a path $\delta \in \mathcal{P}(M)$ and a stage $j \in \mathbb{N}$ such that $M, \delta, j \models \varphi$. We say a formula φ is *satisfied at the initial state w_0* in M, written $M, w_0 \models \varphi$, if $M, \delta, 0 \models \varphi$ for all $\delta \in \mathcal{P}(M)$.

3   Bisimulation and Definability of GDL

In this section, we first define the notion of bisimulation equivalence for GDL, and prove the invariance result of GDL-formulas. Then we present a characterization of the definability of GDL in terms of k-bisimulation.

3.1   Bisimulation and Invariance for GDL

We consider two types of bisimulation for ST-models. The first one is inspired by the notion of bisimulation in [5,6] defined as follows:

Definition 5. *Let $M = (W, w_0, T, L, U, g, \pi)$ and $M' = (W', w'_0, T', L', U', g', \pi')$ be two ST-models (based on the same game signature). We say M and M' are bisimulation-equivalent (bisimilar, for short), written $M \simeq M'$, if there is a binary relation $Z \subseteq W \times W'$ s.t. $w_0 Z w'_0$, and for all states $w \in W$ and $w' \in W'$ with wZw', we have*

1. *All the following hold:*
   (a) $\pi(w) = \pi'(w')$;

(b) $w = w_0$ *iff* $w' = w_0'$;
(c) $w \in T$ *iff* $w' \in T'$;
(d) $L_r(w) = L_r'(w')$ *for any* $r \in N$;
(e) $w \in g(r)$ *iff* $w' \in g'(r)$ *for any* $r \in N$.

2. *For every* $d \in D$ *and* $u \in W$, *if* $U(w,d) = u$, *then there is* $u' \in W'$ *s.t.* $U'(w',d) = u'$ *and* uZu';

3. *For every* $d \in D$ *and* $u' \in W'$, *if* $U'(w',d) = u'$, *then there is* $u \in W$ *s.t.* $U(w,d) = u$ *and* uZu'.

When Z is a bisimulation linking two states w in M and w' in M', we say that w and w' are *bisimilar*, written $M, w \simeq M', w'$. In particular, if $M \simeq M'$, then their initial states are bisimilar, i.e., $M, w_0 \simeq M', w_0'$. In the following, for convenience we denote Condition *(a)-(e)* in Definition 5 as *the local properties* of a state.

With path-based semantics, we define the second type of bisimulation, called *path bisimulation* as follows:

Definition 6. *Consider two ST-models* $M = (W, w_0, T, L, U, g, \pi)$ *and* $M' = (W', w_0', T', L', U', g', \pi')$. *Let* $\delta \in \mathcal{P}(M)$ *and* $\delta' \in \mathcal{P}(M')$, *we say* δ *and* δ' *are* *bisimilar, written* $M, \delta \simeq M', \delta'$, *iff* *for every* $j \geq 0$ *and* $r \in N$, *the local properties hold for* $\delta[j]$ *and* $\delta'[j]$, *and* $\theta_r(\delta, j) = \theta_r(\delta', j)$.

This asserts that two paths are bisimilar if all the corresponding states satisfy the same local properties, and each agent takes the same action at every stage.

It turns out that with the deterministic property, the two types of bisimulation are equivalent. Formally, we have the following result.

Lemma 1. *Given two ST-models* M *and* M', $M \simeq M'$ *iff*

1. *for every* $\delta \in \mathcal{P}(M)$, *there is* $\delta' \in \mathcal{P}(M')$ *such that* $M, \delta \simeq M', \delta'$, *and*
2. *for every* $\delta' \in \mathcal{P}(M')$, *there is* $\delta \in \mathcal{P}(M)$ *such that* $M, \delta \simeq M', \delta'$.

That is, M *is bisimilar to* M' *iff each path that can be developed in one model can also be induced in the other.*

Let us now turn to the logical characterization of bisimulation equivalence. We have the invariance result of GDL-formulas under path-bisimulation.

Proposition 1. *Let* M *and* M' *be two ST-models. For every* $\delta \in \mathcal{P}(M)$ *and* $\delta' \in \mathcal{P}(M')$, *the following are equivalent.*

1. $M, \delta \simeq M', \delta'$
2. $(M, \delta, j \models \varphi$ *iff* $M', \delta', j \models \varphi)$ *for any* $j \in \mathbb{N}$ *and* $\varphi \in \mathcal{L}_{GDL}$.

This result asserts that bisimulation equivalence and the invariance of GDL-formulas match on ST-models. On the one hand, this result justifies that the notion of bisimulation equivalence is natural and appropriate for GDL. On the other hand, two bisimilar ST-models cannot be distinguished by GDL language. This allows us to show the failure of bisimulation-equivalence easily. That is, *two ST-models are not bisimulation-equivalent if there is a GDL-formula that holds in one model and fails in the other.* For instance, let us consider the two ST-models depicted in Fig. 2, where $N = \{r\}$ and $\Phi = \emptyset$. Formula *initial* $\land \bigcirc^2 does(r,c)$ is satisfied in M, but unsatisfied in M'. This leads to $M \not\simeq M'$.

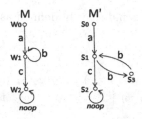

Fig. 2. M and M' are not bisimulation-equivalent.

3.2 k-Bisimulation and Definability of GDL

To show that a property of ST-models is definable in GDL, it suffices to find a defining formula. However, showing that a property is not definable in GDL is not so straightforward. It is well known that the expressive power of basic modal logic with respect to Kripke semantics can be completely characterized in terms of k-bisimulation [6]. In this section, we provide an analogous characterization result for GDL.

Here we consider the definability of properties that are satisfied at the initial state of an ST-model. For $\varphi \in \mathcal{L}_{GDL}$, let $\|\varphi\|$ be the set of all ST-models that satisfy φ at the initial state. i.e., $\|\varphi\| := \{M \mid M, w_0 \models \varphi\}$. The concept of the *definability* is specified as follows:

Definition 7. *A class $\mathcal{M}$ of ST-models is GDL-definable, if there is a formula* $\varphi \in \mathcal{L}_{GDL}$ *s.t.* $\mathcal{M} = \|\varphi\|$.

Similar to [6], we define the concept of *k-bisimulation* as follows:

Definition 8. *Let* $M = (W, w_0, T, L, U, g, \pi)$ *and* $M' = (W', w'_0, T', L', U', g', \pi')$ *be two ST-models. We say M and M' are k-bisimilar, written $M \simeq_k M'$, if there exists a sequence of binary relations $Z_k \subseteq Z_{k-1} \cdots \subseteq Z_0$ s.t. for any* $w \in W$, $w' \in W'$ *and* $i \in \{0, \cdots, k-1\}$,

1. $w_0 Z_k w'_0$
2. *If $w Z_0 w'$, then the local properties hold for w and w';*
3. *If $w Z_{i+1} w'$ and $U(w, d) = u$, then there is $u' \in W'$ s.t. $U'(w', d) = u'$ and $u Z_i u'$;*
4. *If $w Z_{i+1} w'$ and $U'(w', d) = u'$, then there is $u \in W$ s.t. $U(w, d) = u$ and $u Z_i u'$.*

The intuition is that if two ST-models are k-bisimilar, then their initial states w_0 and w'_0 bisimulate up to depth k. Clearly, if $M \simeq M'$, then $M \simeq_k M'$ for all $k \in \mathbb{N}$. We say a class $\mathcal{M}$ of ST-models is *closed under k-bisimulations* if for all ST-models M and M', if $M \in \mathcal{M}$ and $M \simeq_k M'$ then $M' \in \mathcal{M}$.

Before providing the characterization of the definability of GDL, we need some additional notions and results. The depth of next operators for a formula

$\varphi \in \mathcal{L}_{GDL}$, written $deg_N(\varphi)$, is inductively defined as follows:

$$deg_N(\varphi) = \begin{cases} 0, & \text{for } \varphi \text{ is } \bigcirc -\text{free} \\ deg_N(\psi), & \text{for } \varphi = \neg\psi \\ Max\{deg_N(\varphi_1), deg_N(\varphi_2)\}, & \text{for } \varphi = \varphi_1 \wedge \varphi_2 \\ deg_N(\psi) + 1, & \text{for } \varphi = \bigcirc\psi \end{cases}$$

Definition 9. *Let M, M' be two ST-models and $k \in \mathbb{N}$. We say M and M' are k-equivalent, written $M \equiv_k M'$, if at the initial states, they satisfy the same GDL-formulas of degree at most k, i.e., $\{\varphi \in \mathcal{L}_{GDL} \mid deg(\varphi) \leq k$ and $M, w_0 \models \varphi\} = \{\psi \in \mathcal{L}_{GDL} \mid deg(\psi) \leq k$ and $M', w_0' \models \psi\}$.*

We use the fact that *for every ST-model M and every $k \in \mathbb{N}$ there is a formula that completely characterizes M up to k-equivalence.* With action operator and path-based semantics, the way to construct the k-th characteristic formula of an ST-model is non-standard. We need take the following steps.

1. Redefine the set of *atomic propositions*, written *Atm*, as follows: $Atm = \Phi \cup \{initial, terminal\} \cup \{wins(r), legal(r, a) \mid r \in N, a \in \mathcal{A}\}$.
2. Encode the atomic propositions through a valuation V rather than through separate relations or functions. For every $w \in W$, let $V(w) = \{p \in \Phi \mid p \in \pi(w)\} \cup \{initial \mid w = w_0\} \cup \{terminal \mid w \in T\} \cup \{wins(r) \mid w \in g(r)\} \cup \{legal(r, a) \mid a \in L_r(w)\}$. Note $V(w)$ is finite since N, $\mathcal{A}$ and Φ are all finite.
3. For each path $\delta := w_0 \xrightarrow{d_1} w_1 \xrightarrow{d_2} \cdots \xrightarrow{d_j} \cdots$ in M, induce a *trace* $V(\delta) = V(w_0) \cdot does(d_1) \cdot V(w_1) \cdots does(d_j) \cdot V(w_j) \cdots$. Let φ_δ^k be the syntactical representation of δ up to depth k, i.e., $\varphi_\delta^k := (\bigwedge V(\delta[0]) \wedge does(d_1)) \wedge \bigcirc(\bigwedge V(\delta[1]) \wedge does(d_2)) \wedge \cdots \wedge \bigcirc^k(\bigwedge V(\delta[k]) \wedge does(d_{k+1}))$.
4. Define the *k-th characteristic formula Γ_M^k of M* as the disjunctions of all the syntactical representations of paths in M up to depth k, i.e.,

$$\Gamma_M^k := \bigvee_{\delta \in \mathcal{P}(M)} \varphi_\delta^k.$$

Note that Γ_M^k is well-formed as M is finite-branching and all paths are bounded to depth k. It is easy to check that $deg(\Gamma_M^k) = k$ and $M, w_0 \models \Gamma_M^k$.

To illustrate this idea, let us consider the ST-model M depicted in Fig. 3, where $N = \{r\}$, $\Phi = \emptyset$, $T = \{w_{22}, w_{23}\}$ and $g(r) = \{w_{23}\}$. Then the 2-th characteristic formula of M is $\Gamma_M^2 = \varphi_{\delta_1}^2 \vee \varphi_{\delta_2}^2 \vee \varphi_{\delta_3}^2$, where

$\varphi_{\delta_1}^2 = initial \wedge \bigwedge_{i=1}^2 legal(r, a_i) \wedge does(r, a_1) \wedge \bigcirc(\bigwedge_{i=1}^2 legal(r, b_i) \wedge does(r, b_1)) \wedge \bigcirc^2(legal(r, c) \wedge does(r, c))$,

$\varphi_{\delta_2}^2 = initial \wedge \bigwedge_{i=1}^2 legal(r, a_i) \wedge does(r, a_1) \wedge \bigcirc(\bigwedge_{i=1}^2 legal(r, b_i) \wedge does(r, b_2)) \wedge \bigcirc^2(terminal \wedge legal(r, noop) \wedge does(r, noop))$, and

$\varphi_{\delta_3}^2 = initial \wedge \bigwedge_{i=1}^2 legal(r, a_i) \wedge does(r, a_2) \wedge \bigcirc(legal(r, b_3) \wedge does(r, b_3)) \wedge \bigcirc^2(wins(r) \wedge terminal \wedge legal(r, noop) \wedge does(r, noop))$.

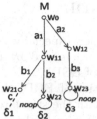

Fig. 3. Characteristic formula Γ_M^2.   **Fig. 4.** A non-definable property.

The following lemma shows that the characteristic formula Γ_M^k captures the essence of k-bisimulation.

Lemma 2. *Let M, M' be two ST-models, and $k \in \mathbb{N}$. The following are equivalent.*

1. $M \simeq_k M'$
2. $M \equiv_k M'$
3. $M', w_0' \models \Gamma_M^k$

This result asserts that (i) k-bisimulation coincides with k-equivalence on ST-models, and (ii) two ST-models are k-bisimilar if and only if for any path developed in one model, its k-th prefix can also be developed in the other.

We are now in the position to provide a characterization for the definability of GDL with respect to k-bisimulation.

Theorem 1. *A class $\mathcal{M}$ of ST-models is GDL-definable iff there is $k \in \mathbb{N}$ s.t. $\mathcal{M}$ is closed under k-bisimulations.*

This theorem indicates that *exactly the properties of ST-models that are closed under k-bisimulation for some $k \in \mathbb{N}$ are definable in GDL*. This provides a feasible approach to test the non-definability of GDL. We can show, for instance, that GDL can express that a player r will win in i steps, i.e., $\bigcirc^i wins(r)$, but it cannot express that *a player has a winning strategy* in general. Indeed for an arbitrary $k \in \mathbb{N}$, we can always construct two ST-models depicted in Fig. 4, where $N = \{r\}$, $\Phi = \emptyset$, $w_{k+1} \in g(r)$ and $s_i \notin g'(r)$ for all $i \in \{0, \cdots, k+1\}$. It is easy to check that $M \simeq_k M'$, but player r has a winning strategy in M while she does not have in M'. By a slight change of M and M' with $w_{k+1} \in T$ and $s_{k+1} \notin T'$, we obtain another GDL-undefinable property that *a game will always reach a terminal state*. It is worth noting that GDL is a lightweight language for describing game rules and specifying game properties, compared to other strategic logics such as ATL [1] and Strategy Logic [7] which can express those properties. This is the price paid for the low complexity of GDL.

4   Bisimulation and Definability of EGDL

In this section, we first define a notion of bisimulation especially designed for EGDL, and then prove that it can be logically characterized by EGDL. We finally provide a characterization of the definability of EGDL.

Let us first introduce the language and semantics of EGDL in [12]. The language of EGDL is obtained by extending GDL with the standard epistemic operators. A formula $\varphi \in \mathcal{L}_{EGDL}$ is defined by the following BNF:

$$\varphi ::= \ p \mid initial \mid terminal \mid legal(r,a) \mid wins(r) \mid$$

$$does(r,a) \mid \neg\varphi \mid \varphi \wedge \psi \mid \bigcirc\varphi \mid \mathsf{K}_r\varphi \mid \mathsf{C}\varphi$$

where $p \in \varPhi$, $r \in N$ and $a \in \mathcal{A}$.

Besides the GDL-components, the formula $\mathsf{K}_r\varphi$ is read as "agent r knows φ", and $\mathsf{C}\varphi$ as "φ is common knowledge among all the agents in N".

The semantics of EGDL is based on epistemic state transition models. An *epistemic state transition (ET)* model is obtained by associating the state transition model with an equivalence relation $R_r \subseteq W \times W$ for each agent r, indicating the states that are indistinguishable for r. To interpret epistemic formulas, the equivalence relation over states is generalized to paths: *two paths $\delta, \lambda \in \mathcal{P}(M)$ are imperfect recall equivalent for agent r at stage $j \in \mathbb{N}$, written $\delta \approx_r^j \lambda$, iff$\delta[j]R_r\lambda[j]$.*

The semantics of EGDL is obtained by adding the following interpretation clauses to Definition 4:

$$M, \delta, j \models \mathsf{K}_r\varphi \text{ iff for all } \lambda \approx_r^j \delta, \ M, \lambda, j \models \varphi$$
$$M, \delta, j \models \mathsf{C}\varphi \ \text{ iff for all } \lambda \approx_N^j \delta, \ M, \lambda, j \models \varphi$$

where $\approx_N^j$ is the transitive closure of $\bigcup_{r \in N} \approx_r^j$.

4.1   Bisimulation Equivalence for ET-models

Different from ST-models, there are two dimensions in ET-models: the temporal and the epistemic dimension. The epistemic relation is actually determined by the stage-path pair, since path equivalence requires not only the corresponding states are undistinguishable, but also the states are reached at the same stage. The state-based bisimulation for GDL fails to capture the latter. Based on above analysis, we define a new notion of path-based bisimulation between ET-models.

Definition 10. *Let $M = (W, w_0, T, \{R_r\}_{r \in N}, L, U, g, \pi)$, $M' = (W', w_0', T', \{R_r'\}_{r \in N}, L', U', g', \pi')$ be two ET-models, $\delta \in \mathcal{P}(M)$, $\delta' \in \mathcal{P}(M')$ and $j \in \mathbb{N}$. We say M, δ, j and M', δ', j are (m,n)-bisimilar, written $M, \delta, j \underset{m}{\overset{n}{\leftrightarrow}} M', \delta', j$, if we have*

1. *The base case*
   (a) *the local properties hold for $\delta[j]$ and $\delta'[j]$.*
   (b) *$\theta_r(\delta, j) = \theta_r(\delta', j)$ for any $r \in N$.*

2. *If $m > 0$, $M, \delta, j + 1 \underset{m-1}{\leftrightarrow^n} M', \delta', j + 1$.*
3. *If $n > 0$, then for all $o \in N \cup \{N\}$,*
   (a) *for any $\lambda \in \mathcal{P}(M)$, $\delta \approx_o^j \lambda$, then there is $\lambda' \in \mathcal{P}(M')$ s.t. $\delta' \approx_o^j \lambda'$ and $M, \lambda, j \underset{m}{\leftrightarrow^{n-1}} M', \lambda', j$;*
   (b) *for any $\lambda' \in \mathcal{P}(M')$, $\delta' \approx_o^j \lambda'$, then there is $\lambda \in \mathcal{P}(M)$ s.t. $\delta \approx_o^j \lambda$ and $M, \lambda, j \underset{m}{\leftrightarrow^{n-1}} M', \lambda', j$.*

This recursively asserts that two paths from ET-models are (m, n)-bisimilar iff (i) at the current stage, they can not be distinguished, i.e., the same local properties hold and each agent takes the same action (Condition 1), and (ii) at the next stage, they also bisimulate each other from both the temporal and epistemic perspectives. This is specified by Condition (2) and Condition (3). Specifically, Condition (2) takes care of the temporal dimension: the depth m of path bisimulation is reduced stage by stage until 0, and Condition (3) deals with the epistemic dimension: for each path indistinguishable from δ, there is a path indistinguishable from δ' that bisimulates it. The number of such bisimilar pairs at stage j is specified by the parameter n.

With this, we define the concept of (m, n)-bisimulation over ET-models as follows:

Definition 11. *Let M and M' be two ET-models. We say M is globally (m,n)-similar to M', written $M \propto_m^n M'$, if for every $\delta \in \mathcal{P}(M)$, there is $\delta' \in \mathcal{P}(M')$ such that $M, \delta, 0 \underset{m}{\leftrightarrow^n} M', \delta', 0$.*

We say M and M' are (m,n)-bisimilar, written $M \underset{m}{\leftrightarrow^n} M'$, if $M \propto_m^n M'$ and $M' \propto_m^n M$.

This asserts that two ET-models are (m, n)-bisimilar iff for every path in one model there is a path in the other such that their initial states are (m, n)-bisimilar. In particular, we say two ET-models M, M' are *path-based bisimilar*, written $M \leftrightarrow M'$, if $M \underset{m}{\leftrightarrow^n} M'$ for all $m, n \in \mathbb{N}$. Finally, a class $\mathcal{M}$ of ET-models is *closed under global (m, n)-simulations* if for all ET-models M and M', if $M \propto_m^n M'$ and $M' \in \mathcal{M}$ then $M \in \mathcal{M}$.

To obtain the characterization results, we need some additional notions. For any EGDL-formula $\varphi \in \mathcal{L}_{EGDL}$, the depth of next operators, written $deg_N(\varphi)$, is defined as for GDL-formulas except for $\varphi \in \{K_r\psi, C\psi\}$, $deg_N(\varphi) = deg_N(\psi)$. The depth of epistemic operators, written $deg_E(\varphi)$, is inductively defined as follows:

$$deg_E(\varphi) = \begin{cases} 0, & \text{for } \varphi \text{ is } K_r, C - \text{free} \\ deg_E(\psi), & \text{for } \varphi \in \{\neg\psi, \bigcirc\psi\} \\ Max\{deg_E(\varphi_1), deg_E(\varphi_2)\}, & \text{for } \varphi = \varphi_1 \wedge \varphi_2 \\ deg_E(\psi) + 1, & \text{for } \varphi \in \{K_r\psi, C\psi\} \end{cases}$$

Let EGDL(m, n) denote the set of all formulas with the depth of next operators and epistemic operators at most m, n, respectively, i.e, EGDL$(m, n) = \{\varphi \in \mathcal{L}_{EGDL} \mid deg_N(\varphi) \leq m$ and $deg_E(\varphi) \leq n\}$. Then we have the following logical characterization result for (m, n)-bisimilar paths.

Proposition 2. *Let M, M' be two ET-models. For every $\delta \in \mathcal{P}(M)$, $\delta' \in \mathcal{P}(M')$ and $j \in \mathbb{N}$, the following are equivalent.*

1. $M, \delta, j \leftrightarrowtail_m^n M', \delta', j$
2. $(M, \delta, j \models \varphi$ iff $M', \delta', j \models \varphi)$ for all $\varphi \in EGDL(m, n)$

This result asserts that (m, n)-bisimulation coincides with the indistinguishability of EGDL-formulas for paths. In particular, this also holds for the initial states.

Similarly, we say two ET-models M and M' are *(m,n)-equivalent*, written $M \equiv_m^n M'$, if at the initial states, they satisfies the same EGDL(m, n)-formulas, i.e., $\{\varphi \in \mathcal{L}_{EGDL(m,n)} \mid M, w_0 \models \varphi\} = \{\psi \in \mathcal{L}_{EGDL(m,n)} \mid M', w_0' \models \psi\}$. Then the following shows that (m, n)-bisimulation is logically characterized by EGDL(m, n).

Theorem 2. *Let M and M' be two ET-models. Then $M \leftrightarrowtail_m^n M'$ iff $M \equiv_m^n M'$.*

This asserts (m, n)-bisimulation equivalence and the invariance of EGDL(m, n)-formulas coincides over ET-models. This result not only justifies that the notion of (m, n)-bisimulation is appropriate for EGDL, but also provides a feasible way to verify the failure of (m, n)-bisimulation. For instance, consider two ET-models M_1, M_2 depicted in Fig. 5, where $N = \{r\}$, $\Phi = \emptyset$. The dotted line denotes the indistinguishability relation of agent r. Notice that the reflexive loops are omitted. Formula $\bigcirc K_r(does(r, b) \rightarrow \neg \bigcirc^2 terminal)$ is not satisfied at w_0 of M_1, but holds at s_0 of M_2. Then $M_1 \not\equiv_3^1 M_2$. This leads to $M_1 \not\leftrightarrowtail_3^1 M_2$.

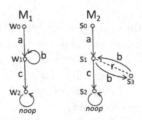

Fig. 5. M_1 and M_2 are not $(3, 1)$−bisimilar.

In particular, we have the following result about the characterization of bisimilar ET-models in terms of EGDL. Recall that $M \leftrightarrowtail M'$ if $M \leftrightarrowtail_m^n M'$ for all $m, n \in \mathbb{N}$.

Proposition 3. *Let M and M' be two ET-models. Then $M \leftrightarrowtail M'$ iff they satisfy the same EGDL-formulas.*

4.2   Logical Characterization of Definability of EGDL

Let us present the characterization of the definability of EGDL in terms of global (m, n)-simulations. Note that the notion of the *definability* of EGDL is defined the same as GDL in Definition 7.

Theorem 3. *A class $\mathcal{M}$ of ET-models is EGDL-definable iff there are $m, n \in \mathbb{N}$ such that $\mathcal{M}$ is closed under global (m, n)-simulations.*

This theorem asserts that *exactly the properties of ET-models that are closed under global (m,n)-simulations for some $m, n \in \mathbb{N}$ are definable in EGDL*. Similar to Theorem 1, this provides a feasible way to verify the non-definability of EGDL. For instance, we can show that EGDL cannot express that a player knows that she has a winning strategy.

We end this section with the following results that the expressivity of GDL is characterized by a special case of (m, n)-bisimulation with $n = 0$.

Proposition 4. *Let M and M' be two ET-models. Then the following are equivalent.*

1. *$M \underset{m}{\leftrightarrow}^0 M'$ for any $m \in \mathbb{N}$*
2. *they satisfy the same GDL formulas.*

Proposition 5. *A class $\mathcal{M}$ of ET-models is GDL-definable iff there is $m \in \mathbb{N}$ s.t. $\mathcal{M}$ is closed under global $(m, 0)$-simulations.*

These results indicates that path-based bisimulation provide a different way to characterize the expressivity of GDL. The first result shows that $(m, 0)$-bisimulation and the invariance of GDL-formulas match over ET-models, and the last characterizes the definability of GDL under $(m, 0)$-bisimulations. This indicates that without considering bisimulation for epistemic relations, $(m, 0)$-bisimulation actually boils down to m-bisimulation for ST-models.

5   Conclusion

In this paper, we have used a bisimulation approach to investigate the expressive power of GDL and EGDL. Specifically, we have defined notions of bisimulations for GDL and EGDL, and obtained the logical characterizations, respectively. We have also shown that a special case of path-based bisimulation can be used to characterize the expressivity of GDL. These results provide a feasible tool to identify the expressive power of game description languages. Finally, it is worth mentioning that bisimulation is a generic approach to identify the expressivity of a logic. Yet it is also sensitive to the logic it applies. Special techniques have to be developed for specific logics. With action operator and path-based semantics, the notions of bisimulation for GDL and EGDL are actually not a trivial and standard generalization of that for modal logic.

Directions of future research are manifold. We intend to explore the van Benthem Characterization Theorem for GDL and EGDL [3]. More recently, GDL has been extended to GDL-II and GDL-III for representing and reasoning about imperfect information games in GGP [19, 20]. We plan to study the expressivity of these languages and compare them with EGDL. Last but not least, bisimulation equivalence provides a natural yet overly strict criterion on game equivalence. It would be interesting to investigate different types of game equivalence in GGP [4, 9, 16, 22].

Acknowledgments. Guifei Jiang acknowledges the support of the National Natural Science Foundation of China (No. 61806102), the Fundamental Research Funds for the Central Universities, and the Major Program of the National Social Science Foundation of China (No. 17ZDA026). Laurent Perrussel acknowledges the support of the ANR project AGAPE ANR-18-CE23-0013.

References

1. Alur, R., Henzinger, T.A., Kupferman, O.: Alternating-time temporal logic. J. ACM **49**(5), 672–713 (2002)
2. van Benthem, J.: Modal correspondence theory. Ph.D. thesis, University of Amsterdam (1977)
3. van Benthem, J.: Correspondence theory. In: Gabbay, D., Guenthner, F. (eds.) Handbook of Philosophical Logic. Synthese Library (Studies in Epistemology, Logic, Methodology, and Philosophy of Science), vol. 165, pp. 167–247. Springer, Dordrecht (1984). https://doi.org/10.1007/978-94-009-6259-0_4
4. van Benthem, J., Bezhanishvili, N., Enqvist, S.: A new game equivalence and its modal logic. In: Proceedings Sixteenth Conference on Theoretical Aspects of Rationality and Knowledge (TARK 2017), pp. 57–74 (2017)
5. Blackburn, P., De Rijke, M., Venema, Y.: Modal Logic. Cambridge University Press, Cambridge (2002)
6. Blackburn, P., van Benthem, J., Wolter, F.: Handbook of Modal Logic, vol. 3. Elsevier, Amsterdam (2006)
7. Chatterjee, K., Henzinger, T.A., Piterman, N.: Strategy logic. Inf. Comput. **208**(6), 677–693 (2010)
8. Genesereth, M., Love, N., Pell, B.: General game playing: overview of the AAAI competition. AI Mag. **26**(2), 62–72 (2005)
9. Goranko, V.: The basic algebra of game equivalences. Stud. Logica **75**(2), 221–238 (2003)
10. Grädel, E., Otto, M.: The freedoms of (guarded) bisimulation. In: Baltag, A., Smets, S. (eds.) Johan van Benthem on Logic and Information Dynamics. OCL, vol. 5, pp. 3–31. Springer, Cham (2014). https://doi.org/10.1007/978-3-319-06025-5_1
11. Hennessy, M., Milner, R.: Algebraic laws for nondeterminism and concurrency. J. ACM (JACM) **32**(1), 137–161 (1985)
12. Jiang, G., Zhang, D., Perrussel, L., Zhang, H.: Epistemic GDL: a logic for representing and reasoning about imperfect information games. In: Proceedings of the 25th International Joint Conference on Artificial Intelligence (IJCAI 2016), pp. 1138–1144 (2016)
13. Lorini, E., Schwarzentruber, F.: A path in the jungle of logics for multi-agent system: on the relation between general game-playing logics and seeing-to-it-that logics. In: Proceedings of the 16th Conference on Autonomous Agents and Multi-Agent Systems (AAMAS 2017), pp. 687–695 (2017)
14. Love, N., Hinrichs, T., Haley, D., Schkufza, E., Genesereth, M.: General game playing: game description language specification. Stanford Logic Group Computer Science Department Stanford University (2006). http://logic.stanford.edu/reports/LG-2006-01.pdf
15. Park, D.: Concurrency and automata on infinite sequences. In: Deussen, P. (ed.) GI-TCS 1981. LNCS, vol. 104, pp. 167–183. Springer, Heidelberg (1981). https://doi.org/10.1007/BFb0017309

16. Pauly, M.: Logic for social software. Ph.D. thesis, University of Amsterdam (2001). ILLC Dissertation Series 2001-10
17. Reiter, R.: The frame problem in the situation calculus: a simple solution (sometimes) and a completeness result for goal regression. Artif. Intell. Math. Theory Comput. Pap. Honor John McCarthy **27**, 359–380 (1991)
18. Ruan, J., van Der Hoek, W., Wooldridge, M.: Verification of games in the game description language. J. Logic Comput. **19**(6), 1127–1156 (2009)
19. Thielscher, M.: A general game description language for incomplete information games. In: Proceedings of the 24th AAAI Conference on Artificial Intelligence (AAAI 2010), pp. 994–999 (2010)
20. Thielscher, M.: GDL-III: a description language for epistemic general game playing. In: Proceedings of the 26th International Joint Conference on Artificial Intelligence (IJCAI 2017), pp. 1276–1282 (2017)
21. Zhang, D., Thielscher, M.: Representing and reasoning about game strategies. J. Philos. Logic **44**(2), 203–236 (2015)
22. Zhang, H., Liu, D., Li, W.: Space-consistent game equivalence detection in general game playing. In: Cazenave, T., Winands, M.H.M., Edelkamp, S., Schiffel, S., Thielscher, M., Togelius, J. (eds.) CGW/GIGA -2015. CCIS, vol. 614, pp. 165–177. Springer, Cham (2016). https://doi.org/10.1007/978-3-319-39402-2_12

A Strategy-Proof Model-Based Online Auction for Ad Reservation

Qinya Li, Fan Wu$^{(\boxtimes)}$, and Guihai Chen

Shanghai Jiao Tong University, Shanghai, China
qinyali@sjtu.edu.cn, {fwu,gchen}@cs.sjtu.edu.cn

Abstract. Ad reservation market is an important part of the Internet advertising industry. Advertisers expect to reserve ad slots in advance, while auctioneers need a mechanism for allocating ad slots and maximizing profits. We propose SMAR, which is a Strategy-proof Model-based online Auction for ad Reservation, to meet their needs. SMAR allows the cancelation policy. It means auctioneers can revoke the reservation and resell ad slots to advertisers with higher bids. SMAR achieves both incentive compatibility and individual rationality. We implement SMAR and compare it with offline VCG and other related works. The results show SMAR has a better performance in both social welfare and revenue.

Keywords: Ad reservation · Auction theory · Strategy proofness · Game theory · Mechanism design

1 Introduction

More and more advertisers (agents or bidders) prefer to use Internet systems to buy advertising display. Many Internet-based ads are sold via instantaneous offline auction, such as the GSP auction (*e.g.*, Google). For traditional offline ads, to bring satisfying advertising effect, advertisers often expect to *reserve* ad slots in advance, while auctioneers are willing to sell out goods in advance due to the uncertainty and unpredictability of the market. Both of them need an effective mechanism to help them make deals for the ad reservation trades.

In this paper, we present SMAR, which is a strategy-proof model-based online auction for ad reservation with a cancellation policy. It means auctioneers can revoke the previous reservation and resell ad slots to bidders with higher bids. We model the arrival pattern and the bid valuation distribution of bidders, and assume auctioneers can reliably know the above information about bidders.

This work was supported in part by the National Key R&D Program of China 2018YFB1004703, in part by China NSF grant 61672348 and 61672353, in part by Supported by the Open Project Program of the State Key Laboratory of Mathematical Engineering and Advanced Computing 2018A09, and in part by Alibaba Group through Alibaba Innovation Research Program. The opinions, findings, conclusions, and recommendations expressed in this paper are those of the authors and do not necessarily reflect the views of the funding agencies or the government.

© Springer Nature Switzerland AG 2019
A. C. Nayak and A. Sharma (Eds.): PRICAI 2019, LNAI 11670, pp. 612–617, 2019.
https://doi.org/10.1007/978-3-030-29908-8_48

Accepted bidders need to pay for entrance fees based on their survival probabilities. It can prevent malicious bidders from bidding and earning compensations. We achieve a strategy-proof mechanism based on the sealed-bid secondary price auction [4]. We implement SMAR and compare it with offline Vickrey-Clarke-Groves (VCG) and COMP [1]. The results show SMAR has a better performance in both social welfare and revenue than others.

2   Preliminaries

2.1   Auction Model

Given an online auction with a trusted auctioneer and potential agents $\mathbb{N} = \{1, 2, \cdots, n\}$. We divide time into $|T|$ time slots with equal length time spans, i.e. $\mathbb{T} = \{1, 2, \cdots, T\}$. There are g ad slots at T. Each agent bids for at most one unit of slots. $\theta_i = (a_i, v_i)$ is agent i's type. $a_i \in \mathbb{T}$ is her arrival time and v_i is her value for the slot. $\theta_i' = (a_i', v_i')$ is her reported type. Agents will cheat on their true types. Based on the heart-beat scheme [3], agents only cheat the arrival time by a late report, i.e., $a_i' \in [a_i, T]$. Let $\mathbb{N}^t = \{i \in N | a_i' \leq t\}$ be the set of agents who have reported arrivals by time t. Clearly, $\mathbb{N}^0 = \emptyset$. $\mathbb{N}^T$ is the set of agents who have attended the auction. $\mathbb{N}^{t-1} \subseteq \mathbb{N}^t \subseteq \mathbb{N}, \forall t \in \mathbb{T}$. $\Delta \mathbb{N}^t = \mathbb{N}^t \setminus \mathbb{N}^{t-1}$ is the set of agents whose reported time is exactly t. Once the agent reports her type, the auctioneer needs to give an immediate response: reject or accept.

During the auction, we maintain a temporary winner set $\mathbb{W}^t \subseteq \mathbb{N}^t$ ($\mathbb{W}^0 = \emptyset$). Each agent needs at most one ad slots, i.e., $|\mathbb{W}^t| \leq g$. Due to the cancellation policy, $\mathbb{W}^t$ does not always contain $\mathbb{W}^{t-1}$. Let $\mathbb{A}^t = \bigcup_{k=1}^t \mathbb{W}^k$ be the set of agents who have been accepted till time t. Clearly, $\mathbb{W}^t \subseteq \mathbb{A}^t \subseteq \mathbb{N}^t, \forall t \in \mathbb{T}$.

Definition 1. *For an agent i with type $\theta_i' = (a_i', v_i')$, there are four possible states: (1) accepted iff $i \in \mathbb{W}^{a_i'}$; (2) rejected iff $i \notin \mathbb{W}^{a_i'}$; (3) survive iff $i \in \mathbb{W}^T$; (4) cancelled iff $i \in \mathbb{A}^T \setminus \mathbb{W}^T$.*

If agent i is accepted, we charge her for an entrance fee p_i^{ent}. It can prevent malicious agents from earning compensations. The accepted agent may be cancelled due to the arrival of agents with higher bids, and get a compensation p_i^{comp}. Lastly, at time T, surviving agents will charge a final price p_i and get ad slots.

Let $\theta' = (\theta_i', \theta_{-i}')$ be the bid profile of agents in $\mathbb{N}^T$. The utility of agent i is:

$$u_i(\theta_i, \theta') = \begin{cases} p_i^{comp} - p_i^{ent}, & i \text{ is cancelled,} \\ v_i - p_i - p_i^{ent}, & i \text{ survives,} \\ 0, & \text{otherwise.} \end{cases} \tag{1}$$

$\mathbf{Pr}_i^{sv} = \mathbf{Pr}_i^{sv}(\theta_i', \theta_{-i}')$ is the survival probability of i. It is related to a_i', θ_i' and other bidders' bid profile θ_{-i}'. The expected utility of an accepted agent i is:

$$E(u_i(\theta_i, \theta_i', \theta_{-i}')) = \left(1 - \mathbf{Pr}_i^{sv}(\theta_i', \theta_{-i}')\right) \times p_i^{comp} + \mathbf{Pr}_i^{sv}(\theta_i', \theta_{-i}') \times (v_i - p_i) - p_i^{ent} \tag{2}$$

Here, agents hope to maximize their utilities or expected utilities, while the auctioneer wants to maximize the social welfare.

Definition 2 (Social Welfare). *The social welfare in our auction model is the summation of the valuations of the whole survivors, i.e.,* $SW = \sum^{i \in \mathbb{W}^T} v_i$.

Due to the existence of entrance fees and compensations, the auctioneer's revenue R and expected revenue $E(R)$ are respectively defined as:

$$R = \sum_i^{i \in \mathbb{W}^T} p_i + \sum_j^{j \in \mathbb{A}^T} p_j^{ent} - \sum_k^{k \in \mathbb{A}^T \setminus \mathbb{W}^T} p_k^{comp} \tag{3}$$

$$E(R) = \sum_i^{i \in \mathbb{A}^T} (p_i^{ent} - (1 - \mathbf{Pr}_i^{sv}) p_i^{comp} + \mathbf{Pr}_i^{sv} p_i) \tag{4}$$

Algorithm 1. RunOneTime

Input: current time t, T, g, model parameters: λ, μ, σ, arriving agent set $\Delta \mathbb{N}^t$, previous winner set $\mathbb{W}^{t-1}$, previous revenue rev;
Output: Current winner set $\mathbb{W}^t$, current revenue rev;
1 $\mathbb{W}^t = \mathbb{W}^{t-1}$;
2 **for** $\forall i \in \Delta \mathbb{N}^t$ **do**
3 $\quad j^* = \underset{i \in \mathbb{W}^t}{argmin}(getBidValue(i))$;
4 $\quad$ **if** $|\mathbb{W}^t| < g$ **then**
5 $\quad\quad \mathbb{W}^t = \mathbb{W}^t \cup \{i\}$;
6 $\quad$ **else if** $v_i' > v_{j*}'$ **then**
7 $\quad\quad \mathbb{W}^t = (\mathbb{W}^t \setminus \{j^*\}) \cup \{i\}$;

8 **for** $\forall i \in \mathbb{W}^t \setminus \mathbb{W}^{t-1}$ **do**
9 $\quad \mathbf{Pr}_i^{sv} = \text{CalculateSurvivalProb}; rev = rev + p_i^{ent}$;
10 **for** $\forall i \in \mathbb{W}^{t-1} \setminus \mathbb{W}^t$ **do**
11 $\quad rev = rev - p_i^{comp}$;
12 **return** $\mathbb{W}^t$, rev;

2.2 Model-Based Assumptions

SMAR is designed via a model-based approach [3], assuming the auctioneer can establish accurate models of bidders' arrival pattern and bid distribution. A *poisson distribution* is used to model the arrival pattern. The number of new agents during each time follows a poisson distribution with the average rate λ, *i.e.*, $N_t \sim P(\lambda)$, $N_t = |\Delta \mathbb{N}^t|$. A *normal distribution* is used to model the distribution of agent's bid, *i.e.*, $v_i \sim N'(\mu, \sigma^2), \forall i \in \mathbb{N}$. Assume all bids are different. The bid value is positive and the minimum gap between two value is one. So, the normal distribution should be rounded to transform a consecutive manner into a discrete one, *i.e.*, $N'(\mu, \sigma^2)$. The rounding is $y' = \lfloor y + 0.5 \rfloor$, *e.g.*, the 3.4 is rounded to 3 and 3.6 is rounded to 4.

3   Auction Design

We try to design a strategy-proof direct revelation mechanism for ad reservation. Agents can report their types without cheating and the utility is non-negative.

Definition 3 (Strategy-Proof Direct Revelation Mechanism [2]). *A direct revelation mechanism is strategy-proof, iff it satisfies incentive-compatibility and individual-rationality.*

We utilize the above probability models to compute the survival probability of agents and allocate slots. If an agent is accepted, she pays for an entrance fee.

$$p_i^{ent} = p_i^{comp}(1 - \mathbf{Pr}_i^{sv}). \tag{5}$$

If an accepted bidder is cancelled, the individual rationality also should be guaranteed. Thus, according to Eq. 2, her utility u_j^c can be

$$u_j^c = p_j^{comp} - p_j^{ent} = \mathbf{Pr}_i^{sv} \times p_j^{comp} > 0 \tag{6}$$

Algorithm 2. Mechanism Overview

Input: Length of the auction T, inventory size g, model parameters: λ, μ and σ;
Output: Final winner set $\mathbb{W}^T$, revenue rev;
1   $\mathbb{W} = \emptyset; \mathbb{N} = \emptyset; rev = 0$;
2   **for** $t \leftarrow 1\ to\ T$ **do**
3   $\quad \Delta\mathbb{N} = \text{CollectArrivingBidder}; \mathbb{N} = \mathbb{N} \cup \Delta\mathbb{N}; \mathbb{W}, rev = \text{RunOneTime}$;
4   $\mathbb{W}^T = \mathbb{W}; \mathbb{N}^T = \mathbb{N}; p_c = \max\{v_i' | i \in \mathbb{N}^T \setminus \mathbb{W}^T\}$;
5   **for** $\forall i \in \mathbb{W}$ **do**
6   $\quad rev = rev + p_c$;
7   **return** $\mathbb{W}^T, rev$;

Next, to achieve incentive compatibility and individual rationality, we propose a policy *Probabilities as Fees (PolicyF)* to define p_i^{comp}, p_i^{ent}, and p_i .

$$p_i^{ent} = 1 - \mathbf{Pr}_i^{sv}; \quad p_i^{comp} = 1; \quad p_i = p_c, \tag{7}$$

Due to the limitation of space, the the strategy-proof of *PolicyF* is omitted.

3.1   Mechanism Overview

We propose two algorithms in the mechanism. The allocation is inspired by the sealed-bid secondary price auction [4]. Algorithm 1 is used to determine the temporary winners and revenue in each period, while Algorithm 2 is used to determine the final winners and revenue. The main idea is that we simulate a "secondary price" auction in an incremental manner. In Algorithm 2, during each time span, it collects the new bidders by *CollectArrivingBidder* and puts bidders set into Algorithm 1. After T time, it calculates p_c and further charges all winners. p_c is *critical payment* that means if the agent wants to win, she has to bid above p_c.

$$p_c = \max\{v'_i | i \in \mathbb{N}^T \setminus \mathbb{W}^T\} \tag{8}$$

3.2   Probability Calculation

The survival probability $\mathbf{Pr}_i^{sv}$ is used to determine i's entrance fee by *PolicyF*. Given an accepted agent with bid $\theta'_i = (a'_i, v'_i)$, we have $\mathbb{E}_i = \{j \in \mathbb{N} | a'_j > a'_i \wedge v'_j > v'_i\}$ and $E_i = |\mathbb{E}_i|$. E_i is the number of future bidders who bid above i. r_i is the rank of i in current winner set $\mathbb{W}^{a'_i}$, i.e., $r_i = \left|\{j \in \mathbb{W}^{a'_i} | v'_j < v'_i\}\right|$.

Theorem 1. *For $a'_i \in \mathbb{T}$ and $v'_i > 0$, denote*

$$pr_i = \frac{1}{\sigma\sqrt{2\pi}} \int_{v'_i + 0.5}^{+\infty} e^{-\frac{(x-\mu)^2}{2\sigma^2}} dx \quad and \quad \lambda_i = (T - a'_i) pr_i \lambda. \tag{9}$$

The E_i follows the Poisson Distribution of average rate λ_i, i.e., $E_i \sim P(\lambda_i)$. So,

$$\mathbf{Pr}_i^{sv} = \sum_{n=0}^{r_i - 1} \mathbf{Pr}[E_i = n]. \tag{10}$$

3.3   Revenue

Our mechanism achieves the same expected revenue as the offline VCG under *PolicyF*. Based on Eqs. 4 and 7, it is $E(R) = \sum_i^{i \in \mathbb{A}^T} \mathbf{Pr}_i^{sv} p_c$. Since p_c is the critical payment, the expected revenue here has the same form as the expected revenue of the offline VCG. Thus, we achieve the optimal expected revenue.

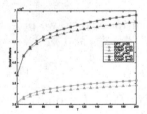

Fig. 1. Social welfare

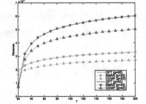

Fig. 2. Revenue comparison

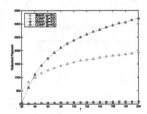

Fig. 3. Refund comparison

4   Evaluation

We compare SMAR with the offline VCG (called OPT) and COMP [1]. T is from 20 to 200. Given $\lambda = 2, \mu = 1000, \sigma = 400$, and generate bidders via the model-based approach. The ad slots g are 20 and 40. For COMP, we config it with $\alpha = 0.1$ and $\gamma = \max(\alpha + \sqrt{\alpha^2 + \alpha}, \alpha/(1-\alpha)) \approx 0.43$. $\alpha \in [0, 1]$ controls the compensation rate. COMP is not truthful, but we assume bidders bid truthfully.

We compare mechanisms from the social welfare, revenue and refunds. Figure 1 shows the results of social welfare. It increases over the auction time. Because more bidders attend the auction, the auctioneer has more chance to capture bidders with high bids. The social welfare in $g = 40$ is less than twice of that in $g = 20$, e.g., in $T = 200$. Because a larger g allows the auctioneer to accept the relatively low bid, which causes the reduction of the average social welfare.

Figure 2 shows the results of revenue. When T is short, a large g may have a low revenue, which is because p_c is close to zero. With the increase of T, the larger g generates the higher revenue. SMAR is with a higher revenue than COMP. Because we pay less compensation to canceled bidders and capture all high bids, while COMP will reject some high bids to guarantee its competitive ration.

Figure 3 shows the results of the refund. Refunds measure the total compensations that the auctioneer pays cancelled bidders. Here, it increases with T. COMP pays more compensations to canceled bidders. Even if SMAR has the same refund as COMP, the revenue of COMP is still less than SMAR. Because SMAR can get entrance fees from accepted bidders and capture all high bids.

5   Conclusion

We have proposed SMAR. It can achieve the optimal social welfare and average revenue. We have implemented it through the simulation. The results have shown SMAR performs better than others in the social welfare, revenue, and refunds.

References

1. Constantin, F., Feldman, J., Muthukrishnan, S., Pál, M.: An online mechanism for ad slot reservations with cancellations. In: Proceedings of the 20th SODA, pp. 1265–1274. Society for Industrial and Applied Mathematics (2009)
2. Mas-Colell, A., Whinston, M.D., Green, J.R., et al.: Microeconomic Theory, vol. 1. Oxford University Press, New York (1995)
3. Nisan, N., Roughgarden, T., Tardos, E., Vazirani, V.V.: Algorithmic Game Theory. Cambridge University Press, Cambridge (2007)
4. Vickrey, W.: Counterspeculation, auctions, and competitive sealed tenders. J. Finan. **16**(1), 8–37 (1961)

Maximum Satisfiability Formulation for Optimal Scheduling in Overloaded Real-Time Systems

Xiaojuan Liao[1(✉)], Hui Zhang[1], Miyuki Koshimura[2], Rong Huang[3], and Wenxin Yu[4]

[1] Chengdu University of Technology, Chengdu, Sichuan, China
liao_xiaojuan@126.com
[2] Kyushu University, Fukuoka, Japan
[3] Donghua University, Shanghai, China
[4] Southwest University of Science and Technology, Mianyang, Sichuan, China

Abstract. In real-time systems where tasks have timing requirements, once the workload exceeds the system's capacity, missed deadlines may incur system overload. Finding optimal scheduling in overloaded real-time systems is critical in both theory and practice. To this end, existing works have encoded scheduling problems as a set of first-order formulas that might be tackled by the Satisfiability Modulo Theory (SMT) solver. In this paper, we move one step forward by formulating the scheduling dilemma in overloaded real-time systems as a Maximum Satisfiability (MaxSAT) problem. In the MaxSAT formulation, scheduling features are encoded as hard constraints and the task deadlines are considered soft ones. An off-the-shelf MaxSAT solver is employed to satisfy as many deadlines as possible, provided that all the hard constraints are met. Our experimental results show that our proposed MaxSAT-based method found optimal scheduling significantly more efficiently than previous works.

Keywords: Scheduling problem · MaxSAT encoding ·
Real-time system · Overloaded system

1 Introduction

Real-time systems, which are designed to handle tasks with completion deadlines, play an important role in a variety of modern applications, such as robotics [2], pacemakers [14], chemical plants [33], telecommunications [6], and multimedia systems [1]. Under ideal circumstances, a real-time system completes all of its

Supported by National Natural Science Foundation of China (Grant Number 61806171), JSPS KAKENHI (Grant Numbers JP17K00307, JP19H04175), Ministry of Education in China Project of Humanities and Social Sciences (Grant Number 17YJCZH260). We thank Dr. Zhuo Cheng for his constructive advice.

© Springer Nature Switzerland AG 2019
A. C. Nayak and A. Sharma (Eds.): PRICAI 2019, LNAI 11670, pp. 618–631, 2019.
https://doi.org/10.1007/978-3-030-29908-8_49

tasks before their deadlines expire. However, in reality, unanticipated emergency conditions may occur causing the workload to exceed the system's capacity, leading to missed deadlines [5]. Such a phenomenon is called *overload*. When overload happens, real-time systems may be dramatically degraded, and catastrophes might occur. Therefore, designing a suitable scheduling strategy that is resistant to system overload is critical to maintain a system's stability.

To date, numerous works have devoted to dynamically detecting and mitigating possible overload in real-time systems. Typical methods involve modulating either the task attributes [30] or system workload [26], exploiting the congestion and feedback mechanisms [7,8,17]. These works, in which the scheduler receives tasks that arrive over time, and must schedule tasks without any knowledge of the future, fall under the rubric of *online scheduling* [23]. Another line of research is to design *off-line scheduling* algorithms to optimally solve the problem, provided that all data are known beforehand[1]. Optimization algorithms can, on one hand, benefit real-time systems where the scheduled application is executed many times [25], on the other hand, serve as a testbed for suboptimal algorithms [3]. The first optimal algorithm for minimizing the number of late jobs on a single machine was presented by Moore [29], working on a special case where the ready times of tasks were the same and no precedence relation was specified. Later, Graham et al. [15] classified the scheduling problems according to the machine environment, job characteristics and optimality criteria. Based on the elaborate specification on the scheduling properties, immense amounts of works boomed. A classic way of finding the optimal scheduling is the exploration of dynamic programming algorithms [4,22], following which a direct combinatorial algorithm was put forward to reduce the time complexity [31]. Another way is to formalize the scheduling problems as a class of generalized problems, such as the mixed integer linear programming [32], constraint programming [20], Satisfiability Modulo Theory (SMT) [9–11,21,25], and Boolean Satisfiability (SAT) problem [3,12,18,24,27]. Motivated by the significant progress in solving these generalized problems, the formalized scheduling problem could be addressed efficiently with the corresponding solving algorithms.

As a pioneering work in satisfiability formalization, Crawford et al. [12] first encoded scheduling problems into a SAT problem, paving the way for subsequent works. Based on Crawford encoding, Koshimura et al. [18] solved six types of open job-shop scheduling problems. Liu et al. [24] presented a SAT-based optimization framework to address the task graph scheduling on multiprocessor systems. These works, in which tasks are not constrained with any specific deadlines, are infeasible for real-time systems, let alone the overloaded situation in real-time systems. To achieve optimal scheduling in overloaded real-time systems, Cheng et al. [9] modeled the overloaded scheduling problem as a set of SMT problems to maximize the total number of completed tasks that can be completed by their deadlines. Subsequently, they extended the model by achieving other objectives

---

[1] Note that an off-line algorithm does not contradict with a real-time system. An off-line scheduling allows a scheduler to make decision based on the total knowledge of the problem, while a real-time system assigns each task with a specific deadline.

[10]. SMT-based scheduling was sufficiently flexible because it handled various objectives with very few changes in the adaption procedure. However, their SMT method suffers from redundant constraints and successive calls to the solver that negatively impact its problem solving's efficiency.

In this paper, we seek to achieve optimal scheduling for real-time systems by presenting a Maximum Satisfiability (MaxSAT) formulation, which is an extended version of SAT. Motivated by the MaxSAT feature that satisfies as many constraints as possible by eliminating the unsatisfied ones, we recast the overloaded scheduling problem as a partial MaxSAT problem and solve it with a state-of-the-art MaxSAT solver. In the MaxSAT formulation, task deadlines are treated as soft constraints, and scheduling features are encoded as hard ones. Then an off-the-shelf MaxSAT solver is employed to meet as many soft constraints as possible, provided that all the hard constraints are met. Instead of repeatedly summoning the solver, our MaxSAT-based scheduling finds the optimal solution by just running the MaxSAT solver once. We experimentally demonstrate the superiority of the MaxSAT formulation and show that regardless of the degree of overload changes, the MaxSAT-based method always outperforms SMT both in terms of execution time and percentage of completed problem instances.

2   Scheduling Model

We adhere to the definition of scheduling problems in previous works [9,10]. For convenience, the notations used in the model are summarized in Table 1.

Table 1. Notations and descriptions in scheduling model

Notation	Description
Γ	Finite set of real-time tasks
τ_l	Task in Γ, where l is its index
$\tau_k \prec \tau_l$	τ_l relies on τ_k
r_l	Ready time of τ_l
c_i	Execution time of τ_l
d_l	Deadline of τ_l
f_i^l	i^{th} fragment of τ_l
q_l	Index of last fragment of τ_l
c_i^l	Execution time of f_i^l
s_i^l	Start execution time of f_i^l

A real-time system is comprised of a finite set of real-time tasks $\{\tau_1, \ldots, \tau_n\}$ that are waiting to be executed. The set is denoted by Γ, i.e., $\Gamma = \{\tau_1, \ldots, \tau_n\}$. All the tasks request a uniprocessor for execution when they arrive in the system.

Each task τ_l can be represented by a 3-tuple $\tau_l = (r_l, c_l, d_l)$, where l is a task's index, r_l is the ready time, i.e., the earliest time at which τ_l can start, c_l is the required execution time, and d_l is the deadline, i.e., the time by which τ_l must be completed. Naturally, $r_l + c_l \leq d_l$. A successfully scheduled task is expected to be finished before its deadline; otherwise, it is worthless to the system. We adopt this so-called *firm-deadline* model [16, 26].

In practical systems, tasks usually have dependency relations. For example, since task τ_l may require the computed result of τ_k, τ_l cannot start until τ_k is finished. Such a dependency relation between tasks is written as $\tau_k \prec \tau_l$. To allow preemption, which indicates that a running task may be interrupted and resumed at a later time, each task $\tau_l \in \Gamma$ is defined as a sequence of indivisible fragments $\langle f_1^l, \ldots, f_{q_l}^l \rangle$. Symbol c_i^l denotes the required execution time of f_i^l. Clearly, $\sum_{i=1}^{q_l} c_i^l = c_l$. For $1 \leq i \leq q_l$, f_{i+1}^l can start to run only after f_i^l is completed. Let s_i^l be the starting execution time of f_i^l, and $s_i^l \geq r_l$ and $s_{i+1}^l \geq s_i^l + c_i^l$.

A system is defined as *overloaded* if no scheduling algorithm can meet the deadlines of all the tasks that have been submitted to it. This paper focuses on designing an offline method to tackle scheduling problems in an overloaded real-time system with a uniprocessor. The *scheduling objective* is to maximize the total number of tasks executed to completion before their deadlines.

3   MaxSAT Formulation for Task Scheduling

In this section, we provide the first ever MaxSAT formulation for solving the task scheduling problem in overloaded real-time systems. A MaxSAT instance consists of a number of constraints that need to be managed by the MaxSAT solver. To formulate all the necessary constraints that characterize the scheduling model, we introduce the following three boolean variables, which are derived from a previous work [18]:

- $sa_{i,t}^l$, which is true if f_i^l starts at time t or later;
- $eb_{i,t}^l$, which is true if f_i^l ends by time t or before;
- $pr_{i,j}^{l,k}$, which is true if f_i^l precedes f_j^k.

Each fragment f_i^l is associated with the following four kinds of time points:

- *Earliest Start Time (EST)* of f_i^l is denoted by EST_i^l, where $EST_i^l = r_l + \sum_{u=1}^{i-1} c_u^l$. Any fragment f_i^l should start at or after EST_i^l, indicating that f_i^l cannot be started before all its previous fragments are finished.
- *Latest Start Time (LST)* of f_i^l is denoted by LST_i^l, where $LST_i^l = d_l - \sum_{u=i}^{q_l} c_u^l$. If f_i^l fails to start before LST_i^l, τ_l cannot end by its deadline d_l, and so this task becomes worthless to the system.
- *Earliest Completion Time (ECT)* of f_i^l is denoted by ECT_i^l, where $ECT_i^l = r_l + \sum_{u=1}^{i} c_u^l$. No fragment f_i^l can be completed before ECT_i^l. Particularly, $ECT_i^l = EST_i^l + c_i^l$.

– *Latest Completion Time (LCT)* of f_i^l is denoted by LCT_i^l, where $LCT_i^l = d_l - \sum_{u=i+1}^{q_l} c_u^l$. If f_i^l fails to end by LCT_i^l, τ_l cannot end by its deadline d_l, and the whole task becomes worthless to the system. Particularly, $LCT_i^l = LST_i^l + c_i^l$.

3.1 Encoding Scheduling Features as Hard Clauses

This subsection introduces several constraints with which to characterize the scheduling model, including task features and fragment features. Logical implication $a \rightarrow b$ is equivalent to $\neg a \vee b$ in classical logic. The set of all the clauses are identified as a conjunction of the following clauses:

(C1) $\forall \tau_l \in \Gamma$, $f_i^l \in \tau_l$, f_i^l starts at or after EST_i^l:

$$sa_{i,EST_i^l}^l; \tag{1}$$

(C2) $\forall \tau_l \in \Gamma$, $\forall f_i^l, f_{i+1}^l \in \tau_l$, f_i^l precedes f_{i+1}^l:

$$pr_{i,i+1}^{l,l}; \tag{2}$$

(C3) $\forall \tau_k, \tau_l \in \Gamma$, $\forall f_i^k \in \tau_k$, $\forall f_j^l \in \tau_l$, if $k \neq l$, $\tau_k \nprec \tau_l$, $\tau_l \nprec \tau_k$, $EST_i^k < LCT_j^l$ and $EST_j^l < LCT_i^k$, and then f_i^k and f_j^l may require the processor at the same time. In this condition, f_i^k precedes f_j^l or f_j^l precedes f_i^k:

$$pr_{i,j}^{k,l} \vee pr_{j,i}^{l,k}; \tag{3}$$

(C4) $\forall \tau_k, \tau_l \in \Gamma$, if $\tau_k \prec \tau_l$, and then the last fragment of τ_k must precede the first fragment of τ_l. That is, if $\tau_k \prec \tau_l$, then $f_{q_k}^k$ precedes f_1^l:

$$pr_{q_k,1}^{k,l}; \tag{4}$$

Further, if τ_k fails to be completed by its deadline, τ_l cannot even start at LST_1^l:

$$\neg eb_{q_k,d_k}^k \rightarrow sa_{1,LST_1^l+1}^l; \tag{5}$$

(C5) $\forall \tau_l \in \Gamma$, $\forall f_i^l \in \tau_l$, if τ_l is completed by its deadline, then each fragment f_i^k of τ_l should finish by LCT_i^l:

$$eb_{q_l,d_l}^l \rightarrow eb_{i,LCT_i^l}^l; \tag{6}$$

(C6) $\forall \tau_l \in \Gamma$, $\forall f_i^l \in \tau_l$, if f_i^l starts at or after time t, then it starts at or after time $t-1$, where t varies in $[EST_i^l + 1, LST_i^l + 1]$:

$$sa_{i,t}^l \rightarrow sa_{i,t-1}^l \quad \left(EST_i^l + 1 \leq t \leq LST_i^l + 1 \right); \tag{7}$$

(C7) $\forall \tau_l \in \Gamma$, $\forall f_i^l \in \tau_l$, if f_i^l ends by t, then it ends by time $t+1$, where t varies in $[ECT_i^l - 1, LCT_i^l - 1]$:

$$eb_{i,t}^l \rightarrow eb_{i,t+1}^l \quad \left(ECT_i^l - 1 \leq t \leq LCT_i^l - 1 \right); \tag{8}$$

(C8) $\forall \tau_l \in \Gamma$, $\forall f_i^l \in \tau_l$, if f_i^l starts at or after time t, then it cannot end before time $t + c_i^l - 1$, where t varies in $[EST_i^l, LST_i^l + 1]$:

$$sa_{i,t}^l \rightarrow \neg eb_{i,t+c_i^l-1}^l \quad (EST_i^l \leq t \leq LST_i^l + 1);\tag{9}$$

(C9) $\forall \tau_k, \tau_l \in \Gamma$, $\forall f_i^k \in \tau_k$, $\forall f_j^l \in \tau_l$, if f_i^k starts at or after time t and f_j^l follows f_i^k, then f_j^l cannot start until f_i^k is finished. That is, for each $pr_{i,j}^{k,l}$ asserted by (C2)–(C4), one clause is generated:

$$sa_{i,t}^k \wedge pr_{i,j}^{k,l} \rightarrow sa_{j,t'}^l;\tag{10}$$

where t varies in $[EST_i^k, LST_i^k + 1]$ and

$$t' = \begin{cases} LST_j^l + 1 & \text{if } t + c_i^k > LST_j^l, \\ EST_j^l & \text{if } t + c_i^k < EST_j^l, \\ t + c_i^k & \text{Otherwise.} \end{cases}$$

This formula reveals the following facts. First, if f_i^k ends after LST_j^l (i.e., $t + c_i^k > LST_j^l$), then f_j^l cannot start at or before LST_j^l. Second, if f_i^k finishes before EST_j^l (i.e., $t + c_i^k < EST_j^l$), then f_j^l starts at or after EST_j^l. Otherwise, f_j^l must start at or after time f_i^k finishes, i.e., $t + c_i^k$.

Up to this point, we have encoded the scheduling features as propositional boolean formulas that can be converted to a set of clauses. Since the scheduling features are intrinsic properties inherent in the tasks and their fragments, such clauses are specified as *hard*, indicating that all of them must absolutely be satisfied. For convenience, we refer to the set of hard clauses introduced in (C1)–(C9) as $\mathcal{C}$.

3.2   Encoding Scheduling Objectives as Soft Clauses

A task is said to be successfully executed if and only if all of its fragments are completed before its deadline. Since all the fragments of a task run sequentially, this constraint can be defined as the task's last fragment that must be completed before its deadline. Correspondingly, the basic scheduling objective, i.e., maximizing the total number of tasks that are completed by their deadlines, is directly encoded as the following clause:

(O) The last fragment of task τ_l must be completed by its deadline:

$$eb_{q_l,d_l}^l \quad (1 \leq l \leq n).\tag{11}$$

The clause introduced in (O) is referred to as $\mathcal{O}$. After encoding, the scheduling objective turns to satisfying as many clauses in $\mathcal{O}$ as possible. Such clauses are labeled *soft*. Conjuncted with $\mathcal{C}$, the problem is then $\{\mathcal{C}, \mathcal{O}\}$. This leads to a partial MaxSAT problem, which tries to find an assignment of variables to satisfy all the hard clauses in $\mathcal{C}$ and the maximum number of soft clauses in $\mathcal{O}$.

3.3   A Pedagogical Example

Consider a simple scheduling problem to describe how the MaxSAT formulation works. Assume a set of real-time tasks $\Gamma = \{\tau_1, \tau_2, \tau_3\}$. Their ready times, execution times, and deadlines are respectively defined as $\tau_1 = (0, 2, 3)$, $\tau_2 = (0, 1, 1)$, $\tau_3 = (1, 1, 2)$. Suppose that τ_1 has two fragments, $\langle f_1^1, f_2^1 \rangle$, and τ_2 and τ_3 each have one, denoted by f_1^2 and f_1^3. The execution time of each fragment is 1. The four critical time points of each fragment are summarized in Table 2.

Table 2. Critical time points of each fragment in three tasks

Task	Fragment	EST	LST	ECT	LCT
τ_1	f_1^1	0	1	1	2
	f_2^1	1	2	2	3
τ_2	f_1^2	0	0	1	1
τ_3	f_1^3	1	1	2	2

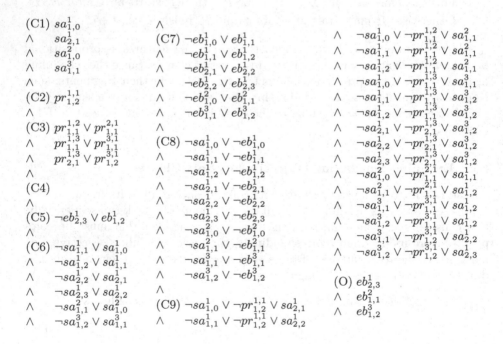

Fig. 1. MaxSAT formulation for exemplified problem

The MaxSAT formulation applied to the scheduling problem is shown in Fig. 1. Constraint (C1) states that each fragment starts at or after its EST. If we consider τ_1 that consists of two fragments, f_1^1 and f_2^1, then $sa_{1,0}^1$ and $sa_{2,1}^1$ should both be assigned as true, indicating that f_1^1 starts at or after time 0 and f_2^1 starts at or after time 1. Constraints (C2) and (C3) work together to specify the execution sequences of the fragments. (C2) forces all the fragments in a single task to be executed sequentially, and (C3) guarantees no overlap of the execution times of any two fragments in different tasks. In the three tasks, only τ_1 has more than one fragment, and hence constraint (C2) only applies to τ_1, ensuring that f_1^1 precedes f_2^1. Constraint (C3) applies to pairs of fragments f_i^k and f_j^l ($k \neq l$) that satisfy $EST_i^k < LCT_j^l$ and $EST_j^l < LCT_i^k$. If both conditions are met, then f_i^k and f_j^l may simultaneously occupy the processor, thus we need to decide in what order to execute them. Constraint (C3) solves this ordering dilemma, which states that either one can precede the other. Consider f_1^1 and f_1^2. As seen in Table 2, $EST_1^1 < LCT_1^2$ and $EST_1^2 < LCT_1^1$. Hence we must explicitly specify that f_1^1 precedes f_1^2 or f_1^2 precedes f_1^1; otherwise, the execution time of these two fragments may overlap. Note that constraint (C3) does not apply to all the fragment pairs. Take f_1^2 and f_1^3 as an example. It is clear that $EST_1^2 < LCT_1^3$, and $EST_1^3 = LCT_1^2$. This means that f_1^3 always starts no earlier than f_1^2. Therefore, no clauses need to be asserted. Constraint (C4) applies to a situation where tasks have dependency relations.

Now consider τ_1, which consists of two fragments f_1^1 and f_2^1. If f_2^1 ends by the deadline of τ_1, then naturally f_1^1 should end by its LST. This is guaranteed by (C5). Constraints (C6)–(C9) are a collection of coherence conditions [12] on the introduced variables for all the fragments of all the tasks. Finally, constraint (O) gives the problem's objective, i.e., completing the last fragment of each task by its deadline.

All of these constraints are conjuncted with $\wedge$ to form a partial MaxSAT problem in CNF, where clauses (C1)–(C9) are declared *hard* and those in (O) are *soft*. Then the CNF formula is input to a MaxSAT solver. The solver's output includes the maximum number of satisfied soft clauses as well as the corresponding assignment of all the boolean variables from which the optimal scheduling table can be obtained.

4   SMT Formulation and Variants

To evaluate the performance of our MaxSAT formulation, we compared it with the previous SMT formulation [9,10] and summarize it in Fig. 2. The logical implication operator is indicated as $\Rightarrow$. Implication $A \Rightarrow B$ is equivalent to $\neg A \vee B$.

The constraints in (C1)–(C5) are explained as follows. Constraint (C1) ensures that each task starts at or after its ready time. Constraint (C2) guarantees that the series of fragments in a task are executed sequentially. Constraint (C3) guarantees no overlap of the execution times of any two fragments. Constraint (C4) states that if τ_l relies on the computed results of τ_k, then τ_l must start after τ_k has been completed. Finally, constraint (C5) defines the firm-deadline model, i.e., each task must be completed by its deadline. All the constraints need to be satisfied by the SMT solver. Clearly, in the case of a system

objective $Max(|\Gamma'|)$ *where* $\Gamma' \subseteq \Gamma$
st:

(C1) $\forall \tau_l \in \Gamma'$ $s_i^l \geq r_l$

(C2) $\forall \tau_l \in \Gamma', \forall f_i^l, f_j^l \in \tau_l, j > i,\; s_j^l \geq s_i^l + c_i^l$

(C3) $\forall \tau_k, \tau_l \in \Gamma',\, k \neq l,$ $\left(s_i^k \geq s_j^l + c_j^l\right) \vee \left(s_j^l \geq s_i^k + c_i^k\right)$
            $\forall f_i^k \in \tau_k, \forall f_j^l \in \tau_l$

(C4) $\forall \tau_k, \tau_l \in \Gamma',\, \tau_k \prec \tau_l,$ $\left(s_1^l \geq s_{q_k}^k + c_{q_k}^k\right) \wedge \left(s_{q_k}^k + c_{q_k}^k > d_k \Rightarrow s_1^l = +\infty\right)$

(C5) $\forall \tau_l \in \Gamma'$ $s_{q_l}^l + c_{q_l}^l \leq d_{q_l}$

Fig. 2. Previously proposed SMT formulation [9,10]

overload, since not all the tasks can be completed by their deadlines, not all the constraints in (C5) can be satisfied when $|\Gamma'| = |\Gamma|$. In previous work [9,10], the SMT formulation was solved iteratively to find optimal $|\Gamma'|$ for the scheduling problem, where $|\Gamma'|$ was determined through a binary search for satisfaction between $[0, |\Gamma|]$. Optimal $|\Gamma'|$ satisfies both the following conditions:

(1) The SMT solver returns a non-empty model with $|\Gamma'|$.
(2) The SMT solver returns an empty model with $|\Gamma'| + 1$.

To avoid repeatedly calling the SMT solver, we update the encoding [9,10] by claiming that completing a task before its deadline is a soft constraint. That is, we make $\Gamma' = \Gamma$ in Fig. 2 and declare the constraints in (C5) to be *soft*. In this way, the solver can satisfy as many constraints as possible, given that all the constraints in (C1)–(C4) are met. The optimization problem can be tackled by a high-performance theorem prover named Z3. In Z3, the soft constraints are specifically claimed by the following prefix expression:

$$\forall \tau_l \in \Gamma, \quad (assert-soft \;\; (\leq \;\; (+ \;\; s_{q_l} \;\; c_{q_l}) \;\; d_l)),$$

where s_{q_l} is the start time of $f_{q_l}^l$, c_{q_l} is the execution time of $f_{q_l}^l$, and d_l is the deadline of $f_{q_l}^l$. Thus, the multiple calls of the SMT solver are eliminated. A tutorial for Z3 optimization is available [28].

Furthermore, redundant constraints exist in (C3) of the SMT formulation (Fig. 2). For example, consider a problem with $\Gamma = \{\tau_1, \tau_2\}$, where $\tau_1 = (0, 1, 2)$ and $\tau_2 = (2, 1, 4)$. According to (C3), the following constraint is asserted: $\left(s_1^1 \geq s_1^2 + 1\right) \vee \left(s_1^2 \geq s_1^1 + 1\right)$, expressing that f_1^1 starts after f_1^2 or f_1^2 starts after f_1^1. However, note that f_1^1 never starts after f_1^2 since τ_1's deadline is no later than the ready time of τ_2. Therefore, the constraints asserted in (C3) are redundant. In addition, if the dependency relations of the two tasks are specified, (C3) should never be imposed on them. Considering such redundancy, we modified the assertion in (C3) by adding more conditions:

(C3*) $\forall \tau_k, \tau_l \in \Gamma',\, k \neq l,\, \forall f_i^k \in \tau_k,$
        $\forall f_j^l \in \tau_l,\, \tau_k \not\prec \tau_l,\, \tau_l \not\prec \tau_k,$ $\left(s_i^k \geq s_j^l + c_j^l\right) \vee \left(s_j^l \geq s_i^k + c_i^k\right).$
        $EST_i^k < LCT_j^l,\, EST_j^l < LCT_i^k$

In summary, in the updated SMT formulation, we added more conditions to the constraints in (C3) and declared the constraints in (C5) to be soft.

Other encodings were kept the same as those described in Fig. 2. By applying more refined constraints and eliminating multiple calls to the solver, SMT-based scheduling can achieve higher efficiency than the previous one. The updated SMT and MaxSAT formulations will be investigated in the experiments described in the next section.

5 Experiments

This section compares the performances of the proposed MaxSAT and the updated SMT formulation [9,10]. The main metrics used for evaluation are the scheduling time spent generating optimal scheduling tables.

We follow a previous method of creating scheduling problems [10] and summarizes it as follows. We created tasks based on uniform distribution with arriving rate λ, which represents the number of tasks that arrive per 100 time units. Clearly, a larger λ indicates that more tasks are arriving in the system during a specific period of time, thus causing more serious overload. In our experiments, λ is assigned 1, 5, 10, 15, and 20 to represent various degrees of system overload. For each λ, a set of scheduling problems is generated, where #tasks ranges from 50 to 500. For each task τ_l, execution time c_l ranges from 1 to 13, and the number of fragments in τ_l, denoted by nf_l, ranges from 1 to 3. The value of deadline d_l is calculated by formula $d_l = r_l + sf_l * c_l$, where sf_i is a slack factor that reflects the deadline's tightness. We assume that for each task τ_l, sf_l ranges from 1 to 4. For each fixed λ and #tasks, 100 problem instances are generated. For convenience, the parameter settings are listed in Table 3.

Table 3. Parameter settings in experiments

Parameter	Description	Value settings
λ	Arriving rate	$\{1, 5, 10, 15, 20\}$
#tasks	Number of tasks in an instance	$[50, 500]$
c_l	Execution time of τ_l	$[1, 13]$
nf_l	Number of fragments in τ_l	$[1, 3]$
sf_l	Slack factor of τ_l	$[1, 4]$
d_l	Deadline of τ_l	$d_l = r_l + sf_l * c_l$

In the following, we conducted tests on a 3.4-GHz Intel E3-1230 processor with 8-GB RAM and experimentally evaluated the performance of the SMT and MaxSAT-based scheduling methods. The selected solver for MaxSAT was QMaxSAT [19], which is a satisfiability-based solver that uses the CNF encodings of cardinality constraints. By contrast, the solver for SMT was Z3 [13], which is a high-performance theorem prover chosen by a previous work [10].

To reveal the degree of overload with various λ, we examined the percentage of tasks that were successfully executed by their deadlines under optimal scheduling. The statistics are shown in Table 4. As λ rises, more tasks arrive in the system per 100 time units, and thus the degree of overload becomes more severe. When $\lambda = 1$, the percentage of successfully completed tasks reaches 99.96%. In comparison, when $\lambda = 20$, the percentage decreases to 66.46%, suggesting that the overloaded situation becomes so serious that only 66.46% tasks were completed by their deadlines.

Table 4. Percentage of completed tasks under optimal scheduling

λ	1	5	10	15	20
Percentage (%)	99.96	95.88	87.52	76.66	66.46

Next we investigated the performances of the SMT and MaxSAT-based methods with various λ. Figure 3 depicts the average computation times for generating optimum scheduling by these two methods. For each instance and solver, we set a time limit of 300 s. Each data point is the average computation time of the solved problem instances. A number with an arrow in the figures denotes the percentage of instances that were successfully solved within the time limit by the corresponding solver, and it is omitted if the solver managed to solve all 100 instances. When the percentage of the solved instances drops to zero, the corresponding curve is omitted because the average computation time has become unpredictable. Clearly, as seen from Fig. 3, with the increase of λ, both SMT and MaxSAT consumed longer time to solve the scheduling problems to optimality. For each λ, the MaxSAT-based method significantly outperformed SMT in terms of both execution time and percentage of successfully solved instances. When $\lambda = 1$ (Fig. 3(a)), the MaxSAT-based method is around five times more efficient than SMT, and when $\lambda = 5$ (Fig. 3(b)), the superiority of MaxSAT becomes even more remarkable since it requires merely one tenth of the time taken by SMT. MaxSAT's advantage continues to expand when λ gets larger. For example, when $\lambda = 10$ and #tasks = 400, MaxSAT completed all the instances within an average of 0.2 s, but there is one instance that SMT failed to solve within the time limit and the rest were completed within around six seconds. As λ and #tasks increase, the percentage of instances that were successfully solved by SMT continued to decrease. Figures 3(d) and (e) represent such significantly fluctuating declines. In particular, when $\lambda = 15$, as #tasks rose from 50 to 500, the percentage of successfully solved instances by SMT fell from 97% to 63%, and the average solving time increased from 2 to 120 seconds. When $\lambda = 20$ and #tasks reached 400, none of the 100 instances were solved by SMT. In contrast, the MaxSAT-based method always solved every instance within the time limit with just moderate average computation time increases.

All our results point to the fact that MaxSAT is a better solution for scheduling tasks in overloaded real-time systems than the SMT formulation.

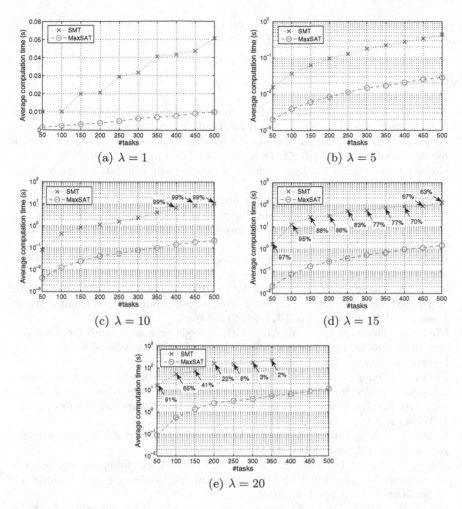

Fig. 3. Average computation times with varied λ

6 Conclusion

This paper presented a novel optimal scheduling approach in overloaded real-time systems that is based on a compact MaxSAT formulation of scheduling problems. In the MaxSAT formulation, scheduling features are encoded as a set of hard clauses, and the scheduling objective is encoded as a set of soft clauses. In this way, the optimal scheduling problem turns to a partial MaxSAT problem, which can be readily solved by an off-the-shelf MaxSAT solver. To evaluate MaxSAT formulation's performance, we performed an extensive experimental evaluation and compared the MaxSAT-based method with the best known SMT formulation for the scheduling problem. Our experimental results concluded that, no matter how the degree of overload and the number of tasks varied, MaxSAT always significantly outperformed SMT in all cases.

References

1. Abeni, L., Buttazzo, G., Superiore, S., Anna, S.: Integrating multimedia applications in hard real-time systems. In: IEEE Real-time Systems Symposium (1998)
2. Aguero, C.E., et al.: Inside the virtual robotics challenge: simulating real-time robotic disaster response. IEEE Trans. Autom. Sci. Eng. 12(2), 494–506 (2015)
3. Gorbenko, A., Popov, V.: Task-resource scheduling problem. Int. J. Autom. Comput. 9(4), 429–441 (2012)
4. Baptiste, P.: An O (n4) algorithm for preemptive scheduling of a single machine to minimize the number of late jobs. Oper. Res. Lett. 24(4), 175–180 (1999)
5. Baruah, S., Haritsa, J.: Scheduling for overload in real-time systems. IEEE Trans. Comput. 46(9), 1034–1039 (1997)
6. Baulier, G.D., et al.: Real-time event processing system for telecommunications and other applications (2002)
7. Cheng, Z., Zhang, H., Tan, Y., Lim, A.O.: DPSC: a novel scheduling strategy for overloaded real-time systems. In: IEEE International Conference on Computational Science and Engineering, pp. 1017–1023 (2014)
8. Cheng, Z., Zhang, H., Tan, Y., Lim, A.O.: Greedy scheduling with feedback control for overloaded real-time systems. In: IFIP/IEEE International Symposium on Integrated Network Management, pp. 934–937 (2015)
9. Cheng, Z., Zhang, H., Tan, Y., Lim, Y.: Scheduling overload for real-time systems using SMT solver. In: IEEE/ACIS International Conference on Software Engineering, Artificial Intelligence, Networking and Parallel/distributed Computing, pp. 189–194 (2016)
10. Cheng, Z., Zhang, H.: SMT-based scheduling for overloaded real-time systems. IEICE Trans. Inf. Syst. E100–D(5), 1055–1066 (2017)
11. Cheng, Z., Zhang, H., Tan, Y., Lim, Y.: SMT-based scheduling for multiprocessor real-time systems. In: IEEE/ACIS International Conference on Computer and Information Science, pp. 1–7 (2016)
12. Crawford, J.M., Baker, A.B.: Experimental results on the application of satisfiability algorithms to scheduling problems. In: Twelfth AAAI National Conference on Artificial Intelligence, pp. 1092–1097 (1994)
13. de Moura, L., Bjørner, N.: Z3: an efficient SMT solver. In: Ramakrishnan, C.R., Rehof, J. (eds.) TACAS 2008. LNCS, vol. 4963, pp. 337–340. Springer, Heidelberg (2008). https://doi.org/10.1007/978-3-540-78800-3_24
14. Er, S.B., Smith, R.E.: Method and apparatus for monitoring and displaying lead impedance in real-time for an implantable medical device (1999)
15. Graham, R.L., Lawler, E.L., Lenstra, J.K., Kan, A.H.G.R.: Optimization and approximation in deterministic sequencing and scheduling: a survey. Ann. Discret. Math. 5(1), 287–326 (1979)
16. Haritsa, J., Carey, M., Livny, M.: On being optimistic about real-time constraints. In: ACM Principles of Database Systems Symposium, pp. 331–343. ACM (1990)
17. Khalilzad, N.M., Nolte, T., Behnam, M.: Towards adaptive hierarchical scheduling of overloaded real-time systems. In: IEEE International Symposium on Industrial Embedded Systems, pp. 39–42 (2011)

18. Koshimura, M., Nabeshima, H., Fujita, H., Hasegawa, R.: Solving open job-shop scheduling problems by SAT encoding. IEICE Trans. Inf. Syst. **E93–D**(8), 2316–2318 (2010)
19. Koshimura, M., Zhang, T., Fujita, H., Hasegawa, R.: QMaxSAT: a partial Max-SAT solver. J. Satisf. Boolean Model. Comput. **8**, 95–100 (2012)
20. Kuchcinski, K.: Constraints-driven scheduling and resource assignment. ACM Trans. Des. Autom. Electron. Syst. **8**(3), 355–383 (2003)
21. Kumar, P., Chokshi, D.B., Thiele, L.: A satisfiability approach to speed assignment for distributed real-time systems. In: Design, Automation and Test in Europe Conference and Exhibition, pp. 749–754 (2013)
22. Lawler, E.L.: A dynamic programming algorithm for preemptive scheduling of a single machine to minimize the number of late jobs. Ann. Oper. Res. **26**(1), 125–133 (1990)
23. Leung, J.Y.T.: Online scheduling. In: Handbook of Scheduling, pp. 328–371. Chapman and Hall/CRC (2004)
24. Liu, W., Gu, Z., Xu, J., Wu, X., Ye, Y.: Satisfiability modulo graph theory for task mapping and scheduling on multiprocessor systems. IEEE Trans. Parallel Distrib. Syst. **22**(8), 1382–1389 (2011)
25. Malik, A., Walker, C., O'Sullivan, M., Sinnen, O.: Satisfiability modulo theory (SMT) formulation for optimal scheduling of task graphs with communication delay. Comput. Oper. Res. **89**, 113–126 (2018)
26. Marchand, M., Chetto, M.: Dynamic scheduling of periodic skippable tasks in an overloaded real-time system. In: IEEE/ACS International Conference on Computer Systems and Applications, pp. 456–464. IEEE (2008)
27. Metzner, A., Herde, C.: RTSAT- an optimal and efficient approach to the task allocation problem in distributed architectures. In: IEEE International Real-Time Systems Symposium, pp. 147–158 (2006)
28. Microsoft, R.: Z3-optimization (2018). https://www.rise4fun.com/Z3/tutorial/optimization
29. Moore, J.M.: An n job, one machine sequencing algorithm for minimizing the number of late jobs. Manag. Sci. **15**(1), 102–109 (1968)
30. Tres, C., Becker, L.B., Nett, E.: Real-time tasks scheduling with value control to predict timing faults during overload. In: IEEE International Symposium on Object and Component-Oriented Real-Time Distributed Computing, pp. 354–358 (2007)
31. Vakhania, N.: Scheduling jobs with release times preemptively on a single machine to minimize the number of late jobs. Oper. Res. Lett. **37**(6), 405–410 (2009)
32. Venugopalan, S., Sinnen, O.: ILP formulations for optimal task scheduling with communication delays on parallel systems. IEEE Trans. Parallel Distrib. Syst. **26**(1), 142–151 (2014)
33. Xenos, D.P., Cicciotti, M., Kopanos, G.M., Bouaswaig, A.E.F., Kahrs, O., Martinez-Botas, R., Thornhill, N.F.: Optimization of a network of compressors in parallel: real time optimization (RTO) of compressors in chemical plants - an industrial case study. Appl. Energy **144**(5), 51–63 (2015)

A Cognitive Model of Human Bias in Matching

Rakefet Ackerman[1], Avigdor Gal[1], Tomer Sagi[2], and Roee Shraga[1(✉)]

[1] Technion – Israel Institute of Technology, Haifa, Israel
ackerman@ie.technion.ac.il, avigal@technion.ac.il,
shraga89@campus.technion.ac.il
[2] University of Haifa, Haifa, Israel
tsagi@is.haifa.ac.il

Abstract. The *schema matching* problem is at the basis of integrating structured and semi-structured data. Being investigated in the fields of databases, AI, semantic Web and data mining for many years, the core challenge still remains the ability to create quality matchers, automatic tools for identifying correspondences among data concepts (*e.g.*, database attributes). In this work, we investigate human matchers behavior using a new concept termed *match consistency* and introduce a novel use of cognitive models to explain human matcher performance. Using empirical evidence, we further show that human matching suffers from predictable biases when matching schemata, which prevent them from providing consistent matching.

Keywords: Schema matching · Data integration · Human-in-the-loop

1 Introduction

Schema matching is at the basis of integrating structured and semi-structured data. The schema matching task revolves around providing correspondences between concepts describing the meaning of data in various heterogeneous, distributed data sources, such as SQL and XML schemata, entity-relationship diagrams, ontology descriptions, interface definitions, and forms format [28].

Schema matching research originated in the database community [28] and has been a focus for other disciplines as well, from artificial intelligence [10,20], to semantic web [17] to data mining [18]. Schema matching research has been going on for more than 30 years now, focusing on designing high quality matchers, automatic tools for identifying correspondences among database attributes. Initial heuristic attempts (*e.g.*, COMA [11]) were followed by theoretical grounding (*e.g.*, see [5,16]).

Recently, the information explosion (a.k.a Big Data) has provided many novel sources for data and with them the need for efficient and effective integration. Crowd-sourcing has allowed pay-as-you-go frameworks for data integration (*e.g.*, [21,35]), to make flexible use of human input in the matching process.

© Springer Nature Switzerland AG 2019
A. C. Nayak and A. Sharma (Eds.): PRICAI 2019, LNAI 11670, pp. 632–646, 2019.
https://doi.org/10.1007/978-3-030-29908-8_50

A basic tenet of the matching process, present from its inception, is that an algorithmic matcher provides a set of definite (true or false) correspondences to be then validated by a human expert. Human validation of algorithmic results assumes the superiority of human matchers over algorithms, which may be naïve, partially because different human matchers may have different opinions and may differ in the way they match schemata [31]. The emergence of crowd-based solutions has not changed this assumption, but merely extended the validation phase to include additional individuals.

A popular contemporary trend involves developing human-level AI. We believe it is equally important to understand human's strengths and predictable biases when determining the appropriate sharing of responsibility with the machine. Hence, in this work we focus on analyzing human's performance in matching. The central new concept in this work is *match consistency*, which we use, aided by cognitive principles, to show that human behavior in matching vary along consistency dimensions, namely temporal, consensuality, and control (as defined in this work). Given a set of human matchers, we assess their abilities, much like traditional models do for algorithms. Additionally, however, human matchers have biases that we can detected and accounted for when making use of human matching.

We present theoretical analyses, using cognitive models, of human matchers strengths and biases (Sect. 3) as well as empirical results on match consistency (Sect. 4) to support our framework. Additionally, Sect. 2 presents background on matching and metacognition. We review of related work in Sect. 5 and conclude in Sect. 6.

2   Background

We next present a formal matching model (Sect. 2.1) and models for human involvement in matching (Sect. 2.2).

2.1   Schema Matching Model

Let S, S' be two schemata with attributes $\{a_1, a_2, \ldots, a_n\}$ and $\{b_1, b_2, \ldots, b_m\}$, respectively. A matching process matches S and S' by aligning their attributes using *matchers* that utilize matching cues such as attribute names, instance data, and schema structure (see surveys *e.g.*, [6] and books *e.g.*, [16]). A matcher's output is conceptualized as a similarity matrix $M(S, S')$ (M for short), having entry $m_{i,j}$ (typically a real number in $[0, 1]$) represent a degree of similarity between $a_i \in S$ and $b_j \in S'$. A *match*, denoted σ, between S and S' is a subset of M's entries.

Matching is a stepped process of applying algorithms, rules, and constraints. Matchers can be separated into *first-line matchers – 1LMs*, which are applied directly to the problem, returning a similarity matrix, and *second-line matchers – 2LMs*, which are applied to the outcome of matchers, receiving similarity matrices and returning a similarity matrix.

Table 1. A similarity matrix example

$S_1 \longrightarrow$ $\downarrow S_2$	1 cardNum	2 city	3 arrivalDay	4 checkIn Time
1 clientNum	**0.84**	0.32	0.32	0.30
2 city	0.29	**1.00**	0.33	0.30
3 checkInDay	0.34	0.33	**0.35**	**0.64**

Example 1 (Matchers). To illustrate the variety of available matchers, consider three 1LMs. Term [16] compares attribute names to identify syntactically similar attributes (*e.g.*, using edit distance and soundex). WordNet uses abbreviation expansion and tokenization methods to generate a set of related words for matching attribute names. Token Path [27] integrates node-wise similarity with structural information by comparing the syntactic similarity of full paths from root to a node.

Example 2 (Similarity Matrices). Table 1 provides an example of an outcome of a matching process between fragments of two reservation systems' schemata, one (S_1) with four attributes and the other (S_2) with three attributes, conceptualized in a similarity matrix. S_1 consists a CardNum attribute with long data-type and a city attribute, which contains some example instances (city names). Attributes may be independent of other attributes or composable, creating compound attributes. *E.g.*, ArrivalDay and CheckInTime attributes can be composed to a compound arrival day/time attribute. S_2 has clientNum, city, and checkInDay attributes.

2.2   Human Involvement in Matching Models

Human schema matching is a complex decision making process, which involves a series of interrelated tasks. Each attribute in one schema is examined to decide whether and which attributes from the other schema correspond. Humans either validate an algorithmic result or locate a candidate attribute unassisted. Human matchers may choose to rely upon superficial information such as string similarity of attribute names (*e.g.*, qty is similar to quantity) or explore additional information such as data-types, instances, and position within the schema hierarchy. The decision whether to explore additional information relies upon self-monitoring of confidence.

Most of the works in schema matching over the years assume that an algorithmic matching system provides a set of definite (true or false) correspondences to be then validated by a human expert who can provide the ultimate matching. Human validation is typically prohibitively large. High matching costs and limited expert availability spawned research into crowd sourcing usage by breaking the matching task into small-sized tasks, suitable for unskilled workers with minimal compensation (pay-as-you-go approach) [23]. McCann *et al.* proposed methods to validate algorithmic matchers, ranging from direct match validation (*e.g., does ccost match pcost?*) to constraint validation (*e.g., does bDate < 2007* *always hold?*) [26]. Zhang *et al.* provided tools for validation task selection [35], Bozovic and Vassalos used feedback to tune matcher weights [8], and Hung *et al.* suggested methods to select conflicting matches in a network of schemata [22]. Sagi and Gal proposed the *Expert Sourcing* model [29], which we follow in this work, in which knowledgeable humans examine substantial parts of the matching task or the entire task (*e.g.*, for small-medium scaled schemata).

3   Match Consistency

Historically, humans (relative) strong matching abilities put them as final decision makers, disregarding biases that affect their ability to provide accurate matches. To capture the impact of human biases on matching, we present a formal notion of a consistent matcher, and use cognitive models to explore human matcher variability.

3.1   Consistent Match Definition

Matchers are typically measured using a global matching evaluation measure, *e.g.*, precision and recall. Such a measure evaluates the similarity matrix a matcher generated against some reference matrix. Given a similarity matrix M of $n \times m$ entries and an evaluation measure E, we define matcher consistency with respect to a consistency dimension $D = \{d_1, d_2, \ldots, d_k\}$ using a dimension function $F_D : \{m_{i,j} \mid m_{i,j} \in M\} \longrightarrow D$ that maps each entry in M into a value in D. In Sect. 3.3 we present a classification of dimensions and give four examples.

Given a similarity matrix M of $n \times m$ entries, a dimension $D = \{d_1, d_2, \ldots, d_k\}$ induces a partition $M_1, M_2, \ldots, M_k$ over M such that $M_l = \{m_{i,j} \in M \mid F_D(m_{i,j}) = d_l\}$. We apply the evaluation measure E over each partition $E(M_1), E(M_2), \ldots, E(M_k)$ and define match consistency using coefficient of variation as follows:

Definition 1 (Match Consistency). *Let M be an $n \times m$ similarity matrix and $D = \{d_1, \ldots, d_k\}$ a consistency dimension. Let $E \in [0, 1]$ be a random variable, with an expected value of $\mu(E)$ and a standard deviation of $sd(E)$, representing an evaluation of partitions $M_1, \ldots, M_k$ over M. MC is a* match consistency *measure of M wrt D, computed as follows:*

$$MC(M, D, E) = 1 - \frac{sd(E)}{\mu(E)} \tag{1}$$

A higher MC value should correlate with increasingly consistent match performance across the partitions induced by D. Coefficient of variation, which was chosen as a best practice measure of data consistency (see [33]), achieves this correlation through its standard deviation component. A higher standard deviation increases the coefficient of variation and reduces the value of MC. This is in line with our understanding of consistency, where a lower standard deviation means a more consistent match. The second parameter is the average performance of the different partitions. Here, we take into account not only consistency but also our desire to achieve a good match, which entails an overall high evaluation measure.

3.2   Self Monitoring of Performance

Cognitive psychology has been examining factors impacting humans when performing knowledge intensive tasks [4]. The metacognitive approach, traditionally applied for learning and answering knowledge questions [7], highlights the role of subjective confidence in regulating efforts while performing challenging tasks.

Metacognition research was recently applied to reasoning and decision making tasks [2]. It suggests that online monitoring of subjective confidence regulates the cognitive effort invested in each task (*e.g.*, identifying a correspondence). The *Discrepancy Reduction Theoretical Framework (DRTF)* explains learning effort investment by suggesting that people set a target knowledge level as a stopping criterion. They continue to invest time and effort, while subjectively monitoring their confidence level, until meeting the stopping criterion [7].

Metacognitive models use three basic components of effort regulation measures, which we use for the matching task:

(1) **Subjective confidence:** Human matchers report matching confidence as their performance monitoring.
(2) **Invested time:** Elapsed time from selection of a term to the final matching decision is used as an objective measure that presumably reflects the metacognitive control decision to either continue or terminate a task, based on the ongoing monitoring of the chance of success.
(3) **Objective performance evaluation:** We use the well accepted precision and recall to evaluate performance.

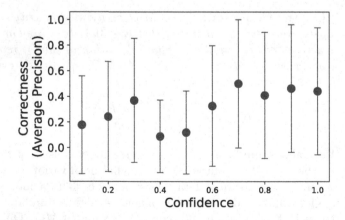

Fig. 1. Correctness by confidence, partitioned into buckets of 0.1

By way of motivation, we provide an illustration (Fig. 1) of the relationship between human confidence in matching and correctness (in terms of precision) based on our experiments (see Sect. 4). It is clear that human subjective confidence cannot serve as a good predictor to matching correctness. In this work we show how human biases affect confidence levels via consistency dimensions.

3.3 Consistency Dimensions

Consistency dimensions can be classified as continuous or discrete and may be performed using individual or collective matchers, as illustrated in Fig. 2. We introduce four consistency dimensions, namely (local and global) temporal, consensuality, and control, as examples to the full set of dimension possibilities.

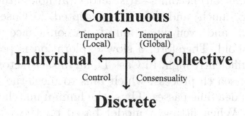

Fig. 2. Consistency dimensions

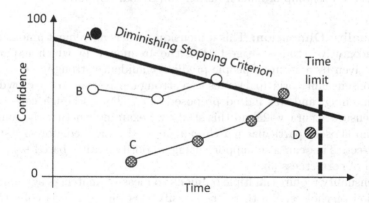

Fig. 3. DCM with hypothetical confidence ratings for four items and a self-imposed time limit (adapted from [1]).

(Local and Global) Temporal Dimension: This dimension is continuous (can be discretized into buckets to fit Definition 1) with both individual and collective variations.

The motivation to analyze the temporal dimension for human biases is rooted in the *Diminishing Criterion Model (DCM)* [1], a DRTF-based model (see Sect. 3.2) that models a common bias in human confidence judgment. DCM stipulates that the stopping criterion of a DRTF model is relaxed over time.

Thus, a human matcher is more willing to accept a low confidence level after investing some time and effort on finding a correspondence.

Figure 3 illustrates hypothetical confidence ratings while performing a schema matching task. Each dot in the figure represents a possible solution to a matching decision (*e.g.*, attributes a_i and b_j correspond), and its associated confidence, which changes over time. The search for a solution starts at time $t = 0$ and the first dot for each of the four examples represents the first solution a human matcher reaches.

As time passes, human matchers continuously evaluate their confidence. In case A, the matcher has a sufficiently high confidence after a short investigation, thus decides to accept it right away. In case B, a candidate correspondence is found quickly but fails to meet the sufficient confidence level. As time passes, together with more comprehensive examination, the confidence level (for the same or a different solution) becomes satisfactory (although the confidence value itself does not change much) and thus it is accepted. In Case C, no immediate candidate stands out, and even when found, its confidence is too low to pass the confidence threshold. Therefore, a slow exploration is performed until the confidence level is sufficiently high. In Case D, an unsatisfactory correspondence is found after a long search process, which fails to meet the stopping criterion before the individual deadline passes. Thus, the human matcher decides to reject the correspondence. When fitting a model based on the temporal dimension we can address human matchers individually, fitting a model for each human matcher separately (Local temporal), or collectively by fitting a general DCM model based on a group of human matchers (Global temporal).

Consensuality Dimension: This dimension models agreement among matchers. Metacognitive studies suggest that the frequency in which a particular answer is given by a group of people predicts confidence strongly [24].

The consensuality principal serves as a strong motivation to use crowd sourcing for matching, and was indeed proposed, *e.g.*, [35]. Although consensuality does not ensure accuracy [25], in this study we examine whether the number of people who chose a particular match can be used as a predictor of its chance to be correct. This can also support using majority voting based solutions as indication of correctness [3].

Consensuality requires multiple opinions to measure matchers agreement and a repetition of choices. We therefore classify this dimension as collective and discrete.

Control Dimension: Control analyzes the consistency of human matchers when assisted by a result of an algorithmic solution. This dimension is discrete (binary, in fact). In this work we consider control as an individual dimension, although it can be easily extended, using a general model for assisted/non-assisted matchers, to be collective.

Metacognitive control decisions are the regulatory decisions people take, given a self-assessment of their chance for success [7]. In the context of this study, the use of algorithmic output for helping the matcher in her task is taken as a control decision.

Variability in this dimension may be attributed to the predicted tendency of humans who do not use system suggestions to be more engaged in the task and recruit more mental effort than those who use suggestions as a way to ease their cognitive load (see [32]). Shraga *et al.* showed that human matchers who rely on algorithmic support are likely to follow the algorithm suggested [31].

4   Empirical Evaluation

The experiments analyze match consistency along consistency dimensions. Results indicate variability along consistency dimensions with varying trends of correctness.

4.1   Experiment Setup

Dataset and evaluation measures are presented next.

Human Matching Dataset: The dataset contains match results of 106 human matchers, all Industrial Engineering undergraduates who studied logistics and database management courses. Participants were briefed in schema matching prior to the task. Four pilot participants completed the task prior to the study to ensure its coherence and instruction legibility. Participants were trained on a pair of small schemata (9–12 attributes) from the *Thalia* dataset[1] prior to the main task.

The main schema matching task was chosen from the *Purchase Order* dataset [11]. The schemata used are medium size, with 142 and 46 attributes, and with high information content (labels, data types, and instance examples). Correspondences are of differing difficulty levels, with both easy matches and complex relationships, which may yield low precision and recall, even when using the strongest of matchers. Potentially, a maximum number of 6,532 correspondences are possible **per human matcher**, by (impossibly) evaluating each and every pair of attributes. In reality, each matcher chose to evaluate 51 correspondences on average, creating a dataset of ∼5,600 human matcher's correspondences (1,229 distinct correspondences). A reference match for evaluation was compiled by domain experts over the years in which this dataset has been used for testing.

A side-by-side view of the two schemata and a dynamic match table were provided. The system records the time it takes for a matcher to determine on a correspondence. Match confidence was inserted by participants directly into the match table as a value between 0 and 1, displayed as a percentage.

Participants were randomly assigned to one of four conditions, differing by the algorithmic support provided. **No suggestions** (0), where participants perform the task with no algorithmic assistance; **limited suggestions** (1a), where

---

[1] www.cise.ufl.edu/research/dbintegrate/thalia/howto.html.

participants are allowed a limited (8 clicks) use of a lifesaver button. A counter of the suggestions used vs. remaining is presented; **unlimited suggestions** (1b), where participants are allowed an unlimited use of a lifesaver button; and **validate algorithmic result** (2), where algorithmic suggestions are pre-entered, letting participants validate, override, or complete them. The latter represents the classic "humans as validators" approach. The algorithmic matcher we used to create suggestions was Term (see Example 1) with typical performance (F1 $\approx$ 0.5) for automatic schema matchers on difficult instances.

To analyze the control dimension, we further separated participants into two groups. The first contains those participants who did not have suggestions (condition 0) or did not use the suggestion (from conditions 1a and 1b). The second contains those who actively requested suggestions from conditions 1a and 1b, in addition to the participants from condition 2.

Duplicate ratings for the same correspondence were removed, taking the latest. Out of the 106 participants, 6 were discarded due to technical faults, leaving 100 valid results. Elapsed time outliers (over 2 standard deviations from the mean of each participant) were removed due to the sensitivity of our measures to outliers, which may occur due to methodical pauses, unrelated to the matching task.

We created a group of the top 10% performing human matchers, considered as performance idealization of humans as validators, to show that even they suffer from biases and therefore are non-distinguishable from others a-priori.

Evaluation Measures: Let M^e be a reference matrix, such that $m_{i,j} = 1$ whenever the correspondence (a_i, b_j) is part of the reference match and $m_{i,j} = 0$ otherwise. The precision (P) and recall (R) evaluation measures are defined as follows:

$$P(\sigma) = \frac{|\sigma \cap M^{e+}|}{|\sigma|}, R(\sigma) = \frac{|\sigma \cap M^{e+}|}{|M^{e+}|} \tag{2}$$

where M^{e+} represent non-zero entries of M^e and recalling that σ is a subset of M's entries. The F1 measure, $F(\sigma)$, is calculated as the harmonic mean of $P(\sigma)$ and $R(\sigma)$.

Given a consistency dimension $D = \{d_1, d_2, \ldots, d_k\}$, precision and recall can be defined similarly per value d_i, by replacing σ with $\sigma \cap M_i$ (see Sect. 3.1). To compute match consistency we use Eq. 1 by estimating μ as the sampled average and sd as the sampled standard deviation over the evaluation measure of choice (*e.g.*, precision).

To analyze human matcher confidence we use metacognitive measures of calibration and resolution, based upon *performance monitoring*.

$$Calibration(\sigma) = \overline{\sigma} - P(\sigma), Resolution(\sigma) = \gamma(\sigma, M^{e+}) \tag{3}$$

Table 2. Resolution, (P)recision, (R)ecall, and (F)1 of matchers.

Matcher	Resolution	Sig. (p-value)	P	R	F
Term	0.63	0.045	0.35	0.80	**0.48**
TokenPath	0.72	0.140	0.25	0.86	0.33
WordNet	0.94	0.035	0.31	**0.87**	0.44
Human matchers	0.16 (SD=0.46)	0.001	**0.63**	0.36	0.45
Top-10 human matchers	0.58 (SD=0.57)	0.104	0.91	0.60	0.71

where $\bar{\sigma}$ is a user average confidence and $\gamma(\cdot, \cdot)$ is GK-Gamma correlation [19].

Positive calibration is interpreted as *overconfidence* and negative calibration as *under-confidence*. *Resolution* measures the extent to which confidence discriminates between correct and incorrect correspondences. GK-Gamma ranges in $[-1, 1]$ where scores of 1 and -1 indicate perfect resolution and 0 indicates no resolution. Negative resolution scores are interpreted as identifying good results as bad and vice-versa.

4.2 Results

We present a confidence analysis and empirical evaluation along consistency dimensions as evidence for human matching biases. Experiments show that human matchers are, in general, overconfident with low ability to distinguish correct from incorrect correspondences. In terms of consistency, results demonstrate significance variability along all dimensions with varying trends of correctness.

Confidence Analysis: We begin with a metacognitive evaluation, examining calibration and resolution (Sect. 4.1) in a schema matching setting. Average calibration (over participants) for match decisions was .26. 45 participants had over .3 calibration and 8 had negative calibration. Overall, the calibration levels demonstrate a right skewed distribution, interpreted as overconfidence, which was reported in the literature as a well-established human tendency [13].

Resolution results are given in Table 2. To compute matches, a 0.5 threshold was applied over the results. Human matchers, as a group, have significant, but low positive resolution (.16) with high variance. Only 25 (31%) had significant resolution (.68 average resolution within the group). Of those, 23 had significant positive resolution (average positive resolution was .79) and 2 had significant negative resolution (average negative resolution was $-.57$).

Note that performing the same calculation with algorithmic matchers yields much better resolution (see Table 2). Comparing human and algorithmic matching, overall the former has better precision, while the latter has better recall. However, even matching algorithms with comparatively fair F1 scores such as TokenPath, demonstrate high resolution. This serves as empirical evidence that the traditional view of "humans as validators", may not be suitable for matching.

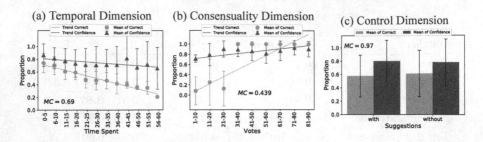

Fig. 4. Confidence (Blue) and correctness (Red) by dimension (Color figure online)

(Local and Global) Temporal Dimension: We validate that the DCM (Sect. 3.3) reflects human matchers behavior by showing the predicted association between elapsed time and reported confidence. We show evidence of temporal bias and our ability to use elapsed time as a predictor of human matching performance. Support for this model would manifest itself via negative correlation between elapsed time and confidence per participant (local) and all participants (global).

Experimental results support the DCM model both locally and globally. A collective negative mean slope of −.23 suggests that on average, confidence decreases with time, which supports global temporal dimension. Zooming in on individual confidence reports, mean slope varied in $[−.274, −.213]$ with 40% individual matchers having significant (negative) correlations. A single-sample t-test was used to reject the null hypothesis of the slopes being random-noise. Also, a one-way ANOVA test was used to reject the null hypothesis of all participants sharing the same confidence mean ($F_{1,80} = 23.6$, p-value $< 10^{-5}$) emphasizing the need for a local model. To support the DCM self-imposed time limit, we followed [34] and found a significant curvilinear relation between time and confidence, reflecting the combination of **two** stopping criteria, which are unique to the DCM [1].

With correlated confidence and elapsed time, we now validate the use of DCM in matching, by examining the accuracy of human matchers as predicted by elapsed time. We tested the correlation between elapsed time and participant chance of providing correct matches. We partitioned the elapsed time into buckets, each of 5 s (0–5, 6–10, *etc.*) and examined an aggregative temporal behavior of human matchers. With $MC = 0.42$, the temporal dimension exhibits high variability among the dimension buckets. Correctness was computed as the precision within the bracket's time frame (Sect. 4.1). Mean slope was found to be −.54 (statistically significant with $p_{val} < 10^{-5}$). This serves as evidence that time spent is predictive of matching (in)correctness.

Figure 4a compares correctness with confidence, by showing precision (red) and mean of confidence across all human matchers (blue), partitioned according to elapsed time. For each measure we also include a linear trend-line and error bars for each time bucket. As discussed before, as time passes, less decisions made by humans are correct and there is a decline in human confidence. We also note that confidence consistently receives higher values than correctness proportion, which reflects the overall overconfidence, as reported above. The difference between the two becomes more prominent as time elapses, which is a classic finding in metacognitive literature, called hard-easy effect [9]. Note that error bars show variance in the way confidence is determined.

Offline examination of the top 10 human matchers reveals a slight (statistically insignificant) improvement in consistency. Accuracy is not available in real-life scenarios and therefore cannot be used to identify the best human matchers a-priori.

Consensuality Dimension: Next, we validate that the agreement level among matchers is correlated with self-reported confidence, and show evidence to its impact on human matching performance (Sect. 3.3).

We partitioned the number of votes for correspondence into increasing agreement levels (0–5, 6–10, *etc.*). For each level we computed the average confidence of correspondences and proportion of correctly matched correspondences out of all correspondences that were determined within the level. Figure 4b (similarly to Fig. 4a) presents confidence (blue) and correctness (red), partitioned by agreement level (number of votes). For each measure we also include a linear trend-line and error bars.

Overconfidence is demonstrated in lower agreement levels, while for higher levels the human matchers underestimate correctness. The error bars illustrate a significant variance in lower agreement levels and becomes negligible at higher levels, possibly as a result of correspondences that are easier to detect in levels where consensus is higher.

We also tested the correlation between level of agreement among participants and participant chance of providing correct matches. Although recent studies suggest that consensuality does not ensure accuracy [25], mean slopes for accuracy was found to be .13 (statistically significant with a p-value $< 10^{-5}$), showing that consensus among matchers is predictive of matching correctness. Consistency was measured at $MC = 0.36$, which is indicative of high variability. Here we see an improvement among the top 10 human matchers probably because they agree more among themselves on correct matches (evaluated only 95 correspondences compared to 1,229 overall).

Control Dimension: Finally, we show the impact of availability of algorithmic correspondences on human matching performance. To evaluate the performance of human matchers we compare the self-reported confidence and objective performance of participants by the control condition.

Figure 4c presents a comparison between participants who used (actively or passively) suggestions (left side) and those who did not (right side). A statistically significant (Pearson) correlation (p-value $< 10^{-5}$) was found between a binary variable indicating the use of a suggestion given a correspondence and a binary variable indicating whether this correspondence is a part of the reference match. Clearly, the human matcher is overconfident, regardless of the algorithmic assistance. Yet, results show better performance of participants who did not use the system's suggestions versus those who did. This can be explained by the fact that human matchers with machine support are more likely to behave as suggested [31], because of shallower processing than without this opportunity (see [32]). It is worth noting that with $MC = 0.85$, the control dimension demonstrates a more consistent pattern than the other two dimensions. Offline examination shows that the top 10 human matchers exhibit better consistency but show larger difference in confidence levels. The matchers assisted among the top 10 are much more confident (.81 compared to non assisted confidence of .69) but also live up to the expectations, achieving high accuracy levels (.86).

5   Related Work

Section 2.2 outlined the main effort in human involvement in matching. We now focus on demonstrating the contribution of this work on the background of state-of-the-art.

Using humans to answer schema matching validation questions was first proposed in [26]. This work was later extended [21,35] by using crowd sourcing to reduce uncertainty. Sarasua *et al.* suggested mechanical turks to validate matching by providing context information [30]. A recent work [35] also acknowledged the fact that the crowd is not always correct, associating probabilities to answers based on the question hardness (hard-easy effect as addressed in this paper) and worker's trustworthiness, which are estimated empirically, based on [15]. We take the observation that humans are not perfect a step further, analyzing cognitive biases that make human evaluation error prone.

Schema matching and ontology alignment [14] are closely related research areas, both aiming at finding matches between concepts. The two vary in their matching objects (schemata *vs.* ontologies), matching refinement (equivalence *vs.* richer semantics such as inclusion), and the underlying mathematical tools (*e.g.*, similarity matrix analysis *vs.* logic). To date, little work was devoted to the role of human matchers in either research areas. Nevertheless, a recent work in ontology alignment have acknowledged the fact that humans (users) can make mistakes [12]. Although it addresses cognitive oriented issues, *e.g.*, cognitive load, their aim is to avoid them. Further, they propose to collect confidence as a future work, which we collected and showed it may be unreliable (overconfidence). Our research insights can be readily applied to ontology alignment.

6   Conclusions and Future Work

This work introduces match consistency as a measure of human matching variability along potential bias dimensions. We view match consistency as a powerful tool to analyze human matching behavior. In future work we intend to identify additional dimensions, beyond the dimensions identified in this work, namely temporal, consensuality, and control. Our empirical evaluations serve as proof-of-concept that validate the important roles of humans as participants in the matching process, and less so as validators. Therefore, future work will involve collaboration models, supporting both human and algorithmic matchers, jointly performing schema matching considering humans biases.

References

1. Ackerman, R.: The diminishing criterion model for metacognitive regulation of time investment. J. Exp. Psychol.: Gen. **143**, 1349 (2014)
2. Ackerman, R., Thompson, V.: Meta-reasoning: monitoring and control of thinking and reasoning. TiCS **21**, 607–617 (2017)
3. Raykar, V.C., et al.: Supervised learning from multiple experts: whom to trust when everyone lies a bit. In: ICML (2009)
4. Barsalou, L.W.: Cognitive Psychology: An Overview for Cognitive Scientists. Psychology Press, New York (2014)
5. Bellahsene, Z., Bonifati, A., Rahm, E. (eds.): Schema Matching and Mapping. Springer, Berlin (2011). https://doi.org/10.1007/978-3-642-16518-4
6. Bernstein, P.A., Madhavan, J., Rahm, E.: Generic schema matching, ten years later. PVLDB **4**, 695–701 (2011)
7. Bjork, R.A., Dunlosky, J., Kornell, N.: Self-regulated learning: beliefs, techniques, and illusions. Ann. Rev. Psychol. **64**, 417–444 (2013)
8. Bozovic, N., Vassalos, V.: Two phase user driven schema matching. In: ADBIS (2015)
9. Brewer, N., Wells, G.L.: The confidence-accuracy relationship in eyewitness identification: effects of lineup instructions, foil similarity, and target-absent base rates. J. Exp. Psychol.: Appl. **12**, 11 (2006)
10. De Una, D., Rümmele, N., Gange, G., Schachte, P., Stuckey, P.J.: Machine learning and constraint programming for relational-to-ontology schema mapping. In: IJCAI (2018)
11. Do, H.H., Rahm, E.: COMA: a system for flexible combination of schema matching approaches. In: VLDB (2002)
12. Dragisic, Z., Ivanova, V., Lambrix, P., Faria, D., Jiménez-Ruiz, E., Pesquita, C.: User validation in ontology alignment. In: Groth, P., Simperl, E., Gray, A., Sabou, M., Krötzsch, M., Lecue, F., Flöck, F., Gil, Y. (eds.) ISWC 2016. LNCS, vol. 9981, pp. 200–217. Springer, Cham (2016). https://doi.org/10.1007/978-3-319-46523-4_13
13. Dunning, D., Heath, C., Suls, J.M.: Flawed self-assessment implications for health, education, and the workplace. Psychol. Sci. Public Interest **5**, 69–106 (2004)
14. Euzenat, J., Shvaiko, P.: Ontology Matching. Springer, New York (2007). https://doi.org/10.1007/978-3-540-49612-0
15. Franklin, M.J., Kossmann, D., Kraska, T., Ramesh, S., Xin, R.: CrowdDB: answering queries with crowdsourcing. In: SIGMOD (2011)
16. Gal, A.: Uncertain Schema Matching. Morgan & Claypool Publishers, San Rafael (2011)
17. Gal, A., Roitman, H., Sagi, T.: From diversity-based prediction to better ontology & schema matching. In: WWW (2016)
18. Gal, A., Roitman, H., Shraga, R.: Heterogeneous data integration by learning to rerank schema matches. In: ICDM (2018)

19. Goodman, L.A., Kruskal, W.H.: Measures of association for cross classifications. J. Am. Stat. Assoc. **49**, 732–764 (1954)
20. Halevy, A.Y., Madhavan, J.: Corpus-based knowledge representation. In: IJCAI (2003)
21. Hung, N.Q.V., Nguyen, T.T., Miklós, Z., Aberer, K., Gal, A., Weidlich, M.: Pay-as-you-go reconciliation in schema matching networks. In: ICDE (2014)
22. Hung, N.Q.V., Tam, N.T., Miklós, Z., Aberer, K.: On leveraging crowdsourcing techniques for schema matching networks. In: Meng, W., Feng, L., Bressan, S., Winiwarter, W., Song, W. (eds.) DASFAA 2013. LNCS, vol. 7826, pp. 139–154. Springer, Heidelberg (2013). https://doi.org/10.1007/978-3-642-37450-0_10
23. Jeffery, S.R., Franklin, M.J., Halevy, A.Y.: Pay-as-you-go user feedback for dataspace systems. In: SIGMOD (2008)
24. Koriat, A.: Subjective confidence in one's answers: the consensuality principle. J. Exp. Psychol.: Learn. Memory Cognit. **34**, 945–959 (2008)
25. Koriat, A.: When reality is out of focus: can people tell whether their beliefs and judgments are correct or wrong? J. Exp. Psychol.: Gen. **147**, 613 (2018)
26. McCann, R., Shen, W., Doan, A.: Matching schemas in online communities: a web 2.0 approach. In: ICDE (2008)
27. Peukert, E., Eberius, J., Rahm, E.: AMC-a framework for modelling and comparing matching systems as matching processes. In: ICDE (2011)
28. Rahm, E., Bernstein, P.A.: A survey of approaches to automatic schema matching. VLDBJ **10**, 334–350 (2001)
29. Sagi, T., Gal, A.: In schema matching, even experts are human. towards expert sourcing in schema matching. In: IIWeb (2014)
30. Sarasua, C., Simperl, E., Noy, N.F.: CrowdMap: crowdsourcing ontology alignment with microtasks. In: ISWC (2012)
31. Shraga, R., Gal, A., Roitman, H.: What type of a matcher are you?: coordination of human and algorithmic matchers. In: HILDA@SIGMOD (2018)
32. Sidi, Y., Shpigelman, M., Zalmanov, H., Ackerman, R.: Understanding metacognitive inferiority on screen by exposing cues for depth of processing. Learn. Instr. **51**, 61–73 (2017)
33. Simonsen, J.C.: Coefficient of variation as a measure of subject effort. Arch. PM&R **76**, 516–520 (1995)
34. Undorf, M., Ackerman, R.: The puzzle of study time allocation for the most challenging items. Psychon. Bull. Rev. **24**, 2003–2011 (2017)
35. Zhang, C., Chen, L., Jagadish, H., Zhang, M., Tong, Y.: Reducing uncertainty of schema matching via crowdsourcing with accuracy rates. TKDE (2018). https://www.computer.org/csdl/journal/tk/5555/01/08533346/17D45XreC6p

Multi-Agent Systems

Adaptive Incentive Allocation for Influence-Aware Proactive Recommendation

Shiqing Wu[1(✉)], Quan Bai[2], and Byeong Ho Kang[2]

[1] Auckland University of Technology, Auckland, New Zealand
shiqing.wu@aut.ac.nz
[2] University of Tasmania, Hobart, Australia
{quan.bai,byeong.kang}@utas.edu.au

Abstract. Most recommendation systems are designed for seeking users' demands and preferences, whereas impotent to affect users' decisions for realizing the system-level objective. In this light, we intend to propose a generic concept named 'proactive recommendation', which focuses on not only maintaining users' satisfaction but also realizing system-level objectives. In this paper, we claim the proactive recommendation is crucial for the scenario where the system objectives are required to realize. To realize proactive recommendation, we intend to affect users' decision-making by providing incentives and utilizing social influence between users. We design an approach for discovering the influential users in an unknown network, and a dynamic game-based mechanism that allocates incentives to users dynamically. The preliminary experimental results show the effectiveness of the proposed approach.

Keywords: Proactive recommendation · Incentives allocation ·
Agent-based modeling · Unknown network

1 Introduction

In recent years, recommendation systems have become increasingly popular and been widely applied in different domains [15]. Although most recommendation approaches are able to discover users' demands and recommend proper items to users [1,6], these approaches are too passive to realize system-level objectives. For example, a traditional recommendation system may be unable to persuade people to take public transportation for commuting if they prefer driving. In this case, it is necessary to propose a new approach, which aims to not only maintain users' satisfaction but also realize system-level objectives, i.e., proactive recommendation. Different from traditional recommendation approaches, in the proactive recommendation, each user is considered as a 'participator' to be coordinated for realizing the system goals.

In scenarios where proactive recommendation systems are required, it is possible that items which are beneficial to the system cannot attract users [20].

© Springer Nature Switzerland AG 2019
A. C. Nayak and A. Sharma (Eds.): PRICAI 2019, LNAI 11670, pp. 649–661, 2019.
https://doi.org/10.1007/978-3-030-29908-8_51

To relieve such conflicts between users and the system, the system needs to embed with effective methods for increasing users' preference towards the beneficial item. In some areas, researchers have proved that providing incentives is an effective strategy for affecting users' behaviors to realize certain system objectives [14,16]. However, the limitations of these approaches still cannot be ignored. First, a naive incentive allocation strategy may not be effective. Since benefits brought from users are variable towards the status of the whole environment, the process of incentives allocation needs to be adaptive as well. Second, incentives provided to users are restricted by the budget amount, so that the number of successfully affected users is also limited. To increase the number of affected users, we consider utilizing a factor which existed between users for affecting more users, i.e., social influence.

In general, users can be influenced by their neighbors or other influential users in the same social network [2]. Namely, by leveraging the power of social influence, it is possible to affect users in selecting system-beneficial items by providing fewer incentives. However, the influence between two users may differ corresponding to diverse items or topics. For instance, a famous blogger who focuses on commenting movies may cause higher influence on his fans' choices of movies for watching, but hardly generates affect in music selection. Hence, regarding a specific topic, the topology of the network should be unknown, and it is necessary to discover the influential users via a series of learning processes rather than directly utilizing some existing knowledge (e.g., the followers of a blogger) to assess their influential ability.

In this paper, we systematically elaborate and formulate the proactive recommendation, which tries to engage users by incentives in selecting system-beneficial items and affecting other users' decisions. Meanwhile, we propose an Agent-based Decisions Making (**ADM**) model. In the ADM, users are modeled as autonomous agents, and each user would select the item with the highest preference as the final decision. Furthermore, we present the Influence Probability Estimation (**IPE**) algorithm for estimating the influential relationship between users, and the Dynamic Game-based Incentive Allocation (**DGIA**) algorithm for calculating the values of incentives providing to users. The experimental results demonstrate that: (1) the IPE algorithm is effective in discovering influential users; and (2) the combination of IPE and DGIA algorithms outperforms other incentive allocation algorithms since it can affect more users given a same budget and time span.

The rest of the paper is organized as follows: Sect. 2 reviews literature related to this research work. Section 3 introduces the formal definitions and problem description. Section 4 presents the proposed model and algorithms. In Sect. 5, the experimental results are presented to evaluate the performance of the proposed model. The conclusion and future work of this paper are presented in Sect. 6.

2   Related Work

Recommendation systems have been deployed widely in many applications over the decades [4,13]. Among these systems, the Collaborative Filtering and

Content-based Filtering approaches [5,12] are widely used and perform effectively in mining users' demands and satisfying their requirements [1,4]. In general, the main objective of these approaches is to predict the user's preference and improve the utilization of data, which results that these approaches can only passively satisfy users' demands, and almost unable to affect users to select items for realizing the system-level objective. By contrast, the main objective of the proactive recommendation is to not only maintain users' satisfaction but also realize the system-level objective.

Individuals tend to take the action when they recognize that they would be rewarded from the action [8]. Inspired by this, to realize the specific system-level objective, some approaches that allocate incentives to users for affecting their decision-making have been proposed. Sengvong and Bai propose an approach trying to affect users' decisions on transportation modes during commute [14]. Their approach determines the value of incentives based on the utility difference between users and the system. Singla et al. attempt to solve the balance problem of sharing-bike by incentivizing users [16]. In addition, some researchers model this incentive allocation problem as a Multi-Armed Bandit (MAB) problem, and propose strategies for solving it by considering a limited budget [3,17–19]. However, the effectiveness of these approaches is limited when the budget is insufficient, as they attempt to incentivize users' directly and the budget would be run out shortly. By contrast, besides providing incentives, the proposed approach considers utilizing social influence between users to affect users' decision-making.

3   Problem Formulation

3.1   Proactive Recommendation

The main challenge of realizing proactive recommendation is how to effectively affect users' decisions, and encourage them to take system-beneficial items or actions. Hence, it is necessary to explore factors altering users' decisions making. In many fields, preference plays a crucial role in affecting users' decision-making process. A phenomenon frequently occurs is that a user would choose A rather B if he has a preference for A over B. Furthermore, a user's preference towards the same item may vary due to the external reason [11]. Inspired by this, in the proactive recommendation, we consider utilizing social influence and incentives to affect users' decisions.

Figure 1 compares the proposed model with existing incentive-based approaches. Figure 1(a) shows the model which is used in most existing incentive-based systems, where green nodes imply to affected users and the red node denotes the user fails to be affected. These systems tend to provide incentives to all users simultaneously, and are capable to affect most users when the budget is sufficient. However, once the budget is insufficient or the number of users becomes too large, the performance would fall down as the allocated incentive to each user may be insufficient to affect the user anymore. Figure 1(b) describes our model which considers social influence between users, where each directed link implies to one influential relationship, e.g. v_1 is able to affect v_2 and v_3.

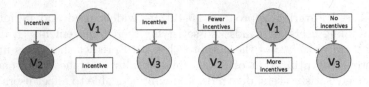

(a) Traditional Incentive Model   (b) The ADM Model

Fig. 1. Comparison of two incentive models

The proposed strategy prefers to provide influential users with more incentives, and allocate fewer incentives to the other users. Users are more possible to affect due to the simultaneous impact of incentives and social influence for other users. Furthermore, adopting such a strategy can save budget to an extent degree for further allocation.

3.2 Formal Definitions

Before introducing the proposed approach, we first give formal definitions and notations which would be frequently mentioned in this paper.

Definition 1. *An action,* $a_m \in A$, *denotes an available option that users can choose, where* $A = \{a_1, ..., a_m\}$ *is a set of finite action options. We regard* a^* *as the action that the system expects users to choose.*

Definition 2. *A user (user agent),* v_i, *is defined as a vertex in a directed social network* $G = (V, E)$, *where* $V = \{v_1, ..., v_n\}$ *represents a set of user agents, and* $E = \{e_{ij}|\{v_i, v_j\} \subseteq V\}$ *denotes a set of edges. Each directed edge* e_{ij} *represents the influence from* v_i *to* v_j, *and the weight* w_{ij} *of each edge denotes the influence value. Concurrently, towards each* a_m, v_i *has a preference degree* u_{v_i,a_m}, *where* $u_{v_i,a_m} \in [0,1]$. *At each time step* t, v_i *would choose an action* $\pi_{v_i,t}$, *where* $\pi_{v_i,t} \in A$. *In this paper, we consider that* v_i *is activated at time step* t *if* $\pi_{v_i,t} = a^*$.

The neighbors of v_i can be classified into two groups, i.e., $N(v_i)^{in}$ and $N(v_i)^{out}$. $N(v_i)^{in}$ denotes users who affect v_i, i.e., $N(v_i)^{in} = \{v_j|e_{ji} \in E\}$. $N(v_i)^{out} = \{v_j|e_{ij} \in E\}$ denotes users who are affected by v_i. At each time step t, v_i may receive an influence $k_{v_i,a_m,t}$ from $N(v_i)^{in}$, where a_m denotes a particular action option. We regard idg_{v_i} as the influential degree of v_i, which indicates the v_i's influential ability to whole network. Furthermore, each v_i can obtain the information regarding $\pi_{v_j,t-1}$ of his neighbors.

Definition 3. *Incentive,* $r_{v_i,t}$ *denotes the reward that the system provides to* v_i *for incentivizing him to choose* a^* *at time step* t. *The value of* $r_{v_i,t}$ *is constrained by the remaining budget* B_t. v_i *only obtains the incentive after he performing* a^*, *and the value of the incentive would be deducted from the* B_t.

Definition 4. *The **influence probability matrix**, X, indicates a matrix describing the influence relationship between users of the social network, where the size of X is depended on the number of users, i.e., $|V| \times |V|$. Each $x_{ij} \in X$ represents an estimated influence probability that v_i successfully affects v_j, where $x_{ij} \in [0,1]$. If $x_{ij} = 0$, it implies v_i impossibly affects v_j, conversely, v_i necessarily affect v_j if $x_{ij} = 1$.*

Definition 5. *The **estimated probability**, $p_{v_i,t}$, describes the probability that v_i accepts an incentive at a time step t, where $p_{v_i,t} \in [0,1]$. $p_{v_i,t}$ is determined by user's action $\pi_{v_i,t}$ and ω_{v_i}, where ω_{v_i} is the probability that v_i takes a^* without receiving incentives and social influence.*

3.3   Problem Description

Given a limited time span $t \in [0,n]$ and a finite budget B at each time step, the major objective of this study is to find out a strategy to incentivize users as many as possible. The *Global Activated Users Percentage (GAUP)* at time step t is represented as μ_t, which indicates the percentage of the number of activated users in the social network, i.e., $\pi_{v_i,t} = a^*$. In other words, GAUP can reflect the completion degree of the system objective. The effectiveness trend of allocation approach can be reflected by GAUP at each time step as well. μ_t can be determined by using Eq. (1).

$$\mu_t = \frac{|\{v_i | \pi_{v_i,t} = a^*, v_i \in V\}|}{|V|} \tag{1}$$

The objective of our approach is to maximize μ_t eventually. Furthermore, the *Global Influenced Activation Coverage (GIAC)* is taken into consideration as well. The GIAC is represented by using notion φ_t, indicating the ratio of activated users who are driven by the social influence. In most existing models, a user is incentivized successfully when the value of the incentive exceeds the user's preference difference. Namely, we can consider v_i is affected due to the social influence if the value of the incentive is less than his preference difference. Suppose $a' = \arg\max_{a_m \in A} u_{v_i,a_m}$ denotes v_i's highest preference, then the preference difference σ_{v_i} can be formulated by using Eq. (2).

$$\sigma_{v_i} = u_{v_i,a'} - u_{v_i,a^*} \tag{2}$$

Subsequently, the GIAC φ_t can be formulated by using Eq. (3).

$$\varphi_t = \frac{|\{v_i | \pi_{v_i,t} = a^*, r_{v_i,t} < \sigma_{v_i}, v_i \in V\}|}{|V|} \tag{3}$$

4   Proactive Recommendation Model

4.1   The Agent-Based Decisions Making (ADM) Model

The ADM model is a decentralized decision-making model which inherits the advantages of Agent-based Modeling. As mentioned previously, each user agent makes a decision not only based on his preference towards the action but also the influence from others and the offered incentive. However, each user agent may have a different position in a networked society, and cause different influence on other agents. It implies that a influential node can play a more pivotal role in affecting others' decisions in society.

Inspired by the Linear threshold model proposed by Granoveteer [7], the total influence generated by neighbors of a focal user has a restriction, i.e. $\sum_{v_j \in N(v_i)^{in}} w_{ji} \leq 1$. Moreover, towards a particular action a_m, we assume that the total influence $k_{v_i,a_m,t}$ exerted on the focal user v_i at time step t is aggregated by the influence caused by users of $N(v_i)^{in}$ who choose a_m at time step $t-1$. $k_{v_i,a_m,t}$ can then be formulated by Eq. (4),

$$k_{v_i,a_m,t} = \sum_{v_j \in N(v_i)^{in}} W(w_{ji}, a_m) \tag{4}$$

where $W(w_{ji}, a_m)$ is a judgment function, formulated by Eq. (5).

$$W(w_{ji}, a_m) = \begin{cases} w_{ji}, & \pi_{v_j,t-1} = a_m \\ 0, & \pi_{v_j,t-1} \neq a_m \end{cases} \tag{5}$$

Different from the linear threshold model, users make decisions not only based on the influence they receive but also their preferences. It implies that setting a random 'threshold' as a constant for users is insufficient to describe the scenario we are tackling. Hence, we assume that v_i needs to reconsider his decision at every time step and always chooses the action with the highest final user preference, as described in the Eq. (6), where $u^*_{v_i,a_m,t}$ denotes the final user preference towards a_m.

$$\pi_{v_i,t} = \arg\max_{a_m \in A} u^*_{v_i,a_m,t} \tag{6}$$

$u^*_{v_i,a_m,t}$ can be calculated by using Eq. (7), where u_{v_i,a_m} denotes v_i's preference towards a_m, $k_{v_i,a_m,t}$ denotes the influence that focal user v_i received from $N(v_i)^{in}$ at time t, and $r_{v_i,t}$ denotes the incentive allocating to v_i for selecting action a^*.

$$u^*_{v_i,a_m,t} = \begin{cases} u_{v_i,a_m} + k_{v_i,a_m,t} + r_{v_i,t}, & a_m = a^* \\ u_{v_i,a_m} + k_{v_i,a_m,t}, & a_m \neq a^* \end{cases} \tag{7}$$

4.2   The Influence Probability Estimation (IPE) Algorithm

Users' behaviors and the influence between users can be captured in the proposed ADM. Stands on the perspective of the system agent, the optimal incentive allocation implies to affecting more users with less budget. To achieve this objective, it is more important to engage influential users. Since only users' actions and preferences can be observed by the system, and the influential relationships between users are unknown, we adopt a behavior-based learning approach to estimate and learn the influence relationship between users. The proposed approach determines whether two users have an influential relationship based on

---

Algorithm 1. The IPE Algorithm

Input: V, X, t
Output: $idg_{v_i}, \forall v_i \in V$

1 **for** *each user agent v_i at time step $t (v_i \in V)$* **do**
2   Waiting for v_i taking action;
3   **for** *each user agent v_j $(v_j \in V \backslash \{v_i\})$* **do**
4 | Update x_{ji} using Eq. (8);
5 **for** *each user agent v_i $(v_i \in V)$* **do**
6      $sum := 0$;
7   **for** *each user agent v_j $(v_j \in V)$* **do**
8 | $sum = sum + x_{ij}$;
9   Calculate idg_{v_i} using Eq. (9);

---

their actions at time step t and $t-1$, i.e., $\pi_{v_j,t}$ and $\pi_{v_i,t-1}$. In the estimation, we consider user v_j is influenced by user v_i if v_i takes an action at time $t-1$ and v_j takes the same action at time t. The estimation of the relationship is possibly not accurate unless experiencing a long-term learning. For example, if ten users all take a_1 at time $t-1$, and then v_i takes action a_1 at time step t as well, then we can only assume a_i is affected by these ten users. With the continuous process of learning, the estimated influential relationship would become accurate.

We utilize the influential probability matrix X to record the influence probabilities among users. The initial values of each element in X are set to 0, implying that the influential relationship between users in the network is unknown to the system. According to the fact in real society, if a neighbor's action is always consistent with a user's own action at last time, then we can consider the neighbor is easier affected by the user [21]. Hence, the value of x_{ij} can be determined by using Eq. (8), where β denotes a coefficient which adjusts the variety speed of x_{ij}. If $\pi_{v_j,t} = \pi_{v_i,t-1}$, then x_{ij} would increase; otherwise, x_{ij} would decrease. Namely, x_{ij} will be very close to 1 if v_i is always successfully affected by v_j; similarly, x_{ij} will be close to 0 if v_j cannot be affected by v_i. Moreover, x_{ij} should be 1 or 0 if x_{ij} exceeds 1 or is less than 0 after updating.

$$x_{ij} = \begin{cases} x_{ij} + \beta \cdot (e^{(x_{ij}-1)} - \beta), & \pi_{v_j,t} = \pi_{v_i,t-1} \\ x_{ij} - \beta \cdot (e^{(-x_{ij})} - \beta), & \pi_{v_j,t} \neq \pi_{v_i,t-1} \end{cases}, x_{ij} \in [0,1] \qquad (8)$$

The influential degree idg_{v_i} reflects the ability of v_i affecting global users in the network, which can be determined by using Eq. (9), where $|V|$ denotes the number of total users in the system.

$$idg_{v_i} = \frac{\sum\limits_{v_j \in N(v_i)^{out}} x_{ij}}{|V|} \qquad (9)$$

The sketch of the IPE algorithm is described in Algorithm 1. The inputs include the user agents set V, the influence probability matrix X and the particular time step t; the output is the influential degree of all user agents, i.e.,

idg_{v_i}, where $\forall v_i \in V$. Lines 1–2 replicate all users in the society take actions simultaneously at time t. Once v_i takes action, the system would calculate x_{ji} in lines 3–4. Lines 5–9 aim to calculate the influential degree of all users. As the process requires to iterate all users, the complexity is $O(n^2)$, where n denotes the number of users in the system.

4.3   Dynamic Game-Based Incentive Allocation (DGIA) Algorithm

To effectively allocate incentives to users, we design a Dynamic Game-based Incentive Allocation (DGIA) algorithm, which considers affecting easily affected users in priority. The DGIA consists of two parts, the Probability Learning Phase, and the Incentives Allocation Phase. The former phase aims to learn the probability $p_{v_i,t}$ that v_i taking the action a^*, and the latter phase is to allocate the incentives to users.

Probability learning phase aims to learn the probability of a user accepting the incentive at each time step. In the dynamic games with incomplete information, as the decision maker cannot grasp the probability of one event, he would make a decision according to his beliefs in general. Such beliefs actually are probability distribution supported by the decision maker's knowledge or experience, and the beliefs would be updated by considering relevant evidence or background (i.e. the result of the event). In the proactive recommendation, the system cannot make a correct allocation of incentives with the incomplete knowledge about users. Thus, inspired by Bayes theorem [9], the probability that v_i accepts the incentive at each time step can be updated corresponding to $\pi_{v_i,t}$. For convenient explanation, we regard ω_{v_i} as the probability of v_i taking a^* based on u_{v_i,a^*}, and $p_{v_i,t-1}$ and $p_{v_i,t}$ as the prior and posterior probability, respectively. As described in Eq. (10), the calculation of $p_{v_i,t}$ is determined by $\pi_{v_i,t}$, where γ is the attenuation coefficient. ω_{v_i} can calculated by using Eq. (11).

$$p_{v_i,t} = \begin{cases} \frac{p_{v_i,t-1}}{p_{v_i,t-1}+\omega_i \cdot (1-p_{v_i,t-1})}, & \pi_{v_i,t} = a^* \\ \gamma \cdot p_{v_i,t-1}, & \pi_{v_i,t} \neq a^* \end{cases} \tag{10}$$

$$\omega_i = \frac{u_{v_i,a^*}}{\sum_{a_m \in A} u_{v_i,a_m}} \tag{11}$$

Incentives allocating phase aims to provide proper incentives to users. To guarantee the completion of the system objective, the value of an incentive towards a particular user should be adjustable according to the user's feature and the status of environment. Hence, the value $r_{v_i,t}$ should be adjustable by the GAUP μ_t, influential degree idg_{v_i}, the probability $p_{v_i,t-1}$ and the preference difference σ_{v_i}, as described in Equation Eq. (12).

$$r_{v_i,t} = (1 - p_{v_i,t-1}) \cdot ((\sigma_{v_i})^{\mu_t-1} + (idg_{v_i})^{\mu_t-1}) \tag{12}$$

Overall, it can be seen that the value of an incentive is determined by following these rules: (1) The user who has large dissatisfaction to a^* may gains

---

Algorithm 2. The DGIA Algorithm

    Input: V, B, μ_{t-1}, t

    Output: μ_t

1 Initialize $B_t = B$;

2 for *each user agent v_i at time step $t (v_i \in V)$* **do**

3 $r_{v_i,t} := 0$;

4   Calculate $r_{v_i,t}$ using Eq. (12);

5   **if** $B_t < r_{v_i,t}$ **then**

6 $\mid$ $r_{v_i,t} = B_t$;

7   Allocate $r_{v_i,t}$ to v_i and observe $\pi_{v_i,t}$;

8   **if** $\pi_{v_i,t} == a^*$ **then**

9 $\mid$ $B_t = B_t - r_{v_i,t}$;

10   Calculate $p_{v_i,t}$ using Eq. (10) according to $\pi_{v_i,t}$;

11 $sum := 0$;

12 for *each user agent v_i $(v_i \in V)$* **do**

13   **if** $\pi_{v_i,t} == a^*$ **then**

14 $\mid$ $sum = sum + 1$;

15 $\mu_t := sum/|V|$;

16 Sort V in descending order of the sum of $p_{v_i,t}$ and idg_{v_i};

---

more incentives; (2) The user who has strong influential ability to others may receive more incentives; (3) The user who is easily activated would receive fewer incentives; (4) Users would receive fewer incentives when the completion degree of system objective is higher. In addition to these four rules, to maximize the efficiency of the use of budget, the system prioritizes to provide incentives to users who are easily accepting a^* and has a strong influential ability.

The DGIA algorithm is shown by Algorithm 2. The inputs include user agents set V, the budget amount per time step B, the last time GAUP μ_{t-1}, and current time steps t; the output is current GAUP μ_t. Line 1 initializes B_t for allocation in time step t. Lines 3–7 calculate the value of an incentive and allocate the incentive to v_i. Lines 8–9 update B_t based on v_i's behaviors. Line 10 updates the probability that v_i accepts incentives. After all users making decisions, the GAUP of current time step would be calculated in Lines 11–15. Line 16 sorts the order of v_i. Note if DGIA is deployed without IPE, this sort function would be executed only based on $p_{v_i,t}$. As the worst-case time complexity of Algorithm 2 is mainly determined by the loops started from Line 2 and Line 12, respectively. Therefore, the complexity is only $O(n)$.

5 Experimental Studies

5.1 Experimental Settings

To evaluate the performance of the proposed approach, we conducted experiments in comparison with other incentive allocation approaches. In the experiments, we create a number of user agents to represent users, and each of them has own neighbors and preferences to actions. Only user agents' preference and their decisions at each time steps are capable captured by the system. We utilize **Ego-Facebook**[1] [10] dataset to establish a social network, which contains 10 anonymized ego-networks, 4039 users and 88234 edges. The average number of edges between users is 21.6. As the public dataset only contains nodes and edges, we assign a random weight for each edge before the experiment, and the sum of weights of edges from $v_j \in N(v_i)^{in}$ to v_i cannot exceed 1, i.e., $\sum_{v_j \in N(v_i)^{in}} w_{ji} \leq 1$. Each user agent has to make a decision at every time step, and his preference towards each action is assigned a random value from 0 and 1 as well.

To evaluate the performance of the proposed approach, we set the time span is 150, the coefficient $\beta = 0.1$, and the attenuation coefficient $\gamma = 0.8$. We also set the number of action options is 2, i.e., select a^* or not. Furthermore, $x_{i,j}$ and μ_0 are both initialized as 0, and $p_{v_i,0}$ of each v_i is initialized as 0.5 at the beginning. Concurrently, two major metrics are utilized for the evaluation process, i.e., GAUP and GIAC. The GAUP has been formulated in Eq. (1), and the GIAC has been formulated in Eq. (3), respectively. We compare the proposed DGIA-IPE approach with following approaches:

- **Uniform Allocation** provides fixed and uniform incentives to all users. The value of an incentive is determined by the budget amount and the number of users.
- **ϵ-first** is an approach based on budgeted Multi-Armed Bandit approach, which splits the budget into two parts, for exploration and exploitation, respectively [17]. They used a uniform pull policy for exploration, i.e., the number of times for incentivizing each user is the same, and the reward-cost ratio ordered greedy algorithm for exploitation. In this experiment, we utilized the same settings for ϵ-first MAB.
- **DBP-UCB** is proposed for engaging users in the system to participate in the bike re-positioning process [16]. DBP-UCB is a dynamic pricing mechanism, which can determine the incentives from a finite price list. Hence, according to the features of ADM model, we set the price list as $\{0, 0.25, 0.5, 0.75, 1, 1.25, 1.5\}$.
- **DGIA** is a part of the proposed approach, which allocates incentives without considering the influential degree of users.

5.2   Experimental Results

We conducted a series of experiments under different budget constraints to evaluate the performance of the proposed approach, and demonstrate two representative experimental results when $B \in \{80, 240\}$. The budget would be reloaded at the beginning of each time step. Figure 2 compares the GAUP of approaches under different budget constraints. When $B = 80$, DGIA outperforms other approaches obviously, whereas DGIA-IPE performs worse at the beginning and

---

[1] https://snap.stanford.edu/data/ego-Facebook.html.

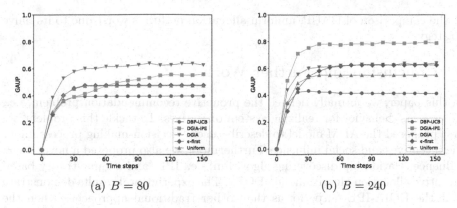

Fig. 2. Comparison of GAUP under different budget constraints

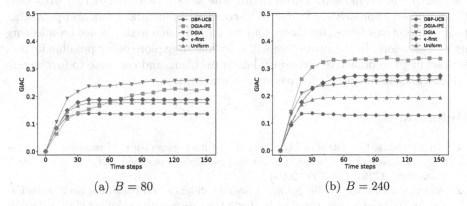

Fig. 3. Comparison of GIAC under different budget constraints

becomes better than other three approaches gradually. The possible reason is that insufficient budget affects the performance of the influence estimation as well as incentive allocation, since DGIA-IPE obviously outperforms other approaches and the GAUP of which reaches around 0.8 when $B = 240$. By contrast, applying uniform allocation can affect limited users' behaviors and performs worse than other approaches. The performance of ϵ-first is same as DBP-UCB when $B = 80$. However, ϵ-first performs as good as DGIA when the budget is sufficient, and outperforms DBP-UCB.

Then, we also compare the GIAC of approaches under the same budget constraints, which reflects the percentage of activated users who are driven by social influence. As described in Fig. 3, though the GIAC of DGIA-IPE is not good as that of DGIA approach with the insufficient budget, it performs obviously better than other approaches when the budget is sufficient. By contrast, DBP-UCB and ϵ-first affect users mainly based on incentives since their trends of the GIAC are stable at a very low stage. However, to our surprise, the GIAC of ϵ-first is similar as that of DGIA and better than DBP-UCB when the budget is sufficient. Akin to the comparison of GAUP, uniform allocation performs worst due to its naive strategy.

6   Conclusion and Further Work

In this paper, we formally defined the proactive recommendation problem, i.e., affect users' behavior for realizing system objectives. To tackle this problem, we first proposed the ADM model to describe users' decision-making process under the incentives and social influence. Furthermore, we also proposed a novel social influence relationship discovering algorithm, i.e., IPE, and a game-theory based incentive allocation algorithm, i.e., DGIA. The experimental results demonstrate that the DGIA-IPE outperforms than other traditional approaches when the budget is sufficient, whereas the DGIA is the most effective approach when the budget is insufficient. Furthermore, the GIAC of DGIA-IPE proves that the proposed approach is capable of discovering influential users and providing proper incentives to engage them, and social influence makes sense on affecting users' behaviors. In the future, we will continue to explore other possible factors for tackling the proactive recommendation problem, and continue to investigate the mechanism for more effective incentive allocation.

References

1. Adomavicius, G., Tuzhilin, A.: Toward the next generation of recommender systems: a survey of the state-of-the-art and possible extensions. IEEE Trans. Knowl. Data Eng. **17**(6), 734–749 (2005)
2. Axsen, J., Orlebar, C., Skippon, S.: Social influence and consumer preference formation for pro-environmental technology: the case of a uk workplace electric-vehicle study. Ecol. Econ. **95**, 96–107 (2013)
3. Biswas, A., Jain, S., Mandal, D., Narahari, Y.: A truthful budget feasible multi-armed bandit mechanism for crowdsourcing time critical tasks. In: Proceedings of the 2015 International Conference on Autonomous Agents and Multiagent Systems, pp. 1101–1109 (2015)
4. Bobadilla, J., Ortega, F., Hernando, A., Gutiérrez, A.: Recommender systems survey. Knowl.-Based Syst. **46**, 109–132 (2013)
5. Breese, J.S., Heckerman, D., Kadie, C.: Empirical analysis of predictive algorithms for collaborative filtering. In: Proceedings of the Fourteenth Conference on Uncertainty in Artificial Intelligence, pp. 43–52 (1998)
6. Felfernig, A., et al.: Persuasive recommendation: serial position effects in knowledge-based recommender systems. In: de Kort, Y., IJsselsteijn, W., Midden, C., Eggen, B., Fogg, B.J. (eds.) PERSUASIVE 2007. LNCS, vol. 4744, pp. 283–294. Springer, Heidelberg (2007). https://doi.org/10.1007/978-3-540-77006-0_34
7. Granovetter, M.: Threshold models of collective behavior. Am. J. Sociol. **83**(6), 1420–1443 (1978)
8. Homans, G.C.: Social Behavior: Its Elementary Forms. Harcourt Brace Jovanovich, San Diego (1974)
9. Koch, K.-R.: Bayes' theorem. Bayesian Inference with Geodetic Applications. LNES, vol. 31, pp. 4–8. Springer, Heidelberg (1990). https://doi.org/10.1007/BFb0048702

10. Leskovec, J., Mcauley, J.J.: Learning to discover social circles in ego networks. In: Pereira, F., Burges, C.J.C., Bottou, L., Weinberger, K.Q. (eds.) Advances in Neural Information Processing Systems, vol. 25, pp. 539–547. Curran Associates, Inc., New York (2012)
11. Mohan Raj, P., et al.: Brand preferences of newspapers-factor analysis approach. Res. J. Econ. Bus. Stud. 5(11), 17–26 (2016)
12. Pazzani, M.J.: A framework for collaborative, content-based and demographic filtering. Artif. Intell. Rev. 13(5–6), 393–408 (1999)
13. Ricci, F., Rokach, L., Shapira, B.: Recommender systems: introduction and challenges. In: Ricci, F., Rokach, L., Shapira, B. (eds.) Recommender Systems Handbook, pp. 1–34. Springer, Boston, MA (2015). https://doi.org/10.1007/978-1-4899-7637-6_1
14. Sengvong, S., Bai, Q.: Persuasive public-friendly route recommendation with flexible rewards. In: 2017 IEEE International Conference on Agents (ICA), pp. 109–114 (2017)
15. Shani, G., Gunawardana, A.: Evaluating recommendation systems. In: Ricci, F., Rokach, L., Shapira, B., Kantor, P.B. (eds.) Recommender Systems Handbook, pp. 257–297. Springer, Boston (2011). https://doi.org/10.1007/978-0-387-85820-3_8
16. Singla, A., Santoni, M., Bartók, G., Mukerji, P., Meenen, M., Krause, A.: Incentivizing users for balancing bike sharing systems. In: Proceedings of the Twenty-Ninth AAAI Conference on Artificial Intelligence, pp. 723–729 (2015)
17. Tran-Thanh, L., Chapman, A., Munoz De Cote Flores Luna, J.E., Rogers, A., Jennings, N.R.: Epsilon-first policies for budget-limited multi-armed bandits. In: Proceedings of the Twenty-Fourth AAAI Conference on Artificial Intelligence, pp. 1211–1216 (2010)
18. Tran-Thanh, L., Chapman, A.C., Rogers, A., Jennings, N.R.: Knapsack based optimal policies for budget-limited multi-armed bandits. In: Proceedings of the Twenty-Sixth AAAI Conference on Artificial Intelligence, pp. 1134–1140 (2012)
19. Tran-Thanh, L., Stein, S., Rogers, A., Jennings, N.R.: Efficient crowdsourcing of unknown experts using bounded multi-armed bandits. Artif. Intell. 214, 89–111 (2014)
20. Wu, S., Bai, Q., Sengvong, S.: GreenCommute: an influence-aware persuasive recommendation approach for public-friendly commute options. J. Syst. Sci. Syst. Eng. 27(2), 250–264 (2018)
21. Yu, C., Zhang, M., Ren, F., Luo, X.: Emergence of social norms through collective learning in networked agent societies. In: Proceedings of the 2013 International Conference on Autonomous Agents and Multi-Agent Systems, pp. 475–482 (2013)

Incentivizing Long-Term Engagement Under Limited Budget

Shiqing Wu[1(✉)] and Quan Bai[2]

[1] Auckland University of Technology, Auckland, New Zealand
shiqing.wu@aut.ac.nz
[2] University of Tasmania, Hobart, Australia
quan.bai@utas.edu.au

Abstract. In recent years, more and more systems have been designed to affect users' decisions for realizing certain system goals. However, most of these systems only focus on affecting users' short-term or one-off behaviors, while ignoring the maintenance of users' long-term engagement. In this light, we intend to design a novel approach which focuses on incentivizing users' long-term engagement. In this paper, inspired by the use of Markov Decision Process (MDP), we first formally model the process of a user's decision-making under long-term incentives. Subsequently, we propose the MDP-based Incentive Estimation (MDP-IE) approach for determining the value of an incentive and the requirement of obtaining that incentive. Experimental results demonstrate that the proposed approach can effectively sustain users' long-term engagement. Furthermore, the experiments also demonstrate that incentivizing users' long-term engagement is more beneficial than one-off or short-term approaches.

Keywords: Incentive allocation · Long-term engagement · Agent-based modeling · Markov Decision Process

1 Introduction

In recent years, an increasing number of systems have been developed to affect users' behaviors to realize certain system objectives [3,16,18]. Leveraging the power of incentives, such systems show great performance in affecting users' behaviors and achieving desired system objectives. However, these approaches only focus on affecting users' short-term or one-off behaviors [10,12,19]. In general, incentivizing users' long-term engagement is typically more beneficial. From a business perspective, though affecting users' short-term behaviors can bring sudden profits, relatively, the cost of maintaining such effects long-term could be prohibitive. For example, it is common that shopping malls would hold large discount events on special days for attracting customers, but such events cannot be sustained for a long time due to their cost. By contrast, incentivizing users' long-term engagement is dominant in maintaining the choices of customers as it

© Springer Nature Switzerland AG 2019
A. C. Nayak and A. Sharma (Eds.): PRICAI 2019, LNAI 11670, pp. 662–674, 2019.
https://doi.org/10.1007/978-3-030-29908-8_52

emphasizes long-term effect and sustainable business goals [7]. More specifically, promoting users' long-term engagement is helpful in cultivating users' habits and increasing brand awareness then continually produces profits; whereas purely applying short-term strategies only generates ephemeral successes which cannot last [9]. Motivated by this background, in this paper, we aim to design a novel incentive allocation strategy, for incentivizing users' long-term engagement.

To successfully incentivize users' long-term engagement, some problems must be considered and tackled. A major problem is how to design a proper pricing policy to maximize the benefits to both users and the system. Overpricing would lead to the inefficient use of the budget, whereas underpricing may hardly attract users' attention [11]. The second problem is how to determine the requirement that a user is able to obtain the incentive. A simple strategy is to provide users a long-term incentive, i.e., the user can obtain the incentive when he continually engages and meets a specific requirement [8]. However, such strategies are insufficient to attract users who only pay attention to the short-term benefits. In other words, from the users' perspective, their long-term behaviors would be affected by not only the value of an incentive but also their attitude towards the long-term incentive. Most existing approaches provide a short-term or one-off incentive and determine the values of incentive only based on the status of users and the system [3,10–12,17]. Obviously, these approaches are insufficient for supporting to incentivize users' long-term engagement, and it is necessary to design a novel strategy for this purpose.

Different from these approaches, in this study, we take users' preferences as well as their attitude towards long-term incentives into consideration when designing the strategy of allocating incentives. To better investigate this problem, we adopt Agent-based Modeling (ABM) [2] to model users, where each user is modeled as an anonymous agent and chooses behaviors following their own preferences. Meanwhile, inspired by the Markov Decision Process (MDP) [1], we propose an Agent-based Long-term Decisions Making (ALDM) model which describes the process of users' decision-making under long-term incentives. Furthermore, we also introduce the MDP-based Incentive Estimation (MDP-IE) algorithm for determining the value of an incentive and the requirement of obtaining that. The experimental results demonstrate that: (1) incentivizing users' long-term engagement is more beneficial than using one-off incentives; and (2) the proposed approach outperforms other incentive allocation approaches under the same budget constraints and time span. To summarize, the contributions of this study are as follows:

- We formally modeled the process of a user's decision-making under long-term incentives while taking the user's attitude towards the long-term incentive into consideration, which is significantly different from traditional incentive-based models.
- We proposed a novel approach which can explore the minimum value of an incentive, and effectively incentivize users' long-term engagement.

The remainder of this paper is organized as follows. Section 2 reviews the literature related to this study. Section 3 introduces the problem description and formal definitions. Section 4 presents the proposed approach for incentivizing users' long-term engagement. We demonstrate the experimental results for evaluating the performance of the proposed approach in Sect. 5 and conclude this paper in Sect. 6.

2   Related Work

Providing incentives is an effective strategy in affecting users' decision-making in many fields, since users tend to take an action when they recognize that they would be rewarded from such an action [4]. Inspired by this, a number of research works related to the incentive problem have been conducted in recent years. Some researchers propose diverse pricing mechanisms for providing effective incentives [3,11,18]. Sengvong and Bai [10] consider affecting users' decisions on selections of transportation modes by providing reward based on the utility difference between the system and users. To affect more users' decision-making under a budget restriction, Wu et al. [17] take the social influence into consideration when determining the value of incentives. Singla et al. [12] attempt to solve the balance problem of sharing-bike by incentivizing users. They determine whether to incentivize a user according to the user's location and the status of nearby bike stations, and their proposed approach DBP-UCB can effectively solve this problem. On the other hand, the cost of incentivizing a user is unknown in advance, and It requires exploration to determine an effective incentive. To address this problem, some researchers regard such a pricing problem as a Multi-Armed Bandit (MAB) problem. They model each price option as a lever, and find out the optimal lever via a series of exploration and exploitation [12,16]. However, these approaches only focus on affecting users' one-off behaviors, whereas our objective is to incentivize users' long-term engagement.

To obtain a long-term incentive, the user needs to choose a specific action continually to satisfy the condition of incentives. Such a process of user' decision-making can be modeled by using Markov Decision Process (MDP) [1], which provides a mathematical framework and has been utilized for solving the optimization problem in many fields [5,6,13]. Besides, users' attitude towards long-term incentives is essential to determine the incentive, since a user who only focuses on short-term profit is difficult to be incentivized due to a long-term incentive. Hence, to determine the condition that a user may obtain the incentive, we also need to explore users' attitude towards long-term incentives.

3   Preliminaries

3.1   Problem Description

The main goal of this research is to design an effective incentive allocation strategy for incentivizing users' long-term engagement. This strategy can be adopted in scenarios where the system expects users to take a specific behavior continually, e.g, encourage users to take a bus daily. Suppose we are going to incentivize

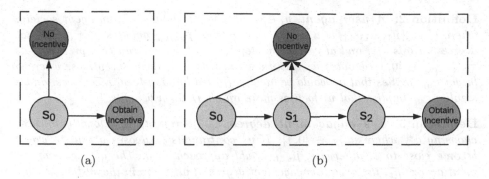

Fig. 1. The process of a user obtaining the incentive. (Color figure online)

a user to take the bus for the commute, the process of the user's decision making can be described by Fig. 1, where each link denotes an action option (e.g., driving or taking a bus), and each yellow node represents a state that the user at (e.g., s_2 denotes the user has continually taken a bus two times). As we can see from Fig. 1(a), the user can obtain an immediate incentive when he takes the bus if we want to incentivize his one-off behaviors. Whereas if we provide the user with a long-term incentive, as shown in Fig. 1(b), the user only obtains an incentive when he continually takes three times bus, otherwise, he would obtain nothing.

Apparently, the process of a user's decision-making in Fig. 1(b) can be modeled as a MDP. The user may hardly to be incentivized when the condition is too difficult since the user cannot obtain incentives in the middle unless the system provides very attractive incentives. Meanwhile, the value of an incentive cannot be infinite due to the constraint of the budget. Hence, the value of an incentive provided to each user should have an **upper bound** and a **lower bound**, where the upper bound denotes the maximum incentive that the system can provide, and the lower bound denotes the minimum incentive to successfully incentivize the user. Furthermore, in addition to the value of incentives, the condition of a user obtaining the incentive also needs to be considered carefully. We will introduce the proposed approach in Sect. 4.

3.2 Formal Definition

Before introducing the proposed approach, we first give formal definitions and notations which would be frequently mentioned in this paper.

Definition 1. *An action, $a_m \in A$, denotes an available option that the system provides to users for choosing, where $A = \{a_1, ..., a_m\}$ is a finite set of all action options. Each a_m has a binary state b_{a_m}, i.e., $b_{a_m} \in \{0, 1\}$, where 1 implies a_m is a beneficial action to the system, and 0 implies conversely. For easier explanation, we use a^* to represent the action that the system expects users to choose.*

Definition 2. *A **user agent**, $u_i \in U$, denotes a member in an agent society, where $U = \{u_1, ..., u_n\}$ is a set of user agents. p_{u_i, a_m} denotes u_i's preference degree towards a_m, and at each time step t, u_i's behavior can be represented as $\pi_{u_i, t}$. $\gamma_{u_i} \in [0, 1]$ denotes u_i's attitude towards long-term incentives, where a larger γ_{u_i} implies that u_i would be highly affected by long-term incentives, and a smaller γ_{u_i} implies that u_i focuses more on short-term benefits.*

Definition 3. *u_i's **engagement degree**, $\mu_{u_i, t}$, reflects u_i's loyalty towards choosing a^*, where $\mu_{u_i, t} \in [0, 1]$. If u_i continuously chooses a^*, $\mu_{u_i, t}$ would become close to 1, otherwise, $\mu_{u_i, t}$ would approach to 0. The average engagement degree, $\overline{\mu_t}$, the overall engagement degree of all users in the system at time step t.*

$\mu_{u_i, t}$ is formulated by using Eq. (1), where $\mu_{u_i, t-1}$ denotes u_i's engagement degree at last time step, $b_{\pi_{u_i, t}}$ implies to the state of $\pi_{u_i, t}$, and β is a fixed coefficient for adjusting the engagement degree.

$$\mu_{u_i, t} = \mu_{u_i, t-1} + \beta \cdot (b_{\pi_{u_i, t}} - \mu_{u_i, t-1}) \tag{1}$$

Subsequently, average engagement degree of all users $\overline{\mu_t}$ can be calculated by using Eq. (2), where $|U|$ denotes the number of user agents in the system.

$$\overline{\mu_t} = \frac{\sum\limits_{u_i \in U} \mu_{u_i, t}}{|U|} \tag{2}$$

Definition 4. *An **incentive**, $R(d_{u_i})$, is used to promote u_i to choose a^*, where d_{u_i} denotes the number of times that u_i needs to choose a^* continually. c_{u_i} denotes the number of times that u_i has already chose a^* continually, and u_i can only obtain the incentive when $c_{u_i} = d_{u_i}$. $R(d_{u_i})$ is restricted by the current remaining budget B, i.e., $R(d_{u_i}) \leq B$. The upper bound and the lower bound of $R(d_{u_i})$ can be represented as $R^{upper}(d_{u_i})$ and $R^{lower}(d_{u_i})$, and $R(d_{u_i}) \in [R^{lower}(d_{u_i}), R^{upper}(d_{u_i})]$.*

Definition 5. *The **influence of an incentive** towards u_i can be represented as $I(c_{u_i}, d_{u_i}, R(d_{u_i}))$. It varies according to the difference between c_{u_i} and d_{u_i}.*

Inspired by MDP, $I(c_{u_i}, d_{u_i}, R(d_{u_i}))$ can be calculated by using Eq. (3). The influence of an incentive would be minimum when u_i has not chose a^* yet (i.e., $c_{u_i} = 0$), and becomes larger with the decreasing difference between c_{u_i} and d_{u_i}. For example, when $c_{u_i} = 2$ and $d_{u_i} = 3$, the influence of the incentive becomes the largest, since u_i would obtain that incentive after choosing a^*.

$$I(c_{u_i}, d_{u_i}, R(d_{u_i})) = \gamma_{u_i}^{(d_{u_i} - c_{u_i} - 1)} \cdot R(d_{u_i}) \tag{3}$$

3.3   Agent-Based Long-Term Decisions Making (ALDM) Model

The ALDM model describes the process of a user' decision-making under long-term incentives. In this model, benefited from ABM, each user is modeled as

---

Algorithm 1. The process of a user's decision-making under the ALDM

    Initialize: $c_{u_i} := 0$;

1 **for** *each time step t $(t = 1, ..., T)$* **do**

2   **if** $c_{u_i} == 0$ **then**

3 | Receive an offer of incentive $R(d_{u_1})$;

4    $temp := p_{u_i,a^*}$;

5   Calculate $I(c_{u_i}, d_{u_i}, R(d_{u_1}))$ using Eq. (3);

6   Update p_{u_i,a^*} using Eq. (5);

7    $\pi_{u_i,t} := \arg\max\limits_{a_m \in A} p_{u_i,a_m}$;

8    $p_{u_i,a^*} = temp$;

9   **if** $\pi_{u_i,t} == a^*$ **then**

10 | $c_{u_i} + +$;

11   **if** $c_{u_i} == d_{u_i}$ **then**

12 | $c_{u_i} = 0$;

13   Receive the incentive;

14   **else**

15 | $c_{u_i} = 0$;

---

an autonomous agent, which behaves based on own preference and the influence of the incentive. In the real world, a user would choose a more preferable item among several provided items. Hence, we assume that u_i would take a_m with the highest p_{u_i,a_m}, as described in Eq. (4), where $\pi_{u_i,t}$ denotes the action that u_i takes at time step t.

$$\pi_{u_i,t} = \arg\max_{a_m \in A} p_{u_i,a_m} \tag{4}$$

Meanwhile, p_{u_i,a^*} can be reinforced by an incentive, which is described in Eq. (5), where $I(c_{u_i}, d_{u_i}, R(d_{u_i}))$ denotes the influence of $R(d_{u_i})$.

$$p_{u_i,a^*} = p_{u_i,a^*} + I(c_{u_i}, d_{u_i}, R(d_{u_i})) \tag{5}$$

The process of each u_i's decision-making can be described in Algorithm 1. Suppose each u_i would take an action at each time step $t \in [1, T]$. In the beginning, c_{u_i} is first initialized as 0, representing u_i has not taken a^* yet. Lines 2–3 provide an offer of $R(d_{u_i})$ to u_i when $c_{u_i} = 0$. The approach for determining $R(d_{u_i})$ and d_{u_i} would be introduced in the next section. Lines 4–8 update p_{u_i,a^*} with considering the influence of the incentive $I(c_{u_i}, d_{u_i}, R(d_{u_i}))$. Among them, Line 7 updates u_i's behavior $\pi_{u_i,t}$, and line 8 restores p_{u_i,a^*}. Lines 9–15 update c_{u_i} according to $\pi_{u_i,t}$.

4 Incentive Estimation Approach

4.1 The Bounds for the Incentive

To incentivize users' long-term engagement effectively, the value of an incentive must be appropriate determined to ensure, since overpricing would lead to the inefficient use of the budget and underpricing may cause failure of attracting users [11]. To avoid happening of these two situations, we formulate a range for the value of incentives, aiming to ensure the budget would not be spent infinitely, and prevent the system to provide the user an insufficient incentive.

We first set the **upper bound** for an incentive, which is determined by two factors, i.e., the average engagement degree $\overline{u_{t-1}}$ and the required number d_{u_i}. The system would provide more incentives for attracting more users' engagement when $\overline{u_{t-1}}$ is small, and conversely when it becomes higher. Meanwhile, the user who needs to take long-term beneficial behaviors would receive more incentives. Hence, the upper bound for an incentive is formulated in Eq. (6), where d_{u_i} denotes the number of times that u_i needs to choose a^* continually for obtaining the incentive, and $ln(d_{u_i})$ denotes an extra bonus.

$$R^{upper}(d_{u_i}) = (1 - \overline{u_{t-1}}) \cdot (d_{u_i} + ln(d_{u_i})) \tag{6}$$

According to Eq. (4), u_i would choose an action with the highest p_{u_i,a_m} at each time step. Suppose $a' = \arg\max_{a_m \in A} p_{u_i,a_m}$. To ensure that u_i can be successfully incentivized, the influence of $R(d_{u_i})$ must exceed the **preference gap**, i.e., the difference between $p_{u_i,a'}$ and p_{u_i,a^*}. The preference gap g_{u_i} can be formulated in Eq. (7).

$$g_{u_i} = p_{u_i,a'} - p_{u_i,a^*} \tag{7}$$

Subsequently, based on Eq. (3), the **lower bound** for an incentive can be calculated by using Eq. (8), where γ_{u_i} denotes u_i's attitude towards a long-term incentive and d_{u_i} denotes the required number of a^* that u_i needs to take continually for obtaining this incentive.

$$R^{lower}(d_{u_i}) = \frac{g_{u_i}}{\gamma_{u_i}^{(d_{u_i}-1)}} \tag{8}$$

However, since g_{u_i} and γ_{u_i} of each user are unknown in advance, we have to explore and estimate the value of these two variables via a series of explorations. In this case, we propose the MDP-based Incentive Estimation (MDP-IE) algorithm, which would be introduced in the next subsection.

4.2   The MDP-Based Incentive Estimation (MDP-IE) Algorithm

The objective of the MDP-IE approach is to incentivize u_i with exploring the minimum of R_{u_i} as well as the optimal d_{u_i}. As we mentioned in Eq. (8), $R(d_{u_i})^{lower}$ is determined by g_{u_i} and γ_{u_i}. Hence, to explore values of these two unknown variables, we first attempt to incentivize u_i' one-off behaviors for a period, i.e., $d_{u_i} = 1$. We can utilize Bisection method to determine $R(1)$, and then estimate the bounds for g_{u_i} based on u_i' behavior, i.e., the upper bound $g_{u_i}^+$ and lower bound $g_{u_i}^-$, and g_{u_i} satisfies that $g_{u_i} \in (g_{u_i}^-, g_{u_i}^+]$. After this period of exploration, we can obtain a relatively accurate $g_{u_i}^+$ which is close to g_{u_i}. Then, we can explore the estimated γ_{u_i}' by providing $R(d_{u_i})(d_{u_i} > 1)$, and adjust γ_{u_i}'

Algorithm 2. The MDP-IE Algorithm

Initialize : Budget B; exploration time t^{exp};
average engagement degree $\overline{\mu_0} := 0$; adjusting rate ω;
for *each* $u_i \in U$ **do**
    $d_{u_i} := 1$; $R(d_{u_i}) := 0.0$; $c_{u_i} := 0$; $\mu_{u_i,0} := 0.0$;
    $g_{u_i}^+ := 1.0$; $g_{u_i}^- := 0.0$ $\gamma_{u_i}' := 0.5$;

1   **for** *each time step* $t \in [1, T]$ **do**
2   **for** *each request by* $u_i \in U$ *at time step* t **do**
3   **if** $c_{u_i} == 0$ *and* $B > 0$ **then**
4   **if** $t > t^{exp}$ *and* $d_{u_i} \neq 1$ **then**
5   **if** $\gamma_{u_i}' < {}^{d_{u_i}-1}\sqrt{\frac{1}{d_{u_i}}}$ **then**
6                 $d_{u_i} - -$;
7   **else if** $\gamma_{u_i}' > \frac{d_{u_i}}{d_{u_i}+1}$ **then**
8                 $d_{u_i} + +$;
9   **else**
10               $d_{u_i} = 1$;
11   Calculate $R^{upper}(d_{u_i})$ using Eq. (6);
12   Calculate $R(d_{u_i})$ using Eq. (9);
13            $R(d_{u_i}) = \min(R^{upper}(d_{u_i}), R(d_{u_i}))$;
           Output: Offer u_i the incentive $R(d_{u_i})$;
           Feedback: Observe u_i's behavior $\pi_{u_i,t}$;
14   **if** $\pi_{u_i,t} == a^*$ **then**
15   **if** $c_{u_i} == d_{u_i}$ **then**
16               $d_{u_i}++$;
17   **if** $d_{u_i} > 1$ **then**
18                 $\gamma_{u_i}' = \min(\gamma_{u_i}' \cdot (1 + \omega), 1.0)$;
19   **else**
20                 $g_{u_i}^+ = \min(g_{u_i}^+, R(d_{u_i}))$;
21               $B- = R(d_{u_i})$;
22   **else**
23   **if** $d_{u_i} > 1$ **then**
24               $\gamma_{u_i}' = \max(\gamma_{u_i}' \cdot (1 - \omega), 0.0)$;
25   **else**
26               $g_{u_i}^- = \max(g_{u_i}^-, R(d_{u_i}))$;
27   Update $\mu_{u_i,t}$ using Eq. (1);
28   Update $\overline{\mu_t}$ using Eq. (2);

---

based on u_i's behavior. The value of $R(d_{u_i})$ can be formulated by using Eq. (9).

$$R(d_{u_i}) = \begin{cases} 0.5 \cdot (g_{u_i}^+ + g_{u_i}^-), \, d_{u_i} = 1 \\ \frac{g_{u_i}^+}{(\gamma_{u_i}')^{(d_{u_i}-1)}}, \qquad d_{u_i} > 1 \end{cases} \qquad (9)$$

The procedure of MDP-IE approach is described in Algorithm 2. In the beginning, we first initialize the overall budget B, the exploring time t^{exp} for incentivizing users' one-off behaviors, and the adjusting rate ω for adjusting γ_{u_i}'.

Meanwhile, the variables of each user should be initialized as well. In this algorithm, $\min(\cdot)$ and $\max(\cdot)$ denote functions returning the minimum and maximum of arguments, respectively. At each time step t, each user agent would interact with the system and then choose an action to behave; while the system provides each u_i an incentive $R(d_{u_i})$ according to the knowledge about the user, and then updates the corresponding variables based on u_i's behavior. Lines 4–10 determine d_{u_i}. Lines 11–13 calculate the incentive by comparing $R^{upper}(d_{u_i})$ and $R(d_{u_i})$. Then lines 14–21 update the corresponding variables when u_i chooses a^* and $c_{u_i} = d_{u_i}$, and lines 22–26 update the variables when u_i fails to choose a^* continually. Finally, line 27 calculates u_i's engagement degree, and line 28 calculates the average engagement degree. The computational complexity is $O(n)$.

The reasons we determine d_{u_i} and adjust γ'_{u_i} are presented as follows:

Lemma 1. *Given a known g_{u_i} and a known γ_{u_i}, and suppose $d_{u_i} = n, (n > 1)$. When $R(d_{u_i})^{upper} > R(d_{u_i})^{lower}$, incentivizing u_i's long-term engagement is more economic than incentivizing his one-off behavior n times if $\gamma_{u_i} > 0.5$.*

proof. Since g_{u_i} and γ_{u_i} are known, we can incentivize u_i's long-term engagement with $R^{lower}(d_{u_i})$, and one-off behavior with g_{u_i}, respectively. Then the inequality can be presented by using Eq. (10). Obviously, $\sqrt[n-1]{\frac{1}{n}}$ is a monotonically increasing function. Let $n = 2$, the lemma is proofed.

$$\frac{g_{u_i}}{\gamma_{u_i}^{(n-1)}} < n \cdot g_{u_i} \Rightarrow \frac{1}{\gamma_{u_i}^{(n-1)}} < n \Rightarrow \sqrt[n-1]{\frac{1}{n}} < \gamma_{u_i} \quad (10)$$

Lemma 2. *Suppose g_{u_i} and γ_{u_i} are known. If $R^{upper}(d_{u_i}) > R^{lower}(d_{u_i})$, set $d_{u_i} = n + 1$ is more economic than $d_{u_i} = n$ when $\gamma_{u_i} > \frac{n}{n+1}, (n > 1)$.*

proof. When $n > 2$, this lemma can be easily proofed by using Eq. (11). Then combined with Lemma 1, we can determine d_{u_i} based on γ'_{u_i} when γ_{u_i} is unknown, i.e., $\sqrt[n-1]{\frac{1}{n}} \leq \gamma'_{u_i} < \frac{n}{n+1}$.

$$\frac{g_{u_i}}{n \cdot \gamma_{u_i}^{(n-1)}} > \frac{g_{u_i}}{(n+1) \cdot \gamma_{u_i}^n} \Rightarrow \gamma_{u_i} > \frac{n}{n+1} \quad (11)$$

Lemma 3. *Suppose we have explored a $g_{u_i}^+$ and a γ'_{u_i}, and it always satisfies that $R(d_{u_i}) < R^{upper}(d_{u_i})$. If $R(d_{u_i})$ can incentivize u_i whereas $R(d_{u_i} + 1)$ cannot, then $\gamma_{u_i} < \gamma'_{u_i}$.*

proof. Since $R(d_{u_i})$ can incentivize u_i whereas $R(d_{u_i} + 1)$ cannot, we can obtain the inequality as described in Eq. (12). Then the lemma is proofed.

$$\frac{g_{u_i}}{\gamma_{u_i}^{(d_{u_i}-1)}} \leq \frac{g_{u_i}^+}{(\gamma'_{u_i})^{(d_{u_i}-1)}} \leq \frac{g_{u_i}^+}{(\gamma'_{u_i})^{(d_{u_i})}} < \frac{g_{u_i}}{\gamma_{u_i}^{(d_{u_i})}} \Rightarrow \frac{1}{\gamma_{u_i}} > \frac{1}{\gamma'_{u_i}} \Rightarrow \gamma_{u_i} < \gamma'_{u_i} \quad (12)$$

Hence, we increase γ'_{u_i} for reducing the value of the next incentive when u_i obtains $R(d_{u_i})$; and reduce γ'_{u_i} when he fails to be awarded.

5   Experimental Results

5.1   Experiment Setup

Before we start demonstrating the experimental results, the detailed setup would be introduced. In the experiments, we adopt 200 user agents to represent the users in the real world. p_{u_i,a_m} and γ_{u_i} of each user are randomly generated from 0 to 1. Each user agent has to choose an action at every time step and is not allowed to acquire any information about others. We also set β is 0.1 for adjusting u_i's engagement degree $\mu_{u_i,t}$. In this study, unless stated otherwise, the total number of time steps is 500, the default budget amount B is 7000, the number of all available action options is 8, and the number of a^* is 1. For the MDP-IE approach, we set the exploring time t^{exp} is 10 and the adjusting rate ω is 0.1. Besides, we also restrict that d_{u_i} cannot exceed 10.

To better evaluate the performance of the proposed approach, we compare the proposed approach with following approaches:

- **Optimal:** this approach can always provide a minimum incentive $R^{lower}(d_{u_i})$ with the optimal d_{u_i} to users as the information about users is known to the system. Namely, the performance of this approach is the most optimal.
- **One-off incentive:** the one-off incentive strategy implies that the system aims to incentivize users' one-time behaviors. Similar to the optimal approach, we assume this approach can obtain information about g_{u_i} in advance as well.
- **fKUBE+DBP-UCB:** fKUBE [15] and DBP-UCB [12] are two UCB-based algorithms for solving the budgeted Multi-Armed Bandit problem. Since these two approaches cannot individually determine $R(d_{u_i})$ and d_{u_i} at the same time, we utilize the fKUBE for determining d_{u_i}, and the DBP-UCB for pricing, respectively. For the DBP-UCB, we use the arithmetic progression to set the pricing options from 0.5 to 5.0, where the difference is 0.5.
- **ϵ-first+DBP-UCB:** ϵ-first [14] would split the budget into two parts for exploration and explication, respectively. In this experiment, we set $\epsilon = 0.1$ and use uniform policy for exploration. Akin to the combination of fKUBE and DBP-UCB, we utilize ϵ-first for choosing the requirement and DBP-UCB for pricing. We use the same pricing options for DBP-UCB as the combination of fKUBE and DBP-UCB.

5.2   Performance Evaluation

As we can observe from Fig. 2, MDP-IE can outperform other approaches even if the budget is not sufficient. Meanwhile, the maximum average engagement degree and the effective time of MDP-IE are close to the optimal result. It implies that the MDP-IE is effective in exploring the minimum $R(d_{u_i})$ as well as appropriate d_{u_i} for incentivizing user's long-time engagement. By contrast, while two combined MAB approaches both give a pretty close performance comparing with the MDP-IE under insufficient budget, the MDP-IE obviously superior

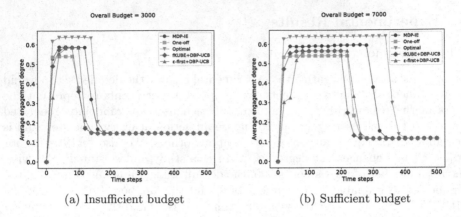

(a) Insufficient budget   (b) Sufficient budget

Fig. 2. Comparison of average engagement degree under different budget constraints.

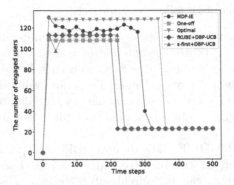

Fig. 3. Comparison of the number of engaged users at each time step, where $B = 7000$.

to MAB-based approaches when the budget is sufficient. Meanwhile, the effectiveness of MDP-IE lasts longer. This is due to the reason that MAB-based approaches incentivize users with too many incentives, and the budget runs out in advance. On the other hand, the one-off incentive approach can incentivize users shortly, but the effect cannot last. Furthermore, it also reflects that some users cannot be attracted by the one-off incentive as the other three approaches can all obtain a higher average engagement degree. A possible reason is that the extra bonus provided in long-term incentives attract users while one-off incentives are insufficient to incentivize them.

Then we also evaluate the performance of approaches by comparing the number of engaged users at each time step when the budget is sufficient. As described in Fig. 3, the proposed MDP-IE can successfully incentivize users longer times than the other three approaches. In addition, the proposed approach can incentivize more users in the initial phase. The possible reason leading to that is the system would spend more budget on exploring the bounds for g_{u_i} during that period. Furthermore, the proposed approach incentivizes around 120 users at each time step, whereas MAB-based approaches perform worse than the proposed approach. Meanwhile, under the same budget constraints, the effectiveness of MAB-based approaches end faster, which implies that MAB-based approaches fail to use of budget efficiently. Concurrently, the comparison between the proposed approach and one-off strategy indicates that some users who cannot be incentivized by one-off incentives are able to be incentivized by long-term incentives.

6   Conclusions and Future Works

In this paper, we proposed a novel strategy of incentivizing users' long-term engagement, which targets to keep users' long-term engagement as much as possible with a limited budget. The long-term decisions making model, i.e., the ALDM, presented in this paper can describe the change of users' preference due to the long-term incentives, and model users' behavior patterns as well. Furthermore, based on the Markov Decision Process, we also proposed the MDP-IE approach which is capable of exploring users' maximum preference difference and attitude towards long-term incentives. Then, to incentivize users effectively and efficiently, we determine the minimum value of incentives as well as requirements of obtaining those incentives. To evaluate the performance of MDP-IE, we compared it with four approaches. Given the same budget and time span, the MDP-IE can perform almost as good as the optimal result and outperform the other three approaches. The experimental results also prove that incentivizing users' long-term engagement is more beneficial than incentivizing users' one-off behaviors.

In future work, we will continue to enrich the details of the ALDM model by adding more external factors. Concurrently, we will attempt more forms of requirements, e.g., establish a kind of sequential requirement which requires users to take diverse actions. Furthermore, we will keep investigating the algorithms which are more effective for incentivizing users' long-term engagement.

References

1. Bellman, R.: A Markovian decision process. J. Math. Mech. **6**, 679–684 (1957)
2. Bonabeau, E.: Agent-based modeling: methods and techniques for simulating human systems. Proc. Nat. Acad. Sci. **99**(3), 7280–7287 (2002)
3. Gan, X., Wang, X., Niu, W., Hang, G., Tian, X., Wang, X., Xu, J.: Incentivize multi-class crowd labeling under budget constraint. IEEE J. Sel. Areas Commun. **35**(4), 893–905 (2017)
4. Homans, G.C.: Social Behavior: Its Elementary Forms. Harcourt Brace Jovanovich, San Diego (1974)
5. Iversen, E.B., Morales, J.M., Madsen, H.: Optimal charging of an electric vehicle using a Markov decision process. Appl. Energy **123**, 1–12 (2014)
6. Ksentini, A., Taleb, T., Chen, M.: A Markov decision process-based service migration procedure for follow me cloud. In: 2014 IEEE International Conference on Communications (ICC), pp. 1350–1354 (2014)

7. Li, W., Bai, Q., Zhang, M., Nguyen, T.D.: Automated influence maintenance in social networks: an agent-based approach. IEEE Trans. Knowl. Data Eng. (2018). https://doi.org/10.1109/TKDE.2018.2867774
8. Liu, Y.: The long-term impact of loyalty programs on consumer purchase behavior and loyalty. J. Mark. **71**(4), 19–35 (2007)
9. Launch Marketing: What are your short-and long-term marketing strategies (2015)
10. Sengvong, S., Bai, Q.: Persuasive public-friendly route recommendation with flexible rewards. In: 2017 IEEE International Conference on Agents, pp. 109–114 (2017)
11. Singla, A., Krause, A.: Truthful incentives in crowdsourcing tasks using regret minimization mechanisms. In: Proceedings of the 22nd International Conference on World Wide Web, pp. 1167–1178 (2013)
12. Singla, A., Santoni, M., Bartók, G., Mukerji, P., Meenen, M., Krause, A.: Incentivizing users for balancing bike sharing systems. In: Proceedings of the Twenty-Ninth AAAI Conference on Artificial Intelligence, pp. 723–729 (2015)
13. Terefe, M.B., Lee, H., Heo, N., Fox, G.C., Oh, S.: Energy-efficient multisite offloading policy using markov decision process for mobile cloud computing. Pervasive Mob. Comput. **27**, 75–89 (2016)
14. Tran-Thanh, L., Chapman, A., de Cote, E.M., Rogers, A., Jennings, N.R.: Epsilon-first policies for budget-limited multi-armed bandits. In: Proceedings of Twenty-Fourth AAAI Conference on Artificial Intelligence, pp. 1211–1216 (2010)
15. Tran-Thanh, L., Chapman, A., Rogers, A., Jennings, N.R.: Knapsack based optimal policies for budget-limited multi-armed bandits. In: Proceedings of Twenty-Sixth AAAI Conference on Artificial Intelligence, pp. 1134–1140 (2012)
16. Truong, N.V., Stein, S., Tran-Thanh, L., Jennings, N.R.: Adaptive incentive selection for crowdsourcing contests. In: Proceedings of the 17th International Conference on Autonomous Agents and Multiagent Systems, pp. 2100–2102 (2018)
17. Wu, S., Bai, Q., Sengvong, S.: GreenCommute: an influence-aware persuasive recommendation approach for public-friendly commute options. J. Syst. Sci. Syst. Eng. **27**(2), 250–264 (2018)
18. Yu, H., Miao, C., Chen, Y., Fauvel, S., Li, X., Lesser, V.R.: Algorithmic management for improving collective productivity in crowdsourcing. Sci. Rep. **7**, 12541 (2017). https://doi.org/10.1038/s41598-017-12757-x
19. Zhao, D., Li, B., Xu, J., Hao, D., Jennings, N.R.: Selling multiple items via social networks. In: Proceedings of the 17th International Conference on Autonomous Agents and Multiagent Systems, pp. 68–76 (2018)

A Compromising Strategy Based on Constraint Relaxation for Automated Negotiating Agents

Shun Okuhara(✉) and Takayuki Ito

Nagoya Institute of Technology, Gokiso, Showa-ku, Nagoya, Japan
okuhara@itolab.nitech.ac.jp

Abstract. This paper presents a compromising strategy based on constraint relaxation for automated negotiating agents. Automated negotiating agents have been studied widely and are one of the key technologies for the future society where multiple heterogeneous agents are collaborately and competitively acting in order to help humans perform daily activities. For example, driver-less cars will be common in the near future. Such autonomous cars will need to cooperate and also compete with each other in traffic situations. A lot of studies including international competitions have been made on negotiating agents. A principal issue is that most of the proposed negotiating agents employ an ad-hoc conceding process, where basically they are adjusting a threshold to accept their opponents' offers. Because merely a threshold is adjusted, it is very difficult to show how and what the agent conceded even after agreement has been reached. To address this issue, we describe an explainable concession process we propose using a constraint relaxation process. In the process, an agent changes its belief that it should not believe a certain constraint so that it can accept its opponent's offer. We also describe three types of compromising strategies we propose. Experimental results demonstrate that these strategies are efficient.

Keywords: Automated negotiating agents · Compromise · Agreement

1 Introduction

Automated negotiating agents have been studied widely in the area of multiagent systems [4,7,8,12–16,19]. Heterogeneous, intelligent and autonomous systems (agents) like self-driving cars have been achieved in actual societies. In such societies, conflicts may occur among multiple agents. Thus it is required to have a social mechanism that forces such agents to reach an agreement to resolve such conflicts through automated negotiation. Many researchers are working on automated negotiating agents in the field of multiagent systems. In particular, international workshops and international competitions have been held since around 2010. Multi-agent systems are one of the most important technological advancements that have been made to address the needs of the next generation.

© Springer Nature Switzerland AG 2019
A. C. Nayak and A. Sharma (Eds.): PRICAI 2019, LNAI 11670, pp. 675–687, 2019.
https://doi.org/10.1007/978-3-030-29908-8_53

The automated negotiation competition ANAC (Automated Negotiating Agents Competition) has been held since 2010 as a testbed for automatic negotiation agent research. ANAC adopts a multi-issue utility model and an alternating-offer protocol. A lot of negotiating agents have been proposed because ANAC changes and extends the rules of negotiations every year. However, there are several drawbacks and problems that the ANAC competition could not focus on. One of them is how to explain the compromise process. In negotiations, agents cannot reach an agreement if they consider only their own profits and interests. Therefore, the compromise strategy is essential to reach an agreement. Most of the existing automated negotiating agents adopt *ad-hoc* compromising processes that only adjust their thresholds to accept the opponent's offer. This has made it difficult to explain how the compromise was achieved in the negotiation. It is important for automated negotiating agents to interact with actual human beings in real society because they need to explain how and why they compromised. For example, if your self-driving car stopped suddenly in the middle of congested traffic, you would want to know why it did so. To address this problem, we propose a compromise process based on constraint relaxation. A constraint is a basic unit of utility. In other words, we define the utility space of an agent as a set of constraints that satisfy the issue values and the argument for them. When a constraint is satisfied, the agent gets a utility value for this constraint. For example, the issues involved in buying a car include the car's color, price, and type. These issues are linked by certain constraints. Thus, there could be a constraint that says if the type is sports car, then the color should be red. Also, if the type is sedan, then color is white. Constraints generate values if they are satisfied, but do not generate any values if they are not satisfied. In the work we report here, we assumed *shared* issues and *individual* issues. In other words, we can say that agents agreed if they have the same issue value for shared issues. For individual issues, each agent can choose issue values to make their utility as high as possible. An agent faces a tradeoff between maximizing its own utility by satisfying the constraint as much as possible while keeping the share value to be the same value as that of the opponent agent. In order to solve this tradeoff, agents perform to compromise. In the compromise process for the strategy we propose, the agent removes constraints one by one from the set of its own constraints. Then, it tries to change the constraints' most preferable issue-value of the shared issue. If the agent can change the issue-value to one that is the same as the opponent's, then they can reach an agreement. Removing constraints is called "constraint relaxation." Specifically, we assume that the agent has a believed constraint set (IN) and an unbelieved (OUT) constraint set. In the initial state, it is assumed that all constraints are IN, and that in the constraint relaxation process, agents move certain constraints from IN to OUT. Various strategies are enabled when the agent moves constraints from IN to OUT. The four methods we propose are:

(1) Relaxation of constraints based on value,
(2) Random constraint relaxation,
(3) Constraint relaxation based on distance,

(4) Constraint relaxation based on value and distance.

Experimental results we obtained demonstrate that methods (1), (3) and (4) are able to obtain social surpluses significantly higher than the (2) random constraint relaxation. The remainder of this paper is organized as follows. In Sect. 2, we describe the automatic negotiation agent and negotiation protocol. In Sect. 3, we describe a compromise algorithm we propose that is based on the newly proposed constraint relaxation. In Sect. 4, we describe and discuss experimental results. In Sect. 5, we clarify the difference between our methods and related research work. Finally, we summarize our paper in Sect. 6.

2   Automated Negotiating Agents

2.1   Utility Hyper-graph

An agent has a complex utility space [11]. A variety of representations have been proposed for complex utility spaces [1,20,21]. In the work we report in this paper, we used hypergraph-based representations [9,10] to focus upon dependency between issues (nodes). A hypergraph is a mathematical representation in which an edge can join multiple nodes. We call a utility space using a hypergraph a "utility hyper-graph," in which nodes are issues and edges are constraints. The utility space U_i of agent i is represented by hypergraph (I, C), wherein $I_i \in I$ is an issues set (node), and C is a constraint set (edge). Each Ii issue has an issues value (Issue Value) within a predetermined range D_i. For example, one issue (color) when purchasing a car has an issue value within a range of "red," "blue," and "green." Constraint $C_j \in C$ is represented by $(v_{C_j}, \phi_{C_j}, \delta_{C_j})$, where v_{C_j} represents the value of constraint C_j and ϕ_{C_j} is a set of issues wherein constraint C_j is joined. Consequently, δ_{C_j} is a set of ranges where $\delta_{C_j} = \{range_{C_j}(I_i) : I_i \in \Phi_{C_j}\}$. The conditions under which constraint C_j is satisfied are as follows. The value assumed by issues I_i is x_{I_i}. If C_j is satisfied, then an agent having C_j obtains the value thereof v_{C_j}.

$$C_j = \begin{cases} satisfy & if \ \ x_{I_i} \in range_{C_j}(I_i) \ \ \forall I_i \in \phi_{C_j} \\ unsatisfy & otherwise \end{cases}$$

Figure 1 shows an example of an agent's utility graph and issues shared.

Here, two agents who have their own utility graph share three issues. Each of the agents has constraints that link issues. The issue takes an issue value. A constraint is satisfied if the issues linked by this constraint have issue values within the predefined ranges. When a constraint is satisfied, the agent obtains a value from this satisfied constraint.

Assumption 1. *A constraint that is difficult to satisfy has a higher value.*

We made the following assumptions in accordance with Assumption 1:

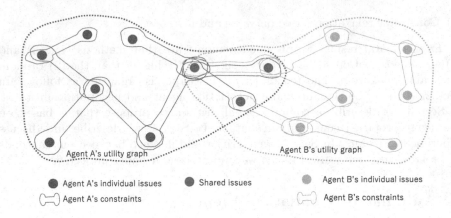

Fig. 1. Sharing issue and utility graph.

- Constraints with a wider issue-value range ($range_{C_j}$) are easier to satisfy and the values are lower. On the other hand, constraints having a narrower value range are more difficult to satisfy and the values are higher.
- Agreement has higher priority, so constraints related to the shared issues have more value than individual constraints.

2.2   Negotiation Protocol

In order to focus only on the compromise algorithms, we propose a simple negotiation protocol. We propose a simultaneous repeated offer protocol. In this protocol, each agent proposes its own offer to the opponent. If both agents can accept the offers, then they reach an agreement. If not, both agents revise their offers by compromising, and then propose again. This repeats until one of the agents cannot compromise anymore.

The Algorithm 1 is the concrete definition.

Algorithm 1. Simultaneous repeated offer protocol.

1: **repeat**
2:     Each agent finds an optimal issue value assignment that maximizes its own utility
3:     Each agent simultaneously proposes the issue value for the shared issue as an offer
4:     **Judging agreement:**
5:     **if** Both agents offer the same issue value for the shared issues **then**
6:         then they reached an agreement
7:     **else**
8:         Each agent performs the compromise process (refer to the next section).
9:     **end if**
10: **until** one of the agents cannot continue, i.e., no constraint can be relaxed or when the prescribed number of iterations is reached.

By performing the compromise process, the agent modifies and revises its utility space so that the agent can compromise with the opponent to reach agreement.

In this protocol, in each round, each agent makes an optimal proposal based on its own utility space. In the field of automated negotiations, the alternating of offered protocols [22] is well known and has been employed very much. However, an agent's strategy changes depending on which agent will give the first proposal. Therefore, we adopted a simple simultaneous repeated offer protocol. Extending it to alternating offer protocols is a subject for future work.

3   Explainable Compromise Process Based on Constraint Relaxation

3.1   Explainable Compromise Process

In this section, we show the compromise process based on constraint relaxation. Constraint relaxation is reducing the sum of allowable utilities (value) by reducing the number of satisfiable constraints.

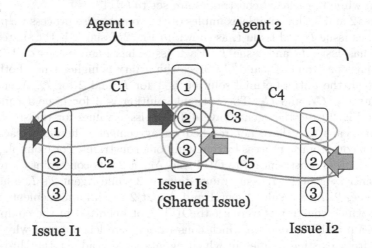

Fig. 2. Example of agreement based on constraint relaxation 1: Initialization.

In the existing research, no explanation has been provided regarding how the agreement is achieved with these values when making a compromise by ad-hoc threshold adjustment. In this research, unsatisfiable constraints are relaxed, i.e., removed, for compromising. Because we can ascertain which constraints are removed in the compromising process, we can understand how the agent compromised and which constraints were relaxed for agreement. In order to achieve this explainable compromising process, we classified the constraints into believed

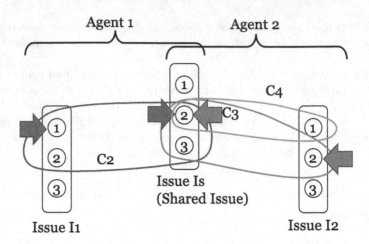

Fig. 3. Example of agreement based on constraint relaxation 2: Agreement by relaxation.

(IN) constraints and non-believed (OUT) constraints. Initially, all constraints are set to IN, while the relaxed constraints are set to OUT.

Figures 2 and 3 show simple examples of the compromise process we propose. Agent 1 has Issue I_1 and Issue I_s as shown in Fig. 2. Issue I_s is the shared issue. Agent 2 has Issue I_2 and Issue I_s. Each issue has issue values of 1, 2 or 3. Agent 1 has constraints C_1 and C_2. Since the utility is higher when both issues are satisfied, the initial optimal solution is 1 for I_1 and 2 for I_s. Agent 2 has constraints C_3, C_4, and C_5. The optimal solution is 3 for Issue I_s and 2 for I_2. In the Fig. 2 case, the agents have different issue values for the shared issue I_s, which means they have not reached an agreement. Therefore, each agent performs a compromise process by removing one constraint. For example, in this case Agent 1 sets constraint C_1 to OUT and Agent 2 sets constraint C_5 to OUT. Consequently, Agent 1's I_s issue value stays at 2 while Agent 2's I_s issue value also becomes 2. As a result, Agent 1 and Agent 2 reach an agreement. Since it is known which constraints were set to OUT (not believed) in the compromise, it becomes possible to explain which constraints were left out and why. This is different from existing studies in which agents simply adjust the threshold of acceptance.

3.2   Compromising Strategies

We propose the following four strategies. All initial constraints are IN and all initial relaxed constraints are OUT.

Random: One of the constraints in IN is randomly selected and pushed into OUT.

min: The lowest value constraint is selected from IN constraints and pushed into OUT.

distance: The constraint that has the longest distance from the shared issue is selected in IN and pushed into OUT. Here, the distance is the number of connected constraints from the shared issue.

min + distance: The constraint with the least value among the constraints most distant from the shared issue is selected from IN and pushed into OUT.

4   Experiment

4.1   Experiment Setting

We performed an experiment to compare the performances of the proposed compromising strategies. Our experimental setting included the following parameters:

- There are two agents.
- One issue can take up to 10 values.
- There is a single shared issue.
- Each agent has x issues.
- Each constraint includes at least one issue.

In other words, each issue is always included in one or more constraints. The number of constraints that include an issue is y. We employed a multistart local search approach as a search method to find the optimal solution. Graph structures based on constraints and issues are assigned randomly. Our experimental setting implies a situation where there are a lot of issues and all of them are subject to a number of constraints. However, the number of constraints is small.

4.2   Preliminary Results and Discussion

In this preliminary experiment, we obtained results for several settings. The results obtained for three settings are shown in Figs. 4, 5 and 6. The graphs in the three figures compare the social welfares for min, random, distance, and distance+min.

In Fig. 4, each agent has 10 issues ($x = 10$) and there is one constraint that includes each of them ($y = 1$). In Fig. 5, each agent has 10 issues ($x = 10$) and there are two constraints that include each of them ($y = 2$). In these two cases, a Tukey-Kramer HSD showed that the social welfare results for the min, distance, and distance+min categories were significantly higher than those for the random category.

In Fig. 6, each agent has 20 issues ($x = 20$) and there are two constraints that include each of them ($y = 2$). In this case, in this particular experiment, the social welfare results for the min and distance+min categories were higher than those for the random category. However, these results are not statistically significant.

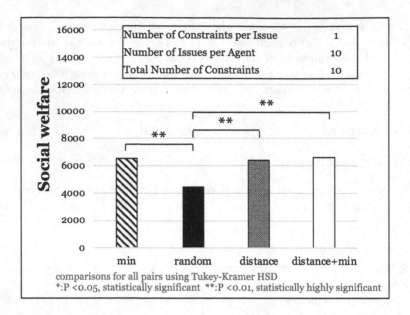

Fig. 4. Experimental results for $x = 10$ and $y = 1$

When there were more than 20 issues per agent, we were unable to get stable experiment results. Namely, it was difficult to obtain results showing a significant difference in the drawing method. This is because there are more than 20 points per agent, the number of solutions exceeds 10^{20} and considerable calculation is required to search the optimal solution. Developing a scalable method will be one of the most important subjects for future work. Also, the graph structure currently given to the agent is randomly given. Developing an optimization strategy based on the structure of the graph will also be a subject for future work.

5   Related Work

In this section, we describe the differences between our study and related work. In the field of automated negotiation research, the compromise process was first proposed by Klein et al. [18]. His main argument is that it is reasonable for the agent to gradually compromise at the Pareto front in simple negotiations where the issues are independent and the utility space is linear in each issue. However, if the issues are interdependent, the process is not simple because utility space is complicated, which makes the agent unable to find the Pareto front easily. To address this problem, Klein et al. proposed an SA-based agreement point search protocol (implicitly assuming compromising). In addition, Faratin et al. [5] analyzed various compromise functions.

The ANAC Competition [3] has been held annually since 2010. It is common for ANAC agents to adopt a method for estimating and presenting proposals that can be statistically accepted from the opponent's offers and accepting the

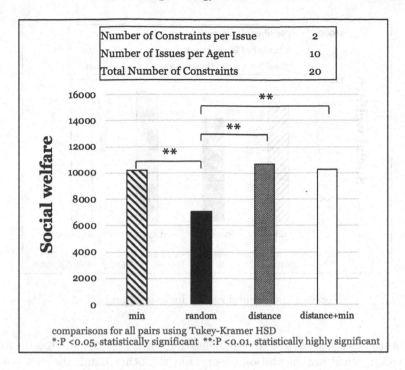

Number of Constraints per Issue	2
Number of Issues per Agent	10
Total Number of Constraints	20

comparisons for all pairs using Tukey-Kramer HSD
*:P <0.05, statistically significant **:P <0.01, statistically highly significant

Fig. 5. Experimental results for $x = 10$ and $y = 2$

proposal by adjusting the threshold considering the time discount utility. For example, AgentK [17], the winning agent of ANAC 2010, estimates the opponent utility space and the attitude (hostile or compromising) towards agreement from the opponent's offer history. If the partner seems to be a compromiser, concession is made, and if the partner is hostile, it will not concede more than a certain threshold. The above is the strategy that pioneered ANAC's basic concession strategy. Fawkes [2], the winning agent of ANAC 2013, estimates optimal concessions using discrete wavelet prediction based on an opponent's offer history. Most existing studies have focused on how to adjust the threshold so that the opponent's offer can be accepted. The threshold is a kind of upper limitation utility with which the agent can accept the opponent's offer. However, these studies give no explanation about how to achieve the threshold value. Thus, they do not explain why the agent compromises. This is a real problem because if your self-driving car compromises, you will not be able to obtain any explanation about the compromise. Also, as far as the authors know, no research has been done that assesses how compromising can be explained for an automated negotiation agent that assumes multi-argument utility functions.

Sycara has published a series of studies [24–26] that proposed negotiation and compromising processes that are explainable because they use case-based reasoning. The point is that they defined compromise and persuasion in the form of logical arguments within the framework of case-based reasoning. Sycara's

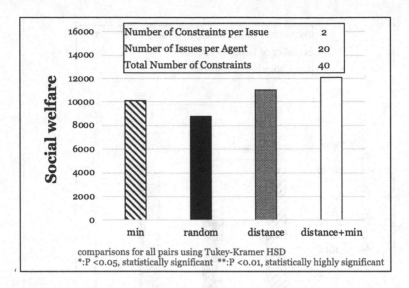

Fig. 6. Experimental results for $x = 20$ and $y = 2$

series of studies is also related to argumentation theory [23, 27] and has developed into mathematical argumentation theory. On the other hand, the viewpoint of this research is focusing on how to construct an explainable compromise process based on the utility function that can be handled numerically.

The Distributed Constraint Satisfaction/Optimization Problem (DCSP/DCOP) [28] has been one of the major topics in multiagent research. Because our model is based on constraints, it is closely related to DCSP/DCOP. The main difference is that our model focuses on negotiation situations where agents are basically trying to maximize their own individual utilities, but they compromise because they need to make an agreement. This is because if they cannot reach an agreement, there is no utility. In DCSP/DCOP, however, agents basically do not consider their own individual utilities. The main focus is on constraint satisfaction or optimization with distributed cooperative agents.

Wakaki et al. [27] published a paper about a DTMS (Distributed Truth Maintenance System) in which they proposed a classification of *consistency* in multiagent environments. They classified the distributed consistency concept into Inconsistent, Local-Consistency, Local-and-Shared-Consistency, and Global Consistency categories. In this study, an agreement means that each agent has its internal consistency while they have a consistent shared issue-value, which is the Local-and-Shared-Consistency. The compromising method we propose is one of the methods for obtaining Local-and-Shared-Consistency. However, the constraint graph we use represents utility space. On the other hand, a DTMS does not express preferences.

6 Conclusion

We have developed and in this paper propose an explainable compromise process for automatic negotiation agents. Most existing automatic negotiation compromising processes are ad-hoc adjustments of a threshold to accept an opponent's offers. However, our proposed method enables an explanation to be obtained by eliminating constraints one by one.

The following are our contributions: (1) The novel explainable compromise process we developed is based on a utility graph structured with constraints and issues. (2) For automatic multi-issue negotiation, we developed a new model that distinguishes between shared issues and personal issues. (3) For the compromise process, we developed a constraint relaxation process based on distance and value and demonstrated its effectiveness.

As a subject for future work, we should attempt to develop a more sophisticated compromising process. For example, it should be possible to create a process that can find possible combinations of the fewest constraints to be relaxed so that agents can change their alternatives.

References

1. Aydogan, R., Baarslag, T., Hindriks, K., Jonker, C., Yolum, P.: Heuristics for using CP-nets in utility-based negotiation without knowing utilities. Knowl. Inf. Syst. **45**, 357–388 (2015). https://doi.org/10.1007/s10115-014-0798-z
2. Baarslag, T.: What to bid and when to stop. Master's thesis, Delft University of Technology (2014)
3. Baarslag, T., et al.: The first international automated negotiating agents competition. Artif. Intell. J. (AIJ) (2012, to appear)
4. Bai, Q., Ren, F., Fujita, K., Zhang, M., Ito, T. (eds.): Multi-agent and Complex Systems. SCI, vol. 670. Springer, Singapore (2017). https://doi.org/10.1007/978-981-10-2564-8
5. Faratin, P., Sierra, C., Jennings, N.R.: Negotiation decision functions for autonomous agents. Int. J. Robot. Auton. Syst. **24**(3–4), 159–182 (1998). http://eprints.ecs.soton.ac.uk/2117/
6. Fioretto, F., Pontelli, E., Yeoh, W.: Distributed constraint optimization problems and applications: a survey. J. Artif. Intell. Res. **61**, 623–698 (2018)
7. Fujita, K., Ito, T., Zhang, M., Robu, V. (eds.): Next Frontier in Agent-Based Complex Automated Negotiation. SCI, vol. 596. Springer, Tokyo (2015). https://doi.org/10.1007/978-4-431-55525-4
8. Fukuta, N., Ito, T., Zhang, M., Fujita, K., Robu, V. (eds.): Recent Advances in Agent-based Complex Automated Negotiation. SCI, vol. 638. Springer, Cham (2016). https://doi.org/10.1007/978-3-319-30307-9
9. Hadfi, R., Ito, T.: Modeling complex nonlinear utility spaces using utility hypergraphs. In: Torra, V., Narukawa, Y., Endo, Y. (eds.) MDAI 2014. LNCS (LNAI), vol. 8825, pp. 14–25. Springer, Cham (2014). https://doi.org/10.1007/978-3-319-12054-6_2
10. Hadfi, R., Ito, T.: On the complexity of utility hypergraphs. In: Fukuta, N., Ito, T., Zhang, M., Fujita, K., Robu, V. (eds.) Recent Advances in Agent-based Complex Automated Negotiation. SCI, vol. 638, pp. 89–105. Springer, Cham (2016). https://doi.org/10.1007/978-3-319-30307-9_6

11. Ito, T., Hattori, H., Klein, M.: Multi-issue negotiation protocol for agents: exploring nonlinear utility spaces. In: Proceedings of 20th International Joint Conference on Artificial Intelligence (IJCAI 2007), pp. 1347–1352 (2007)
12. Ito, T., Hattori, H., Zhang, M., Matsuo, T.: Rational, Robust, and Secure Negotiations in Multi-agent Systems. SCI, vol. 89. Springer, Berlin (2008). https://doi.org/10.1007/978-3-540-76282-9
13. Ito, T., Zhang, M., Robu, V., Fatima, S., Matsuo, T.: Advances in Agent-Based Complex Automated Negotiations. SCI, vol. 233. Springer, Berlin (2009). https://doi.org/10.1007/978-3-642-03190-8
14. Ito, T., Zhang, M., Robu, V., Fatima, S., Matsuo, T.: New Trends in Agent-Based Complex Automated Negotiations. SCI, vol. 383. Springer, Berlin (2011). https://doi.org/10.1007/978-3-642-24696-8
15. Ito, T., Zhang, M., Robu, V., Fatima, S., Matsuo, T., Yamaki, H.: Innovations in Agent-Based Complex Automated Negotiations. SCI, vol. 319. Springer, Berlin (2010). https://doi.org/10.1007/978-3-642-15612-0
16. Ito, T., Zhang, M., Robu, V., Matsuo, T.: Complex Automated Negotiations: Theories, Models, and Software Competitions. SCI, vol. 435. Springer, Berlin (2013). https://doi.org/10.1007/978-3-642-30737-9
17. Kawaguchi, S., Fujita, K., Ito, T.: Compromising strategy based on estimated maximum utility for automated negotiation agents competition (ANAC-10). In: Mehrotra, K.G., Mohan, C.K., Oh, J.C., Varshney, P.K., Ali, M. (eds.) IEA/AIE 2011. LNCS (LNAI), vol. 6704, pp. 501–510. Springer, Heidelberg (2011). https://doi.org/10.1007/978-3-642-21827-9_51
18. Klein, M., Faratin, P., Sayama, H., Bar-Yam, Y.: Negotiating complex contracts. Group Decis. Negot. **12**(2), 58–73 (2003)
19. Marsa-Maestre, I., Lopez-Carmona, M.A., Ito, T., Zhang, M., Bai, Q., Fujita, K. (eds.): Novel Insights in Agent-based Complex Automated Negotiation. SCI, vol. 535. Springer, Tokyo (2014). https://doi.org/10.1007/978-4-431-54758-7
20. Robu, V., La Poutré, H.: Constructing the structure of utility graphs used in multi-item negotiation through collaborative filtering of aggregate buyer preferences. In: Ito, T., Hattori, H., Zhang, M., Matsuo, T. (eds.) Rational, Robust, and Secure Negotiations in Multi-Agent Systems. SCI, vol. 89, pp. 147–168. Springer, Berlin (2008). https://doi.org/10.1007/978-3-540-76282-9_9
21. Robu, V., Somefun, D.J.A., Poutré, J.L.: Modeling complex multi-issue negotiations using utility graphs. In: AAMAS 2005: Proceedings of the Fourth International Joint Conference on Autonomous Agents and Multiagent Systems, pp. 280–287. ACM, New York (2005)
22. Rubinstein, A.: Perfect equilibrium in a bargaining model. Econom.: J. Econom. Soc. **50**(1), 97–109 (1982)
23. Sierra, C., Jennings, N.R., Noriega, P., Parsons, S.: A framework for argumentation-based negotiation. In: Singh, M.P., Rao, A., Wooldridge, M.J. (eds.) ATAL 1997. LNCS, vol. 1365, pp. 177–192. Springer, Heidelberg (1998). https://doi.org/10.1007/BFb0026758
24. Sycara, K.P.: Resolving goal conflicts via negotiation. In: Proceedings of Fifth National Conference on Artificial Intelligence, pp. 245–250 (1988)
25. Sycara, K.P.: Argumentation: planning other agents' plans. In: Proceedings on International Joint Conference on Artificial Intelligence (IJCAI 89), pp. 517–523 (1989)
26. Sycara-Cyranski, K.: Arguments of persuasion in labor mediation. In: Proceedings of International Joint Conference on Artificial Intelligence (IJCAI 85), pp. 294–296 (1985)

27. Wakaki, T., Nitta, K.: Mathematical discussion. Tokyo Denki University Press (2017). (Japanese)
28. Yokoo, M., Durfee, E.H., Ishida, T., Kuwabara, K.: The distributed constraint satisfaction problem: formalization and algorithms. IEEE Trans. Knowl. Data Eng. **10**(5), 673–685 (1998). https://doi.org/10.1109/69.729707

Semantics of Opinion Transitions in Multi-Agent Forum Argumentation

Ryuta Arisaka[(⊠)] and Takayuki Ito

Nagoya Institute of Technology, Nagoya, Japan
ryutaarisaka@gmail.com, ito.takayuki@nitech.ac.jp

Abstract. There are online forums such as changemyview where a user may submit his/her views on a subject matter, against which other users argue to try to change the opinions of his/hers. To measure the quality of such discussion, one useful criterion is how influential a given topic is to participating users' opinion changes, as may be measured by the change (if any) in the proportion of supporting-objecting-mixed opinions by users. In this work, we incorporate the notion of agency into a previously proposed argumentation framework for issue-based information systems, QuAD, and formulate semantics of opinion transitions by newly considering agent-wise evaluation of QuAD initial scores.

1 Introduction

Consider the following argumentation example modelled off a thread[1] in changemyview, which is an online forum allowing a user to submit his/her views on a certain subject matter, against which other users may argue to try to change the opinions of his/hers.

1. **Proposition argument P**: There should exist a system to ensure politicians admit to their blatant lies.
2. **User e_1 supporting P**: Agreed. It should be a community that will decide if a politicians lies are blatant or serious enough to warrant retraction. If they find such a lie, politicians are legally required to take back their false claims or they will be removed from office.
3. **User e_2 objecting e_1**: We already have a system to hold politicians accountable for their lies: voting. Also, I think this would be a very problematic system to implement. Politicians would just make vague statements that can not be proven one way or the other. Saying, "Border crossings are a threat to national security" or "Medicare for all would be bad for America" is a personal opinion statement and can not be true or false.
4. **User e_1 concurring with e_2 in part**: True, vague statements might be a problem. As for voting, I feel that it is too slow to call them out on their lies.

---

[1] "CMV: There should exist a system to ensure politicians admit to their blatant lies" on reddit.com.

© Springer Nature Switzerland AG 2019
A. C. Nayak and A. Sharma (Eds.): PRICAI 2019, LNAI 11670, pp. 688–703, 2019.
https://doi.org/10.1007/978-3-030-29908-8_54

5. **User e_3 objecting e_2 in part**: I don't think voting is actually a good enough deterrent. Like, how can a voter know that Trump is lying to them when they only watch Fox News? Fox isn't going to report that. Basically, voting to oust liars only works if the voters believe they're liars.

Assume the following arguments:

P: There should exist a system to ensure that politicians admit to their blatant lies.

a_1: Agreed. It should be a community that will decide if a politician's lies are blatant or serious enough to warrant retraction.

a_2: Politicians will be legally required to take back their serious false claims, or he/she will be removed from office.

a_3: We already have a system to hold politicians accountable for their lies: voting.

a_4: It is hard to implement the proposed system because politicians would just make vague statements which cannot be proven to be either true or false.

a_5: It is true that vague statements might be a problem.

a_6: As for voting, I feel it is too slow to call them out on their lies.

a_7: I do not think voting is a good enough deterrent. Voting to oust liars only works if the voters believe they are liars.

$\boxed{A}$ shows the attack/support relations among them, and also which non-proposition arguments belong to which agent (user).

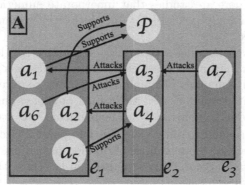

P is the proposition argument for this discussion, to which e_1 provides two supportive opinions a_1 and a_2. Later on in the discussion, e_1 revises his view on a_2, concurring by a_5 that vague statements which may be neither true nor false is problematic to the idea embedded in a_2. Thus, comparing one of e_1's initial opinions about P, i.e. a_2, and his/her later opinion a_5 in response to e_2's a_4, we see that an opinion transition, as far as a_2 is concerned, has occurred for e_1 with respect to P.

Interestingly, this discussion initiated by e_1 is not just about e_1's opinions. e_2 by his/her argument has expressed his views against P, and e_3 has likewise expressed his/her views on P, if indirectly, through objection to one of e_2's counterarguments, namely a_3. Indeed, a view of any participants to this discussion can be supporting or objecting P, and just as e_1, their opinions on P can also transition.

Concerning the condition of an opinion change, it is worth noting that an agent's opinion change does not entail from his/her opinion merely getting attacked by another agent, such as e_2's a_3 in $\boxed{A}$ attacked by e_1's a_6 and e_3's a_7, since e_2 may have simply dismissed the counter-arguments as nothing important to give a thought to. This can be contrasted to e_1 acknowledging a_4 with a_5, where it is obvious that e_1 actually supports a_4 against his/her former opinion a_2.

1.1 Semantics of Agents' Opinion Transitions

Given the purport of changemyview being that agents mutually act upon their opinions of proposition arguments, one useful criterion to judge the discussion quality is whether a proposition argument has been sufficiently influential for altering agents' initial opinions of the proposition arguments. This can be measured for example by the change (if any) in the proportion of supporting-objecting-mixed opinions of a proposition argument, or, if it matters which specific agents have changed their opinions, by the change in the number of agents who altered their opinions of a proposition argument. By setting forth some threshold value n, we can then accept those proposition arguments with the change greater than n.

For the acceptability of arguments, a Dung-like bipolar argumentation theory [12] allows inference of acceptable arguments from attack/support relations among them [11]. There are studies that incorporate numerical information to argumentation theory, e.g. [6,14,16,21], in which finer grading of acceptability semantics is achieved.

However, while a Dung-like acceptability semantics [13] looks at one argumentation graph for outputting a set of sets of nodes as acceptable arguments, the above-described semantics of opinion transitions require us to firstly (A) know both their initial opinions and their final opinions, and secondly to (B) juxtapose the two, which is not exactly in the realm of Dung acceptability semantics. In fact, while a_3 (and a_4) in $\boxed{A}$ are initial opinions of e_2's of P, in which sense they are relevant to the semantics of opinion transitions we envisage, a_3 is not relevant to a Dung acceptability semantics in the sense that it is not going to be an acceptable argument.

Towards the direction of dealing with quantitative judgement of arguments' strengths in a forum argumentation with an acyclic argumentation structure, there is a line of study [9,27] that incorporates ideas in argumentation theory into issue-based information systems [20], with an emphasis put on inference of final scores on arguments from initial scores given to them. However, currently, the theory QuAD proposed in [9] does not presume the notion of agency and agent-wise inference, which does not allow us to simply make use of the theory for our purpose of obtaining opinion changes per agent.

To fill the technical gap, we propose a multi-agent forum argumentation as a multi-agent-aware QuAD, and characterise the novel semantics of opinion transitions within.

1.2 Related Work

Change of Opinions. Change of belief is a key topic in belief revision [1,19]. A belief is a logical expression, and so a clear distinction of whether a belief opposes previously held beliefs is naturally obtained. Similar is true of logic-based argumentation expressing arguments as logical expressions [3,15,25], in which an attack relation between arguments is inferrable from logical contradiction between the two expressions. In the setting, it is in fact possible to apply belief revision techniques [5]. However, in general, an attack is not limited to strictly logical contradiction. Particularly in rhetoric argumentation, if one so chooses, he/she may attack any argument with any argument irrespective of logical (ir)relevance. What constitutes an opinion change in argumentation is indeed a question that must be answered. Nevertheless, since an opinion must be of something, if, say, an agent supports a proposition argument initially, like e_1 supporting P with a_2 in $\boxed{A}$, and if the same agent later on supports a counter-argument to a_2, like e_1 supporting a_4 (a counter-argument to a_2) with a_5, then we see that e_1 has changed his/her view as regards P. Similarly, when an agent who has expressed an opinion supportive of a proposition argument, and later on attacks another agent's opinion supportive of his/her opinion, we again see that the agent's opinion has changed.

Abstract Multi-Agent Argumentation and Forum Argumentation. There are several theories of abstract multiagent argumentation [4,7,10,28], some dedicated to agent argumentation under incomplete information. While they study acceptability statuses of a given argumentation graph treating all nodes equally, in this work we assume proposition arguments as are distinct from other arguments to be put forward to attack or support them, directly or indirectly. The differentiation is from classifications of issue-based information systems (IBIS) [20], which, in addition to proposition arguments (called positions in IBIS) and other arguments (called arguments in IBIS), also has issues. QuAD [9,27] is an argumentation theory that takes into account the distinction. The aim of QuAD is to derive, through argumentation theoretic recursive aggregation, final scores of given arguments as indications of the strength of support from the initial scores given to them. In comparison, our goal is to infer from the initial and final opinions of agents' (and from some threshold values) which proposition arguments are sufficiently influential in altering agents' views. Since the inference considers the relation between two different states and since we also consider multi-agent local scopes with agent-wise opinion change, the theory of QuAD is not immediately applicable. QuAD with voting from users [26] embeds a set of users as a parameter. However, a user in their framework is not for defining agents' local scopes, while ours is. For theoretical generalisations of QuAD and related argumentation formalisms such as social argumentation [21], there is a study [24] that measures both quality and acceptability of an opinion through aggregation of votes and numerical values for a chosen set of criteria. Still, our focus differs in that we deal with the influence of a proposition argument primarily through participating agents' opinion changes. There are other studies [2,8] that detail which properties QuAD and other related formalisms with numerical judgements satisfy.

In the rest, we will: go through Dung abstract argumentation, its preferred set characterisation and QuAD (in Sect. 2); and develop a multi-agent forum argumentation model as a multi-agent adaptation of QuAD, illustrate our score inference and opinion change semantics in examples, and also study how the score inference may be characterised by Dung-like preferred sets (in Sect. 3), before drawing conclusions.

2   Technical Preliminaries

Let $\mathcal{A}$ be a class of abstract entities that we understand as arguments, whose member is referred to by a or p with or without a subscript and a superscript, and whose subset is referred to by A with or without a subscript and with or without a superscript.

2.1   Dung Abstract Argumentation

Dung abstract argumentation considers an argumentation as a graph, with nodes as arguments and edges as attacks. The purpose of Dung theory is to judge which set(s) of arguments are acceptable in the graph. The semantics is called acceptability semantics.

Specifically, a Dung argumentation is defined to be a tuple (A, R) with $R \subseteq A \times A$. $a_1 \in A$ is said to attack $a_2 \in A$ if and only if, or iff, $(a_1, a_2) \in R$. $A_1 \subseteq A$ is said to *defend* $a_x \in A$ iff every $a_y \in A$ attacking a_x is attacked by at least one member of A_1. A_1 is said to be: *conflict-free* iff no member of A_1 attacks a member of A_1; admissible iff it is conflict-free and defends all the members of A_1; complete iff A_1 is admissible and includes every argument defended by A_1; and *preferred* iff A_1 is a maximal complete set. The set of all preferred sets of (A, R) is called the preferred semantics of (A, R).

Later studies based on this Dung theory extend it one way or another, such as by considering more than one binary relation and by considering several types of arguments, as does QuaD below.

2.2   QuAD: Quantitative Argumentation Debates

Let $\mathcal{A}^p, \mathcal{A}^a, \mathcal{A}^s$ be a class of proposition arguments, that of non-proposition attacking arguments and that of non-proposition supporting arguments. They are such that $\mathcal{A} = \mathcal{A}^p \cup \mathcal{A}^a \cup \mathcal{A}^s$, and that they are disjoint. We refer to: a finite subset of $\mathcal{A}^p$ by A^p; that of $\mathcal{A}^a$ by A^a; and that of $\mathcal{A}^s$ by A^s, each with or without a subscript.

A QuAD framework is a tuple (A^p, A^a, A^s, R, f) with: $R \subseteq (A^a \cup A^s) \times (A^p \cup A^a \cup A^s)$; and $f : (A^p \cup A^a \cup A^s) \rightarrow [0, 1]$. $(A^p \cup A^a \cup A^s, R)$ is assumed acyclic, i.e. there exists no $a \in (A^p \cup A^a \cup A^s)$ such that $(a, a) \in R^+$. Here R^+ is the transitive closure of R. Also, if $(a_1, a_2), (a_1, a_3) \in R$, then a_2 is a_3. f associates some numerical value, called initial score, to $A^p \cup A^a \cup A^s$ for certain quantitative judgement. $a_1 \in (A^p \cup A^a \cup A^s)$ is said to be connected to $a_2 \in (A^p \cup A^a \cup A^s)$ iff either a_1 is a_2, or $(a_1, a_2) \in R^+$, or else $(a_2, a_1) \in R^+$. For every $a \in (A^a \cup A^s)$, there exists some $a_x \in A^p$ such that a is connected to a_x. In this paper, we further assume that $A^p \neq \emptyset$ because our semantics of opinion transitions are with respect to member(s) of A^p.

The following characterisation of attack and support derives from [11], with a slight difference that we include the Dung-attack as a special case of (in)direct-attack instead of supported-attack. Assume $a_1, a_2 \in (A^p \cup A^a \cup A^s)$. a_1 is said to:

- support a_2 iff $(a_1, a_2) \in R$ and $a_1 \in A^s$.[2]
- (in)direct-attack a_2 iff either both $(a_1, a_2) \in R$ and $a_1 \in A^a$, or else there exists some $a_3 \in (A^p \cup A^a \cup A^s)$ such that a_1 (in)direct-attack a_3 and that a_3 supports a_2. a_1 may be simply said to attack a_2 when $(a_1, a_2) \in R$ and $a_1 \in A^a$ both hold.
- supported-attack a_2 iff there exists some sequence $a_3, \ldots, a_{k+3} \in (A^p \cup A^a \cup A^s)$ such that a_1 supports a_3, that a_{k+3} attacks a_2, and that a_{i+3} supports a_{i+4} for all $0 \leq i \leq k - 1$.

$A_1 \subseteq (A^p \cup A^a \cup A^s)$ is said to be conflict-free [11] iff, for no $a_1, a_2 \in A_1$, a_1 (in)direct-attack or supported-attack a_2. Characterisation of defence in this paper differs from that in [11], the definition of which we will therefore postpone until Sect. 3.

3   Multi-Agent Forum Argumentation Frameworks and Opinion Transition Semantics

As per our introduction, we focus on formulation of the semantics that judge which sets of proposition arguments have sufficiently influenced agents opinion changes. To this end, we extend QuAD into: $(A^p, A^a, A^s, R, f, E, f_\mathbf{E})$ where E is a finite set of abstract entities we understand as agents and $f_\mathbf{E}$ is a surjective function: $A^a \cup A^s \rightarrow E$. Intuitively, when $f_\mathbf{E}(a) = e$, a is in the scope of e, i.e. e expresses a non-proposition argument a. We call such a tuple MQuAD (Multi-agent QuAD).

3.1   Initial and Last Opinions by an Agent

Since we are interested in formulating semantics of opinion transitions for $a \in A^p$ based on agents' initial and last opinions of a, we define what constitutes initial or last opinions. Intuitively, since the graph $(A^p \cup A^a \cup A^s, R)$ is acyclic by definition, for every $a \in A^p$, we have sequence(s) $a, a_1, a_2, \ldots$ such that

---

[2] "and" instead of "and" is used in this paper when the context in which it appears strongly indicates truth-value comparisons. It follows the semantics of classical logic conjunction.

$(a_1, a), (a_2, a_1), \ldots \in R$. Thus, any non-proposition argument a_k that appears first as a non-proposition argument by e in those sequences should be considered e's initial opinion of a; and, similarly, any non-proposition argument a_k that appears for the last time as a non-proposition argument by e in those sequences should be considered e's last opinion of a. Let us make this intuition formal, and obtain the set of all initial and last opinions by an agent.

Definition 1 (Forward and backward connected sets). *Let $C(A)$ denote the set of all $a \in (A^p \cup A^s \cup A^a)$ connected to some member of A, and let $\delta : 2^{(A^p \cup A^a \cup A^s)} \rightarrow 2^{(A^p \cup A^a \cup A^s)}$ be such that $\delta(A_1) = A_1 \cup \{a \in (A^p \cup A^a \cup A^s) \mid \exists a_x \in A_1.(a_x, a) \in R^+\}$. For any $A \subseteq (A^p \cup A^a \cup A^s)$, we say that $\delta(A)$ is A's forward connected set, and that $C(A) \backslash \delta(A)$ is A's backward connected set.*

Definition 2 (Agent's initial and last opinion sets). *We say that $A \subseteq (A^a \cup A^s)$ is e's initial opinion set for $a_1 \in A^p$ iff $A \subseteq C(\{a_1\})$ and for every $a \in A$ and for every $a_x \in \delta(\{a\})$, $f_E(a_x) = e$ iff a is a_x. We say in particular that $A \subseteq (A^a \cup A^s)$ is e's maximal initial opinion set for $a_1 \in A^p$ iff there exists no e's initial opinion set that is strictly larger than A.*

We say that $A \subseteq (A^a \cup A^s)$ is e's last opinion set for $a_1 \in A^p$ iff $A \subseteq C(\{a_1\})$ and $f_E(a) = e$ for every $a \in A$ and $f_E(a) \neq e$ for every $a \in (C(A) \backslash \delta(A))$. We say in particular that $A \subseteq (A^a \cup A^s)$ is e's maximal last opinion set for $a_1 \in A^p$ iff there exists no e's last opinion set that is strictly larger than A.

Example 1 (Initial and last opinion sets). $(A^p, A^a, A^s, R, f, E, f_E)$ for $\boxed{A}$ *is such that:*

$$A^p \equiv \{p\}. \qquad A^a \equiv \{a_3, a_4, a_6, a_7\}. \qquad A^s \equiv \{a_1, a_2, a_5\}.$$
$$R \equiv \{(a_1, p), (a_2, p), (a_3, a_1), (a_4, a_2), (a_5, a_4), (a_6, a_3), (a_7, a_3)\}.$$
$$E \equiv \{e_1, e_2, e_3\}. \qquad f_E(a_{1,2,5,6}) = e_1, \; f_E(a_{3,4}) = e_2, \; f_E(a_7) = e_3.$$

At this point, assume some f.

e_1's maximal initial opinion set for p is $\{a_1, a_2\}$. To see that it is an initial opinion set for p, it suffices to verify the conditions given in Definition 2. First of all, clearly, $\{a_1, a_2\} \subseteq C(\{p\}) \equiv \{p, a_{1,\ldots,7}\}$. Secondly, $\delta(\{a_1\}) = \{p, a_1\}$ and $\delta(\{a_2\}) = \{p, a_2\}$. We have $f_E(a_{1,2}) = e$ but $f_E(p) \neq e$, as required. To verify that it is maximal, suppose we add either of a_5 and a_6 into the set, then we see that $\delta(\{a_5\})$ (similarly for $\delta(\{a_6\})$) is $\{a_5, a_4, a_2, p\}$, and $f_E(a_2) = e$ as well as $f_E(a_5) = e$, which does not satisfy the second condition. Such a set is not e_1's initial opinion set for p. Suppose we add either of a_3, a_4 and a_7 into the set, then it is immediate that $f_E(a_{3,4,5}) \neq e$, and the second condition is again not satisfied. Such a set is not e_1's initial opinion set for p.

e_1's maximal last opinion set for p, on the other hand, is $\{a_5, a_6\}$. To see that it is e_1's last opinion set for p, it suffices to verify the conditions given in Definition 2. Firstly, we have $\{a_5, a_6\} \subseteq C(\{p\})$. Secondly, we have $f_E(a_{5,6}) = e_1$. For the third condition, note we have $C(\{a_5, a_6\}) = \{p, a_{1,2,3,4,5,6}\}$, and also: $\delta(\{a_5, a_6\}) = C(\{a_5, a_6\})$ in this example. Therefore, $C(\{a_5, a_6\}) \backslash \delta(\{a_5, a_6\}) = \emptyset$. Thus, vacuously, $f_E(a) \neq e$ for every $a \in \emptyset$. Verification that it is maximal is done in a similar manner to the verification that $\{a_1, a_2\}$ is e's maximal initial opinion set for p.

Similarly, $\{a_3, a_4\}$ is e_2's maximal initial and last opinion sets for p. Unlike for e_1, here $C(\{a_3, a_4\}) \setminus \delta(\{a_3, a_4\}) = \{a_7\}$ which is not empty. However, clearly $f_{\mathbf{E}}(a_7) \neq e_2$, satisfying the third condition for $\{a_3, a_4\}$ being e_2's maximal last opinion set for p.

As a matter of fact, we have:

Proposition 1 (Uniqueness). *For every $e \in E$, there exists only one maximal initial and last opinion set for any $p \in A^p$, as given by a set union of all e's initial or last opinion sets for p.*

Proof. Suppose two initial (last) sets A_1, A_2 by $e \in E$ of $p \in A^p$. Suppose, by way of showing contradiction, that $A_1 \cup A_2$ is not an initial (last) set by e of p, then there exists at least one $a_x \in A_1 \cup A_2$ that appears in $\delta(A_1 \cup A_2) \setminus (A_1 \cup A_2)$. Refer both to Definitions 2 and 1 to detect contradiction.   □

3.2   Evaluation of Opinion Sets

Now that we can tell which sets are maximal initial and last opinion sets of each agent for each proposition argument, we set about learning whether $e \in E$ is supporting, objecting, or having a mixed opinion of a given $p \in A^p$. For this purpose, we define that f satisfy the following simple initial scoring classifying a single argument into either a supportive or objecting opinion of a proposition argument. By default, each proposition argument is considered supportive of itself. We note that the initial scores are often regarded as given beforehand of discussion in the relevant literature; however, derivation of a truly objective score out of the context of the discussion is not feasible, thus we understand the initial scores as numerical values that will be used for inference.

Definition 3 (Supportive/objecting initial scoring). *We define f to be such that:*

1. *$f(a_1) = 1$ iff either of the following holds.*
   (a) *$a_1 \in A^p$.*
   (b) *there is some $a \in (A^p \cup A^a \cup A^s)$ such that $(a_1, a) \in R$, and that either of the following holds.*
       i. *$a_1 \in A^s$ and $f(a) = 1$.*
       ii. *$a_1 \in A^a$ and $f(a) = 0$.*
2. *$f(a_1) = 0$ iff there is some $a \in (A^p \cup A^a \cup A^s)$ such that $(a_1, a) \in R$, and that either of the following holds.*
   (a) *$a_1 \in A^s$ and $f(a) = 0$.*
   (b) *$a_1 \in A^a$ and $f(a) = 1$.*

$f(a) = 1$ means that a is directly or indirectly supporting $p \in A^p$ that it is connected to. Meanwhile, $f(a) = 0$ means that a is directly or indirectly objecting $p \in A^p$ that it is connected to.

For the evaluation of an agent's maximal initial and last opinion sets, if each initial opinion $a \in (A^a \cup A^s)$ of $p \in A^p$ by $e \in E$ is supportive of p, i.e. $f(a) = 1$ for all such a, then it is clear that e is initially supportive of p. Similarly, if each one of them is objecting p, i.e. $f(a) = 0$, then clearly e is initially objecting p. In case some of e's initial opinions of p are supportive, while the others are objecting, then we judge that e's opinion of p is mixed. We reflect the three-valued judgement in the following definition. We emphasise that the judgement is per agent, since every agent that expresses any opinion of $p \in A^p$ has its own set of initial and last opinion sets.

Definition 4 (Opinion set evaluation). *Let* eval : $2^{(A^p \cup A^a \cup A^s)} \times E \to [0,1]$ *be such that:*

1. *eval(A_1, e_1) is undefined if either A_1 is empty or else A_1 is not e_1's initial or last opinion set.*
2. *Otherwise, we have:*
   (a) *eval$(A_1, e_1) = 1$ if, for every $a \in A_1$, $f(a) = 1$.*
   (b) *eval$(A_1, e_1) = 0$ if, for every $a \in A_1$, $f(a) = 0$.*
   (c) *eval$(A_1, e_1) = 0.5$, otherwise.*

Thus, for every $e \in E$, we can apply eval to e's maximal initial and last opinion sets to see whether e is supportive or objecting p, or neither.

Example 2 (Opinion set evaluation). (Continued from Example 1) For our example $\boxed{A}$, we assume that f is as has been defined in Definition 3. Any $a \in (A^p \cup A^a \cup A^s)$ with $f(a) = 1$ is shown with a border around the circle in $\boxed{B}$, while any $a \in (A^p \cup A^a \cup A^s)$ with $f(a) = 0$ is shown without any border in $\boxed{B}$.

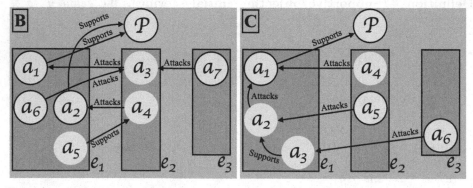

For the evaluation of e_1's maximal initial and last opinion sets of p, eval$(\{a_1, a_2\}, e_1) = 1$, and also eval$(\{a_5, a_6\}, e_1) = 0.5$, so e_1 initially was supportive of p, but progression of the forum argumentation let him question some supportive premises of p, namely of a_2, by the end. Meanwhile,

$\mathsf{eval}(\{a_3, a_4\}, e_2) = 0$, and $\mathsf{eval}(\{a_7\}, e_3) = 1$, and so initially e_2 is supportive of, and e_3 is objecting, p. However, because $\{a_3, a_4\}$ and $\{a_7\}$ are at the same time the maximal last opinion sets of e_2, and respectively of e_3, we see that opinions by the two agents did not change throughout the forum argumentation.

3.3 Opinion Set Evaluation Characterisation with Dung-like Preferred Sets

While the semantics of QuAD-based argumentation theory is intended to be more quantitative than Dung-like acceptability semantics, it is still of interest to identify the relation between the evaluation we have defined and Dung-based characterisation. Here, we show that eval in Definition 4 can be equivalently expressed with preferred sets.

Let us define the notion of defence to begin with:

Definition 5 (Defence). *Let* $\sigma : A^p \cup A^a \cup A^s \to 2^{(A^p \cup A^a \cup A^s)}$ *be such that* $\sigma(a)$ *is the least set that satisfies all the following.*

- $a \in \sigma(a)$.
- $a_1 \in \sigma(a)$ *if* $a_1 \in A^s$ *and there exists some* $a_2 \in \sigma(a)$ *such that* $(a_1, a_2) \in R$.

We say that $A_1 \subseteq (A^p \cup A^a \cup A^s)$ *defends* $a \in (A^p \cup A^a \cup A^s)$ *iff, for every* $a_x \in (A^a \cup A^s)$ *(in)direct-attacking* a, *there exists some* $a_y \in A_1$ *such that* a_y *supported-attacks or (in)direct-attacks some* $a_z \in \sigma(a_x)$.

Next, we define a new concept of up-to-preferred set which is such that preferredness, i.e. (1) conflict-freeness, (2) defendedness, (3) containedness of all that are defended, and (4) maximality among all those sets that satisfy (1) + (2) + (3) (See Sect. 2), holds under some subset of $(A^p \cup A^a \cup A^s)$, that is, up to the subset:

Definition 6 (Up-to-preferred sets). *We say that* $A_1 \subseteq (A^p \cup A^a \cup A^s)$ *is complete up to* $A_2 \subseteq (A^p \cup A^a \cup A^s)$ *iff* $A_1 \subseteq A_2$ *and* A_1 *is both conflict-free and including every* $a \in (A^p \cup A^a \cup A^s) \cap A_2$ *that it defends in another* $MQuAD$:$(A^p \cap A_2, A^a \cap A_2, A^s \cap A_2, R \cap (A_2 \times A_2), f', E', f'_E)$ *where:*

- f' *is* $((A^p \cup A^a \cup A^s) \cap A_2) \to [0, 1]$ *satisfying Definition 3.*
- $E' = \{e \in E \mid \exists a \in ((A^a \cup A^s) \cap A_2).f_E(a) = e\}$.
- f'_E *is* $((A^a \cup A^s) \cap A_2) \to E'$ *such that if* $f'_E(a) = e$ *then* $f_E(a) = e$.

We say that $A_1 \subseteq (A^p \cup A^a \cup A^s)$ *is preferred up to* $A_2 \subseteq (A^p \cup A^a \cup A^s)$ *iff* A_1 *is a maximal complete set up to* A_2.

The motivation of an up-to-preferred set is to group together arguments that get 1 via f, and group together those that get 0 via f when $A_1 \subseteq (A^p \cup A^a \cup A^s)$ is preferred up to $\delta(A_x)$ where A_x is some maximal initial or last opinion set of some agent of some $p \in A^p$.

Example 3 (Defence and up-to-preferred sets). Consider the MQuAD in $\boxed{C}$. As preferredness of a set requires that it includes all that it defends, the defence in Definition 5 should allow us to have (or not have) all that get the same value via f.

Let us begin with e_1's opinion sets. e_1's maximal initial opinion set for p is $\{a_1\}$, clearly. Hence, we should be able to obtain that $\{p, a_1\}$ is preferred up to $\delta(\{a_1\})$, to judge that a_1, belonging to the same set with p, is supportive of p. But, indeed, $\{p, a_1\}$ is preferred up to $\delta(\{a_1\}) = \{p, a_1\}$, as $\{p, a_1\}$ involves no member of A^a. For e_1's maximal last opinion set for p, which is $\{a_3\}$, we have that $\delta(\{a_3\}) = \{a_3, a_2, a_1, p\}$. There is only one subset of $\delta(\{a_3\})$, i.e. $\{a_2, a_3\}$ which is preferred up to $\delta(\{a_3\})$. Now, because the set does not contain p, we like to conclude that both a_2 and, in particular, $a_3 \in \{a_3\}$ are objecting p, which is indeed the case, for $f(a_2) = f(a_3) = 0$ in $\boxed{C}$.

Let us now look at the opinion sets of e_2 and e_3, to see why defence is as defined in Definition 5. e_2's maximal initial (and last) opinion set for p is $\{a_4, a_5\}$. We have $\delta(\{a_4, a_5\}) = \{a_4, a_5, a_2, a_1, p\}$. There are two preferred sets up to $\delta(\{a_4, a_5\})$, one is $\{a_4\}$, the other is $\{a_5, a_1, p\}$. We would like to conclude from the first set that a_4 is objecting p because it does not contain p, which is indeed the case. From the second set, we would like to conclude in particular that a_5 is supportive of p, which is indeed the case. Similarly for e_3, we obtain that $\{a_6, a_1, p\}$ is the (only one) preferred set up to $\delta(\{a_6\})$, and as we expect, a_6 is supportive of p.

Now, were the definition of defence as given in [11]: $A_1 \subseteq (A^p \cup A^a \cup A^s)$ defends $a \in (A^p \cup A^a \cup A^s)$ iff, for every $a_x \in (A^a \cup A^s)$ (in)direct-attacking or supported-attacking a, there exists some $a_y \in A_1$ such that a_y supported-attacks or (in)direct-attacks a_x, we would obtain for e_3 that $\{a_6\}$ is the only one preferred set up to $\delta(\{a_6\})$, as, $\{p, a_1, a_6\}$ would defend neither a_1 nor p, for none of them attacks a_2 which is (in)direct-attacking $a_1 \in \{p, a_1, a_6\}$. This motivated Definition 5. In passing, we note that our support interpretation does not fall into any of the 3 typical support interpretations (deductive, necessary, evidential) given in [12]. Given, however, it is not our intent to advertise a new interpretation of support, we refer an interested reader to [12] for the technical detail.

For the equivalence proof, the following two observations come in handy.

Lemma 1. *Suppose $A_1 \subseteq A^p \cup A^a \cup A^s$, if A_1 is not a subset of any preferred set up to $\delta(A_1)$, then at least one of the following holds true. (1) A_1 is not a subset of $\delta(A_1)$. (2) A_1 is not conflict-free. (3) A_1 does not defend some member of A_1.*

Lemma 2. *Let $h : A^p \cup A^a \cup A^s \to 2^{(A^p \cup A^a \cup A^s)}$ be such that, for any $a_1, a_2 \in A^p \cup A^a \cup A^s$, if $a_1 \in \sigma(a_2)$ or $a_2 \in \sigma(a_1)$, then $h(a_1) = h(a_2)$, and vice versa. For any $F \equiv (A^p, A^a, A^s, R, f, E, f_E)$, let $H(F)$ denote a Dung argumentation (A', R') where $A' \equiv \bigcup_{a \in A^p \cup A^a \cup A^s} \{h(a)\}$, and where $R' \subseteq A' \times A'$ is such that $(h(a_1), h(a_2)) \in R'$ iff some $a_x \in h(a_1)$ attacks some $a_y \in h(a_2)$. Then, all the following hold good.*

1. *For every $a' \in A'$, there exists some $n \in \{0, 1\}$ such that, for every $a \in A^p \cup A^a \cup A^s$, if $h(a) = a'$, then $f(a) = n$.*
2. *$A'_1 \subseteq A'$ is conflict-free in $H(F)(\equiv (A', R'))$ iff $\{a \in A^p \cup A^a \cup A^s \mid h(a) \in A'_1\}$ is conflict-free in F.*

Theorem 1 (Equivalence). *For $e \in E$, suppose that A_1 is e's maximal initial or last opinion set for $p \in A^p$. We have: (1) eval$(A_1, e) = 1$ iff $A_1 \cup \{p\}$ is a subset of a preferred set up to $\delta(A_1)$; (2) eval$(A_1, e) = 0$ iff A_1 is a subset of a preferred set A_2 up to $\delta(A_1)$ with $p \notin A_2$; (3) eval$(A_1, e) = 0.5$ iff otherwise.*

Proof. For one direction, assume firstly that eval$(A_1, e) = 1$. We show that $A_1 \cup \{p\}$ is a subset of a preferred set A_2 up to $\delta(A_1)$. Suppose, by way of showing contradiction, that there exists no preferred set up to $\delta(A_1)$ such that $A_1 \cup \{p\}$ is its subset. Then, by Lemma 1, (1) $A_1 \cup \{p\} \not\subseteq \delta(A_1)$, (2) $A_1 \cup \{p\}$ is not conflict-free, or (3) $A_1 \cup \{p\}$ does not defend at least one member of $A_1 \cup \{p\}$. (1) is clearly not the case. Suppose (2) holds true, then there exist some $a_1, a_2 \in A_1 \cup \{p\}$ such that a_1 (in)direct-attacks or supported-attacks a_2. Then, by Lemma 2, $f(a_1) \neq f(a_2)$, and therefore, eval$(A_1, e) \neq 1$, contradiction. Finally for (3), suppose that $A_1 \cup \{p\}$ does not defend $a_x \in A_1 \cup \{p\}$. Then there exists some $a_y \in \delta(A_1)$ such that a_y attacks a_x. However, no $a \in A^p \cup A^a \cup A^s$ such that $(a, a_x) \in R$ is in $\delta(A_1)$ by Definition 1, contradiction. All the other cases are proved likewise through Lemmas 1 and 2. $\qquad\square$

3.4   Semantics of Opinion Transitions

We define two semantics for opinion transitions that occur among agents in multi-agent forum argumentation. For one of them that we term A^p-centred opinion transition semantics, only the change in the proportion of supporting-objecting-mixed-opinionated agents matters, while for the other that we term E-centred opinion transition semantics, the number of agents who have changed their opinions is relevant.

Let G denote the class of functions $A^p \to \mathbb{N}$.

Definition 7 (E-centred opinion transition semantics). *Let init(e, p) and last(e, p) denote e's maximal initial opinion set and respectively e's maximal last opinion set for p. Let $\varpi : A^p \to \mathbb{N}$ be such that $\varpi(p) = |\{e \in E \mid$ eval$(init(e, p), e)$ is defined and eval$(init(e, p), e) \neq$ eval$(last(e, p), e))\}|$.*

Let $\alpha : G \to 2^{A^p}$ be such that $\alpha(g) = \{p \in A^p \mid g(p) < \varpi(p)\}$. For any $g \in G$, we say that $\alpha(g)$ is E-centred opinion transition semantics with respect to g.

Definition 8 (A^p-centred opinion transition semantics).
Let ss, os, ms : $A^p \times \{init, last\} \to \mathbb{N}$ be such that:

- *ss$(p, init) = |\{e \in E \mid$ eval$(init(e, p), e) = 1\}|$.*
- *os$(p, init) = |\{e \in E \mid$ eval$(init(e, p), e) = 0\}|$.*
- *ms$(p, init) = |\{e \in E \mid$ eval$(init(e, p), e) = 0.5\}|$.*
- *ss$(p, last) = |\{e \in E \mid$ eval$(last(e, p), e) = 1\}|$.*

- $os(p, last) = |\{e \in E \mid eval(last(e, p), e) = 0\}|$.
- $ms(p, last) = |\{e \in E \mid eval(last(e, p), e) = 0.5\}|$.

Let $\beta : G \to 2^{A^p}$ be such that $\beta(g) = \{p \in A^p \mid g(p) < (\max(0, ss(p, init) - ss(p, last)) + \max(0, os(p, init) - os(p, last)) + \max(0, ms(p, init) - ms(p, last)))\}$. For $g \in G$, we say $\beta(g)$ is A^p-centred opinion transition semantics with respect to g.

We illustrate the difference in detail:

Example 4 (Opinion transition semantics). Let us first illustrate with $\boxed{D}$ that the two opinion transition semantics are not identical. Assume $g \in G$ is such that $g(p) = 1$.

Since $eval(init(e_1, p), e_1) \neq eval(last(e_1, p), e_1)$ and $eval(init(e_2, p), e_2) \neq eval(last(e_2, p), e_2)$ and $eval(init(e_3, p), e_3) = eval(last(e_3, p), e_3)$, we have that $\varpi(p) = 2$. Therefore, $\alpha(g) = \{p\}$. On the other hand, $ss(p, init) - ss(p, last) = 0$ and $os(p, init) - os(p, last) = 0$ and $ms(p, init) - ms(p, last) = 0$. Hence, $\beta(g) = \emptyset$.

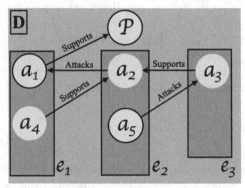

That is, while both e_1 and e_2 change their opinions, whose changes are reflected in E-centred opinion transition semantics, when we count the numbers of initially supporting/objecting/mixed-opinionated agents for p, and compare them with the numbers of supporting/objecting/mixed-opinionated agents for p at the end, we see no change in the proportion. So A^p-centred opinion transition semantics which considers only the proportion does not include p.

Meanwhile, for our running example $\boxed{A}$, when $g \in G$ is such that $g(p) = 0$, we obtain that $\alpha(g) = \beta(g) = \{p\}$, since only e_1 changes its opinions, and the proposition also changes by 1 agent.

Clearly:

Proposition 2 (Subsumption). *For any $g \in G$, we have $\beta(g) \subseteq \alpha(g)$, but not necessarily $\alpha(g) \subseteq \beta(g)$.*

4   Conclusion

We considered multi-agent-aware QuAD, and formulated two semantics of opinion transitions. They judge which proposition arguments are influential in changing agents' opinions, one by monitoring the change in the proportion of supporting-objecting-mixed-opinionated agents towards them, and another by monitoring the change in the number of agents whose opinions have altered. The threshold values can be determined from required percentage changes. In Theorem 1, we showed that evaluation of agents' initial and last opinion sets (via eval) can be equivalently characterised by up-to-preferred sets, identifying a link to Dung notations.

By keeping track of agents' opinion changes in a forum argumentation, we can learn which proposition arguments are most persuasive. The knowledge is beneficial for better understanding 'critical' topics in a forum argumentation.

With the rising popularity of online argumentation, it has become important that we be able to handle large-scale forum argumentations involving hundreds of agents who may generate thousands of comments [17,18,22]. For those, the overhead to conduct annotations and to determine precise initial scores of opinions in the continuous scale of $[0, 1]$ is rather steep. More discrete initial scoring and agent-wise inferences such as considered in this work seem to make a less expensive, and yet - given the improving performance of argumentation mining techniques to detect support and attack links, e.g. [23] - still effective alternative for analysing online forum argumentations. There are many representative large-scale consensus support systems around, into one of which we are currently implementing our method for empirically evaluating its scalability.

Acknowledgement. We thank anonymous reviewers for suggestions on future work.

References

1. Alchourrón, C.E., Gärdenfors, P., Makinson, D.: On the logic of theory change: partial meet contraction and revision functions. J. Symb. Logic **50**, 510–530 (1985)
2. Amgoud, L., Ben-Naim, J.: Weighted bipolar argumentation graphs: axioms and semantics. In: IJCAI, pp. 5194–5198 (2018)
3. Amgoud, L., Besnard, P.: Bridging the gap between abstract argumentation systems and logic. In: Godo, L., Pugliese, A. (eds.) SUM 2009. LNCS (LNAI), vol. 5785, pp. 12–27. Springer, Heidelberg (2009). https://doi.org/10.1007/978-3-642-04388-8_3
4. Arisaka, R., Bistarelli, S.: Defence outsourcing in argumentation. In: COMMA, pp. 353–360 (2018)
5. Arisaka, R., Satoh, K.: Balancing rationality and utility in logic-based argumentation with classical logic sentences and belief contraction. In: Baldoni, M., Chopra, A.K., Son, T.C., Hirayama, K., Torroni, P. (eds.) PRIMA 2016. LNCS (LNAI), vol. 9862, pp. 168–180. Springer, Cham (2016). https://doi.org/10.1007/978-3-319-44832-9_10
6. Arisaka, R., Satoh, K.: Coalition formability semantics with conflict-eliminable sets of arguments. In: AAMAS, pp. 1469–1471 (2017)
7. Arisaka, R., Satoh, K., van der Torre, L.: Anything you say may be used against you in a court of law. In: Pagallo, U., Palmirani, M., Casanovas, P., Sartor, G., Villata, S. (eds.) AICOL 2015-2017. LNCS (LNAI), vol. 10791, pp. 427–442. Springer, Cham (2018). https://doi.org/10.1007/978-3-030-00178-0_29

8. Baroni, P., Rago, A., Toni, F.: From fine-grained properties to broad principles for gradual argumentation: a principled spectrum. Int. J. Approximate Reason. **105**, 252–286 (2019)

9. Baroni, P., Romano, M., Toni, F., Aurisicchio, M., Bertanza, G.: Automatic evaluation of design alternatives with quantitative argumentation. Argument Comput. **6**(1), 24–49 (2015)

10. Brewka, G., Eiter, T.: Argumentation context systems: a framework for abstract group argumentation. In: Erdem, E., Lin, F., Schaub, T. (eds.) LPNMR 2009. LNCS (LNAI), vol. 5753, pp. 44–57. Springer, Heidelberg (2009). https://doi.org/10.1007/978-3-642-04238-6_7

11. Cayrol, C., Lagasquie-Schiex, M.C.: On the acceptability of arguments in bipolar argumentation frameworks. In: Godo, L. (ed.) ECSQARU 2005. LNCS (LNAI), vol. 3571, pp. 378–389. Springer, Heidelberg (2005). https://doi.org/10.1007/11518655_33

12. Cayrol, C., Lagasquie-Schiex, M.-C.: Bipolarity in argumentation graphs: towards a better understanding. In: Benferhat, S., Grant, J. (eds.) SUM 2011. LNCS (LNAI), vol. 6929, pp. 137–148. Springer, Heidelberg (2011). https://doi.org/10.1007/978-3-642-23963-2_12

13. Dung, P.M.: On the acceptability of arguments and its fundamental role in nonmonotonic reasoning, logic programming, and n-person games. Artif. Intell. **77**(2), 321–357 (1995)

14. Dunne, P.E., Hunter, A., McBurney, P., Parsons, S., Wooldridge, M.: Weighted argument systems: basic definitions, algorithms, and complexity results. Artif. Intell. **175**(2), 457–486 (2011)

15. Gabbay, D.M., Garcez, A.S.D.: Logical modes of attack in argumentation networks. Stud. Logica. **93**(2), 199–230 (2009)

16. Gabbay, D.M., Rodrigues, O.: An equational approach to the merging of argumentation networks. J. Logic Comput. **24**(6), 1253–1277 (2014)

17. Introne, J., Laubacher, R., Olson, G., Malone, T.W.: The climate CoLab: large scale model-based collaborative planning. In: CTS, pp. 40–47 (2011)

18. Ito, T.: Towards agent-based large-scale decision support system: the effect of facilitator. In: HICSS (2018)

19. Katsuno, H., Mendelzon, A.O.: On the difference between updating a knowledge base and revising it. In: Belief Revision. Cambridge University Press (1992)

20. Kunz, W., Rittel, H.W.J., Messrs, W., Dehlinger, H., Mann, T., Protzen, J.J.: Issues as elements of information systems. Technical report, University of California (1970)

21. Leite, J., Martins, J.: Social abstract argumentation. In: IJCAI, pp. 2287–2292 (2011)

22. Malone, T.W., Klein, M.: Harnessing collective intelligence to address global climate change. Innov. Technol. Gov. Global. **2**(3), 15–26 (2007)

23. Menini, S., Cabrio, E., Tonelli, S., Villata, S.: Never retreat, never retract: argumentation analysis for political speeches. In: AAAI, pp. 4889–4896 (2018)

24. Patkos, T., Bikakis, A., Flouris, G.: A multi-aspect evaluation framework for comments on the social web. In: KR, pp. 593–596 (2016)

25. Prakken, H., Sartor, G.: Argument-based extended logic programming with defeasible priorities. J. Appl. Non-class. Logics **7**, 25–75 (1997)

26. Rago, A., Toni, F.: Quantitative argumentation debates with votes for opinion polling. In: An, B., Bazzan, A., Leite, J., Villata, S., van der Torre, L. (eds.) PRIMA 2017. LNCS (LNAI), vol. 10621, pp. 369–385. Springer, Cham (2017). https://doi.org/10.1007/978-3-319-69131-2_22

27. Rago, A., Toni, F., Aurisicchio, M., Baroni, P.: Discontinuity-free decision support with quantitative argumentation debates. In: KR, pp. 63–73 (2016)
28. Rienstra, T., Perotti, A., Villata, S., Gabbay, D.M., van der Torre, L.: Multi-sorted argumentation. In: Modgil, S., Oren, N., Toni, F. (eds.) TAFA 2011. LNCS (LNAI), vol. 7132, pp. 215–231. Springer, Heidelberg (2012). https://doi.org/10.1007/978-3-642-29184-5_14

Learning Individual and Group Preferences in Abstract Argumentation

Nguyen Duy Hung[1(✉)] and Van-Nam Huynh[2]

[1] Sirindhorn International Institute of Technology, Khlong Nueng, Thailand
hung.nd.siit@gmail.com
[2] Japan Advanced Institute of Science and Technology, Nomi, Japan
huynh@jaist.ac.jp

Abstract. In **A**bstract **A**rgumentation, given the same AA framework rational agents accept the same arguments unless they reason by different AA semantics. Real agents may not do so in such situations, and in this paper we assume that this is because they have different preferences over the confronted arguments. Hence by reconstructing their reasoning processes, we can learn their hidden preferences, which then allow us to predict what else they must accept. Concretely we formalize and develop algorithms for such problems as learning the hidden preference relation of an agent from his expressed opinion, by which we mean a subset of arguments or attacks he accepted; and learning the collective preferences of a group from a dataset of individual opinions. A major challenge we addressed in this endeavor is to represent and reason with "answer sets" of preference relations which are generally exponential or even infinite.

1 Introduction

Argumentation is a form of reasoning unifying other forms such as non-monotonic and defeasible reasoning. Much of its recent development rests on Dung's Abstract Argumentation framework (AAF) [11] defined simply as a pair (Arg, Att) of a set of arguments Arg and a binary attack relation Att. For an illustration, Fig. 1 shows an AAF from the following story line[1].

President Trump nominated Judge Kavanaugh to supreme court on July 9, 2018 [18], *arguing that: Judge Kavanaugh should be confirmed because has an excellent judicial record and temperament...(Argument T in Fig. 1).*

The Senate Judiciary Committee then began Judge Kavanaugh's confirmation hearing. At the end of the hearing, Judge Kavanaugh was accused of sexually assaulting Dr Ford thirty-six years prior at a party. Dr Ford testified that:

F : ...I was pushed ...100% certain that it was Kavanaugh who attacked me...[2]

Judge Kavanaugh had two lines of defense. The first uses argument K rebutting F.

K : ...I'm not questioning that Dr. Ford may have been sexually assaulted by some person in some place at some time, but I have never done this to her or to anyone...

---

[1] This AAF is used in all running examples throughout the paper.
[2] The unexpressed conclusion of this argument is that Judge Kavanaugh is not qualified to be a Justice. Hence F attacks T.

© Springer Nature Switzerland AG 2019
A. C. Nayak and A. Sharma (Eds.): PRICAI 2019, LNAI 11670, pp. 704–717, 2019.
https://doi.org/10.1007/978-3-030-29908-8_55

The second line uses argument B, which is supported by a FBI supplemental investigation afterwards:

B : ...No one can corroborate Dr Ford's testimony...eyewitnesses named by Dr Ford denied any memory of the party whatsoever...

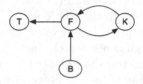

Fig. 1. AAF depicted by graph

An intuitive but powerful concept in AAF is the acceptability of an argument, namely that an argument A is acceptable wrt a set of arguments S iff any argument attacking A is attacked by S. Semantics of AA are defined by extending this concept, and widely viewed as representing reasoning standards possibly adopted by rational agents. Hence an underlying assumption of this view is that given the same AAF, rational agents accept the same arguments unless they reason by different AA semantics. However, real agents may not do so in such situations. Illustratively, in Judge Kavanaugh's confirmation, the US senate voted almost along party lines: 48 democrats voted NO, 49 republicans and 1 democrat voted YES, and 2 republicans abstained. Note that at least in the eyes of the public, these senators arrived at their decisions by the same reasoning standard, from the same knowledge (arguably) represented by the AAF in Fig. 1 for which all semantics produce the same set of acceptable arguments: $\{T, B, K\}$. In this paper we assume that different audiences may arrive at different acceptable arguments because they have different preferences over the confronted arguments, even though they reason by the same reasoning standard[3]. So a senator who voted NO might have preferred F to both B and K, and hence rejected attacks (B, F) and (K, F). Since preferences are the only thing that individuates different audiences, differences in opinions have to be traceable to differences in preferences. Thus, an agent's hidden preferences can be learned by reconstructing how the agent reasoned to arrive at his opinion.

It is worth noting that argument preferences have been used in Preference-based Argumentation frameworks (PAFs) [2,3,5,6], however the interest there is to reason from (argument) preferences to (argument) acceptances. Concretely, a PAF $(\mathcal{A}rg, \mathcal{A}tt, >_P)$ assumes an input preference relation $>_P$, and uses this to remove such an attack $(A, B) \in \mathcal{A}tt$ that $B >_P A$ (B is preferred to A according to P) to arrive at a standard AAF from which arguments are evaluated by AA semantics. Our interest is to reason from acceptances to preferences. In particular, the problem of individual preference learning takes as input an incomplete PAF $(\mathcal{A}rg, \mathcal{A}tt, \_)$ and a set $\mathcal{O} \subseteq \mathcal{A}rg \cup \mathcal{A}tt$ of arguments and attacks accepted

---

[3] We focus on the grounded semantics but our approach can be extended to others.

by an agent, and returns an "answer set" $\mathcal{Q}(\mathcal{O})$ consisting of all such preference relations $>_P$ that wrt PAF $(\mathcal{A}rg, \mathcal{A}tt, >_P)$, the elements of $\mathcal{O}$ are all acceptable. From an answer set $\mathcal{Q}(\mathcal{O})$, we can then predict whether the agent must also accept another set $\mathcal{A} \subseteq \mathcal{A}rg \cup \mathcal{A}tt$, by checking whether the elements of $\mathcal{A}$ are all acceptable in PAF $(\mathcal{A}rg, \mathcal{A}tt, >_P)$ for any $>_P$ in $\mathcal{Q}(\mathcal{O})$. A major challenge in this endeavor is to represent and reason with answer sets ranging from empty set (e.g. $\mathcal{Q}(\{F, K\}) = \emptyset$), to the whole set of all preference relations over $\mathcal{A}rg$ (e.g. $\mathcal{Q}(\{B\})$), and so exponential or even infinite. A direction one might pursue is to assume that arguments are characterized by attributes with values from certain domains and translate $\mathcal{Q}(\mathcal{O})$ into a set of preference relations over those values, hoping that there are not too many relations of this kind. Unfortunately, since we are working on AA which ignores any internal structures of arguments, this direction is not possible. So in this paper we start with the development of a compact representation for answer sets based on so called *preference states*. Learning individual preferences is reduced to the derivation of a finite set of preference states $\mathcal{Q}_{\mathcal{O}}$ that compactly represents a possibly infinite answer set $\mathcal{Q}(\mathcal{O})$. Our computational structure developed for this purpose is aptly called *preference derivations* which, as suggested in the previous paragraph, infer the agent's preferences by reconstructing how he reasoned to arrive at his opinion. Group preference learning generalizes individual preference learning. In particular, group preference learning takes as input an opinion dataset $\mathcal{D} = \{\mathcal{O}_1^{c_1}, \ldots, \mathcal{O}_n^{c_n}\}$ (where $\mathcal{O}_i \subseteq \mathcal{A}rg \cup \mathcal{A}tt$ and $c_i \geq 1$ the count of $\mathcal{O}_i$) instead of an individual opinion. For example, the above senate vote can be represented by $\mathcal{D}_1 = \{\{F, (F, T)\}^{48}, \{T\}^{50}, \{\}^2\}$, where $\{F, (F, T)\}$ consisting of argument F and attack (F, T) to represent a NO vote, and $\{T\}$ representing a YES vote, while $\{\}$ representing an abstention. The opinion counts could reflect the opinion percentages, as those in dataset $\mathcal{D}_2 = \{\{F\}^{52}, \{K\}^{38}, \{\}^{10}\}$ representing a public poll by CNN during at the senate voting time: when asked who to believe on the sexual accusation, 52% believes Ford, 38% believes Kavanaugh and 10% has no opinion [1]. Group preference learning computes the *collective preferences* of the whole group, and certainly needs to *fuse* the answer sets for individual opinions in the dataset appropriately. It turns out that there are different ways this fusion can be defined, resulting in different types of answers. In this paper, we define and compute two types called *ideal* and *optimal* answers.

Preferences draw much attention from many AI areas: recommender systems, automatic planning, social choice, etc (see [16] for a review). However most relevant to our work are studies of preferences in AA. Adding preferences over arguments allows for more expressivity, for example to express that some arguments promote more important values [6], or some are constructed from more prioritized beliefs [7]. Instead of having a binary preference relation $>_P$, one may consider the case in which arguments can express preferences between other arguments as in [4,10,13,15]. In essence, we can say that this line of work focuses on the semantics of argumentation with preferences. Learning or eliciting preferences over arguments has been largely unexplored except [14] which tackles a special case of the individual preference learning addressed in this paper. In particular, their algorithm can learn from only a conflict-free set of arguments in a finite framework. Additionally, their algorithm aims to explicitly enumerate all preference relations (transitivity is not guaranteed), and this is precisely what we want to avoid. Note that in AA we must not assume any internal structure of arguments and this contrasts our work with others that work on specific models of structured arguments, e.g. assumption-based arguments [12], or learn preferences of different kinds, e.g. preferences between defaults [8,9]. The rest of this paper is structured as follows. Section 2 presents the background, then *preference-based dispute derivations*, which can compute the grounded semantics of PAF. However their main purpose is to facilitate the development of preference derivations in Sect. 3, which deals with the learning of individual preferences. Section 4 generalizes this problem to the learning of group preferences, defining and computing so called *ideal* and *optimal* answers. Finally Sect. 5 concludes the paper (Due to the lack of space, proofs are given sketchily or even omitted).

2   Background

2.1   Argumentation Frameworks

Abstract Argumentation framework (AAF) is a pair $\mathcal{F} = (\mathcal{A}rg, \mathcal{A}tt)$ of a set $\mathcal{A}rg$ of arguments and an attack relation $\mathcal{A}tt \subseteq \mathcal{A}rg \times \mathcal{A}rg$. $S \subseteq \mathcal{A}rg$ attacks $A \in \mathcal{A}rg$ iff $(B, A) \in \mathcal{A}tt$ for some $B \in S$. $A \in \mathcal{A}rg$ is acceptable wrt to S iff S attacks every argument attacking A. S is *conflict-free* iff S does not attack itself; *admissible* extension iff S is conflict-free and each argument in S is acceptable wrt S; *complete* extension iff S is admissible and contains every arguments acceptable wrt S; *grounded* extension iff S is the least complete set. A is acceptable under semantics *sem* (e.g. *gr* stands for the *grounded*), denoted $\mathcal{F} \vdash_{sem} A$, if A is in a *sem* extension. $\mathcal{F}$ is said to be finitary if for any $A \in \mathcal{A}rg$, the set of arguments with directed paths to A (in the graph of $\mathcal{A}tt$) is finite.

Preference-based Argumentation framework (PAF) is a triple $(\mathcal{A}rg, \mathcal{A}tt, >_P)$ where $(\mathcal{A}rg, \mathcal{A}tt)$ is an AAF and $>_P \subseteq \mathcal{A}rg \times \mathcal{A}rg$ is transitive and asymmetric with $A >_P B$ meaning that A is (strictly) preferred to B. The *AA reduction* of PAF $\mathcal{P}$ is AAF $\mathcal{P}_\downarrow = (\mathcal{A}rg, \mathcal{A}tt_\downarrow)$ where $\mathcal{A}tt_\downarrow = \mathcal{A}tt \setminus \{(B, A) \mid A >_P B\}$. $S \subseteq \mathcal{A}rg$ is an extension of $\mathcal{P}$ under *sem* if S is an extension of AAF $\mathcal{P}_\downarrow$ under *sem*. For an argument A, $\mathcal{P} \vdash_{sem} A$ iff $\mathcal{P}_\downarrow \vdash_{sem} A$. In this paper, we restrict ourselves to PAFs containing finitary AAFs (for short, finitary PAFs).

2.2   Computing AAF and PAF Semantics

A proof that $\mathcal{F} \vdash_{sem} A$ can be represented by a dispute between two antagonistic parties, Proponent and Opponent. Proponent starts the dispute by putting forwards A then two parties alternate in attacking each other's previous arguments. For example, in [17], a dispute is constructed by a so called *simple dispute derivation* defined as a finite sequence of tuples the form $\langle P_i, O_i \rangle$ with: $P_i \subseteq \mathcal{A}rg$ contains arguments presented by Proponent but not yet attacked by Opponent;

$O_i \subseteq \mathcal{A}rg$ contains arguments presented by Opponent but not yet counter-attacked by Proponent. To ensure that the constructed dispute is a sound proof under the grounded semantics, the *ground dispute derivation* [17] enforces the acyclicity of graph $G_i \subseteq \mathcal{A}tt$ made from the attacks actually used by the parties. Readers are referred to [17] for further details, however the above basics suffice for our development of *preference-based dispute derivations* below.

To check if PAF $\mathcal{P} \vdash_{sem} A$, one can first reduce $\mathcal{P}$ to AAF $\mathcal{P}_\downarrow$ then check if $\mathcal{P}_\downarrow \vdash_{sem} A$ by AAF proof procedures. For efficiency the reduction should be done just enough to answer a given query $\mathcal{P} \vdash_{sem} A$. Concretely, suppose that at some step i of a dispute derivation, Proponent selects an attack $(B, A) \in O_i$ of Opponent. If A is preferred to B, then the reduction is triggered to remove the attack (B, A). Otherwise, Proponent should select some attack (C, B) to attack argument B and in this case B must not be preferred to C. So a *preference-based dispute derivation* is defined as a sequence of tuples of the form $\langle P_i, O_i, SP_i, SO_i, G_i \rangle$ with: P_i, G_i are defined exactly as in the ground dispute derivation of [17]; $O_i \subseteq \mathcal{A}tt$ now contains attacks presented by Opponent but not yet counter-attacked by Proponent; and SP_i is the set of arguments presented by Proponent up to step i (so $P_i \subseteq SP_i$), while SO_i is the set of attacks presented by Opponent and already counter-attacked by Proponent.

Definition 1. *Given a selection function, a **preference-based dispute derivation** is a sequence* $\langle P_0, O_0, SP_0, SO_0, G_0 \rangle \ldots \langle P_i, O_i, SP_i, SO_i, G_i \rangle \ldots$ *where*

1. *$P_i, SP_i \subseteq \mathcal{A}rg$; $O_i, SO_i \subseteq \mathcal{A}tt$; and $G_i \subseteq \mathcal{A}rg \times \mathcal{A}rg$ is a graph over $\mathcal{A}rg$.*
2. *At each step i, an element X is selected from P_i or O_i.*
   (a) *If X is an argument selected from P_i, then: $P_{i+1} = P_i \setminus \{X\}$; $O_{i+1} = O_i \cup \{(Y, X) \mid (Y, X) \in \mathcal{A}tt\}$; $SP_{i+1} = SP_i$; $SO_{i+1} = SO_i$; and $G_{i+1} = G_i$.*
   (b) *If X is an attack (B, A) selected from O_i, then:*
      i. *If A is preferred to B, then: $P_{i+1} = P_i$; $O_{i+1} = O_i \setminus \{X\}$; $SP_{i+1} = SP_i$; $SO_{i+1} = SO_i$; and $G_{i+1} = G_i$.*
      ii. *Otherwise, $B \notin SP_i$ and there exists some attack $(C, B) \in \mathcal{A}tt \setminus (SO_i \cup O_i)$ such that B is not preferred to C, and: $P_{i+1} = P_i \cup \{C\}$ if $C \notin SP_i$, otherwise $P_{i+1} = P_i$; $O_{i+1} = O_i \setminus \{X\}$; $SP_{i+1} = SP_i \cup \{C\}$; $SO_{i+1} = SO_i \cup \{X\}$; and $G_{i+1} = G_i \cup \{(C, B), (B, A)\}$ is acyclic.*

A preference-based dispute derivation **for a set arguments** $S \subseteq \mathcal{A}rg$ begins with a tuple $\langle S, \emptyset, S, \emptyset, \emptyset \rangle$. It is **successful** if it ends with a tuple $\langle \emptyset, \emptyset, SP_n, SO_n, G_n \rangle$. Theorem 1, which can viewed as restarting Theorems 2 and 8 of [17] and hence borrows the proofs thereof, says that preference-based dispute derivations represent a sound, complete, and terminating PAF proof procedure.

Theorem 1. *Let S be a finite set of arguments in a finitary PAF $\mathcal{P} = (\mathcal{A}rg, \mathcal{A}tt, P)$.*

1. *If there is a successful preference-based dispute derivation for S, then $\mathcal{P} \vdash_{gr} A$ for each $A \in S$.*

2. *If $\mathcal{P} \vdash_{gr} A$ for each $A \in \mathcal{S}$, then for any selection function there is a successful preference-based dispute derivation for $\mathcal{S}$.*
3. *There are no infinite preference-based dispute derivations for $\mathcal{S}$.*

3   Learning Individual Preferences

Individual preference learning takes as input an incomplete PAF framework $(\mathcal{A}rg, \mathcal{A}tt, \_)$ and a set $\mathcal{O} \subseteq \mathcal{A}rg \cup \mathcal{A}tt$ of arguments and attacks accepted by a rational agent (see the table below for illustrations), and returns an "answer set" $\mathcal{Q}(\mathcal{O})$ consisting of all such preference relations $>_P$ that wrt the complete PAF $(\mathcal{A}rg, \mathcal{A}tt, >_P)$, the elements of $\mathcal{O}$ are all acceptable. Answers are restricted to transitive and asymmetric relations, instead of any binary relations over $\mathcal{A}rg$.

Opinion $\mathcal{O}$	Reading
$\{T\}$	Approve Kavanaugh's confirmation (YES vote)
$\{F, (F, T)\}$	Disapprove Kavanaugh's confirmation (NO vote)
$\{K\}$	Believe Kavanaugh's testimony
$\{F\}$	Believe Ford's testimony

3.1   Preference States

Our compact representation for answer set is based on preference states.

Definition 2. *1. A preference statement is either a positive one of the form $(A > B)$ stating that A is preferred to B, or a negative one of the form $\neg(A > B)$ stating that A is not preferred to B, where $A, B \in \mathcal{A}rg$.*
2. A preference state $Q = Q^+ \cup Q^-$ contains a set Q^+ (resp. Q^-) of positive (resp. negative) preference statements satisfying two constraints:
(a) Q^+ defines a transitive, asymmetric relation over $\mathcal{A}rg$: if $(A > B), (B > C) \in Q^+$ then $(A > C) \in Q^+$; if $(A > B) \in Q^+$ then $(B > A) \notin Q^+$.
(b) Q^+ and Q^- are consistent: If $(A > B) \in Q^+$ then $\neg(A > B) \notin Q^-$.
The set of all preference states is denoted by $\mathcal{Q}$.

Definition 3. *For $Q, Q_1, Q_2 \in \mathcal{Q}$, we write $Q = Q_1 \uplus Q_2$ if Q^+ coincides with the transitive closure of $Q_1^+ \cup Q_2^+$ and $Q^- = Q_1^- \cup Q_2^-$.*

Obviously for any pair $Q_1, Q_2 \in \mathcal{Q}$, there exists at most one state Q that $Q_1 \uplus Q_2 = Q$. In cases such Q does not exist, we write $Q_1 \uplus Q_2 = null$.

Definition 4. *Let Q be a preference state. The PAF generated by Q is $\mathcal{P}_Q = (\mathcal{A}rg, \mathcal{A}tt, >_Q)$ where $A >_Q B$ iff $(A > B) \in Q$. An argument is acceptable by Q iff it is acceptable in PAF $\mathcal{P}_Q$. An attack is acceptable by Q iff it belongs to the AA reduction of PAF $\mathcal{P}_Q$.*

Now let's restate the individual preference learning problem.

Definition 5. *For $\mathcal{O} \subseteq \mathcal{A}rg \cup \mathcal{A}tt$, an answer of PreferenceDerivation($\mathcal{O}$)* **problem** *is such a preference state Q that the elements of $\mathcal{O}$ are all acceptable by Q. The set of all these answers is denoted by $\mathcal{Q}(\mathcal{O})$.*

Several answer sets $\mathcal{Q}(\mathcal{O})$ are shown below.

Opinion $\mathcal{O}$	Answer set $\mathcal{Q}(\mathcal{O})$
$\{B\}, \{\}$	$\mathcal{Q}$ (the set of all preference states)
$\{K\}$	$\{Q \in \mathcal{Q} \mid (F > B) \notin Q \text{ or } (K > F) \in Q\}$
$\{T\}$	$\{Q \in \mathcal{Q} \mid (T > F) \in Q \text{ or } (F > B) \notin Q \text{ or } (K > F) \in Q\}$
$\{F\}$	$\{Q \in \mathcal{Q} \mid (F > K), (F > B) \in Q\}$
$\{F, (F,T)\}$	$\{Q \in \mathcal{Q} \mid (F > K), (F > B) \in Q \text{ and } (T > F) \notin Q\}$
$\{K, F\}$	$\{\}$

We shall show that any (possibly infinite) answer set $\mathcal{Q}(\mathcal{O})$ has a finite core part $\mathcal{Q}_{\mathcal{O}} \subseteq \mathcal{Q}(\mathcal{O})$ that can be "completed" to obtain $\mathcal{Q}(\mathcal{O})$. Function $Compl(.)$ defined in Definition 7 serves this purpose. Intuitively, the completion of $\mathcal{Q}_{\mathcal{O}}$ is obtained by "lifting" each preference state $Q \in \mathcal{Q}_{\mathcal{O}}$ to a set of preference states $\lceil Q \rceil$.

Definition 6. *For a preference state Q, $\lceil Q \rceil \triangleq \{Q' \in \mathcal{Q} \mid Q'_{\uparrow} \supseteq Q\}$ where $Q'_{\uparrow} \triangleq Q'^{+} \cup \{\neg(A > B) \mid (A > B) \notin Q'^{+}\}$.*

That is, $\lceil Q \rceil$ contains Q' just in case the "Clark completion" $Q'_{\uparrow}$ of Q' implies Q. It is worth noting that $Q = \emptyset$ then $\lceil Q \rceil = \mathcal{Q}$.

Definition 7. *For a set of preference states $\mathcal{R} \subseteq \mathcal{Q}$, the* **completion** *of $\mathcal{R}$, denoted $Compl(\mathcal{R})$, is the set $\bigcup\{\lceil Q \rceil \mid Q \in \mathcal{R}\}$.*

3.2 Preference Derivations

Intuitively, a preference derivation for an agent's expressed opinion $\mathcal{O} \subseteq \mathcal{A}rg \cup \mathcal{A}tt$ simulates a dispute in which Proponent impersonates the agent to defend the elements of $\mathcal{O}$ from attacks by Opponent, and a neutral observer examines the moves of Proponent to determine the preferences of the agent. Like in preference-based dispute derivations, Opponent can keep bringing up all possible counter-arguments to whatever presented by Proponent. However *while preference-based dispute derivations use preferences to constraint Proponent's moves, preference derivation derive preferences from Proponent's moves.* Suppose that Proponent needs to counter an attack (B, A) by Opponent against an argument A presented previously by Proponent. There are several avenues for Proponent, as captured by cases of Definition 9.

- Case 2.a: If Proponent has preferred A to B (i.e. $(A > B) \in Q_i$), then the attack (B, A) is simply disregarded.
- Case 2.b.i: Otherwise, Proponent has an option to state that $(A > B)$ provided that this statement does not contradict with his currently revealed preferences, i.e. $Q_i \uplus \{(A > B)\}$ does not yield a null value.
- Case 2.b.ii: If Proponent does not state that $(A > B)$, then he needs to select an argument C to attack B, such that (C, B) is not used by Opponent before, i.e. $(C, B) \in Att \setminus (SO_i \cup O_i)$. The selection of C reveals a piece of Proponent's preferences, $\neg(B > C)$, because otherwise C does not really attack B. Hence $\neg(B > C)$ is added into Q_i to obtain Q_{i+1}.

Definition 8. *1. A **preference derivation** using a selection function sl is a sequence of pairs $(T_0, Q_0), \ldots, (T_i, Q_i), (T_{i+1}, Q_{i+1}) \ldots$ where for each $i \geq 0$, two conditions below holds:*

(a) T_i is a tuple of the form $\langle P_i, O_i, SP_i, SO_i, G_i \rangle$ as defined in Definition 1 and Q_i is a preference state as defined in Definition 2.

(b) $(T_{i+1}, Q_{i+1}) \in Follow(T_i, Q_i, sl)$ where Follow is a ternary function defined by Definition 9^4.

*2. A **preference derivation for** $\mathcal{O} \subseteq Arg \cup Att$ is a finite preference derivation starting with $(T_0, Q_0) = (\langle \mathcal{O} \cap Arg, \emptyset, \mathcal{O} \cap Arg, \emptyset, \emptyset \rangle, \{\neg(B > A) \mid (A, B) \in \mathcal{O} \cap Att\})$ and ending with (T_n, Q_n) of the form $(\langle \emptyset, \emptyset, \_, \_, \_ \rangle, \_)$*

Definition 9. *Given a tuple $T_i = \langle P_i, O_i, SP_i, SO_i, G_i \rangle$ and a selection function sl that selects an element X from either: (1) P_i component of T_i, or (2) O_i component of T_i; $Follow(T_i, Q_i, sl)$ is defined respectively as follows.*

1. If X is an argument selected from P_i, then $Follow(T_i, Q_i, sl)$ consists of only one pair (T_{i+1}, Q_{i+1}) where $T_{i+1} = \langle P_{i+1}, O_{i+1}, SP_{i+1}, SO_{i+1}, G_{i+1} \rangle$ is obtained from T_i as in step 2.a of Definition 1, and $Q_{i+1} = Q_i$.

2. If X is an attack (B, A) selected from O_i, then there are two cases.

(a) If $(A > B) \in Q_i$, then $Follow(T_i, Q_i, sl)$ consists of only one pair (T_{i+1}, Q_{i+1}) where $T_{i+1} = \langle P_i, O_i \setminus \{X\}, SP_i, SO_i, G_i \rangle$ and $Q_{i+1} = Q_i$.

(b) If $(A > B) \notin Q_i$, then $Follow(T_i, Q_i, sl)$ consists of such pairs (T_{i+1}, Q_{i+1}) with $Q_{i+1} \neq null$ that satisfy either conditions below.

  i. $Q_{i+1} = Q_i \uplus \{(A > B)\}$ and $T_{i+1} = \langle P_i, O_i \setminus \{X\}, SP_i, SO_i, G_i \rangle$.

  ii. $Q_{i+1} = Q_i \uplus \{\neg(B > C)\}$ where $(C, B) \in Att \setminus (SO_i \cup O_i)$ and $B \notin SP_i$; and T_{i+1} is obtained from T_i as in step 2.b.ii of Definition 1.

Theorem 2. *For any finite set $\mathcal{O} \subseteq Arg \cup Att$ and any selection function sl,*

1. There are finitely many preference derivations for $\mathcal{O}$, and they are all finite.

2. If $(T_0, Q_0) \ldots (T_n, Q_n)$ is a preference derivation for $\mathcal{O}$, then $\lceil Q_n \rceil \subseteq \mathcal{Q}(\mathcal{O})$.

3. For any $Q \in \mathcal{Q}(\mathcal{O})$, there exists a preference derivation $(T_0, Q_0) \ldots (T_n, Q_n)$ (using sl) for $\mathcal{O}$ such that $Q_\uparrow \supseteq Q_n$.

---

[4] An algorithmic form of function *Follow* can be easily worked out but we skip this.

Proof. Only Property (1) is proved here because the proof helps further reading. First let's visualize each preference derivation using a selection function sl for a set $\mathcal{O} \subseteq Arg \cup Att$ as a branch of a so called *preference derivation tree* whose nodes are labeled by pairs of the form (T_i, Q_i). The root is labeled by the pair (T_0, Q_0) specified in case 2 of Definition 8 (so Q_0 is finite since $\mathcal{O}$ is finite) and the leafs are labeled by pairs (T_n, Q_n) of the form $(\langle \emptyset, \emptyset, \_, \_, \_\rangle, \_)$. The sets of pairs labeling the children of an internal node with label (T_i, O_i) is exactly $Follow(T_i, O_i, sl)$. For an illustration, Fig. 2 shows a preference derivation tree for $\mathcal{O} = \{T\}$. Note that $Follow(T_i, O_i, sl)$ is a singleton set in cases 1 and 2.a (Definition 9). In case 2.b, $Follow(T_i, O_i, sl)$ contains one pair (T_{i+1}, Q_{i+1}) constructed by 2.b.i, and possibly many pairs constructed by 2.b.ii. Since we assume that AAF (Arg, Att) is finitary, the set $Att \setminus (SO_i \cup O_i)$ mentioned in 2.b.ii is finite, and hence $Follow(T_i, O_i, sl)$ must be a finite set. In other words, the preference derivation tree is finite in breath. It also follows from the finitary assumption of AAF (Arg, Att) that each branch of the preference derivation tree is finite.

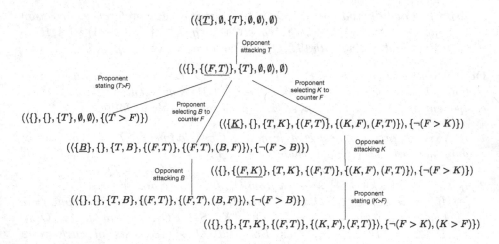

Fig. 2. A preference derivation tree.

It follows immediately from Property (1) of Theorem 2 that:

Lemma 1. *For any finite set $\mathcal{O} \subseteq Arg \cup Att$ and selection function sl, the set of all preference states Q_n occurring in the last pair (T_n, Q_n) of a preference derivation for $\mathcal{O}$ using sl, denoted $\mathcal{Q}_\mathcal{O}^{sl}$, is finite.*

An algorithm computing $\mathcal{Q}_\mathcal{O}^{sl}$ is shown in Algorithm 1. It works by constructing the preference derivation tree for $\mathcal{O}$ in a top-down manner. Line 4 consists of two destructive assignments initializing "variables" T and $\mathcal{Q}_\mathcal{O}^{sl}$. Variable T is the set of frontiers of the preference derivation tree under construction. So initially T contains only the root of the tree. The tree is then expanded by selecting some frontier (T, Q) in $\mathcal{T}$ (line 6) and adding its children. A child (T', Q') with T' having the form $\langle \emptyset, \emptyset, \_, \_, \_ \rangle$ is a leaf node and hence Q' is added into $\mathcal{Q}_{\mathcal{O}}^{sl}$. Otherwise (T', Q') is added to the set of frontiers waiting for further expansions. Different strategies of selecting a frontier at line 6 shall lead to different ways to construct the same tree, ranging from depth-first construction to breath-first construction. Since the final tree is finite both in depth and breath, the algorithm always terminates and returns $\mathcal{Q}_{\mathcal{O}}^{sl}$. For example, $ComputeQ(\{T\}, sl)$ constructs the preference derivation tree shown in Fig. 2. Regardless of selection strategies at line 6, $ComputeQ(\{T\}, sl)$ always returns $\mathcal{Q}_{\{T\}}^{sl} = \{\{(T > F)\}, \{\neg(F > B)\}, \{\neg(F > K), (K > F)\}\}$.

Algorithm 1. $ComputeQ(\mathcal{O}, sl)$

1: **Input:** A set $\mathcal{O} \subseteq \mathcal{A}rg \cup \mathcal{A}tt$; a selection function sl
2: **Output:** $\mathcal{Q}_{\mathcal{O}}^{sl}$ - the set of all preference states Q_n occurring in the last pair (T_n, Q_n) of a preference derivation for $\mathcal{O}$ using sl.
3: $T_0 = \langle \mathcal{O} \cap \mathcal{A}rg, \emptyset, \mathcal{O} \cap \mathcal{A}rg, \emptyset, \emptyset \rangle;\ Q_0 = \{\neg(B > A) \mid (A, B) \in \mathcal{O} \cap \mathcal{A}tt\}$
4: $\mathcal{T} := \{(T_0, Q_0)\};\ \mathcal{Q}_{\mathcal{O}}^{sl} := \{\}$
5: **while** $\mathcal{T} \neq \emptyset$ **do**
6:     select any pair (T, Q) from $\mathcal{T}$
7:     **for each** $(T', Q') \in Follow(T, Q, sl)$ **do**
8:         **if** T' is of the form $\langle \emptyset, \emptyset, \_, \_, \_ \rangle$ **then**
9:             $\mathcal{Q}_{\mathcal{O}}^{sl} := \mathcal{Q}_{\mathcal{O}} \cup \{Q'\}$
10:        **else**
11:            $\mathcal{T} := \mathcal{T} \cup \{(T', Q')\}$
12:        **end if**
13:    **end for**
14: **end while**
15: **return** $\mathcal{Q}_{\mathcal{O}}^{sl}$

Theorem 3 below says that $\mathcal{Q}_{\mathcal{O}}^{sl}$ is also a compact representation of $\mathcal{Q}(\mathcal{O})$ that we are looking for. Since for any selection function sl, $\mathcal{Q}_{\mathcal{O}}^{sl}$ would satisfy our purpose, we can omit sl for readability.

Theorem 3. *For any finite set $\mathcal{O} \subseteq \mathcal{A}rg \cup \mathcal{A}tt$ and any selection function sl, $Compl(\mathcal{Q}_{\mathcal{O}}^{sl}) = \mathcal{Q}(\mathcal{O})$.*

Proof. For each $Q_n \in \mathcal{Q}_{\mathcal{O}}^{sl}$, $\lceil Q_n \rceil \subseteq \mathcal{Q}(\mathcal{O})$ (Theorem 2) and hence $Compl(\mathcal{Q}_{\mathcal{O}}^{sl}) = \bigcup \{\lceil Q_n \rceil \mid Q_n \in \mathcal{Q}_{\mathcal{O}}^{sl}\} \subseteq \mathcal{Q}(\mathcal{O})$. Now consider $Q \in \mathcal{Q}(\mathcal{O})$. Property 3 of Theorem 2 says that there exists $Q_n \in \mathcal{Q}_{\mathcal{O}}^{sl}$ such that $Q_n \subseteq Q_\uparrow$. By Definition 6, $Q \in \lceil Q_n \rceil$ and hence $Q \in Compl(\mathcal{Q}_{\mathcal{O}}^{sl})$. In other words, $Compl(\mathcal{Q}_{\mathcal{O}}^{sl}) \supseteq \mathcal{Q}(\mathcal{O})$.

4   Learning Group Preferences

Learning in this section takes as input an opinion dataset $\mathcal{D} = \{\mathcal{O}_1^{c_1}, \ldots, \mathcal{O}_n^{c_n}\}$ where $\mathcal{O}_i \subseteq \mathcal{A}rg \cup \mathcal{A}tt$ and $c_i \geq 1$ the count of $\mathcal{O}_i$, and returns preference states that fuse the preferences of all agents. It turns out that there are different ways this fusion can be defined, resulting in different types of answers.

Definition 10. *Let $\mathcal{D} = \{\mathcal{O}_1^{c_1}, \ldots, \mathcal{O}_n^{c_n}\}$ be a dataset. An **ideal answer of** PreferenceLearning$(\mathcal{D})$ **problem** is such a preference state Q that $Q \in \mathcal{Q}(\mathcal{O}_i)$ for any $\mathcal{O}_i^{c_i} \in \mathcal{D}$. The set of all ideal answers is $\mathcal{Q}_{id}(\mathcal{D}) \triangleq \mathcal{Q}(\mathcal{O}_1) \cap \cdots \cap \mathcal{Q}(\mathcal{O}_n)$.*

Lemma 2 below says that the ideal answer set $\mathcal{Q}_{id}(\mathcal{D})$ can be computed easily by the algorithms developed in the previous section.

Lemma 2. *For any dataset $\mathcal{D} = \{\mathcal{O}_1^{c_1}, \ldots, \mathcal{O}_n^{c_n}\}$, $\mathcal{Q}_{id}(\mathcal{D}) = \mathcal{Q}(\mathcal{O}_1 \cup \cdots \cup \mathcal{O}_n)$.*

Unfortunately ideal answers do not always exist, e.g. for our sample datasets in the introduction: $\mathcal{Q}_{id}(\mathcal{D}_1) = \mathcal{Q}_{id}(\mathcal{D}_2) = \emptyset$. In such cases we want to satisfy as many individual opinions as possible, by considering so called *optimal answers*.

Definition 11. *An **optimal answer** of PreferenceLearning$(\mathcal{D})$ problem is such a preference state $\hat{Q}$ that maximizes objective function $f_{\mathcal{D}}(Q) = sum\{c \mid \mathcal{O}^c \in \mathcal{D} \text{ and } Q \in \mathcal{Q}(\mathcal{O})\}$. The set of all optimal answers is denoted by $\mathcal{Q}_{op}(\mathcal{D})$.*

Let's revisit our sample datasets. Obviously $\mathcal{Q}_{op}(\mathcal{D}_1) = \mathcal{Q}(\{T\})$; and each $Q \in \mathcal{Q}_{op}(\mathcal{D}_1)$ belongs to $\mathcal{Q}(\{T\})$ and $\mathcal{Q}(\{\})$, but not $\mathcal{Q}(\{F, (F, T)\})$, i.e. $f_{\mathcal{D}_1}(Q) = sum\{50, 2\}$. Similarly $\mathcal{Q}_{op}(\mathcal{D}_2) = \mathcal{Q}(\{F\})$. Each $Q \in \mathcal{Q}_{op}(\mathcal{D}_2)$ belongs to $\mathcal{Q}(\{F\})$ and $\mathcal{Q}(\{\})$, but not $\mathcal{Q}(\{K\})$, i.e. $f_{\mathcal{D}_2}(Q) = sum\{52, 10\}$. Lemma 3 says that an ideal answer is also optimal. The reverse also holds if there exists an ideal answer.

Lemma 3. *For any PreferenceLearning$(\mathcal{D})$ problem, $\mathcal{Q}_{id}(\mathcal{D}) \subseteq \mathcal{Q}_{op}(\mathcal{D})$. Further if $\mathcal{Q}_{id}(\mathcal{D}) \neq \emptyset$ then $\mathcal{Q}_{op}(\mathcal{D}) = \mathcal{Q}_{id}(\mathcal{D})$.*

Computing optimal answers is by definition an optimization problem but we can either harness existing optimization algorithms or develop tailored ones. For lack of space let us follow the first approach using an implementation of objective function $f_{\mathcal{D}}(Q)$ in Algorithm 2, whose correctness is ensured by Theorem 4.

Theorem 4. *If $Compute\mathcal{Q}(.)$ maps $\mathcal{D} = \{\mathcal{O}_1^{c_1}, \ldots, \mathcal{O}_n^{c_n}\}$ to $\mathcal{PD} = \{\mathcal{Q}_{\mathcal{O}_1}^{c_1}, \ldots, \mathcal{Q}_{\mathcal{O}_n}^{c_n}\}$, then $f_{\mathcal{D}}(Q) = sum\{c \mid Q \in Compl(\mathcal{Q}_{\mathcal{O}}) \text{ for some } \mathcal{Q}_{\mathcal{O}}^c \in \mathcal{PD}\}$.*

Algorithm 2. $f_{\mathcal{D}}(Q)$

1: **Context:** A dataset $\mathcal{D} = \{\mathcal{O}_1^{c_1}, \ldots, \mathcal{O}_n^{c_n}\}$ and its mapping $\mathcal{PD} = \{\mathcal{Q}_{\mathcal{O}_1}^{c_1}, \ldots, \mathcal{Q}_{\mathcal{O}_n}^{c_n}\}$
    by function $Compute\mathcal{Q}(.)$.
2: **Input:** A preference state Q.
3: **Output:** Value of $f_{\mathcal{D}}(Q)$
4: sum : $= 0$
5: **for each** $\mathcal{Q}_{\mathcal{O}}^c \in \mathcal{PD}$ **do**
6:     **if** $Member(Q, \mathcal{Q}_{\mathcal{O}}) == True$ **then**
7:         sum := sum + c
8:     **end if**
9: **end for**
10: **return** sum

Note that Algorithm 2 requires an implementation of function $Member/2$, which is called in Line 6. This function receives as input two parameters: a preference state $Q \in \mathcal{Q}$ and a set $\mathcal{R} \subseteq \mathcal{Q}$, then check whether $Q \in Compl(\mathcal{R})$. It is important to do the check without computing the completion of $\mathcal{R}$ and the implementation shown in Algorithm 3 satisfies this requirement. It works as follows: $Q \in Compl(\mathcal{R})$ iff there exist $Q' \in \mathcal{R}$ such that $Q \in \lceil Q' \rceil$ or equivalently $Q' \subseteq Q_\uparrow$. To check if $Q' \subseteq Q_\uparrow$, we first check if $Q'^+ \subseteq Q^+$ (line 5) then verify whether for each $\neg(B > A) \in Q'^-$, $(B > A) \notin Q^+$ (lines 7–12). This verification is equivalent to verifying that $Q'^- \subseteq Q_\uparrow^- = \{\neg(B > A) \mid (B > A) \notin Q^+\}$, but it does not requires an explicit construction of the set $Q_\uparrow^-$.

Algorithm 3. $Member(Q, \mathcal{R})$

1: **Input:** $Q \in \mathcal{Q}$ and $\mathcal{R} \subseteq \mathcal{Q}$
2: **Output:** True if $Q \in Compl(\mathcal{R})$; False otherwise.
3: **for each** $Q' \in \mathcal{R}$ **do**
4:     /* check if $Q' \subseteq Q_\uparrow$ */
5:     **if** $Q'^+ \subseteq Q^+$ **then**
6:         foundQ' := True
7:         **for each** $\neg(B > A) \in Q'^-$ **do**
8:             **if** $(B > A) \in Q^+$ **then**
9:                 foundQ' := False
10:                 **break**
11:             **end if**
12:         **end for**
13:         **if** foundQ' **then**
14:             **return** True
15:         **end if**
16:     **end if**
17: **end for**
18: **return** False

---

5   Conclusions

In this paper we formalize and develop algorithms for learning the hidden preferences over arguments of rational agents. We address both the learning of individual preferences and of group preferences. A major challenge in this endeavor is to represent and reason with "answer sets" of preference relations which are in general exponential or even infinite. Hence at the heart of our work are a representation called *preference states* and a computational structure called *preference derivations* which derive preference states by reconstructing how the agents reasoned to arrive at their opinions. Learning or eliciting preferences over arguments has been largely unexplored in the current literature except the work of [14], which tackles a special case of the individual preference learning problem addressed in this paper. In particular, their algorithm can learn from only a conflict-free set of arguments in a finite framework. Additionally, their algorithm aims to explicitly enumerate all answers (transitivity is not guaranteed), and this is precisely what we want to avoid. The authors also do not address group preference learning. Our contributions can be extended in several directions. First, one may want to drop the assumption that preferences over arguments are the only thing that individuates different audiences. Second, one may want to consider agents that reason by other semantics. In group preference learning, one might want to define other types of answers, or develop special optimization algorithms rather harnessing the available ones. It is also interesting to explore applications of the developed algorithms, for example, in public opinion analyses.

Acknowledgment. This paper is based upon work supported in part by the Asian Office of Aerospace R&D (AOARD) (Grant No. FA2386-17-1-4046), and the US Office of Naval Research Global (ONRG, Grant No. N62909-19-1-2031).

References

1. CNN poll conducted by SSRS. http://cdn.cnn.com/cnn/2018/images/10/08/rel9a.-.kavanaugh.pdf. Accessed Oct 2018
2. Amgoud, L., Cayrol, C.: A reasoning model based on the production of acceptable arguments. Ann. Math. AI **34**(1–3), 197–215 (2002)
3. Amgoud, L., Cayrol, C., Lagasquie-Schiex, M.C., Livet, P.: On bipolarity in argumentation frameworks. Int. J. Intell. Syst. **23**(10), 1062–1093 (2008)
4. Baroni, P., Cerutti, F., Giacomin, M., Guida, G.: Encompassing attacks to attacks in abstract argumentation frameworks. In: Sossai, C., Chemello, G. (eds.) ECSQARU 2009. LNCS (LNAI), vol. 5590, pp. 83–94. Springer, Heidelberg (2009). https://doi.org/10.1007/978-3-642-02906-6_9
5. Bench-Capon, T.: Persuasion in practical argument using value-based argumentation frameworks. J. Logic Comput. **13**(3), 429–448 (2003)
6. Bench-Capon, T., Atkinson, K.: Abstract argumentation and values. In: Simari, G., Rahwan, I. (eds.) Argumentation in Artificial Intelligence, pp. 45–64. Springer, Boston (2009). https://doi.org/10.1007/978-0-387-98197-0_3
7. Benferhat, S., Dubois, D., Prade, H.: Argumentative inference in uncertain and inconsistent knowledge bases. In: Uncertainty in AI. Morgan Kaufmann (1993)
8. Brewka, G.: Preferred subtheories: an extended logical framework for default reasoning. In: IJCAI 1989, pp. 1043–1048 (1989)
9. Brewka, G.: Reasoning about priorities in default logic. In: AAAI, pp. 940–945 (1994)
10. Brewka, G., Woltran, S.: Abstract dialectical frameworks. In: Proceedings of KR 2010, pp. 102–111. AAAI Press (2010)
11. Dung, P.M.: On the acceptability of arguments and its fundamental role in non-monotonic reasoning, logic programming and n-person games. AIJ **77**(2), 321–357 (1995)
12. Dung, P.M., Kowalski, R.A., Toni, F.: Assumption-based argumentation. In: Simari, G., Rahwan, I. (eds.) Argumentation in Artificial Intelligence, pp. 199–218. Springer, Boston (2009). https://doi.org/10.1007/978-0-387-98197-0_10
13. Hanh, D.D., Dung, P.M., Hung, N.D., Thang, P.M.: Inductive defense for sceptical semantics of extended argumentation. J. Logic Comput. **21**(1), 307–349 (2010)

14. Mahesar, Q., Oren, N., Vasconcelos, W.W.: Computing preferences in abstract argumentation. In: Miller, T., Oren, N., Sakurai, Y., Noda, I., Savarimuthu, B.T.R., Cao Son, T. (eds.) PRIMA 2018. LNCS (LNAI), vol. 11224, pp. 387–402. Springer, Cham (2018). https://doi.org/10.1007/978-3-030-03098-8_24
15. Modgil, S.: Reasoning about preferences in argumentation frameworks. AIJ **173**(9–10), 901–934 (2009)
16. Pigozzi, G., Tsoukiàs, A., Viappiani, P.: Preferences in artificial intelligence. Ann. Math. Artif. Intell. **77**(3), 361–401 (2016)
17. Thang, P.M., Dung, P.M., Hung, N.D.: Towards a common framework for dialectical proof procedure in abstract argumentation. J. Logic Comput. **19**(6), 1071–1109 (2009)
18. Wikipedia. Page name: Brett kavanaugh supreme court nomination. Accessed July 2019

Epistemic Argumentation Framework

Chiaki Sakama[1](✉) and Tran Cao Son[2]

[1] Wakayama University, Wakayama 640 8510, Japan
sakama@wakayama-u.ac.jp
[2] New Mexico State University, Las Cruces, NM 88003, USA
tson@cs.nmsu.edu

Abstract. The paper introduces the notion of an *epistemic argumentation framework* (EAF) as a means to integrate the beliefs of a reasoner with argumentation. Intuitively, an EAF encodes the beliefs of an agent who reasons about arguments. Formally, an EAF is a pair of an argumentation framework and an *epistemic constraint*. The semantics of the EAF is defined by the notion of an ω-*epistemic labelling set*, where ω is complete, stable, grounded, or preferred, which is a set of ω-labellings that collectively satisfies the epistemic constraint of the EAF. The paper shows how EAF can represent different views of reasoners on the same argumentation framework. It also includes representing preferences in EAF and multi-agent argumentation. Finally, the paper discusses the complexity of the problem of determining whether or not an ω-epistemic labelling set exists.

Keywords: Argumentation framework · Epistemic information · Multiple agents · Preferences

1 Introduction

Rational agents often claim that they make their decision based on their knowledge and beliefs when facing alternative and conflicting choices. Consider two examples:

- On January 15, 2019, British Prime Minister's Theresa May suffered a humiliating defeat in the vote on the Brexit deal; 432 Members of Parliament (MPs) voted against the deal while 202 were for it.[1] The MPs who voted against the deal believe that the deal is bad for Britain. Those who voted for the deal believe that the deal is the best that Britain can get.
- In the US presidential election, a voter selects one candidate from a set of candidates (often only two candidates). Everyone claims that he/she has made the "right choice".

---

[1] *"Brexit vote"*, Jan. 15th, 2019. washingtonpost.com.

© Springer Nature Switzerland AG 2019
A. C. Nayak and A. Sharma (Eds.): PRICAI 2019, LNAI 11670, pp. 718–732, 2019.
https://doi.org/10.1007/978-3-030-29908-8_56

In each scenario above, an agent (an MP or a voter) listens to various arguments, which either support or reject a potential decision, and then opts for one among the possibilities, which he/she believes is the right choice. In each situation, the arguments supporting/against a choice, their counter-arguments, etc. can be easily encoded in an *abstract argumentation framework* (AF) introduced in [12]. For instance, $AF = (\{(a)ccept, (r)eject\}, \{(a,r),(r,a)\})$, having two arguments mutually attacking each other, represents (in its most condensed form) the AF that the MPs have for making their choice about the Brexit's deal. Given arguments made by each agent in each scenario, an argumentation semantics of the corresponding AF provides the result of rational reasoning. The stable semantics of the above AF supports two alternative choices, while the ground semantics of the AF supports "no decision". As such, it would likely result in the unanimous choice by all agents who participate in argumentation and claim that they are rational.

The above discussion raises the question "how to express an agent's opinion for supporting an argument among conflicting arguments in the outcome of an AF?" Arguably, there are two possibilities: the agent modifies the AF so that the new AF supports his/her choice or the agent is simply biased towards his/her conclusion. In the first case, nothing other than the agent's beliefs could influence his/her choice of arguments and/or attacks that lead to the new AF, which ultimately leads to his/her conclusion. In this approach, a modified AF represents objective evidences and subjective beliefs indistinguishably. If one merges objective evidences (normally invariant) and subjective beliefs (possibly variant) in a single AF, however, it must be revised whenever an agent changes its own belief. Moreover, it would become hard to distinguish subjective beliefs from objective evidences in a personally customized AF. In this respect, it is desirable to have a mechanism that can distinguishably represent subjective beliefs (or biases) of agents as well as objective evidences as an AF.

In the second case, biases, reflecting beliefs of agents, could be viewed as agents' preferences. Furthermore, there is a huge amount of literature in AF on dealing with preferences in argumentation. It is therefore instructive to consider whether previously developed approaches to dealing with preferences would be sufficient to capture biases. In most approaches in abstract AF, the key idea is to extend an AF with a syntactic component that records the preferences such as a preference relation among arguments or an attack relation between arguments and attacks, and then define a new semantics for this extended AF (detailed discussion is in Sect. 4). Approaches to dealing with preferences have thus far only considered biases/preferences between arguments (e.g., prefer an argument over another one) or preferences between arguments and attacks. However, it is difficult to apply those approaches to represent preferences in a complicated situation. Suppose the following scenario: a person, who goes to a restaurant, has a preference on the combination of food and drink: white wine for fish and red wine for meat. However, the person wants no red wine other than French one, so he/she will take white wine for meat if French red wine is unavailable. It is hard to specify such conditional preference using preference relations among individual arguments. Then we represent preferences as a formula over epistemic literals.

In this paper, we propose an approach to incorporate agents' *beliefs* into an argumentation framework (AF). Specifically, we propose an extension of AF, called *epistemic argumentation framework* (EAF). EAF introduces the third component to an AF, *an epistemic constraint*, that represents the belief of an agent given an AF. We study formal properties of EAF and show that it can be used in representing preferences and decision making in multiagent environments. We also investigate computational complexity and discuss related issues. The rest of the paper is organized as follows. Section 2 reviews basic notions of argumentation frameworks used in this paper. Section 3 introduces epistemic argumentation frameworks and addresses its applications. Section 4 discusses related issues and Sect. 5 concludes the paper. Due to space limit, we omit proofs of propositions, which will be provided in the full paper.

2   Argumentation Framework

This paper uses *(abstract) argumentation frameworks* introduced by [12].

An *argumentation framework* (AF) is a pair (Ar, att) where Ar is a (finite) set of *arguments* and $att \subseteq Ar \times Ar$. We write $a \rightarrow b$ (say, a *attacks* b) iff $(a, b) \in att$. We say that a *indirectly attacks* b if there is a finite sequence $x_0, ..., x_{2n+1}$ $(n \geq 1)$ such that $a = x_0$ and $b = x_{2n+1}$ and for each $0 \leq i \leq 2n$, $(x_i, x_{i+1}) \in att$.

For the semantics of AFs, we use the labelling-based semantics [10]. A *labelling* of (Ar, att) is a (total) function $\mathcal{L} : Ar \rightarrow \{ \text{in, out, und} \}$. When $\mathcal{L}(a) = \text{in}$ (resp. $\mathcal{L}(a) = \text{out}$ or $\mathcal{L}(a) = \text{und}$) for an argument $a \in Ar$, it is written as $\text{in}(a)$ (resp. $\text{out}(a)$ or $\text{und}(a)$). In this case, the argument a is said to be *accepted* (resp. *rejected* or *undecided*) in $\mathcal{L}$. Given $AF = (Ar, att)$ and a labelling $\mathcal{L}$, define $\text{in}(\mathcal{L}) = \{ x \mid \mathcal{L}(x) = \text{in for } x \in Ar \}$, $\text{out}(\mathcal{L}) = \{ x \mid \mathcal{L}(x) = \text{out for } x \in Ar \}$, and $\text{und}(\mathcal{L}) = \{ x \mid \mathcal{L}(x) = \text{und for } x \in Ar \}$. A labelling $\mathcal{L}$ of (Ar, att) is also represented as a set $S(\mathcal{L}) = \{ \lambda(x) \mid \mathcal{L}(x) = \lambda \text{ for } x \in Ar \}$. We say that $\lambda(x)$ represents the *justification state* of $x \in Ar$.

A labelling $\mathcal{L}$ of $AF = (Ar, att)$ is a *complete labelling* if for each argument $a \in Ar$, it holds that:

- $\mathcal{L}(a) = \text{in}$ iff $\mathcal{L}(b) = \text{out}$ for every $b \in Ar$ such that $(b, a) \in att$.
- $\mathcal{L}(a) = \text{out}$ iff $\mathcal{L}(b) = \text{in}$ for at least one $b \in Ar$ such that $(b, a) \in att$.
- $\mathcal{L}(a) = \text{und}$, otherwise.

Let $\mathcal{L}$ be a complete labelling of AF. Then,

- $\mathcal{L}$ is a *stable labelling* iff $\text{und}(\mathcal{L}) = \emptyset$.
- $\mathcal{L}$ is a *grounded labelling* iff $\text{in}(\mathcal{L}) \subseteq \text{in}(\mathcal{L}')$ for any complete labelling $\mathcal{L}'$ of AF.
- $\mathcal{L}$ is a *preferred labelling* iff there is no complete labelling $\mathcal{L}'$ of AF such that $\text{in}(\mathcal{L}) \subset \text{in}(\mathcal{L}')$.

We often abbreviate complete, stable, grounded, and preferred labelling as *co, st, gr,* and *pr*, respectively.

3   Epistemic Argumentation Framework

3.1   Epistemic Labelling Set

Given $AF = (Ar, att)$, define $\mathcal{A}_{AF} = \{\, \text{in}(a), \text{out}(a), \text{und}(a) \mid a \in Ar \,\}$. An *epistemic atom* over AF is of the form $\mathbf{K}\,\varphi$ or $\mathbf{M}\,\varphi$ where φ is a propositional formula over $\mathcal{A}_{AF}$. An *epistemic literal* is an epistemic atom or its negation. An *epistemic formula* (over $\mathcal{A}_{AF}$) is a propositional formula constructed over epistemic literals together with $\top$ (true) and $\bot$ (false). Intuitively, $\mathbf{K}\,\varphi$ (resp. $\mathbf{M}\,\varphi$) states that the agent believes that φ is *true* (resp. *possibly true*).[2] We will use epistemic formulas to represent the epistemic side of an agent given an AF.

Let φ be a propositional formula over $\mathcal{A}_{AF}$ and $\mathcal{L}$ be a labelling over AF. Then $S(\mathcal{L})$ is considered an interpretation of φ. We say that φ is true in $\mathcal{L}$, denoted by $\mathcal{L} \models \varphi$, if φ is interpreted to be true under $S(\mathcal{L})$.

Definition 1 (satisfaction). A set SL of labellings *satisfies* an epistemic formula φ, denoted by $SL \models \varphi$, if one of the following conditions holds:

(i) $\varphi = \top$,
(ii) $\varphi = \mathbf{K}\,\psi$ and $\mathcal{L} \models \psi$ for every $\mathcal{L} \in SL$,
(iii) $\varphi = \mathbf{M}\,\psi$ and $\mathcal{L} \models \psi$ for some $\mathcal{L} \in SL$,
(iv) $\varphi = \neg\psi$ and $SL \not\models \psi$,
(v) $\varphi = \varphi_1 \wedge \varphi_2$ and $(SL \models \varphi_1$ and $SL \models \varphi_2)$,
(vi) $\varphi = \varphi_1 \vee \varphi_2$ and $(SL \models \varphi_1$ or $SL \models \varphi_2)$.

An epistemic formula φ is *consistent* if there exists a (non-empty) set SL of labellings such that $SL \models \varphi$; otherwise, φ is *inconsistent*. Some basic properties are addressed.

Proposition 1. *Let SL be a set of labellings. For any propositional formula φ and ψ over $\mathcal{A}_{AF}$,*

(i) $SL \models \neg\mathbf{M}\,\varphi$ iff $SL \models \mathbf{K}\,\neg\varphi$,
(ii) $SL \models \neg\mathbf{K}\,\varphi$ iff $SL \models \mathbf{M}\,\neg\varphi$,
(iii) $SL \models \mathbf{M}\,(\varphi \vee \psi)$ iff $SL \models \mathbf{M}\,\varphi \vee \mathbf{M}\,\psi$,
(iv) $SL \models \mathbf{K}\,(\varphi \wedge \psi)$ iff $SL \models \mathbf{K}\,\varphi \wedge \mathbf{K}\,\psi$.

Definition 2 (epistemic argumentation framework). An *epistemic argumentation framework* (EAF) is a triple (Ar, att, φ) where $AF = (Ar, att)$ is an argumentation framework and φ is an epistemic formula (called an *epistemic constraint*).

---

[2] By the meaning, it might be better to write $\mathbf{B}\varphi$ rather than $\mathbf{K}\,\varphi$, but we use $\mathbf{K}$ because we implement it using epistemic logic programs in which $\mathbf{K}$ and $\mathbf{M}$ are used (see Sect. 5).

Intuitively, an EAF (Ar, att, φ) represents the view of an agent who, given $AF = (Ar, att)$, believes that φ is true. So, an EAF consists of two different types of information: an objective evidence AF and a subjective belief φ of an agent. We also refer to an EAF by (AF, φ) whenever it is clear from the context what AF refers to.

Example 1. In the introductory example, consider an AF with the set of arguments $\{ (f)ish, (m)eat, (w)hite, (r)ed, (u)navailable \}$ and the set of attacks $\{(f, m), (m, f), (w, r), (r, w), (r, u), (u, r) \}$.

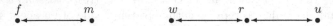

Then, some EAFs are defined as follows:

- $EAF_1 = (AF, \mathbf{M}\mathrm{in}(r))$ represents the view of an agent who believes that r is possibly accepted.
- $EAF_2 = (AF, \mathbf{K}\mathrm{in}(w) \vee \mathbf{K}\mathrm{in}(r))$ represents the view of an agent who believes that either w or r should be accepted.
- $EAF_3 = (AF, \mathbf{K}(\mathrm{in}(m) \wedge \neg\mathrm{in}(u) \rightarrow \mathrm{in}(r)) \wedge \mathbf{K}(\mathrm{in}(f) \rightarrow \mathrm{in}(w)))$ represents the view of an agent whose belief is given by the statement: "*if m is accepted and u is unaccepted, then r should be accepted; and if f is accepted then w should be accepted.*"

Next we define the semantics of an EAF.

Definition 3 (epistemic labelling set). Let $EAF = (AF, \varphi)$ and $\omega \in \{co, st, gr, pr\}$. A set SL of labellings is an ω-*epistemic labelling set* of (AF, φ) if (i) each $\mathcal{L} \in SL$ is an ω-labelling of AF, and (ii) SL is a $\subseteq$-maximal set of ω-labellings of AF that satisfies φ. An EAF possibly has multiple ω-epistemic labelling sets.

Intuitively, an ω-epistemic labelling set is a collection of ω-labellings that reflects the belief of an agent. In particular, $EAF = (AF, \top)$ has the unique ω-epistemic labelling set that coincides with the set of ω-labellings of AF. In what follows, we assume $\omega \in \{co, st, gr, pr\}$ unless stated otherwise. By definition, EAF always has an ω-epistemic labelling set (possibly as an empty set).

Proposition 2. $EAF = (AF, \bot)$ *has the ω-epistemic labelling set $\emptyset$.*

Our primary interest is an EAF that has non-empty ω-epistemic labelling sets.

Example 2. Consider the EAFs of Example 1 under the stable semantics. First, AF in the EAFs has four stable labellings:

$$L_1 = \{ \mathrm{in}(f), \mathrm{out}(m), \mathrm{out}(w), \mathrm{in}(r), \mathrm{out}(u) \},$$
$$L_2 = \{ \mathrm{out}(f), \mathrm{in}(m), \mathrm{out}(w), \mathrm{in}(r), \mathrm{out}(u) \},$$
$$L_3 = \{ \mathrm{in}(f), \mathrm{out}(m), \mathrm{in}(w), \mathrm{out}(r), \mathrm{in}(u) \},$$

$$L_4 = \{\, \text{out}(f),\ \text{in}(m),\ \text{in}(w),\ \text{out}(r),\ \text{in}(u)\,\}.$$

This implies EAF_1 has a unique stable epistemic labelling set $\{L_1, L_2, L_3, L_4\}$; EAF_2 has two stable epistemic labelling sets $\{L_1, L_2\}$ and $\{L_3, L_4\}$; and EAF_3 has a unique stable epistemic labelling set $\{L_2, L_3, L_4\}$. Suppose that it turns that French red wine is unavailable. The situation is represented by

$$EAF_4 = (AF,\ \mathbf{K}(\text{in}(m) \wedge \neg\text{in}(u) {\rightarrow} \text{in}(r)) \wedge \mathbf{K}(\text{in}(f) {\rightarrow} \text{in}(w)) \wedge \mathbf{K}\text{in}(u)).$$

Then EAF_4 has a unique stable epistemic labelling set $\{L_3, L_4\}$.

As shown in the above example, EAF can represent belief change of an agent by revising an epistemic constraint without modifying AF. The revised EAF then produces new epistemic labelling sets that reflect new belief states of an agent. In Example 2, EAF_4 introduces an additional constraint $\mathbf{K}\text{in}(u)$ to EAF_3, which results in eliminating L_2 from the stable epistemic labelling set of EAF_3. For two epistemic formulas φ_1 and φ_2, we say that φ_1 is *stronger* than φ_2 if $\varphi_1 \models \varphi_2$ (in the sense of classical logic). Introducing a stronger constraint to EAF eliminates elements of SL in general.

Proposition 3. *Let $EAF_1 = (AF, \varphi_1)$ and $EAF_2 = (AF, \varphi_2)$ be two EAFs such that φ_1 is stronger than φ_2. Then, for each ω-epistemic labelling set SL_1 of EAF_1 there exists some ω-epistemic labelling set SL_2 of EAF_2 such that $SL_1 \subseteq SL_2$.*

In argumentation frameworks, stable, grounded, or preferred labellings are complete labellings. In epistemic argumentation frameworks, a similar result holds.

Proposition 4. *Let (AF, φ) be an EAF. If a non-empty set SL of labellings is a stable, grounded, or preferred epistemic labelling set of (AF, φ), then $\mathcal{L} \in SL$ is an element of a complete epistemic labelling set of (AF, φ).*

We next consider a sufficient condition for the uniqueness of ω-epistemic labelling sets.

Lemma 5. *Let φ be a conjunction of epistemic literals over $\mathcal{A}_{AF}$. If two sets of labellings SL_1 and SL_2 satisfy φ (i.e., $SL_1 \models \varphi$ and $SL_2 \models \varphi$), then $SL_1 \cup SL_2 \models \varphi$.*

Using the lemma, we can prove the next result.

Proposition 6. *Let (AF, φ) be an EAF such that φ is a conjunction of epistemic literals. Then (AF, φ) has a unique ω-epistemic labelling set.*

Assume that φ is a DNF in which each disjunct is a conjunction of epistemic literals. Due to Proposition 1, we can assume that each disjunct in φ is of the form $\mathbf{K}\,\psi_0 \wedge \mathbf{M}\,\psi_1 \wedge \cdots \wedge \mathbf{M}\,\psi_n{}^3$ where ψ_i $(0 \leq i \leq n)$ is a propositional formula over $\mathcal{A}_{AF}$, which will be denoted by $EC(\psi_0; \psi_1, \ldots, \psi_n)$. We can prove:

---

[3] $\neg\mathbf{M}\,\psi$ (resp. $\neg\mathbf{K}\,\psi$) is converted to $\mathbf{K}\,\neg\psi$ (resp. $\mathbf{M}\,\neg\psi$), and $\mathbf{K}\,\psi_1 \wedge \mathbf{K}\,\psi_2$ is converted to $\mathbf{K}\,(\psi_1 \wedge \psi_2)$.

Lemma 7. *Let SL be a set of labellings such that $SL \models EC(\psi_0; \psi_1, \ldots, \psi_n)$. Then, for each $i = 1, \ldots, n$, there exists some $\mathcal{L} \in SL$ such that $\mathcal{L} \models \psi_0 \wedge \psi_i$.*

Proposition 8. *Let $\varphi = \bigvee_{j=1}^{k} EC(\psi_j; \psi_1^j, \ldots, \psi_{n_j}^j)$ $(k \geq 1)$ be an epistemic formula. Then, $EAF = (AF, \varphi)$ has a non-empty ω-epistemic labelling set if there exists an integer j $(1 \leq j \leq k)$ such that for each $1 \leq i \leq n_j$, AF has an ω-labelling $\mathcal{L}$ and $\mathcal{L} \models \psi_j \wedge \psi_i^j$.*

Each AF semantics imposes some specific condition on every argument, e.g., the stable semantics allows no argument to be undecided, while the grounded semantics keeps controversial arguments undecided. EAF is useful for selecting intended labellings from the set of all possible labellings.

Example 3. Consider the AF in Example 1. Since the availability of French red wine is unknown before visiting a restaurant, an agent wants to keep the argument u undecided. The situation is specified as the epistemic constraint $\varphi = \mathbf{K}und(u)$. Then (AF, φ) has the single preferred epistemic labelling set $\{\{\mathrm{in}(f), \mathrm{out}(m), \mathrm{und}(w), \mathrm{und}(r), \mathrm{und}(u)\}, \{\mathrm{out}(f), \mathrm{in}(m), \mathrm{und}(w), \mathrm{und}(r), \mathrm{und}(u)\}\}$.

3.2 Representing Preference

Preference among arguments can be specified in EAF as follows. Let $\succeq$ be a pre-order (i.e., reflexive and transitive) relation over $Ar \times Ar$ such that $(x, y) \in \succeq$ implies that x indirectly attacks y or vice versa. $x \succeq y$ means that an argument x is at least as preferred as y. We write $x \succ y$ if $x \succeq y$ and $y \not\succeq x$.

Definition 4 (preference over arguments). Given $AF = (Ar, att)$ and a preorder relation $\succeq \subseteq Ar \times Ar$, define $EAF = (AF, \varphi_A)$ where

$$\varphi_A = \bigwedge_{x \succ y} \mathbf{K}(\mathrm{in}(y) \supset \mathrm{in}(x)).$$

Intuitively speaking, φ_A represents that an argument x should be accepted whenever another argument y of lower preference is accepted. Note that the preference is specified as $x \succ y$ but not as $x \succeq y$ in φ_A. When both $x \succeq y$ and $y \succeq x$ exist, there is no reason to prefer one of them. In this case, the conjunct involved x and y in φ_A is $\top$.

Proposition 9. *Let $EAF = (AF, \varphi_A)$ be an EAF defined as above. Then, for any ω-epistemic labelling set SL of EAF, there is no $\mathcal{L} \in SL$ such that $\mathrm{in}(x) \notin \mathcal{L}$ and $\mathrm{in}(y) \in \mathcal{L}$ for any $x \succ y$.*

Example 4. Consider $AF = (\{a, r\}, \{(a, r), (r, a)\})$ with $r \succ a$. Then $EAF = (AF, \varphi_A)$ with $\varphi_A = \mathbf{K}(\mathrm{in}(a) \supset \mathrm{in}(r))$ has the unique stable epistemic labelling set $\{\{\mathrm{in}(r), \mathrm{out}(a)\}\}$, and the unique complete epistemic labelling set $\{\{\mathrm{in}(r), \mathrm{out}(a)\}, \{\mathrm{und}(r), \mathrm{und}(a)\}\}$.

In Example 4, the complete epistemic labelling set contains $\{\text{und}(r), \text{und}(a)\}$. This can be eliminated by introducing the constraint $\varphi_A = \mathbf{K}(\text{in}(a) \vee \text{und}(a) \supset \text{in}(r))$.

Preference over arguments is generalized to preference over justification states of arguments as follows. A pre-order relation $\sqsupseteq$ over justification states of arguments is a collection of elements of the form $\lambda(x) \sqsupseteq \mu(y)$ where $\lambda, \mu \in \{\text{in}, \text{out}, \text{und}\}$, meaning that $\lambda(x)$ is at least as preferred as $\mu(y)$ for arguments x and y. We write $\lambda(x) \sqsupset \mu(y)$ if $\lambda(x) \sqsupseteq \mu(y)$ and $\mu(y) \not\sqsupseteq \lambda(x)$.

Definition 5 (preference over justification states). Given $AF = (Ar, att)$ and a preorder relation $\sqsupseteq \subseteq \mathcal{A}_{AF} \times \mathcal{A}_{AF}$, define $EAF = (AF, \varphi_J)$ where

$$\varphi_J = \bigwedge_{\lambda(x) \sqsupset \mu(y)} \mathbf{K}(\mu(y) \supset \lambda(x)).$$

φ_J states that if the justification state $\lambda(x)$ is preferred to $\mu(y)$ for $x, y \in Ar$, then $\mathcal{L} \models \mu(x)$ implies $\mathcal{L} \models \lambda(x)$ for any $\mathcal{L} \in SL$ where SL is any ω-epistemic labelling set of EAF.

By definition, Definition 4 is considered a special case of Definition 5 with $\mu = \lambda = \text{in}$.

Proposition 10. *Let $EAF = (AF, \varphi_J)$ be an EAF defined as above. Then, for any ω-epistemic labelling set SL of EAF, there is no $\mathcal{L} \in SL$ such that $\lambda(x) \notin \mathcal{L}$ and $\mu(y) \in \mathcal{L}$ for any $\lambda(x) \sqsupset \mu(y)$. In particular, $\mu(y) \notin \mathcal{L}$ for any $\mathcal{L} \in SL$ if $x = y$.*

Example 5. Suppose that in Example 4, an MP prefers keeping the decision undecided if possible. This is represented by $\sqsupset = \{(\text{und}(x), \text{in}(x)), (\text{und}(x), \text{out}(x)) \mid x \in \{a, r\}\}$ which is translated to $\varphi_J = \bigwedge_{x \in \{a,r\}} \mathbf{K}(\text{in}(x) \supset \text{und}(x)) \wedge \mathbf{K}(\text{out}(x) \supset \text{und}(x))$. Then $EAF = (AF, \varphi_J)$ has the unique complete epistemic labelling set $\{\{\text{und}(r), \text{und}(a)\}\}$. Furthermore, $\emptyset$ is the stable epistemic labelling set, since there is no choice to make a and r undecided.

In this way, EAF enables us to specify preference over not only arguments but also justification states of arguments. Furthermore, it could also be useful to introduce preferences among epistemic formulas. For instance, we could write $\mathbf{K}\lambda(x) > \mathbf{K}\mu(x)$ for some argument x to indicate that we prefer SL_1 over SL_2 whenever $SL_1 \models \mathbf{K}\lambda(x)$ and $SL_2 \models \mathbf{K}\mu(x)$ for two arbitrary ω-epistemic labelling sets SL_1 and SL_2. We leave such extensions for future work.

3.3   Multiple Agents

Suppose that two agents share $AF = (\{a, r\}, \{(a, r), (r, a)\})$. If they have the same belief represented by the epistemic constraint $\varphi = \mathbf{K}\text{in}(a)$, the EAF (AF, φ) has the single epistemic complete labelling set $\{\{\text{in}(a), \text{out}(r)\}\}$ and the agents agree on accepting a. On the other hand, if two agents have conflicting beliefs $\varphi_1 = \mathbf{K}\mathrm{in}(a)$ and $\varphi_2 = \neg\mathbf{K}\mathrm{in}(a)$ respectively, then they do not agree on accepting a or r. In this section, we assume multiple agents who share the same AF while having different beliefs in general. The situation is represented by the collection of EAFs (AF, φ_i) $(1 \leq i \leq n)$. First, we define two different types of agreements.

Definition 6 (agreement). Let $AF = (Ar, att)$ and $EAF_1 = (AF, \varphi_1)$, ..., $EAF_n = (AF, \varphi_n)$ $(n \geq 1)$. Then $EAF_1, \ldots, EAF_n$ *credulously agree on* $\lambda(a)$ for $a \in Ar$ where $\lambda \in \{\mathrm{in}, \mathrm{out}, \mathrm{und}\}$ *under* ω-*epistemic labelling* if each EAF_i $(i = 1, \ldots, n)$ has an ω-epistemic labelling set SL_i such that $SL_i \models \mathbf{M}\lambda(a)$. In contrast, $EAF_1, \ldots, EAF_n$ *skeptically agree on* $\lambda(a)$ *under* ω-*epistemic labelling* if for any ω-epistemic labelling set SL_i of EAF_i $(i = 1, \ldots, n)$ $SL_i \models \mathbf{K}\lambda(a)$.

The above definition characterizes two different situations (credulous or skeptical) in which agents reach an agreement on $\lambda(a)$. For simplicity reasons, Definition 6 assumes that different agents employ the same ω-epistemic labelling, but the definition is easily extended to a case in which agents employ different ω-labellings.

Proposition 11. *Let* $AF = (Ar, att)$ *and* $EAF_1 = (AF, \varphi_1)$, ..., $EAF_n = (AF, \varphi_n)$ $(n \geq 1)$. *Then,* $EAF_1, \ldots, EAF_n$ *skeptically agree on* $\lambda(a)$ *for* $a \in Ar$ *under* ω-*epistemic labelling iff* EAF_i *and* $EAF_i' = (AF, \varphi_i \wedge \mathbf{K}\,\lambda(a))$ $(i = 1, \ldots, n)$ *have the same* ω-*epistemic labelling sets.*

Proposition 12. *Let* $AF = (Ar, att)$ *and* $EAF_1 = (AF, \varphi_1)$, ..., $EAF_n = (AF, \varphi_n)$ $(n \geq 1)$. *If* $EAF_1, \ldots, EAF_n$ *credulously agree on* $\lambda(a)$ *for* $a \in Ar$ *under* ω-*epistemic labelling, then* $(AF, \varphi_1 \vee \cdots \vee \varphi_n)$ *has an* ω-*epistemic labelling set* SL *such that* $SL \models \mathbf{M}\lambda(a)$. *Conversely, if* $(AF, \varphi_1 \wedge \cdots \wedge \varphi_n)$ *has an* ω-*epistemic labelling set* SL *such that* $SL \models \mathbf{M}\lambda(a)$, *then* $EAF_1, \ldots, EAF_n$ *credulously agree on* $\lambda(a)$ *under* ω-*epistemic labelling.*

We next show that EAF can be used for formalizing majority voting. In the presence of $EAF_i = (AF, \varphi_i)$ $(1 \leq i \leq n)$, define:

$$M_\psi^\omega = \{\, i \mid EAF_i \text{ has an } \omega\text{-epistemic labelling set } SL \text{ s.t. } SL \models \mathbf{M}\,\psi \,\},$$
$$N_\psi^\omega = \{\, i \mid \text{for each } \omega\text{-epistemic labelling set SL of } EAF_i, \, SL \models \mathbf{K}\,\psi \,\}.$$

Definition 7 (majority voting). Let $AF = (Ar, att)$ and $EAF_i = (AF, \varphi_i)$ for $(1 \leq i \leq n)$. For $a \in Ar$, $\lambda(a)$ is *credulously* (resp. *skeptically*) *adopted by majority voting under* ω-*epistemic labelling* iff the cardinality of the set $M_{\lambda(a)}^\omega$ (resp. $N_{\lambda(a)}^\omega$) is greater than the cardinality of the set $M_{\mu(a)}^\omega$ (resp. $N_{\mu(a)}^\omega$) where $\lambda, \mu \in \{\mathrm{in}, \mathrm{out}, \mathrm{und}\}$ and $\lambda \neq \mu$.

When $|M_{\lambda(a)}^\omega| = n$ (resp. $|N_{\lambda(a)}^\omega| = n$) in Definition 7, EAF_1, ..., EAF_n credulously (resp. skeptically) agree on $\lambda(a)$.

Algorithm 1. Existence(EAF, ω)

Input: ω, $EAF = (AF, \varphi)$.
Output: **true** if EAF has a (non-empty) ω-epistemic labelling set; **false**
      otherwise.

1 Convert to DNF: $\varphi = \vee_{j=1}^{k} EC(\psi_j; \psi_1^j, \ldots, \psi_{n_j}^j)$

2   where $EC(\psi; \psi_1, \ldots, \psi_k) = \mathbf{K}\psi \wedge \bigwedge_{i=1}^{k} \mathbf{M}\psi_i$

3 **for** $j = 1$ *to* k **do**

4 | $num\_labelling := 0$

5 | **for** $i = 1$ *to* n_j **do**

6 | | **if** $\mathrm{D}(\omega, AF, \psi_j \wedge \psi_i^j) = $ **true then**

7 | | | $num\_labelling := num\_labelling + 1$

8 | **if** $num\_labelling = n_j$ **then return true**

9 **return false**

Example 6. Consider $AF = (\{a, r\}, \{(a, r), (r, a)\})$ and three EAFs: $EAF_1 = (AF, \mathbf{K} \, in(a))$, $EAF_2 = (AF, \neg\mathbf{M} \, und(a))$, and $EAF_3 = (AF, \mathbf{K} \, und(a))$. Then $in(a)$ is credulously adopted by majority voting under the complete epistemic labelling, while it is not skeptically adopted.

3.4 Complexity

We assume that the readers are familiar with the well-known notations in computational complexity (e.g., P-c, NP-c, coNP-c, etc.). Let $\omega \in \{gr, st, co, pr\}$ and $EAF = (AF, \varphi)$. Due to Proposition 8, we can check for the existence of a non-empty ω-epistemic labelling set using Algorithm 1, assuming the existence of a procedure $\mathrm{D}(\omega, AF, \psi)$ that determines the existence of an ω-labelling $\mathcal{L}$ of AF such that $\mathcal{L} \models \psi$.

In essence, Algorithm 1 shows that checking whether EAF has a non-empty ω-epistemic labelling set can be reduced to checking whether a labelling $\mathcal{L}$ of AF satisfies a formula over $\mathcal{A}_{AF}$. In line 1 we assume that φ has at most k disjuncts, and each contains at most p conjuncts, where p and k are polynomial in the size of the AF and refer to φ as a (k, p)-DNF.[4] Under this assumption, Algorithm 1 will call $\mathrm{D}(\omega, AF, \psi)$ at most $k \times p$ times. Consider the following decision problem:

$\mathbf{Exists}_{\omega}^{(k,p)}$: Given an $AF = (Ar, att)$ and a (k, p)-DNF epistemic formula φ over $\mathcal{A}_{AF}$, does (AF, φ) have a non-empty ω-epistemic labelling set?

The above discussion gives us the next result.

Proposition 13. $\mathbf{Exists}_{\omega}^{(k,p)}$ *is P-c for* $\omega = gr$ *and NP-c for* $\omega \in \{co, st, pr\}$.

---

[4] The DNF of a formula φ might have exponential number of disjuncts in general, however, it would be a rare case that belief of an agent is expressed by an exponential formula.

The proof of the above results relies on the following facts: (i) the grounded labelling of AF can be computed in polynomial time and is unique; and given a labelling $\mathcal{L}$ and a propositional formula ψ over $\mathcal{A}_{AF}$, (ii) checking whether there exists an ω-labelling satisfying a formula is NP-c for $\omega \in \{st, co, pr\}$ (by the result $\mathtt{Cred}_\sigma$ in [13, Table 1] or in [14]); (iii) checking whether a given labelling $\mathcal{L}$ satisfies a propositional formula over $\mathcal{A}_{AF}$ is polynomial.

4   Related Work

EAF could be viewed as an approach to limiting the set of extensions (or labellings) of an argumentation framework for semantical consideration and this is similar, at least in the spirit, to argumentation with preferences and probabilistic argumentation. By introducing epistemic constraints, it is similar to works focusing on a reasoner's belief. The key difference between EAF and the other approaches can be summarized as follows.

Constrained argumentation frameworks (CAF) proposed in [11] are syntactically similar to EAF. Both are of the form $\langle A, R, C \rangle$ where (A, R) is an AF and C is a propositional formula (over A) in a CAF whilst it is an epistemic formula (over $\mathcal{A}_{AF}$) in an EAF. The key distinction between CAF and EAF lies in the use of the constraint. In CAF, C is imposed on extensions of the AF leading to a new set of extensions of the original AF. In contrast, φ does not change the labellings of the original AF in an EAF (AF, φ). Another extension of Dung's AF is *abstract dialectical framework* (ADF) [9] where each argument has an associated acceptance condition expressed by a propositional formula over the existing arguments. In EAF individual arguments do not have acceptance conditions, while epistemic constraints specify beliefs concerning which arguments are to be (un)accepted in the final outcome.

Probabilistic argumentation as proposed in [16,17] focuses on the uncertainty of arguments rather than reasoners' beliefs. This approach represents the beliefs of agents by a probability assignment to arguments [16] or an epistemic labelling [17]. It provides methods for computing epistemic extensions of an AF which contain arguments with probability greater than a certain threshold or assigning labels to arguments in accordance to the probability of the labelling, i.e., it merges an objective evidence and subjective beliefs in a single framework, which is in contrast to our approach. Moreover, it differs from EAFs significantly as its extensions might not correspond to any type of extensions of the original AF. On the other hand, there would be a connection between probabilistic argumentation and EAF. For instance, we consider that for each $EAF = (AF, \varphi)$ and ω, there would exist a probabilistic distribution P with respect to AF with the property that x is believed wrt P $(P(x) > 0.5)$ then $in(x)$ is skeptically entailed by every ω-labelling set of EAF. We believe that the inverse could be true as well. We leave the precise formulation and proof of this interesting problem for future work. Recent work in this direction has introduced *epistemic attack semantics* that considers extended probability distribution, which assigns degrees of belief to arguments and attacks [22] which is then further investigated in dynamic setting [18]. Whether formulas in EAF could sufficiently model this type of extension is an open question that we intend to pursuit as well.

Argumentation with preferences or priorities has been studied extensively in recent years. Preference over arguments is introduced as a preorder relation over arguments in [2–4,19], while a new attack relation that ranges from arguments to attacks is used in [20]. Our representation of preferences is close to the approach employing a preorder but there are differences from them. For instance, given $AF = (\{a, b\}, \{(a, b)\})$ with the preference $a \preceq b$, Kaci and van der Torre [19] provide its semantics by extensions of $AF_1 = (\{a, b\}, \{\})$, and Amgoud and Vesic [3] convert AF to $AF_2 = (\{a, b\}, \{(b, a)\})$. As such, the structure of the original argumentation graph is changed, and as a result, extensions of the preference-based AF are not extensions selected from those of AF. Wakaki [23] introduces *preference-based AF* (PAF) which, as we do, selects extensions based on preference relation over arguments. Our representation of preference in EAF is different from PAF in the sense that EAF can represent preference over not only arguments but justification states. *Value-based argumentation framework* (VAF) [6] represents preference in AF by assigning values to arguments. In VAF acceptable arguments may change depending on the order of values. Arguments acceptable irrespective of any value order are called *objectively acceptable* and those acceptable for some order are called *subjectively acceptable*. In EAF justification states of arguments change depending on epistemic constraints, so the effect of epistemic constraints in EAF is similar to the effect of value in VAF. On the other hand, VAF may produce extensions that are not those of the original AF, while EAF produces labellings that are also labellings of the original AF. Airiau *et al.* [1] consider the problem such that given a profile of argumentation frameworks $(AF_1, \ldots, AF_n)$, one for each agent, can this profile be explained in terms of a single master argumentation framework, an association of arguments with values, and a profile of preference orders over values $(\succeq_1, \ldots, \succeq_n)$, one for each agent? Their approach represents individual views of a common AF by preference orders over values, which is in contrast with our approach in which individual views are encoded by epistemic formulas over arguments. Visser *et al.* [24] introduce an *epistemic argumentation framework* for reasoning about preferences with uncertain information. They provide languages and inference schemes for instantiated AFs, which is in contrast with our framework for abstract argumentation.

Schwarzentruber *et al.* [21] introduce a logical framework for reasoning about arguments owned by agents and their knowledge about other agents' arguments. They introduce epistemic logics to represent belief state of agents in dialogues and define Kripke semantics. For instance, they represent that "an agent 1 believes that there exists an argument about global warming (gw) owned by an agent 2" by the formula: $B_1(\langle U \rangle(gw \wedge ownedby(2)))$. Our approach is different from theirs in two ways: first EAF is an extension of AF and we do not use modal logic based on Kripke structures. Second, our primary interest in this paper is to represent an agent's own beliefs, and we do not consider reasoning about beliefs of other agents. Finally, we note that an EAF realizes meta-level reasoning about arguments in abstract argumentation frameworks. In this sense, it could be viewed as a kind of *meta-level arguments* discussed in [7].

5 Conclusion and Future Work

An epistemic argumentation framework introduces belief of agents to argumentation frameworks. A unique feature of EAF is that it can represent arguments and attacks as objective evidence in AF, while at the same time, it can encode subjective beliefs of individual agents by epistemic constraints over the outcome. By separating objective knowledge and subjective beliefs, individual agents could produce different conclusions based on their biases toward a common AF. Such a situation happens, for instance, in a court case where jurors share the same open AF while could reach different conclusions based on their biases. Moreover, the separation has an advantage that an individual agent can easily revise his/her belief without changing the structure of an AF.

This paper addresses declarative aspects of EAFs. From the procedural viewpoint, a system for computing epistemic labelling sets is built on top of *answer set solvers* [8]. More precisely, suppose an EAF (AF, φ) where φ is a CNF $\varphi = \psi_1 \wedge \cdots \wedge \psi_n$ in which $\psi_i \, (1 \leq i \leq n)$ is a disjunction of simple epistemic literals of the form $\mathbf{E}\lambda(x)$ or $\mathbf{E}\neg\lambda(x)$ where $\mathbf{E} \in \{\mathbf{M}, \mathbf{K}\}$ and $\lambda \in \{\mathtt{in}, \mathtt{out}, \mathtt{und}\}$. In this case, the EAF is transformed to an *epistemic logic program* [15] Π and ω-epistemic labelling sets are computed by *world views* of Π. We will address the issue in the full paper.

In this paper, we focus on representing an agent's own belief in EAFs. On the other hand, EAF could be extended to reasoning about beliefs of other agents and representing an agent's own belief based on beliefs of other agents. This type of belief contains a constraint such that "$\mathbf{K}_1\mathtt{in}(a) \supset \mathbf{M}_2\mathtt{in}(a)$" (if an agent 1 supports the acceptance of an argument a then an agent 2 would not argue against it). EAF is used for characterizing several problems in argumentation. For instance, the *enforcement* [5] of an argument a in AF is captured as finding an EAF $(AF', \mathbf{M}\,\mathtt{in}(a))$ having a non-empty ω-epistemic labelling set where AF' is an expansion of AF. We introduce EAF for complete, stable, grounded, or preferred semantics, but the framework is extended to other semantics such as semi-stable, stage, ideal, etc. Those issues are left for future work.

Acknowledgement. This work is supported by JSPS KAKENHI Grant Numbers JP17H00763 and JP18H03288. The authors also acknowledge the partial support from the NSF grants HRD-1345232, IIS-1812628, and OIA-1757207.

References

1. Airiau, S., Bonzon, E., Endriss, U., Maudet, N., Rossit, J.: Rationalisation of profiles of abstract argumentation frameworks: characterisation and complexity. J. Artif. Intell. Res. **60**, 149–177 (2017)

2. Amgoud, L., Cayrol, C.: On the acceptability of arguments in preference-based argumentation. In: Proceedings of the 14th Conference Uncertainty in Artificial Intelligence, pp. 1–7 (1998)
3. Amgoud, L., Vesic, S.: On the role of preference in argumentation frameworks. In: Proceedings of the 22nd IEEE International Conference on Tools with Artificial Intelligence, pp. 219–222 (2010)
4. Amgoud, L., Vesic, S.: A new approach for preference-based argumentation frameworks. Ann. Math. Artif. Intell. **63**(2), 149–183 (2011)
5. Baumann, R., Brewka, G.: Expanding argumentation frameworks: enforcing and monotonicity results. In: Proceedings of the 3rd International Conference on Computational Models of Argument, Frontiers in AI and Applications, vol. 216, pp. 75–86. IOS Press (2010)
6. Bench-Capon, T.J.M.: Value-based argumentation frameworks. In: Proceedings of the 9th International Workshop on Non-Monotonic Reasoning, pp. 443–454 (2002)
7. Boella, G., Gabbay, D.M., van der Torre, L., Villata, S.: Meta-argumentation modelling I: methodology and techniques. Stud. Logica **93**(1), 297–355 (2009)
8. Brewka, G., Eiter, T., Truszczynski, M.: Answer set programming at a glance. Commun. ACM **54**(12), 92–103 (2011)
9. Brewka, G., Woltran, S.: Abstract dialectical frameworks. In: Proceedings of the 12th International Conference on Principles of Knowledge Representation and Reasoning, pp. 102–111 (2010)
10. Caminada, M.W.A., Gabbay, D.M.: A logical account of formal argumentation. Stud. Logica **93**(2–3), 109–145 (2009)
11. Coste-Marquis, S., Devred, C., Marquis, P.: Constrained argumentation frameworks. In: Proceedings of the 10th International Conference on Principles of Knowledge Representation and Reasoning, pp. 112–122 (2006)
12. Dung, P.M.: On the acceptability of arguments and its fundamental role in non-monotonic reasoning, logic programming and n-person games. Artif. Intell. **77**, 321–357 (1995)
13. Dvořák, W.: On the complexity of computing the justification status of an argument. In: Modgil, S., Oren, N., Toni, F. (eds.) TAFA 2011. LNCS (LNAI), vol. 7132, pp. 32–49. Springer, Heidelberg (2012). https://doi.org/10.1007/978-3-642-29184-5_3
14. Dvořák, W., Dunne, P.E.: Computational problems in formal argumentation and their complexity. In: Handbook of Formal Argumentation, pp. 631–688. College Publications (2018)
15. Gelfond, M.: Strong introspection. In: Proceedings of the 9th National Conference on Artificial Intelligence (AAAI), pp. 386–391 (1991)
16. Hunter, A.: A probabilistic approach to modelling uncertain logical arguments. J. Approx. Reason. **54**, 47–81 (2013)
17. Hunter, A., Thimm, M.: Probabilistic reasoning with abstract argumentation frameworks. J. Artif. Intell. Res. **59**, 565–611 (2017)
18. Hunter, A., Polberg, S., Potyka, N.: Updating belief in arguments in epistemic graphs. In: Proceedings of the 16th International Conference on Principles of Knowledge Representation and Reasoning, pp. 138–147 (2018)
19. Kaci, S., van der Torre, L.: Preference-based argumentation: arguments supporting multiple values. J. Approx. Reason. **48**, 730–751 (2008)
20. Modgil, S.: Reasoning about preferences in argumentation frameworks. Artif. Intell. **173**(9–10), 901–934 (2009)

21. Schwarzentruber, F., Vesic, S., Rienstra, T.: Building an epistemic logic for argumentation. In: del Cerro, L.F., Herzig, A., Mengin, J. (eds.) JELIA 2012. LNCS (LNAI), vol. 7519, pp. 359–371. Springer, Heidelberg (2012). https://doi.org/10.1007/978-3-642-33353-8_28
22. Thimm, M., Polberg, S., Hunter, A.: Epistemic attack semantics. In: Proceedings of the 7th International Conference on Computational Models of Argument, Frontiers in AI and Applications, vol. 305, pp. 37–48. IOS Press (2018)
23. Wakaki, T.: Preference-based argumentation built from prioritized logic programming. J. Logic Comput. **25**(2), 251–301 (2015)
24. Wietske, V., Hindriks, K.V., Jonker, C.M.: Argumentation-based qualitative preference modelling with incomplete and uncertain information. Group Decis. Negot. **21**(1), 99–127 (2012)

Modeling Convention Emergence by Observation with Memorization

Chin-wing Leung$^{(\boxtimes)}$, Shuyue Hu, and Ho-fung Leung

The Chinese University of Hong Kong, Hong Kong, China
{cwleung,syhu,lhf}@cse.cuhk.edu.hk

Abstract. Convention emergence studies how global convention arises from local interactions among agents. Traditionally, the studies on convention emergence are conducted by means of agent-based simulations, whereas very few studies are based on model-based approaches. In this paper, we employ model-based approach to study the convention emergence by observation with memorization in a large population under social learning. In particular, we derive the recurrence equations of the population dynamic, which is the evolution of action distribution over time, under the external majority (EM) strategy. The recurrence equations precisely predict the behaviour of the multi-agent system at any time point, which is verified with the agent-based simulations. Based on the recurrence equations, We prove the converge behavior under various situations and work out the optimal memory length under different number of actions. Finally, we show that the EM strategy outperforms other popular strategies such as Q-learning and Highest Cumulative Reward (HCR) in convergence speed under social learning, even in very large convention space.

Keywords: Multiagent system · Convention emergence · Model-based approach

1 Introduction

Social convention or norm is a common action chosen among the whole population. A typical example is to decide to drive on the left side or right side of the road. Usually it does not matter which action is taken, as long as the action is coordinated by everyone in society. Convention is an effective way to facilitate coordination among agents, which can reduce the potential for conflict and help ensure agents to achieve their goal in an efficient manner [13,22].

The research into convention emergence studies how global convention arises from local interactions among agents, where local interaction is usually treated as an pairwise interaction in a society [6,19,21,23]. The study of convention emergence has started a few decades ago [6,21,23]. One of the research topics is to design a mechanism to facilitate the efficient and stable emergence of norm, which improves coordination in an agent society. There have been a large number

© Springer Nature Switzerland AG 2019
A. C. Nayak and A. Sharma (Eds.): PRICAI 2019, LNAI 11670, pp. 733–745, 2019.
https://doi.org/10.1007/978-3-030-29908-8_57

of studies on norm emergence concerning different agent interaction models, agent learning strategies and network topologies [2].

One of the seminal works of studying convention emergence problem is done by Shoham and Tennenholtz [21]. They propose the external majority (EM) approach and show that it is an effective agents' strategy for convention emergence under sequential pairwise interaction, where only one pair of agents are interacting with each other at a time. The idea of EM is to choose the majority action in the memory, where the memory stores the opponents' actions that the agent has encountered in the past.

The study is conducted through agent-based simulation. As mentioned in the paper [21], it is desirable if we can develop a mathematical theory to explain and predict the result. As opposed to the agent-based simulations, the model-based approaches aim to derive recurrence/differential equations to describe the population dynamics. The major advantage of the model-based approach is that it enables the researchers to explain the dynamical system by a mathematical/statistical process, and thus provides an insight on how the system is evolved in a mathematical description. Another advantage of the model-based approach is that it enables us to provide a proof of convergence through the derived equations [2, 17].

More recently, Sen and Airiau study the convention emergence problem under the social learning framework [19]. Their experimental results show that Q-learning is the most effective strategy among various strategies they have compared. Since then, the reinforcement learning-based approach has been widely adopted in the research in convention emergence [1, 16, 27, 28, 31].

Note that there has not been a research applying the EM strategy onto social learning, yet we identify that the observation with memorization type strategy such as EM is indeed very effective under social learning, it seems the EM strategy has been overlooked by others.

Motivated by the above issues, we decide to develop a model to describe the population dynamics of the EM strategy under social learning in large population. We also study the convergence behavior of the EM strategy under finite and infinite population, and provide the converging conditions on the memory length. We then work out the optimal memory length based on the equations under different number of actions. Finally, we show that the EM strategy outperforms other popular strategies such as Q-learning and Highest Cumulative Reward (HCR) in convergence speed under social learning, which is confirmed in very large convention space.

The rest of this paper is organized as follows. The related work of our study will be introduced in Sect. 2. In Sect. 3, we illustrate and model the population dynamics for the EM strategies under social learning. In Sect. 4, we compare the convergence speed of above strategies with other strategies which are widely used in the literature. Finally, conclusion will be given in Sect. 5.

2   Related Research

In 1992, Shoham and Tennenholtz study the convention emergence problem and propose the EM strategy [21]. For an agent using EM as its strategy, the agent maintains a memory to store the opponent actions it has encountered. For each interaction it has participated, the agent observes and remember the opponent action, and then the agent updates its strategy according to the majority action in the memory. Two years later, the same authors model the agent interaction as a 2-player coordination game and propose the Highest Cumulative Reward (HCR) strategy [22], and the HCR strategy is later shown to be equivalent to EM in the case of 2 actions [23].

The above works has motivated large number of subsequent studies in convention emergence. Walker et al. study the effect on convention emergence different variation of EM [29]. Kittock, Delgado et al. study the population dynamics using EM and HCR under different network topologies [6,7,13]. Epstein et al. study the majority rule with a variable observation radius under a ring network [8]. Urbano et al. study the optimal memory length of EM on different networks [25].

In 2007, Sen and Airiau propose the social learning framework [19], which is a new kind of agent interaction model. In social learning, for each round of iteration, every agents are paired up to play games and update their strategies simultaneously.

Afterward, lots of works are proposed under the setting of social learning. Mukherjee, Villatoro et al. study the effect on convention emergence in the existance of network topology [16,26]. Villatoro et al. study the effect on convergence time under various factors, such as the neighbourhood size and the memory size of past rewards [27,28]. Chen et al. study a hybrid approach of observation and Q-learning under the social learning framework [3].

On the other hand, Urbano et al. propose another agent strategy based on force [24] and Mihaylov et al. propose the Win-Stay-Lose-probability-Shift (WSLpS) strategy for the convention emergence [14]. Yu et al. propose the collective learning framework [31]. Shibusawa and Hao et al. further improve the efficiency of emergence under collective learning framework [10,20].

We note that all the above works are studied via agent-based simulations.

In 2009, Parunak develops an equation based model [17] to study the convergence under a simplified version of the collective cognitive convergence (CCC) model in [18]. They derive the recurrence equation of the dynamics on the proportion of agents interested in certain topics. Later, Brooks et al. come up with a model-based approach based on the bias revision strategy under sequential pairwise interaction [2]. They derive the recurrence equation on the dynamics of the action distribution, and calculate the converged value and time to converge.

We also note that there are number of studies [4,9,11,15] which relate to majority/probabilistic pooling on graph. In their setting, every agents will look at their neighbors' actions and update their actions based on majority/probabilistic rule in every iteration.

3   External Majority Under Social Learning

We adopt the social learning framework [19]. Consider a population consisting of N agents, where N tends to infinity. Each agent is able to perform d actions $(a_1, ..., a_d)$. Let $\boldsymbol{p}^{(t)} = (p_1^{(t)}, ..., p_d^{(t)})$ be the *action distribution* at time t, which is the probability distribution of actions in the population at time t. The agents interact among themselves repeatedly. For each iteration, every agents are randomly paired up to interact simultaneously. In each interaction, both agents will select their actions to perform based on their strategies. The goal of an agent is to select the same action as their opponent, without communication in advance. Based on the outcome of the interaction, the agents will update their choices according to their strategies, the updated action distribution becomes $\boldsymbol{p}^{(t+1)}$. The process continues until the same action is chosen by whole population. We note that the population dynamics we discuss is referred to the trajectory of action distribution $\boldsymbol{p}^{(t)}$ over time.

Consider all agents will adopt the same External Majority (EM) strategy [21], with a slight modification. For an agent using EM, it will choose to perform the most frequent action that it has encountered in previous M rounds, where M is the memory length. In case of tie, the agent will randomly choose one action among the most frequent actions. We note that in the original paper [21], the agent will stick to the current action in case of tie, however, it will cause some confusion when the number of action is larger than 2.[1]

3.1   Dynamics of Action Distribution

As all agents are homogeneous, we can focus on one single agent, and then apply to the whole population. At time t, the agent is randomly assigned an opponent. It then observes the opponent's action. When N is large, such process could be described as drawing a sample from the multinoulli distribution with parameter $\boldsymbol{p}^{(t)}$. Let a_j be the observed action at time t, we write $\boldsymbol{X}^{(t)} = \boldsymbol{e}_j = (0, ..., 1, ..., 0)$ be the observed sample. $\boldsymbol{e}_j$ is the basis vector in $\mathbb{R}^d$, where the jth entry of $\boldsymbol{e}_j$ is equal to 1, otherwise 0. We have $\boldsymbol{X}^{(t)} \in \{\boldsymbol{e}_1, ..., \boldsymbol{e}_d\} \sim \text{Multinoulli}(\boldsymbol{p}^{(t)})$. Let $\boldsymbol{Y}^{(t)} = \sum_{i=0}^{M-1} \boldsymbol{X}^{(t-i)} = (Y_1^{(t)}, ..., Y_d^{(t)})$ be the sum of all samples in the memory, where $\boldsymbol{Y}^{(t)}$ is the sum of independent but not identical multinoulli random variable, which follows the poisson multinomial distribution (PMD) [5]. Let $\xi_{\boldsymbol{k}} = Pr(\boldsymbol{Y}^{(t)} = \boldsymbol{k})$ be the probability mass function (pmf) of $\boldsymbol{Y}^{(t)}$, where

---

[1] Consider the following example. Let the number of action be 3 and memory length be 5. Consider an agent with following observation sequence $(a_1, a_1, a_1, a_2, a_2, a_3, a_3)$. We can see that the action a_1 changes from majority to minority over time. However, the "stick to the current action" strategy will choose a_1 even it is a minority action, since a_2, a_3 are in the tie. However, it is not a reasonable choice. In general, when an action is changing from majority to minority, and other actions become majority and in a tie at the same time, then the "stick to the current action" strategy will choose a minority action. Therefore we make the modification in the case of tie to yield a reasonable outcome.

$\boldsymbol{k}$ is a d-dimensional vector with non-negative elements sum to M, that is $\boldsymbol{k} \in [0, ..., M]^d$ such that $\sum_j k_j = M$, $k_j \geq 0$. The pmf has no explicit form, but the probability can be calculated by the method of enumeration. However, the computation complexity grows exponentially when M is large, as we have to consider d^M combinations for all possible $\boldsymbol{k}$. To deal with this issue, we generalize the algorithm proposed by Hong [12]. The result is summarized by Theorem 1.

Theorem 1. *Let* $\boldsymbol{Y}^{(t)} = \sum_{i=0}^{M-1} \boldsymbol{X}^{(t-i)}$ *as* $\boldsymbol{Y} = \sum_{i=1}^{M} \boldsymbol{X}_i$, *and* $\boldsymbol{p}^{(t)}, ..., \boldsymbol{p}^{(t-M+1)}$ *as* $\boldsymbol{p}_1, ..., \boldsymbol{p}_M$, *then we have*

$$\xi_{\boldsymbol{k}} = Pr(\boldsymbol{Y} = \boldsymbol{k})$$

$$= \frac{1}{(M+1)^d} \sum_{l_1=0}^{M} \cdots \sum_{l_d=0}^{M} e^{-i\frac{2\pi}{M+1}(l_1 k_1 + ... + l_1 k_d)} x_l \tag{1}$$

where $x_l = \prod_{i=1}^{M}[\sum_{j=1}^{d} p_{ij} e^{i\frac{2\pi}{M+1}l_j}]$, $\boldsymbol{l} \in [0, ..., M]^d$.

The idea of the proof is to apply a d-dimensional Fourier transform to the characteristic function of the PMD and obtain the pmf.

To update its strategy, the agent will randomly choose an action among the majority actions in the memory $A_{EM} = \{a_j : j \in argmax_{j'}\{Y_{j'}^{(t)}\}\}$. For an agent, the probability of choosing action a_j at time $t+1$ is

$$p_j^{(t+1)} = \sum_{b=1}^{d} \frac{1}{b} Pr(j \in A_{EM} \wedge |A_{EM}| = b) \tag{2}$$

where $|A_{EM}|$ is the number of majority actions in the memory, and $j = 1, ..., d$. The explicit form of Eq. (2) depends on the memory length M. For example, for $M = 3$, Eq. (2) becomes

$$
\begin{aligned}
p_j^{(t+1)} =& Pr(Y_j^{(t)} = 3) + Pr(Y_j^{(t)} = 2) + \frac{1}{3} Pr(Y_j^{(t)} = 1, Y_k^{(t)} \leq 1 \; \forall k \neq j) \\
=& [p_j^{(t-0)} p_j^{(t-1)} p_j^{(t-2)}] + [p_j^{(t-0)} p_j^{(t-1)}(1 - p_j^{(t-2)}) + p_j^{(t-0)} p_j^{(t-2)}(1 - p_j^{(t-1)}) \\
& + p_j^{(t-1)} p_j^{(t-2)}(1 - p_j^{(t-0)})] + \frac{1}{3}[p_j^{(t-0)} \sum_{k \neq j} \sum_{l \neq k, j} p_k^{(t-1)} p_l^{(t-2)} \\
& + p_j^{(t-1)} \sum_{k \neq j} \sum_{l \neq k, j} p_k^{(t-0)} p_l^{(t-2)} + p_j^{(t-2)} \sum_{k \neq j} \sum_{l \neq k, j} p_k^{(t-0)} p_l^{(t-1)}] \\
=& ... \\
=& \frac{1}{3}(p_j^{(t-0)} + p_j^{(t-1)} + p_j^{(t-2)}) + \frac{1}{3}(p_j^{(t-0)} p_j^{(t-1)} + p_j^{(t-0)} p_j^{(t-2)} + p_j^{(t-1)} p_j^{(t-2)}) \\
& - \frac{1}{3}(p_j^{(t-0)} \sum_k p_k^{(t-1)} p_k^{(t-2)} + p_j^{(t-1)} \sum_k p_k^{(t-0)} p_k^{(t-2)} + p_j^{(t-2)} \sum_k p_k^{(t-0)} p_k^{(t-1)})
\end{aligned}
\tag{3}
$$

for $t \geq 3$. The derivation requires us to expand each term directly.

For larger M, the explicit form requires tedious calculation. In general, for any M, the action distribution $\boldsymbol{p}^{(t+1)} = (p_1^{(t+1)}, ..., p_d^{(t+1)})$ can be calculated numerically by Algorithm 1. The idea is to consider every possible observations and aggregate the probabilities in a bottom up manner.

Algorithm 1. Calculate $\boldsymbol{p}^{(t+1)}$ under EM

Input: $\boldsymbol{p}^{(t)}, ..., \boldsymbol{p}^{(t-M+1)}$
Output: $\boldsymbol{p}^{(t+1)}$

1: initialize $p_j^{(t+1)} = 0$ for all j
2: **for** each observation $\boldsymbol{k}$ in the set of possible observations $\boldsymbol{Y}^{(t)}$, $\{\boldsymbol{k} \in [0, ..., M]^d :$
        $\sum_j k_j = M, k_j \geq 0\}$ **do**
3:      obtain the pmf $\xi_{\boldsymbol{k}}$ from equation (1)
4:      find the set of majority actions in $\boldsymbol{k}$, $A_{EM} = \{a_j : j \in argmax_{j'}\{k_{j'}\}\}$
5:      find the number of majority actions $b = |A_{EM}|$
6:      **for** every action $a_j \in A_{EM}$ **do**
7:          $p_j^{(t+1)} += \frac{1}{b}\xi_{\boldsymbol{k}}$
8:      **end for**
9: **end for**

As all agents are homogeneous, the recurrence Eq. (2) is applied to every agent in the population, hence the whole population. Therefore Eq. (2) is a recurrence equation to describe the population dynamics of the convention emergence. The trajectory of action distribution can be calculated numerically using Algorithm 1, given the initial distribution $\boldsymbol{p}^{(0)}$.

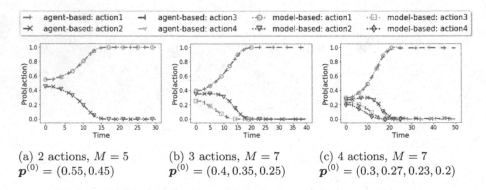

(a) 2 actions, $M = 5$ (b) 3 actions, $M = 7$ (c) 4 actions, $M = 7$
$\boldsymbol{p}^{(0)} = (0.55, 0.45)$ $\boldsymbol{p}^{(0)} = (0.4, 0.35, 0.25)$ $\boldsymbol{p}^{(0)} = (0.3, 0.27, 0.23, 0.2)$

Fig. 1. Population dynamics using EM: model-based vs agent-based

Figure 1 presents a comparison of the population dynamics obtained using model-based approach versus agent-based simulations. The population dynamics under different number of actions (d), initial distributions $(\boldsymbol{p}^{(0)})$ and memory sizes (M) are plotted. The dotted line is obtained by Algorithm 1, and the dash line is obtained from agent-based simulations. For the agent-based simulations, the results are averaged over 100 simulations using 10,000 agents per simulation. We can see that the majority action in $\boldsymbol{p}^{(0)}$ always become the final convention. We also note that the results from our model almost match the results from agent-based simulations, confirming that the model can explain the population dynamics well.

3.2 Emergence Behaviour with Finite and Infinite Number of Agents

The model we formulate in Sect. 3.1 is based on the assumption that N tends to infinity. In agent-based simulations, we usually simulate using a large number of agents in the society but the number cannot be infinity. We expect the emergence behavior will be the same as N is large, except some particular situations. One obvious example is the case where the initial action distribution $\boldsymbol{p}^{(0)}$ is uniformly distributed. In agent-based simulations, we can observe that convention will emerge if the number of iterations is large enough. This is due to the randomness on the agents pairing processing and actions selecting processing injected to the system, which causes the action distribution becoming slightly non-uniform over time. As $N \to \infty$, the randomness is averaged out and the action distribution will keep uniformly distributed over time.

Following the discussion, we study the convergence behaviour of convention emergence in the cases when $N < \infty$ and $N \to \infty$. The following theorems concludes our findings.

Theorem 2. *For $N < \infty$ but sufficiently large ($N > 4$, $N > d$), convention will emerge iff $M \geq 2$.*

The idea of the proof is similar to the proof of convention emergence in previous papers [2,14]. We consider each possible action profile of agents $(\alpha_1, ..., \alpha_N)$ as a state in a Markov Chain, and then we show that there is non-zero probability for any non-absorbing states reaching the absorbing states (states of convention emergence).

As $N \to \infty$, the above idea will not work as the number of states is infinite. We have to consider the explicit form of the recurrence Eq. (2) to draw the conclusion.

Theorem 3. *As $N \to \infty$, $M \leq 2$, convention will not emerge.*

We expend $p^{(t)}$ directly and show $p_j^{(t)} = p_j^{(t-1)}$ $\forall t$ to conclude the result.

Theorem 4. *As $N \to \infty$, $M = 3$, convention will emerge iff the initial distribution has single mode.*

The idea of the proof is to consider Eq. (3) and show the probability of choosing a majority action is increasing in general. We have not yet obtain a general proof for $M > 3$, but we hypothesize that the more the agents can remember, the more likely the convention can emerge.

Theorem 5. *For $M = 3$, the uniformly distributed action distribution is not stable.*

The idea of the proof is to consider $\boldsymbol{x} = [\boldsymbol{p}^{(t-1)\top}, \boldsymbol{p}^{(t-2)\top}, \boldsymbol{p}^{(t-3)\top}]^\top$ as a dynamical system. By showing the uniformly distributed situation is an operating point, we can consider the linearized state space model and show that the system matrix has at least one positive eigenvalue.

From the above theorems, we can summarize two major differences on the emergence behaviour between $N < \infty$ and $N \to \infty$. The first difference is when $N < \infty$, convention will emerge if $M \geq 2$, whereas for $N \to \infty$, convention will not emerge when $M = 2$. This is due to the randomness of agents pairing processing and actions selecting processing of the system. In fact, we have performed an agent-based simulations with $M = 2$ over 1000 iterations and observe that the action distribution over time will vary around the initial distribution without exhibiting any emerging behaviour. However, as time goes to infinity, by Theorem 2, there must be a time point that the action distribution converges to a single action (absorbing state) and stays afterward. The second difference is examplified by the example we discuss at the begining of Sect. 3.2, when $N < \infty$, convention will emerge if we have multiple majorities in the initial action distribution (including the uniformly distributed case), whereas for $N \to \infty$, convention will not emerge in such cases. We show in Theorem 5 that the uniformly distributed situation is not stable. When $N < \infty$, due to randomness, the action distribution for the majority actions must become slightly uneven, and the action distribution never return to the stationary point.

3.3   Identification of Optimal Memory Length

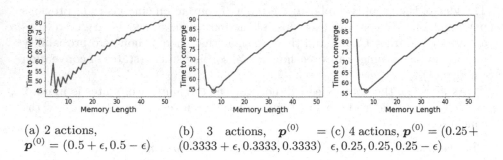

(a) 2 actions, $\boldsymbol{p}^{(0)} = (0.5 + \epsilon, 0.5 - \epsilon)$

(b) 3 actions, $\boldsymbol{p}^{(0)} = (0.3333 + \epsilon, 0.3333, 0.3333)$

(c) 4 actions, $\boldsymbol{p}^{(0)} = (0.25 + \epsilon, 0.25, 0.25, 0.25 - \epsilon)$

Fig. 2. Optimal memory length using EM

The recurrence Eq. (2) allows us to study the optimal memory length from a theoretical perspective. We consider the randomness in agent-based simulations and make the initial distribution to be slightly uneven by adding and subtracting a small ϵ. We then study the time to converge under different memory lengths (from 3 to 50). The time to converge is defined as the first time when the proportion of agents choosing majority action has passed a threshold L $(t = min_{t'}\{max\{p_j^{(t')}\} > L\})$. Figure 2 presents the results of our finding. We set $L = 0.9999$, and $\epsilon = 0.0001$, the circles indicate optimal points for the speed of convergence. Generally, the curve is in a "V" shape. It is because the accuracy of identifying the majority action increases as the agent considers more samples. However, when the memory length further increases, the old samples start to deteriorate the estimation, as the action distribution keeps changing over time. In case of 2, 3 and 4 actions, the corresponding optimal memory lengths are 5, 7 and 7 respectively. We can expect that as the number of actions increase, the optimal memory lengths will increase accordingly, as we more samples in order to identify the majority action.

In the case of 2 actions, we can see that the time to converge is oscillating between odd and even memory lengths. It is because the speed of convergence is slowed down as agents have to deal with the tie situation when memory length is even. When number of actions increases, the pattern disappears as the tie situation may happen even when memory length is odd.

4   Experiment on Convergence Speed

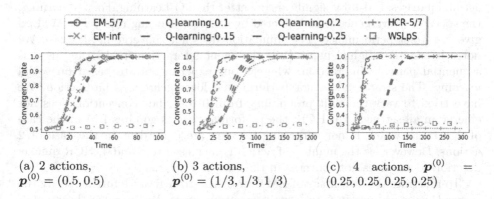

(a) 2 actions, $p^{(0)} = (0.5, 0.5)$

(b) 3 actions, $p^{(0)} = (1/3, 1/3, 1/3)$

(c) 4 actions, $p^{(0)} = (0.25, 0.25, 0.25, 0.25)$

Fig. 3. Comparison on speed of convergence

One major concern in convention emergence is the speed of convergence. In this section we compare the EM strategy with various popular methods, including Q-Learning [30], HCR [22] and WSLpS [23], under social learning. We conduct the experiments using agent-based simulations. For Q-Learning and HCR, as payoff is required for the learning, the standard $(1, -1)$ coordination game is considered, agents receive $+1$ payoff if they choose same action, and -1 payoff if their actions does not match. Let the convergence rate at time t be the proportion of agents choosing the majority action at time t $(rate_t = max_j\{p_j^{(t)}\})$.

For every strategies, we conduct 100 simulations using 10,000 agents. The initial action is uniformly distributed. In the first experiment, we conduct the simulations for the cases of 2, 3, and 4 actions, the maximum number of iterations are 100, 200, 300 respectively. The convergence rate at each time step is recorded, and the average convergence rate is calculated. The strategies and their settings are described as below:

EM: The memory length L is set to be 5 in the case of 2 actions, and 7 in the case of 3 and 4 actions, which are the optimal memory lengths obtained in Sect. 3.3.

EM-inf: EM with unlimited memory length.

Q-Learning: The learning rates vary from 0.1 to 0.25, and the exploration rate ϵ is set to 0.2, which is the common setting.[2]

HCR: We use the same memory length as we used in EM[3], note that in case of 2 actions, HCR is shown to be equivalent to EM.

WSLpS: The shift probability α is set to 0.5, as suggested in the paper [14].

The results for 2, 3 and 4 actions are presented in Fig. 3. Overall, the EM gives the best performance. It can establish conventions within 50 iterations in all cases. It confirms our intuition that observation with memorization is a very effective way to identify the majority, which is the key to the quick convergence. The EM-inf and Q-Learning methods give moderate performance: their convergence speed are close in the case of 2 actions, but when the number of action increases, EM-inf is significantly better than Q-Learning. For Q-Learning, the speed of convergence is not affected by different learning rates. The WSLpS give the worst performance. The simulation does not converge in all cases. We note that in the original paper [14], the agent interaction model they used is sequential pairwise interaction, whereas our experiments are based on social learning. The strategy does not perform well for synchronous interaction. We have tried to vary the shift probability from 0 to 1, but convention does not emerge in all settings. For HCR, the performance is as good as EM in the case of 2 actions, which is not surprising as they are equivalent in the case of 2 actions. However, as the number of action is increased to 3 and 4, HCR quickly deteriorate and does not converge in the case of 4 actions.

To illustrate the advantage of EM over other strategies, we further study the convention emergence in very large convention space. We increase the number of actions to 10, 100, 1000 and repeat the same experiments. As other strategies do not converge even in the case of 4 actions, we only compare the performance of EM with Q-learning. For EM, we first keep the memory length to be 7. As we have mentioned in Sect. 3.3, the optimal memory length is increasing when the number of action increase, therefore we also conduct the experiments with larger memory size. Starting from 10, we add the memory length by 5 and conduct the

---

[2] We measure the convergence rate based on agents' policy, as the actual actions are affected by exploration.

[3] In case of tie, the agent will randomly choose one action among the actions with highest cumulative reward. The changes is made for the same reason as in Sect. 3.

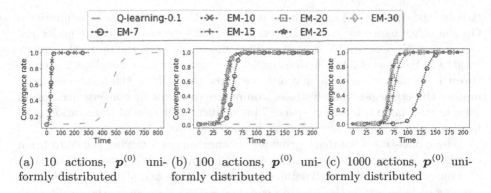

(a) 10 actions, $p^{(0)}$ uni- (b) 100 actions, $p^{(0)}$ uni- (c) 1000 actions, $p^{(0)}$ uniformly distributed   formly distributed   formly distributed

Fig. 4. Convention convergence in large convention space

simulation, and we stop increasing the memory length once the performance has started to deteriorate. For Q-learning, as we have mentioned before, the learning rates do not affect the performance by a lot, we fix the learning rates to be 0.1. As before, we conduct 100 simulations using 10,000 agents, and report the average convergence rate.

The simulations results are presented in Fig. 4. We can see that the performance of EM is very good in all the cases. When the memory length is 7, the convention emerged within 100 iterations in the setting of 10 and 100 actions. Even when the number of action is up to 1000, the convention can emerge within 200 iterations. For Q-learning, we can see that the convergence speed has deteriorated to around 800 iterations in the case of 10 actions, and the convention does not emerged in the case of 100 actions, even we have conducted the simulations up to 1000 iterations. We do not perform the experiment for Q-learning under 1000 actions, since it fails to converged in the setting of 100 actions.

Focus back on EM, we can see that the performance is indeed better for larger memory length as the number of action increases. The best memory lengths are 7, 15, 25 in the case of 10, 100, 1000 actions. The performance difference are hard to notice in the plot, but we can compare with the actual result. For example in the setting of 10 actions, EM with memory length 7 out-perform EM with memory length 10 by 1 iteration. We can see that with the best memory lengths, the convention can emerged within 100 iteration even when the number of actions has went up to 1000. The second set of experiments confirm that EM is very effective in social learning, and the time to converge is very robust against increase in convention space.

5   Conclusion

We develop a model to describe the population dynamics (action distribution over time) of EM strategy under social learning in large population. In particular, we derive the recurrence equations for the population dynamics and an algorithm to solve the equations. The results match well with the agent-based simulations. On the other hand, we study the condition of emergence behaviour under the case where $N \to \infty$ and $N < \infty$. In addition, we work out the optimal memory lengths in the case of 2 to 4 actions under the model we have developed. Finally, through the agent-based simulations, we show that EM is the fastest strategy among the strategies that we have compared in terms of convergence speed, even in very large convention space. This confirm that idea of observation with memorization is indeed very effective in social learning.

The model-based approach provides another way of studying the convention emergence problem apart from agent-based simulations. It provides the description on how the population dynamics evolves over time and allow us to study the emergence behavior in the limiting case $(N \to \infty)$. Moreover, the result from model-based approach does not suffered from simulation error. As for future work, we would like to develop a model-based approach for other strategies under social learning, such as Q-Learning, so that the comparison of convergence speed can be done in theoretical manner.

References

1. Airiau, S., Sen, S., Villatoro, D.: Emergence of conventions through social learning. Auton. Agent. Multi-Agent Syst. **28**(5), 779–804 (2014)
2. Brooks, L., Iba, W., Sen, S.: Modeling the emergence and convergence of norms. In: IJCAI, pp. 97–102 (2011)
3. Chen, B., Yu, C., Diao, Q., Liu, R., Wang, Y.: Social or individual learning? An aggregated solution for coordination in multiagent systems. J. Syst. Sci. Syst. Eng. **27**(2), 180–200 (2018)
4. Cruise, J., Ganesh, A.: Probabilistic consensus via polling and majority rules. Queueing Syst. **78**(2), 99–120 (2014)
5. Daskalakis, C., Kamath, G., Tzamos, C.: On the structure, covering, and learning of poisson multinomial distributions. In: 2015 IEEE 56th Annual Symposium on FOCS, pp. 1203–1217. IEEE (2015)
6. Delgado, J.: Emergence of social conventions in complex networks. Artif. Intell. **141**(1–2), 171–185 (2002)
7. Delgado, J., Pujol, J.M., Sangüesa, R.: Emergence of coordination in scale-free networks. Web Intell. Agent Syst. Int. J. **1**(2), 131–138 (2003)
8. Epstein, J.M.: Learning to be thoughtless: social norms and individual computation. Comput. Econ. **18**(1), 9–24 (2001)
9. Gärtner, B., Zehmakan, A.N.: Majority model on random regular graphs. In: Bender, M.A., Farach-Colton, M., Mosteiro, M.A. (eds.) LATIN 2018. LNCS, vol. 10807, pp. 572–583. Springer, Cham (2018). https://doi.org/10.1007/978-3-319-77404-6_42
10. Hao, J., Sun, J., Chen, G., Wang, Z., Yu, C., Ming, Z.: Efficient and robust emergence of norms through heuristic collective learning. ACM Trans. Auton. Adapt. Syst. (TAAS) **12**(4), 23 (2017)
11. Hassin, Y., Peleg, D.: Distributed probabilistic polling and applications to proportionate agreement. Inf. Comput. **171**(2), 248–268 (2001)
12. Hong, Y.: On computing the distribution function for the poisson binomial distribution. Comput. Stat. Data Anal. **59**, 41–51 (2013)

13. Kittock, J.E.: Emergent conventions and the structure of multi-agent systems. In: Proceedings of the 1993 Santa Fe Institute CSSS, vol. 6, pp. 1–14. Citeseer (1993)
14. Mihaylov, M., Tuyls, K., Nowé, A.: A decentralized approach for convention emergence in multi-agent systems. Auton. Agent. Multi-Agent Syst. 28(5), 749–778 (2014)
15. Mossel, E., Neeman, J., Tamuz, O.: Majority dynamics and aggregation of information in social networks. Auton. Agent. Multi-Agent Syst. 28(3), 408–429 (2014)
16. Mukherjee, P., Sen, S., Airiau, S.: Norm emergence under constrained interactions in diverse societies. In: Proceedings of the 7th AAMAS, vol. 2, pp. 779–786. International Foundation for Autonomous Agents and Multiagent Systems (2008)
17. Parunak, H.: A mathematical analysis of collective cognitive convergence. In: Proceedings of the 8th AAMAS, vol. 1, pp. 473–480. International Foundation for Autonomous Agents and Multiagent Systems (2009)
18. Parunak, H., Belding, T.C., Hilscher, R., Brueckner, S.: Modeling and managing collective cognitive convergence. In: Proceedings of the 7th AAMAS, vol. 3, pp. 1505–1508. International Foundation for Autonomous Agents and Multiagent Systems (2008)
19. Sen, S., Airiau, S.: Emergence of norms through social learning. In: IJCAI, vol. 1507, p. 1512 (2007)
20. Shibusawa, R., Sugawara, T.: Norm emergence via influential weight propagation in complex networks. In: ENIC 2014 European, pp. 30–37. IEEE (2014)
21. Shoham, Y., Tennenholtz, M.: Emergent conventions in multi-agent systems: initial experimental results and observations. In: Proceedings of the 3rd KR, pp. 225–231 (1992)
22. Shoham, Y., Tennenholtz, M.: Co-learning and the evolution of social activity. Technical report, Department of Computer Science, Stanford University, CA (1994)
23. Shoham, Y., Tennenholtz, M.: On the emergence of social conventions: modeling, analysis, and simulations. Artif. Intell. 94(1–2), 139–166 (1997)
24. Urbano, P., Balsa, J., Antunes, L., Moniz, L.: Force versus majority: a comparison in convention emergence efficiency. In: Hübner, J.F., Matson, E., Boissier, O., Dignum, V. (eds.) COIN -2008. LNCS (LNAI), vol. 5428, pp. 48–63. Springer, Heidelberg (2009). https://doi.org/10.1007/978-3-642-00443-8_4
25. Urbano, P., Balsa, J., Ferreira, P., Antunes, L.: How much should agents remember? The role of memory size on convention emergence efficiency. In: Lopes, L.S., Lau, N., Mariano, P., Rocha, L.M. (eds.) EPIA 2009. LNCS (LNAI), vol. 5816, pp. 508–519. Springer, Heidelberg (2009). https://doi.org/10.1007/978-3-642-04686-5_42
26. Villatoro, D., Sabater-Mir, J., Sen, S.: Social instruments for robust convention emergence. In: IJCAI, vol. 11, pp. 420–425 (2011)
27. Villatoro, D., Sen, S., Sabater-Mir, J.: Topology and memory effect on convention emergence. In: IEEE/WIC/ACM International Joint Conferences on WI-IAT, WI-IAT 2009, vol. 2, pp. 233–240. IEEE (2009)
28. Villatoro, D., Sen, S., Sabater-Mir, J.: Exploring the dimensions of convention emergence in multiagent systems. Adv. Complex Syst. 14(02), 201–227 (2011)
29. Walker, A., Wooldridge, M.: Understanding the emergence of conventions in multi-agent systems. In: ICMAS, vol. 95, pp. 384–389 (1995)
30. Watkins, C.J., Dayan, P.: Q-learning. Mach. Learn. 8(3–4), 279–292 (1992)
31. Yu, C., Zhang, M., Ren, F., Luo, X.: Emergence of social norms through collective learning in networked agent societies. In: Proceedings of the 2013 AAMAS, pp. 475–482. International Foundation for Autonomous Agents and Multiagent Systems (2013)

Breaking Deadlocks in Multi-agent Reinforcement Learning with Sparse Interaction

Toshihiro Kujirai[✉] and Takayoshi Yokota

Tottori University, 4-101, Koyama-cho Minami, Tottori 680-8550, Japan
tkujiraski@gmail.com, yokota@eecs.tottori-u.ac.jp

Abstract. Although multi-agent reinforcement learning (MARL) is a promising method for learning a collaborative action policy that will enable each agent to accomplish specific tasks, the state-action space increased exponentially. Coordinating Q-learning (CQ-learning) effectively reduces the state-action space by having each agent determine when it should consider the states of other agents on the basis of a comparison between the immediate rewards in a single-agent environment and those in a multi-agent environment. One way to improve the performance of CQ-learning is to have agents greedily select actions and switch between Q-value update equations in accordance with the state of each agent in the next step. Although this "GPCQ-learning" usually outperforms CQ-learning, a deadlock can occur if there is no difference in the immediate rewards between a single-agent environment and a multi-agent environment. A method has been developed to break such a deadlock by detecting its occurrence and augmenting the state of a deadlocked agent to include the state of the other agent. Evaluation of the method using pursuit games demonstrated that it improves the performance of GPCQ-learning.

Keywords: Reinforcement learning · Multi-agent · Sparse interaction · Fully cooperative · Deadlock

1 Introduction

Multi-agent reinforcement learning (MARL) is a promising method for learning a collaborative action policy that will enable each agent to accomplish specific tasks (Bloembergen et al. 2015; Vlassis 2007). Each agent tries to learn an optimal action policy, one that maximizes the expected cumulative rewards, while sharing the environment with other agents. Agents that learn their action policy by considering the states and actions of other agents are called joint-action learners. Those that learn it independently are called independent learners (Claus and Boutilier 1998).

If each agent shares the same reward for a task, i.e., a fully cooperative task, independent learners can sometimes learn a collaborative action policy without considering the states and actions of other agents because a random exploration strategy may enable them to learn collaborative actions coincidently (Lauer and Riedmiller 2000; Sen et al. 1994). While joint-action learners may perform better because they take information about other agents into account, they suffer an exponential increase in

© Springer Nature Switzerland AG 2019
A. C. Nayak and A. Sharma (Eds.): PRICAI 2019, LNAI 11670, pp. 746–759, 2019.
https://doi.org/10.1007/978-3-030-29908-8_58

the state-action space for learning, thereby reducing the learning speed and increasing the cost of communication and the cost of estimating information about other agents (Tan 1993).

In many real-world tasks, agents behave independently most of the time and sometimes must behave cooperatively. For example, consider a task involving multiple robots working together to move a heavy box to a specific position. A rational approach is for them to independently approach the box and then cooperatively move it in the same direction to the final position. Each should decide its actions taking other agents' positions and actions into account only when the robots are close to each other. The basic idea of MARL with sparse interaction is to reduce the state-action space by considering information about other agents only when necessary because a smaller state-action space makes the learning process more efficient. This means that identifying when cooperative actions are required is a key function in MARL with sparse interaction. Melo and Veloso (2009) reported a method in which a pseudo-action, COORDINATE, is added to the action space of each agent. The agents learn when they should consider other agents by estimating the Q-value for the COORDINATE action for each state. Hauwere et al. (2010, 2011) proposed the coordinating Q-learning (CQ-learning) concept. Each agent determines when it should consider the state of other agents by comparing the immediate rewards in a single-agent environment with those in a multi-agent environment. In CQ-learning, the state-action space is partially augmented when an agent detects a difference in the immediate rewards using Student's t-test. Kujirai and Yokota (2018, 2019) reported three methods for improving the performance of CQ-learning: greedily selecting actions (GCQ-learning), switching between Q-value updating equations on the basis of the state of each agent in the next step (PCQ-learning), and their combination (GPCQ-learning). Evaluation using several maze games validated their effectiveness, especially that of GPCQ-learning.

We previously observed that agents using GPCQ-learning sometimes fall into a deadlock if there is no difference in the immediate rewards between a single-agent environment and a multi-agent environment. We have now developed a method for breaking the deadlock by detecting its occurrence and augmenting the state of a deadlocked agent to include the state of the other agent. Evaluation using pursuit games demonstrated that it improves the performance of GPCQ-learning.

The reminder of this paper is organized as follows. Section 2 gives an overview of MARL and discusses related work. Section 3 discusses MARL with sparse interaction and the deadlock caused by GPCQ-learning. Section 4 presents our method for breaking the deadlock. Section 5 describes our evaluation and compares the performance of our proposed method with those of existing methods including GPCQ-learning. Section 6 concludes this paper with a summary of the key points.

2 Multi-agent Reinforcement Learning

2.1 MDP and Reinforcement Learning

A Markov decision process (MDP) is formalized as a problem in which an agent optimizes its action policy by maximizing the expected cumulative reward resulting from the actions it takes in its environment. The MDP is defined as a tuple (S, T, R, π), where S stands for the state space of the agent, $T(= p(s'|s, a))$ and $R(= r(s, a, s'))$ stand for the transition probability matrix and immediate reward matrix for the combinations of state s, action a, and next state s', and $\pi(= p(a|s))$ stands for the action policy of the agent. An optimal policy, i.e., one that maximizes the expected cumulative reward, is described as π^*.

Reinforcement learning is one method for iteratively estimating π^*. Q-learning (Watkins 1989, 1992) is a typical reinforcement learning method. Instead of estimating π^*, Q-learning estimates the optimal Q-value Q^* by updating the Q-value using (1), wherein α_t indicates the learning rate and γ indicates the discount rate.

$$Q(s, a) \leftarrow (1 - \alpha_t)Q(s, a) + \alpha_t[r(s, a) + \gamma max_{a'}Q(s, a)] \tag{1}$$

2.2   Extended Multi-agent Systems

As shown in Table 1, an MDP can be extended for multi-agent systems in at least four ways. A natural extension is multi-agent MDP (MMDP), in which agents share all the system states (full observability) and rewards. The agents share all their states and actions and obtain the same rewards from the environment as a result of their joint actions (Boutilier 1996).

Table 1. Extended multi-agent systems.

	Full observability	Full joint observability
Shared rewards	MMDP	DEC-MDP
Independent rewards	MG	DEC-MG

Another extension is decentralized MDP (DEC-MDP), in which each agent can observe only its own states, and the agents obtain the same rewards (Melo and Veloso 2011). If they can know the complete state of the environment by sharing their observations, the agents are said to have full joint observability.

These two extensions are called fully cooperative games because the agents obtain the same rewards.

In contrast, in a Markov game (MG) and a decentralized Markov game (DEC-MG), each agent has an independent reward function. This results in a competitive situation (Aras 2004).

2.3   Related Work

The focus here is on fully cooperative games, which have at least one optimal action policy for each agent. The agents can serendipitously learn a cooperative behavior without having any information about the other agents. This is because coincidental actions that lead to high cumulative reward are reinforced (Sen et al. 1994).

For example, assume that an agent randomly selects an action at a certain position in a maze game, and the action results in the agent obtaining a high cumulative reward because the action coincidentally prevents the agent from colliding with another agent. The agent may thereby learn a cooperative action policy without having any information about the other agents. An agent learns a more precise cooperative action policy if it has knowledge of not only its own state-action combinations but also those of other agents. However, the resulting exponential increase in the state-action space and communication cost between agents slows down the learning process.

Figure 1 shows two example maze games in which each agent i tries to find an optimal path from start position S_i to goal G_i. In both games, the goal for each agent is the start position of another agent. They collide if each one simply takes the shortest path. The optimal solution is for one of the agents to take a detour immediately a collision. However, finding this solution requires extensive exploration of potential detours because there are a number of unsuitable detour routes.

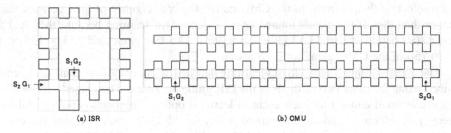

Fig. 1. Example maze games.

Figure 2 shows the average number of steps needed to complete the two games for every 100 episodes using three different learning methods. These methods are straightforward extensions of Q-learning for a multi-agent environment. The first method is independent learning, which is Q-learning itself. Each agent learns its own action policy without having any information about the other agents. The second one is joint-state learning (JSQ-learning), in which each agent always knows the states of the other agents and decides its actions independently on the basis of its own policy. The third is joint-state-action learning (JSAQ-learning), in which one super-agent observes all the states and decides the joint actions for all agents.

As shown in Fig. 2, even the agents using independent learning learned how to avoid collisions and reach their goals unimpeded. In the ISR game, which has a small state-action space, the agents using independent learning converged the fastest although the average number of steps to the goal was the highest because they did not explicitly consider the other agents. In the CMU game, which has a larger state-action space, the agents trained using independent learning had superior performance because 10,000 episodes were not enough for the other methods to learn an optimal policy in the large state-action space.

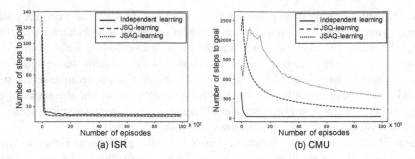

Fig. 2. Learning curves for two maze games.

Policy convergence was slower for the agents trained using JSQ-learning, and the average number of steps to the goal was less than that for the agents trained using independent-learning in the ISR game because these agents consider the other agents when selecting their actions. In the CMU game, the convergence of their policies was slower than that of the agents trained using independent learning because they had a larger state-action space: $43 \times 43 \times 4 = 7396$ in the ISR game and $133 \times 133 \times 4 = 70,756$ in the CMU game for each agent.

The agents trained using JSAQ-learning learned a better policy than the agents trained using the other two methods in the ISR game because the combined state-action space was small enough for each agent to learn an optimal joint-action policy. All the agents had difficulty learning an optimal policy in the CMU game because the combined state-action space was too large ($133 \times 133 \times 4 \times 4 = 283,024$) for the agents to sufficiently explore all the state-action pairs in the limited number of episodes.

3 Multi-agent Reinforcement Learning with Sparse Interaction

3.1 Existing Methods of MARL with Sparse Interaction

In some fully cooperative games, agents can decide their actions without considering any information about the other agents for most states. That is, an independent optimal action policy might be optimal for most states. In the other states, each agent needs information about the other agents to learn an optimal action policy. Therefore, exponential increases in the state-action space and in the communicational cost can be avoided by having each agent consider information about the other agents only when necessary. This type of framework is called decentralized sparse interaction MDP (DEC-SIMDP) (Melo and Veloso 2011) and is a special case of DEC-MDP. Several methods have been proposed for agents to learn their action policy for DEC-SIMDP.

As mentioned in the introduction, Melo and Veloso (2009) reported a method in which a pseudo-action, COORDINATE, is added to the action space of each agent. When COORDINATE is selected as an action by an agent, the agent obtains information about the other agents and behaves in accordance with that information while suffering the penalty of communication cost. Because the Q-value for selecting

COORDINATE can be obtained with Q-learning, the agent can decide when it should consider the other agents. Setting the cost of COORDINATE for a specific task is a difficult issue because, if the cost is too low, the agents will always choose COOR-DINATE, and, if the cost is too high, the agents will seldom choose it.

Hauwere et al. (2010, 2011) proposed a method in which the state of an agent is augmented to include the state of another agent if the two agents are likely to interfere with each other. Each agent behaves in accordance with Q-values learned in advance in a single-agent environment. Each agent can identify potential interference with other agents by comparing the distribution of a state's immediate rewards to those for a single-agent environment. Once the state of an agent is augmented, the agent selects an action on the basis of Q-values corresponding to the augmented joint state when the agent and another agent are in an augmented joint state (Fig. 3).

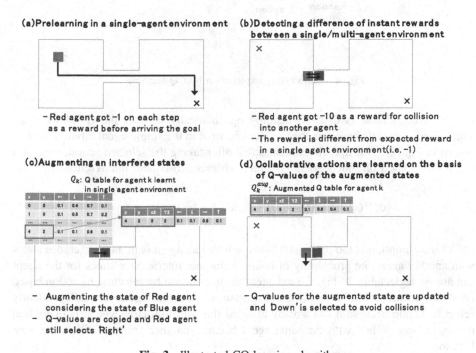

Fig. 3. Illustrated CQ-learning algorithm.

This method is called cooperating Q-learning (CQ-learning). To be more specific, CQ-learning augments the state of agent k s_k, creating augmented joint state $\vec{s_k} = (s_k, s_l)$ that considers the state of agent l. Its Q-values are represented as $Q_k^{aug}(\vec{s_k}, a_k)$ when Student's t-test rejects the hypothesis that the distribution of immediate rewards in the state comes from that of a single-agent environment. This partial augmentation of joint states dramatically reduces the state-action space in sparse interaction tasks compared with JSQ-learning and JSAQ-learning and thereby improves the efficiency and optimality of the learned action policy.

Kujirai and Yokota (2018, 2019) pointed out two issues on CQ-learning and proposed an improved method of CQ-learning, called GPCQ-learning. The first issue on CQ-learning is its unnecessary exploration. An agent using CQ-learning selects its action ε-greedily even it is not in an interfered state. This causes unnecessary exploration and interferences in a multi-agent environment. In addition to that, taking a random action might coincidently prevent interference with another agent, resulting in a lost opportunity for the agent to identify the difference between a single-agent environment and multi-agent environment (Fig. 4)

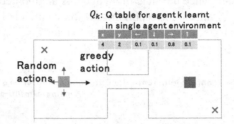

Fig. 4. Unnecessary exploration by CQ-learning.

The second issue is that CQ-learning optimistically updates the Q-values of an augmented joint state based on the Q-values learned in the single-agent environment as shown in Eq. (2). This updating assumes that after taking the selected action, the agent can behave based on independent Q-values without subsequent interference.

$$Q_k^{aug}(\overrightarrow{s_k}, a_k) \leftarrow (1 - \alpha_t)Q_k^{aug}(\overrightarrow{s_k}, a_k) + \alpha_t[r_k + \gamma max_{a'_k}Q_k(s'_k, a'_k)] \tag{2}$$

This assumption is too optimistic because when an agent is in an interference states with another agent the probability of being in another interference states for the agent can not be neglectable. In Fig. 5 a red agent avoid collision by selecting its action based on its augmented Q-values. CQ-learning assumes that the agent can independently select its action because it has already avoided the collision. However, it is likely that the agent may collide with the same agent because another agent is still in the near location.

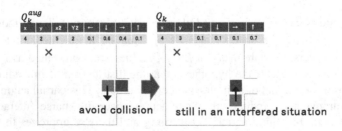

Fig. 5. Example subsequent interferences. (Color figure online)

GPCQ-learning greedily selects an action when an agent is not in an interfered state to avoid unnecessary exploration, and it changes the equation for updating the Q-value of the augmented joint state depending on whether the agent is in an interfered state in the next step in order to avoid optimistic evaluation of the optimal Q-value of the augmented joint state. GPCQ-learning was demonstrated to outperform CQ-learning in several maze games.

3.2   Deadlock Resulting from GPCQ-Learning

In pursuit games, agents (depicted by numbers in Fig. 6) try to move next to a target (depicted by T) in a square field, and the game finishes when all the agents are next to the target. In this paper, the target does not move from the initial position. A state of each agent is represented by a difference in positions between the agent and the target, i.e. $(-6 \leq dx \leq 6, -6 \leq dy \leq 6)$. Actions are *Up*, *Down*, *Left*, and *Right* to move. Rewards are designed as -1 for a movement, -10 for a collision with another agent, and 0 for a movement next to the target and a finish.

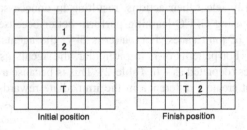

Initial position   Finish position

Fig. 6. Example pursuit game.

First, an agent is trained to learn how to move in order to touch the target in a single-agent environment. The initial positions of the target and the agent are randomly selected in every episode. Then, multiple agents try to find an optimal policy for moving in order to touch the target at the same time in a multi-agent environment. The initial positions of the target and the agents were fixed in seven patterns for evaluation, as shown in Fig. 7. For patterns 1–3, the agents can touch the target by greedily selecting their actions without interference with the other agents. For patterns 4–7, an agent collides with another agent if it greedily selects its action on the basis of Q-values learned in a single-agent environment.

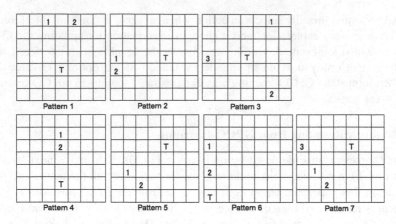

Fig. 7. Seven initial agent/target-position patterns used for evaluation.

For patterns 1–3, the agents using GPCQ-learning find optimal paths resulting in the minimum number of steps to finish while agents using CQ-learning take more steps because they ε-greedily select their actions, resulting in unnecessary exploration. For pattern 4, the agents using GPCQ-learning perform better because they avoid unnecessary augmentation of joint states. For patterns 5–7, the agents using CQ-learning perform much better. For patterns 6–7, in particular, the agents using GPCQ-learning rarely finish the games (depicted as – in Table 2). This is because a repetitive pattern of action-states does not create a difference in the immediate rewards.

Table 2. Comparison of CQ and GPCQ-learning for pursuit games

Patterns	Interference	No. of agents	Min No. of steps	CQ		GPCQ	
				Mean	Std dev.	Mean	Std dev.
1	No	2	4	5.17	1.89	4.00	0.00
2	No	2	4	6.23	8.12	4.00	0.00
3	No	3	4	5.61	13.7	4.00	0.00
4	No	2	4	5.88	1.51	5.14	0.53
5	Yes	2	4	8.53	12.4	274.00	1620.00
6	Yes	2	4	117	219	–	–
7	Yes	3	4	39.5	45.8	–	–

Looking at Fig. 8, we see that agent 1 first greedily selects an action of *upward* in accordance with the prelearned Q-value and detects a difference in immediate rewards because it collides with agent 2 (Fig. 8(a)). It then augments its state with the state of agent 2. For this augmented joint state, agent 1, using GPCQ-learning, decides its action ε-greedily and may select an action *Left* (Fig. 8(b)). After it moves to the left, because it is no longer in an augmented joint state, it greedily selects action *Right* to move back to the previous position in accordance with the prelearned Q-value (Fig. 8 (c)). The reward of selecting action *Right* in a multi-agent environment is the same (i.e. −1) as that in a single-agent environment. Agent 2 also gains the same reward (i.e. 0) as in a single-agent environment because rewards for a touch and finish are the same. Because both rewards are the same as in a single-agent environment, the state of agent 1 is not augmented. Even if agent 1 selects an action of *Right* or *Down*, it may return to the same position because of the same reason. Because any state of the agent is no longer augmented in the situation, it becomes trapped in repetitive movements (i.e. deadlock).

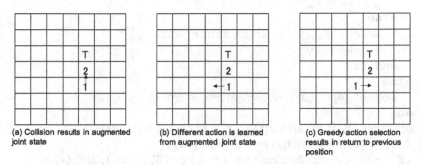

(a) Collision results in augmented joint state

(b) Different action is learned from augmented joint state

(c) Greedy action selection results in return to previous position

Fig. 8. Mechanism of deadlock resulting from GPCQ-learning.

4 Proposed Method

In Algorithm 1, the hatched portions show the differences between CQ-learning and GPCQ-learning and the underlined portion shows the difference between GPCQ-learning and our proposed method. GPCQ-learning selects its action greedily in an unaugmented state (line 13) and changes updating equations based on whether the agent is still in an interfered state in the next step (line 28–29). The proposed method detects repetitive movements between an augmented joint state and a non-augmented state on the basis of states' log and then augments the state of a deadlocked agent to include the state of the other agent, which enables the agent to learn how to avoid the deadlock by updating Q-values.

Although the proposed algorithm can only detect two steps cycle deadlocks, a longer cycle deadlock can be neglectable thanks to the assumption of the sparse interaction.

Algorithm 1: Improved GPCQ-learning algorithm

1: Train Q_k independently first, initialize Q_k^{aug} to zero
   and W_k=empty
2: Set $t=0$
3: **while** true **do**
4:   observe local state $s_k(t)$
5:   **if** $s_k(t)$ is part of a $\overrightarrow{s_k}$ and the info of $\overrightarrow{s_k}$ present
     in the system state s(t) **then**
6:     **if** a set of $\overrightarrow{s_k}$ contains more than two s_k with
       $s_k(t) = s_k$ **then**
7:       Select an agent l randomly from a set of $\overrightarrow{s_k}$
         with $s_k(t) = s_k$
8:       Select $a_k(t)$ in accordance with $Q_k^{aug}(s_k, s_l)$
         ε-greedily
9:     **else**
10:      Select $a_k(t)$ in accordance with Q_k^{aug}
         ε-greedily
11:    **end if**
12: **else**
13:    Select $a_k(t)$ in accordance with Q_k greedily
14: **end if**
15: observe $r_k = R_k(s(t), a(t))$, s'_k from $T(s(t), a(t))$
16: Store $<s_k(t), a_k(t), r_k(t)>$ in $W_k(s_k, a_k)$
17: **if** p-value of Student's t-test $(W_k(s_k, a_k), E(R_k(s_k, a_k))$
     $< p_{th}$ **then**
18:    Store $<s_k(t), a_k(t), s_l(t), r_k(t)>$ in $W_k(s_k, a_k, s_l)$
       for all other agents
19:    **for** all extra information s_l about another agent l
       Present in s(t) **do**
20:      **if** p-value of Student's t-test
         $W_k(s_k, a_k, s_l), E(R_k(s_k, a_k)) < p_{th}$ **then**
21:        augment s_k with s_l to $\overrightarrow{s_k}$ and add it to Q_k^{aug}
22:      **end if**
23:    **end for**
24: **end if**
25: **if** agent k selected a greedy action and agent k
     was/will be in the same augmented joint state
     at $t=t-1/t+1$ **then**
26:    augment s_k with s_l to $\overrightarrow{s_k}$ and add it to Q_{k}^{aug}
27: **if** $s_k(t)$ is part of $\overrightarrow{s_k}$ and information of $\overrightarrow{s_k}$
     is in s(t) **then**
28:    **if** s'_k and s'_l is in part of $\overrightarrow{s_k}$ **then**
29:      $Q_k^{aug}(\overrightarrow{s_k}, a_k) \leftarrow (1 - \alpha_t)Q_k^{aug}(\overrightarrow{s_k}, a_k) +$
         $\alpha_t[r_k + \gamma max_{a'_k}Q_k^{aug}(s'_k, a'_k, s'_l)]$

```
30:    else
31:        Q_k^{aug}(\vec{s_k}, a_k) \leftarrow (1 - \alpha_t)Q_k^{aug}(\vec{s_k}, a_k) +
            \alpha_t[r_k + \gamma max_{a'_k}Q_k(s'_k, a'_k)]
32:    end if
33: else
34:    No need to update Q-value
35: end if
36: t=t+1
37:end while
```

5 Evaluation

We evaluated our proposed learning method in comparison with existing methods: independent learning, JSQ-learning, JSAQ-learning, CQ-learning, and GPCQ-learning. The number of episodes was set to 20,000 for independent learning, JSQ-learning, and JSAQ-learning and 10,000 for CQ-, GPCQ-, and improved GPCQ-learning because CQ-learning and its extensions require prelearning (in this case, 10,000 episodes) in a single-agent environment. In the prelearning, the initial positions of the agents and the target are randomly selected, and ε was set to 0.3 to ensure that the agents could sufficiently explore the environment.

The seven initial agent/target-position patterns shown in Fig. 7 were used for our evaluation. For CQ-, GPCQ-, and improved GPCQ-learning, the length of the window used to calculate the distribution of immediate rewards was set to 20. The threshold of the Student's t-test, p_{th}, was set to 0.01, as was done by Hauwere et al. (2010, 2011).

State-action space of an agent is $13 \times 13 \times 4 = 676$ for an agent using independent learning. If the number of agents is two, the space is $13 \times 13 \times 13 \times 13 \times 4 = 114,244$ for the agents using JSQ-learning, $13 \times 13 \times 13 \times 13 \times 4 \times 4 = 456,976$ for the agents using JSAQ-learning. If the number of agents is three, the space is $13 \times 13 \times 13 \times 13 \times 13 \times 13 \times 4 = 19,307,236$ for the agents using JSQ-learning and $13 \times 13 \times 13 \times 13 \times 13 \times 13 \times 4 \times 4 \times 4 = 308,915,776$ for the agents using JSAQ-learning.

Table 3 shows the number of steps to finish and the standard deviation. For patterns 3 and 7 in which there are three agents, the increase of the state-action space by JSQ-learning and JSAQ-learning clearly reduces the efficiency of finding an optimal action policy. For patterns 1–3, the proposed method, as well as GPCQ-learning, found the optimal paths because there were no interferences between the agents if the agents greedily selected their actions. A slight improvement was obtained for pattern 4. The proposed method substantially outperformed GPCQ-learning for patterns 5–7 while the performance of GPCQ-learning was worst because of deadlocks. For pattern 7, the path found using GPCQ-learning was far from being optimal because there were frequent collisions between the agents in this setting, which is inconsistent with the assumption of sparse interaction. In this case, independent learning, which coincidentally found better paths, performed the best.

Table 3. Evaluation results

Patterns	Independent		JSQ		JSAQ		CQ		GPCQ		Proposed	
	Mean	Std dev.	Mean	Std dev.	Mean	Std dev.	Mean	Std dev.	Mean	Std dev.	Mean	Std dev.
1	5.07	1.56	5.23	2.35	6.91	9.65	5.17	1.89	4.00	0.00	4.00	0.00
2	5.65	1.76	5.38	2.82	7.11	10.5	6.23	8.12	4.00	0.00	4.00	0.00
3	5.02	1.54	6.87	8.14	108	110	5.61	13.7	4.00	0.00	4.00	0.00
4	6.77	2.56	6.69	4.12	7.73	14.2	5.88	1.51	5.14	0.527	5.09	0.41
5	5.11	1.72	5.22	2.47	7.47	10.7	8.53	12.4	274	1620	4.35	0.87
6	6.40	5.07	6.59	4.91	8.54	31.2	117	219	–	–	5.50	1.02
7	8.32	7.27	11.5	11.8	141	119	39.5	45.8	–	–	9.66	3.65

6 Conclusion

We previously observed that agents using GPCQ-learning sometimes fall into a deadlock if there is no difference in the immediate rewards between a single-agent environment and a multi-agent environment.

Our proposed method breaks such a deadlock by detecting them and augmenting the state of a deadlocked agent to include the state of the other agent.

Evaluation against existing five methods, including GPCQ-learning, using seven initial agent/target-position patterns demonstrated that the proposed method outperforms existing methods for most patterns.

References

Bloembergen, D., Tuyls, K., Hennes, D., Kaisers, M.: Evolutionary dynamics of multi-gent learning: a survey. J. Artif. Intell. Res. **53**(1), 659–697 (2015)

Vlassis, N.: A concise introduction to multiagent systems and distributed artificial intelligence. Synth. Lect. Artif. Intell. Mach. Learn. **1**(1), 1–71 (2007)

Claus, C., Boutilier, C.: The dynamics of reinforcement learning in cooperative multiagent systems. In: Proceedings of the 15th National Conference on Artificial Intelligence, pp. 746–752 (1998)

Lauer, M., Riedmiller, M.: An algorithm for distributed reinforcement learning in cooperative multi-agent systems. In: Proceedings of the 17th International Conference on Machine Learning, pp. 535–542 (2000)

Sen, S., Sekaran, M., Hale, J.: Learning to coordinate without sharing information. In: Proceedings of the 12th National Conference on Artificial Intelligence, pp. 426–431 (1994)

Tan, M.: Multi-agent reinforcement learning: independent vs. cooperative agents. In: Proceedings of the 10th International Conference on Machine Learning, pp. 330–337 (1993)

Melo, F., Veloso, M.: Learning of coordination: exploiting sparse interactions in multiagent systems. In: Proceedings of the 8th International Conference on Autonomous Agents and Multiagent Systems, pp. 773–780 (2009)

Hauwere, Y., Vrancx, P., Nowé, A.: Learning multi-agent state space representations. In: Proceedings of the 9th International Conference on Autonomous Agents and Multiagent Systems, pp. 715–722 (2010)

Hauwere, Y.: Sparse interactions in multi-agent reinforcement learning. Ph.D. thesis, Vrije Universiteit Brussel (2011)

Kujirai, T., Yokota, T.: Greedy action selection and pessimistic Q-value updates in cooperative Q-learning. In: Proceedings of the SICE Annual Conference, pp. 821–826 (2018)

Kujirai, T., Yokota, T.: Greedy action selection and pessimistic Q-value updating in multi-agent reinforcement learning with sparse interaction. SICE J. Control Meas. Syst. Integr. **12**(3), 76–84 (2019)

Watkins, C.J.C.H.: Learning from delayed rewards. Ph.D. thesis, Cambridge University (1989)

Watkins, C.J.C.H., Dayan, P.: Q-learning. Mach. Learn. **8**(3–4), 279–292 (1992)

Boutilier, C.: Planning, learning and coordination in multiagent decision processes. In: Proceedings of the 6th Conference on Theoretical Aspects of Rationality and Knowledge, pp. 195–210 (1996)

Melo, F., Veloso, M.: Decentralized MDPs with sparse interactions. Artif. Intell. **175**(11), 1757–1789 (2011)

Aras, R., Dutech, A., Charpillet, F.: Cooperation through communication in decentralized Markov games. In: Proceedings of the International Conference on Advances in Intelligent Systems - Theory and Applications (2004)

Author Index

Printed in the United States
By Bookmasters

Printed in the United States
By Bookmasters